Fundamentals of

Reinforced Cement Concrete Designs

Fundamentals of
Reinforced Cement Concrete Designs

PS Gahlot ME (Str Engg)

Former

Professor, Department of Civil Engineering
National Institute of Technical Teachers Training and Research
Chandigarh

Director/Principal
Yagyavalkya Institute of Technology (YIT)
Rajasthan Institute of Engineering and Technology (RIET)
Jaipur Institute of Engineering and Management (JIEM)
Rajasthan Technical University (RTU), Jaipur, Rajasthan

Deep Gehlot BE, BSc
DGM, Tata Consulting Engineers (TCE) Ltd

CBS

CBS Publishers & Distributors Pvt Ltd

New Delhi • Bengaluru • Chennai • Kochi • Kolkata • Mumbai
Hyderabad • Nagpur • Patna • Pune • Vijayawada

Fundamentals of
Reinforced Cement Concrete Designs

ISBN: 978-93-85915-61-1

Copyright © Authors and Publisher

First Edition: 2016

Published by Satish Kumar Jain and produced by Varun Jain for

CBS Publishers & Distributors Pvt Ltd

4819/XI Prahlad Street, 24 Ansari Road, Daryaganj, New Delhi 110 002, India.

Ph: 23289259, 23266861, 23266867   Website: www.cbspd.com

Fax: 011-23243014   e-mail: delhi@cbspd.com; cbspubs@airtelmail.in.

Corporate Office: 204 FIE, Industrial Area, Patparganj, Delhi 110 092

Ph: 4934 4934   Fax: 4934 4935   e-mail: publishing@cbspd.com; publicity@cbspd.com

Branches

- **Bengaluru:** Seema House 2975, 17th Cross, K.R. Road, Banasankari 2nd Stage, Bengaluru 560 070, Karnataka
  Ph: +91-80-26771678/79   Fax: +91-80-26771680   e-mail: bangalore@cbspd.com
- **Chennai:** 7, Subbaraya Street, Shenoy Nagar, Chennai 600 030, Tamil Nadu
  Ph: +91-44-26680620, 26681266   Fax: +91-44-42032115   e-mail: chennai@cbspd.com
- **Kochi:** Ashana House, No. 39/1904, AM Thomas Road, Valanjambalam, Ernakulam 682 018, Kochi, Kerala
  Ph: +91-484-4059061-62-64-65   Fax: +91-484-4059065   e-mail: kochi@cbspd.com
- **Kolkata:** 6/B, Ground Floor, Rameswar Shaw Road, Kolkata-700 014, West Bengal, India
  Ph: +91-33-22891126, 22891127, 22891128   e-mail: kolkata@cbspd.com
- **Mumbai:** 83-C, Dr E Moses Road, Worli, Mumbai-400018, Maharashtra
  Ph: +91-22-24902340/41   Fax: +91-22-24902342   e-mail: mumbai@cbspd.com

Representatives

- **Hyderabad** 0-9885175004   • **Nagpur** 0-9021734563   • **Patna** 0-9334159340
- **Pune** 0-9623451994   • **Vijayawada** 0-9000660880

Printed at: Swastik Packagings , 506 F.I.E., Patparganj, Delhi - 92

Foreword

Reinforced cement concrete (RCC) is used in almost all structures and hence understanding of RCC designs of various structural elements is most essential for all civil engineers. Design of various elements play a vital role in the safety and stability of any structure. Thus, proper knowledge of fundamentals of RCC designs is necessary for practising engineers for understanding and achieving quality, progress, economy and safety at the construction site. Proper knowledge of fundamentals of RCC design also helps in seeking good employment, fast professional growth and reputation. Most of the structures have many types of RCC elements and hence safe design of any structure requires thorough knowledge of design of various RCC elements.

The book has been developed keeping in mind the limitations of present-day students and basic principles of learning. Each chapter starts with statement of learning objectives to clarify the knowledge explained in the chapter. Provision of objective questions at the end of each chapter facilitates the students for self-evaluation of their learning. This also helps the teacher for making teaching effective. Summary at the end of each chapter brings out key points for easy grasping of important basic knowledge and recall of such knowledge whenever required prior to examination or competitive examination for employment. Each design emphasises provision of IS codal specifications for safety, quality, durability and serviceability of all the structure. Proper design, quality and construction technique of RCC elements affect the strength, stability and life of the structure.

All civil engineers are required to have the basic knowledge of design, quality and construction techniques and hence it is mandatory for all civil engineering students at various levels to understand and master fundamentals of RCC designs.

The authors have done an excellent job to explain the designs of various RCC elements very effectively in their book on *Fundamentals of Reinforced Cement Concrete Designs*. The book meets long standing need for a comprehensive treatment on the important subject of RCC designs. Needless to say, India has already became a global destination, thanks to the maturity of Indian democracy. During this crucial juncture, I am sure, this book will be of immense help to all namely students, teachers and practising field engineers and supervisors at various levels.

Ashok Kumar Basa FIE
President
The Institution of Engineers (India)

20.11.2014

Preface

Design of RCC elements configures the basic foundation course for all civil engineers. Structural engineers require full understanding (mastery) of the fundamentals of RCC designs for efficient, economic and safe design of any RCC structure. Keeping this need in mind, we have written this book on *Fundamentals of Reinforced Cement Concrete Designs* for benefit of students desirous of excelling in their examinations and professional career. We have a long experience in teaching and execution of RCC works and the book has been brought out considering the basic needs of the subject to benefit the budding and practising engineers. We have structured and sequenced the contents adopting the basic principles of learning hierarchy.

The book starts with general requirements of RCC materials specifying their behaviour and limiting properties. Having specified material properties, design procedures are introduced and explained. Next, with the understanding of material's properties and design procedures, the knowledge is applied for the design of actual RCC elements. After dealing with the design of individual elements, design of structures as a whole is dealt. This approach and sequence of learning from concepts to application is quite effective. The book is compiled in seven units according to the theory of learning (concept to principles to application and analysis) for effective learning of design of RCC structures.

Number of solved and unsolved problems along with objective questions in each chapter will facilitate expertise learning of RCC designs. Summary will facilitate practising engineers to recall the specific RCC designs for achieving good quality, economy and progress in design and construction. Summary, problems and objective questions will further facilitate students in mastery learning and seeking good grades in examinations and selection for employment. The book will also help teachers for effective teaching.

Systéme international (SI) units of measurement have been used throughout the book. Latest IS code (IS:456-2000) provisions have been adopted wherever necessary in designs.

PS Gahlot
Deep Gehlot

Contents

Unit I: General Requirements of RCC Designs

Unit II: Limit State Design of Beams

4. Limit State Design of Beam Section for Bending Moment

5. Limit State Design for Shear, Torsion, Bond and Development Length of Reinforcement in Beams

6. Limit States of Serviceability for Bending Elements

Unit III: Limit State Design of Slabs

7. Limit State Design of Slabs

Unit IV: Design of Compression Members (Columns)

8. Limit State Design of Columns

Unit V: Design of Foundations

9. Limit State Design of Footings

Unit VI: Retaining Structure

10. Limit State Design of Retaining Walls

11. Water Retaining Structures (Tanks)

UNIT VII: Miscellaneous Structures

12. Introduction to Prestressed Structural Concrete

13. Design of RCC Staircases

14. Design of RCC Culverts

15. Yield Line Analysis of Slabs

Chapter 1

General Requirements of Reinforced Cement Concrete

LEARNING OBJECTIVES

After the study of *general requirements of reinforced cement concrete*, the learner will understand the basic design requirements and will be able to:

- State the general application of reinforced cement concrete (RCC) in construction
- State special characteristics of constituent materials
- Describe important properties of cement concrete
- Describe important properties of steel reinforcement
- State different material requirements of RCC elements as per IS:456-2000
- Specify concrete covers in different RCC elements in different environments
- Specify different design loads for RCC structures based on occupancy as per IS:875 (Part I, II, III).

1.1 INTRODUCTION

Reinforced cement concrete is a composite material comprising cement concrete and steel reinforcement bars. The composite action is caused by adequate bond between reinforcement and cement concrete. Cement concrete is generally waterproof when properly prepared and compacted, and hence RCC structures remain quite durable. The loads on RCC structures are shared between steel and concrete because of bond. There is compatibility of strains in steel and concrete because of bond. Both concrete and steel reinforcement deform together, i.e. strains in concrete and steel are equal.

In flexural elements, concrete resists compression while steel resists tension. The reinforcement makes the RCC element ductile up to certain loading range. Concrete is strong in compression and takes care of compressive forces. In case of columns, steel also supplements concrete in bearing axial compressive forces in addition to any accidental tensile forces caused by eccentricity and transverse loads.

RCC has established itself as the most important construction material of 20th century. It is extensively used for construction of slabs, beams, columns, walls, footings, retaining walls, water tanks, reservoirs, dams, bridges, piers, electric poles, sleepers, swimming pools, cooling towers, silos, chimneys, tunnels, bunkers, girders, pipes, box culverts, piles, arches, portals, etc.

Development of suitable and reliable methods of design during 20th century has made RCC as the most common construction material. Presently, RCC construction has surpassed all other materials of construction all over the world. During the beginning of 20th century elastic design methods were developed and later in 1950s' ultimate load design methods were adopted. During the later part of 20th century, comprehensive approach of *limit state design* was developed and adopted. Latest IS codes and other country's codes consider limit state design as the most suitable approach. Latest IS:456 code for RCC and IS:800 for steel incorporates mainly limit state design approach.

1.2 MATERIALS FOR REINFORCED CEMENT CONCRETE

Reinforced cement concrete mainly consists of cement concrete and steel reinforcement. The specifications and properties of these materials are covered by various Indian standards (e.g. IS:456). The characteristic strengths of concrete and steel form the main basis of design for RCC. Let us first consider the properties of cement concrete.

1.3 CEMENT CONCRETE

Specifications of cement concrete are covered by IS:456-2000 which incorporates major changes in properties of cement concrete and design procedures. For further understanding of properties of cement concrete, refer to *Quality Management of Cement Concrete Construction* by Gahlot and Gehlot, CBS Publishers, New Delhi.

1.3.1 Grade of Concrete

Concrete grades are specified by its characteristic strength (f_{ck}). Grade of concrete is denoted by letter M followed by its 28 days compressive (crushing) strength specified in N/mm^2 as measured on 150 mm size cubes. Characteristic strength is the mean strength in which not more than 5% samples fall below this strength. Thus, M30 grade means the characteristic strength after 28 days moist curing will be 30 N/mm^2 as measured on 150 mm size cubes. Various grades used in concrete are given in Table 1.1.

Table 1.1: Grades of cement concrete

Grade	M10	M15	M20	M25	M30	M35	M40	M45	M50	M60	M70	M80
Compressive strength– f_{ck} (N/mm^2)	10	15	20	25	30	35	40	45	50	60	70	80

Note: For grades greater than M50, design strengths and other characteristics are considered only after proper verification by actual tests and research.

1.3.2 Tensile Strength of Concrete

Tensile strength of concrete is much smaller than its compressive strength (characteristic strength–f_{ck}). Flexural tensile strength (f_{cr}) is related to its characteristic strength (f_{ck}), expressed in N/mm^2 by the following relation:

$f_{cr} = 0.70 \sqrt{f_{ck}}$ N/mm^2, where f_{ck} is the compressive/characteristic strength of CC expressed in N/mm^2.

1.3.3 Modulus of Elasticity of Concrete

The modulus of elasticity of cement concrete (CC) is primarily influenced by the elastic properties of the aggregate and to a lesser extent by the conditions of curing, age of the concrete, the mix proportions and the type of cement. The modulus of elasticity (E_c) is normally related to the *compressive strength* (f_{ck}) of concrete by the following relation:

$E_c = 5000 \sqrt{f_{ck}}$ (N/mm^2), where E_c is the short term static modulus of elasticity in N/mm^2 and f_{ck} is the characteristic cube compressive strength of concrete in N/mm^2.

1.3.4 Shrinkage of Concrete

Cement concrete shrinks during setting and hardening process. The total shrinkage of concrete depends on the constituents of concrete, size of the member and environmental conditions (temperature and humidity). For a given humidity and temperature, the total shrinkage of concrete is most influenced by the total amount of water present in the concrete at the time of mixing and, to a lesser extent, by the cement content. Generally, for design purposes total shrinkage strain may be taken as 0.0003 unless otherwise specified.

1.3.5 Creep of Concrete

Creep of concrete depends in addition to the factors listed for shrinkage, on the stress in concrete, age at loading and the duration of loading. As long as the stress in concrete does not exceed one-third of its characteristic strength, creep may be assumed to be proportional to the stress. In absence of detailed experimental study, the ultimate creep strain/elastic strain at the given age may be assumed as given in Table 1.2.

Table 1.2: Creep coefficients

Age at loading	7 days	28 days	1 year
Creep coefficient	2.2	1.6	1.1

1.3.6 Coefficient of Thermal Expansion

The coefficient of thermal expansion depends on nature of the cement, the aggregate, the cement content, the relative humidity and the size of the section. The coefficient of thermal expansion for concrete with different aggregates may be taken as given in Table 1.3.

Table 1.3: Coefficient of expansion

Type of aggregate	Quartzite	Sandstone	Granite	Basalt	Limestone
Coefficient of thermal expansion per °C	$1.2–1.3 \times 10^{-5}$	$0.9–1.2 \times 10^{-5}$	$0.7–0.95 \times 10^{-5}$	$0.8–0.95 \times 10^{-5}$	$0.6–0.9 \times 10^{-5}$

1.3.7 Cement Concrete Workability

The concrete mix proportions should be such that the concrete is of adequate workability for the site placing conditions. The workability should be adequate for compaction with the available means for compaction. Suggested values of workability of concrete measured in accordance with IS:1199 are given in Table 1.4.

Table 1.4: Ranges of workability for different placing conditions

Placing conditions	Blinding concrete, shallow sections, pavements using pavers	Mass concrete, lightly reinforced sections, slabs, beams, walls, columns, floors, hand placed pavements, canal lining, footing, etc.	(a) Heavily reinforced, slabs, beams, walls, columns, (b) Slip form work, pumped concrete	Trench fill and *in situ* piling	Tremie concrete
Degree of workability	Very low	Low	Medium	High SCC*	Very high SCC*
Slump (mm)/ Compacting factor	Slump (0–10 mm) Compacting factor (0.75–0.80)	Slump 25–75 mm CF 0.85–0.95	(a) Slump 50–100 mm (b) Slump 75–100 mm	Slump 100–150 mm	Measured by flow

SCC*: Self compacting concrete

1.3.8 Cement Concrete Durability

Durability of concrete is one that performs satisfactorily in the anticipated working environment and exposure conditions. The mix proportions and the type of cement and admixtures used are such that the concrete maintains its integrity and protects the embedded steel reinforcement from corrosion. The durability of concrete is affected by its permeability due to ingress of water, air, carbon dioxide, chlorides, sulphates and other deleterious substances. Impermeability of concrete depends on its constituents and compaction quality.

The durability of concrete is influenced by:

- The **environment**
- The **cover** to the embedded steel
- The **type** and **quality of constituents** (materials)
- The **cement content** and **type**
- The **W/C** ratio
- Full **compaction** and **adequate curing**
- The **shape and size** of the member

Cement concrete is more vulnerable to deterioration due to chemical or climatic attack when it is in thin sections. Thin sections under hydrostatic pressure from one side, partially immersed and at corners and edges of members having lower life but can be enhanced by providing extra cover to steel and also by surface coating to prevent the ingress of water, carbon dioxide or aggressive chemicals. There are 5 type of exposure conditions as given in Table 1.5.

Table 1.5: Environmental exposure conditions

Environment level	Exposure conditions
1. Mild	Concrete surface protected against weather or aggressive conditions.
2. Moderate	Concrete surface sheltered from severe rain, concrete exposed to rain, concrete continuously under water, concrete buried in non-aggressive soil, concrete surface sheltered in coastal area.
3. Severe	Concrete surface exposed to severe rain, alternate wetting and drying, occasional freezing whilst wet or severe condensation, concrete fully immersed in seawater, concrete exposed to coastal area.
4. Very severe	Concrete exposed to seawater, corrosive fumes, freezing whilst wet, aggressive soil or underground water.
5. Extreme	Concrete in tidal zones, contact with liquid/solid aggressive chemicals.

1.3.9 Durability in Exposure to Sulphate Attack

Concrete members, when exposed to sulphate attack, the durability is achieved by using special type of cement, certain minimum quantity of cement and water/cement ratio not more than a certain value as given in Table 1.6.

Table 1.6: Requirements for cement concrete exposed to sulphate attack (IS:456-2000)

S. No.	Concentration of sulphate (expressed as SO_3 concentration)		Type of cement	Minimum cement content (kg/m³)	Maximum W/C ratio by mass
	Total SO_3 in				
	Soil (%)	Groundwater (g/l)			
1	Traces (< 0.2)	Less than 0.3	• OPC (Ordinary PC) • PSC (Portland slag) • PPC (Pozzalanic PC)	280	0.55
2	0.20–0.50	0.30–1.2	• OPC • PSC • PPC	330	0.50
			• Supersulphated • Sulphate resisting PC	310	0.50
3	0.50–1.0	1.2–2.5	• Supersulphated • Sulphate resisting PC	330	0.50
			• PPC • PSC	350	0.45
4	1.0–2.0	2.5–5.0	• Supersulphated • Sulphate resisting PC	370	0.45
5	More than 2.0	More than 5.0	• Sulphate resisting PC • Supersulphated with protective coating	400	0.40

Notes: 1. Cement content is irrespective of the grade of concrete.

2. Cement content given is the minimum and cement content more than this may be used on the basis of various conditions.

3. For very severe conditions and thin sections under hydrostatic pressure on one side only, considerations should be given to a further reduction of W/C ratio.

The protection of the steel reinforcement in concrete against corrosion depends upon an adequate thickness of good quality concrete cover. Nominal concrete cover to meet durability requirements in different exposure conditions is given in Table 1.7 (a) and (b).

Table 1.7(a): Nominal concrete cover to meet durability requirements (IS:456-2000)

Exposure	Mild	Moderate	Severe	Very severe	Extreme
Nominal concrete cover in mm not less than:	20	30	45	50	75

Notes: 1. For main reinforcement up to 12 mm diameter bars for mild exposure the nominal cover may be reduced by 5 mm (i.e. 15 mm).

2. Minimum concrete cover should not deviate from the required by (+) 10 mm to (–) 0 mm.

3. Where concrete grade is M35 and above, for severe and very severe exposure conditions, a reduction of 5 mm may be allowed.

Table 1.7(b): Nominal cover for specified period of fire resistance (IS:456-2000)

Fire resistance (hours)	Beams		Slabs		Ribs		Columns (mm)
	Simply supported (mm)	Conti-nuous (mm)	Simply supported (mm)	Conti-nuous (mm)	Simply supported (mm)	Conti-nuous (mm)	
0.50	20	20	20	20	20	20	40
1.00	20	20	20	20	20	20	40
1.50	20	20	25	20	35	20	40
2.00	40	30	35	25	45	35	40
3.00	60	40	45	35	55	45	40
4.00	70	50	55	45	65	55	40

Notes: 1. The nominal covers given relate to specifically to the minimum member dimensions.

2. Cases below bold line require extra attention.

1.3.10 Mix Design (Mix Proportions), W/C Ratio and Concrete Grade

To achieve durability of cement concrete, we need careful selection of mix proportions and type of materials. Mix design for durability is based on various considerations as given in Tables 1.4–1.7 derived from IS:456-2000.

For 20 mm nominal maximum size of aggregate for different grades of cement concrete for various exposure conditions, minimum cement content and maximum W/C ratio limits are given in Table 1.8 for suitable durability.

Concrete ingredients shall be mixed in a mechanical mixer. The mixer should comply with Indian standards 1791 and 12119. The mixer shall be fitted with water measuring (metering) device. The mixing shall be continued until there is uniform distribution of ingredients and the mass is homogeneous and of uniform colour and consistency.

Table 1.8: Minimum cement content and maximum W/C ratio in RCC (IS:456-2000)

Exposure conditions	Grade of concrete	Minimum cement content (kg/m³)	Maximum W/C ratio by mass
Mild	M20	300	0.55
Moderate	M25	300	0.50
Severe	M30	320	0.45
Very severe	M35	340	0.45
Extreme	M40	360	0.40

Note: Cement content is irrespective of the cement grade. Use of fly ash and blast furnace slag may be considered in respect of cement content and water–cement ratio.

Concrete mix proportions are designed for a *target strength* given by

$$f_t = f_{ck} + ks \qquad \qquad \text{... Eq. (1.1)}$$

where f_{ck} is desired characteristic strength (M), k is statistical constant based on the quality control at site, and s is the standard deviation for the desired grade and quality control.

Standard deviation s for various grades of concrete and normal quality control at site are given in Table 1.9 as per IS:456-2000.

Table 1.9: Standard deviation s for different grades (IS:456-2000)

Grade of concrete	M10	M15	M20	M25	M30	M35	M40	M45	M50	Remarks
Assumed standard deviation s (N/mm²)		3.5		4.0			5.0			The values of s depends on site quality control of batching, mixing, controlled water addition, etc.

1.4 STEEL REINFORCEMENT

Steel reinforcements of different types and strengths are used with cement concrete of different grades. Steel reinforcement is used in the tension zone of flexural members since concrete is weak in tension. Compression members carrying compressive forces, steel reinforcement is used to share part of compressive load and thus enhancing the compressive load carrying capacity apart from resisting accidental tension developed due to eccentricity or lateral loading.

There are four types of steel reinforcements used in structural concrete members which are designated as follows:

(i) Mild steel and medium tensile steel bars conforming to IS:432 (Part 1).

(ii) High strength deformed steel bars conforming to IS:1786.

(iii) Hard drawn steel wire fabric conforming to the specifications of IS:1566.

(iv) Structural steel conforming to IS:2062 (grade A) covering rolled steel sections.

It may be noted that most of the steel used for reinforced cement concrete comprises high strength deformed steel bars (Fe 415) which has much superior bond strength with concrete due to protruding ribs on the surface. The reinforcements used in concrete must be free from all loose mill scale, rust, oil, mud, and other substances since these materials reduce the bond of steel with concrete resulting in poor composite action of RCC.

Most of the steel reinforcements have yield point stress (f_y) or may have 0.20% proof stress if definite yield stress is not exhibited. The most commonly used steel reinforcement is the high strength deformed bars (HYSD) with yield strength of 415 N/mm². Generally steel bars are available in diameters of 5, 6, 8, 10, 12, 16, 18, 20, 22, 25, 28, 32, 36, 40, 45 and 50 mm (most common).

Reinforcement must be cleaned by sand blasting to remove surface coatings which are injurious to bond with concrete. The *modulus of elasticity* of steel shall be taken as 200 kN/mm² (2×10^5 N/mm²). Characteristic yield strength (f_y) of different steels shall be assumed as minimum *yield stress* or 0.20% *proof stress* specified in relevant IS codes.

1.5 REQUIREMENTS AND STRESSES OF STEEL REINFORCEMENT

1.5.1 Detailing

Assembly and placing reinforcement in RCC members are as important as the design of members. Reinforcement of required size and quantity is cut to proper length and placed at the appropriate spacing and depth as per design. The high strength deformed bars (HYSD) are not provided with hooks at ends but plain mild steel bars (MS) are provided with end hooks of specified details. In modern construction practices mild steel plain bars are rarely used. Bar bending schedule must be prepared with the approval of the engineer (designer).

All reinforcements shall be placed and maintained in the position shown in the design drawings by providing proper cover blocks, spacers and supporting bars or benches. Reinforcement should be secured against displacement outside the specified limits. The tolerance limits shall be as under:

(i) For effective depth of 200 mm or less ± 10 mm

(ii) For effective depth of more than 200 mm ± 15 mm

(iii) Actual concrete cover should not deviate + 10 mm, – 0 mm

Nominal concrete covers are specified in Tables 1.7(a) and (b) in accordance with IS:456-2000.

Reinforcing steel of the same type and grade should be used as main reinforcement throughout the member/structure. However, different types and grades of steel can be used for main and secondary reinforcements. Bars may be arranged singly or in pairs or in groups of 3–4 bars bundled and enclosed within stirrups or ties. Bars in bundle must have development length as per the formula:

$$L_d = \frac{\phi \cdot \sigma_s}{4\,\tau_{bd}} \qquad\qquad \text{... Eq. (1.2)}$$

where, ϕ is the nominal diameter of bar

σ_s is the stress in the bar at the section considered

τ_{bd} is the design bond stress of steel and concrete.

Design bond stress in limit state method for plain and deformed bars is given in Table 1.10(a). Permissible bond stress in working stress method is given in Table 1.10(b).

Table 1.10(a): Design bond stress in limit state method (in tension)

Grade of concrete		M20	M25	M30	M35	M40
Design bond stress	Plain bars	1.20	1.40	1.50	1.70	1.90
τ_{bd} (N/mm²)	Deformed bars	1.92	2.24	2.40	2.72	3.04

Note: For bars in compression, the bond stress in tension shall be increased by 25%. It may be noted that the bond stress in deformed bars is 60% more than the plain bars.

Table 1.10(b): Permissible bond stress in tension bars in working stress method

Grade of concrete		M15	M20	M25	M30	M35	M40	M45	M50
Permissible bond stress intension	Plain bars	0.60	0.80	0.90	1.00	1.10	1.20	1.30	1.40
Average τ_{bd} (N/mm²)	Deformed bars	0.96	1.28	1.44	1.60	1.76	1.92	2.08	2.24

Note: For bars in compression, the bond stress is increased by 25% of bars in tension.

It may be noted that bond stress in HYDS bars is 60% more than that for plain bars.

As a general rule, curtailment of tension reinforcement is done so that these bars extend by a length equal to *effective depth* of the member or *12 times the bar diameter* beyond the point where it is no more required *whichever is greater*. Flexural reinforcement shall not be terminated in tension zone.

1/3rd positive moment reinforcement in simply supported members and 1/4th the positive reinforcement in continuous members shall extend along the same face of the member into the support, to a length equal to 1/3rd the development length (L_d).

Nominal diameter of deformed bar is average diameter of equivalent circle having the same area. *Minimum distance between two parallel main bars* shall usually be not less than the greatest of the following:

  (i) The diameter of the bar, if equal diameters are there

  (ii) The diameter of the largest bar

  (iii) 5 mm more than the nominal maximum size of the coarse aggregate used

In case of two layers, the minimum vertical distance between two bars shall be 15 mm, 2/3rd the nominal maximum size of aggregate or the maximum size of bars, whichever is greater.

In beams the horizontal distance between parallel reinforcement bars or groups, near the tension face of the beam shall not be greater than the value given in Table 1.11.

Table 1.11: Clear distance between bars for different steels

f_y (N/mm²)	Percent redistribution to or from section considered				
	−30	−15	0	+15	+30
	Clear distance between bars (mm)				
250	215	260	300	300	300
415	125	155	180	210	235
500	105	130	150	175	195

In slabs, the horizontal distance between parallel main reinforcement bars shall not be more than three times the effective depth of solid or 300 mm whichever is smaller. Horizontal distance between the reinforcement provided against shrinkage and temperature shall not be more than 5 times the effective depth of a solid slab or 450 mm whichever is smaller.

Cover to steel reinforcement is given in Tables 1.7(a) and (b) as required for different exposure conditions and fire resistance.

1.5.2 Reinforcement Requirements in Different Elements

Beams

The minimum area of tension reinforcement shall be not less than that given by

$$A_{st} = \frac{0.85\,bd}{f_y} \qquad \text{... Eq. (1.3)}$$

where, A_{st} = Minimum area of tension steel
  b = Breadth of beam or the breadth of the web of T-beam (mm)
  d = Effective depth (mm)
  f_y = Characteristic strength of reinforcement (N/mm²).

The maximum reinforcement area in tension zone shall not exceed $0.04b \cdot D$.

The maximum area of compression reinforcement shall not exceed $0.04b \cdot D$. Compression reinforcement in beams shall be enclosed by stirrups for lateral restraint.

Where the depth of the web in a beam exceeds 750 mm, side face reinforcement shall be provided along the two faces. Total area of such side face reinforcement shall be not less than 0.10% of the web area and shall be distributed equally on two faces at a spacing not exceeding 300 mm or web thickness whichever is less.

The transverse reinforcement in beams shall be taken around the outermost tension and compression bars. In T-beams lateral ties shall enclose longitudinal bars close to outer face of the flange. Maximum spacing of shear reinforcement stirrups shall not exceed 0.75d for vertical stirrups and d for inclined stirrups at 45°, where 'd' is the effective depth of the section. In no case the spacing shall exceed 300 mm.

Minimum shear reinforcement in the form of stirrups shall be provided such that

$$\frac{A_{sv}}{b \cdot S_v} \geq \frac{0.40}{0.87\,f_y} \qquad \text{... Eq. (1.4)}$$

where, A_{sv} = Total area of stirrup legs effective in shear
  S_v = Stirrup spacing along the length of the member
  b = Breadth of beam or web of flanged beam
  f_y = Characteristic strength of stirrup reinforcement
    (N/mm²) which shall not be taken greater than 415 N/mm².

Torsion reinforcement shall be provided as below:

Rectangular closed stirrups placed perpendicular to the axis of the member. The spacing of the stirrups shall not exceed the least of: x_1, $\dfrac{x_1 + y_1}{4}$, 300 mm, where x_1 and y_1 are respectively the short and long lengths of the legs of the stirrups.

Longitudinal reinforcement shall be placed as close as practicable to the corners of the cross-section and in all cases, there shall be at least one longitudinal bar in each corner of the ties. When cross-sectional dimension of the member exceeds 450 mm, additional longitudinal bars shall be provided to satisfy the minimum reinforcement and spacing requirements.

Minimum reinforcement in either direction in slabs shall not be less than 0.15% (in case of plain bars) and 0.12% (HYSD bars) of the total cross-sectional area. The diameter of reinforcing bars shall not exceed one-eighth of the total thickness of the slab.

For columns, the longitudinal reinforcement shall be not less than 0.80%, nor more than 6% of the gross-sectional area of the column. To avoid congestion of bars at the lap joint, the longitudinal reinforcement may be restricted to 4% only.

Whenever size of the column provided is more than that required as per design, the minimum percent of main steel shall be based on the area of concrete actually required to resist the direct stress (compressive) and not on the actually provided. Minimum number of longitudinal bars shall be four in rectangular and six in circular columns.

Minimum diameter of longitudinal bars (main) be not less than 12 mm (beams and columns).

Circular columns with helical reinforcement shall have minimum 6 longitudinal bars in contact with the helix and spaced equally along the inner circumference. Maximum spacing of the longitudinal bars along the periphery shall not exceed 300 mm.

In case of pedestal where longitudinal reinforcement is not taken into account in strength calculations, a minimum of 0.15% of the cross-sectional area shall be provided.

Generally, the longitudinal bars will be tied laterally to safeguard against *buckling*. If the longitudinal bars are not spaced more than 75 mm on either side, transverse reinforcement need to go round corner and alternate bars for the purpose of lateral support. Diameter of transverse reinforcement (ties) shall not exceed 20 mm.

Pitch of the lateral ties in compression members shall be not more than the least of the following:

  (i) Least lateral dimension of the compression members

 (ii) 16 times the diameter of the smallest main bar

(iii) 300 mm.

Diameter of lateral ties: The diameter of polygonal or lateral ties shall be not less than 1/4th the diameter of the largest longitudinal bars.

Helical reinforcement shall be uniform and helix spaced evenly and shall have 1½ extra turns. Generally, pitch of the helical turns shall be not more than 75 mm nor more than $\frac{1}{6}$ th of the core diameter nor less than 25 mm, nor less than 3 times the diameter of the helix bar.

Expansion joints are provided in the structure when the member is longer than 45 m in plan. Expansion joints are provided in every 25–30 m continuous length. The structures adjacent to the joint should preferably be supported on separate columns or walls but not necessarily separate foundations.

1.5.3 Stresses and Design

Various design stresses and permissible stresses are specified in different Tables 1.12–1.19.

Shear reinforcement is provided as per design shear strength of concrete (τ_c) in limit state design method (Table 1.12). In working stress method, the shear reinforcement is determined by considering permissible shear stress (Table 1.16).

Table 1.12: Limit state design shear strength of concrete (τ_c)

% Steel $\dfrac{100\,A_s}{bd}$	Grade of concrete					
	M15	M20	M25	M30	M35	M40 and above
≤ 0.15	0.28	0.28	0.29	0.29	0.29	0.30
0.25	0.35	0.36	0.36	0.37	0.37	0.38
0.50	0.46	0.48	0.49	0.50	0.50	0.51
0.75	0.54	0.56	0.57	0.59	0.59	0.60
1.00	0.60	0.62	0.64	0.66	0.67	0.68
1.25	0.64	0.67	0.70	0.71	0.73	0.74
1.50	0.68	0.72	0.74	0.76	0.78	0.79
1.75	0.71	0.75	0.78	0.80	0.82	0.84
2.00	0.71	0.79	0.82	0.84	0.86	0.88
2.25	0.71	0.81	0.85	0.88	0.90	0.92
2.50	0.71	0.82	0.88	0.91	0.93	0.95
2.75	0.71	0.82	0.92	0.94	0.96	0.98
3.00 and above	0.71	0.82	0.92	0.96	0.99	1.01

Table 1.13: Maximum limit state design shear stress, $\tau_{c\,max}$ (N/mm²)

Concrete grade	M15	M20	M25	M30	M35	M40 and above
$\tau_{c\,max}$ (N/mm²)	2.5	2.8	3.1	3.5	3.7	4.0

Shear reinforcement shall be provided to carry a shear force equal to $(V_u - \tau_c \cdot b \cdot d)$. V_u is the shear force and τ_c is the design shear stress for the given grade of concrete and % age of steel reinforcement. The strength of shear reinforcement V_{us} shall be calculated as below:

(a) For vertical stirrups: $V_{us} = \dfrac{0.87\,A_{sv} \cdot f_y \cdot d}{S_v}$ ... Eq. (1.5)

(b) For inclined stirrups: $V_{us} = \dfrac{0.87\,f_y \cdot A_{sv} \cdot d}{S_v}(\sin \alpha + \cos \alpha)$ ... Eq. (1.6)

(c) For single bar or single group of bars: $V_{us} = 0.87 f_y \cdot A_{sv} \cdot \sin \alpha$ ... Eq. (1.7) where, S_v = Spacing of the stirrups or bent up bars along the length

A_{sv} = Total cross-sectional area of stirrup legs or bent up bars within a distance S_v

τ_v = Nominal shear stress

τ_c = Design shear strength of the given concrete grade

b = Breadth of the member which for flanged beams shall be taken as the breadth of the web (b_w)

d = Effective depth of the member

f_y = Characteristic strength of the stirrups or bent up bars which shall not be taken greater than 415 N/mm²

α = Angle between the inclined stirrups or bent up bars and the axis of the member, not less than 45°.

For members in direct tension, the design direct tensile stress shall be taken as per Table 1.14.

Tensile stress (direct) shall be calculated as $\dfrac{\text{Force}}{(A_c + m \cdot A_{st})}$ ... Eq. (1.8)

where, A_c is the area of concrete and A_{st} is the area of steel in tension.

Table 1.14: Direct limit state design tensile stress for different concrete grades

Grade of concrete	M10	M15	M20	M25	M30	M35	M40	M45	M50
Tensile stress (Direct) N/mm²	1.20	2.00	2.80	3.20	3.60	4.00	4.40	4.80	5.20

In working stress method, permissible stresses are considered in design and these permissible stresses for different grades of concrete are given in Table 1.15.

Table 1.15: Permissible stresses in concrete (working stress method)

Grade of concrete	Permissible stress in compression (N/mm²)		Permissible bond stress average (N/mm²)	
	Bending (σ_{cbc})	Direct (σ_{cc})	Plain bars in tension	Plain bars in compression
M15	5.00	4.00	0.60	0.75
M20	7.00	5.00	0.80	1.00
M25	8.50	6.00	0.90	1.12
M30	10.00	8.00	1.00	1.25
M35	11.50	9.00	1.10	1.37
M40	13.00	10.00	1.20	1.50
M45	14.50	11.00	1.30	1.62
M50	16.00	12.00	1.40	1.75

Permissible shear stress depends on the existence of tensile reinforcement and hence the values of permissible shear stress for different grades of concrete is given with reference to percent of tensile steel (for working stress method) in Table 1.16.

$$\text{Percent tensile steel} = \frac{100\,A_{st}}{bd}$$

Table 1.16: Working stress permissible shear stress in different grades of concrete

Percent steel $\left(100\dfrac{A_{st}}{bd}\right)$	Permissible shear stress in concrete (τ_c) N/mm^2 Concrete grades					
	M15	M20	M25	M30	M35	M40 and above
≤ 0.15	0.18	0.18	0.19	0.20	0.20	0.20
0.25	0.22	0.22	0.23	0.23	0.23	0.23
0.50	0.29	0.30	0.31	0.31	0.31	0.32
0.75	0.34	0.35	0.36	0.37	0.37	0.38
1.00	0.37	0.39	0.40	0.41	0.42	0.42
1.25	0.40	0.42	0.44	0.45	0.45	0.46
1.50	0.42	0.45	0.46	0.48	0.49	0.49
1.75	0.44	0.47	0.49	0.50	0.52	0.52
2.00	0.44	0.49	0.51	0.53	0.54	0.55
2.25	0.44	0.51	0.53	0.55	0.56	0.57
2.50	0.44	0.51	0.55	0.57	0.58	0.60
2.75	0.44	0.51	0.56	0.58	0.60	0.62
3.00 and above	0.44	0.51	0.57	0.60	0.62	0.63

Maximum shear stress $(\tau_{c\,max})$ in any structural member in working stress method is given in Table 1.17.

Table 1.17: Maximum shear stress $\tau_{c\,max}$ (N/mm^2) in working stress

Concrete Grade	M15	M20	M25	M30	M35	M40 and above
$\tau_{c\,max}$ (N/mm^2)	1.60	1.80	1.90	2.20	2.30	2.50

Working stress design of shear reinforcement is done based on the nominal shear stress τ_v in beams. The nominal shear stress τ_v in beams shall not exceed $\tau_{c\,max}$, given in Table 1.17. When shear stress τ_v exceeds $\tau_{c\,max}$, the beam section is redesigned so that the nominal shear stress τ_v is less than $\tau_{c\,max}$. When the nominal shear stress τ_v in beam is less than τ_c given in Table 1.16, minimum nominal shear reinforcement is provided in the beam. When the shear stress exceeds τ_c, proper shear reinforcement is provided by:

(i) Vertical stirrups at a spacing of S_v, $V_s = \dfrac{\sigma_{sv} \cdot A_{sv} \cdot d}{S_v}$ ... Eq. (1.9)

(ii) Inclined stirrups at a spacing of S_v, $V_s = \dfrac{\sigma_{sv} \cdot A_{sv} \cdot d}{S_v}(\sin\alpha + \cos\alpha)$ ... Eq. (1.10)

(iii) Bent up bars with stirrups at the section, $V_s = \sigma_{sv} \cdot A_{sv} \sin\alpha$ ... Eq. (1.11)

where,

A_{sv} = Total cross-sectional area of stirrup legs or bent up bars within a distance (S_v)

S_v = Spacing of the stirrups or bent up bars along the length (span) of the member

τ_c = Shear strength of the concrete (design shear strength in limit state and permissible shear stress in working stress)

b = Breadth of the member which for flanged beams shall be taken as the breadth of the web (b_w)

d = Effective depth of the section

σ_{sv} = Permissible tensile stress in shear reinforcement which shall not be taken greater than 230 N/mm² while in limit state it shall be taken as 0.87 f_y N/mm²

α = Angle between the inclined stirrups or bent up bars and the axis of the member not less than 45°

Table 1.18: Permissible stresses in steel reinforcement (working stress method)

S. No.	Type of stress in steel reinforcements	Permissible stresses (N/mm²)		
		MS grade I (IS:432)	Medium tensile steel (IS:432)	HYSD Fe 415 (IS:1786)
1	Tensile stress (σ_{st}, σ_{sv}) (a) Diameters up to 20 mm (b) Diameters over 20 mm	 140 130	Half yield stress subject to a maximum 190	 230 230
2	Compressive stress in columns (direct) σ_{sc}	130	130	190
3	Compressive stress in bars in bending elements (beams, slabs) when the compressive resistance of the concrete is taken into account	The calculated compressive stress in the surrounding concrete multiplied by 1.5 times the modular ratio (σ_{cbc} × 1.5 m) or σ_{sc} whichever is lower		
4	Compressive stress in bars in bending elements (beams, slabs) when the compressive resistance of the concrete is not taken into account (a) Up to 20 mm diameter (b) Over 20 mm diameter	 140 130	Half the guaranteed yield stress subject to a maximum of 190	 190 190

Notes: (i) For HYS deformed bars of grade Fe 500, the permissible stress in direct tension and flexural tension shall be 0.55 f_y. The permissible stress for shear and compression reinforcement shall be same as for grade Fe 415.

(ii) For the purpose of standard IS:456, the yield stress of steels for which there is no clearly defined yield point, should be taken to be 0.20% proof stress.

(iii) When MS conforming to Grade II of IS:432 (Part I) is used, the permissible stresses shall be 90% of the permissible stresses in column (3) or reinforcement calculated on the basis of MS Grade I, may be increased by 10%.

(iv) For welded mesh fabric conforming to IS:1566, the permissible value in tension σ_{st} is 230 N/mm^2.

1.6 DESIGN LOADS

Design of any structural member requires the knowledge of the various types and amount of loads. These loads are:

(a) Dead loads of the members/materials (IS:875 – Part I)

(b) Imposed loads (live loads, static or moving loads) – (IS:875 – Part II)

(c) Wind loads and earthquake loads (IS:875 – Part III)

(a) Dead Loads

Dead loads are based on the materials of the member. The dead loads are calculated from unit weights of materials which are specified in IS code 875 – Part I for common building materials as shown in Table 1.19.

Table 1.19: Unit weights of construction materials (IS:875 – Part I – 1987)

S. No.	Construction material	Unit weight
1	Brick masonry	18.85–22.00 kN/m^3
2	Stone masonry	20.40–26.50 kN/m^3
3	Cement concrete (Plain)	22.00–23.50 kN/m^3
4	Reinforced cement concrete (RCC)	22.75–26.5 kN/m^3 (25.00)
5	Cement mortar	20.40 kN/m^3
6	Terrazzo (10 mm thick)	0.24 kN/m^2 (24 kN/m^3)
7	Mastic asphalt (10 mm thick)	0.215 kN/m^2 (21.5 kN/m^3)
8	Brick wall (100 mm thick)	1.91 kN/m^2 (19.1 kN/m^3)
9	Brick wall (200 mm thick)	3.84 kN/m^2 (19.2 kN/m^3)
10	Marble	26.70 kN/m^3
11	Laterite	20.40–23.55 kN/m^3
12	Granite	25.90–27.45 kN/m^3
13	Asbestos cement sheeting	0.118–0.130 kN/m^2
14	Common burnt clay bricks	15.70–18.85 kN/m^3
15	Hollow concrete blocks	1.41 kN/m^3
16	Solid concrete blocks	17.65 kN/m^3
17	Plain concrete (IS:456-2000)	24.0 kN/m^3
18	Reinforced cement concrete (IS:456-2000)	25.0 kN/m^3

(Contd.)

Table 1.19: Unit weights of construction materials (IS:875 – Part I – 1987) (Contd.)

S. No.	Construction material	Unit weight
19	Rubble masonry	20.8 kN/m^3
20	Concrete tile flooring (25 mm thick)	0.50 kN/m^2 (20 kN/m^3)
21	Dry soil	$13.85–18.05 \text{ kN/m}^3$
22	Moist soil	$15.70–19.60 \text{ kN/m}^3$
23	Fine dry sand (FA)	$15.10–15.70 \text{ kN/m}^3$
24	Dry stone aggregate (CA)	$15.70–18.35 \text{ kN/m}^3$
25	Teak wood	6.28 kN/m^3

(b) Imposed (Live Loads)

The imposed loads (also called live loads) are different for different types of floor occupancy in various building structures. These loads are specified in IS:875 – Part II (1987) and are given in Table 1.20.

Table 1.20: Live loads on floors with different occupancy

S. No	Type of structures and type of floor occupancy	Minimum live load (kN/m^2)
1	**Residential buildings**	
(a)	All rooms, kitchen, toilets, etc.	2.0
(b)	Staircases, corridors, passages, stores, balconies, etc.	3.0, or point load of 1.5 kN/m at the edge
(c)	Balcony, staircases liable to overcrowding	5.0
2	**Hostels, hotels, boarding houses, dormitories, etc.**	
(a)	Living rooms, bedrooms, bath and toilets, etc.	2.0
(b)	Kitchen, laundries, lounges, billiards room, indoor games room, corridors, passages, etc.	3.0
(c)	Store rooms	5.0
(d)	Dining rooms, cafeteria, restaurants, etc.	4.0
(e)	Office rooms	2.5
(f)	Balconies	Same as rooms to which these are connected but not less than 4.0 kN/m^2 or 1.5 kN/m at the edge as point line load
(g)	Boiler/plant rooms (to be calculated but not less than)	5.0 or concentrated load of 6.7 kN

(Contd.)

Table 1.20: Live loads on floors with different occupancy (contd.)

S. No	Type of structures and type of floor occupancy	Minimum live load (kN/m²)
(h)	Garages including parking areas for vehicles not exceeding 25 kN gross weight but not less than Garage floors for vehicle not exceeding 40 kN gross weight (heavy weight vehicles)	2.5 5.0–7.5
3	**Educational buildings/institutional buildings**	
(a)	Classrooms, dining rooms, restaurants, kitchens, laboratories, laundries, reading rooms, X-rays, etc.	3.0
(b)	Office lounges, staff rooms, OPD rooms, etc.	2.5
(c)	Dormitories, toilets, bedrooms, dressing rooms, etc.	2.0
(d)	Projection rooms, store rooms, boiler/plant rooms, etc.	5.0
(e)	Libraries and archives, stack areas (2. 2 m height), additional for each metre height beyond 2.2 m height	6.0 + 2.0 kN/m² for each 1 m of height beyond 2.2 m
(f)	Reading rooms without separate storage, fire escapes, boiler/plant rooms, corridors, lobbies, passages, staircases, balconies (same as rooms attached) with minimum	4.0
4	**Assembly buildings**	
(a)	Assembly areas with fixed seats, restaurants, corridors, passages, staircases, balconies, fire escapes, etc.	4.0 (1.50 kN/m point load at the outer edge of balcony)
(b)	Assembly areas without fixed seats, projection room, stage, corridors, passages, wheeled vehicles, trolleys, etc.	5.0
(c)	Boiler/plant rooms including weight of machinery	7.50

(Contd.)

Table 1.20: Live loads on floors with different occupancy (contd.)

S. No	Type of structures and type of floor occupancy	Minimum live load (kN/m²)
5	**Business and office buildings**	
(a)	Rooms for general use with separate storage	2.5
(b)	Rooms without separate storage, corridors, balconies, stair cases, fire escapes, passages, not less than	4.0 (1.5 kN/m point load on outer edge of balcony)
(c)	Banking halls, cafeterias, dining rooms, kitchens, etc.	3.0
(d)	Business computing machine rooms, etc.	3.5
(e)	Record rooms, storage, vaults and strong rooms, boiler/plant rooms, etc.	5.0
6	**Mercantile buildings**	
(a)	Retail shops, corridors, passages, staircases, lobbies, fire escapes, balconies, etc. not less than	4.0 (1.5 kN/m concentrated load on outer edge of balcony
(b)	Whole sale shops to be calculated, but not less than	6.0
(c)	Office rooms	2.5
(d)	Dining rooms, restaurants, cafeteria, kitchen, laundries, etc.	3.0
(e)	Boiler rooms/plant rooms, corridors, passages for vehicular traffic, stair-cases, trolleys, etc. not less than	5.0
7	**Industrial buildings**	
(a)	Work areas without machinery/equipment, etc.	2.5
(b)	Work area with machinery to be calculated but not less than –Light duty	5.0
	–Medium duty	7.0
	–Heavy duty	10.0
(c)	Cafeteria, dining rooms, kitchens, etc.	3.0
(d)	Corridors, passages, staircases, fire escapes	4.0
(e)	Corridors, passages, staircases subjected to machine loads, boiler rooms, etc. to be calculated, but not less than	5.0

(Contd.)

Table 1.20: Live loads on floors with different occupancy (contd.)

S. No	Type of structures and type of floor occupancy	Minimum live load (kN/m²)
8	**Storage buildings**	
(a)	Storage rooms (other than cold storage), warehouses, to be calculated based on the bulk density of materials stored but not less than	2.4 kN/m², per each metre of storage height with a minimum of 7.5 kN/m²
(b)	Cold storage, to be calculated but not less than	5.0 kN/m² per each metre of storage height with a minimum of 15.0 kN/m².
(c)	Corridors, passages, staircases, fireescapes, as per the floor serviced but not less than	4.5 kN/m²
(d)	Corridors, passages subject to loads greater than from crowds, such as wheeled vehicles, trolleys, etc.	5.0 kN/m²
(e)	Boiler rooms and plant rooms....	7.5 kN/m²
9	**Roofs**	
(a)	Flat, curved, or sloping up to 10° (access provide)	1.50 kN/m²
(b)	Access not provided except for maintenance	0.75 kN/m²
(c)	For slopes greater than 10°, for each 1° slope, load shall be (0.75 – 0.001 × slope in excess of 10° up to 20°), and (0.75 – 0.001 × slope in excess of 10° up to 20° – 0.002 × slope in excess of 20°)	

Notes:

For low income housing, the floor loads may be reduced

For unrestricted assembly of persons, the value of live loads may be increased to 4.0 kN/m²

In case of industrial building actual loading if more than 10 kN/m², the design may be based on the actual loading

In case of change of occupancy, the live loads must be calculated for the new occupancy and structural members be strengthened accordingly

The members must be designed for the worst condition of live loads placed in most critical positions

(c) Wind Loads and Earthquake Loads

The wind loads become more critical in case of tall structures. The wind loads acting on the structure depends on the location (geographical zone) of the structure. These loads are specified in IS:875-1987 (Part III) according to zone of the structure location and height above ground level. The wind load on the structure also depends on the shape of the structure and plan dimensions. Relevant wind pressure coefficients are adopted according to the shape of the structure in plan. These coefficients are also specified in IS:875-1987 (Part III). Wind loads on sloping roofs may vary according to the slope of the roof slab.

India is divided in different zones according to intensity of earthquake. Earthquake induces acceleration in the structure due to vibration. This causes horizontal and vertical forces on the structure. According to the location zone, the acceleration coefficients are specified in IS codes. From these coefficients, horizontal and vertical forces caused by earthquake are calculated by multiplying the mass with the respective acceleration coefficients (Fig. 1.1).

$$\text{Force} = \text{mass} \times \text{acceleration coefficient}$$

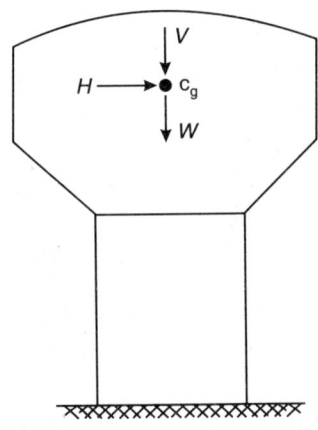

Fig. 1.1 Earthquake forces on the structure

| Horizontal force (H) acting at mass centre of structure | = mass × horizontal acceleration coefficient = $(m \times \alpha_H)$ |
| Vertical force (V) acting at the mass centre of structure | = mass × vertical acceleration coefficient = $(m \times \alpha_V)$ |

These loads shall be considered in design of tall structures, OHSR, multistoreyed buildings, etc.

1.7 ARRANGEMENT OF REINFORCEMENT

(a) Slabs

Minimum reinforcements, cover, and spacing of main and secondary (distribution) bars are shown in Fig. 1.2 as per IS:456-2000.

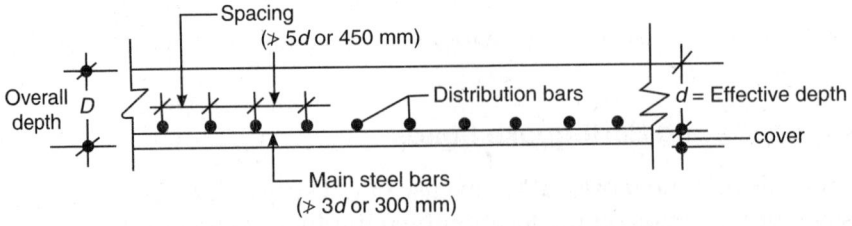

Fig. 1.2 Reinforcement details in RCC slab

Minimum reinforcement:

Plain ms bars ≮ 0.15% of gross area of cross-section

HYSD bars ≮ 12% of gross area of cross-section

Spacing: Main bars ≯ 3d or 300 mm, whichever is smaller

Distribution bars ≯ 5d or 450 mm, whichever is smaller

(b) Beams (Fig.1.3)

b = Breadth (mm)

d = Effective depth (mm)

A_{st} = Main tension reinforcement (mm²)

f_y = Characteristic yield strength of reinforcement (N/mm²)

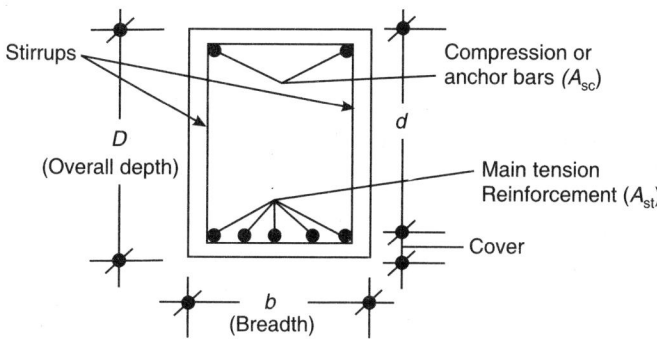

Fig. 1.3 Beam section (simply supported)

As specified earlier, the minimum tension reinforcement (A_{st}) as per IS:456-2000 shall be such that:

$$\frac{A_{st}}{bd} \geq \frac{0.85}{f_y}$$

where, A_{st} = Minimum area of tension steel (mild steel)

b = Breadth of beam or web width in flanged beam

d = Effective depth

f_y = Characteristic yield strength of tension reinforcement.

Minimum tension reinforcement $\dfrac{A_{st}}{b \cdot d}$ ≮ 0.34% for mild steel (f_y = 250 N/mm²),

≮ 0.20% for HYSD bars (f_y = 415 N/mm²)

Maximum reinforcement ≯ 0.04 $b \cdot d$ for both tension and compression to avoid congestion.

Spacing between bars ≮ diameter of the larger bar nor less than nominal size of coarse aggregate plus 5 mm, whichever is greater.

Nominal cover ≮ 25 mm nor less than the diameter of the bar.

When the overall depth of beam exceeds 750 mm, side face reinforcement of 0.10% of web area shall be provided and distributed equally on two faces at a spacing not exceeding 300 mm or web thickness whichever is less.

As specified earlier, the minimum shear reinforcement as stirrups should satisfy

$$\frac{A_{sv}}{b \cdot S_v} \geq \frac{0.40}{0.87 f_y}$$

where,

f_y = Characteristic yield strength (or proof strength) of stirrups (N/mm^2)

b = Breadth (mm)

S_v = Spacing of the stirrups (mm)

A_{sv} = Total area of stirrups legs (2 legs) – (mm^2)

(c) Column Reinforcement (Basic Details)

Generally column sections are square, rectangular or circular in shape. Main longi-tudinal reinforcing bars are provided along the circumference inside lateral ties. Lateral ties are provided to prevent buckling of main bars under compressive loads. The details of minimum and maximum quantity of reinforcement, minimum number of bars and their diameters, cover requirements, etc. are shown in Fig. 1.4. For further details IS:456-2000 may be referred.

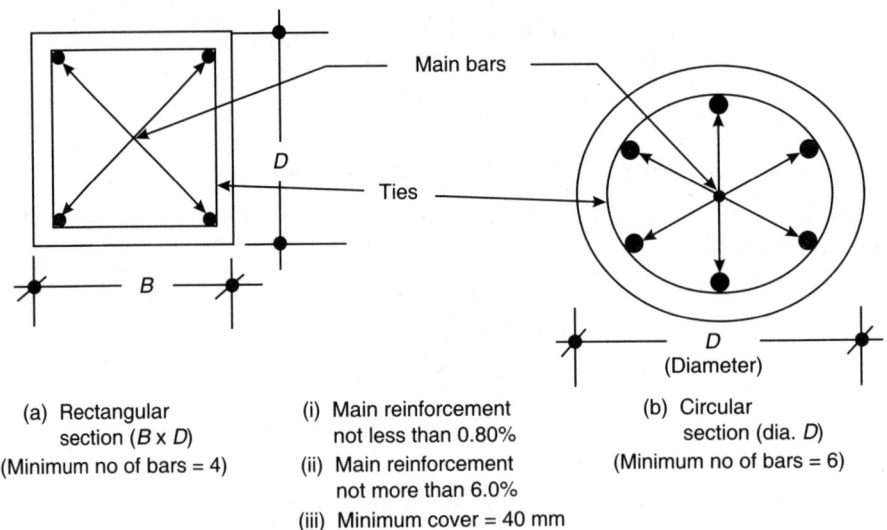

(a) Rectangular
section (B x D)
(Minimum no of bars = 4)

(i) Main reinforcement
not less than 0.80%
(ii) Main reinforcement
not more than 6.0%
(iii) Minimum cover = 40 mm

(b) Circular
section (dia. D)
(Minimum no of bars = 6)

Fig. 1.4 Column sections

Minimum main reinforcement (not less than) 0.80%

Maximum main reinforcement (not more than) 6.0%

Minimum number of main bars: Rectangular – 4

Circular – 6

Diameter of main bars ≮ 12 mm

Minimum cover = 40 mm or diameter of main bars, whichever is greater.

Lateral ties: Diameter ≮ $\frac{1}{4}$ th the diameter of largest main bars nor < 5 mm.

Pitch of lateral ties not more than the least of the following:

(a) Least lateral dimension

(b) 16 times the smallest diameter of the main bars

(c) 300 mm

SUMMARY

In modern time, RCC is the most popular construction material used for large number of structures such as buildings, dams, roads, bridges, retaining walls, water tanks, towers, tunnels, bunkers, swimming pools, industrial buildings, etc. The RCC is used as a construction material due to its excellent characteristics and many advantages.

Reinforced cement concrete is a composite material and monolithic action is based on the bond between concrete and steel. The strains in steel and concrete are compatible due to adequate bond.

Concrete is strong in compression and weak in tension and hence in flexural members steel reinforcements are provided in tension zone. Steel reinforcements are also provided in compression members to supplement the load carrying capacity of the structural members.

Design of RCC structures is based on properties and characteristics of its constituent materials. The most important constituents are cement concrete and steel reinforcement. For suitable design, proper understanding of properties and characteristics of cement concrete and steel reinforcement is essential.

Characteristic strength of concrete (f_{ck}) depends on the grade of concrete (M20, M30, …) which is obtained by suitable concrete mix design. The characteristic strength (f_{ck}) of concrete is defined as compressive crushing strength as measured on 150 mm cubes after 28 days of normal curing. Different properties of concrete depend on the characteristic strength (f_{ck}). Various properties of concrete required for design are specified for different grades of concrete.

Flexural tensile strength $f_{cr} = 0.70 \sqrt{f_{ck}}$ N/mm^2

Modulus of elasticity $E_c = 5000 \sqrt{f_{ck}}$ N/mm^2

Modular ratio $\left(m = \dfrac{E_s}{E_c} \right) = \dfrac{280}{3\sigma'_{cbc}}$, $\sigma'_{cbc} = \dfrac{f_{ck}}{3}$

Various stresses are specified in IS:456-2000 for different grades of concrete.

W/C ratio, workability, quality and quantity of cement and aggregate, concreting operations, and additions of chemical admixtures, etc. play critical role in achieving characteristic strength and durability of concrete. All the factors are considered in concrete mix design for a particular grade (M20, …, M50) and exposure to certain environment. The target strength (f_t) of cement concrete mix shall be calculated from the characteristic strength (f_{ck}) and quality control at site ($f_t = f_{ck} + ks$).

Concrete cover (distance of steel reinforcement from the exposed surface) plays important role for better durability and design of the structural member. Suitable concrete covers under various exposure conditions shall be adopted as specified in IS:456-2000 code of practice.

Stresses in different grades of reinforcement (Fe 250, Fe 415, Fe 500) are also specified in IS:456-2000 code of practice and must be adopted suitably in the design of structural elements. The steel reinforcement must be extended by appropriate development length (L_d) beyond the points of cut off where it may not be required.

Certain minimum required reinforcement shall be provided in the design as per the type of steel and type of structural member as specified in IS:456-2000 code of practice.

Different types of stresses considered in design are specified in IS:code 456-2000 for different grades of concrete and steel. These stresses are different for consideration in limit state method of design and working stress method of design. Partial safety factors for specific materials (concrete and steel) are considered in design.

Steel reinforcement may comprise plain mild steel bars (Fe 250) or high yield strength deformed bars (hot or cold worked Fe 415 and Fe 500 HYSD). Appropriate design stresses in steel are considered according to the type and grade of steel. Before studying design of structural members, we must understand the properties and behaviour of materials (concrete and steel reinforcement). For properties and behaviour, relevant IS codes must be referred. For limit state design of RCC structural members, IS:456-2000 must be referred at every stage.

PRACTICE QUESTIONS

Q. 1.1 Select the correct response given after each statement to complete it correctly and fill in the response sheet provided.

(i) The bond between steel reinforcement and cement concrete results in .... of strains in steel and concrete.

(a) compatibility
(b) inequality
(c) superiority
(d) complexity

(ii) Reinforced cement concrete has become most common construction material because of development of .... methods.

(a) easy construction
(b) rational design
(c) cheap construction
(d) mechanised

(iii) M30 grade of concrete means ....

(a) permissible concrete stress of 30 N/mm²
(b) 30 days concrete stress
(c) characteristic strength of 30 N/mm²
(d) 28 days bond strength of 30 N/mm²

(iv) Flexural tensile strength (f_{cr}) of cement concrete shall be normally .... N/mm² (f_{ck} is the characteristic strength in N/mm²).

(a) $f_{cr} = 5000 \sqrt{f_{ck}}$
(b) $f_{cr} = \dfrac{\sqrt{f_{ck}}}{5.0}$

(c) $f_{cr} = 0.70 f_{ck}$
(d) $f_{cr} = 0.70 \sqrt{f_{ck}}$

(v) The modulus of elasticity (E_c) of cement concrete shall be normally .... (f_{ck} is the characteristic strength in N/mm^2)

(a) $E_c = 50 \sqrt{f_{ck}}$ (b) $E_c = 5000 \sqrt{f_{ck}}$

(c) $E_c = 0.70 \sqrt{f_{ck}}$ (d) $E_c = 700 f_{ck}$

(vi) In absence of detailed investigations, the total shrinkage strain of cement concrete may be assumed as ....

(a) 3×10^{-4} (b) 3×10^4

(c) 0.0002   (d) 0.03

(vii) In absence of detailed investigations, the creep coefficient of cement concrete at 28 days age may be assumed as ....

(a) 3.3   (b) 2.2

(c) 1.6   (d) 1.1

(viii) For adequate compaction and site placing conditions, the .... plays an important role.

(a) characteristic strength   (b) W/C ratio

(c) segregation   (d) workability

(ix) Life of RCC elements gets affected critically by provision of ....

(a) adequate concrete cover to steel   (b) diameter of steel bars

(c) workability of cement concrete   (d) size of coarse aggregate

(x) Extreme environmental conditions are considered when the concrete surface is exposed to ....

(a) rain water   (b) continuously under water

(c) aggressive chemicals   (d) alternate wetting and drying

(xi) Nominal concrete cover for durability consideration in severe exposure conditions will be ....

(a) 20 mm   (b) 30 mm

(c) 45 mm   (d) 75 mm

(xii) Generally the maximum W/C ratio for extreme exposure of M40 grade of concrete shall be ....

(a) 0.55   (b) 0.50

(c) 0.45   (d) 0.40

(xiii) Permissible tensile strength of HYSD bars for design purposes may be taken as .... N/mm^2.

(a) 130   (b) 190

(c) 230   (d) 415

(xiv) Design bond strength of HYSD bars shall be.........................

(a) 60% higher than plain MS bars

(b) 25% higher than plain MS bars

(c) 25% lower than plain MS bars

(d) 10% higher than plain MS bars

    (xv) Maximum spacing of main reinforcement in slabs (effective depth = d) shall be ....
- (a) 450 mm or $5d$, whichever is smaller
- (b) 300 mm or $3d$, whichever is smaller
- (c) equal to overall thickness D
- (d) 16 times the diameter of bars

Response sheet to Q. 1.1 (i to xv)

Question	(i)	(ii)	(iii)	(iv)	(v)	(vi)	(vii)	(viii)	(ix)	(x)	(xi)	(xii)	(xiii)	(xiv)	(xv)
Response (a/b/c/d)															

Chapter 2

Design Philosophies

LEARNING OBJECTIVES

After the study of *design philosophies*, the learner knows the basic design principles and will be able to:

- Describe meaning of design
- List different approaches to design
- Differentiate between various design approaches
- State basic features of: (a) working stress method (b) ultimate load method and (c) limit state design method
- List different limits of collapse of structural elements
- State meaning of serviceability of concrete elements
- Define characteristic strength of concrete in limit state
- State meaning of design load in limit state method
- Explain idealized stress–strain curves for materials.

2.1 INTRODUCTION

The meaning of structural design of cement concrete element is to decide the dimensions of the concrete element and also decide about the quantity and location of steel reinforcement in the member so that the member remains safe against the intended/likely type and quantity of loading caused by type of occupancy. The member must also be safe and durable under the given environmental exposure.

Based on the type and quantity of given loading pattern, the dimensions are determined using the desired approach for the desired shape of the cross-section and exposure conditions. The exposure conditions help in deciding the concrete mix (proportions, type and quantity of cement, use of appropriate admixture, W/C ratio, etc.) and concrete cover to steel.

The objective of design is to provide adequate safety, adequate serviceability (deflection – stiffness and durability) and reasonable economy.

Reinforced cement concrete members can be designed by using analytical approaches based on certain theories or based on experimental investigations on models of full size elements. Most common theoretical approaches for design of RCC are based on:

(i) The working stress method (also known as linear elastic method)

(ii) The ultimate load method (also known as load factor method)

(iii) The limit state method

2.2 WORKING STRESS METHOD (LINEAR ELASTIC THEORY)

The *working stress method* of design of structural concrete elements is based on *linear elastic theory*. It was first developed in Europe during beginning of 20th century and was included in many national codes of practices as theoretical method of design of reinforced concrete (RC) sections. In this theoretical method, the materials are assumed to behave in linear elastic manner in the working stress range. The desired safety is ensured by restricting the stresses in concrete and steel to permissible limits obtained by applying appropriate factor to the characteristic strength (f_{ck} or f_y) of the materials.

$$\text{Permissible (or working) stress in concrete } (\sigma_{cbc}) = \frac{f_{ck}\,(\text{characteristic strength})}{\text{Fs}\,(\text{factor of safety})}$$

$$\text{Permissible stress in steel } (\sigma_{st}) = \frac{f_y\,(\text{characteristic stress or proof stress})}{\text{Fs}\,(\text{factor of safety})}$$

The resulting permissible (or working) stresses under service loads (actual loads carried by the structural elements) will be well within the linear elastic range of the materials.

In the elastic theory for the flexural (bending) members, the important assumptions in simple theory of bending are also adopted in the design of flexural RCC elements. These assumptions are reproduced briefly:

(i) At any point, plane sections before bending will remain plane after bending. This means the elastic strains in concrete (e_c) and steel (e_s) shall be equal at the same point. Also, the stresses will be directly proportional to their elastic moduli, i.e.

$$\text{Strain } (e) = \frac{\text{Stress}}{\text{Modulus of elasticity}}$$

$$e = e_c = e_s \text{ or } \frac{\sigma_{cbc}}{E_c} = \frac{\sigma_s}{E_s} \text{ or } \sigma_s = \frac{E_s \times \sigma_{cbc}}{E_c} = m\sigma_{cbc} \qquad \text{... Eq. (2.1)}$$

Also this means that the strains at certain points above and below NA will be proportional to their respective distances from NA.

(ii) All tensile stresses are resisted by reinforcement and none by concrete, except as otherwise specified.

(iii) The stress–strain relationship of steel and concrete under working loads is a straight line.

(iv) The modular ratio (m) according to IS:456-2000 has the value:

$$m = \frac{E_s}{E_c} = \frac{280}{3\sigma_{cbc}} \qquad \text{... Eq. (2.2)}$$

where, σ_{cbc} is the permissible compressive bending stress in concrete due to bending expressed in N/mm^2 (generally $\sigma_{cbc} = \dfrac{f_{ck}}{3}$ for concrete).

The assumption of linear elastic behavior is considered justifiable since the specified permissible stresses are kept well below the ultimate strength of the material. The ratio of yield (or proof) stress of steel reinforcement or the concrete cube crushing strength to corresponding permissible (or working) stress is usually called the factor of safety. The working stress method of design assumes a factor of safety of about 3 for concrete with respect to crushing strength and a factor of safety of about 1.80 for steel reinforcement with respect to the yield (or proof) strength.

$$\text{Permissible concrete stress } (\sigma_{cbc}) = \frac{\text{Concrete cube strength } (f_{ck} = \sigma_{cu})}{\text{Factor of safety (3)}} \quad \text{... Eq. (2.3)}$$

$$\text{Permissible steel stress } (\sigma_{st}) = \frac{\text{Yield (or proof) stress } (f_y)}{\text{Factor of safety (1.80)}} \quad \text{... Eq. (2.3a)}$$

where, σ_{cbc} = Permissible stress in concrete (N/mm^2)

            σ_{cu} = Crushing stress in concrete at failure (N/mm^2)

            σ_{st} = Permissible stress in steel (N/mm^2)

            f_y = Yield (or proof) stress in steel (N/mm^2)

Reinforced cement concrete is a composite material and its analysis is based on the bond between steel bars and cement concrete. The bond results in strain (deformation) compatibility between steel and concrete. Because of equal strains in steel and concrete at the same point, the stresses in steel and concrete are linearly related, i.e.

$$f_s = \frac{E_s}{E_c} \cdot f_c = mf_c \quad \text{... Eq. (2.4)}$$

where, f_s = Stress in steel

            f_c = Stress in concrete

            m = Modular ratio of steel and concrete $\left(m = \dfrac{E_s}{E_c} \right)$.

Merits of Working Stress Method

    (i) It is very simple in concept and design application.

    (ii) Generally the structures designed by working stress are relatively large in size and hence provide higher safety factor and serviceability (lesser deflections and crack width, etc.) under service loads.

    (iii) Working stress method is only applicable in case of service load condition for analysis of serviceability (i.e. deflection and crack width, etc.), so it is essential to understand working stress method even in other methods of design (such as ultimate load and limit state methods).

Demerits of Working Stress Method

Although, the structures designed by working stress method have performed satisfactorily, yet it has following demerits:

    (i) The working stress method neither indicates the real strength nor provides the actual factor of safety of the structure at failure.

(ii) The modular ratio in working stress design results in larger percentage of compression steel than that given by limit state design, thus making the design uneconomical.

(iii) Because of actual nonlinear stress–strain relationship and creep, the concrete does not have definite modulus of elasticity (E_c).

(iv) The working stress method does not differentiate between different types of loads that may act simultaneously but may have different uncertainties.

The basics of working stress method are explained in Chapter 3.

2.3 ULTIMATE LOAD METHOD

The ultimate load method was developed in 1950 after World War II. This was done to ward off demerits of working stress method and to improve economics of the structure. The method is based on the ultimate strength of reinforced concrete at ultimate load. The ultimate load is obtained by enhancing the service load by some factor (called as load factor) for giving a desired margin of safety. The ultimate load method is also called load factor method. The ultimate load method was introduced as an alternative to working stress method in American and British codes in 1957.

In the ultimate load method, stress condition at the time of collapse of the structure is analysed, and hence nonlinear stress–strain curves of concrete and steel are considered. The safety measure in the design is obtained by the use of proper load factor. In this approach, different load factors can be considered for different types of loads. It may be noted that the safety in strength performance at ultimate collapse load may not guarantee satisfactory serviceability performance at service loads.

Ultimate load factor method utilizes a large reserve of strength in plastic stage (inelastic region) and of ultimate strength of the RC element, the resulting member is very slender and thin. Because of this, excessive deformations and wide cracks occur. Also this method does not take into account the effect of shrinkage and creep. Let us now understand the basic merits and demerits of this approach.

Merits of Ultimate Load Method

(i) The ultimate load method uses actual stress–strain curve in totality representing more realistic safety against collapse load. While working stress method utilises only the straight (linear) portion of the stress–strain curve. In ultimate load method the stress block parameters are defined by the actual stress–strain curve.

(ii) The load factor method gives the more realistic margin of safety against collapse.

(iii) The method allows to use different load factors for different types of loads and the combination thereof.

(iv) The failure load calculated by ultimate load factor matches better with the experimental collapse load.

(v) The method is based on the ultimate strain as the failure criteria.

(vi) The method utilises the reserve of strength in plastic range for analysis of ultimate strength.

Demerits of Ultimate Load Method

(i) This method does not consider the serviceability criteria (deflections and cracking) at ultimate load condition.

(ii) The use of high strength reinforcement (HYSD) and high strength concrete result in thin sections and hence results in higher deflections and wide cracks.

(iii) This method does not take into consideration the effects of creep and shrinkage.

(iv) In the ultimate load method, the distribution of stress at the ultimate load is taken as the distribution at service loads increased by the load factors. This assumption is erroneous because the redistribution of stress occurs with the increase of loading from service loads to ultimate loads.

To conclude, it may be said that the ultimate load method ensures safety at ultimate loads, but it disregards the serviceability criteria at service loads. Therefore, both working stress method and ultimate load method of designs suffer with some weaknesses and hence recently the *limit state method* has been introduced to eliminate the weaknesses of both the previous methods.

2.4 LIMIT STATE METHOD

The working stress method provides structural designs performing satisfactorily at working loads, but it is unrealistic at the ultimate state of collapse. Similarly, the ultimate load method provides realistic safety factor at the collapse load but does not guarantee satisfactory serviceability requirements at service (or working) loads. Therefore, the most efficient method shall include the consideration of ultimate strength of the structure as well as the serviceability and durability requirements at the service loads. The *limit state method* has evolved to include both the aspects of design.

In the limit state method, a structure component is designed for safety against collapse load and checked for serviceability at the working loads. This makes the structure (or component) safe for its intended use. Thus, this method takes care of safety against collapse load and serviceability at working loads.

International code of practice for reinforced concrete was published in 1963, which has introduced the philosophy of limit state method. British code introduced limit state method in 1973 and while the Indian code of practice IS:456-1978 introduced this method.

A limit state is a state of *impending failure* beyond which a structure ceases to perform its intended function satisfactorily, both in terms of safety against collapse load and serviceability at working loads. The objective of this method is to achieve acceptable possibilities so that the structure does not reach a limit state. As per IS:456-2000, various limit states will be considered in design to ensure *safety and serviceability*. The section shall be designed for the most critical limit state and shall be checked for other limit states. Design shall be based on characteristic strengths and other values. The characteristic strengths are based on statistical data available or existing experience. The design values are derived from the characteristic values using partial safety factors for material strengths and loads.

Different limit states (beyond which the structure fails to perform) are as under:

(I) Limit states of collapse:
  (a) Limit state of collapse in flexure
  (b) Limit state of collapse in compression
  (c) Limit state of collapse in compression under uniaxial bending
  (d) Limit state of collapse in compression under biaxial bending
  (e) Limit state of collapse in shear
  (f) Limit state of collapse in bond
  (g) Limit state of collapse in torsion
  (h) Limit state of collapse in tension

(II) Limit states of serviceability:
  (a) Limit state of deflection
  (b) Limit state of cracking
  (c) Other limit states, such as durability, vibration, fire resistance, etc.

Limit State of Collapse

The limit state of collapse of the structure or any component of the structure shall be considered when rupture or buckling of one or more critical sections occur. The resistance to bending, shear, torsion and axial loads or combinations thereof at every section shall not be less than the appropriate value at that section produced by the probable most unfavorable combination of loads using the partial safety factors.

Different limit states shall be explained in detail in subsequent chapters.

Limit State of Serviceability

The limit state of serviceability of any structure or element relate to its performance or behavior at working loads.

Normally, the design of structural members or the structure is based on the analysis of limit states of collapse against ultimate loads and on serviceability limit states of cracking and deflection under service loads.

Durability is ensured by providing concrete with appropriate nominal cover, concrete grade, type of cement, cement content, water–cement ratio and use of suitable admixture depending on the exposure conditions.

2.5 CHARACTERISTIC STRENGTH AND DESIGN STRENGTH

2.5.1 Characteristic Strength

The characteristic strength of material (such as concrete) according to the Indian standard code of practice (IS:456-2000) means that value of the strength below which not more than 5% of strength results fall (i.e. there is probability that the actual strength may be less than the characteristic strength in 5% cases. According to this probability, the statistical constant value (k) may be taken as 1.65 approximately.

Characteristic strength (f_{ck}) = mean strength $(f_m) - k \times$ standard deviation, i.e.

$$f_{ck} = f_m - k \cdot s \qquad \text{... Eq. (2.5)}$$

where,

f_{ck} = Characteristic strength of material

f_m = Mean strength of material

k = Statistical constant (approximately equal to 1.65)

s = Standard deviation for the set of test results according to quality control.

The value of standard deviation $s = \sqrt{\dfrac{\sum(s_i - s_m)^2}{n-1}}$ \qquad ... Eq. (2.6)

where,

s_i = Strength of ith sample

s_m = Mean strength of n samples

n = Number of samples tested (as per IS:456-2000, $n \geq 30$).

Characteristic strength of concrete (f_{ck}) is taken as its compressive (crushing) strength as measured by 150 mm size cubes after 28 days moist curing and expressed in N/mm^2 (MPa). The characteristic strength is also designated by its grade of concrete as M10, M20, M50, M80, etc. These concrete grades are grouped into 3 categories as ordinary (M10–M20), standard (M25–M50) and high strength (M55–M80).

2.5.2 Design Strength

Design strength (f_d) of concrete is given as

$$f_d = \frac{f_{ck}}{\gamma_m} \qquad \text{... Eq. (2.7)}$$

where,

f_{ck} = Characteristic strength of concrete (material)

f_d = Design strength in limit state or load factor methods

γ_m = Partial safety factor for the material.

The partial safety factor for the cement concrete is generally taken as 1.50, according to IS:456-2000 in the limit state method.

2.5.3 Design Load

For the purpose of design, the characteristic load is also enhanced by using partial safety factor appropriate to the nature of loading, i.e.

Design load \qquad\qquad $F_d = F_k \cdot \gamma_f$ \qquad ... Eq. (2.8)

where,

F_d = Design load

F_k = Characteristic load

γ_f = Partial safety factor based on the nature of loading (as specified in IS:456-2000).

2.6 IDEALISED STRESS–STRAIN CURVES FOR MATERIALS

2.6.1 Cement Concrete

The stress–stain curve for cement concrete is prepared with respect to the compressive stress in the cube up to crushing stage. The stress–strain curve for tension is also assumed the same as for the compressive stress. IS:456-2000 specifies idealised stress–strain curves for concrete in cube and for concrete in actual structure. Design curve is also drawn after applying partial safety factor for concrete ($\gamma_m = 1.5$). Thus, for cement concrete there are three curves $\left(\text{one for } f_{ck}, \text{one for } 0.67f_{ck} \text{ for the structure and one}\right.$

idealised design curve for $\left. \dfrac{0.67 f_{ck}}{1.50} = 0.45 f_{ck}\right)$. These curves are shown in Fig. 2.1. The

curve generally remains parabolic upto a strain of 0.002 and then it becomes straight (i.e. there is no increase in stress beyond the strain of 0.002 but the strain increases without increase of stress).

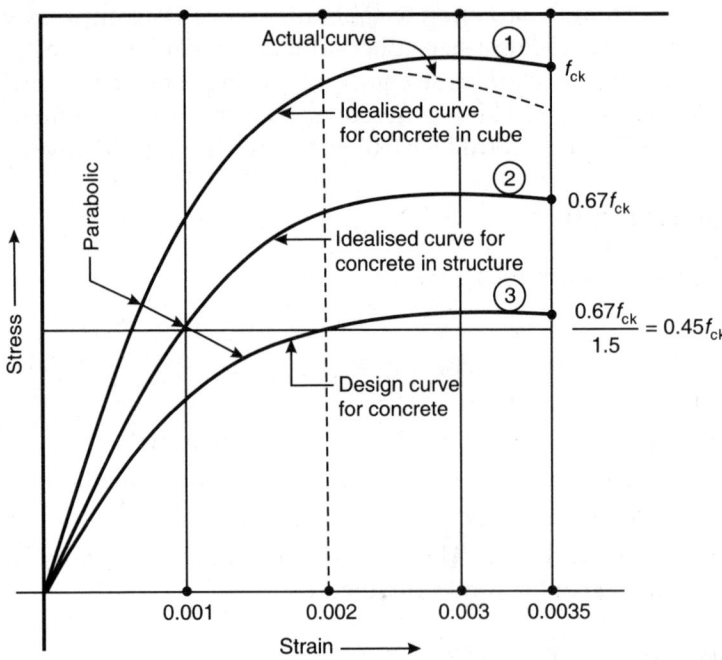

Fig. 2.1 Stress–strain curve for concrete

2.6.2 Mild Steel

The stress–strain curve for mild steel is shown in Fig. 2.2. The modulus of elasticity (E_s) for mild steel and other steels is generally taken as $E_s = \dfrac{\text{Stress}}{\text{Strain}} = 2 \times 10^5 \text{ N/mm}^2$.

The stress–strain relationships for steels in tension and compression are assumed to be the same. In mild steel, stress is proportional to strain up to the yield point and thereafter the strain increases without increase in stress (plastic stage). The design stress in the limit state of steel is equal to $\dfrac{f_y}{\gamma_m}$. The value of γ_m for steel is taken as 1.15, and the design stress in the limit state method will be taken as equal to $\dfrac{f_y}{1.15} = 0.87f_y$.

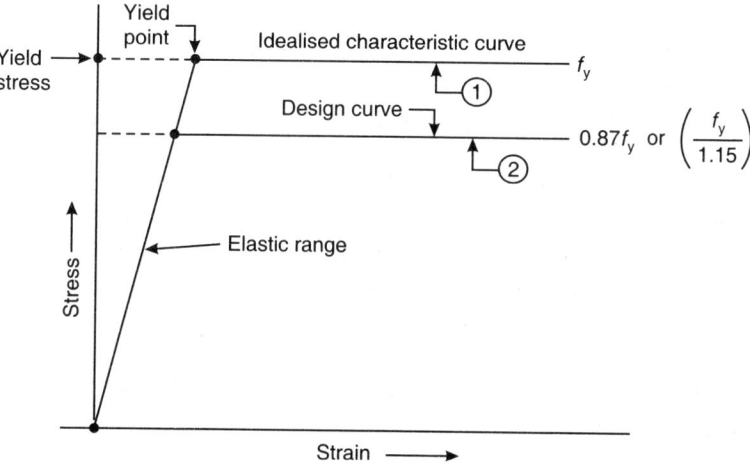

Fig. 2.2 Stress–strain curve for MS

For cold worked steel (Fe 415 and Fe 500 HYSD bars), there is no definite yield stress and the yield stress is considered as 0.2% proof stress. The stress–strain curves are shown in Fig. 2.3 for these type of steels (Fe 415 and Fe 500) respectively. These steels are cold worked and strained beyond the yield point by twisting or stretching. By cold working and twisting these steel bars get higher yield strength.

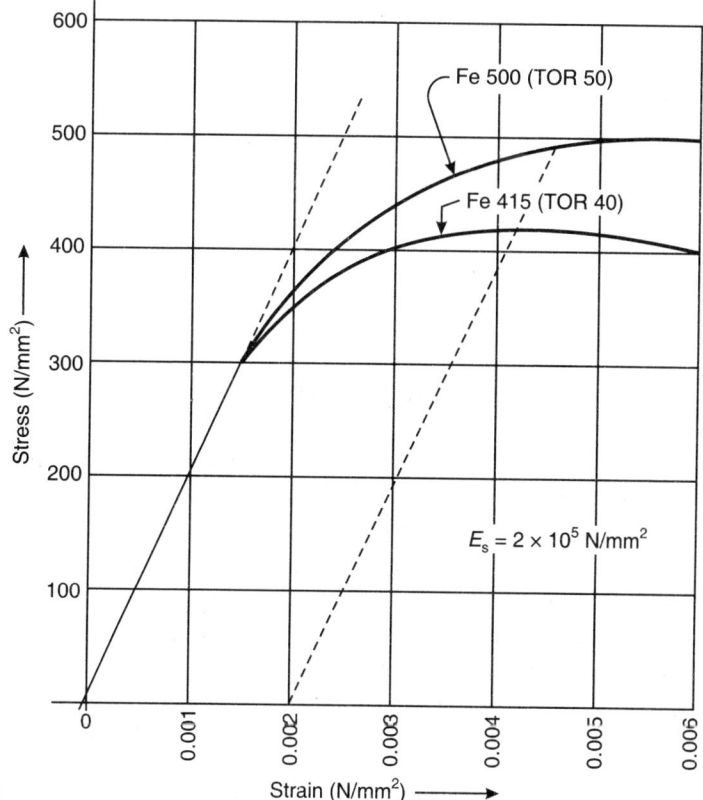

Fig. 2.3 Stress–strain curve for cold twisted deformed bars

Since these HYSD bars are twisted, these are also specified as Fe 415 (TOR 40) and Fe 500 (TOR 50). Deformed bars are provided with lugs or ribs for enhancing their bond strength as shown in Fig. 2.4.

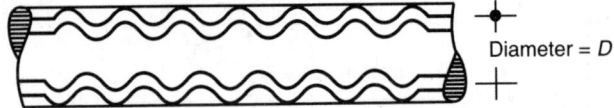

Diameter = D

Fig. 2.4 Deformed bars (TOR steel)

Permissible stresses (in working stress method) for HYSD bars conforming to IS:1786 (Fe 415) is generally taken as 230 N/mm² in tension and 190 N/mm² in compression in columns as well as in flexural compression. Flexural compression in steel is considered as $(1.5\ m)$ times stress in surrounding concrete and this stress should be less than or equal to 190 N/mm².

Characteristic strengths in limit state method and permissible strength in working stress method for steel bars are given in Table 2.1.

Table 2.1: Characteristic and permissible strengths of different steels

Type of reinforcement (steel bars)	Yield stress or 0.2% proof stress (N/mm²)	Characteristic strength (f_y) (N/mm²)	Permissible tensile strength σ_{st} (N/mm²)
Mild steel bars conforming to grade I of IS:432 (Part I) or deformed MS bars of IS:1139	250 Average	250	140 (up to 20 mm ϕ) 130 (over 20 mm ϕ)
High yield strength deformed bars conforming to IS:1109 or Fe 415 grade of IS:1786-1979	415	415	230
High yield strength deformed bars of grade Fe 500 of IS:1786-1979	500	500	275

SUMMARY

Structural design provides dimensions of the member and quantity of steel reinforcement for safety, economy and serviceability of the structural element. The objective of design is to provide adequate safety and serviceability and reasonable economy. The exposure and placement conditions help in deciding mix proportions, admixture, W/C ratio, and concrete cover, etc.

There are 3 approaches to design, viz. (a) working stress method, (b) ultimate load method and (c) limit state method.

The working stress method is based on restricting the stresses in steel and concrete at service loads much lower than respective characteristic strengths of steel $\left(\dfrac{f_y}{1.8}\right)$ and concrete $\left(\dfrac{f_{ck}}{3}\right)$ respectively. Strains in steel (e_s) and concrete (e_c) at the same point in the beam section shall be equal ($e_s = e_c$) due to bond between steel and concrete. Stresses in steel and concrete are related as $\sigma_{st} = m\sigma_{cbc}$ due to strain compatibility ($e_s = e_c$). Working stress method is simple but does not provide real factor of safety at the collapse.

The ultimate load method is based on enhancing the service loads by multiplying with a constant (known as load factor) to obtain collapse load on the structure or element. Thus, the load factor represents the real margin of safety at the collapse. The collapse load is based on the ultimate strength and the load matches quite well with the experimental collapse load on the structural element. The ultimate load method does not provide realistic idea of serviceability of the member at working service loads.

The limit state method of design considers various limits of collapse as well as serviceability of the structure both at the collapse and service load conditions. The limit state method takes into consideration merits of both working stress method and ultimate load method. The limit state method is fully based on IS:456-2000 code. All provisions and stress limits specified in the code must be thoroughly understood and adopted fully in designs of all RCC elements/structures. Characteristic strengths of concrete (f_{ck}) and steel (f_y) are based on the condition that not more than 5% samples may fall below the mean values, i.e. $f_{ck} = f_m - ks_d$

The design strength (f_d) is considered with partial safety factor (γ_m) on characteristic strength (f_{ck} or f_y), i.e.

$$f_d = \frac{f_{ck}}{\gamma_m}$$

where,

       f_d = Design strength

       f_{ck} = Characteristic strength

       γ_m = Partial safety factor for material

            (for concrete γ_m = 1.5 and for steel γ_m = 1.15)

The design load (W_d or F_d) is obtained by enhancing the service load by multiplying with a constant (γ_f) known as load factor.

$$F_d \text{ (or } W_d) = F_k \cdot \gamma_f$$

where,

       F_k = Characteristic or service load (N)

       F_d = Design load (N)

       γ_f = Load factor (also called partial safety factor for loading)

Stress–strain curves are drawn for collapse load, characteristic strength (f_{ck}) and for design strength (f_d). Wherever in steel the yield stress is not well defined, 0.2% (0.002) strain is considered for proof stress. The yield strength of steel is enhanced by twisting steel bars beyond yield strength. The bond between steel and concrete gets enhanced by providing lugs or ribs on the bar surface in case of high yield strength deformed (HYSD) bars. For different grades of steel, values of characteristic strengths (for limit state) and permissible strengths (for working stress) are adopted from Table 2.1.

PRACTICE QUESTIONS

Q. 2.1 Select the correct response given after each statement to complete it correctly and fill in the response sheet provided.

(i) Structural design makes it possible to decide on the .... of the structural member to resist given loads safely.

(a) shape

(b) dimensions

(c) exposure conditions

(d) mix proportions of cement concrete

(ii) Structural design also results in calculation of .... matching with the dimensions of the member.

(a) quantity of reinforcement

(b) concrete cover to reinforcement

(c) quantity of cement

(d) quality of cement

(iii) The objective of structural design is to provide adequate ....

(a) safety

(b) economy

(c) serviceability

(d) economy, safety and serviceability

(iv) In present times, according to IS:456-2000, RCC members are designed by using ....

(a) theoretical approach of working stress

(b) experimental approach of working stress

(c) experimental approach of ultimate load

(d) theoretical approach of limit state

(v) The working stress method is based on ....

(a) linear elastic theory   (b) linear limit states

(c) parabolic elastic theory   (d) ultimate yield line theory

(vi) Reinforced cement concrete beam design is not based on the assumption of ....

(a) plane section before bending remains plane after bending

(b) all tensile stresses are resisted by steel reinforcement

(c) perfect bond between steel and concrete up to service loads

(d) service loads must be less than the self-weight of concrete

(vii) According to IS:456-2000, the modular ratio of steel to concrete in working stress method is taken as ....

(a) $500\sqrt{f_{ck}}$ (b) $0.70\sqrt{f_{ck}}$

(c) $\dfrac{280}{f_{ck}}$ (d) $\dfrac{500}{0.70\,f_{ck}}$

(viii) In working stress method, the permissible stress (in flexure) in compression for M30 grade cement concrete will be .... N/mm².

(a) 30   (b) 10

(c) 3.0   (d) 2.8

(ix) Merit of linear elastic theory for RCC is that it provides ....

(a) real margin of safety at failure stage

(b) linear stress–strain relationship up to failure stage

(c) larger sections and higher safety against collapse

(d) higher economy and logical design

(x) The advantage of working stress method of design of RC members is that it ....

(a) utilises the reserve strength available in plastic stage

(b) uses ultimate load at the failure stage

(c) is simple and uses strength up to proportional limit

(d) allows large deformations without increase in stress

(xi) The advantage of ultimate load method of design of RC member is that it ....

(a) utilises the reserve strength available in plastic stage

(b) does not permit yielding of material

(c) makes higher durability of structure

(d) is simple and assumes straight line relations in stress–strain

(xii) Limit state method of design considers the .... and ....

(a) ultimate load at the collapse, serviceability at working loads

(b) working loads at proportional limit, serviceability at the collapse

(c) ultimate loads at the collapse, serviceability at the collapse

(d) working loads at the proportional limit and serviceability at the proportional limit

(xiii) Limit states of collapse in beam elements refer to ....

(a) collapse in deflection

(b) collapse in crack width

(c) collapse due to durability

(d) collapse due to bending compression or tension

(xiv) For limit states consideration, the load on beams are placed at ....

(a) middle span point of SS beams

(b) most unfavorable locations of the beam

(c) support of simply supported beam

(d) fixed end of cantilever

(xv) Characteristic strength of concrete is determined as its crushing (compressive) strength ....

(a) after 7 days of moist curing and measured on 150 mm cubes

(b) after 28 days of moist curing and measured on 100 mm cubes

    (c) after 28 days of moist curing and measured on 150 mm cubes

    (d) after 90 days of moist curing and measured on 100 mm cylinder

(xvi) Characteristic strength (f_{ck}) of concrete is obtained from mean crushing strength by using equation with usual notations ....

    (a) $f_{ck} = f_m + k \cdot s$ (b) $f_{ck} = f_m$

    (c) $f_{ck} = f_m - k \cdot s$ (d) $f_{ck} = f_m - s$

(xvii) Standard deviation (s) of n samples of concrete is given by ....

$$\left[\begin{array}{l} \text{where,} \qquad s_i = \text{strength of } i\text{th sample} \\ \qquad\qquad s_m = \text{mean strength of } n \text{ samples} \end{array} \right]$$

    (a) $s = \sqrt{\dfrac{\sum (s_i - s_m)^2}{(n+1)}}$ (b) $s = \dfrac{\sum (s_i - s_m)^2}{(n+1)}$

    (c) $s = \dfrac{\sum (s_i - s_m)^2}{(n-1)}$ (d) $s = \sqrt{\dfrac{\sum (s_i - s_m)^2}{(n-1)}}$

(xviii) In limit state, the design strength f_d of material with usual notations is given by ....

$$\left[\begin{array}{l} \text{where,} \qquad f_d = \text{design strength} \\ \qquad\qquad \gamma_m = \text{partial safety factor} \\ \qquad\qquad n = \text{number of samples} \end{array} \right]$$

    (a) $f_d = \dfrac{f_{ck}}{\gamma_m}$ (b) $f_d = \dfrac{\gamma_m}{f_{ck}}$

    (c) $f_d = \gamma_m \cdot f_{ck}$ (d) $f_d = \dfrac{\gamma_m f_{ck}}{(n-1)}$

(xix) In limit state method, the design load (W_d) will be calculated by ....

$$\left[\begin{array}{l} \text{where,} \qquad W_d = \text{design load} \\ \qquad\qquad W_k = \text{characteristic load (service load)} \\ \qquad\qquad \gamma_f = \text{partial safety factor for the nature of loading} \end{array} \right]$$

    (a) $W_d = \gamma_f \cdot W_k$ (b) $W_d = \dfrac{W_k}{\gamma_f}$

    (c) $W_d = \dfrac{W_k}{1.5\,\gamma_f}$ (d) $W_d = \dfrac{W_k \cdot \gamma_f^2}{1.5}$

(xx) In an idealised stress–strain curve for mild steel, the design strength (f_d) of steel is considered as ....

    (a) $f_d = \dfrac{\text{ultimate collapse strength}}{1.80}$ (b) $f_d = \dfrac{\text{yield strength}}{1.15}$

    (c) $f_d = $ yield strength   (d) $f_d = $ ultimate strength

(xxi) In case of HYSD steel bars, equivalent yield strength is considered as proof stress at .... strain.

    (a) 2%   (b) 0.20%

    (c) 0.02   (d) 0.003

(xxii) In cold-worked high yield strength deformed (HYSD) bars, the increase in yield stress is obtained by ....

    (a) twisting bars at a stress higher than yield or proof stress

    (b) twisting bars at a stress much lower than proof stress

    (c) pulling bars at a stress up to yield stress

    (d) pulling bars at a stress much below proof stress

(xxiii) Cold-worked HYSD bars (Fe 415) are also designated as ....

    (a) TOR 40

    (b) TOR 500

    (c) TOR 360 (i.e. 415 × 0.87)

    (d) TOR 275 $\left(\text{i.e. } \dfrac{415}{1.50}\right)$

(xxiv) The bond strength of HYSD bars is improved by .... on the bar surface.

    (a) cold working

    (b) lugs (or ribs)

    (c) strain hardening

    (d) corrosion pitting

(xxv) The permissible compressive stress in HYSD steel bars in beams in compression zone shall be practically considered as ....

    (a) more than 230 N/mm^2

    (b) more than 190 N/mm^2 but less than 230 N/mm^2

    (c) m times $\left(m = \dfrac{E_s}{E_c}\right)$ stress in surrounding concrete

    (d) $1.5m$ times $\left(m = \dfrac{E_s}{E_c}\right)$ stress of surrounding concrete

Response sheet to Q. 2.1 (i to xxv)

Question	(i)	(ii)	(iii)	(iv)	(v)	(vi)	(vii)	(viii)	(ix)	(x)	(xi)	(xii)	(xiii)	(xiv)	(xv)
Response (a/b/c/d)															

Question	(xvi)	(xvii)	(xviii)	(xix)	(xx)	(xxi)	(xxii)	(xxiii)	(xxiv)	(xxv)
Response (a/b/c/d)										

Chapter 3

Working Stress Method

LEARNING OBJECTIVES

After the study of *working stress method* for design of flexural elements of RCC, the learner understands the basic design procedure and will be able to:

- ◉ State basic assumptions for design of reinforced cement concrete elements
- ◉ State fundamental assumptions in straight line theory in design of beams
- ◉ Explain the concept of equivalent concrete area in reinforced concrete element at different stages of loading
- ◉ Explain the concept of neutral axis of beam sections
- ◉ Explain moment of resistance of a beam section and determine various design constants for given grades of concrete and steel
- ◉ Explain the concepts of balanced, under reinforced and over reinforced concrete
- ◉ Explain the procedures of determining (a) moment of resistance, (b) stresses in materials and (c) design of RCC section of flexural members with given data
- ◉ Explain the design of slab spanning in one direction for given loading

3.1 INTRODUCTION

It has already been explained that the plain cement concrete has high compressive strength and relatively very low tensile strength. As explained in structural mechanics, the beam sections are subjected to both compressive and tensile stresses on either side of neutral axis. The concrete beam sections, therefore, need to be reinforced with steel on tension zone to resist bending moments caused in beams.

Consider a simply supported (SS) beam carrying transverse loads as shown in Fig. 3.1(a). A cantilever beam is shown in Fig. 3.1(b).

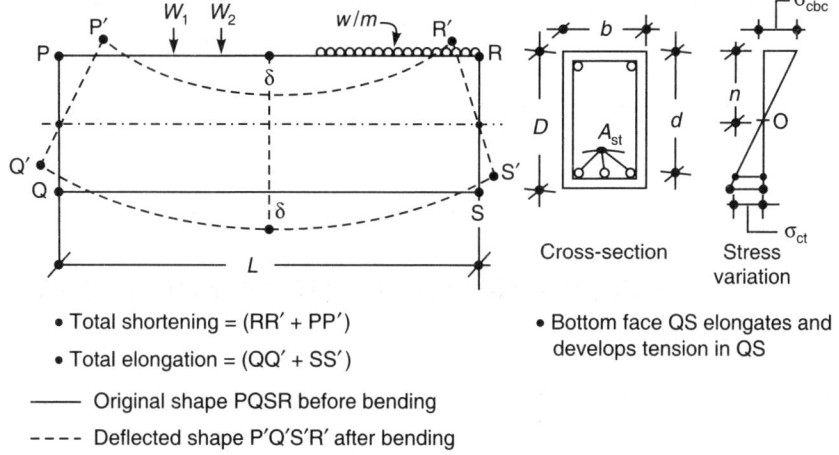

- Total shortening = (RR' + PP')
- Total elongation = (QQ' + SS')

⎯⎯ Original shape PQSR before bending

- - - - Deflected shape P'Q'S'R' after bending

- Bottom face QS elongates and develops tension in QS

Fig. 3.1(a): Simply supported beam carrying transverse loads
(bottom face elongates and develops tension)

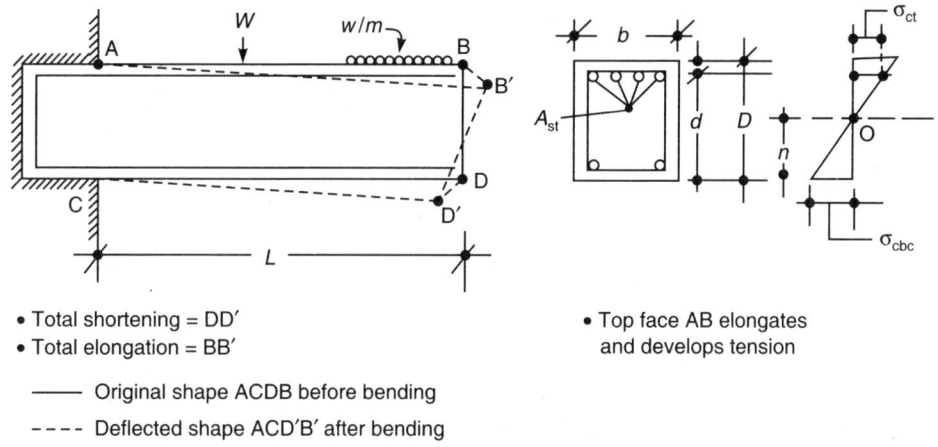

- Total shortening = DD'
- Total elongation = BB'

⎯⎯ Original shape ACDB before bending

- - - - Deflected shape ACD'B' after bending

- Top face AB elongates and develops tension

Fig. 3.1(b): Cantilever beam carrying transverse loads
(top face elongates and develops tension)

When the beam is subjected to transverse loads, it bends. In case of simply supported beam, the lower face of the beam develops tension and it cracks. The concrete becomes ineffective in tension zone and hence steel bars are required to resist this tension. Top face of SS beam develops compression, which concrete is capable of resisting. Thus, top face of SS beam develops compression while bottom face develops tension to be resisted by the steel reinforcement provided.

In cantilever beams, tension develops in top face and hence steel reinforcement is required in top face. The bottom face of cantilever beam develops compression which is resisted by concrete. Thus, in any bending or flexural element, tension is carried by steel reinforcement and compression is carried by concrete. Stirrups are provided for resisting shear while minimum stirrups are provided to resist temperature and shrinkage cracks. To hold the stirrups minimum two anchor bars are provided in compression zone. These anchor bars are accounted in calculations of moment of resistance in case of doubly reinforced sections but anchor bars are not accounted in singly reinforced sections.

In reinforced cement concrete (RCC), the steel reinforcement and cement concrete resist forces by combined action. This combined action is dependent on the following factors:

(a) There is bond between steel bars and concrete.

(b) There is no corrosion in embedded steel bars so that bond remains intact.

(c) Both concrete and steel have almost equal coefficients of thermal expansions so that there is no differential movement.

3.2 ASSUMPTIONS IN LINEAR ELASTIC THEORY OF RC BEAM SECTIONS

Following assumptions are made in linear elastic theory (working stress) of RC beam sections:

(i) At any section of the beam, plane section before bending remains plane section after bending. This means the strain at any point above or below neutral axis remains proportional to the distance from the neutral axis.

(ii) The stress–strain relationships of concrete and steel are straight lines under working loads.

(iii) All tensile forces are resisted by steel reinforcement and tensile resistance of concrete is ignored.

(iv) Concrete resists compressive forces (totally in case of singly reinforced beams and partly in case of doubly reinforced beams).

(v) The modular ratio $\left(m = \dfrac{E_s}{E_c} \right)$ has the value of $\dfrac{280}{3\sigma_{cbc}}$ according to IS:456-2000, where σ_{cbc} is the permissible compressive stress due to bending in concrete expressed in N/mm^2.

(vi) The modulus of elasticity of steel under compression and tension is assumed to be same. Similarly, the modulus of elasticity under compression and tension for concrete is also assumed to be the same.

3.3 EQUIVALENT AREAS OF COMPOSITE SECTIONS

(a) Column Section

Consider a reinforced cement concrete column (Fig. 3.2). When an axial load acts on the column, the steel bars and the concrete deform together and the load carried is shared by both (steel and concrete).

$$P = P_s + P_c = p_s \cdot A_s + A_c \cdot p_c \qquad \text{... Eq. (3.1)}$$

Because of bond between steel and concrete, we have

$$e = e_s = e_c$$

or $\qquad \dfrac{p_s}{E_s} = \dfrac{p_c}{E_c}$ or $p_s = \dfrac{E_s}{E_c} \cdot p_c = m p_c \qquad \text{... Eq. (3.2)}$

From Eqs (3.1) and (3.2), we have

$$P = p_s \cdot A_s + p_c \cdot A_c = m p_c \cdot A_s + p_c \cdot A_c$$

$$p_c = \frac{P}{(mA_s + A_c)} = \frac{P}{A_e}$$

where, A_e (equivalent concrete area) $= (A_c + m \cdot A_s)$

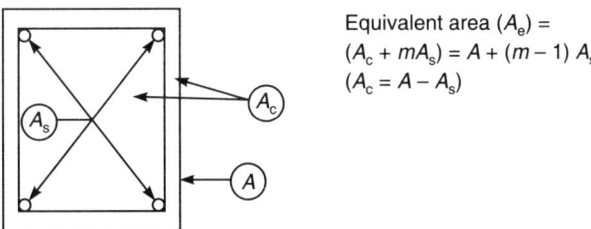

Equivalent area $(A_e) =$
$(A_c + mA_s) = A + (m - 1) A_s$
$(A_c = A - A_s)$

Fig. 3.2: Equivalent area in cross-section of a compression member

Thus, equivalent area of the section subjected to direct compression will be

$$A_e = mA_s + A_c = A + (m - 1) A_s \qquad \dots \text{Eq. (3.3)}$$

where,

A_e = Equivalent concrete area in compression
A_c = Area of concrete = $(A - A_s)$
A_s = Area of steel
A = Area of the composite section

$$m = \frac{E_s \text{ (modulus of elasticity of steel)}}{E_c \text{ (modulus of elasticity of concrete)}} = \text{Modular ratio}$$

Thus, steel area (A_s) represents (mA_s) equivalent concrete area in tension zone, since concrete area in tension does not resist any tension. Similarly, steel area (A_{sc}) in compression zone, the equivalent concrete area according to experimental research is found to be $(1.5 \, m - 1) \, A_{sc}$ and is adopted by IS:456-2000.

(b) Beam Section

(i) Uncracked
(ii) Cracked

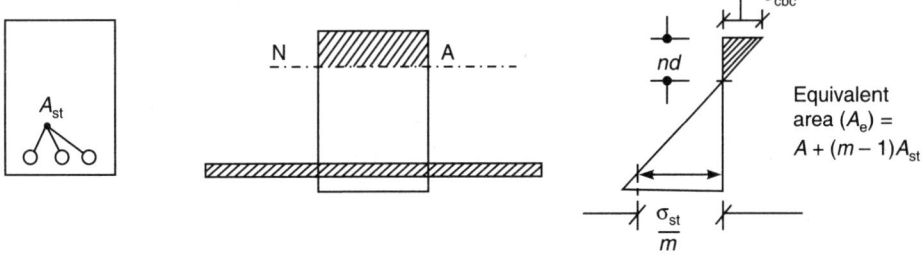

Equivalent
area $(A_e) =$
$A + (m - 1)A_{st}$

Fig. 3.3: Equivalent area A_e in uncracked beam section for tension steel

(i) Uncracked Section

Figure 3.3 shows a RCC beam section having steel reinforcement area A_{st}, A as total area of cross-section and A_c as area of concrete. When the load on the beam is small and the tensile stress in concrete remains less than permissible tensile stress of concrete, it does not crack.

The strains are equal and hence

$$\frac{\sigma_{st}}{E_s} = \frac{\sigma'_{cb}}{E_c} \text{ or } \sigma_{st} = \frac{E_s}{E_c}\cdot\sigma'_{cb} = m\sigma'_{cb} \text{ or } \sigma_{st} = m\sigma'_{cb} \qquad \text{... Eq. (3.4)}$$

i.e. stress in steel is m times stress in surrounding concrete
or stress in surrounding concrete

$$\sigma'_{cb} = \frac{\sigma_{st}}{m} \qquad \text{... Eq. (3.4a)}$$

Load carried by steel (or equivalent concrete area) is $\sigma_{st}\cdot A_{st} = \sigma'_{cb}\cdot A_e$

or $\qquad \sigma'_{cb}\cdot A_e = A_{st}\cdot\sigma'_{st} \text{ or } A_e = A_{st}\cdot\dfrac{\sigma_{st}}{\sigma'_{cb}} = m\cdot A_{st} \left(\text{since } \dfrac{\sigma_{st}}{\sigma'_{cb}} = m\right)$

Net concrete area (uncracked) $= A + (m-1)\,A_{st}$

Net increase in concrete area due to provision of steel (uncracked)

$$= (m-1)\,A_{st} \qquad \text{... Eq. (3.5)}$$

(ii) Cracked Section

If the load on the beam is increased, the tensile stress in concrete in tension zone increases beyond permissible limits of concrete and the concrete in tension zone cracks (Fig. 3.4).

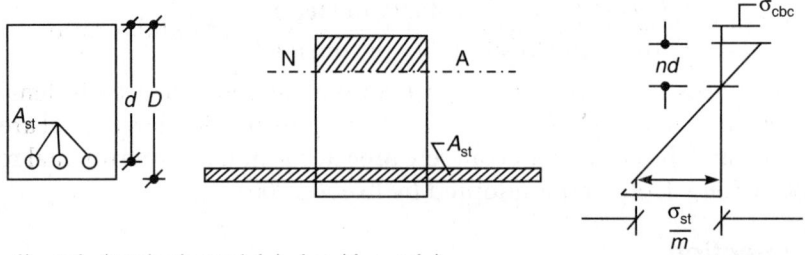

$A'_e = mA_{st}$ (tension in steel A_{st}), $A_e = (A_c + mA_{st})$

Fig. 3.4: Equivalent area A_e in cracked section of beam

Since the concrete in tension zone (below NA in case of SS beam section) cracks, the load carried by steel area is equal to load carried by equivalent concrete area.

$$A_{st}\cdot\sigma_{st} = A_e\cdot\sigma_{cb} \text{ or } A_e = \frac{\sigma_{st}}{\sigma_{cb}}\cdot A_{st} = m\cdot A_{st} \qquad \text{... Eq. (3.6)}$$

Thus, tensile steel area A_{st} is equal to transformed concrete area $A_e = mA_{st}$
(since cracked concrete area is ineffective in tension zone).

where, $\qquad A_e$ = Equivalent area of tension steel

$\qquad\qquad A_{st}$ = Area of tension steel

$\qquad\qquad m$ = Modular ratio

3.4 NEUTRAL AXIS IN BENDING ELEMENTS

Consider Figs 3.1(a) and (b) of any flexural or bending element carrying transverse loads. The fibres of the beam element are subjected to variable stresses (compressive on one side and tensile on the other side). There are some fibres between these compressive and tensile fibres which do not undergo any linear strain or stress. Such fibres between the compressive and tensile fibres carrying zero stress are known as neutral fibres. Line joining all such neutral fibres at a particular section across the cross-section is called neutral axis (NA).

Consider a SS beam of rectangular cross-section as shown in Fig. 3.5(a). The stress variation across the section is shown in Fig. 3.5(b). Consider a small strip of length dx in the span before bending and dx' (or dx'') after bending as shown in Fig. 3.5(c).

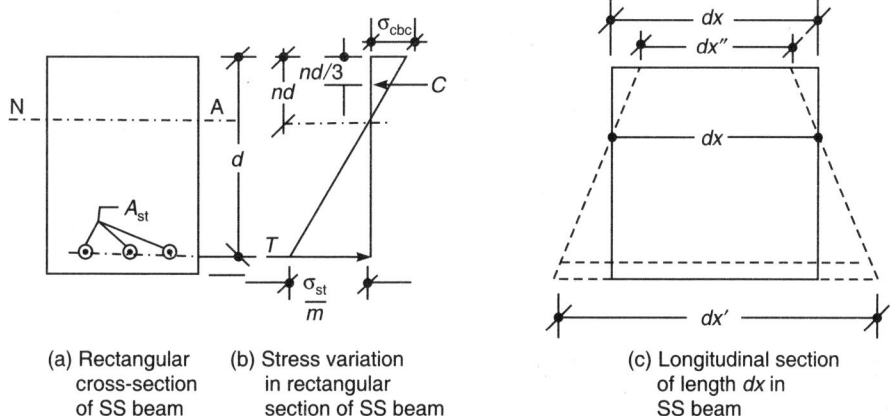

(a) Rectangular cross-section of SS beam

(b) Stress variation in rectangular section of SS beam

(c) Longitudinal section of length dx in SS beam

Fig. 3.5: Stress distribution in rectangular section of SS beam

Let the SS beam considered in Fig. 3.5 be subjected to a moment M under service loads. Abbreviations used are:

σ_{cbc} = Compressive stress developed in concrete in extreme compressive fibre

A_{st} = Area of tensile reinforcement

σ_{st} = Stress in tensile steel reinforcement

d = Effective depth (depth – cover)

b = Width of the member

n = Neutral axis depth factor (coefficient)

nd = Depth of neutral axis from extreme compressive fibre

m = Modular ratio $\left(\dfrac{280}{3\sigma_{cbc}}\right)$

j = Lever arm coefficient $\left(1 - \dfrac{n}{3}\right)$

C = Flexural compressive force in concrete

T = Tensile force in steel reinforcement

M = Moment of resistance of the section

Since the concrete is weak in tension, the concrete below NA is neglected as the portion below NA is subjected to tension. The section is reinforced with steel bars to resist tension developed in tension zone (below NA in SS beams and above NA in cantilever beams). The steel area in tension zone is converted into an equivalent area of concrete. The equivalent area of concrete is calculated as $A_e = mA_{st}$.

This equivalent area of concrete contributes for internal equilibrium of the forces developed in the section (i.e. $C = T$).

Consider stress diagram shown in Fig. 3.5(b), we have

$$\frac{\sigma_{cbc}}{nd} = \frac{\sigma_{st}/m}{(d-nd)} \quad \text{or} \quad \frac{\sigma_{cbc}}{\sigma_{st}/m} = \frac{n}{(1-n)}$$

or $\qquad n(\text{NA coefficient}) = \dfrac{1}{1+\dfrac{\sigma_{st}}{m \cdot \sigma_{cbc}}} = \dfrac{m \cdot \sigma_{cbc}}{(m\sigma_{cbc} + \sigma_{st})}$ ... Eq. (3.7)

Consider moment equilibrium of the forces in the section, we have:

$$M = C\left(d - \frac{nd}{3}\right) = \frac{1}{2}\sigma_{cbc} \cdot b \cdot nd\left(d - \frac{nd}{3}\right)$$

or $\qquad M = \dfrac{1}{2}\sigma_{cbc} \cdot b \cdot n \cdot d^2\left(1 - \dfrac{n}{3}\right) = \dfrac{1}{2}\sigma_{cbc} \cdot n \cdot j \cdot bd^2$

or $\qquad M = Q \cdot b \cdot d^2$ ... Eq. (3.8)

where, $\qquad j = \left(1 - \dfrac{n}{3}\right)$

M = Moment of resistance of the section

$$Q = \frac{1}{2}n \cdot j \cdot \sigma_{cbc}$$

Also $\qquad d = \sqrt{M/Q \cdot b}.$ ... Eq. (3.8a)

M is also calculated from tension side, i.e.

$$M = T\left(d - \frac{nd}{3}\right) = \sigma_{st} \cdot A_{st}\left(d - \frac{nd}{3}\right) = \sigma_{st} \cdot A_{st} \cdot jd$$

or $\qquad A_{st} = \dfrac{M}{\sigma_{st} \cdot jd}$ ... Eq. (3.9)

The design coefficients (constants) n, j and Q are dependent on the permissible stresses in concrete (σ_{cbc}) and steel (σ_{st}). The permissible stresses in different grades of concrete and steel are specified in Tables 21 and 22 of IS:456-2000 and are reproduced in Tables 1.15 and 1.18 respectively for ready reference. Permissible and maximum shear stresses in concrete are specified in Table 23 of IS:456-2000 and are reproduced in Tables 1.16 and 1.17.

The design coefficients for various grades of concrete and steel are given in Table 3.1.

Table 3.1: Design coefficients for various grades of concrete

Grade of concrete	Permissible stress (N/mm²)		Modular ratio m $280/3\sigma_{cbc}$	NA coefficient n	Lever arm coefficient j	Moment coefficient Q
	σ_{cbc}	σ_{st}				
		140		0.404	0.865	0.874
M15	5	230	19	0.292	0.903	0.659
		280		0.253	0.916	0.579

(Contd.)

Table 3.1: Design coefficients for various grades of concrete (Contd.)

Grade of concrete	Permissible stress (N/mm²)		Modular ratio m $280/3\sigma_{cbc}$	NA coefficient n	Lever arm coefficient j	Moment coefficient Q
	σ_{cbc}	σ_{st}				
M20	7	140	13	0.394	0.869	1.198
		230		0.284	0.905	0.900
		280		0.245	0.918	0.787
M25	8.5	140	11	0.400	0.867	1.474
		230		0.289	0.904	1.110
		280		0.250	0.917	0.974
M30	10	140	9	0.391	0.870	1.701
		230		0.281	0.906	1.273
		280		0.243	0.919	1.117
M35	11.5	140	8	0.397	0.868	1.981
		230		0.286	0.905	1.488
		280		0.247	0.918	1.304
M40	13	140	7	0.394	0.869	2.226
		230		0.284	0.905	1.671
		280		0.245	0.918	1.462
Formulae				$n = \left(\dfrac{m\sigma_{cbc}}{m\sigma_{cbc} + \sigma_{st}} \right)$	$j = \left(1 - \dfrac{n}{3} \right)$	$Q = 0.5\sigma_{cbc} \cdot n \cdot j$

3.5 ANALYSIS AND DESIGN OF REINFORCED BEAM SECTIONS (SINGLY REINFORCED)

Analysis and design of beam sections subjected to bending moments pertain to three categories of problems.

(a) **Category I**

Cross-sectional dimensions and area of tensile reinforcement given and stresses developed in the two materials under given loads required to be computed.

(b) **Category II**

Permissible stresses in concrete and steel are specified and it is required to compute the moment of resistance and tensile steel area for the given section.

(c) **Category III**

Permissible stresses in concrete and steel are specified and it is required to compute cross-sectional dimensions and tensile steel area for the given loading (or bending moment).

(a) Problem Category I

When the cross-sectional dimensions and steel area are known and stresses in the two materials are required under given loading (maximum BM), it is important to establish

NA. The NA of the cross-section passes through the centre of gravity of the effective beam section. Consider a rectangular section (breadth b and depth d) and tensile steel reinforcement (A_{st}) as shown in Fig. 3.6.

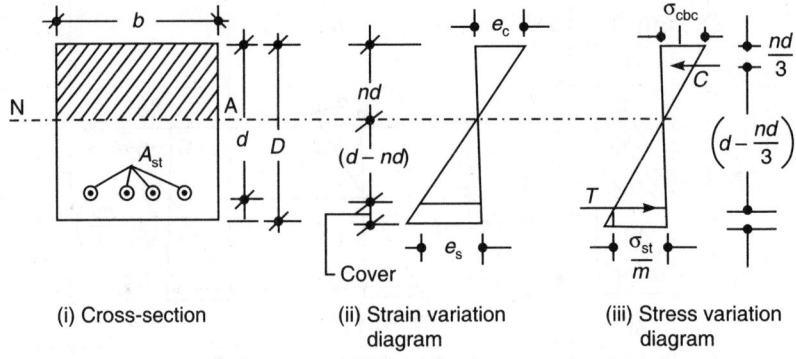

(i) Cross-section   (ii) Strain variation diagram   (iii) Stress variation diagram

Fig. 3.6: Singly reinforced beam section and strain–stress diagrams

Equating the moment of equivalent area in tension zone and moment of concrete area in compression about the NA, we have

$$b \cdot nd \cdot \frac{nd}{2} = (mA_{st})(d - nd)$$

or

$$n^2 \cdot \frac{bd^2}{2} = mA_{st} \cdot d(1 - n)$$

or

$$n^2 \cdot bd + 2m\, A_{st}\, n - 2m\, A_{st} = 0$$

Let

$$\frac{A_{st}}{bd} = p', \text{ we have}$$

$$n^2 + 2n\, mp' - 2mp' = 0, n = \frac{-2mp' \pm \sqrt{4m^2 p'^2 + 8mp'}}{2}$$

or

$$n = -mp' \pm \sqrt{2mp' + (mp')^2} \qquad \text{... Eq. (3.10)}$$

Thus, neutral axis depth coefficient n can be calculated from the known values of A_{st}, b, d and $p' \left(\dfrac{A_{st}}{bd} \right)$ by using Eq. (3.10). From the computed values of n and maximum bending moment at the section, the stresses developed in the section can be computed by the following equations:

$$\frac{\sigma_{cbc}}{2} \cdot b \cdot nd \left(d - \frac{nd}{3} \right) = \text{max BM } (M)$$

Also

$$\sigma_{st} \cdot A_{st} \left(d - \frac{nd}{3} \right) = \text{max BM } (M)$$

i.e.

$$\sigma_{cbc} = \frac{2M}{b \cdot d^2 n \left(1 - \dfrac{n}{3} \right)} = \frac{2M}{njbd^2} \qquad \text{... Eq. (3.11)}$$

and

$$\sigma_{st} = \frac{M}{A_{st} \cdot jd}, \text{ where } j = \left(1 - \frac{n}{3} \right) \qquad \text{... Eq. (3.11a)}$$

(b) Problem Category II

When the cross-sectional dimensions and area of tensile steel are known and moment of resistance is required to be computed for the given permissible stresses in concrete and steel.

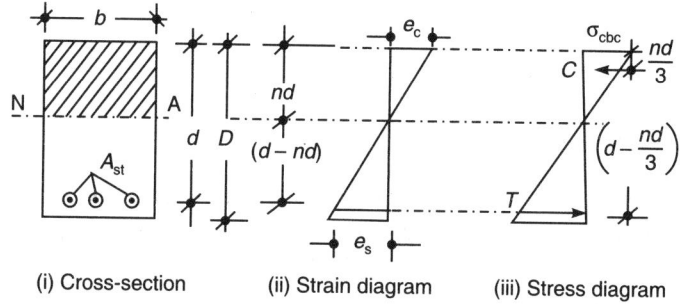

| (i) Cross-section | (ii) Strain diagram | (iii) Stress diagram |

Fig. 3.7: Singly reinforced section and strain–stress diagrams

Consider a rectangular section of breadth b and depth d (overall depth D) and tensile steel (A_{st}) as shown in Fig. 3.7. The NA (neutral axis) lies at CG of the section. Equivalent concrete area of tensile steel is mA_{st}. Let CG be at a distance of nd from the compression face, from Eq. (3.10), we have:

$$n \text{ (actual)} = \left(\sqrt{2mp' + (mp')^2} - mp' \right), \text{ where } p' = \frac{A_{st}}{bd}.$$

If the permissible stresses in concrete and steel develop simultaneously, the section is balanced, in this case $n_a = n_c$.

If the stress in concrete develops equal to permissible (in bending compression) but stress in steel remains less than the permissible, the section is called *over reinforced* and the safe moment of resistance shall be based on compression in concrete. In this case the actual neutral axis lies below the critical neutral axis. This is the case of *over reinforcement* and the *actual NA lies below critical NA* $(n_a > n_c)$. The safe moment of resistance shall be based on the permissible compressive stress in extreme concrete fibre.

If the actual NA lies above the critical NA (i.e. $n_a < n_c$), the section is *under reinforced* and the safe *moment of resistance* (M_r) shall be based on the permissible stress in tensile steel reinforcement intension zone.

Thus, before determining the safe moment of resistance of the section, we must first determine the actual and critical neutral axes to establish if the section is balanced $(n_a = n_c)$ or *over reinforced* $(n_a > n_c)$ or *under reinforced* $(n_a < n_c)$.

For balanced section, the moment of resistance shall be the same with respect to permissible compressive stress in concrete and permissible tensile stress in tensile steel reinforcement.

In case of under reinforced section $(n_a < n_c)$, the safe moment of resistance (M_r) shall be based on the permissible stress in tensile steel reinforcement.

In case of over reinforced section $(n_a > n_c)$, the safe moment of resistance (M_r) shall be based on the permissible compressive stress in extreme concrete fibre.

$$M_r = \frac{1}{2}\sigma_{cbc} \cdot bnd \left(d - \frac{nd}{3} \right) = \frac{1}{2}\sigma_{cbc}\, n \cdot j \cdot bd^2 = Qbd^2 \qquad \dots \text{Eq. (3.12)}$$

$$Q = \frac{1}{2}\,\sigma_{cbc} \cdot n \cdot j$$

with reference to compressive stress in concrete (n = actual NA for over reinforced)

$$M_r = A_{st} \cdot \sigma_{st}\left(d - \frac{nd}{3}\right) = A_{st} \cdot \sigma_{st}\, jd \qquad\qquad \text{... Eq. (3.13)}$$

with respect to tensile stress in steel (for under reinforced).

(c) Problem Category III

When the loading and maximum bending moments along with permissible stresses in concrete and steel are known, we are required to determine the appropriate size (breadth and depth) of the section and quantity of steel reinforcement (A_{st}) in tension zone. Generally, it is designed as balanced beam section or slightly under reinforced so as to avoid sudden brittle failure in compression.

These 3 types of problems shall be explained by illustrative examples.

Example 3.1

A RCC beam section 200 mm × 300 mm (effective) consists of concrete grade M30 having permissible bending compressive stress of 10 N/mm² and steel grade Fe 415 having permissible tensile bending stress of 230 N/mm². If the beam is reinforced with 4 bars of 12 mm on the tension face, calculate the safe moment of resistance of the beam section and also determine if the beam is balanced, under reinforced or over reinforced?

Solution: $\quad b = 200$ mm, $d = 300$ mm, $A_{st} = 4(113) = 452$ mm²

$$\sigma_{cbc} = 10\,\text{N/mm}^2,\ \sigma_{st} = 230\,\text{N/mm}^2,\ m = \frac{280}{3\sigma_{cbc}} = 9.33 = 9$$

$$n_c = \frac{m\sigma_{cbc}}{m\sigma_{cbc} + \sigma_{st}} = \frac{9\times10}{(9\times10+230)} = \frac{90}{320} = 0.281 \text{ (Refer to Table 3.1)}$$

$$p' = \frac{A_{st}}{bd} = \frac{452}{200\times300} = 0.00753,\ mp' = 9 \times 0.00753 = 0.0678$$

From Eq. (3.10), we have

$$n_a = \left(\sqrt{2mp' + (mp')^2} - mp'\right) = \sqrt{2\times0.0678 + (0.0678)^2} - 0.0678$$

$$= \left(\sqrt{0.1356 + 0.004593} - 0.0678\right) = 0.30663$$

$$0.30663 > 0.281,\ \text{i.e. } n_a > n_c \Rightarrow \text{ over reinforced}$$

Safe M_r on the basis of permissible concrete stress (bending compression)

$$M_r = \frac{1}{2}\sigma_{cbc} \cdot b \cdot nd\left(d - \frac{nd}{3}\right),\ n = n_a = 0.3066$$

$$= \frac{1}{2}(10)(200)(300)^2(0.3066)(1 - 0.1022) = 1.37633 \times 18 \times 10^6 \text{ N-mm}$$

$$= 24.774 \times 10^6 \text{ N-mm} = 24.774 \text{ kN-m}$$

M_r (with tensile steel) = $452 \times 230 \times 300 (1 - 0.1022) = 28.00 \times 10^6$ N-mm

$\therefore$ Safe $M_r = 24.774$ kN-m (with reference to compressive concrete stress).

Thus, beam section is over reinforced.

Example 3.2

A RCC beam section of 300 mm × 600 mm is reinforced with 2 bars of 12 mm at an effective cover of 50 mm in tension. The beam section carries a service load moment of 36 kN-m. Concrete of M20 grade and steel of Fe 415 grade have been used. Determine the stresses in concrete and steel.

Solution: $b = 300$ mm, $d = 600 - 50 = 550$ mm, $A_{st} = 2 \times 113 = 226$ mm²

M20 grade, $m = 13$, service moment = 36 kN-m (36×10^6 N-mm)

$$\text{Actual NA, } n_a = \left(\sqrt{2mp' + (mp')^2} - mp' \right), p' = \frac{226}{300 \times 550} = 1.37 \times 10^{-3}$$

$$n_a = \left[\sqrt{35.61 \times 10^{-3} + (17.81 \times 10^{-3})^2} - 17.81 \times 10^{-3} \right] = 0.171$$

$$M = 36 \times 10^6 \text{ N-mm} = 0.5\, \sigma_{cbc}\, (300)\, (0.171)\, 550 \times \left(550 - \frac{0.171 \times 550}{3} \right)$$

$$\sigma_{cbc} = \frac{36 \times 10^6}{0.5 \times 300 \times 0.171 \times 550 \times 518.65} = \frac{36 \times 10^6}{7316854} = 4.92 \text{ N/mm}^2$$

Also $$M = A_{st} \cdot \sigma_{st} \left(d - \frac{nd}{3} \right) = 226 \times 550\ (0.943)\ \sigma_{st}$$

$\therefore$ $$\sigma_{st} = \frac{36 \times 10^6}{226 \times 550\ (0.943)} = 307.13 \text{ N/mm}^2$$

Example 3.3

A rectangular beam section of 300 mm width and 500 mm effective depth is reinforced with 4 bars of 16 mm diameter in tension zone. Determine the stresses in concrete in compression zone and steel in tension zone, when a maximum BM of 50 kN-m acts on the beam section. Concrete of M20 grade is used.

Solution: $M = 50 \times 10^6$ N-mm, $b = 300$ mm, $d = 500$ mm, $\sigma_{cbc} = 7$ N/mm² for M20 CC.

$$m = \frac{280}{3\sigma_{cbc}} = \frac{280}{3 \times 7} = 13.33 = 13 \text{ (as per IS:456)}$$

$$A_{st} = 4\ (201 \text{ mm}^2) = 804 \text{ mm}^2, p = \frac{804}{300 \times 500} = 53.6 \times 10^{-4}, mp = 0.0697$$

$$n = \{(mp)^2 + 2mp\}^{1/2} - mp = (0.00486 + 0.1394)^{1/2} - 0.0697 = 0.31012$$

$$j = \left(1 - \frac{0.31012}{3} \right) = 0.8966$$

$$\sigma_{st} = \frac{50 \times 10^6}{A_{st} \cdot jd} = \frac{50 \times 10^6}{804 \times 0.89663 \times 500} = 138.72 \text{ N/mm}^2$$

$$\sigma_{cbc} = \frac{(\sigma_{st}/m)n}{(1-n)} = \frac{138.72 \times 0.31012}{13(1 - 0.31012)} = \frac{43.02}{8.968} = 4.80 \text{ N/mm}^2$$

Example 3.4

Design a rectangular beam section singly reinforced subjected to a moment of 100 kN-m. Use M25 grade of concrete and Fe 415 grade of steel. Assume depth to breadth ratio of 2.

Solution: For Fe 415 grade steel, permissible tensile stress $\sigma_{st} = 230 \text{ N/mm}^2$

For M25 CC, permissible compressive bending stress $= 8.5 \text{ N/mm}^2$, $m = \dfrac{280}{3(8.5)} = 11$

NA depth coefficient $n = \dfrac{m\sigma_{cbc}}{m\sigma_{cbc} + \sigma_{st}} = \dfrac{11 \times 8.5}{(93.5 + 230)} = 0.289$

Lever arm coefficient $j = \left(1 - \dfrac{n}{3}\right) = 0.904$

Moment of resistance coefficient $Q = 0.5\sigma_{cbc} \cdot n \cdot j = 1.11034$

For singly reinforced sections,

moment of resistance $= 0.5\sigma_{cbc} \cdot n \cdot j \cdot bd^2 = Qbd^2 = 1.11034 \, b \cdot d^2$

$\therefore \qquad b = \dfrac{d}{2}, \quad \therefore 100 \times 10^6 \text{ N-mm} = 1.11034 \left(\dfrac{d}{2}\right)d^2 = \dfrac{1.11034}{2} \cdot d^3$

or $\qquad d^3 = \dfrac{200 \times 10^6}{1.11034} = 180125 \times 10^3, \, d = (180125 \times 10^3)^{1/3} = 56.45 \times 10$

$d = 566 \text{ mm}, \, b = 283 \text{ mm}$

Provide $b = 300 \text{ mm}, d = 570 \text{ mm} (D = 620 \text{ mm})$, cover 50 mm

Check:

$M_r \text{ (compression)} = Qb \cdot d^2 = 1.11034 \times 300 \times 570^2$

$\qquad\qquad\qquad\qquad = 108.22 \times 10^6 > 100 \times 10^6 \text{ N-mm (safe, OK)}$

$A_{st} = \dfrac{M}{\sigma_{st} \cdot jd} = \dfrac{100 \times 10^6}{230 \times 0.904 \times 570} = 844 \text{ mm}^2 \text{ (provide 3 bars of 20 mm diameter)}$

$M_r \text{ (tension)} = \sigma_{st} \cdot A_{st} \cdot jd = 230 \times 3 \times 314 \times 0.904 \times 570$

$\qquad\qquad\qquad = 111.6 \times 10^6 \text{ N-mm} > 100 \times 10^6 \text{ (safe, OK)}$

Example 3.5

A SS RCC beam of span 4 m has a rectangular section of 250 mm width by 500 mm overall depth with effective cover of 50 mm. The beam is reinforced with 3 bars of 16 mm diameter on the tension side. M20 CC and Fe 250 steel are used in the beam construction. Determine the maximum live load uniformly distributed on the entire beam which can be safely allowed.

Solution: Span $= 4$ m, dead load (self-weight) of beam $= 25.0 \times \dfrac{250}{1000} \times \dfrac{500}{1000}$

$\qquad\qquad\qquad\qquad\qquad\qquad\qquad\qquad\qquad = 3.1255 \text{ kN/m}$

Width $b = 250$ mm, depth $D = 500$ mm, effective depth $= 450$ mm

$A_{st} = 3 \times 201 = 603 \text{ mm}^2$

M20 CC, $\sigma_{cbc} = 7 \text{ N/mm}^2$, Fe 250, $\sigma_{st} = 140 \text{ N/mm}^2$,

$$p = \frac{603}{250 \times 450} = 0.00536$$

$$m = \frac{280}{3 \times 7} = 13.3 = 13, \; n_c = \frac{m\sigma_{cbc}}{m\sigma_{cbc} + \sigma_{st}} = \frac{13 \times 7}{(91 + 140)} = 0.394, \; j = 0.869$$

$$Q = 1.198, \; mp = 0.0697, \; (mp)^2 = 0.004855$$

Depth of NA (actual)

$$\frac{b(nd)^2}{2} = m \cdot A_{st}(d - nd) \; \text{or} \; n = \left(\sqrt{(mp)^2 + 2mp} - mp \right)$$

$$n = \left(\sqrt{0.004855 + 0.1394} \right) - 0.0697 = 0.37981 - 0.0697 = 0.31011$$

Actual $\quad j = \left(1 - \dfrac{n}{3} \right) = \left(1 - \dfrac{0.31011}{3} \right) = 0.8966.$

Actual NA is above critical NA ($n < n_c$), the section is under reinforced.

Moment of resistance $M_r = A_{st} \cdot \sigma_{st} \cdot jd = 603 \times 140 \times 0.8966 \times 450 = 34.062 \times 10^6$ N-mm.

Let the total u.d.l. on the beam be w kN/m

$$\text{SS BM (max)} = \frac{w \cdot L^2}{8} = \frac{w \times 4^2}{8} = 2w \text{ kN-m}$$

$$\text{BM} = M_r, \; 2w \times 10^6 = 34.062 \times 10^6, \; w = \frac{34.062}{2} = 17.031 \text{ kN/m}$$

Live load = (permissible load − self load) = 17.031 − 3.125 = 13.906 kN/m

Thus, maximum live load permissible on the beam is 13.90 kN/m

Example 3.6

A simply supported RC beam consists of rectangular section 200 mm wide by 350 mm deep (overall). The section is reinforced with 3 bars of 16 mm ϕ at an effective cover of 50 mm on the tension face. Determine the stress in concrete when the stress in steel is 120 N/mm². Take $m = 18$. Also suggest the grade of concrete considering the stress to be permissible and factor of safety = 3.

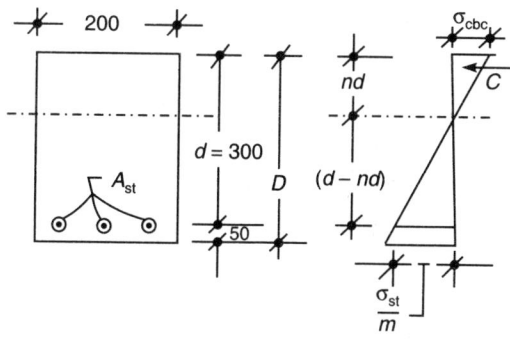

Fig. 3.8: Beam section

Solution: $\quad b = 200$ mm, $D = 350$ mm, $d = 350 - 50 = 300$ mm (Fig. 3.8).

$$m = 18, \; \sigma_{cbc} = ?, \; \sigma_{st} = 120 \text{ N/mm}^2, \; = \frac{\sigma_{st}}{m} = \frac{120}{18} = \frac{20}{3}.$$

$$A_{st} = 3(201) = 603 \text{ mm}^2, p = \frac{603}{200 \times 300} = 0.01005$$

$$mp = 0.1809, (mp)^2 = 0.032725$$

Actual NA

$$n = \sqrt{2mp + (mp)^2} - mp = (0.394525)^{0.50} - 0.1809$$
$$= 0.62811 - 0.1809 = 0.4472$$

From stress diagram

$$\frac{\sigma_{st}}{m} / (d - nd) = \frac{\sigma'_{cbc}}{nd} \text{ or } \frac{120}{18(1-n)} = \frac{\sigma'_{cbc}}{n} \text{ or } \sigma_{cbc} = \frac{20 \times 0.4472}{3(1-0.4472)}$$

$$= \frac{8.944}{1.6584} = 5.393 \text{ N/mm}^2$$

Characteristic strength required = factor of safety × permissible stress,

i.e. $f_{ck} = 3 \times 5.393 = 16.179 \text{ N/mm}^{2.}$

Suitable grade of concrete will be M20 (more than 16.179 N/mm²).

Example 3.7

A rectangular, singly reinforced beam, 300 mm width and 550 mm overall depth is used as a cantilever beam over an effective span of 3 m. The reinforcement comprises 4 bars of 20 mm diameter placed at an effective cover of 50 mm in top face. If the cantilever beam carries a load of 10 kN/m, inclusive of self-weight, determine the stresses developed in steel and concrete. Take $m = 13$.

Solution: $b = 300$ mm, $D = 550$ mm, $d = 550 - 50 = 500$ mm, cantilever span = 3 m,

$$A_{st} = 4 \times 314 = 1256 \text{ mm}^2, w = 10 \text{ kN/m, Max. BM} = \frac{wl^2}{2} = \frac{10 \times 3^2}{2} = 45 \text{ kN-m,}$$

$$M = 45 \text{ kN-m}, m = 13, p = \frac{A_{st}}{bd} = \frac{1256}{300 \times 500} = 0.008373,$$

$$(mp)^2 = 0.0000701$$

Actual NA depth coefficient $= n$

Equating the moments of two areas about NA, we get

$$n = \sqrt{(mp)^2 + 2mp} - mp = \sqrt{0.01185 + 0.2177} - 0.10885 = 0.370264$$

$$j = \left(1 - \frac{n}{3}\right) = \left(1 - \frac{0.370264}{3}\right) = 0.87658$$

$$nd = 0.370264 \times 500 = 185.13 \text{ mm}, jd = 438.3 \text{ mm,}$$

$$(d - nd) = 500 - 185.13 = 314.87$$

M_r (with compressive face) $= 0.5 \sigma_{cbc} \times nd \cdot b \cdot jd$

$$= 0.5 \sigma_{cbc} \times 185.313 \times 300 \times 438.3 \text{ N-mm}$$

Max BM $= M_r \therefore 0.5 \sigma_{cbc} \times 185.313 \times 300 \times 438.3 = 45 \times 10^6 \text{ N-mm}$

$$\sigma_{cbc} = \frac{45 \times 10^6}{12171371.85} = 3.70 \text{ N/mm}^2$$

$$\sigma_{st} = \frac{m\sigma_{cbc} \times (d - nd)}{nd} = 81.75 \text{ N/mm}^2$$

Also $\sigma_{st} = \dfrac{45 \times 10^6}{1256 \times 438.3} = 81.75 \text{ N/mm}^2$

Example 3.8

Design a RCC SS beam section using M20 grade concrete and Fe 415 grade steel, keeping the width of rectangular section equal to 200 mm. Span of the beam is 4 m and it carries u.d.l. of 10 kN/m over the entire span including the self-weight of the beam.

Solution: $b = 200 \text{ mm}, d = ?, w = 10 \text{ kN/m}, l = 4 \text{ m},$

$$\text{Max. BM} = \frac{wl^2}{8} = \frac{10 \times 4^2}{8} = 20 \text{ kN-m},$$

For M20 CC, permissible $\sigma_{cbc} = 7 \text{ N/mm}^2$,
For Fe 415, permissible $\sigma_{st} = 230 \text{ N/mm}^2$,

$$m = \frac{280}{3 \times 7} = 13 \text{ (as per IS:456-2000).}$$

Critical $n_c = \dfrac{m\sigma_{cbc}}{m\sigma_{cbc} + \sigma_{st}} = \dfrac{13 \times 7}{(91 + 230)} = \dfrac{91}{321} = 0.284, j = \left(1 - \dfrac{0.284}{3}\right) = 0.9053$

$Q = 0.5\,\sigma_{cbc} \cdot n \cdot j = 0.90$
$M_r = \text{Max. BM or } 0.90\,b \cdot d^2 = 20 \times 10^6 \text{ N-mm.}$

or $d^2 = \dfrac{20 \times 10^6}{0.90 \times 200} = 111127, d = 333.36 \text{ mm, (say) provide 335 mm and}$

$D = 380 \text{ mm (cover 45 mm)}$

$A_{st} = \dfrac{20 \times 10^6}{230 \times 0.9053 \times 335} = 287 \text{ mm}^2 \text{ (provide 12 mm } \phi \text{ 3 bars,}$

$A_{st} = 3 \times 113 = 339 \text{ mm}^2).$

Beam: 200 mm wide × 380 mm overall depth and 3 bars of 12 diameter in tension face.

3.6 DOUBLY REINFORCED BEAM SECTIONS

We have seen that the balanced section of singly reinforced beam is the most economical from the steel requirement point of view. To increase moment of resistance of the section ($b \times d$) beyond balanced M_r of Qbd^2, we have two options: (i) to increase the quantity of tensile steel making it over reinforced, or (ii) to provide steel reinforcement in compression zone and equivalent extra tension reinforcement. It is observed that over reinforcing the beam section makes it uneconomical because M_r does not increase in the same ratio as increase in the cost of steel. Since M_r of the balanced beam beyond $Q_c \cdot b \cdot d^2$, leads to increase in stress in concrete beyond permissible limits, and hence the first approach of over reinforcing singly reinforced beam section is not the best approach. Whenever there is limitation on cross-sectional dimensions, specially depth (d), we must reinforce the beam section in compression zone in addition to reinforcing in tension zone. Such a section which is reinforced both in tension and compression zones is known as *doubly reinforced beam section*.

Doubly reinforced beams are also provided whenever there is possibility of reversal of stresses due to external loading (e.g. wind pressures, eccentricity on either side, earthquake forces, or any other accidental loadings, etc.).

The steel reinforcement provided in the compression zone is subjected to compressive stress. The concrete in compression zone undergoes creep strains due to continued compressive stress and as a result the strain in concrete goes on increasing with time. This results in increase of strains in compressive steel beyond its own creep strain. Thus, the total compressive strain in compressive zone steel is much higher than the strain in the surrounding concrete due to flexure alone. As per IS:456-2000, the compressive steel stress should be calculated by multiplying the stress in surrounding concrete by $1.5m$ times instead of usually multiplying factor of m times. The permissible stresses in steel in compression zone shall be taken as given in Table 1.18.

Figure 3.9 shows a doubly reinforced beam section, equivalent concrete section and strain–stress variation diagrams. The coefficient of NA depth from top can be determined by equating the moments of the equivalent concrete area in compression and equivalent area of tension steel about the NA.

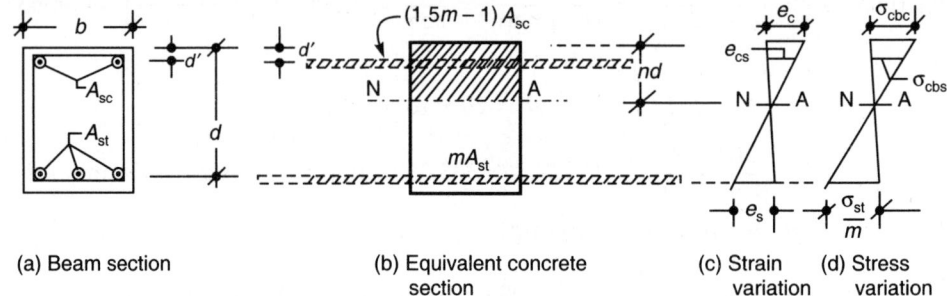

| (a) Beam section | (b) Equivalent concrete section | (c) Strain variation | (d) Stress variation |

Fig. 3.9: Doubly reinforced beam section and strain–stress diagrams

$$b \cdot nd \cdot \frac{nd}{2} + (1.5m - 1) A_{sc} (nd - d') = m A_{st} (d - nd)$$

We may substitute $A_{st} = p \cdot bd$ and $A_{sc} = p'bd$ in above equation, we get

$$0.5\, b \cdot d^2 \cdot n^2 + (1.5m - 1) p'bd (nd - d') = mpbd (d - nd)$$

or
$$n^2 + 2 (1.5m - 1) p' \left(n - \frac{d'}{d} \right) - 2mp (1 - n) = 0 \qquad \text{... Eq. (3.14)}$$

Putting $(1.5m - 1)$ approximately 1.50 m for simplicity and solving for n, we have

$$n = \left[m^2 (p + 1.5p')^2 + 2m \left(p + \frac{1.5\, p'd'}{d} \right) \right]^{1/2} - m\, (p + 1.5p') \qquad \text{... Eq. (3.14a)}$$

The lever arm of the resultant compressive force about tensile force is given as jd (d is the distance of CG of the tensile steel from the top compressive face),

i.e.
$$jd = \left\{ d - \frac{C_c \cdot \dfrac{nd}{3} + C_s \cdot d'}{(C_c + C_s)} \right\}$$

where,
$$C_c = 0.5\, \sigma_{cbc} \cdot b \cdot nd$$

$$C_s = (1.5m - 1)\, A_{sc} \cdot \sigma'_{cbc} = (1.5m - 1)\, p'bd\, \sigma'_{cbc} \left(1 - \frac{d'}{nd}\right)$$

or
$$jd = \left(d - \frac{nd}{3}\right) + \frac{(1.5m-1)p'\left(\dfrac{nd}{3} - d'\right)\left(n - \dfrac{d'}{d}\right)}{mp\,(1-n)} \qquad \text{... Eq. (3.15)}$$

The moment of resistance (M_r) of section with respect to compressive and tensile forces can be determined as under:

$$M = C \cdot jd = \left[0.5\sigma_{cbc} \cdot b \cdot nd + (1.5m - 1)\, A_{sc}\, \sigma'_{cbc} \left(1 - \frac{d'}{nd}\right)\right] jd \qquad \text{... Eq. (3.16)}$$

Alternatively, M_r can also be found by taking moment of compressive forces about the tensile steel, i.e. with usual notations

$$M = M_1 + M_2 = [0.5\sigma_{cbc} b \cdot nd]\left(d - \frac{nd}{3}\right) + (1.5m - 1)\, A_{sc}\, \sigma'_{cbc}\, (d - d')$$

$$\text{(compressive concrete)} + \text{(compressive steel equivalent)}$$

or
$$M = M_1 + M_2 = Qb \cdot d^2 + \sigma'_{cbc}\,(1.5m - 1)\, A_{sc}\,(d - d') \qquad \text{... Eq. (3.16a)}$$

Tensile steel comprises two parts, viz. first part A_{st_1} for the balanced singly reinforced section corresponding to Qbd^2 $\left(\text{i.e. } A_{st} = \dfrac{Qbd^2}{jd\sigma_{st}}\right)$ and the remaining extra M_r (M_2) due to compressive steel equivalent concrete moment

$$= \sigma'_{cbc}\,(1.5m - 1)\, A_{sc}\,(d - d').$$

Equivalent tensile steel area (A_{st_2}) is found by equating moment of compressive steel about the NA and moment of equivalent tensile steel area about the NA, i.e.

$$(1.5m - 1)\, A_{sc}\,(nd - d') = m\, A_{st_2}\,(d - nd)$$

or
$$A_{st_2} = \frac{(1.5m - 1)\, A_{sc}\,(nd - d')}{m(d - nd)} = \frac{(1.5m - 1)\left(n - \dfrac{d'}{d}\right) A_{sc}}{m(1 - n)} \qquad \text{... Eq. (3.17)}$$

Alternatively,

A_{st_2} can be found by $\dfrac{M_2}{\sigma'_{st}\, jd} = \dfrac{(1.5m - 1)(d - d')\sigma'_{cbc}\, A_{sc}}{\sigma_{st}(d - d')}$

$$A_{st_2} = \frac{(1.5m - 1)\, A_{sc}\sigma_{cbc}}{\sigma_{st}} \qquad \text{... Eq. (3.18)}$$

where, d' = Depth of A_{sc} from top (cover)

         σ'_{cbc} = Stress in concrete surrounding compressive steel

         σ'_{st} = Stress in tensile steel at the level it is located

         = σ_{st}, if it is located at the same level as A_{st}, and lever arm

         = $(d - d')$

Total moment of resistance can be found directly by first finding CG of two compressive force components (C_c – due to compression in concrete and C_s – due to compressive steel equivalent concrete) and the distance of CG of compressive forces from the CG of tensile steel bars.

$$M = (C_c + C_s) \times \text{lever arm} = T \times \text{lever arm}$$

Also $\quad T = C_c + C_s,$ considering equilibrium of internal forces in the beam section.

Different types of problems are similar to singly reinforced beams based on whether the section is over reinforced, under reinforced or balanced. If the actual NA is above critical NA (i.e. $n < n_c$), the section is under reinforced in tension and if actual NA is below critical NA (i.e. $n > n_c$) the section is over reinforced in tension. The safe moment of resistance is based on tension if the section is under reinforced. The safe moment of resistance is based on compression if the section is over reinforced. In case of balanced section, moment of resistance in tension and in compression are equal.

Three types of problems will now be illustrated by examples for understanding.

Example 3.9

A rectangular beam section of 300 mm width and 500 mm effective depth is reinforced with 4 bars of 18 mm diameter intension face and 4 bars of 14 mm diameter in compression face at an effective cover of 50 mm. Determine the stresses induced in the top compression fibre of the concrete, compression steel and tension steel when subjected to a moment of 95 kN-m. Concrete of M20 grade and steel of Fe 415 grade are used in the beam.

Solution: Tension steel is assumed to be in one layer, $m = 13$

$$A_{sc} = 4 \times \frac{\pi}{4} (14)^2 = 4(153.94) = 615.75 \text{ mm}^2$$

$$A_{st} = 4 \times \frac{\pi}{4} (18)^2 = 4(254.5) = 1018 \text{ mm}^2$$

$$p' = \frac{615.75 \times 100}{300 \times 500} = 0.4105\% \cong 0.41\%$$

$$p = \frac{1018 \times 100}{300 \times 500} = 0.675\% \cong 0.68\%$$

$$d = 500 \text{ mm}, \ d' \text{ (cover to compression steel)} = 50 \text{ mm}$$

$$\frac{d'}{d} = \frac{50}{500} = 0.10$$

$$j \text{ (lever arm coefficient)} = \left(1 - \frac{n}{3}\right) + (1.5m - 1) \ p'\left(\frac{n}{3} - \frac{d'}{d}\right)\left(n - \frac{d'}{d}\right)$$

$$n = \left\{ m^2 (p + 1.5p')^2 + 2m\left(p + 1.5p' \frac{d'}{d}\right) \right\}^{1/2} - m(p + 1.5p')$$

Putting various values:

$$n = \left\{ 169(0.0068 + 0.00616)^2 + 26\left(0.0068 + 0.00616 \times \frac{1}{10}\right) \right\}^{1/2}$$

$$- 13(0.0068 + 0.00616)$$

$$= [0.02839 + 0.19282]^{1/2} - 0.16848 = 0.47033 - 0.16848 = 0.30185$$

or $\quad n = 0.3019$

$$j = \left(1 - \frac{0.3019}{3}\right) + (1.5 \times 13 - 1)\, 0.004105 \left(\frac{0.3019}{3} - 0.10\right)(0.3019 - 0.10)$$

or $\quad j = 0.8994 + 18.5 \times 0.004105 \times 0.0006333 \times 0.2019 = 0.89941$

$$\sigma_{st} = \frac{95 \times 10^6}{1018 \times 0.89941 \times 500} = 207.5 \ \text{N/mm}^2$$

$$\sigma_{cbc} = \frac{\sigma_{st} \cdot n}{m(1 - n)} = \frac{207.5 \times 0.3019}{13(1 - 0.3019)} = 6.903 \ \text{N/mm}^2$$

$$\sigma_{sc} = 1.5m\, \sigma_{cbc} \left(1 - \frac{d'}{nd}\right) = 90.021 \ \text{N/mm}^2$$

Example 3.10

Determine from first principle the permissible moment of resistance for a doubly reinforced rectangular section of 300 mm width × 500 mm effective depth. The section is reinforced with 4 bars of 18 mm diameter on tension face and 4 bars of 14 mm diameter on the compression face at an effective cover of 50 mm. The beam is constructed using M20 grade CC and Fe 415 grade steel.

Solution: $A_{st} = 4(254.5) = 1018 \ \text{mm}^2$, $A_{sc} = 4(154) = 616 \ \text{mm}^2$, $m = 13$

$$d = 500 \ \text{mm}, \ d' = 50 \ \text{mm}, \ \frac{d'}{d} = \frac{50}{500} = 0.10, \ \sigma_{cbc} \ (\text{M20 CC}) = 7 \ \text{N/mm}^2$$

σ_{st} (Fe 415) $= 230 \ \text{N/mm}^2$.

Critical (balanced) NA coefficient $n_c = \dfrac{m\sigma_{cbc}}{m\sigma_{cbc} + \sigma_{st}} = \dfrac{280}{280 + 3\sigma_{st}} = 0.2887$

Actual NA n by taking moment of equivalent concrete area about NA.

$$(1.5m - 1)\, A_{sc}\,(nd - d') + b \cdot nd \cdot \frac{nd}{2} = m\, A_{st}\,(d - nd)$$

or $\qquad\qquad n^2 + 0.3284n - 0.1917 = 0$

On solving, we get $n = 0.3034 \Rightarrow n_a > n_c$ $(0.3034 > 0.2887)$, beam is over reinforced. Moment of resistance is based on the compression in concrete,

i.e. $\qquad \sigma_{cbc} = 7 \ \text{N/mm}^2, \sigma'_{cbc} = \dfrac{7}{0.3034} \times (0.3034 - 0.10) = 4.693 \ \text{N/mm}^2$

Moment of compressive forces about tensile steel,

$$M = 0.5 \times 7 \times 300\,(0.3034 \times 500)\left(1 - \frac{0.3034}{3}\right)500 + (1.5 \times 13 - 1)\,616\,(500 - 50) \times 4.693$$

$$= 71.59 \times 10^6 + 24.07 \times 10^6 \ \text{N-mm} = 95.66 \times 10^6 \ \text{N-mm} \ (95.66 \ \text{kN-m})$$

Actual stress in tension steel,

$$\sigma_{st} = \frac{M}{A_{st} \cdot jd} = \frac{95.66 \times 10^6}{1018 \times 0.8992 \times 500} = 209.01 \ \text{N/mm}^2$$

where j is found by considering CG of compressive forces.

CG of compressive forces

$$\tilde{y} = \frac{b \cdot nd \dfrac{\sigma_{cbc}}{2} \cdot \dfrac{nd}{3} + (1.5m - 1)A_{sc} \cdot \dfrac{\sigma_{cbc}}{nd}(nd - d')d'}{b \cdot nd \dfrac{\sigma_{cbc}}{2} + (1.5m - 1)A_{sc} \cdot \dfrac{\sigma_{cbc}}{nd}(nd - d')}$$

$$= \frac{(3451933 + 569800)}{(75000 + 11396)} = \frac{4021733}{86396} = 50.42 \text{ mm}$$

Distance from tensile steel $= 500 - 50.42 = 449.58$ mm

$$j = \frac{449.58}{500} = 0.8992 \text{ (lever arm coefficient)}$$

Example 3.11

If tensile steel on the tension face comprises 4 bars of 16ϕ as in Example 3.10, determine M_r. All other data remains unchanged except the tensile steel. Also calculate stresses in steel and concrete.

Solution. $A_{st} = 4(201) = 804 \text{ mm}^2, p = \dfrac{804}{300 \times 500} = 0.00536, p' = 0.00411$

$$d' = 50 \text{ mm}, d = 500 \text{ mm}, = \frac{d'}{d} = \frac{50}{500} = 0.10, m = 13, \left(m = \frac{280}{3\sigma_{cbc}} = 13 \right)$$

M20 CC: $\sigma_{cbc} = 7 \text{ N/mm}^2, \sigma_{st}$ (Fe 415) $= 230 \text{ N/mm}^2$

$$n_c = \frac{m\sigma_{cbc}}{m\sigma_{cbc} + \sigma_{st}} = \frac{91}{91 + 230} = \frac{91}{321} = 0.284$$

Actual NA(n_a) by taking moment of equivalent area,

$$(1.50m - 1)\,616\,(nd - d') + b \cdot nd \cdot \frac{nd}{2} = m \cdot A_{st} \cdot (d - nd)$$

or $\qquad 18.5 \times 616 \left(n - \dfrac{d'}{d} \right) + \dfrac{300}{2}\, n^2 \cdot 500 = 13 \times 804\,(1 - n)$

or $\qquad 75000\,n^2 + 11396n - 439.6 + 10452n - 10452 = 0$

or $\qquad n^2 + 0.29131n - 0.1546 = 0 \Rightarrow n = \dfrac{-0.2913 \pm \sqrt{(0.29131)^2 + 4 \times 0.1546}}{2} = 0.2737$

$n_a < n_c$ (0.2737 < 0.284), so the section is under reinforced and the M_r is based on tensile force in steel.

$$\sigma_{st} = 230, = \frac{\sigma_{st}}{m} = \frac{230}{13} = 17.7 \text{ N/mm}^2, \frac{17.7}{(d - nd)} = \frac{\sigma_{cbc}}{nd} \text{ or } \sigma_{cbc} = 17.7\left(\frac{n}{1 - n} \right)$$

$$= 6.67 \text{ N/mm}^2$$

$$= \frac{\sigma'_{cbc}}{(nd - d')} = \frac{\sigma_{cbc}}{nd}, \sigma'_{cbc} = \frac{6.67}{0.2737}(0.2737 - 0.1) = 4.233 \text{ N/mm}^2$$

Distance of compressive steel from tensile steel $= (d - d') = 500 - 50 = 450$ mm

Distance of centroid of compressive force in concrete $= \left(d - \dfrac{nd}{3} \right)$

$$= 500 - 45.62 = 454.39 \text{ mm}$$

$\sigma_{sc} = 4.233 \times 13 \times 1.5 = 82.54 \, \text{N/mm}^2$

Moment of resistance

$$M = M_1 + M_2 = Q \cdot bd^2 + (1.5m - 1) \, A_{sc} \cdot \sigma'_{cbc} \times 454.39$$

$$= \frac{1}{2} \times 6.67 \, (300 \times 0.2737 \times 500) \, 454.39 + (1.5 \times 13 - 1) \, 616 \times 4.233 \times 450$$

$$= 62.21 \times 10^6 + 21.71 \times 10^6 = 83.92 \times 10^6 \, \text{N-mm} \, (83.92 \, \text{kN-m})$$

$M_r = 83.92 \, \text{kN-m}, \, \sigma_{st} = 230 \, \text{N/mm}^2, \, \sigma_{cbc} = 6.67 \, \text{N/mm}^2$

$\sigma_{sc} = 82.54 \, \text{N/mm}^2$

Example 3.12

A rectangular beam section of 250 mm × 500 mm is used for a simply supported beam of effective span of 5 m. The beam is reinforced with 3 bars of 16 mm diameter on tension face at an effective concrete cover of 50 mm. The section is also reinforced with 2 bars of 16 mm on compression face at an effective cover of 50 mm. Concrete is of M20 grade and steel is of Fe 415 grade. Determine the maximum permissible live load uniformly distributed over the entire span.

Solution: Effective span = 5.0 m, $b = 250$ mm, $D = 500$ mm, $d = 450$ mm,

$$d' = 50 \, \text{mm}, = \frac{d'}{d} = \frac{50}{450} = 0.111, \, A_{st} = 3(201) = 603 \, \text{mm}^2,$$

$A_{sc} = 2(201) = 402 \, \text{mm}^2$

M20 grade CC: $\sigma_{cbc} = 7.0 \, \text{N/mm}^2$,

Fe 415 steel:

$\sigma_{st} = 230 \, \text{N/mm}^2, \, m = 13.$

$$n_c = \frac{m\sigma_{cbc}}{m\sigma_{cbc} + \sigma_{st}} = \frac{13 \times 7}{91 + 230} = \frac{91}{321} = 0.2835, \, n_c \cdot d = 127.6 \, \text{mm}$$

Actual depth of NA,

$$\frac{b \cdot (n_a d)^2}{2} + (1.5m - 1) \, A_{sc} \, (n_a d - d') = m \, A_{st} \, (d - n_a d)$$

or

$$\frac{250 \times 450 \times 450 n_a^2}{2} + 18.5 \times 402 \times 450 \, (n_a - 0.111) = 13 \times 603 \times 450 \, (1 - n_a)$$

or

$$n_a^2 + 0.2719 \, n_a - 0.1541 = 0 \Rightarrow n_a = \frac{-0.2719 \pm \sqrt{0.07343 + 0.6164}}{2}$$

$n_a = 0.27933 < 0.2835 = n_c$, the section is under reinforced.

$n_a d = 125.7 \, \text{mm}$

M_r based on tensile steel

$\sigma_{st} = 230 \, \text{N/mm}^2,$

$$\sigma_{cbc} = \frac{\sigma_{st}}{m} \cdot \frac{n_a d}{(d - n_a d)} = \frac{230 \times 125.7}{13(450 - 125.7)} = 6.86 \, \text{N/mm}^2,$$

$$\sigma'_{cbc} = \frac{6.86 \times 75.7}{125.7} = 4.1313 \, \text{N/mm}^2$$

$\sigma'_{sc} = (1.5m - 1) \, \sigma'_{cbc} = 18.5 \times 4.1313 = 76.43 \, \text{N/mm}^2$

M_r (moment of compressive forces about tensile steel)

$$= 76.43 \times 402 \times (450 - 50) + 250 \times 125.7 \times \frac{6.86}{2} \left(450 - \frac{125.7}{3} \right)$$

$$M = 12.29 \times 10^6 + 43.99 \times 10^6 = 56.28 \times 10^6 \text{ N-mm}$$

Max BM SS beam $(l = 5 \text{ m}) = \dfrac{w(5)^2}{8} \times 10^6 = 56.28 \times 10^6,$

$$w = \frac{56.28 \times 8}{25} = 18.01 \text{ kN/m}$$

Self-weight $= 0.250 \times 0.500 \times 1 \times 25 = 3.130 \text{ kN/m}$

Max live load $= 18.01 - 3.13 = 14.88 \text{ kN/m}$

Live load (u.d.l.) $= 14.88 \text{ kN/m}.$

Example 3.13

Design a rectangular beam section of 300 mm width and 550 mm overall depth with a concrete cover of 50 mm to carry a bending moment of (a) 60 kN-m (b) 100 kN-m. Use M20 grade concrete and Fe 415 grade steel.

Solution: (a) $M = 60$ kN-m, the beam section may be designed as singly reinforced or doubly reinforced depending on its balanced moment of resistance.

For M20 CC and Fe 415 (Table 3.1)

$n_c = 0.284, j = 0.905, Q = 0.90, b = 300 \text{ mm}, D = 550 \text{ mm}, d = 500 \text{ mm}, m = 13.$

$\sigma_{cbc} = 7 \text{ N/mm}^2, \sigma_{st} = 230 \text{ N/mm}^2, M_{bal} = Qbd^2 = 0.90 \times 300 \times (500)^2$

$$= 67.5 \times 10^6 \text{ N-mm}$$

$M < M_{bal}$, hence beam shall be designed as singly under reinforced.

$$A_{st_1} = \frac{M}{\sigma_{st} \cdot jd} = \frac{60 \times 10^6}{230 \times 0.905 \times 500} = 577 \text{ mm}^2 \text{ (provide 3 bars of 16 mm diameter)}$$

$$A_{st} = 603 \text{ mm}^2$$

$$p = \frac{A_{st_1}}{bd} = \frac{603}{300 \times 500} = 0.00402 \ (0.402\%).$$

Actual modified values of n (NA depth coefficient) for $p = 0.00402$, $m = 13$

$$n_a = (m^2 p^2 + 2mp)^{1/2} - mp = (0.0027311 + 0.10452)^{1/2} - 0.0523 = 0.2752$$

Actual $j = \left(1 - \dfrac{n_a}{3} \right) = 0.9083, A_{st} = \dfrac{M}{\sigma_{st} \cdot jd} = \dfrac{60 \times 10^6}{230 \times 0.9083 \times 500} = 574 \text{ mm}^2$

Provide 3 bars 16 mm ϕ $(A_{st} = 603 \text{ mm}^2)$

M (tension) $= A_{st} \cdot \sigma_{st} \cdot jd = 603 \times 230 \times 0.9083 \times 500 = 63 \times 10^6 \text{ N-mm} = 63 \text{ kN-m}$

M (compression) $= \dfrac{1}{2} \sigma_{cbc} \cdot b \cdot nd \cdot jd$

$$= \frac{1}{2} \times 300 \times 0.2752 \times 500 \times 0.9083 \times 500 = 65.61 \times 10^6 \text{ N-mm}.$$

Hence safe permissible $M_r = 63$ kN-m.

(b) $M = 100$ kN-m, M_1 (singly reinforced) as in part (a) $= 63$ kN-m, $A_{st_1} = 603 \text{ mm}^2$ as in part (a).

M_2 (required) = (100 – 63) = 37 kN-m or (37 × 10^6 N-mm)

σ_{cbc} (extreme concrete fibre stress) for 63 × 10^6 N-mm = $7 \times \dfrac{63.0}{65.61} = 6.722 \text{ N/mm}^2$

Let us provide compression steel at 50 mm cover ($d' = 50$ mm).

$$A_{sc} = \frac{M_2}{(1.5m-1)\sigma'_{cbc}(d-d')} = \frac{37 \times 10^6}{18.5 \times \sigma'_{cbc}(450)}$$

$$\sigma'_{cbc} = \frac{\sigma_{cbc}(n_a d - d')}{n_a \cdot d} = 4.28 \text{ N/mm}^2$$

or $A_{sc} = \dfrac{37 \times 10^6}{18.5 \times 4.28 \times 450} = 1038 \text{ mm}^2$ (2 bars 20ϕ plus 2 bars 16ϕ or 4 bars 18ϕ),

$A_{sc} = 1030 \text{ mm}^2$ or 1018 mm^2

$$A_{st_2} = \frac{M_2}{\sigma_{st}(d-d')} = \frac{37 \times 10^6}{230(500-50)} = 358 \text{ mm}^2 \text{ (2 bars of 16 mm}\phi, A_{st} = 402 \text{ mm}^2\text{)}.$$

Total $A_{st} = A_{st_1} + A_{st_2} = (603 + 402) = 1005 \text{ mm}^2$
(5 bars of 16ϕ) or (4 bars of 18ϕ)

$\qquad A_{sc} = 1016 \text{ mm}^2$ (4 bars 18 mmϕ)

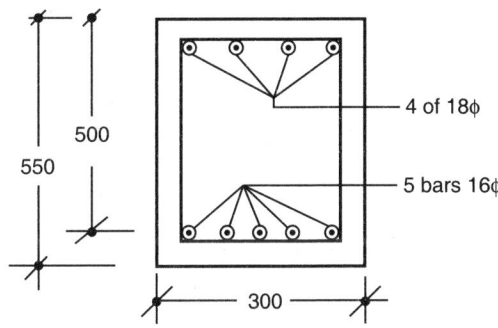

Fig. 3.10: Details of reinforcement (for $M = 100$ kN-m)

Example 3.14

A doubly reinforced rectangular beam is 250 mm wide and 500 mm deep. If the permissible stresses in concrete and steel are 10 N/mm² and 230 N/mm² respectively, determine the steel areas to sustain a maximum bending moment of 120 kN-m. Assume the concrete cover of 40 mm on both faces. Take $m = 9$.

Solution: $\sigma_{cbc} = 10 \text{ N/mm}^2$, $\sigma_{st} = 230 \text{ N/mm}^2$, $m = 9$, $M_r = 120 \times 10^6 \text{ N-mm}$
$\qquad d = 500 - 40 = 460 \text{ mm}$, $d' = 40 \text{ mm}$, $b = 250 \text{ mm}$.

$$n_c \text{ (balance)} = \frac{m\sigma_{cbc}}{m\sigma_{cbc} + \sigma_{st}} = \frac{9 \times 10}{(90 + 230)} = 0.2813, n_c d = 129.4 \text{ mm}$$

$$\sigma'_{cbc} = \frac{\sigma_{cbc}(n_c \cdot d - d')}{n_c \cdot d} = \frac{10 \times (129.4 - 40)}{129.4} = 6.91 \text{ N/mm}^2$$

$$M_r = b \cdot n_c \cdot d \cdot \frac{\sigma_{cbc}}{2}\left(460 - \frac{129.4}{3}\right) + (1.5 \times 9 - 1) \cdot A_{sc} \times 6.91 \,(460 - 40)$$

$$= 120 \times 10^6 \text{ N-mm.}$$

or $\quad 12.5\, A_{sc} \cdot 6.91 \times 420 = 10^6\left[120 - 250 \times 129.4 \times \frac{10}{2} \times \frac{416.9}{10^6}\right] = 10^6 \times 52.6$

$$A_{sc} = \frac{52.6 \times 10^6}{12.5 \times 6.91 \times 420} = 1449 \text{ mm}^2 \text{ (5 bars of 20 mm}\phi \text{ (1570 mm}^2\text{) or 6 bars of 18}\phi$$
(1526 mm²)

By considering internal equilibrium of forces, we have

total tension = total compression

$$A_{st} \cdot \sigma_{st} = b \cdot n_c \cdot d \cdot \frac{\sigma_{cbc}}{2} + (1.5m - 1)\, A_{sc} \cdot \sigma'_{cbc}$$

$$230\, A_{st} = 250 \times 129.4 \times \frac{10}{2} + 12.5 \times 1449 \times 6.91 = 161750 + 125157 = 286907 \text{ N}$$

$$A_{st} = \frac{286907}{230} = 1270 \text{ mm}^2$$

Alternatively:

$$j \text{ (lever arm)} = \left(d - \frac{n_c d}{3}\right) = \left(460 - \frac{129.4}{3}\right) = 416.9 \text{ mm}$$

$$M_1 \text{ (Singly reinforced)} = b \cdot n_c \cdot d \cdot \frac{\sigma_{cbc}}{2}\left(d - \frac{n_c d}{3}\right)$$

$$= 250 \times 129.4 \times \frac{10}{4} \times 416.9 = 67.44 \times 10^6 \text{ N-mm}$$

$$M_2 = 120 \times 10^6 - 67.44 \times 10^6 = 52.66 \times 10^6 \text{ N-mm}$$

$$A_{st_1} = \frac{M_1}{\sigma_{st} \cdot jd} = \frac{67.44 \times 10^6}{230 \times 416.9} = 703.3 \text{ mm}^2$$

$$A_{st_2} = \frac{M_2}{\sigma_{st}(d - d')} = \frac{52.66 \times 10^6}{230 \times 420} = 545 \text{ mm}^2$$

Total $\quad A_{st} = 703.3 + 545.2 = 1248.5 \text{ mm}^2$

Compression steel to balance A_{st_2}

$$A_{sc} = \frac{m(d - nd) \times A_{st}}{(1.5m - 1)(nd - d')} = \frac{9 \times (460 - 129.4) \times 545}{12.5 \times 89.4} = 1451 \text{ mm}^2 \,(A_{sc} > A_{st} \text{ and is un-}$$

economical). Hence by steel beam theory,

$$A_{st} = A_{sc} = \frac{M}{\sigma_{st}(d - dc)} = \frac{120 \times 10^6}{230 \times (460 - 40)} = 1242 \text{ mm}^2$$

Example 3.15

Design a RC beam subjected to a reversal of bending moment of equal magnitude of ±150 kN-m. The permissible stresses in concrete and steel are 7 N/mm² and 230 N/mm² respectively and $m = 14$. Design the rectangular section if $b = 0.5d$, and cover (effective) = 0.1d.

Solution: $\sigma_{cbc} = 7\,\text{N/mm}^2$, $\sigma_{st} = 230\,\text{N/mm}^2$, $m = 14$, $n_c = \dfrac{m\sigma_{cbc}}{(m\sigma_{cbc} + \sigma_{st})} = \dfrac{14 \times 7}{98 + 230}$

$$= 0.299$$

$b = 0.5d$, cover $d' = 0.10d$

Since BM gets reversed, provide tensile steel on both the faces, hence the beam section acts as doubly reinforced beam section and $A_{st} = A_{sc} = A_s$. Equating the moment of equivalent area about NA, we get

$$\frac{b \cdot n^2 d^2}{2} + (1.5m - 1)\, A_{sc}\,(nd - d') = m\, A_{st}\,(d - nd)$$

Putting $A_{st} = A_{sc} = A_s$,

$b = 0.5d$, $d' = 0.1d$, we get

$$\frac{0.5d(0.299)^2 \cdot d^2}{2} + 20\, A_s\,(0.299d - 0.1d) = 14\, A_s\,(d - 0.299d)$$

or $0.02235d^3 + 3.98A_s \cdot d = 9.814A_s \cdot d$

or $0.02235d^2 = 5.834 \cdot A_s,$ $A_s = 0.03831d^2$

$$\sigma'_{cbc} = \frac{\sigma_{cbc}\,(n \cdot d - d')}{n \cdot d} = \frac{7(0.299 - 0.1)d}{0.299d} = 4.66\,\text{N/mm}^2$$

$$A_{sc} = 0.003831d^2, \quad M_r = M = 150 \times 10^6\,\text{N-mm}$$

∴ $0.5d\,(0.299d)\,\dfrac{7}{2}\left(d - \dfrac{0.299d}{3}\right) + (1.5 \times 14 - 1)\,0.003831d^2 \times 4.66 \times (d - 0.1d)$

$$= 150 \times 10^6$$

or $0.4711d^3 + 0.3213d^3 = 150 \times 10^6$

or $0.7924d^3 = 150 \times 10^6$, or $d^3 = \dfrac{150 \times 10^6}{0.7924} = 189.29 \times 10^6$

or $d = 574.17\,\text{mm}$, $b = 0.5 \times 574.17 = 287.1\,\text{mm}$, $d' = 57.42\,\text{mm}$

$A_{st} = A_{sc} = A_s = 0.003831d^2 = 1263\,\text{mm}^2$ (4 bars of 20 mmφ)

Practical dimension of beam 290 mm × 580 mm (effective), cover 55 mm, $A_{st} = 1256\,\text{mm}^2 = A_{sc}$, provide 4 bars of 20 mm in each face at a cover of 55 mm.

3.7 ANALYSIS AND DESIGN OF T-BEAM SECTIONS

3.7.1 Introduction

We have seen that the concrete is strong in compression, but it is not effective in tension. The concrete in beam sections is effectively used to resist forces on compression face while the concrete on tension side does not resist any substantial force and is used only to embed the tension reinforcement. Thus, it suggests that the flexural members should be such that the beam has greater concrete area (or width) in compression face in comparison to the area (or width) in tension face.

In beam–slab construction, a T-beam is more common than a independent rectangular beam construction to support slab (Fig. 3.11). Generally, beam and slab construction is done monolithically and thus beam behaves as T-beam.

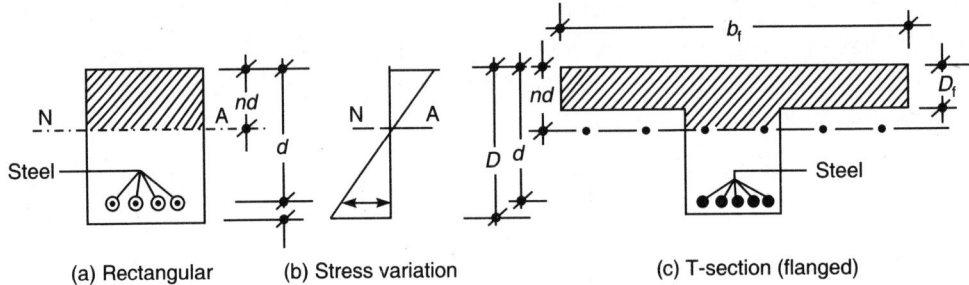

| (a) Rectangular | (b) Stress variation | (c) T-section (flanged) |

Fig. 3.11: Rectangular and flanged T–sections in beams

The slab in beam–slab construction acts as a compression flange of a T-beam while the rectangular section below the slab forms the web of the T-beam. In this system the roof slab is designed to bend over the span AC, CE or BD, DF, etc. (Fig. 3.12 (a) and (b) and the beams bend over span AB, CD or EF, etc. The main reinforcement of slab (positive and negative) is provided at right angles to AB, CD or EF, etc. The beam supports uniformly distributed load transferred from the slab. The slab forms the upper part of the T-beam. The slab bends in the lateral direction while as T-beam flange it bends in the longitudinal direction (AB, CD or EF) and resists compression.

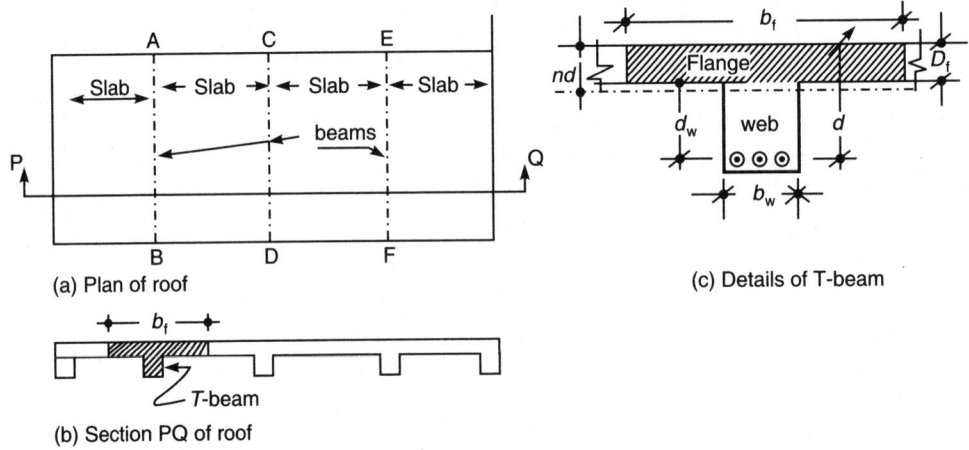

(a) Plan of roof

(b) Section PQ of roof

(c) Details of T-beam

Fig. 3.12: T-beam roof–slab construction

3.7.2 Dimensions of a T-beam

Figure 3.13 shows arrangement of T-beam–slab and cross-section details of T-beam: (i) breadth b_f of flange (ii) thickness of the flange (slab) D_f (iii) breadth of the rib b_w (iv) depth of the rib d_w and (v) depth of T-beam d (top of flange to tensile steel level).

(i) Breadth of the flange (b_f)

Breadth b_f of the flange is that portion of slab which acts monolithically with the beam and which resists the compressive stresses due to bending. According to IS:456-2000, a slab which is assumed to act as a flange of a T-beam (or of a L-beam) shall satisfy the following:

(a) The slab be cast integrally with the web, or the web and the slab shall be effectively bonded together;

(b) If the main reinforcement of the slab is parallel to the beam, transverse reinforcement shall be provided as shown in Fig. 3.13. The transverse reinforcement shall not be less than 60% of the main reinforcement at the mid span of the slab.

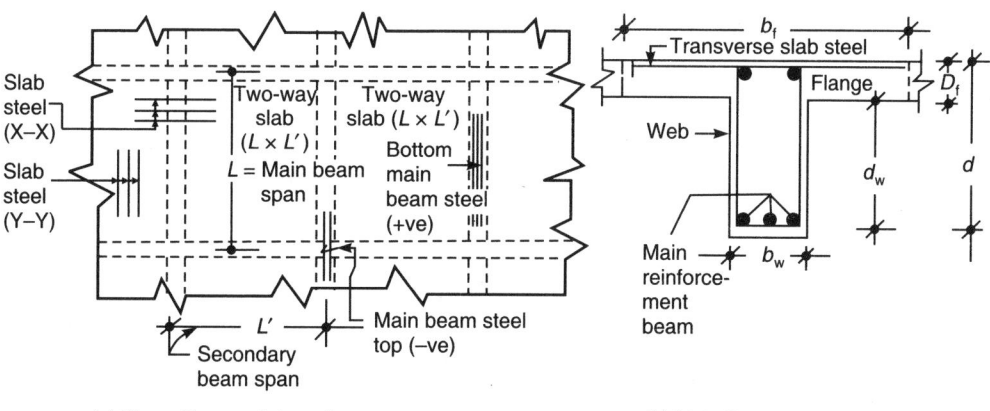

(a) Plan of beam-slab roof   (b) Main T-beam section

Fig. 3.13: Beam–slab arrangement of roof

Generally main span of the slab spans across T-beams, whereby the main reinforcement of the slab (beam flange) runs at right angles to the main span of the beam. This reinforcement helps the slab and the beam in bonding and ensuring the monolithic action between the beam and the slab.

According to IS:456-2000, the effective flange width b_f shall be taken as the least of the following:

(a) $b_f = \dfrac{l_0}{6} + b_w + 6D_f$

(b) $b_f = b_w + \dfrac{1}{2}$ (sum of clear distances between the adjacent beams on either side)

(c) For isolated T-beams, the effective flange width shall be obtained as below, but in no case greater than the actual width (refer to IS:456-2000, page 37).

$$b_f = \frac{l_0}{\left(\dfrac{l_0}{b} + 4\right)} + b_w$$

where, b_f = Effective width of flange

l_0 = Distance between points of zero moments in the beam

b_w = Breadth of web (width of beam rib on tension side)

D_f = Thickness of flange (total thickness of slab)

b = Actual width of flange

Note: For continuous beams and frames, l_0 may be taken as 0.70 times the effective span.

(ii) Thickness of the flange (D_f)

The thickness of the flange is taken equal to the total thickness of slab including concrete cover. D_f is calculated on the basis of the bending of slab between the beams.

(iii) Breadth of the rib or web (b_w)

The breadth of the rib is equal to the width of the tension face of the web to accommodate the required tensile reinforcement. The width of rib should be at least 1/3rd of the depth of the rib (d_w). Architecturally, the width of the rib should be equal to the width of the supporting column.

(iv) Depth of rib or web (d_w)

The depth of rib is the distance between the bottom of flange (slab) and centre of the tensile steel and is dependent on the effective depth (d) of the beam (i.e. the distance between the top of the flange and centre of the tensile reinforcement),

i.e. $$d = d_w + D_f$$

The overall depth (D) of the T-beam is generally assumed as 1/12 to 1/15 of the simply supported span. The ratio of overall depth (D) to the span of continuous beams may be assumed on the basis of loading pattern as:

- Light loading: 1/20 to 1/15 (D/L)
- Medium loading: 1/15 to 1/12 (D/L)
- Heavy loading: 1/12 to 1/10 (D/L)

3.7.3 Position of Neutral Axis (NA)

Neutral axis in a T-beam may be located inside or outside the flange (Fig. 3.14). NA may fall within the flange in case of comparatively thicker slab or flange. NA depth is determined in the same manner as that adopted in case of singly reinforced rectangular sections. The breadth b is replaced with b_f in case of T-beam, when NA lies within the flange, i.e.

$$b_f \cdot \frac{(nd)^2}{2} = mA_{st}(d - nd) \qquad \qquad \text{... Eq. (3.19)}$$

when $nd = D_f$, we have

$$\frac{bf}{2}(D_f)^2 = mA_{st}(d - D_f) \qquad \qquad \text{... Eq. (3.20)}$$

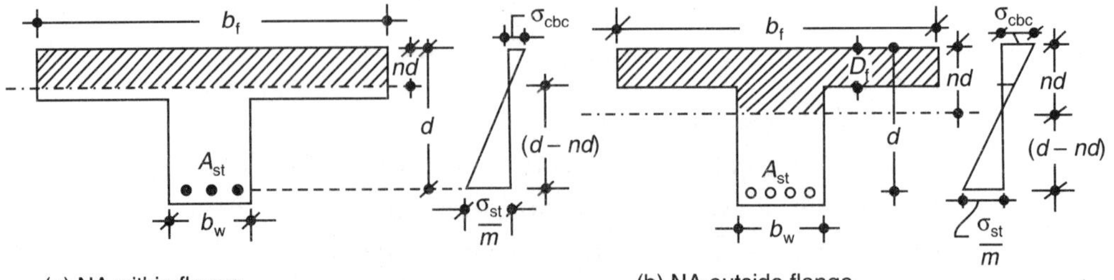

(a) NA within flange   (b) NA outside flange

Fig. 3.14: Location of NA in T-beam sections

In case the NA falls outside the flange, the location of NA can be determined as usual by taking moment of compressive area and tensile area about the NA.

$$b_f \cdot D_f \left(n \cdot d - \frac{D_f}{2} \right) + \frac{b_w \cdot (nd - D_f)^2}{2} = mA_{st} \, (d - nd) \qquad \dots \text{Eq. (3.21)}$$

In Eq. (3.21) the term $\dfrac{b_w \, (nd - D_f)^2}{2}$ is usually very small and it can be neglected. In case this term is neglected the expression for nd becomes

$$b_f \cdot D_f \left(nd - \frac{D_f}{2} \right) = mA_{st} \, (d - nd) \qquad \dots \text{Eq. (3.22)}$$

For given dimensions of T-beam, NA can be determined by using Eqs (3.19), (3.20), (3.21) or (3.22). When nd determined by Eq. (3.19) gives $nd > D_f$, revise the nd values by using Eq. (3.22).

When the maximum permissible stresses in concrete (σ_{cbc}) and steel (σ_{st}) are known, the NA can be determined by using stress diagram:

$$\frac{\sigma_{cbc}}{nd} = \frac{\sigma_{st}}{m(d - nd)} \quad \text{or} \quad \frac{n}{(1-n)} = \frac{m\sigma_{cbc}}{\sigma_{st}} \qquad \dots \text{Eq. (3.23)}$$

3.7.4 Lever Arm and Moment of Resistance

Figure 3.15 shows a T-beam along with stress diagram and CG of forces for a case when NA lies outside the flange. The stress at the bottom level of flange is say σ_{cb_1}. CG of forces $\tilde{y}$ from the top of the flange is given as:

$$\tilde{y} = \frac{\left\{ \sigma_{cbc} \cdot \dfrac{D_f}{2} \cdot \dfrac{D_f}{3} \right\} + \left(\sigma_{cb_1} \dfrac{D_f}{3} D_f \right)}{\left(\sigma_{cbc} \cdot \dfrac{D_f}{2} \right) + \left(\sigma_{cb_1} \cdot \dfrac{D_f}{2} \right)} = \frac{(\sigma_{cbc} + 2\sigma_{cb_1})}{(\sigma_{cbc} + \sigma_{cb_1})} \cdot \frac{D_f}{3} \qquad \dots \text{Eq. (3.24)}$$

But from stress diagram, in Fig. 3.15, we have

$$\sigma_{cb_1} = \frac{(nd - D_f)}{nd} \cdot \sigma_{cbc} \qquad \dots \text{Eq. (3.25)}$$

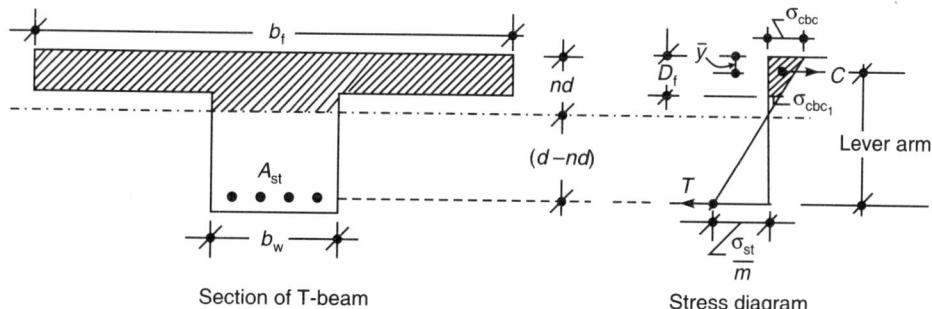

Section of T-beam   Stress diagram

Fig. 3.15: T-beam and lever arm of couple (C or T × lever arm)

Putting value of σ_{cb_1} in Eq. (3.24), we have

$$\tilde{y} = \frac{1 + 2\dfrac{(nd - D_f)}{nd}}{\left(1 + \dfrac{nd - D_f}{nd} \right)} \times \frac{D_f}{3} = \frac{(3nd - 2D_f)}{(2nd - D_f)} \cdot \frac{D_f}{3} \qquad \dots \text{Eq. (3.26)}$$

$$\text{Lever arm} = (d - \tilde{y}) = d - \frac{D_f}{3} \cdot \left(\frac{3nd - 2D_f}{2nd - D_f} \right) \qquad \text{... Eq. (3.27)}$$

If the actual NA is above the critical NA, the section is under reinforced, the stress σ_{st} in steel reaches its maximum permissible value σ_{st} before the stress σ_{cbc} reaches its maximum permissible stress of compression in concrete. If the actual NA falls below its critical NA value, the section is over reinforced (Fig. 3.16).

The moment of resistance of the T-beam is found by multiplying the total resultant compression and the lever arm (Fig. 3.16).

When the NA falls within the flange,

$$M_r = b_f \cdot nd \cdot \frac{\sigma_{cbc}}{2} \left(d - \frac{nd}{3} \right) \qquad \text{... Eq. (3.28)}$$

When the NA lies in the web, the moment of resistance is given by

$$M_r = b_f \cdot D_f \frac{\sigma_{cbc} + \sigma_{cbc_1}}{2} \times \text{lever arm} \qquad \text{... Eq. (3.28a)}$$

Neglecting the small compression of the rib,

or $\qquad M_r = \frac{b_f \cdot D_f \cdot \sigma_{cbc}}{2} \left\{ 1 + \frac{nd - D_f}{nd} \right\} \left\{ d - \frac{D_f}{3} \left(\frac{3nd - 2D_f}{2nd - D_f} \right) \right\} \qquad \text{... Eq. (3.28b)}$

Also $\qquad M_r = \sigma_{st} \cdot A_{st} \left\{ d - \frac{D_f}{3} \left(\frac{3nd - 2D_f}{2nd - D_f} \right) \right\} \qquad \text{... Eq. (3.29)}$

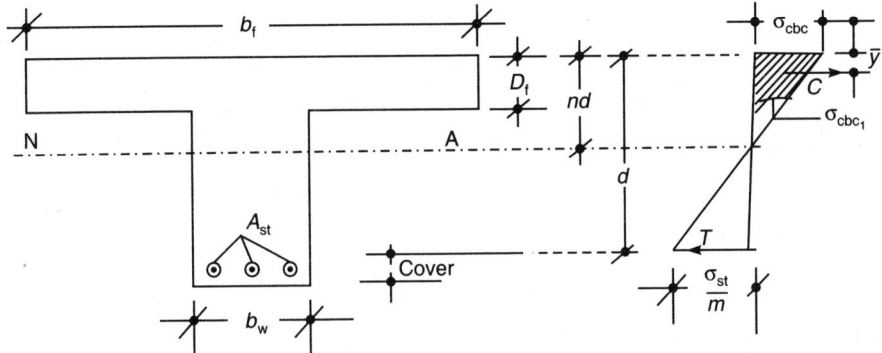

Fig. 3.16: T-beam section and stress diagram

When the rib is also considered, total moment of resistance is the sum of moment of resistance of the flange and the web (rib) portion above NA, i.e.

$$M_r = M_{rf} + M_{rb}$$

$$= \frac{b_f \cdot D_f \cdot \sigma_{cbc}}{2} \left\{ 1 + \frac{nd - D_f}{nd} \right\} \left\{ d - \frac{D_f}{3} \left(\frac{3nd - 2D_f}{2nd - D_f} \right) \right\}$$

$$+ b_w \cdot (nd - D_f) \cdot \frac{\sigma_{cbc_1}}{2} \left\{ d - nd + \frac{2}{3} (nd - D_f) \right\} \qquad \text{... Eq. (3.30)}$$

3.7.5 Balanced Depth of T-Beam

Balanced or critical depth factor of NA of T-beam is found by the same equation as

$$n = \frac{m\sigma_{cbc}}{(m\sigma_{cbc} + \sigma_{st})}, \text{ with usual notations.}$$

In case of T-beam, the depth of balanced section is determined by ignoring the small compressive force offered by web portion and considering compression of flange only. Compressive stress at the level of flange bottom,

$$\sigma_{cbc_1} = \frac{\sigma_{cbc}}{nd}(nd - D_f) = \sigma_{cbc}\left(1 - \frac{D_f}{nd}\right)$$

$$\text{Compressive force of flange} = b_f \cdot D_f\left(\frac{\sigma_{cbc} + \sigma_{cbc_1}}{2}\right) = \frac{b_f \cdot D_f}{2}\left\{\sigma_{cbc} + \sigma_{cbc}\left(1 - \frac{D_f}{nd}\right)\right\}$$

or
$$C = \frac{b_f \cdot D_f \cdot \sigma_{cbc}}{2}\left\{1 + 1 - \frac{D_f}{nd}\right\} = \frac{b_f \cdot D_f \cdot \sigma_{cbc}}{2}\left\{2 - \frac{D_f}{nd}\right\} = \frac{b_f \cdot D_f \cdot \sigma_{cbc}(2nd - D_f)}{2nd}$$

i.e.
$$C = b_f \cdot D_f \cdot \sigma_{cbc}\frac{(2nd - D_f)}{2nd} \qquad \text{... Eq. (3.31)}$$

$$\text{Lever arm} = \left\{d - \frac{D_f}{3}\left(\frac{3nd - 2D_f}{2nd - D_f}\right)\right\}$$

∴
$$M_r = C \times \text{lever arm} = b_f \cdot D_f \cdot \sigma_{cbc}\left(\frac{(2nd - D_f)}{2nd}\right)\left\{d - \frac{D_f}{3}\frac{(3nd - 2D_f)}{2nd - D_f}\right\}$$

Lever arm in T-beams generally varies between 0.87d to 0.95d and for approximation it may be assumed as 0.90d. For this approximation,

$$M_r = b_f \cdot D_f \cdot \sigma_{cbc}\left(\frac{2nd - D_f}{2nd}\right)(0.90d)$$

or
$$M_r = 0.45b_f \cdot D_f \cdot \sigma_{cbc}\left(\frac{2nd - D_f}{n}\right) \qquad \text{... Eq. (3.32)}$$

Thus, the unknown depth d can be found from known values of M_r, b_f, D_f, σ_{cbc} and n. From this calculated value of d other values of tension steel may be calculated for the design of T-beam. If necessary, more accurate values of d may be calculated by considering the accurate location of NA and rib compression, etc.

3.7.6 Economic Depth of T-beam

In T-beam sections, most of the compressive area of concrete is provided by the flange over a small depth (D_f) and hence the critical depth (d) is also small. Thus, it requires large area of tension steel. The cost of steel is several times (r) more than that of concrete and thus making T-beam construction uneconomical. Since the flange concrete is fixed from the slab consideration, let us consider concrete of web and tension steel for economical depth calculations.

Let R_s = Cost of steel per cubic mm (steel cost $R_s = r$ times concrete cost R_c)

           R_c = Cost of concrete per cubic mm

$$r = \frac{R_s \text{ (cost of steel)}}{R_c \text{ (cost of concrete)}} \text{ per unit volume}$$

           d_c = Depth of concrete cover to steel

Volume of concrete per unit length in web $= b_w (d - D_f + d_c) \times 1$

Cost of concrete in the rib per mm length $= b_w (d - D_f + d_c) R_c$

Also, area of steel $A_{st} = \dfrac{M}{\sigma_{st} \cdot jd}$, where $\sigma_{st} = $ permissible stress of steel.

Total cost (Q) of T-beam per mm length is given as

$$Q = R_c \cdot b_w (d - D_f + d_c) + (rR_c)\dfrac{M}{\sigma_{st} \cdot jd} = R_c \left[b_w (d - D_f + d_c) + \dfrac{r \cdot M}{\sigma_{st} \cdot jd} \right]$$

In this equation b_w, D_f and d_c are fixed dimensions, whereas d is variable and affects the total cost Q. Hence, for the cost Q to be minimum, we have $\dfrac{dQ}{dd} = 0$,

or
$$\dfrac{d}{dd} \left\{ R_c \, b_w (d - D_f + d_c) + \dfrac{r \cdot M}{\sigma_{st} \cdot jd} \right\} = 0$$

or
$$R_c \left[b_w (1) - \dfrac{M \cdot r}{\sigma_{st} \cdot j \cdot (d^2)} \right] = 0, \text{ or } b_w \cdot \sigma_{st} \cdot jd^2 = M \cdot r, \text{ since } D_f, d_c \text{ are fixed}$$

or
$$d = \sqrt{\dfrac{M \cdot r}{\sigma_{st} \cdot b_w \cdot j}} \qquad \text{... Eq. (3.33)}$$

If
$$\sigma_{st} = 140 \text{ N/mm}^2, j = 0.9, d = \sqrt{\dfrac{M \cdot r}{126 b_w}} \qquad \text{... Eq. (3.33a)}$$

If
$$\sigma_{st} = 230 \text{ N/mm}^2, j = 0.9, d = \sqrt{\dfrac{M \cdot r}{207 b_w}} \qquad \text{... Eq. (3.33b)}$$

The width b_w of web is kept minimum to accommodate tension steel. If necessary, steel may be provided in two layers which may reduce the effective depth.

3.7.7 Shear and Development Length

According to IS:456-2000, the nominal shear stress $\tau_v = \dfrac{V}{b_w \cdot d}$ ... Eq. (3.34)

where, $\tau_v = $ Nominal shear stress (N/mm^2)

             $V = $ Shear force (N)

             $b_w = $ Width of web

             $d = $ Depth of T-beam (effective)

For the bond, IS:456-2000 recommends the development length as $L_d = \dfrac{\phi \cdot \sigma_{st}}{4 \tau_{bd}}$

                                                               ... Eq. (3.35)

where, $L_d = $ Development length (mm)

             $\sigma_{st} = $ Permissible tensile stress in steel (N/mm^2)

             $\tau_{bd} = $ Permissible bond stress in steel (N/mm^2)

              $\phi = $ Bar diameter (mm)

Types of problems in T-beams are similar to rectangular sections, like:

- Determination of moment of resistance of a given T-section
- Determination of stresses in a given T-section subjected to a given bending moment
- Design of a section to resist a given bending moment (or loading and span) of a T-beam.

Example 3.16

An isolated T-beam is simply supported over a span of 8 m. The dimension of the T-beam are: Breadth of flange 750 mm, total thickness of the flange 125 mm, over all depth of the beam 450 mm, breadth of the rib 300 mm, effective cover to tensile reinforcement 50 mm and tensile reinforcement of 4 bars of 20 mm. Determine the moment of resistance of the beam if steel of Fe 415 grade and concrete of M20 grade are used.

Solution: $\sigma_{cbc} = 7\,\text{N/mm}^2$, $\sigma_{st} = 230\,\text{N/mm}^2$ (Fe 415), span $l_0 = 8000$ mm

For M20 CC: $b_w = 300$ mm, $D_f = 125$ mm, D (overall) $= 450$ mm,

$$d \text{ (effective)} = 450 - 50 = 400 \text{ mm}, \; A_{st} = 4\left(\frac{\pi}{4} \times 20^2\right) = 1256 \text{ mm}^2$$

$$m = \frac{280}{3\sigma_{cbc}} = \frac{280}{21} = 13.3 = 13$$

Effective flange width in an isolated T-beam,

$$b_f = \frac{l_0}{\left(\dfrac{l_0}{b} + 4\right)} + b_w = \frac{8000}{\left(\dfrac{8000}{750} + 4\right)} + 300 = \left(\frac{8000 \times 750}{11000} + 300\right) = 845.5 \text{ mm}$$

(actual available $b_f = 750$ mm)

Thus, $b_f = 750$ mm

Let NA lies in the flange, equating the moments of equivalent areas about the NA,

$$\frac{750}{2}(nd)^2 = 13\,(1256)\,(400 - nd)$$

or $\qquad 375(nd)^2 + 16328nd - 6531200 = 0,$

or $\qquad nd^2 + 43.54nd - 17416.5 = 0$

$$nd = \frac{-43.54 \pm \sqrt{43.54^2 + 4 \times 17416.5}}{2} = \frac{-43.54 \pm 267.51}{2}$$

$$= 112.0 \text{ mm} < 125 \text{ mm}$$

Hence, NA lies within the flange.

Critical NA $nd_c = \dfrac{m\sigma_{cbc}(400)}{m\sigma_{cbc} + \sigma_{st}} = \dfrac{13 \times 7 \times 400}{13 \times 7 + 230} = \dfrac{91 \times 400}{321} = 113.4 \text{ mm} (>112 \text{ mm})$

Since the actual NA (112 mm) lies above the critical NA (113.4 mm), the section is under reinforced and hence M_r is based on the tensile face and σ_{cbc} (actual)

$$= \frac{\sigma_{st} \times nd}{m(d - nd)} = \frac{230}{13} \times \frac{112}{(400 - 112)} = 6.88 \text{ N/mm}^2$$

Thus $\quad M_r = (b_f \cdot nd) \dfrac{\sigma_{cbc}}{2}\left(d - \dfrac{nd}{3}\right) = 750 \times 112 \times \dfrac{6.88}{2}\left(400 - \dfrac{112}{3}\right)$

$\qquad\qquad = 104.797123 \text{ N-mm} \times 10^6$

Also $\quad M_r = A_{st} \cdot \sigma_{st} \cdot jd = 1256 \times 230 \times \dfrac{1088}{3} = 104770000 \text{ N-mm}$

Hence $\quad M_r = 104.797123 \text{ kN-m}$

Example 3.17

As isolated T-beam has a flange width of 1200 mm, flange thickness of 100 mm and effective depth of 400 mm. The web width is 250 mm and is reinforced with 5 bars of 20 mm mild steel (Fe 250) in tension zone. Determine the moment of resistance of the section, if the permissible stresses in concrete and steel are 5.0 N/mm² and 140 N/mm² respectively. Assume $m = 19 \left(\dfrac{280}{3 \times 5} = 18.67 = 19\right)$. Neglect the compression in the web.

The beam is simply supported over a span of 4 m.

Solution: $L_0 = 4000$ mm, $b_f = 1200$ mm, $D_f = 100$ mm, $d = 400$ mm, $b_w = 250$ mm,
$\qquad A_{st} = 1570$ mm².

The effective $b_f = \dfrac{l_0}{\left(\dfrac{l_0}{b}\right)+4} + b_w = \dfrac{4000}{\left(\dfrac{4000}{1200}\right)+4} + 250 = 795.5$ mm

$\qquad\qquad\qquad\qquad\qquad\qquad\qquad = 796$ mm (< 1200 mm actual)

Thus, effective width of the flange for T-beam section $= 796$ mm

Assuming the NA to fall within the flange, we have

$$b_f \cdot \dfrac{nd^2}{2} = m \cdot A_{st}\,(d - nd) \text{ or } \dfrac{796}{2} \times nd^2 = 19\,(1570)\,(400 - nd)$$

or $\qquad\qquad nd^2 + 74.95\, nd - 29980 = 0$

$nd = \dfrac{-74.95 \pm \sqrt{74.95^2 + 4 \times 29980}}{2} = \dfrac{-74.95 + 354.31}{2} = 139.70 \text{ mm } (> 100 \text{ mm } D_f)$

Therefore, NA falls in the web.

Revision of NA

Neglecting the compression of the web and equating the moments of equivalent areas about the NA,

$$796 \times 100 = \left(nd - \dfrac{100}{2}\right) = 19 \times 1570\,(400 - nd)$$

or $\qquad 109430nd = \dfrac{19 \times 1570 \times 400}{39800000}$ or $109430nd = 15912000 = 145.4$ mm

$\therefore \qquad\qquad nd = 145.4$ mm

$$nd_c \text{ (critical NA)} = \dfrac{m\sigma_{cbc} \times 400}{(m\sigma_{cbc} + \sigma_{st})} = \dfrac{19 \times 5 \times 400}{(19 \times 5 + 140)} = \dfrac{95 \times 400}{235} = 161.7 \text{ mm}$$

Actual NA is above critical NA and hence the section is under reinforced and the actual stress in steel reaches to 140 N/mm² first.

$$\sigma_{st} = 140 \, N/mm^2, \, m = 19$$

Thus, $$\sigma_{cbc} = \frac{140 \times 145.4}{19 \times (400 - 145.4)} = 4.21 \, N/mm^2$$

$$\sigma'_{cbc} = \frac{(145.4 - 100)4.21}{145.4} = 1.314 \, N/mm^2$$

Hence, CG of compressive forces $\tilde{y}$ (from top)

$$\tilde{y} = \frac{(\sigma_{cbc} + 2\sigma'_{cbc})}{(\sigma_{cbc} + \sigma'_{cbc})} \cdot \frac{D_f}{3} = \frac{(4.21 + 2 \times 1.314)}{(4.21 + 1.314)} \times \frac{100}{3} = 41.26 \, mm$$

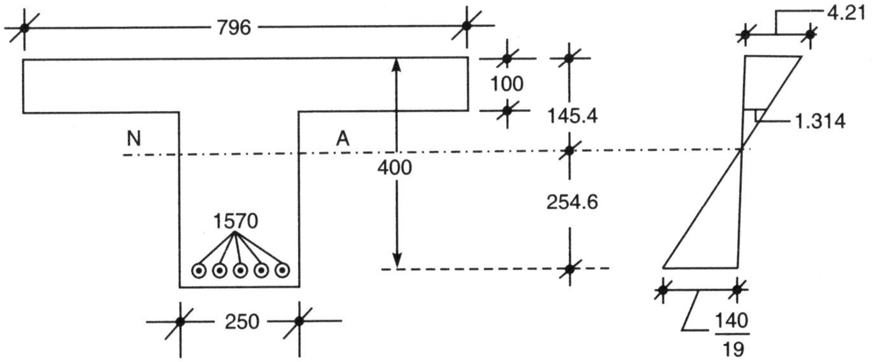

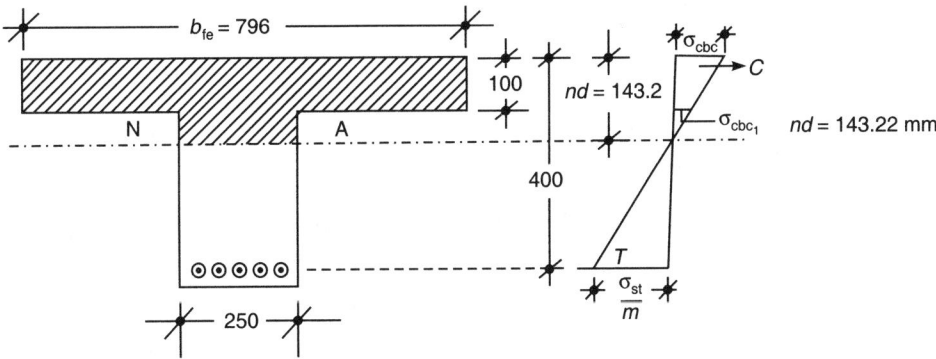

Fig. 3.17: Details of T-section and stress variation (N/mm²)

Lever arm $(jd) = (d - \tilde{y}) = (400 - 41.26) = 358.74 \, mm, \left(j = \frac{358.74}{400} = 0.897 \right)$

Moment of resistance $(M_r) = b_f D_f \left(\frac{\sigma_{cbc} + \sigma'_{cbc}}{2} \right) jd$

or $M_r = 796 \times 100 \left(\frac{4.21 + 1.314}{2} \right) 358.74 = 78870855 \, N\text{-}mm \, (78.87 \, kN\text{-}m)$

Alternatively:

$M_r = A_{st} \cdot \sigma_{st} \cdot jd = 1570 \times 140 \times 358.74 = 78860000 \, N\text{-}mm \, (78.86 \, kN\text{-}m)$

Example 3.18

Determine the moment of resistance of an isolated beam as in Example 3.17 by considering compressive force of the web also.

Fig. 3.18: Details of T-section and stress variation

Solution: $m = 19$, $A_{st} = 1570$ mm², b_{fe} (effective) $= 796$ mm $< b_f = 1200$ mm, $b_w = 250$ mm, $l_0 = 4000$ mm, $D_f = 100$ mm, $d = 400$ mm, $\sigma_{st} = 140$ N/mm², $\sigma_{cbc} = 5$ N/mm² (permissible)

Equating the moments of equivalent areas about the NA, we have

$$796 \times 100 \left(nd - \frac{100}{2} \right) + 250 \, (nd - 100) \left(\frac{nd - 100}{2} \right) = 19 \times 1570 \, (400 - nd)$$

or $\qquad 79600nd - 79600 \times 50 + 125 \, (nd - 100)^2 = 11932000 - 29830nd$

or $\qquad 109430nd - 15912000 + 125 \, (nd^2 - 200nd + 10000) = 0$

or $\qquad 125nd^2 + 84430nd - 14662000 = 0$

or $\qquad nd^2 + 675.44nd - 117296 = 0$

∴ $\qquad nd = 143.22$ mm $< nd_c = 161.7$ mm

The section is under reinforced.

Hence, $\sigma_{st} = 140$ N/mm² (actual), $nd = 143.22$ mm

$$\sigma_{cbc} \text{ (actual)} = \frac{140(143.22)}{19 \times (400 - 143.22)} = 4.11 \text{ N/mm}^2 \text{ (actual)}$$

$$\sigma_{cbc_1} = \frac{4.11}{143.22} \, (143.22 - 100) = 1.24 \text{ N/mm}^2 \text{ (actual)}$$

$\tilde{y} = $ (CG of compression)

$$= \frac{b_f \cdot D_f \left(\dfrac{\sigma_{cbc} + \sigma_{cbc_1}}{2} \right) \dfrac{D_f}{3} \left\{ \dfrac{2\sigma_{cbc_1} + \sigma_{cbc}}{\sigma_{cbc_1} + \sigma_{cbc}} \right\} + b_w \, (nd - D_f) \dfrac{\sigma_{cbc_1}}{2} \left\{ D_f + \dfrac{nd - D_f}{3} \right\}}{b_f \cdot D_f \left\{ \dfrac{\sigma_{cbc} + \sigma_{cbc_1}}{2} \right\} + b_w \, (nd - D_f) \cdot \dfrac{\sigma_{cbc_1}}{2}}$$

$$\tilde{y} = \frac{\dfrac{796 \times 100}{2} (4.11 + 1.24) \dfrac{100}{3} \left\{ \dfrac{2.48 + 4.11}{4.11 + 1.24} \right\} + 250(43.22) \dfrac{1.24}{2} \left\{ 100 + \dfrac{43.22}{3} \right\}}{79600 \left\{ \dfrac{4.11 + 1.24}{2} \right\} + 250 \, (43.22) \dfrac{1.24}{2}}$$

$$= \frac{8742733 + 766421}{212930 + 6699} = \frac{1640654}{219629} = 7.47 \text{ mm from top}$$

Lever arm $jd = (d - \tilde{y}) = 400 - 7.47 = 392.53$ mm

∴ $M_r = A_{st} \cdot \sigma_{st} \cdot jd = 1570 \times 140 \times 392.53 = 86278094$ N-mm (86.28 kN-m)

Alternatively, $M_r = \left\{ b_f \cdot D_f \left(\dfrac{\sigma_{cbc} + \sigma_{cbc_1}}{2} \right) + b_w \, (nd - D_f) \dfrac{\sigma_{cbc_1}}{2} \right\} jd$

$$M_r = \left\{ 79600 \left(\frac{4.11 + 1.24}{2} \right) + 250 \, (43.22) \frac{1.24}{2} \right\} 392.53$$

$$= \{212930 + 6699\} \, 392.53 = 86220000 \text{ N-mm (86.22 kN-m)}$$

$M_r = 86.22$ kN-m

Example 3.19

An isolated T-beam has flange width of 1200 mm, thickness of 100 mm and effective depth 400 mm. The web width is 250 mm and is reinforced with 5 bars of 25 mm diameter of Fe 415 steel. The concrete is of M20 grade. Permissible stresses for steel $\sigma_{st} = 230$ N/mm², concrete $\sigma_{cbc} = 7$ N/mm², $m = 13$. Determine the maximum u.d.l. which a cantilever beam of 3 m span can carry safely including its own weight. Neglect the compression in the web portion. The beam is so placed that the flange is under compression.

Solution: b_f (actual) = 1200 mm, b_w = 250 mm, d = 400 mm, D_f = 100 mm

$$\text{Projection span } l_0 = 3 \text{ m}, A_{st} = 5 \times \frac{\pi}{4} (25)^2 = 2454 \text{ mm}^2, m = 13, \sigma_{st} = 230 \text{ N/mm}^2$$

$$\sigma_{cbc} = 7 \text{ N/mm}^2$$

Effective width of flange for T-section is least of:

(i) $b_f = \dfrac{l_0}{\dfrac{l_0}{b}+4} + b_w = \dfrac{3000}{\dfrac{3000}{1200}+4} + 250 = 712$ mm

(ii) $b_f = \dfrac{l_0}{6} + b_w + 6D_f = \dfrac{3000}{6} + 250 + 6 \times 100 = 1350$ mm

(iii) $b_f = b_w + \dfrac{1}{2}$ (sum of clear distances to the adjacent beams on either side) not applicable

Hence, $b_{fe} = 712$ mm

Let the NA be in the web and neglect the compression in the web. Taking the moments of equivalent areas about NA,

$$712 \times 100 \left(nd - \frac{100}{2} \right) = 13 \times 2454 (400 - nd)$$

or $\qquad (71200 + 31902)nd = 12760800 + 3560000)$

$$nd \text{ (actual)} = \frac{16320800}{103102} = 158.3 \text{ mm}$$

$$nd_c = \frac{m\sigma_{cbc} \cdot d}{m\sigma_{cbc} + \sigma_{st}} = \frac{13 \times 7 \times 400}{13 \times 7 + 230} = 113.4 \text{ mm} < nd_a, \text{ over reinforced}$$

Hence, stress in concrete develops first at permissible level at 7 N/mm².

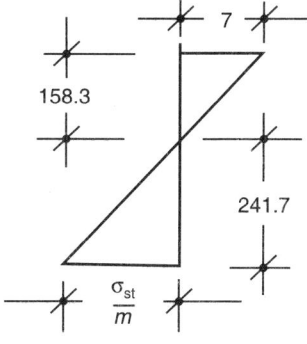

Fig. 3.19: Stress variation across T-section

$$\text{Stress in steel} = \frac{m \times 241.7 \times 7}{158.3} = 139 \text{ N/mm}^2$$

$$\sigma'_{cbc} = \frac{7 \times 58.3 \times 7}{158.3} = 2.58 \text{ N/mm}^2$$

$$\bar{y} = (\text{CG of compression}) = \frac{\sigma_{cbc} + 2\sigma'_{cbc}}{\sigma_{cbc} + \sigma'_{cbc}} \times \frac{D_f}{3} = \frac{7 + 5.16}{7 + 2.58} \times \frac{100}{3} = 42.3 \text{ mm}$$

Lever arm $= (d - \bar{y}) = (400 - 42.3) = 357.7 \text{ mm}$

$$M_r = b_f \cdot D_f \cdot \left(\frac{\sigma_{cbc} + \sigma'_{cbc}}{2} \right) jd = 71200 \left(\frac{7 + 2.58}{2} \right) 357.7 = 122000000 \text{ N-m (122 kN-m)}$$

Alternatively:

$M_r = A_{st} \cdot \sigma_{st} \cdot jd = 2454 \times 139 \times 357.7 = 122013616 \text{ N-mm (122.14 kN-m)}$

Let the u.d.l. on the cantilever T-beam (including self-weight) $= w$ kN/m, span $= 3$ m.

$$\text{Max BM (cantilever)} = \frac{w \times (3)^2}{2} = 4.5w \text{ kN-m} = 122 \text{ kN-m}$$

$$\therefore \qquad w \text{ (u.d.l.)} = \frac{122}{4.5} = 27.11 \text{ kN/m (including self-weight of beam).}$$

Example 3.20

The flange of a T-beam is 150 mm thick and 1500 mm wide. The web of the beam is 300 mm and the depth upto the centre of tensile steel is 700 mm. Tensile reinforcement consists of 6 bars of 25 mm diameter arranged in one layer. If the beam section is subjected to a BM of 300 kN-m, find the stresses developed in concrete and steel if $m = 13$, and simply supported beam span is 8 m.

Solution: $m = 13$, $A_{st} = 6 \times 491 = 2946 \text{ mm}^2$, $l_0 = 8000 \text{ mm}$, $D_f = 150 \text{ mm}$,

$$d = 700 \text{ mm, Actual } b_f = 1500 \text{ mm, effective } b_f = \frac{l_0}{\dfrac{l_0}{b} + 4} + 300 = \frac{8000}{\dfrac{8000}{1500} + 4} + 300$$

$$= 1157.14 \text{ mm} \cong 1157 \text{ mm}$$

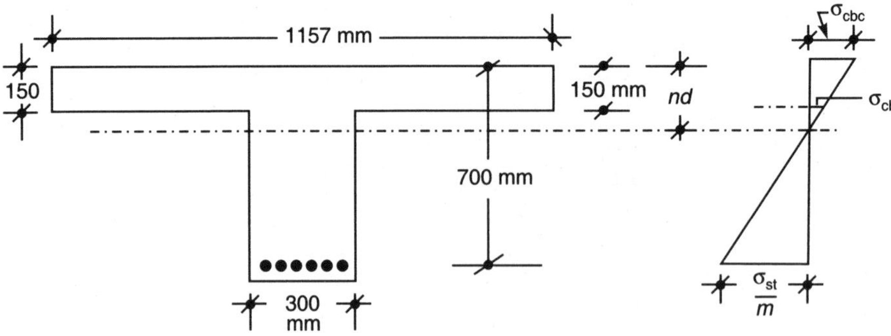

Fig. 3.20: T-section and stress variation

Thus, effective $b_f = 1157 \text{ mm}$, $D_f = 150 \text{ mm}$

Neglecting the compression in web and equating the moments of equivalent areas about NA, we have

$$1157 \times 150 \left(nd - \frac{150}{2} \right) = 13 \times 2946 (700 - nd)$$

or $\qquad nd\ (173550 + 38298) = (26808600 + 13016250) = 39824850$

$$nd = \frac{(39824850)}{(211848)} = 188.0 \text{ mm} > 150 \text{ mm} \ (D_f)$$

If σ_{cbc} is the maximum stress in concrete flange top,

$$\sigma_{cbc_1} = \frac{\sigma_{cbc} \cdot (188 - 150)}{188} = 0.202\sigma_{cbc}$$

$$\tilde{y} = \frac{D_f}{3}\left(\frac{2\sigma_{cbc_1} + \sigma_{cbc}}{\sigma_{cbc_1} + \sigma_{cbc}}\right) = \frac{150}{3} \times \frac{1.404}{1.202} = 58.4 \text{ mm},$$

lever arm $= 700 - \tilde{y} = 641.6$ mm

$$M_r = b_f\,D_f \cdot \left(\frac{\sigma_{cbc} + \sigma_{cbc_1}}{2}\right)jd = 1157 \times 150 \left(\frac{1.202}{2}\right)\sigma_{cbc} \times 641.6 = 66921157.7\sigma_{cbc}$$

$\therefore \quad 300 \times 10^6 = 66921157.7\sigma_{cbc}, \ \sigma_{cbc} = 4.482 \text{ N/mm}^2 \sim 4.48 \text{ N/mm}^2$

$$\sigma_{st} = \frac{M_r}{A_{st} \cdot jd} = \frac{300 \times 10^6}{2946 \times 641.6} = 158.72 \text{ N/mm}^2$$

Example 3.21

A reinforced concrete T-beam has a flange 1200 mm wide and 120 mm thick. The effective depth of beam is 450 mm and width of web is 300 mm. The value of $m = 13$, and permissible stresses in concrete and steel are respectively 7 N/mm² and 230 N/mm². The beam is simply supported over a span of 8 m. Determine the tensile reinforcement and M_r of the beam section.

Solution: $l_0 = 8$ m, $d = 450$ mm, $b_f = 1200$ mm, $D_f = 120$ mm, $b_w = 300$ mm, $m = 13$.

$\qquad \sigma_{st} = 230 \text{ N/mm}^2, \ \sigma_{cbc} = 7 \text{ N/mm}^2$

Effective flange width $b_f = \dfrac{l_0}{\dfrac{l_0}{b}+4} + b_w = \dfrac{8000}{\dfrac{8000}{1200}+4} + 300 = 1050$ mm

Critical NA $= nd_c$

$$nd_c = \frac{m\sigma_{cbc} \times d}{m\sigma_{cbc} + \sigma_{st}} = \frac{13 \times 7 \times 450}{13 \times 7 + 230} = 0.2835 \times 450 = 127.6 \text{ mm}$$

Equating the moments of equivalent areas about NA and neglecting the compressive area of web

$$1050 \times 120\left(127.6 - \frac{120}{2}\right) = 13\,A_{st}\,(450 - 127.6)$$

$$A_{st} = \frac{1050 \times 120 \times 67.6}{13 \times 322.4} = 2033 \text{ mm}^2 \text{ (balanced section)}$$

$$M_r = b_f \cdot D_f \cdot \left(\frac{\sigma_{cbc} + \sigma'_{cbc}}{2}\right) \text{ lever arm}$$

$$\sigma'_{cbc} = \left(\frac{nd_c - D_f}{nd_c}\right) \times \sigma_{cbc} = \left(\frac{127.6 - 120}{127.6}\right) \times 7 = 0.42 \text{ N/mm}^2$$

$$\tilde{y} = \left(\frac{\sigma_{cbc} + 2\sigma'_{cbc}}{\sigma_{cbc} + \sigma'_{cbc}}\right) \cdot \frac{D_f}{3} = \left(\frac{7+0.84}{7+0.42}\right) \times \frac{120}{3} = 42.26 \text{ mm}$$

Lever arm $= (d - \tilde{y}) = (450 - 42.26) = 407.74 \text{ mm}$

$$M_r = A_{st} \cdot \sigma_{st}\, jd = b_f \cdot D_f \left(\frac{\sigma_{cbc} + \sigma'_{cbc}}{2}\right) jd = 1050 \times 120 \times \left(\frac{7+0.42}{2}\right) \times 407.74$$

$$= 190.6 \times 10^6 \text{ N-mm} = 190.6 \text{ kN-m}$$

Example 3.22

A T-beam roof consists of 120 mm thick slab cast monolithically with 300 mm wide web (beams) which are spaced 4 m centre to centre. The super-imposed load including slab and web self-weight is 5 kN/m^2. The effective span of simply supported beam is 6.0 m. Determine the depth and area of tensile steel in intermediate beams if the concrete of M20 grade and steel of Fe 415 grade are used in the beam. Consider $m = \dfrac{280}{3 \times 7} = 13$. Find the actual M_r of the section.

Solution: $D_f = 120$ mm, $b_f = $ (least of $\dfrac{l_0}{6} + b_w + 6D_f$ and c/c of beams)

$$b_f \text{ (effective)} = \frac{6000}{6} + 300 + 6(120) = 2020 \text{ mm or } 4000 \text{ c/c}, b_f = 2020 \text{ mm}$$

Effective d is determined from economic depth criteria by assuming lever arm (jd) and cost of steel to concrete ratio $'r'$. Thus criteria are based on maximum stresses in bending, shear and limiting deflection, etc. these criteria shall be discussed in later chapters.

Assume $j = 0.90$, and r (cost ratio) say $= 60$ for preliminary calculations.

Load on beam = weight of 4 m slab span $= 4 \times 5 = 20$ kN/m on beam

$$w = 20 \text{ kN/m, BM (max)} = \frac{20}{8}(L^2) = \frac{5}{2}(6 \times 6) = 90 \text{ kN-m } (90 \times 10^6 \text{ N-mm})$$

$$\text{Economic depth } d = \sqrt{\frac{r.M}{\sigma_{st} \cdot j \cdot b_w}} = \sqrt{\frac{60 \times 90 \times 10^6}{230 \times 0.9 \times 300}} = 295 \text{ mm (provide 300 mm)}$$

Lever arm $= 0.9 \times 300 = 270$ mm

$$A_{st} = \frac{M}{\sigma_{st} \times 270} = \frac{90 \times 10^6}{230 \times 270} = 1450 \text{ mm}^2 \text{ (provide 5 bars of 20 mm, } A_{st} = 1570 \text{ mm}^2)$$

Now by considering actual dimensions, we have to check the stresses in steel and concrete, which must be within safe permissible limits.

$$d = 300 \text{ mm, } b_f = 2020 \text{ mm, } b_w = 300 \text{ mm, } A_{st} = 1570 \text{ mm}^2,$$

Equating the moments of the equivalent areas about NA, we have

$$2020 \times 120 \left(nd - \frac{120}{2}\right) = 13(1570)(300 - nd)$$

Neglecting web area $242400 (nd - 60) = 20410 (300 - nd)$

$$nd = \frac{20667000}{262810} = 78.64 \text{ mm } (< 120 \text{ mm, i.e. NA in flange})$$

Thus, nd (actual) lies within the flange.

$$M_r = b_f \cdot nd \cdot \frac{\sigma_{cbc}}{2} \times \left(d - \frac{nd}{3} \right) = \frac{2020 \times 78.64}{2} \sigma_{cbc} \left(300 - \frac{78.64}{3} \right)$$

Coefficient of lever arm $= 273.79/300 = 0.912$

$$90 \times 10^6 = 1010 \times 78.64 \times 273.79 \, \sigma_{cbc},$$

σ_{cbc} (actual) $= 4.14 \, \text{N/mm}^2 \, (< 7 \, \text{N/mm}^2)$, OK

Also $\qquad 90 \times 10^6 = (1570)\sigma_{st} \times 273.79$

σ_{st} (actual) $= 209.4 \, \text{N/mm}^2 \, (< 230 \, \text{N/mm}^2)$, OK

Thus, effective depth $= 300 \, \text{mm}$ (overall $D = 350 \, \text{mm}$, with cover $= 50 \, \text{mm}$)

Area of tensile steel $(A_{st}) = 1570 \, \text{mm}^2$ (5 bars of 20 mm)

Stresses in concrete and steel are: $\sigma_{cbc} = 4.14 \, \text{N/mm}^2 \, (< 7.0 \, \text{N/mm}^2)$

$$\sigma_{st} = 209.4 \, \text{N/mm}^2 \, (< 230 \, \text{N/mm}^2)$$

Actual BM capacity (under reinforced) $= M = A_{st} \cdot \sigma_{st} \cdot jd$

$$= 1570 \times 230 \times 273.79 = 98.87 \, \text{kN-m}$$

Alternatively, $M = b_f \cdot nd \cdot \dfrac{\sigma_{cbc}}{2} \times jd = 2020 \times 78.64 \times \dfrac{7}{2} \times 273.79 = 152.22 \, \text{kN-m}$

Thus, safe moment of resistance $= 98.87 \, \text{kN-m}$ at a stress of $\sigma_{st} = 230 \, \text{N/mm}^2$ in steel.

SUMMARY

The concrete elements subjected to bending moment develop flexural compression on one face and flexural tension on the other face. Concrete is weak in tension and hence steel reinforcement is provided to resist tension on the tension face. The reinforced concrete beams offer composite action due to bond between the concrete and the steel reinforcement.

For simplified analysis and design of RCC beams, the important assumptions are:

(i) Plane sections before bending remain plane after bending

(ii) The stress–strain diagrams for concrete and steel remain straight line up to working loads

(iii) Concrete resists compression in compression zone while steel reinforcement resists tension in tension zone

(iv) Modular ratio $E_s/E_c \, (m) = \dfrac{280}{3\sigma_{cbc}}$, (as per IS:456-2000)

(v) Moduli of elasticity of concrete and steel under compression and tension shall be considered the same.

The equivalent concrete area for steel areas (A_{sc} and A_{st}) shall be:

$\qquad A_{ec} = A + (m-1) A_{sc}$ in compression zone

$\qquad A_{et} = m \, A_{st}$ in tension zone (since concrete is ineffective in tension)

Neutral axis (NA) represents a line joining fibres of zero stress strain. The internal compression and tension forces developed due to external loading always remain in equilibrium (i.e. $C = T$). The NA passes through the centroid of the section.

When the permissible compressive stress in extreme concrete fibre and the permissible tensile stress in extreme steel bars develops simultaneously, the section is balanced and NA is called critical. The critical NA depth coefficient (n_c) is given by

$$n_c = \frac{m\sigma_{cbc}}{(m\sigma_{cbc} + \sigma_{st})}$$

The actual NA depth coefficient (n_a) can be found by equating moment of equivalent compressive area about NA and moment of equivalent tensile area of the section about NA, i.e.

$$b \cdot n_a \cdot d \cdot \frac{n_a \cdot d}{2} = m\, A_{st}\,(d - n_a \cdot d), \text{ where } n_a = \text{actual NA depth factor}$$

If $A_{st} = p \cdot bd$, we have $n_a = -mp \pm \sqrt{2mp + (mp)^2}$

Moment of resistance, M_r = external BM (M), i.e.

$$b \cdot n_a \cdot d \left(d - \frac{n_a \cdot d}{3} \right) \times \frac{\sigma_{cbc}}{2} = M = A_{st} \cdot \sigma_{st} \cdot jd, \text{ where } j = \left(1 - \frac{n_a}{3} \right)$$

$$M = Qbd^2 = \frac{1}{2}\sigma_{cbc} \cdot n_a \left(1 - \frac{n_a}{3} \right) bd^2 = A_{st} \cdot \sigma_{st} \cdot jd$$

Whenever: (i) $n_a > n_c$, the section shall be over reinforced

$$(M \text{ is based on concrete compression, } M = \frac{b}{2} \cdot nd\sigma_{cbc} \cdot jd)$$

(ii) $n_a = n_c$, the section shall be balanced and

$$(M = \frac{b \cdot nd}{2}\sigma_{cbc} \cdot jd \text{ or } A_{st} \cdot \sigma_{st} \cdot jd)$$

(iii) $n_a < n_c$, the section shall be under reinforced

$$(M \text{ is based on steel tension, } M = A_{st} \cdot \sigma_{st} \cdot jd)$$

Stress in steel and concrete are related as follows:

$$\frac{\sigma_{st}}{m(d - n_a \cdot d)} = \frac{\sigma_{cbc}}{n_a \cdot d} \text{ or } \frac{\sigma_{st}}{\sigma_{cbc}} = \frac{m(d - n_a \cdot d)}{n_a \cdot d} = \frac{m(1 - n_a)}{n_a}$$

If the external BM is more than the critical moment of resistance of the given section, the moment of resistance of the section is enhanced either by increasing the depth of the section or if there is restriction on increasing the depth, steel reinforcement is also provided in the compression face. Stress in compression steel is based on the strain in concrete at the level of compression steel (stress $\sigma_{sc} = E_s \cdot e_{sc} = mE_c \cdot e_{sc} = m\sigma'_{cbc}$). As per IS:456-2000, the internal equilibrium equation is given as:

$$b \cdot nd \cdot \frac{nd}{2} + (1.5m - 1)\, A_{sc} \cdot (nd - d') = m\, A_{st}\,(d - nd)$$

$$jd = \left(d - \frac{nd}{3} \right) + (1.5m - 1)\, p' \left(\frac{nd}{3} - d' \right)\left(n - \frac{d'}{d} \right)$$

where, $p' = \dfrac{A_{sc}}{bd}$ and $p = \dfrac{A_{st}}{bd}$

Total M_r of doubly reinforced beam $M = M_1 + M_2$

or
$$M = Qbd^2 + \sigma'_{cbc} (1.5m - 1) A_{sc}(d - d')$$

$$A_{st_1} = \frac{Qbd^2}{jd \cdot \sigma_{st}}, \text{ for balanced singly reinforced portion.}$$

$$M_2 = \sigma'_{cbc} (1.5m - 1) A_{sc} (d - d'),$$

$$A_{st_2} = \frac{M_2}{\sigma_{st} \cdot jd} = \frac{(1.5m - 1) A_{sc} \sigma'_{cbc}(d - d')}{\sigma_{st} \cdot jd}$$

where, σ'_{cbc} = Concrete stress at the level of compression steel

d' = Cover to compression steel, A_{sc} = Compression steel

or
$$A_{st_2} = \frac{(1.5m - 1) A_{sc} \sigma'_{cbc}}{\sigma_{st}}, \text{ since } jd = (d - d')$$

A_{st_2} bars are placed in level with A_{st_1} bars.

Also by equilibrium of internal forces, we have

Total tensile force (T) = Total compressive forces $(C_c + C_s)$

In roof slab–beam construction, beams are cast monolithically with the slab. The slab acts as the flange of the T-beam and the beam below slab acts as web. In T-beam action, the flange provides maximum area of concrete in compression using the concrete most efficiently. Web area generally remain in tension zone and provides adequate space to accommodate tension steel bars and connect flange slab and beam monolithically.

According to IS:456-2000, the effective flange width b_f is given as least of:

$$b_f = \left(\frac{l_0}{6} + b_w + 6D_f\right) \text{ or } [b_w + \frac{1}{2}(\text{clear distance between adjacent beams})]$$

For isolated beams, the effective flange width (b_f) shall be obtained as

$$b'_f (\text{effective}) = \left[\frac{l_0}{\left(\frac{l_0}{b_f} + 4\right)} + b_w\right], \text{ but not greater than actual } b_f.$$

l_0 in continuous beams may be taken as $0.70l$

where, l = effective span

Generally overall depth (D) in T-beam is assumed to be:

1/20 to 1/15 of span for light loading

1/15 to 1/12 of span for medium loading

1/12 to 1/10 of span for heavy loading.

In case of T-beam section, the NA may lie within the flange (slab), or outside the flange in web. For approximate design calculations, the compression in web (being small) is generally neglected without severely affecting the design. Neutral axis is calculated by different equations based on its location w.r.t. flange.

(a) When NA lies within the flange:

$$\frac{b'_f (nd)^2}{2} = mA_{st} (d - nd), b'_f \text{ is effective flange width}$$

$$M_r = b'_f \, nd \cdot \frac{\sigma_{cbc}}{2}\left(d - \frac{nd}{3}\right)$$

(b) When NA lies outside the flange and compression in web is neglected:

$$b'_f \, D_f \left(nd - \frac{D_f}{2}\right) = m A_{st} \, (d - nd)$$

CG of compressive forces from top of flange $(\tilde{y})$

$$\tilde{y} = \frac{\left\{\sigma_{cbc} \cdot \frac{D_f}{2} \cdot \frac{D_f}{3}\right\} + \left\{\sigma_{cb_1} \cdot \frac{D_f}{2} \cdot \frac{2D_f}{3}\right\}}{\left(\sigma_{cbc} \cdot \frac{D_f}{2}\right) + \left(\sigma_{cb_1} \cdot \frac{D_f}{2}\right)} = \frac{(\sigma_{cbc} + 2\sigma_{cb_1})}{(\sigma_{cbc} + \sigma_{cb_1})} \cdot \frac{D_f}{3}$$

where, $\quad \sigma_{cb_1} = \left(\dfrac{nd - D_f}{nd}\right) \cdot \sigma_{cbc},$ and $\tilde{y} = \dfrac{3nd - 2D_f}{2nd - D_f} \cdot \dfrac{D_f}{3}$

Lever arm $(jd = d - \tilde{y}) = \left[d - \dfrac{D_f}{3} \dfrac{(3nd - 2D_f)}{(2nd - D_f)} \right]$

$$M_r = b_f \cdot D_f \cdot \frac{\sigma_{cbc}}{2} \left\{ 1 + \frac{nd - D_f}{nd} \right\} \left[d - \frac{D_f}{3} \frac{(3nd - 2D_f)}{(2nd - D_f)} \right]$$

Also $\qquad M_r = \sigma_{st} \cdot A_{st} \left[d - \dfrac{D_f}{3} \dfrac{(3nd - 2D_f)}{(2nd - D_f)} \right]$

Value of j may be assumed $= 0.90$ (generally j varies from 0.87 to 0.95)

$$M_r = 0.45 \, b_f \cdot D_f \cdot \sigma_{cbc} \left(\frac{2nd - D_f}{n}\right)$$

Economical depth of T–beam section

$$d_e = \sqrt{\frac{r \cdot M}{\sigma_{st} \cdot b_w}}, \text{ where } r \text{ is ratio of cost of steel to that of CC.}$$

While design requires to satisfy equilibrium of internal forces, various equations can be used for the design and analysis.

PRACTICE QUESTIONS

(I) Objective Questions

Q. 3.1 Select the correct response given after each statement to complete it correctly and fill in the response sheet provided.

(i) The combined effective action of RCC section does not depend on
....

(a) the bond between steel and concrete

(b) the corrosion of the embedded steel

(c) the coefficient of expansion of steel and concrete

(d) the grades of steel and concrete.

(ii) Following assumption is not made in case of elastic theory (working stress method) of RCC beams: ....

    (a) plane section before bending remains plane section after bending

    (b) the stress–strain relationship of concrete and steel are straight lines under service loads

    (c) the moduli of elasticity of steel and concrete are equal

    (d) the concrete resists basically compressive force and the steel resists tensile force.

(iii) The equivalent concrete area of a uncracked RCC column section will be A_e = ... (with usual notations of steel area A_{st}, column area A and modular ratio m).

    (a) $(A + m\,A_{sc})$

    (b) $\{A + (m - 1)\,A_{sc}\}$

    (c) $(mA + A_{sc})$

    (d) $(m - 1)\,A + A_{sc}$

(iv) In an uncracked RCC column section, the strain in steel will be .... the strain in concrete.

    (a) more than

    (b) less than

    (c) equal to

    (d) m times (where m is modular ratio)

(v) According to IS:456-2000, the modular ratio of steel to concrete (m) shall be taken as ....

    (a) $m = \dfrac{280}{3\sigma_{cbc}}$, where σ_{cbc} is permissible stress of concrete

    (b) $m = \dfrac{280\sigma_{st}}{\sigma_{cbc}}$, where σ_{st} is permissible stress of steel

    (c) $m = \dfrac{5000\sqrt{f_{ck}}}{E_c}$, where f_{ck} is characteristic strength of concrete and E_s is elastic modulus of steel

    (d) $m = \dfrac{E_c}{280\,\sigma_{cbc}}$, where E_s and σ_{cbc} are as specified in options (a) and (c) above

(vi) In a rectangular singly reinforced beam, the moment of resistance of the section in working stress method will be M_r = ...with usual notations.

    (a) $\sigma_{cbc}\,j \cdot n \cdot b \cdot d^2$

    (b) $\dfrac{1}{2}\sigma_{cbc}\,n \cdot j \cdot bd^2$

    (c) $\dfrac{1}{3}\sigma_{cbc}\,n \cdot j \cdot bd$

    (d) $\dfrac{1}{2}\sigma_{st}\,n(1 - n)\,b \cdot d$

(vii) According to working stress method, the moment coefficient in a singly reinforced rectangular RCC beam will be Q = ... with usual notations.

(a) $\dfrac{1}{\left(1+\dfrac{\sigma_{st}}{m\sigma_{cbc}}\right)}$

(b) $\left(1-\dfrac{n_c}{3}\right)$

(c) $\dfrac{280}{\sigma_{cbc}}$

(d) $0.5 \cdot \sigma_{cbc} \cdot n_c \cdot \left(1-\dfrac{n_c}{3}\right)$

(viii) In working stress method, if the actual NA is above critical NA, the section will be .... and the moment of resistance will be based on ....

   (a) under reinforced, compression in concrete
   (b) over reinforced, compression in concrete
   (c) under reinforced, tension in steel
   (d) over reinforced, tension in steel

(ix) In working stress method, if the actual NA lies below the critical NA, the section will be .... and the moment of resistance will be based on ....

   (a) under reinforced, tension in steel
   (b) over reinforced, compression in concrete
   (c) under reinforced, compression in concrete
   (d) over reinforced, tension in steel

(x) In working stress method, if the actual NA coincides with the critical NA, the section will be known as .... and the moment of resistance (M_r) of a singly reinforced beam section will be based on ....

   (a) under reinforced, only tension in steel
   (b) over reinforced, only tension in concrete
   (c) balanced, either tension in steel or compression in concrete
   (d) balanced, either tension in concrete or compression in steel

(xi) Whenever there is limit to depth of rectangular beam section and the required moment of resistance (M_r) is much higher, the design of the section is best done by ....

   (a) increasing the breadth of the beam section
   (b) increasing the area of tensile steel
   (c) decreasing the area of tensile steel
   (d) providing steel in compression face also.

(xii) A doubly reinforced rectangular beam section has steel reinforcement ....

   (a) double the steel quantity required for singly reinforced section
   (b) double the steel quantity needed for a beam of double the size
   (c) both in compression face as well as in tension face
   (d) both for side faces of the beam section

(xiii) Effective equivalent concrete area of compressive steel (A_{sc}) in a doubly reinforced beam section with tensile steel (A_{st}) and modular ratio (m) will be considered according to IS:456-2000 as ....

(a) $(1.5m - 1) A_{sc}$ (b) $(m - 1) A_{sc}$

(c) $(1.5m - 1) A_{st}$ (d) $m (A_{sc} + A_{st})$

(xiv) In a doubly reinforced rectangular beam section ($b \times d$ eff.) and reinforced with tensile steel (A_{st}) and compressive steel (A_{sc}), the NA with usual notations is found by the equation ....

(a) $\dfrac{b \cdot nd^2}{2} + (m - 1) A_{sc} (nd - d') = 1.5m A_{st} (d - nd)$

(b) $\dfrac{b \cdot nd^2}{3} + (1.5m - 1) A_{sc} (nd - d') = 1.5m A_{st} (d - nd)$

(c) $\dfrac{b \cdot nd^2}{2} + (1.5m - 1) A_{sc} (nd - d') = m A_{st} (d - nd)$

(d) $b \cdot nd^2 + (m - 1) A_{sc} (nd - d') \dfrac{1}{2} = m A_{st} (d - nd)$

(xv) In a doubly reinforced rectangular effective beam section ($b \times d$) reinforced with compression steel (A_{sc}) at an effective concrete cover (d') and tensile steel (A_{st}), the moment of resistance (M_r) with usual notations is given by $M_r =$.........

(a) $Qb \cdot d^2 + \sigma_{cbc} m \cdot A_{sc} (d - d')$

(b) $Qb \cdot d^3 + (m - 1) A_{sc} (d)$

(c) $Qb \cdot d^3 + (m - 1) A_{sc} \dfrac{nd - d'}{nd} \sigma_{cbc} (d)$

(d) $Qb \cdot d^2 + (1.5m - 1) A_{sc} \dfrac{nd - d'}{nd} \sigma_{cbc} (d - d')$

(xvi) In a slab-beam roof construction of a hall, the slab acts as the flange of the T-beam. According to IS:456-2000, the effective flange width (b_f) of an intermediate beam of an effective span (l_0) shall be ....

(a) least of centre to centre distance between adjacent beams and $\left(\dfrac{l_0}{6} + b_w + 6D_f \right)$

(b) greater of centre to centre distance between adjacent beams and $\left(\dfrac{l_0}{6} + b_w + 6D_f \right)$

(c) equal to centre to centre distance between adjacent beams

(d) equal to 12 times the depth (D) of the beam

(xvii) If the main reinforcement of the flange slab is parallel to the beam, the transverse reinforcement according to IS:456-2000 shall be ....

(a) more than the main reinforcement

(b) equal to the main reinforcement

(c) not less than 60% of the main reinforcement

(d) less than 40% of the main reinforcement

(xviii) For isolated T-beams, the effective flange width (b_f) with usual notations shall be b'_f ....

    (a) equal to the actual flange width (b_f)

    (b) equal to $\left\{ \dfrac{b_f \cdot l_0}{(l_0 + 4b_f)} + b_w \right\}$

    (c) greater of $\left\{ \dfrac{b_f \cdot l_0}{(l_0 + 4b_f)} + b_w \right\}$ and centre to centre span of the beam

    (d) least of centre to centre distance between the beams and $\left\{ \dfrac{l_0}{6} + b_w + 6D_f \right\}$

(xix) When the NA in a T-beam lies just at the bottom of the flange slab, the tensile steel can be found by equation ....

    (a) $b_f \cdot D_f \left\{ nd - \dfrac{D_f}{2} \right\} + \dfrac{b_w}{2} (nd - D_f)^2 = m A_{st} (d - nd)$

    (b) $b_f \cdot D_f \left\{ nd - \dfrac{D_f}{3} \right\} = m A_{st} (d - nd)$

    (c) $\dfrac{b_f \cdot D_f^2}{2} = m A_{st} (d - D_f)$

    (d) $\dfrac{b_f \cdot nd^2}{2} = m A_{st} \left(d - \dfrac{nd}{3} \right)$

(xx) Moment of resistance (M_r) of an over reinforced T-beam when the compression in the web in neglected, with usual notations is given by M_r = ...

    (a) $b_f \cdot D_f \cdot \dfrac{\sigma_{cbc}}{2} \left\{ 1 + \dfrac{nd - D_f}{nd} \right\} \left\{ d - \dfrac{D_f}{3} \left(\dfrac{3nd - 2D_f}{2nd - D_f} \right) \right\}$

    (b) $b_f \cdot D_f \cdot \dfrac{\sigma_{cbc}}{2} \left\{ 1 + \dfrac{nd - D_f}{nd} \right\} \left\{ d - \dfrac{D_f}{3} \left(\dfrac{3nd - 2D_f}{2nd - D_f} \right) \right\} +$

    $b_w (nd - D_f) \dfrac{\sigma_{cbc}}{2} \left\{ d - nd + \dfrac{2}{3} (nd - D_f) \right\}$

    (c) $0.45 b_f \cdot D_f \cdot \sigma_{cbc} \left(\dfrac{2nd - D_f}{n} \right)$

    (d) $b_w \cdot D_f \cdot \dfrac{\sigma_{cbc}}{2} \left\{ 1 + \dfrac{nd - D_f}{nd} \right\} \left\{ d - \dfrac{D_f}{3} \left(\dfrac{3nd - 2D_f}{2nd - D_f} \right) \right\}$

Response sheet to Q. 3.1 (i to xx)

Question	(i)	(ii)	(iii)	(iv)	(v)	(vi)	(vii)	(viii)	(ix)	(x)
Response (a/b/c/d)										

Question	(xi)	(xii)	(xiii)	(xiv)	(xv)	(xvi)	(xvii)	(xviii)	(xix)	(xx)
Response (a/b/c/d)										

(II) Numerical Questions

Q. 3.2 A rectangular beam section 300 mm wide and 500 mm deep is reinforced in tension zone with 2 bars of 12 mm diameter at an effective cover of 50 mm. The section is subjected to a service load causing maximum BM of 40 kN-m. M20 grade concrete and Fe 415 HYSD grade steel are used in beam section. Determine the stresses in concrete and steel. Also state if the RC section is balanced, under reinforced or over reinforced.

[**Hint:** $nd = 123.34$ mm, $nd_c = 127.8$ mm, section under reinforced, $\sigma_{cbc} = 5.29$ N/mm², $\sigma_{st} = 432.86$ N/mm² (unsafe as more than 230 N/mm²)]

Q. 3.3 Calculate the moment of resistance of a rectangular RCC beam section 300 mm wide and 500 mm effective depth and reinforced with 3 bars of 16 mm diameter in tension zone. Concrete of M25 grade and steel of Fe 415 HYSD grade are used in the beam section.

[**Hint:** Under reinforced, $M_r = 63.42$ kN-m, $nd_a = 128.22$ mm]

Q. 3.4 Calculate the M_r of a RCC rectangular beam section 300 mm × 500 mm (eff.) and reinforced with 6 bars of 20 mm diameter. Concrete of M25 grade and steel of Fe 415 HYSD are used. Also, find the actual stresses in steel and concrete at maximum permissible moment of resistance.

[**Hint:** Design constants: $\sigma_{st} = 230$ N/mm², $\sigma_{cbc} = 8.5$ N/mm², $m = 11$, $n_c = 0.288$, $j = 0.90$, $Q = 1.10$, actual $nd_a = 202.68$ mm (over reinforced), $M_r = 111.75$ kN-m, $\sigma_{st} = 137.16$ N/mm², $\sigma_{cbc} = 8.5$ N/mm²]

Q. 3.5 A simply supported RCC one way slab of 4.0 m effective span is reinforced with 10 mm diameter bars of Fe 415 HYSD steel spaced at 200 mm centre to centre at an effective depth of 180 mm. Using M20 grade concrete, determine the maximum permissible live load on the slab if the self-weight of slab and finishes are 5.5 kN/m².

[**Hint:** Consider 1 m strip. Design constants: $\sigma_{st} = 230$ N/mm², $\sigma_{cbc} = 7.0$ N/mm², $nd_c = 0.284d$ (51.12 mm), $nd_a = 38.2$ mm, under reinforced, $M_r = 15.2$ kN-m, $w_1 = 2.1$ kN/m² (in addition to $w_d = 5.5$ kN/m²]

Q. 3.6 A RCC rectangular beam section of 250 mm × 500 mm (depth) is reinforced with 3 bars of 16 mm diameter on the tension face and 2 bars of 16 mm diameter on compression face at an effective covers of 50 mm on each face. Concrete of M15 grade and steel of Fe 250 grade are used. Determine the maximum permissible live load on the beam.

[**Hint:** $\sigma_{cbc} = 5$ N/mm², $\sigma_{st} = 140$ N/mm², $m = 19$, $nd_a = 142$ mm, $d = 450$ mm,
$M_r = 34.01$ kN-m, $w_d = 3$ kN/m, $w_1 = 14.0$ kN/m]

Q. 3.7 A rectangular beam section of 300 mm width and 500 mm effective depth is reinforced with 4 bars of 18 mm diameter in tension face and 2 bars of 14 mm diameter in compression face at an effective cover of 50 mm. Determine the stresses induced in the top compression fibre of the concrete, compression steel and tension steel when the beam is subjected to a moment of 100 kN-m. Concrete of M20 grade and steel of Fe 415 HYSD grade are used.

[**Hint:** $\sigma_{cbc} = 7$ N/mm², $\sigma_{st} = 230$ N/mm², $b = 300$ mm, $d = 500$ mm, $d' = 50$ mm,

$A_{st} = 1018$ mm², $A_{sc} = 308$ mm², $p' = 0.0020502$, $p = 0.0068$, $m = 13$, $\dfrac{d'}{d} = 0.10$.

$n \text{ (coefficient)} = \left\{ m^2 (p+1.5p')^2 + 2m \left(p+1.5p' \cdot \dfrac{d'}{d} \right) \right\}^{\frac{1}{2}} = m(p+1.5p') = 0.32026$

$j \text{ (lever arm coefficient)} = \left(1 - \dfrac{n}{3} \right) + (1.5m-1) p' \left(\dfrac{n}{3} - \dfrac{d'}{d} \right) \left(n - \dfrac{d'}{d} \right) = 0.8933$

$\sigma_{st} = \dfrac{100 \times 10^6}{1018 \times 0.8933 \times 500} = 220$ N/mm²

$\sigma_{cbc} = \dfrac{\sigma_{st} \cdot n}{m(1-n)} = 7.97$ N/mm² (> 7.0 N/mm², unsafe)

$\sigma_{sc} = 106.9$ N/mm²]

Q. 3.8 A rectangular beam section of 300 mm width and 500 mm effective depth is reinforced with 4 bars of 16 mm diameter in tension face and 4 bars of 14 mm diameter in the compression face at an effective cover of 50 mm. M20 grade CC and Fe 415 steel are used in the beam. Determine from the first principle M_r and actual stresses in the concrete and steel for this M_r.

[**Hint:** $\sigma_{cbc} = 7$ N/mm², $\sigma_{st} = 230$ N/mm², $m = 13$, $A_{st} = 804$ mm², $A_{sc} = 616$ mm²,
$nd_c = 0.2835d$, $nd_a = 0.2737d$ (under reinforced), $\sigma_{st} = 230$ N/mm²
$\sigma_{cbc} = 6.67$ N/mm², $\sigma_{sc} = 82.5$ N/mm², $M = Qb \cdot d^2 + (1.5m-1) A_{sc} \cdot \sigma'_{cbc} \times 454.39$
$= 83.92$ kN-m]

Q. 3.9 A rectangular beam section of 250 mm × 500 mm is used for simply supported beam of 5 m effective span. The beam is reinforced with 3 bars of 16 mm diameter on tension face at an effective cover of 50 mm. The section is also reinforced with 2 bars of 16 mm on compression face at an effective cover of 50 mm. Concrete of M20 grade and steel of Fe 415 grade are used. Determine the maximum permissible live load uniformly distributed over the entire span.

[**Hint:** $nd_c = 127.6$ mm, $d = 450$ mm, $b = 250$ mm, $nd_a = 125.7$ mm (under reinforced)
$\sigma_{st} = 230$ N/mm², $M_r = 56.28$ kN-m, w (live load) = $\{18.01 - 313 (w_d)\}$
$= 14.88$ kN/m]

Q. 3.10 Design a rectangular beam section of 300 mm width and 500 mm effective depth and 50 mm concrete covers to carry a BM (a) 60 kN-m (b) 100 kN-m. Use M20 CC and Fe 415 steel.

[**Hint:** $b = 300$ mm, $d = 500$ mm, $\sigma_{cbc} = 7$ N/mm^2, $\sigma_{st} = 230$ N/mm^2, $m = 13$

Design constant: $n_c = 0.284$, $j = 0.905$, $Q = 0.90$, M_r (balanced) = 68 kN-m

(a) BM = 60 kN-m $< M_r$ of 68 kN-m singly reinforced, 3 bars 16ϕ ($A_{st} = 603$), $M_r = 63$ kN-m (> 60 kN-m)

(b) BM = 100 kN-m $> M_r$ of 68 kN-m, design doubly reinforced section

$M_1 = 63$ kN-m [(as in (a)], M_2 (required) = $100 - 63 = 37$ kN-m

$$A_{sc} = \frac{M_2}{(1.5m-1)\sigma'_{cbc}(d-d')} = 1038 \text{ mm}^2 \text{ (2 bars of 20 mm plus 2 bars of}$$

16 mm), $A_{sc} = 1030$ mm^2

$$A_{st_2} = \frac{M_2}{\sigma_{st}(d-d')} = \frac{37 \times 10^6}{230\,(450)} = 358 \text{ mm}^2 \text{ (2 bars of 16 mm diameter)}$$

Total $A_{st} = A_{st_1} + A_{st_2} = 603 + 402 = 1005$ mm^2 (5 bars of 16ϕ)]

Q. 3.11 Design the steel reinforcement for a doubly reinforced rectangular beam of 250 mm $\times$ 500 mm with effective concrete covers of 40 mm on both the faces to carry a maximum BM = 120 kN-m. Permissible stresses are $\sigma_{cbc} = 10$ N/mm^2, $\sigma_{st} = 230$ N/mm^2, $m = 9$.

[**Hint:** $nd_c = 129.4$ mm, $M_r = b \cdot nd_c \cdot \dfrac{\sigma_{cbc}}{2}\left(460 - \dfrac{129.4}{3}\right) + (1.5 \times 9 - 1)\,A_{sc} \cdot \sigma'_{cbc}(460 - 40)$

$= 120 \times 10^6$ N-mm

$A_{sc} = 1450$ mm^2 (5 bars of 20 mm, $A_{sc} = 1570$ mm^2)

$$A_{st} = \left[b \cdot nd_c \frac{\sigma_{cbc}}{2} + 1450 \times \sigma'_{cbc}(1.5 \times 9 - 1)\right]\frac{1}{230} = 1270 \text{ mm}^2$$

(4 bars of 20 mm ϕ, $A_{st} = 1256$ mm^2)]

Q. 3.12 Design a RC rectangular beam section with its $b = 0.5d$ and cover $\dfrac{d}{10}$. The beam is subjected to reversal of stresses to a maximum of $\pm$ 200 kN-m, permissible stresses are $\sigma_{cbc} = 7$ N/mm^2, $\sigma_{st} = 230$ N/mm^2, $m = 14$.

[**Hint:** $n_c = 0.299$, $\dfrac{b \cdot nd^2}{2} + (1.5m - 1)\,A_{sc}\left(nd - \dfrac{d}{10}\right) = 14 \cdot A_{st}\,(d - nd)$,

$A_{st} = A_{sc} = A_s = 0.003831d^2$, $\sigma_{cbc} = 4.66$ N/mm^2, $d = 632$ mm, $b = 316$ mm

$= 1530$ mm^2 (5 bars of 20 mm diameter on each face).

$D = 680$ mm with concrete cover of 45 mm and width $b = 320$ mm]

Q. 3.13 An isolated T-beam is simply supported over a span of 6 m. The dimensions of the T-beam are width of flange 750 mm, total thickness of the flange slab 125 mm, overall depth of the beam 450 mm, breadth of the web 300 mm, effective cover to tensile steel 50 mm and tensile reinforcement of 4 bars of 20 mm diameter. Determine the moment of resistance of the beam section, if steel of Fe 415 grade and concrete of M20 grade are used.

[**Hint:** $d = 400$ mm, b_f (eff) = 750 mm, $nd = 112$ mm, $nd_c = 125$ mm, under reinforced as $nd_c = 113.4$ mm, $M_r = 104.80$ kN-m

Q. 3.14 An isolated T-beam has a flange width of 1500 mm, thickness of 150 mm and effective depth of 500 mm. The web width is 350 mm and is reinforced with

6 bars of 20 mm of Fe 415 HYSD grade in tension zone. Determine the moment of resistance of the section, if the permissible stress in concrete and steel are respectively 7 N/mm² and 230 N/mm². Assume $m = 13$. Neglect the compression of the web. The beam is simply supported over a span of 6 m.

[**Hint:** $nd = 128.6 (< 141.74\ nd_c)$, under reinforced within the flange. $jd = 457.13$ mm, $M_r = 198.1$ kN-m]

Q. 3.15 In Question 3.14, if the thickness of the flange reduces to 100 mm instead of 150 mm, and all other data remains the same, determine the M_r of the section considering compressive force in the web also.

[**Hint:** $nd = 130.72$ mm $(< 141.74$ mm $= nd_c)$, under reinforced.

$jd = (500 - 43.58) = 456.42$, $\sigma_{st} = 230$ N/mm², $\sigma_{cbc} = 6.263$ N/mm²,

$\sigma'_{cbc} = 1.472$ N/mm², $\tilde{y} = 43.58$ mm $M_r = 197.78$ kN-m]

Q. 3.16 An isolated T-beam has flange width of 1200 mm, thickness of 100 mm and effective depth 400 mm. The web width is 300 mm and is reinforced with 5 bars of 25 mm diameter of Fe 415 steel. The concrete of M20 grade have been used. Determine the maximum u.d.l. which a simply supported beam of 6 m span can carry safely including its own weight. Neglect the compression in the web portion. The beam is so placed that the flange is under compression.

[**Hint:** b_f (eff) $= 762$ mm, $nd = 153.30$ mm, $nd_c = 113.4$ mm, over reinforced

$\sigma_{cbc} = 7$ N/mm², $\sigma_{st} = 146.44$ N/mm², $\sigma'_{cbc} = 2.434$ N/mm², $\tilde{y} = 41.93$ mm,

$jd = 358.07$ mm, $M_r = 128.7$ kN-m, w (u.d.l.) $= 28.6$ kN/m]

Q. 3.17 The flange of a T-beam is 120 mm thick and 1200 mm wide. The web of the beam is 300 mm wide and the depth up to the centre of the tensile steel is 600 mm. Tensile reinforcement consists of 6 bars of 25 mm diameter. If the beam section is subjected to a maximum BM of 300 kN-m, find the stresses in the materials if $m = 13$, and SS beam span is 6 m.

[**Hint:** $nd = 194$ mm, $nd_c = 170.1$ mm, $\sigma_{cbc} = 6.82$ N/mm², over reinforced

$\tilde{y} = 51.05$ mm, $jd = 548.95$ mm, $\sigma_{st} = 185.55$ N/mm²]

Q. 3.18 A RCC T-beam has flange of 1200 mm width and 120 mm slab thickness. The effective depth of beam is 500 mm and web width is 300 mm. If $m = 13$, permissible stresses in concrete and steel are 7 N/mm² and 230 N/mm² respectively and the beam is simply supported over a span of 7.20 m, determine the tensile reinforcement and the moment of resistance of the section.

[**Hint:** b_f (effective) $= 1020$ mm, $nd_c = 127.6$ mm, $\sigma_{cbc} = 7$ N/mm², $\sigma_{st} = 230$ N/mm² (max),

$$\sigma'_{cbc} = \frac{7(127.6 - 120)}{127.6} = \frac{7.6 \times 7}{127.6} \text{ or } \sigma'_{cbc} = 0.42 \text{ N/mm}^2$$

$$\tilde{y} = \frac{120}{3}\left(\frac{7.84}{7.42}\right) = 42.26 \text{ mm, Lever arm } jd = 407.74 \text{ mm}$$

$A_{st} = 1974$ mm² (4 bars of 25 mm diameter $A_{st} = 1964$ mm²)

$M_r = 185.1$ kN-m]

Q. 3.19 A T-beam slab roof consists of 120 mm thick slab cast monolithically with 300 mm wide web (beam) which are spaced at 5 m centre to centre. The super-imposed load including self-weight of slab and web is 5 kN/m². The effective span of simply supported beam is 8 m. Determine the depth and area of tensile steel in intermediate beams if the concrete of M30 grade and steel of Fe 415 grade are used in the beam construction. Consider $m = 9$, $\sigma_{cbc} = 10$ N/mm², $\sigma_{st} = 230$ N/mm².

[**Hint:** b_f (eff) = 2353 mm, assume $jd = 0.9d$, cost ratio $r = 60$ (say)

Load on beam $w = 5 \times (5 \times 1) = 25$ kN/m, BM (max) = 200 kN-m.

$$\text{Economic depth} = \sqrt{\frac{r \cdot m}{\sigma_{st}\, j \cdot b_w}} = \sqrt{\frac{60 \times 200 \times 10^6}{230 \times 0.9 \times 300}} = 440 \text{ mm}$$

Provide $D = 500$ mm, $d = 450$ mm

Lever arm = $0.9 \times 450 = 405$ mm, $A_{st} = \dfrac{M}{\sigma_{st} \cdot jd} = 2147$ mm² (Provide 5 bars of 25 mm)

Design can be rechecked with actual dimensions ($A_{st} = 2455$ mm²)

$d = 450$ mm, $b_f = 2353$ mm, $b_w = 300$ mm, $A_{st} = 2455$ mm²,

Actual $nd = 88.3$ mm < 120 mm, NA lies within the flange.

$$\text{Lever arm} = \left(450 - \frac{88.3}{3} \right) = 420.6 \text{ mm} \left(j = \frac{420.6}{450} = 0.935 \right)$$

$\sigma_{st} \times 2455 \times 420.6 = 200 \times 10^6$, $\sigma_{st} = 193.7$ N/mm² < 230 N/mm², OK

$\dfrac{2353}{2} \times 88.3 \times \sigma_{cbc} \times 420.6 = 200 \times 10^6$, $\sigma_{cbc} = 4.58$ N/mm² < 10 N/mm²

Thus, the section is OK, M_r capacity = 237.5 kN-m, OK]

UNIT – II Limit State Design of Beams

Chapter 4

Limit State Design of Beam Section for Bending Moment

LEARNING OBJECTIVES

After the study of *limit state design of beam section for bending moment*, the learner understands the basic principles and procedure of limit state design for different RCC beam sections and will be able to:

- State various assumptions in limit states for safety and serviceability of RCC flexural members
- Differentiate between working stress, ultimate load and limit state methods of design of flexural RCC members
- Explain the concept of characteristic strength of materials
- Explain the concept of partial safety factor in limit state design
- Explain the assumptions in the limit state of collapse in flexure
- Explain the stress–strain relationships of RCC materials
- Explain the stress–block parameters according to IS:456-2000
- Explain the variation of moment of resistance of a beam with respect to percent (p_t) of tensile steel
- Explain the limit state design for singly reinforced rectangular beam sections
- Explain the limit state design for doubly reinforced rectangular beam sections
- Explain the limit state design for singly reinforced and doubly reinforced T- and L-beam sections.

4.1 INTRODUCTION

The working stress method of design gives satisfactory performance of the structure at working loads, but it becomes unrealistic at the ultimate state of collapse. Similarly, the ultimate load method does not represent the realistic evaluation of safety factor at the service loads. Therefore, the limit state method is an ideal method which provides a realistic safety margin against collapse and ensures serviceability at the working loads. Considering these shortcomings of the common method of working stress and ultimate collapse load method, a more realistic method of limit state design was developed in 1950. *Limit state method of design* incorporates the salient features of both working stress and ultimate collapse load methods.

Thus, the limit state design method was adopted by many countries and incorporated in their design standard codes. In India also, the limit state design method was fully incorporated in to code IS:456-2000 and presently the design of RCC structures is totally based on IS:456-2000 provisions. The limit state method of design is oriented towards the simultaneous satisfaction of safety against collapse and guarantee against satisfactory serviceability at service loads. The limit state method is a judicious combination of the working stress method and the ultimate load method. The acceptable limit of safety and serviceability requirements, before failure occurs is called a *limit state*.

4.2 LIMIT STATES OF SAFETY AND SERVICEABILITY

The safe and satisfactory design must ensure the achievement of an acceptable probability that the specified life of any structure is not curtailed prematurely by the attainment of an unsatisfactory limit state which covers the various forms of failure.

Some of the *limit states of failure* are stated below:
- Failure of members due to flexure, shear, torsion or combination
- Failure by fatigue due to repeated loads
- Failure due to bond and anchorage of steel bars
- Failure due to elastic instability of members
- Failure due to impact, earthquake, wind or fire, etc.
- Failure due to the destructive chemicals, corrosion of steel, etc.

The limit state of collapse (or failure) depends on the ultimate strength.

Various limits of serviceability are given below:
- Excessive deflections affecting the false ceiling causing discomfort
- Excessive local damage leading to cracking or spalling of concrete affecting appearance of the structure. The limit state of excessive deflections and crack width is applicable at service loads and is estimated on the basis of elastic analysis (working stress method)

Thus, in this method two types of limit states are considered in design as:
 (i) Limit state of collapse (strength limit)
(ii) Limit state of serviceability

4.3 CHARACTERISTIC AND DESIGN VALUES WITH PARTIAL SAFETY FACTORS

Different limit states should be considered in design to ensure adequate degree of safety and serviceability. The structural member should be designed on the basis of the most critical limit state and then checked for other limit states. For this the design should be based on characteristic values of loads, which consider the variations in the loads likely to be supported by the member. The design values are derived from the characteristic load values by applying partial safety factors. The partial safety factors of a structure are based on probability theory due to large number of variables involved. There is no rational method to find the safety factors accurately and hence partial safety factors are assumed.

(i) Characteristic Material Strength

The characteristic strength refers to that value of strength of the material below which not more than 5% of the results are expected to fall. The characteristic strength of concrete is denoted by f_{ck} (N/mm^2) and the values for different grades of concrete are already specified in Chapter 1. Present Indian standards for reinforcing steel bars specify characteristic strength value as its yield strength (f_y) or 0.2% proof stress in relevant code.

(ii) Design Strength Values

The design strength of the material (f_d) is given by

$$f_d = \frac{f_{ck}}{\gamma_m} \qquad \text{... Eq. (4.1)}$$

where, f_{ck} = Characteristic strength of material (f_{ck} for concrete and f_y for steel)

γ_m = Partial safety factor appropriate to the material

(iii) Partial Safety Factor (γ_m)

Partial safety factor for material (γ_m) in limit state of collapse shall be taken as 1.5 for concrete and 1.15 for steel (i.e. γ_{mc} = 1.5, and γ_{ms} = 1.15). The values given in IS: code 456-2000 have already considered γ_{mc} = 1.5 and γ_{ms} = 1.15 in various tables and equations for the limit state. γ_{mc} is higher for concrete, since there are greater chances of variation of strength of concrete due to improper compaction, improper mixing and batching, inadequate curing, and variations in properties of ingredients. Since the steel reinforcement is manufactured in the factory, the chances of variations in strength are much less, and hence a lower value of γ_{ms} = 1.15 has been adopted.

Thus, design stress in steel: $f_{yd} = \dfrac{f_y}{1.15} \cong 0.87 f_y$

Compressive strength in concrete shall be taken as $f_c = \dfrac{f_{ck}}{1.5} = 0.67 f_{ck}$ and as per IS:456-2000, a further partial safety factor of γ_{mc} = 1.5, shall be taken and hence the design strength $f_{dc} = \dfrac{0.67 f_{ck}}{1.5} = 0.447$. Partial safety factor for materials for serviceability consideration shall be taken as (deflection and local damage) γ_m = 1.0.

(iv) Design Loads

Superimposed loads are also variable and depends on the loading pattern and type of structure. The term characteristic load means that value of load which has 95% probability of not being exceeded during the lifetime of the structure. Various IS codes (IS:875-Part 1–4) specify the loads for different structures and situations. These loads shall be adopted as characteristic loads.

From these characteristic loads, the design loads are calculated by applying partial safety load factor using the following equation:

Design load

$$(F_d) = F \cdot (\gamma_f) \qquad \qquad \text{...Eq. (4.2)}$$

where, F = Characteristic load on the structure

γ_f = Partial safety load factor appropriate to the nature of loading and the type of limit state considered

F_d = Design load for the structure

The partial safety load factor shall be considered as given below in Table 4.1 (Table 18 of IS:456-2000, page 68).

Table 4.1: Value of partial safety load factor (γ_f)

Load combination	Limit state of collapse			Limit states of serviceability		
	DL	IL	WL	DL	IL	WL
DL + IL	1.5	1.5	1.0	1.0	1.0	–
DL + WL	1.5 or	–	1.5	1.0	–	1.0
DL + IL + WL	1.2	1.2	1.2	1.0	0.8	0.8
	(0.90* for stability)					

* consideration against overturning

Note: DL = Dead load, IL = Imposed load, WL = Wind load, EL = Earthquake load

For considering earthquake effects, substitute EL for WL.

It may be noted that the design strength of material is reduced in the ratio of partial safety factor or material (γ_m).

$$f_d \text{ (design strength)} = \frac{f}{\gamma_m}$$

while the design load (F_d) is obtained by multiplying the characteristic load with the partial safety load factor (γ_f), i.e.

$$F_d = F \times \gamma_f$$

4.4 ASSUMPTIONS IN LIMIT STATE OF COLLAPSE IN FLEXURE

Design for the limit state of collapse in bending is based on the following assumptions:

(i) Plane sections normal to the axis before bending remain plane after bending.

(ii) Maximum strain in concrete at the outermost compression fibre is taken as 0.0035 in bending.

(iii) The tensile strength of the concrete is neglected.

(iv) The maximum strain in the tensile reinforcement in the section at failure shall not be less than $\left(\dfrac{f_y}{1.15\,E_s} + 0.002 \right)$

where, f_y = Characteristic strength of steel

E_s = Modulus of elasticity of steel

(v) The stresses in the reinforcement are derived from the stress–strain curve for the type of steel used for reinforcement and the partial safety factor (γ_{ms}) for steel shall be assumed as 1.15 (Figs 4.3 and 4.4).

(vi) The modulus of elasticity (E) of steel shall be assumed same both in axial compression and tension but its value is assumed higher (1.5 times) in case of flexural compression.

(vii) The relationship between the compressive stress and the strain in concrete may be assumed so that it results in the strength prediction in agreement with the test results. The partial safety factor (γ_{mc}) shall be 1.50 (Fig. 4.1).

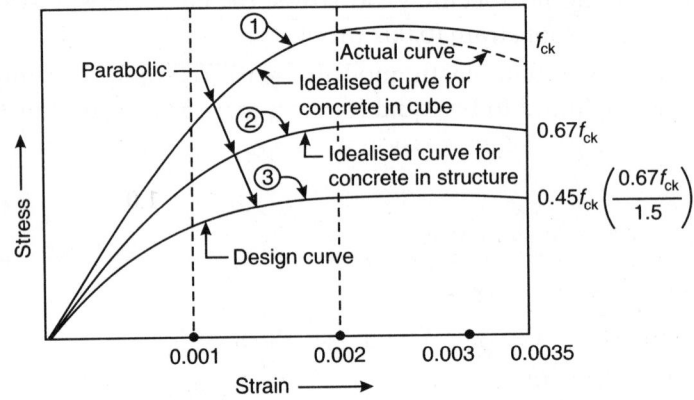

Fig. 4.1: Stress–strain curve for concrete

Assumption (i) has been verified by tests and it implies linear variation of strain across the depth of the concrete beam section as shown in Fig. 4.2.

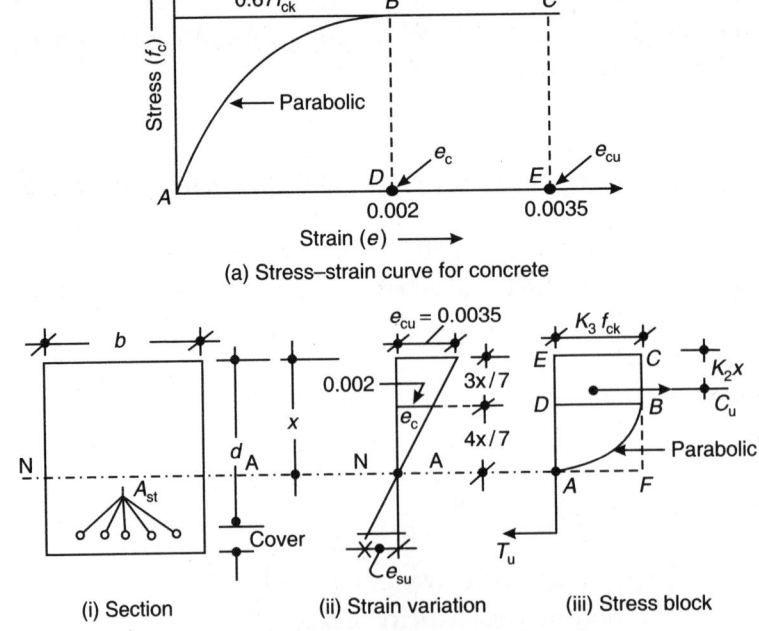

(a) Stress–strain curve for concrete

(i) Section   (ii) Strain variation   (iii) Stress block

(b) Strain–stress variations across the section

Fig. 4.2: Strain and stress variations for concrete

Assumption (ii) regarding maximum strain in concrete when the concrete section reaches the maximum moment capacity depends on many factors viz. rate of loading, time of loading, grade of concrete, percentage of reinforcement, shape of cross-section, etc. IS:456-2000 code adopts a conservative general value of e_{cu} equal to 0.0035.

Assumption (iii) states that the tensile strength of concrete is very small and ignoring the same does not make any substantial difference in the design.

Assumption (iv) is intended to ensure ductile failure of the section, i.e. the section undergoes large deformation before failure.

Assumption (v) refers to stress–strain curves for the specific type of steel. For cold twisted steel bars having no specific yield stress, the characteristic stress f_y is taken corresponding to 0.2% proof strain (0.002 proof strain).

Assumption (vi) states that elastic modulus with reference to compression and tension are practically found to be equal for steel in axial stresses, but E_s is assumed 1.50 times higher in flexural compression in concrete and hence equivalent concrete area is taken as $1.5m\,A_{sc}$ for compressive steel A_{sc}.

Assumption (vii) refers to compressive stress distribution in concrete section which may be assumed rectangular, parabolic, trapezoidal or any other shape but the prediction of strength remains in close agreement with the test results. The strain variation and stress variation diagrams are assumed as shown in Figs 4.1 and 4.2 respectively.

The stress–strain curve for concrete shown in Fig. 4.1 (refer to Fig. 21, page 69 of IS:456-2000) forms the basis of the design in limit state method. The compressive strength of concrete in the structure is considered as $0.67f_{ck}$ as in curve 2 in Fig. 4.1. The 0.67 factor accounts for difference in cube strength and strength of concrete in the structure. In this curve, initial portion is parabolic while at a strain of 0.002 (0.20% strain) the stress remains constant with increasing load up to a strain of 0.0035 at the time of failure. The curve 3 in Fig. 4.1 refers to design curve using a partial safety factor of $\gamma_m = 1.5$ for concrete. Thus, maximum compressive stress in concrete for design

$$= \frac{0.67\,f_{ck}}{1.5} = 0.45\,f_{ck}.$$

The stress–strain curve for mild steel is shown in Fig. 4.3, while the stress–strain curves for Fe 415 and Fe 500 steels are shown in Figs 4.4(a) and (b). Curves 1 in Figs 4.3 and 4.4 represent characteristic strengths while curves 2 represent design strengths in respective steels.

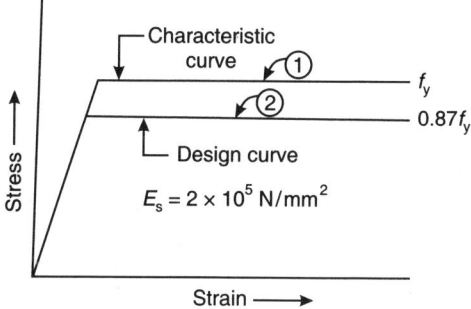

Fig. 4.3: Stress–strain curve for mild steel

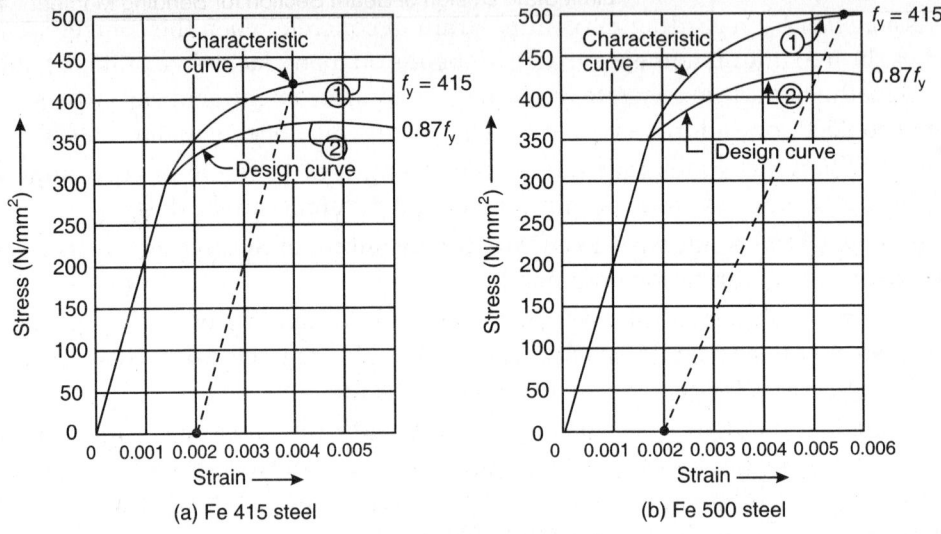

Fig. 4.4: Stress–strain curves

In limit state method design strength for steel $f_{yd} = \dfrac{f_y}{1.15} = 0.87\, f_y$

Thus, for mild steel: $f_{yd} = 0.87 \times 250 \cong 218\ \text{N/mm}^2$

For Fe 415 steel: $f_{yd} = 0.87\, f_y = 0.87 \times 415 \cong 361\ \text{N/mm}^2$

For Fe 500 steel: $f_{yd} = 0.87\, f_y = 0.87 \times 500 \cong 435\ \text{N/mm}^2$

4.5 STRESS BLOCK PARAMETERS

Indian standard code IS:456-2000 developed theory for determining ultimate flexural moment resistance of RCC sections based on strain compatibility. A limiting strain in concrete (0.0035) forms the criterion for crushing of concrete in flexural compression.

The stress–strain curve of concrete under compression is generally obtained from cube of concrete subjected to longitudinal compressive loading. In actual flexural member, the stress–strain vary across the depth of the section while in case of cube crushing test the stress–strain remain fairly uniform. The maximum flexural compression in a beam section and the cube crushing strength differs. The IS code recommends the compressive strength of concrete in the structure equal to $0.67\%f_{ck}$. The stress–strain curve for concrete is shown in Fig. 4.2(a) which is similar to Fig. 4.1 curve 2. Figure 4.2(b) (ii) shows the variation of strain across the depth while Fig. 4.2(b) (iii) shows the variation of stress across the depth of the section. The stress curve ABCEDA in Fig. 4.2(b) (iii) is similar to Fig. 4.2(a). The stress from A to B is parabolic and straight line (linear) variation from B to C at a constant stress of $0.67f_{ck}$.

The total compressive force C_u and its location below the top fibre can be expressed in terms of stress block factors k_1, k_2 and k_3 as under:

k_1 = Shape factor, defined as the ratio between the stress block area ABCED and area of rectangle AFCE, i.e.

$$k_1 = \frac{\text{Area of ABCE}}{\text{Area of AFCE}} = \frac{\text{Area of ABCE}}{\text{Area} (x \times \text{CE})} \qquad \text{... (Fig. 4.2(b) (iii))}$$

Ultimate strain in concrete $e_{cu} = \text{AE} = 0.0035$, the strain $\text{AD} = e_c$, after which concrete yields at constant stress of $0.67f_{ck}$ and is equal to 0.002 for all grades of concrete.

Ratio $\qquad \dfrac{\text{AD}}{\text{AE}} = \dfrac{e_c}{e_{cu}} = \dfrac{0.0020}{0.0035} = \dfrac{4}{7}$,

(since $e_c = 0.002$, and after plastic flow $e_{cu} = 0.0035$ at failure) (Fig. 4.2(b) (ii)).

Also $\qquad \dfrac{\text{DE}}{\text{AE}} = \dfrac{(0.0035 - 0.002)}{(0.0035)} = \dfrac{0.0015}{0.0035} = \dfrac{3}{7}$

Thus, $\text{AD} = \dfrac{4}{7}\,\text{AE}$, and $\text{ED} = \dfrac{3}{7}\,\text{AD}$

Area ABCE = Area ABD + Area BCED

$$= \frac{2}{3}\,\text{AD} \times \text{BD} + (\text{DE} \times \text{CE}) = \frac{2}{3} \times \frac{4}{7}\,\text{AE} \times \text{CE} + \frac{3}{7}\,\text{AE} \times \text{CE}$$

$$= \frac{8}{21}\,\text{AE} \times \text{CE} + \frac{3}{7}\,\text{AE} \times \text{CE} = \frac{17}{21}\,(\text{AE} \times \text{CE}) = \text{Area AFCE} \times \frac{17}{21}$$

Thus, $k_1 = \dfrac{17}{21} = 0.8095 \cong 0.81$

Let the resultant compressive force be located at a distance $k_2 x$ below the top compression fibre where, x is the depth of the NA.

$$k_2\,\text{AE} = \frac{(\text{Area ABD})x_1 + (\text{Area BCED})x_2}{\text{Area ABCE}}$$

$$k_2\,\text{AE} = \frac{\dfrac{8}{21}\left(\text{DE} + \dfrac{3}{8}\,\text{AD}\right) + \dfrac{9}{21}\left(\dfrac{1}{2}\,\text{ED}\right)}{\dfrac{17}{21}} = \frac{\left[\dfrac{8}{21}\left(\dfrac{3}{7} + \dfrac{3}{8} \times \dfrac{4}{7}\right) + \dfrac{9}{21}\left(\dfrac{1}{2} \times \dfrac{3}{7}\right)\right]\text{AE}}{\dfrac{17}{21}}$$

$$k_2 = \frac{\dfrac{36}{7} + \dfrac{27}{14}}{17} = \frac{99}{238} = 0.416 \cong 0.42$$

Hence, compressive force (C) is located at a depth of $k_2 \cdot x(0.416 = 0.42x)$ below the top compression fibre. Let the maximum ordinate EC of the stress block be $k_3 \cdot f_{ck}$. This is the maximum compressive stress in flexure before failure. IS:456-2000 recommends the value of $k_3 = 0.67$. Thus, the compressive force C_u in concrete = $b \times$ area ABCE of stress block (Fig. 4.2(b) (iii)),

$$C_u = k_1,\, x \cdot \text{EC} \times b = k_1 \cdot k_3 f_{ck} \cdot bx,$$

$$M_u = C_u(d - k_2 \cdot x) = f_{ck.}\, k_1 \cdot k_3\, bd^2 \cdot \frac{x}{d}\left(1 - k_2\,\frac{x}{d}\right)$$

$$M_u = f_{ck} \cdot k_1\, k_3 \cdot bd^2 \cdot \left(\frac{x}{d}\right)\left(1 - k_2\,\frac{x}{d}\right) \qquad \text{...Eq. (4.3)}$$

4.6 DESIGN PARAMETERS OF STRESS BLOCK (IS:456-2000)

The design parameters can be derived from the general expressions for C_u and M_u found in article, as in Fig. 4.5 by incorporating partial safety factors, γ_{mc} and γ_{ms} for concrete and steel. For concrete stress–strain curve 3 (Fig. 4.1) and for steel stress–strain curves (2) (Figs 4.3 and 4.4) shall be adopted. Based on these curves and appropriate partial safety factors, the strain and stress block diagrams are shown in Fig. 4.5 for a singly reinforced beam section as recommended by IS:456-2000.

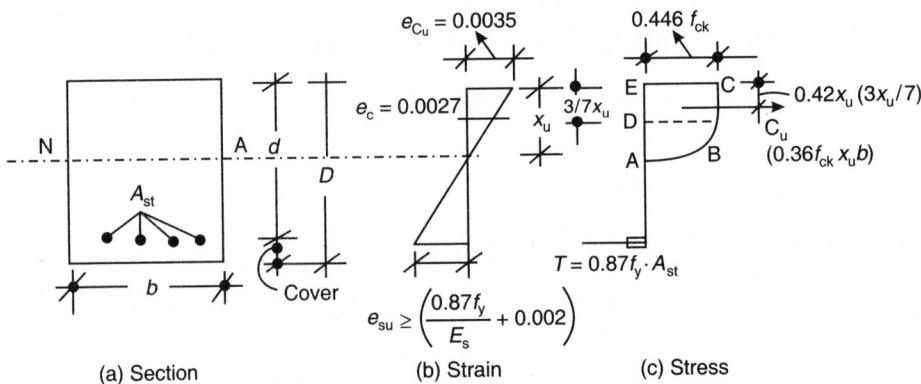

Fig. 4.5: Strain and stress diagrams for design

The maximum stress CE (Fig. 4.5(c)) = $\dfrac{k_3 f_{ck}}{\gamma_{mc}} = \dfrac{0.67 f_{ck}}{1.5} = 0.446 f_{ck} \cong 0.45 f_{ck}$

Factor $k_2 = 0.416 \cong 0.42$, and depth of NA is x_u. The strain e_{cu} in concrete at the time of collapse = 0.0035.

The maximum stress in steel is limited to $\dfrac{f_y}{\gamma_{ms}} = \dfrac{f_y}{1.15} = 0.87 f_y$

The area of stress block for compression in concrete is given by

$$C_u = \text{Area ABCE} = k_1 \cdot x_u \times \dfrac{k_3 f_{ck}}{\gamma_{mc}}$$

$k_1 = 0.8095$, $k_3 = 0.67$, and $\gamma_m = 1.50$, gives $C_u = \dfrac{0.8095 \times 0.67}{1.5} x_u f_{ck}$

i.e. $\qquad C_u = 0.36 \, x_u f_{ck}$ (per unit width) $\qquad\qquad$ ... Eq. (4.4)

Total compressive force for a rectangular beam section of width b will be
$C_u = 0.36 \cdot x_u \cdot b \cdot f_{ck}$ $\qquad\qquad\qquad\qquad\qquad\qquad$ ... Eq. (4.4a)

The maximum strain at failure e_{cu} in concrete = 0.0035.

The code IS:456-2000 requires that at failure, the strain in steel reinforcement (e_{su}) should not be less than $\left(\dfrac{0.87 f_y}{E_s} + 0.002\right)$, thus keeping a stress of $0.87 f_y$ in steel. This limits the depth x_u of NA to its maximum value. The depth x_u of the NA is given by

$$\dfrac{x_u}{e_{cu}} = \dfrac{(d - x_u)}{e_{su}} \quad \text{or} \quad x_u = \dfrac{e_{cu} \cdot d}{e_{cu} + e_{su}} \qquad\qquad \text{... Eq. (4.5)}$$

Substituting the minimum value of $e_{su} = \left(0.002 + \dfrac{0.87\,f_y}{E_s}\right)$ and $e_{cu} = 0.0035$, we have

$$\frac{x_{u\,(max)}}{d} = \frac{0.0035}{\left(0.0035 + 0.002 + \dfrac{0.87\,f_y}{E_s}\right)} = \frac{0.0035\,E_s}{\left(0.0055\,E_s + 0.87\,f_y\right)} \qquad \text{... Eq. (4.6)}$$

Putting $E_s = 2 \times 10^5\,\text{N/mm}^2$ for steels, we have

$$\frac{x_{u\,(max)}}{d} = \frac{0.0035 \times 2 \times 10^5}{\left(0.0055 \times 2 \times 10^5 + 0.87\,f_y\right)} = \frac{700}{\left(1100 + 0.87\,f_y\right)}$$

i.e.
$$\frac{x_{u\,(max)}}{d} = \frac{700}{\left(1100 + 0.87\,f_y\right)} \qquad \text{... Eq. (4.6a)}$$

The limiting values of the depth of NA for different grades of steel are given in Table 4.2.

Table 4.2: NA depth coefficients (maximum)

f_y (N/mm^2)	250 (MS)	415 (Fe 415)	500 (Fe 500)
$\dfrac{x_{u\,(max)}}{d}$	0.53(0.5313)	0.48(0.4791)	0.46(0.4560)

Note: The limiting values of $\dfrac{x_{u\,(max)}}{d}$ given in Table 4.2 ensures ductile failure of the beam section.

4.7 ANALYSIS OF SINGLY REINFORCED RECTANGULAR SECTIONS

In a singly reinforced rectangular beam section, the total compressive force C_u is given by Eq. (4.4a) as

$$C_u = 0.36 x_u \cdot b \cdot f_{ck}$$

The total tensile force (T_u) in steel reinforcement is given as

$$T_u = 0.87\,f_y \cdot A_{st}$$

For equilibrium of the beam section, we have $T_u = C_u$

i.e.
$$0.36\,f_{ck}\,x_u \cdot b = 0.87\,f_y \cdot A_{st}$$

$$\frac{x_u}{d} = \frac{0.87\,f_y\,A_{st}}{0.36\,f_{ck}\,b \cdot d} = 2.417\left(\frac{f_y}{f_{ck}}\right)\cdot\left(\frac{A_{st}}{bd}\right) = 2.417 p_t \cdot \frac{f_y}{f_{ck}}$$

i.e.
$$\frac{x_u}{d} = 2.417 p_t \cdot \frac{f_y}{f_{ck}} = 2.417\,\frac{f_y}{f_{ck}}\cdot\frac{A_{st}}{bd} \qquad \text{... Eq. (4.7)}$$

where $p_t = \left(\dfrac{A_{st}}{bd}\right)$ is the ratio of steel reinforcement to concrete area.

The limiting value of $\dfrac{x_{u\,(max)}}{d}$ is obtained from Eq. (4.6a) or Table 4.2.

According to the values of $\left(\dfrac{x_u}{d}\right)$ and $\left(\dfrac{x_{u\,(max)}}{d}\right)$, three different cases may arise:

(i) **Case I:** $\left(\dfrac{x_u}{d} < \dfrac{x_{u\,(max)}}{d}\right)$, i.e. $\left(\dfrac{x_u}{d}\right)$, is less than the limiting value $\left(\dfrac{x_{u\,(max)}}{d}\right)$ repre-

senting *under reinforced* section.

In this case, the quantity of steel reinforcement is less than the limiting value of reinforcement.

$$p_{t\,lim} = \frac{1}{2.417} \cdot \frac{f_{ck}}{f_y} \cdot \left(\frac{x_{u\,(max)}}{d}\right) = 0.414 \frac{f_{ck}}{f_y} \cdot \left(\frac{x_{u\,(max)}}{d}\right) \qquad \ldots \text{Eq. (4.8)}$$

$p_{t\,lim}$ for different grades of concrete and different grades of steel are given below in Table 4.3.

Table 4.3: Value of $p_{t\,lim}$ (%)

Grade of concrete	Value of $100\,p_{t\,lim}$ for different steels		
(f_{ck}) N/mm²	$f_y = 250$ N/mm²	$f_y = 415$ N/mm²	$f_y = 500$ N/mm²
M15	1.317	0.718	0.571
M20	1.756	0.957	0.762
M25	2.194	1.197	0.952
M30	2.633	1.437	1.143
M35	3.072	1.676	1.330
M40	3.511	1.915	1.524

$$p_{t\,lim} = 0.414 \frac{f_{ck}}{f_y} \cdot \frac{x_{u\,(max)}}{d}; \qquad\qquad \frac{x_{u\,(max)}}{d} \text{ from Table 4.2} \left(\frac{700}{1100 + 0.87\,f_y}\right)$$

The moment of resistance for beam sections with under reinforced sections:

$$M_u = T_u\,(d - 0.416\,x_u), \text{ where } T_u = 0.87\,f_y \cdot (p_t \cdot bd)$$

$$\therefore \quad M_u = 0.87\,f_y \cdot A_{st}\,(d - 0.416x_u) = 0.87\,f_y \cdot A_{st}\,d \left(1 - 0.416\,\frac{x_u}{d}\right) \qquad \ldots \text{Eq. (4.9)}$$

Putting value of $\dfrac{x_u}{d}$ from Eq. (4.7) as $\dfrac{x_u}{d} = 2.417\,p_t \cdot \dfrac{f_y}{f_{ck}}$

$$M_u = 0.87\,f_y \cdot A_{st} \cdot d \left(1 - 0.416 \times 2.417 \frac{A_{st}}{bd} \cdot \frac{f_y}{f_{ck}}\right), \text{ since } p_t = \frac{A_{st}}{bd}$$

or $\qquad M_u = 0.87 f_y \cdot A_{st}\,d \left\{1 - \dfrac{A_{st}}{bd} \cdot \dfrac{f_y}{f_{ck}}\right\}$, since $0.416 \times 2.417 \cong 1.0$

i.e. $\qquad M_u = 0.87 f_y \cdot A_{st} \cdot d \left\{1 - \dfrac{A_{st}}{bd} \cdot \dfrac{f_y}{f_{ck}}\right\} \qquad\qquad \ldots \text{Eq. (4.9a)}$

(ii) Case II: $\left(\dfrac{x_u}{d} = \dfrac{x_{u\,(max)}}{d} \right)$

This is the case of balanced section in which the steel reinforcement attains its yield stress at the same time when the concrete attains its ultimate (collapse) strain. The stress block is fully developed as shown in Fig. 4.5(c). The moment of resistance M_u of the balanced section develops its limiting value $M_{u\,lim}$. The value of $M_{u\,lim}$ is given as:

$$M_{u\,lim} = 0.36\,f_{ck} \cdot x_{u\,lim} \cdot b \cdot (d - 0.416\,x_{u\,lim})$$

or $\qquad M_{u\,lim} = 0.36\,f_{ck} \cdot \dfrac{x_{u\,lim}}{d} \left(1 - 0.416\,\dfrac{x_{u\,lim}}{d} \right) bd^2 \qquad \dots \text{Eq. (4.10)}$

(iii) Case III: $\left(\dfrac{x_u}{d} > \dfrac{x_{u\,(max)}}{d} \right)$

In this case, the beam section has tensile reinforcement more than the minimum steel required for the balanced section so that the steel reinforcement does not yield and the concrete in extreme compression fibres starts failing at a limiting strain of 0.0035. Thus, compression fibres in concrete fails first attaining the limiting (ultimate) strain of 0.0035 while the steel does not yield. The failure or collapse is sudden and does not give any warning before failure. Such a failure is also known as brittle and such a design is avoided. The code IS:456-2000 recommends that if the value of $\dfrac{x_u}{d}$ obtained is more than limiting value , the section should be redesigned.

In real life, the design should be balanced or under reinforced. The notion of providing more steel reinforcement than the required quantity for the balanced section to enhance the safety level is totally wrong and dangerous. Hence, over reinforcement must be avoided in RCC design.

4.8 MOMENT OF RESISTANCE (M_u) (As Per IS:456-2000)

(i) Limiting depth of NA $\dfrac{x_{u\,(max)}}{d} = \dfrac{700}{\left(1100 + 0.87\,f_y\right)}$

(ii) Depth of NA $\dfrac{x_u}{d} = \dfrac{0.87\,f_y}{0.36\,f_{ck}} \cdot \dfrac{A_{st}}{bd} = 2.417\,\dfrac{f_y}{f_{ck}} \cdot \dfrac{A_{st}}{bd}$

Compare $\left(\dfrac{x_u}{d} \right)$ with $\dfrac{x_{u\,(max)}}{d}$

If $\dfrac{x_u}{d} < \dfrac{x_{u\,lim}}{d}$, then under reinforced and

$$M_u = 0.87\,f_y \cdot A_{st} \cdot d \left(1 - \dfrac{A_{st}}{bd} \cdot \dfrac{f_y}{f_{ck}} \right)$$

(iii) If $\dfrac{x_u}{d} = \dfrac{x_{u\,lim}}{d}$, the section is balanced and the moment of resistance is then given as :

$$M_u = 0.36 f_{ck} \cdot \frac{x_{u\,lim}}{d}\left(1 - 0.416 \frac{x_{u\,lim}}{d}\right) bd^2$$

(iv) If $\frac{x_u}{d} > \frac{x_{u\,lim}}{d}$, the section is over reinforced. The moment of resistance M_u shall be limited to $M_{u\,lim}$. Over reinforced sections are uneconomical and dangerous since these sections fail suddenly without giving any warning. IS:456-2000 recommends redesign of over reinforced sections.

Limiting Moment of Resistance Coefficient (R_u)

In a balanced section, the steel reinforcement attains its yield stress simultaneously with concrete reaching its limiting strain (0.0035) in extreme compression fibre. The stress block is fully developed as shown in Fig. 4.5(c). At this point, the moment of resistance M_u reaches its limiting value $M_{u\,lim}$ which can be found by taking the moment of compressive force in concrete about the centre of tensile steel, i.e.

$$C_u = (b \cdot x_{u\,lim})(0.36 f_{ck}) = 0.36 f_{ck}\, b \cdot x_{u\,lim}$$

Lever arm $\quad \alpha = (d - 0.416\, x_{u\,lim})$

$$M_u = 0.36 f_{ck} \cdot \frac{x_{u\,lim}}{d}\left(1 - \frac{0.416 x_{u\,lim}}{d}\right) bd^2$$

or $\qquad \dfrac{M_u}{bd^2} = 0.36 f_{ck} \cdot \dfrac{x_{u\,lim}}{d}\left(1 - 0.416 \dfrac{x_{u\,lim}}{d}\right) = R_{u\,lim}$ (say) $\qquad$ ... Eq. (4.11)

From Table 4.2, the value of $\frac{x_{u\,lim}}{d}$ for different steels are given as

Fe 250 (MS): $\frac{x_{u\,lim}}{d} = 0.5313\ (0.53)$

Fe 415: $\frac{x_{u\,lim}}{d} = 0.4791\ (0.48)$

Fe 500: $\frac{x_{u\,lim}}{d} = 0.4560\ (0.46)$

Substituting these values in Eq. (4.11), we get values of $R_{u\,lim}$ (Table 4.4).

Table 4.4: Values of $R_{u\,lim}$ (N-mm/mm³) for different grades of steel and concrete

Grade of concrete f_{ck} (N/mm²)	Grade of steel reinforcement		
	Fe 250 (MS) $R_{u\,lim} = 0.1489 f_{ck}$	Fe 415 HYSD $R_{u\,lim} = 0.1381 f_{ck}$	Fe 500 HYSD $R_{u\,lim} = 0.1330 f_{ck}$
M15	2.234	2.072	1.995
M20	2.978	2.762	2.660
M25	3.723	3.453	3.325
M30	4.467	4.143	3.990
M35	5.212	4.834	4.665
M40	5.956	5.524	5.320

$$M_{u\ lim} = R_{u\ lim} \cdot bd^2 \qquad \qquad \text{... Eq. (4.12)}$$

$R_{u\ lim}$ can be taken from the Table 4.4 or can be computed from Eq. (4.11), i.e.

$$R_{u\ lim} = 0.36 f_{ck} \cdot \frac{x_{u\ lim}}{d} \left(1 - 0.416 \frac{x_{u\ lim}}{d}\right)$$

$\dfrac{x_{u\ lim}}{d}$ can be taken from the Table 4.2 or can be calculated by Eq. (4.6a)

as
$$\frac{x_{u\ lim}}{d} = \frac{700}{\left(1100 + 0.87 f_y\right)}$$

It may be noted that R_u of limit state is generally more than $1.5Q$ of working stress method and hence limit state method gives economical design with smaller concrete section. It may be further noted that balanced beam section gives smallest section. Over reinforced section with higher quantity of steel reinforcement becomes costly and does not use steel or concrete optimally. Since steel is costly, under reinforced section (with p_t less than limiting) is always desirable in limit state. Code also recommends under reinforced designs.

4.9 DESIGN OF BEAM SECTION (SINGLY REINFORCED)

Design of beam section involves determinations of sectional dimensions (width b and depth d) and the area of steel reinforcement in tension A_{st}. Singly reinforced sections are such that M_u is borne fully within the provided section. The process of design is explained below:

I. Design Cross-Sectional Dimensions (b and d) and Tensile Steel (A_{st})

Step (i): Find the limiting depth of NA with reference to limiting strain of 0.0035 in concrete at the extreme compression fibre.

$$\left(\frac{x_u}{d}\right)^{lim} = \frac{0.0035}{\left(0.0055 + \dfrac{0.87 f_y}{E_s}\right)} = \frac{700}{\left(1100 + 0.87 f_y\right)}$$

Step (ii): Assume suitable $\dfrac{d}{b}$ ratio in accordance with given situation

$\left(\dfrac{d}{b} \text{ is generally taken between 1.5 to 3, if not specified}\right)$

Step (iii): Find d from the moment of resistance [Eq. (4.10)]

$$M_{u\ lim} = 0.36 f_{ck} \cdot \frac{x_{u\ lim}}{d} \left(1 - 0.416 \frac{x_{u\ lim}}{d}\right) bd^2 = R_u \cdot bd^2$$

Values of R_u can be chosen from the Table 4.4 with reference to the given grades of concrete and steel.

Step (iv): Knowing d and $\dfrac{d}{b}$ ratio, b can be computed and then from these values of b and d, A_{st} or $p_{t\ lim}$ can be computed from Eq. (4.9) or Table 4.3.

$$M_u = 0.87 f_y \, A_{st} \cdot d \left(1 - 0.416 \frac{x_{u\,lim}}{d} \right)$$

Alternatively, Eq. (4.8) for $p_{t\,lim} = 0.414 \frac{f_{ck}}{f_y} \cdot \frac{x_{u\,lim}}{d}$

$$A_{st} = p_{t\,lim} \cdot bd$$

II. Design to Find Tensile Steel Area When *b* and *d* are Known

In this case, first it is checked whether the given M_u, *b* and *d* requires a singly reinforced or doubly reinforced section.

Step (i):   Find $M_{u\,lim}$ corresponding to *b* and *d* using given grades of concrete and steel from Eq. (4.10), i.e

$$M_{u\,lim} = 0.36 f_{ck} \frac{x_{u\,lim}}{d} \left(1 - 0.416 \frac{x_{u\,lim}}{d} \right) bd^2$$

Step (ii):   If $M_u = M_{u\,lim}$, the section is balanced and the area of steel can be computed from Eq. (4.8), i.e.

$$p_{t\,lim} = 0.414 \frac{f_{ck}}{f_y} \cdot \frac{x_{u\,lim}}{d}$$

and $\qquad A_{st} = p_{t\,lim} \, b \cdot d$

Step (iii): If $M_u > M_{u\,lim}$, the beam shall be doubly reinforced which will be dealt later.

Step (iv):   If $M_u < M_{u\,lim}$, the section shall be under reinforced and singly reinforced.

Determine $\frac{x_u}{d}$ for the given M_u, i.e.

$$M_u = 0.36 f_{ck} b \cdot x_u (d - 0.416 \, x_u)$$

This gives quadratic equation in terms of x_u which can be solved.

Step (v):   Find A_{st} by: $M_u = 0.87 f_y \, A_{st} (d - 0.416 \, x_u)$, for under reinforced section

$$A_{st} = \frac{M_u}{0.87 \, f_y \left(d - 0.416 x_u \right)}$$

Alternatively, $\qquad p_t = 0.414 \frac{f_{ck}}{f_y} \cdot \frac{x_u}{d}$

$$A_{st} = p_t \cdot bd$$

Steps (iv) and (v) can be avoided and A_{st} can be found directly by solving quadratic

equation as $M_u = 0.87 f_y \cdot A_{st} \cdot d \left(1 - \frac{A_{st}}{bd} \cdot \frac{f_y}{f_{ck}} \right)$

or $\qquad A_{st} = \frac{0.5 f_{ck}}{f_y} \left[1 - \sqrt{1 - \frac{4.6 M_u}{f_{ck} \cdot bd^2}} \right] \cdot bd \qquad$ ... Eq. (4.13)

Different types of problems will be explained through solved examples.

Example 4.1

Determine the moment of resistance of a singly reinforced concrete beam of 200 mm width and 400 mm effective depth and is reinforced with 3 bars of 16 mm HYSD Fe 415 steel. Concrete used in the beam construction is of M20 grade.

Solution: $b = 200$ mm, $d = 400$ mm, $A_{st} = 3 \times 201 = 603$ mm^2, $f_{ck} = 20$ N/mm^2,
$f_y = 415$ N/mm^2

$$\left(\frac{x_u}{d}\right) = \frac{0.87 \, f_y \cdot A_{st}}{0.36 \, f_{ck} \cdot bd} = \frac{0.87 \times 415 \times 603}{0.36 \times 20 \times 200 \times 400} = 0.378$$

$\left(\dfrac{x_u}{d}\right)_{lim}$ for Fe 415 grade steel from the Table 4.2 $\cong 0.48$

The section is under reinforced $\left(\dfrac{x_u}{d} < \dfrac{x_{u\,lim}}{d}\right)$

$$\left\{\frac{x_{u\,lim}}{d} = \frac{700}{\left(1100 + 0.87 \, f_y\right)}\right\}$$

M_u shall be based on tension reinforcement for under reinforced sections.

$$M_u = 0.87 \, f_y \cdot A_{st} \, d \left(1 - \frac{A_{st} \cdot f_y}{bd \, f_{ck}}\right) = 0.87 \times 415 \times 603 \times 400 \left(1 - \frac{603 \times 415}{200 \times 400 \times 20}\right)$$

$$= 73.465 \times 10^6 \text{ N-mm (73.465 kN-m)}$$

Example 4.2

A reinforced concrete beam of rectangular section 200 mm × 550 mm deep is reinforced with 6 bars of 20 mm diameter at an effective cover of 50 mm. Using M20 grade of concrete and Fe 415 HYSD steel bars, compute the safe moment of resistance of the section.

Solution: $b = 200$ mm, $d = 550$ (mm) $- 50$ (mm) $= 500$ mm, $f_{ck} = 20$ N/mm^2,
$f_y = 415$ N/mm^2, $A_{st} = 6 \times 314 = 1884$ mm^2.

$$\frac{x_{u\,lim}}{d} = 0.48 \text{ (from Table 4.2) or } \frac{x_{u\,lim}}{d} = \frac{700}{\left(1100 + 0.87 \, f_y\right)} = 0.48$$

$$\frac{x_u}{d} = \frac{0.87 \, f_y \cdot A_{st}}{0.36 \, f_{ck} \cdot bd} = \frac{0.87 \times 415 \times 603}{0.36 \times 20 \times 200 \times 500} = 0.945 > \frac{x_{u\,lim}}{d} = 0.48$$

The section is over reinforced.

M_u will be based on compression fibres, i.e.

$$M_u \text{ (safe)} = 0.138 \, f_{ck} \times bd^2 = 0.138 \times 20 \times 200 \times 500^2$$
$$= 138 \times 10^6 \text{ N-mm} = 138 \text{ kN-m.}$$

Example 4.3

A reinforced concrete beam section 300 mm wide and 490 mm effective depth and reinforced with 7 bars of 16 mm diameter of Fe 415 grade of steel M20 grade of concrete is used, compute the moment of resistance of the section.

Solution: $b = 300$ mm, $d = 490$ mm, $A_{st} = 7 \times 201 = 1407$ mm^2, $f_{ck} = 20$ N/mm^2,
$f_y = 415$ N/mm^2.

Depth of NA, $\dfrac{x_u}{d} = \dfrac{0.87 \, f_y \cdot A_{st}}{0.36 \, f_{ck} \cdot bd} = \dfrac{0.87 \times 415 \times 1407}{0.36 \times 20 \times 300 \times 490} = 0.48, \; \dfrac{x_{u\,lim}}{d} = 0.48$

Thus, the section is balanced and hence

$$M_u = 0.138 \, f_{ck} \, bd^2 = 0.138 \times 20 \times 300 \, (490)^2 = 198.8 \times 10^6 \text{ N-mm (198.8 kN-m)}$$

From tension consideration,

$$M_u = 0.87 f_y\, A_{st} \cdot d \left(1 - \frac{A_{st} \cdot f_y}{bd \cdot f_{ck}}\right) = 0.87 \times 415 \times 1407 \times 490 \left(1 - \frac{1407 \times 415}{300 \times 490 \times 20}\right)$$

$$= 248918700\,(1 - 0.1986) = 199.5 \times 10^6 \text{ N-mm (199.5 kN-m)}$$

As the values of M_r w.r.t. compression and tension are almost equal, hence the section is balanced and safe $M_r\,(M_u) = 198.8$ kN-m.

Example 4.4

Determine the area of tension reinforcement required for a singly reinforced beam of 300 mm breadth and 600 mm effective depth to resist a factored moment of 200 kN-m if concrete is of M25 grade and steel of Fe 415 grade.

Solution: $b = 300$ mm, $d = 600$ mm, $f_{ck} = 25$ N/mm^2, $f_y = 415$ N/mm^2,

$$M = 200 \times 10^6 \text{ N-mm.}$$

For Fe 415, $M_{u\,lim} = 0.138\, f_{ck} \cdot bd^2 = 0.138 \times 25 \times 300\,(600)^2$

$$= 372.6 \times 10^6 \text{ N-mm} > M_r \text{ of } 200 \times 10^6 \text{ N-mm}$$

The section is under reinforced.

$\therefore$ $$M_u = 0.87\, f_y\, A_{st}\, d \left(1 - \frac{A_{st} \cdot f_y}{bd \cdot f_{ck}}\right) = 0.87 \times 415\, A_{st} \times 600 \left(1 - \frac{415 A_{st}}{300 \times 600 \times 25}\right)$$

or $\qquad 200 \times 10^6 = 216630\, A_{st}\,(1 - 9.22 \times 10^{-5}\, A_{st})$

or $\qquad A_{st}^2\,(19.9733) - 216630\, A_{st} + 200 \times 10^6 = 0$

or $\qquad A_{st}^2 - 10846\, A_{st} + 10013368 = 0$

$$A_{st} = \frac{10846 \pm \sqrt{(10846)^2 - 4(10013368)}}{2} = \frac{10846 \pm 8808}{2} = 1019 \text{ or } 9827 \text{ mm}^2$$

Alternatively, $$A_{st} = \frac{0.5 f_{ck} \cdot bd}{f_y}\left[1 - \sqrt{1 - \frac{4.6 M_u}{f_{ck}\, bd^2}}\right]$$

$$A_{st} = 1020 \text{ mm}^2$$

Thus, minimum steel reinforcement $= 1019$ mm^2 (provide 4 bars of 18 mm diameter or 5 bars of 16 mm diameter.

Example 4.5

Design a rectangular RC beam section of 200 mm width to bear a factored moment of 250 kN-m. Concrete of grade M20 and steel of Fe 415 grade are used in the beam construction.

Solution: $b = 200$ mm, $d = ?$, $f_{ck} = 20$ N/mm^2, $f_y = 415$ N/mm^2, $M_u = 250 \times 10^6$ N-mm, $A_{st} = ?$

Design the section as balanced or under reinforced.

$$M_{u\,lim} = 0.138\, f_{ck} \cdot bd^2 \text{ or } 250 \times 10^6 = 0.138 \times 20 \times 200\,(d)^2$$

or $$d = \frac{\sqrt{250 \times 10^6}}{0.138 \times 20 \times 200} = 673 \text{ mm (provide } d = 675 \text{ mm)}$$

From Table 4.2: $\left(\dfrac{x_{u\,lim}}{d}\right) = 0.48 = \dfrac{0.87 A_{st} \cdot f_y}{0.36 f_{ck} \cdot bd}$ (equating tension and compression for balanced section) or $\qquad A_{st} = \dfrac{0.48 \times 0.36 \times 20 \times 200 \times 675}{0.87 \times 415} = 1292 \text{ mm}^2$

Provide effective depth 675 mm and tensile reinforcement 7 bars of 16ϕ
(or 4 bars of 20 mm diameter, $A_{st} = 1256 \text{ mm}^2$)

$$M_u = 0.87 \times 415 \times 1256 \times 675 \left(1 - \dfrac{1256 \times 415}{200 \times 675 \times 20}\right)$$

$$= 306098190 \ (1 - 0.19305) \cong 247 \text{ kN-m}.$$

Thus, provide beam section of 200 mm × 680 mm and tensile steel of 4 bars of 20 mm.

Example 4.6

Design a balanced singly reinforced concrete beam section for a service moment of 60 kN-m. The width of the beam is limited to 200 mm. Use M20 CC and Fe 415 steel bars.

Solution: $M = 60 \times 10^6$ N-mm, $b = 200$ mm, $f_{ck} = 20 \text{ N/mm}^2$,

$\qquad M_u = 1.5 \times 60 \times 10^6 = 90 \times 10^6$ N-mm (assuming $\gamma_f = 1.5$)

For balanced beam section: $\dfrac{x_{u\,lim}}{d} = \dfrac{700}{(1100 + 0.87 \times 415)} = 0.48$

$M_{u\,lim} = 0.36 f_{ck} \cdot \dfrac{x_{u\,lim}}{d} = \left(1 - 0.416 \dfrac{x_{u\,lim}}{d}\right) bd^2$

$90 \times 10^6 = 0.36 \times 20 \times 0.48 \ (1 - 0.416 \times 0.48) \times d^2 \times 200, \ d^2 = 162695$

$\qquad d = 404$ mm (provide $d = 410$ mm)

A_{st} from Eq. (4.13): $A_{st} = \dfrac{0.5 f_{ck}}{f_y}\left[1 - \sqrt{1 - \dfrac{4.6 M_u}{f_{ck} bd^2}}\right] \cdot bd$

$$A_{st} = \dfrac{0.5 \times 20}{415}\left[1 - \sqrt{1 - \dfrac{4.6 \times 90 \times 10^6}{20 \times 200 \times 410^2}}\right] 200 \times 410$$

$A_{st} = 751 \text{ mm}^2$ (3 bars of 18ϕ say, $A_{st} = 764 \text{ mm}^2$).

Example 4.7

Design a singly reinforced beam of 8 m span (SS) to carry 16 kN/m superimposed load over the entire span. The beam construction uses M30 CC and Fe 500 steel. Assume suitable self-weight of beam.

Solution: $b = ?, d = ?, A_{st} = ?, w_1 = 16 \text{ kN/m}, w_d = ?, L = 8 \text{ m}, f_{ck} = 30 \text{ N/mm}^2$,

$\qquad f_y = 500 \text{ N/mm}^2$

Let $\dfrac{d}{b} = 2$, i.e. $b = 0.5d$, self-weight (assume) $= 0.3 \times 0.6 \times 125 = 4.5 \text{ kN/m}$

Total u.d.l $w = w_1 + w_d = 16 + 4.5 = 20.5 \text{ kN/m}$, Load factor $= 1.50$, ultimate design load $w_u = 1.5 \times 20.5 = 30.75 \text{ kN/m} \cong 31 \text{ kN/m}$,

Max BM (M_u) $= \dfrac{w_u L^2}{8} = \dfrac{31}{8} (8)^2 = 248 \text{ kN/m} \ (248 \times 10^6 \text{ N-mm})$

$M_{u\,lim} = 0.36 \times 30 \times 0.46 \ (1 - 0.416 \times 0.46) \ (0.5d) \ (d^2) = 2.00866 d^3$

$$d^3 = \frac{248 \times 10^6}{2.00866} = 123465287, d = 498 \text{ mm}, b = 249 \text{ mm}$$

Provide 250 mm × 500 mm (effective) and 250 mm × 550 mm overall

$$p_{t\,\text{lim}} = 0.414 \frac{f_{ck}}{f_y} \cdot \frac{x_{u\,\text{lim}}}{d} = 0.414 \frac{30}{500} \times 0.46 = 0.01143 \ (1.143\%)$$

$A_{st} = 0.01143 \times 250 \times 500 = 1429 \text{ mm}^2$ (6 bars of 18 mm or 5 bars of 20 mm)

$A_{st} = 1526 \text{ mm}^2$ or 1570 mm^2

Example 4.8

A RCC beam of rectangular section 250 mm × 400 mm effective depth is reinforced with 4 bars of 16 mm diameter in tension zone. The beam is constructed using M25 grade concrete and Fe 500 grade steel. Consider $m = 11$, find the actual stresses in concrete and steel when the beam is subjected to a total factored BM of 100 kN-m.

Solution: $f_{ck} = 25 \text{ N/mm}^2, f_y = 500 \text{ N/mm}^2, M_u = 100 \times 10^6 \text{ N-mm}, A_{st} = 4 \times 201 = 804 \text{ mm}^2,$ $b = 250 \text{ mm}, d = 400 \text{ mm}, m = 11$

For the given steel and concrete $\dfrac{x_{u\,\text{lim}}}{d} = 0.456 \ (0.46)$ (Table 4.2)

Equating tensile and compressive forces considering internal equilibrium, we have

$$0.36 f_{ck} \cdot b \cdot x_u = 0.87 f_y \cdot A_{st} \text{ or } \frac{x_u}{d} = \frac{0.87 \times 500 \times 804}{0.36 \times 25 \times 250 \times 400} = 0.389$$

From Table 4.2: $\dfrac{x_{u\,\text{lim}}}{d} = \dfrac{700}{1100 + 0.87 \times 500} = 0.456 \ (\cong 0.46)$

Actual $\dfrac{x_u}{d} < \dfrac{x_{u\,\text{lim}}}{d}$, hence the section is under reinforced.

Moment of resistance shall be with reference to steel and hence

$$M_u = 100 \times 10^6 = f_s \cdot A_{st} \cdot d \left\{ 1 - 0.416 \frac{x_u}{d} \right\}, \text{ where } f_s = \text{actual stress in steel}$$

or $\qquad 10^8 = f_s \times 804 \times 400 = \{1 \times 0.416 \times 0.389\} = 804 \times 400 \times 0.8382 f_s$

$\therefore \qquad f_s = \dfrac{10^8}{804 \times 400 \times 0.8382} = 371 \text{ N/mm}^2 \text{ (steel stress)}, (< 0.87 f_y, \text{ i.e. } 435 \text{ N/mm}^2)$

Equating tensile force with compressive force, we have

$$f_s \cdot A_s = f_c \cdot x_u \cdot b \ (0.36), \text{ or } f_c \ (0.389 \times 400) \ 250 \times 0.36 = 371 \times 804$$

or $\qquad f_c = \dfrac{371 \times 804}{0.36 \times 0.389 \times 400 \times 250} = 21.3 \text{ N/mm}^2 \text{ (concrete stress)}$

Strain in steel $\dfrac{e_s}{(d - x_u)} = \dfrac{0.0035}{x_u}$, or $e_s = \dfrac{0.0035 \times 400(1 - 0.389)}{0.389 \times 400} = 0.0055$

Ultimate moment of resistance (M_u) w.r.t. compression is given by

$$M_u = C \times \text{lever arm} = 0.36 f_{ck} \cdot b \cdot x_u \ (d - 0.416 x_u) = 100 \text{ kN-m}$$

Ultimate M_r (M_u) w.r.t. tension is given by

$M_u = T \times$ lever arm $= 0.87 f_y \cdot A_{st} (d - 0.41\ 6x_u) = 87.01$ kN-m, while 117.26 kN-m at $f_y = 500$ N/mm^2

Example 4.9

Determine ultimate moment of resistance of a rectangular beam section of 300 mm wide and 500 mm effective depth. The beam is reinforced with 6 bars of 25 mm diameter. Concrete grade of M30 and steel of Fe 415 grade are used in the construction of the beam.

Solution: $f_{ck} = 30$ N/mm^2, $f_y = 415$ N/mm^2, $b = 300$ mm, $d = 500$ mm, $A_{st} = 6 \times 491 = 2946$ mm^2.

$$M_u = 0.36 f_{ck} \cdot b \cdot x_u(d - 0.416 x_u), \text{ where } x_u = \frac{f_s \cdot A_{st}}{0.36\ f_{ck} \cdot b}, f_s = 0.87 f_y$$

$$x_u = \frac{0.87 \times 415 \times 2946}{0.36 \times 30 \times 300} = 328.3 \text{ mm}, x_{u\,lim} = \frac{700 \times 500}{(1100 + 0.87\ f_y)} = 0.479 \times 500 = 239.5 \text{ mm}$$

$x_u > x_{u\,lim},$ the section is over reinforced.

Strain in steel $e_s = \dfrac{e_c(d - x_u)}{x_u} = \dfrac{0.0035(500 - 328.3)}{328.3} = 0.0018305$

$f_s = 0.87 f_y$ for balanced section and under reinforced
  $< 0.87 f_y$ for over reinforced sections.
$f_s = e_s \cdot E_s = 0.0018305 \times 2 \times 10^5 = 366.1$ N/mm^2

Hence $M_u = 0.36 f_{ck} \cdot b \cdot x_u(d - 0.416 x_u) = 0.36 \times 30 \times 300 \times 328.3 (500 - 328.3)$
        $= 182.64 \times 10^6$ N-mm.

Also from tension criteria, $M_u = 0.87 f_y \cdot A_{st} (d - 0.416\ x_u)$
or $M_u = 0.87 \times 415 \times 2946 (500 - 328.3) = 182.63 \times 10^6$ N-mm.

Example 4.10

Determine the actual stresses in concrete and steel if a rectangular singly reinforced beam of 300 mm width × 500 mm effective depth size is reinforced with 6 bars of 25 mm and subjected to a BM = 350 kN-m. Concrete of M30 grade and steel of Fe 415 grade are used in the constriction of beam.

Solution: $f_{ck} = 30$ N/mm^2, $f_y = 415$ N/mm^2, $b = 300$ mm, $d = 500$ mm, $A_{st} = 2946$ mm^2.

$$x_u = \frac{f_s \cdot A_{st}}{0.36\ f_{ck} \cdot b} = \frac{0.87 \times 415 \times 2946}{0.36 \times 30 \times 300} = 328.3 \text{ mm},$$

$x_{u\,lim} = 0.479 \times 500 = 239.5$ mm
$x_u > x_{u\,lim},$ beam is over reinforced.

$$e_s = \frac{0.0035 (500 - 328.3)}{328.3} = 0.0018305$$

$f_s = e_s \times E_s = 0.0018305 \times 2 \times 10^5 = 366.1$ N/mm^2
$M_u = $ (over reinforced) $= 0.36 f_{ck} \cdot b \cdot x_u(d - 0.416\ x_u) = 386.58 \times 10^6$ N-mm
For stress $350 \times 10^6 = 0.36 f_c \cdot b \cdot x_u(d - 0.416\ x_u), f_c = 27.16$ N/mm^2
Thus, $f_c = 27.16$ N/mm^2, $f_s = 366.1$ N/mm^2.

4.10 DOUBLY REINFORCED RECTANGULAR SECTIONS

When there is restriction to increase depth (d) of a singly reinforced concrete beam section to resist bending moment more than the capacity of singly reinforced beam section with the restricted depth (d), doubly reinforced beam shall be provided. Figure 4.6 shows a doubly reinforced rectangular beam section. A beam is called doubly reinforced when steel reinforcement is provided in both compressive and tensile faces. The triangular distribution of strain across the beam section along the depth is shown in Fig. 4.6(b), while non-linear stress distribution is shown in Fig. 4.6(c). The total compressive force in concrete (C_c) and compressive force (C_{sc}) in compression steel (A_{sc}) are shown in Fig 4.6.

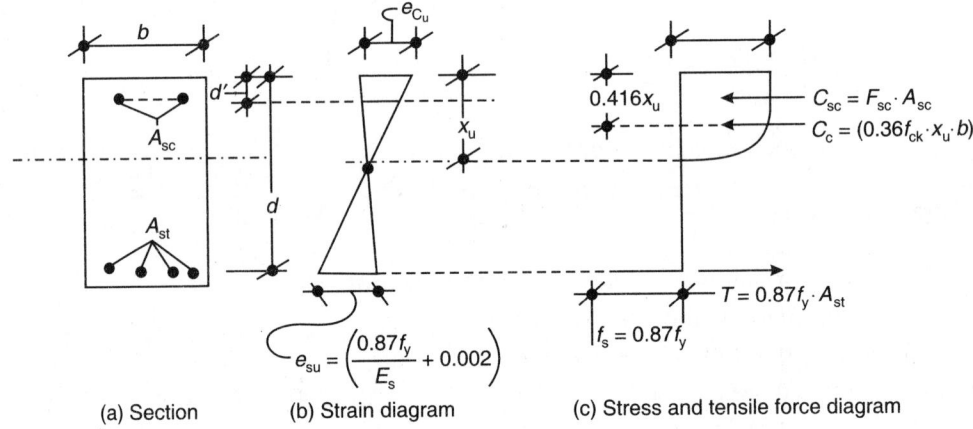

| (a) Section | (b) Strain diagram | (c) Stress and tensile force diagram |

Fig. 4.6: Doubly reinforced rectangular beam section

The ultimate moment of resistance (M_u) of section shall depend on whether the section is a balanced, under reinforced or over reinforced. The type of section can be found by finding the depth of neutral axis (x_u) and then comparing with the critical depth of NA ($x_{u\,lim}$) corresponding to the balanced section. The depth of NA (x_u) is found by considering equilibrium of internal forces (tensile force T = total compressive force C) as given below:

$$C = T$$

$$0.36\,f_{ck} \cdot b \cdot x_u + (f_{sc} - f_{cc})\,A_{sc} = 0.87 f_y\,A_{st}$$

or
$$x_u = \frac{\{0.87\,f_y \cdot A_{st} - (f_{sc} - f_{cc})A_{sc}\}}{0.36\,f_{ck} \cdot b} \qquad \text{... Eq. (4.14)}$$

where, f_{sc} = Stress in compression steel corresponding to strain e_{sc}

$$e_{sc} = \frac{0.0035\,(x_u - d')}{x_u} \qquad \text{... Eq. (4.15)}$$

For mild steel: $f_{sc} = 0.87 f_y = e_{sc} \times E_s$, where $E_s = 2 \times 10^5\,\text{N/mm}^2$

For cold worked steel:

$$f_{sc} = e_{sc} \cdot E_s \text{, for } e_{sc} \le \frac{0.696\,f_y}{E_s} \text{, and for } e_{sc} > \frac{0.696\,f_y}{E_s} \text{, the value of } f_{sc} \text{ is obtained from}$$

Table 4.5.

f_{cc} = Stress in concrete corresponding to strain e_{sc},

$$e_{sc} = \frac{0.0035\,(x_u - d')}{x_u}$$

$f_{cc} = 0.446 f_{ck}\,(e_{sc} - 250 e_{sc}^2),$ for $e_{sc} < 0.002$ ... Eq. (4.16)

$\quad = 0.446 f_{ck}$ for $e_{sc} \geq 0.002$ ... Eq. (4.16a)

It may be observed that the values of x_u depend on e_{sc} and e_{sc} depends on x_u [Eq. (4.15)]. Thus, the value of x_u shall be determined by iterative process till the value of x_u converges as explained below:

Step (i):   Assume $x_u = x_{u\,lim} = \dfrac{0.0035d}{\left(0.0055 + \dfrac{0.87\,f_y}{E_s}\right)} = \dfrac{700d}{\left(1100 + 0.87\,f_y\right)}$

(Taking $E_s = 2 \times 10^5 \, \text{N/mm}^2$)

Step (ii): Compute $e_{sc} = \dfrac{0.0035\,(x_u - d')}{x_u}$

Step (iii): Find the value of f_{sc} from Table 4.5 corresponding to the value of e_{sc} obtained in step (ii).

Step (iv): Substitute the value of f_{sc} in Eq. (4.14) and compute the value of modified

$$x_u = \frac{0.87\,f_y \cdot A_{st} - \left(f_{sc} - f_{cc}\right) A_{sc}}{0.36\,f_{ck} \cdot b}$$

Step (v): Repeat step (ii) to Step (v) till x_u converges.

The method shall be explained by examples. Different cases arise according to the value of the depth of NA (x_u) in comparison to the value of $x_{u\,lim}$.

Case I : If $x_u < x_{u\,lim}$, then section shall be under reinforced

Case II : If $x_u = x_{u\,lim}$, then section shall be balanced.

Case III : If $x_u > x_{u\,lim}$, then section shall be over reinforced

Table 4.5: Important points of stress–strain curve for cold worked deformed bars (HYSD)

Stress level	Fe 415 steel: $f_y = 415 \, \text{N/mm}^2$		Fe 500 steel: $f_y = 500 \, \text{N/mm}^2$	
	Strain	Stress (N/mm²)	Strain	Stress (N/mm²)
$0.80 \times (0.87 f_y)$	0.00144	288.7	0.00174	347.8
$0.85 \times (0.87 f_y)$	0.00163	306.7	0.00195	369.6
$0.90 \times (0.87 f_y)$	0.00192	324.8	0.00226	391.3
$0.95 \times (0.87 f_y)$	0.00241	342.8	0.00277	413.0
$0.975 \times (0.87 f_y)$	0.00276	351.8	0.00312	423.9
$1.0 \times (0.87 f_y)$	0.00380	360.9	0.00417	434.8

Note: Design strength $f_{yd} = 0.87,$ $f_y = \dfrac{f_y}{\gamma_m} = \dfrac{f_y}{1.15} = 0.87 f_y$

Linear interpolation may be made for intermediate values.

Case I: Under Reinforced Section (when $x_u < x_{u\,lim}$)

In this case, the failure occurs in tension face as the limiting stress develops in tensile reinforcement. The moment of resistance is given by:

$$M_u = 0.87 f_y \cdot A_{st} \times \alpha \qquad \qquad \text{... Eq. (4.17)}$$

where, α is the lever arm, i.e. the distance between the centre of tensile steel and CG of resultant compressive force.

Moment of resistance will also be found by taking moment of compressive forces about the centre of tensile reinforcement.

$$M_u = 0.36 f_{ck} x_u\, b\, (d - 0.416\, x_u) + f_{sc} A_{sc}\, (d - d')$$

or

$$M_u = 0.36 f_{ck} \frac{x_u}{d} \left(1 - 0.416 \frac{x_u}{d} \right) b d^2 + f_{sc} \cdot A_{sc} \cdot (d - d') \qquad \text{... Eq. (4.18)}$$

M_u found by both the Eqs (4.17) and (4.18) will be equal. The value of lever arm α can be found by finding the distance $\tilde{y}$ of CG of the compressive forces from the top compression fibre as

$$\tilde{y} = \frac{0.36 f_{ck} \times x_u \times b \times 0.416 x_u + f_{sc} \times A_{sc} \times d'}{0.36 f_{ck} \times x_u \times b + f_{sc} \times A_{sc}} \qquad \text{... Eq. (4.19)}$$

Lever arm
$$\alpha = (d - \tilde{y}) = \left[d - \frac{\{0.36 f_{ck} \times x_u \times b \times 0.416 x_u + f_{sc} \times A_{sc} \times d'\}}{(0.36 f_{ck} \times x_u \times b + f_{sc} \times A_{sc})} \right]$$
$$\text{... Eq. (4.19a)}$$

Equation (4.19) have been derived neglecting the loss of concrete occupied by compression steel (A_{sc}). These equations can be modified as under if this loss of concrete due to A_{sc} is taken into account:

$$M_u = 0.36 f_{ck} \times x_u \times b\, (d - 0.416 x_u) + (A_{sc} \times f_{sc} - 0.446 f_{ck} A_{sc})\, (d - d') \qquad \text{... Eq. (4.20)}$$

where, $\frac{x_u}{d}$ can be determined on the basis of equation

$C_u = 0.36 f_{ck} \times x_u \times b + f_{sc} \times A_{sc}$, which will be modified as under:

$$0.36 f_{ck} \times x_u \cdot b + (f_{sc} - 0.446 f_{ck})\, A_{sc} = 0.87 f_y A_{st} \qquad \text{... Eq. (4.21)}$$

Normally the term $0.446 f_{ck} \times A_{sc}$ is very small and can be neglected without causing any appreciable effect.

Case II: Balanced Section (when $x_u = x_{u\,lim}$)

The yield stress in the tension reinforcement reaches to limiting value at the same instant when the compression fibre of concrete reaches the ultimate strain of 0.0035. Thus, M_u reaches $M_{u\,lim}$. For balanced sections, the value of ultimate moment of resistance M_u will be equal to $M_{u\,lim}$, i.e.

$$M_u = M_{u\,lim} = 0.36 f_{ck} \frac{x_{u\,lim}}{d} \left(1 - 0.416 \frac{x_{u\,lim}}{d} \right) b d^2$$
$$+ (f_{sc} - 0.446 f_{ck}) \times A_{sc}\, (d - d') \qquad \text{... Eq. (4.22)}$$

Case III: Over Reinforced Section (when $x_u > x_{u\,lim}$)

In case $x_u > x_{u\,lim}$, the section will be over reinforced and the value of x_u found above will not be valid, x_u will be found by considering equilibrium of internal forces as given below:

Total compressive force (C) = Total tension (T)

$$0.36 f_{ck} \cdot b \cdot x_u + (f_{sc} - f_{cc}) A_{sc} = f_s \cdot A_{st} = 0.87 f_y \cdot A_{st} \qquad \text{... Eq. (4.23)}$$

or $\qquad x_u = \dfrac{\{f_s \cdot A_{st} - (f_{sc} - f_{cc}) A_{sc}\}}{0.36 f_{ck} \cdot b} = \dfrac{0.87 f_y \cdot A_{st} - (f_{sc} - f_{cc}) A_{sc}}{0.36 f_{ck} \cdot b}$

i.e $\qquad x_u = \dfrac{0.87 f_y \cdot A_{st} - (f_{sc} - f_{cc}) A_{sc}}{0.36 f_{ck} \cdot b} \qquad \text{... Eq. (4.24)}$

where, $\quad f_s$ = Stress in tensile steel corresponding to strain

$$e_s = \frac{0.0035\,(d - x_u)}{x_u}$$

f_{sc}, f_{cc} = Stresses in compression steel and concrete respectively corresponding to

strain $e_{sc} = \dfrac{0.0035\,(x_u - d')}{x_u}$.

The value of x_u can be determined by solving the above equations.

(i) Assume $x_u = x_{u\,lim} = \dfrac{700 d}{\left(1100 + 0.87 f_y\right)}$

(ii) Find $e_{sc} = \dfrac{0.0035\,(x_u - d')}{x_u}$ and $f_{sc} = E_s \times e_{sc} = \dfrac{2 \times 10^5 \times 0.0035\,(x_u - d')}{x_u}$

(iii) Now find modified $x_u = \dfrac{0.87 f_y \cdot A_{st} - (f_{sc} - f_{cc}) A_{sc}}{0.36 f_{ck} \cdot b}$, $\qquad (f_{cc} \cong 0.446 f_{ck})$

(iv) Repeat steps (ii) and (iii) till the value of x_u converges.

In over reinforced beam sections, failure occurs when strain inextreme concrete fibres reaches its limiting value of 0.0035, while the strain in tensile steel remain below the yield strain e_{su}. The moment of resistance

$$M_u = 0.36 f_{ck} \cdot x_{u\,lim} \cdot b \cdot (d - 0.416 x_{u\,lim}) + (f_{sc} - 0.446 f_{ck}) A_{sc} (d - d')$$

4.11 DESIGN OF DOUBLY REINFORCED BEAM SECTIONS

Figure 4.7 shows a doubly reinforced beam section and its equivalent singly reinforced section plus a doubly reinforced section with steel beam couple formed by tensile steel A_{st_2} and A_{sc} at a distance of d' from the top concrete fibre as:

$$A_{st_2} \times 0.87 f_y \times (d - d') = f_{sc} \times A_{sc}(d - d')$$

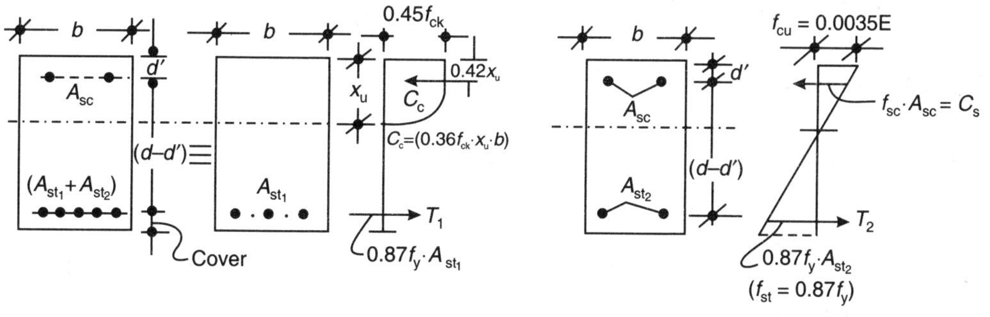

Fig. 4.7: Doubly reinforced section ($b \times d_{eff}$)

Steel beam theory is simple method of analyzing doubly reinforced beam section. This method is not accurate, but provides fairly good results for practical purposes without introducing errors of any practical importance. In this text, we shall use only steel beam theory for doubly reinforced beam sections.

In doubly reinforced beam sections, there are two components, viz:

(i) Singly reinforced section with restricted depth (d) having the limiting moment of resistance: $M_{u \, lim_1} = R_u \cdot bd^2$

(ii) A steel beam with tension steel (A_{st_2}) and compression steel (A_{sc}) at a distance (d') from top compression fibre and at ($d - d'$) from the centre of tensile steel (A_{st_2}).The moment of couple of steel beam

$$M_{u_2} = \{M_u \text{ (applied)} - M_{u \, lim_1}\} = \{M_u - R_u \cdot bd^2\} \qquad \qquad \text{... Eq. (4.25)}$$

where,

$M_{u \, lim} = M_r$ offered by singly reinforced beam as per IS:456-2000, i.e.

$= 0.148 f_{ck} \cdot bd^2$, for Fe 250 grade steel

$= 0.138 f_{ck} \cdot bd^2$ for Fe 415 HYSD grade steel

$= 0.133 f_{ck} \cdot bd^2$ for Fe 500 grade steel

$M_{u_2} = M_r$ offered by the steel beam neglecting the effect of concrete, i.e.

$= f_{sc} A_{sc} (d - d')$, moment couple of steel beam

f_{sc} = The stress in the compression steel corresponding to the strain reached by it when the extreme concrete fibre strain reaches 0.0035

A_{sc} = Area of compression steel

d = Effective depth of the rectangular beam section up to the CG of the tension steel

d' = Distance of the compression steel CG from the extreme compression fibre of concrete

A_{st_1} = Part of tension steel equivalent to singly reinforced beam section

A_{st_2} = Part of tensile steel equivalent to steel beam theory M_r

$A_{st} = A_{st_1} + A_{st_2}$

The design and analysis of doubly reinforced beams shall be explained through simple examples by following steps:

Step (i): $M_{u \, lim}$ (singly reinforced) $= R_u \, bd^2$

where, $R_u = M_r$ coefficient 0.148f_{ck}, 0.138f_{ck}, 0.133f_{ck}, for MS 250, Fe 415 and Fe 500 respectively.

Step (ii): Compute $A_{st_1} = \dfrac{0.36 f_{ck} \cdot bx_{u \, lim}}{0.87 f_y}$, assuming $f_{st} = 0.87 f_y$

Step (iii): Compute $A_{st_2} = (A_{st} - A_{st_1})$, where A_{st} = total tensile steel,

Step (iv): Compute $A_{sc} = \dfrac{0.87 f_y \cdot A_{st_2}}{f_{sc}}$ and $f_{sc} = \left\{ \dfrac{0.0035 \left(x_{u \, lim} - d'\right)}{x_{u \, lim}} \right\} E_s$

Step (v): Compute ultimate total M_r of the section as

$M_u = M_{u \, lim} + f_{sc} A_{sc} (d - d')$

Example 4.11

A doubly reinforced cement concrete beam constructed with M20 concrete and Fe 415 steel has 250 mm width and 500 mm effective depth. The beam section has 2 bars of 12 mm diameter in compression face with an effective cover 40 mm and 4 bars of 20 mm diameter in tensile face respectively. Determine the flexural strength of the beam section according to IS:456-2000.

Solution: $b = 250$ mm, $d = 500$ mm, $d' = 40$ mm, $A_{st} = 4 \times 314 = 1256$ mm^2

$$A_{sc} = 2 \times 113 = 226 \text{ mm}^2, f_{ck} = 20 \text{ N/mm}^2, f_y = 415 \text{ N/mm}^2$$

$$\frac{x_{u\,lim}}{d} = 0.48 \text{ for Fe 415}, x_{u\,lim} = 0.48 \times 500 = 240 \text{ mm}$$

$$f_{sc} = e_{sc} \times E_s = \frac{0.0035\,(d-d')}{240} \times 2 \times 10^5 = 583 \text{ N/mm}^2, \text{ but} \not> 0.87 \times 415$$

$$= 361 \text{ N/mm}^2$$

$$A_{st_2} = \frac{f_{sc} \cdot A_{sc}}{0.87\,f_y} = \frac{361 \cdot A_{sc}}{0.87 \times 415}, A_{sc} = A_{st_2} = 226 \text{ mm}^2$$

$$A_{st_1} = (A_{st} - A_{st_2}) = 1256 - 226 = 1030 \text{ mm}^2$$

Actual NA (revised): x_u by equilibrium of internal forces for singly reinforced section $C_1 = T_1$

$$0.36\,f_{ck} \cdot b \cdot x_u = 0.87\,f_y\,A_{st_1}$$

or $x_u = \dfrac{0.87 \times 415 \times 1030}{0.36 \times 20 \times 250} = 206.6$ mm < 240 mm (x_u assumed $= x_{u\,lim}$ initially); section is under reinforced.

Moment of resistance,

$$M_u = 0.87 f_y \cdot A_{st_1}\,(d - 0.42 x_u) + f_{sc} \cdot A_{sc}\,(d - d'), \text{ taking } f_{sc} = 0.87\,f_y$$

$$= 0.87 \times 415 \times 1030\,(500 - 0.42 \times 206.6) + 0.87 \times 415 \times 226\,(500 - 40)$$

$$= (153.67 + 37.53)10^6 = 191.2 \times 10^6 \text{ N-mm} = 191.2 \text{ kN-m}$$

Example 4.12

A simply supported beam of 6 m effective span carries a total factored u.d.l. of 60 kN/m including self-weight. The size is limited to 250 mm width and 450 mm overall depth. Effective concrete cover of 50 mm is permitted. If concrete of M25 is used, determine the tensile and compressive steel areas of Fe 415 HYSD grade.

Solution: $b = 250$ mm, $D = 450$ mm, cover $(d') = 50$ mm, $d = 400$ mm, $f_{ck} = 25$ N/mm^2

$$f_y = 415 \text{ N/mm}^2, w_u = 60 \text{ kN/m}, L = 6 \text{ m}$$

$$M_u \text{ (SS beam)} = \frac{w_u \cdot L^2}{8} = \frac{60 \times 6 \times 6}{8} = 270 \text{ kN-m } (270 \times 10^6 \text{ N-mm})$$

For Fe 415: $\dfrac{x_{u\,lim}}{d} = 0.48$, $x_{u\,lim} = 0.48 \times 400 = 192$ mm

$M_{u\,lim} = 0.138\,f_{ck} \cdot bd^2 = 0.138 \times 25 \times 250\,(400)^2 = 138 \times 10^6$ N-mm (138 kN-m) (singly reinforced)

$$A_{st_1} = \frac{138 \times 10^6}{0.87 \times 415\,(400 - 0.42 \times x_{u\,lim})} = 1197 \text{ mm}^2$$

Max strain in compressive steel $= \dfrac{0.0035\,(192-50)}{192} = 0.0026$

$f_{sc} = 0.0026 \times E_s = 0.0026 \times 10^5 \times 2 = 520$ but $\not> 0.87 \times 415 = 361\ \text{N/mm}^2$

$\therefore$ f_{sc} allowed $= 361\ \text{N/mm}^2$

$$A_{sc} = \frac{M_{u2}}{f_{sc}\,(d-d')} = \frac{(270-138)10^6}{361\,(400-50)} = \frac{132 \times 10^6}{361 \times 350} = 1045\ \text{mm}^2$$

$$A_{st_2} = \frac{f_{sc} \times A_{sc}}{0.87\,f_y} = \frac{361 \times 1045}{361} = 1045\ \text{mm}^2$$

Total $A_{st} = A_{st_1} + A_{st_2} = 1197 + 1045 = 2242\ \text{mm}^2$ (6 bars 22 mm ϕ, $A_{st} = 2280\ \text{mm}^2$)

Compressive steel $A_{sc} = 1045\ \text{mm}^2$ (3 bars of 22 mm ϕ, $A_{sc} = 1140\ \text{mm}^2$)

Example 4.13

Determine the ultimate moment of resistance for a rectangular concrete beam section of 300 mm width and 500 mm effective depth. Consider M20 grade CC and Fe 415 steel grade and effective cover of 50 mm both in case of compression steel as well as tensile reinforcement. The beam section is reinforced with 4 bars of 25 mm diameter in tension face and 4 bars of 16 mm diameter in compression face.

Solution: $f_{ck} = 20\ \text{N/mm}^2$, $f_y = 415\ \text{N/mm}^2$, $b = 300\ \text{mm}$, $d = 500\ \text{mm}$, $d' = 50\ \text{mm}$

$A_{sc} = 4 \times 201 = 804\ \text{mm}^2$, $A_{st} = 4 \times 491 = 1964\ \text{mm}^2$

$$\frac{x_{u\lim}}{d} = \frac{700}{\left(1100 + 0.87\,f_y\right)} = 0.48 \text{ for Fe 415 steel}$$

$x_u \lim = 0.48 \times 500 = 240\ \text{mm}$

Assume $x_u = x_u \lim$ (240 mm) to start iterative process to find x_u (actual).

By considering equilibrium of internal forces, total compression will be equal to tension,

i.e. $0.36\,f_{ck} \cdot b \cdot x_u + A_{sc}\,(f_{sc}-f_{cc}) = 0.87 \times 415 \times A_{st}$

or $x_u = \dfrac{0.87 \times 415 \times 1964 - 804\,(f_{sc}-f_{cc})}{0.36 \times 20 \times 300} = \dfrac{709102 - 804\,(f_{sc}-f_{cc})}{0.36 \times 20 \times 300}$

or $x_u = 328.29 - 0.3722\,(f_{sc}-f_{cc})$ (1)

Thus, x_u depends on f_{sc} and f_{cc} which depend on e_{sc} · e_{sc} depends on x_u in turn, x_u will be found by iterative process till it converges.

e_{sc} (for assumed $x_u = 240\ \text{mm}$) $= \dfrac{0.0035\,(240-50)}{240} = 0.00277$

From stress–strain curves or Table 4.5 for Fe 415, we have

$f_{sc} = 352\ \text{N/mm}^2$, $f_{cc} = 15.83\ \text{N/mm}^2$

Substituting these in Eq. (1) for x_u, we get

x_u (revised) $= 328.29 - 0.3722\,(352 - 15.83) = 203.2\ \text{mm}$

e_{sc} (revised) $= \dfrac{0.0035\,(203.2-50)}{203.2} = 0.00264$

f_{sc} (Table 4.5) $= 348.7\ \text{N/mm}^2$, $f_{cc} = 15.1\ \text{N/mm}^2$

x_u (revised) $= 328.29 - 0.3722\,(348.7 - 15.1) = 204.1$ mm

$x_u = 205.0$ mm (say the final value), $e_{sc} = 0.00265, f_{sc} = 349$ N/mm$^2, f_{cc} = 15$ N/mm^2

$x_u < x_{u\,lim}$ (240), the section is under reinforced.

Therefore, $M_u = 0.36\,f_{ck}\,b \cdot x_u(d - 0.416.x_u) + A_{sc}\,(349 - 15)\,(500 - 50)$

$$= 183.64 \times 10^6 + 120.84 \times 10^6 = 304.48 \text{ kN-m.}$$

This method is quite laborious and time consuming, and hence students can adopt the steel beam theory.

Example 4.14

Solve Example 4.13 by using steel beam theory as per IS:456-2000.

Solution: $f_{ck} = 20$ N/mm$^2, f_y = 415$ N/mm$^2, b = 300$ mm, $d = 500$ mm, $d' = 50$ mm

$A_{sc} = 804$ mm$^2, A_{st} = 1964$ mm$^2, M_u = ?$

$x_u\,lim = 0.48d = 0.48 \times 500 = 240$ mm

or $\quad f_{sc}$ (stress in compression steel) $= e_{sc} \cdot E_s = \dfrac{0.0035\,(x_{u\,lim} - d')}{x_{u\,lim}} \times 2 \times 10^5$

or $\quad f_{sc} = \dfrac{0.0035 \times 190}{240} \times 2 \times 10^5 = 554.17$ N/mm^2 but not more than 0.87×415 or

$$(361 \text{ N/mm}^2)$$

$\therefore \quad A_{st_2} = \dfrac{f_{sc} \cdot A_{sc}}{0.87\,f_y} = \dfrac{0.87\,f_{y\,(804)}}{0.87\,f_y} = 804$ mm^2

$A_{st_1} = (A_{st} - A_{st_2}) = 1964 - 804 = 1160$ mm^2

Actual NA x_u by equating internal forces in singly reinforced section $0.36\,f_{ck} \cdot b \cdot x_u = 0.87\,f_y \cdot A_{st_1}$

or $x_u = \dfrac{361 \times 1160}{(0.36 \times 20 \times 300)} = 193.87$ mm $< x_{u\,lim}$ (240 mm); the section is under reinforced.

$$x_u = 194 \text{ mm}$$

Moment of resistance M_u (neglecting loss of concrete area due to A_{sc}),

$M_u = 0.87\,f_y \cdot A_{st_1}\,(d - 0.416\,x_u) + f_{sc} \cdot A_{sc}\,(d - d')$

$\quad = 361 \times 1160\,(500 - 80.70) + 0.87 \times 415 \times 804\,(500 - 50)$

$\quad = (175.6 + 130.6)10^6 = 306.2 \times 10^6$ N-mm $= 306.2$ kN-m

Thus, the approximate method result is very near the exact method result ($306.2 \cong 304.5$ kN-m).

Example 4.15

Design a rectangular beam section subjected to an ultimate moment of 300 kN-m. Consider concrete of grade M30 and steel of grade Fe 415. Assume effective depth to breadth ratio of 2 and singly reinforced beam to provide about 60% flexural strength.

Solution: $\quad d = 2b, M_1 = \dfrac{60}{100} \times 300 = 180$ kN-m (180×10^6 N-mm),

$M_2 = 0.4 \times 300 = 120$ kN-m (120×10^6 N-mm)

$\dfrac{x_u\,lim}{d} = \dfrac{0.0035}{\left(0.0055 + \dfrac{0.87\,f_y}{E_s}\right)} = \dfrac{700}{(1100 + 361)}$ or $x_u\,lim = 0.48d$

$$M_1 = 0.36 f_{ck} \left(\frac{d}{2}\right) x_{u_1} (d - 0.42 x_{u_1}) = 0.36 \times \frac{30}{2} (d) (0.48d) (d - 0.42 \times 0.48d)$$

$$180 \times 10^6 = 2.592 d^3 (1 - 0.2016) = 2.069453 d^3, \ d^3 = 86.979515 \times 10^6$$

$$d = 100 \sqrt[3]{86.98} = 4.43 \times 100 = 443 \text{ mm (provide } d = 450 \text{ mm)}$$

$$b = \frac{d}{2} = \frac{450}{2} = 225 \text{ mm}, \ D = d + \text{cover} = 450 + 50 = 500 \text{ mm}$$

$\therefore$ $x_u \lim = 0.48 \times 450 = 216 \text{ mm}$

$M_{u1 \lim}$ (singly reinforced) $\cong 180$ kN-m approximately

$$A_{st1} = \frac{M_{u1}}{0.87 f_y (d - 0.42 \times 216)} = \frac{180 \times 10^6}{361 (450 - 90.72)} = 1388 \text{ mm}^2$$

$$M_{u1 \lim} = 0.36 f_{ck} \cdot b \cdot x_{u1} (450 - 0.42 \times 216) = 188.6 \times 10^6 \text{ N-mm}$$

$\therefore$ $M_{u2} = (300 - 188.6)10^6 = 111.4 \times 10^6 \text{ N-mm}$

Using steel beam theory and neglecting loss of concrete area due to compression steel, we have

$$f_{sc} \cdot A_{sc} (d - d') = 111.4 \times 10^6 \text{ N-mm}, \ A_{sc} = \frac{111.4 \times 10^6}{0.87 \times 415 (450 - 50)} = 771 \text{ mm}^2$$

Total $A_{st} = 1388 + 771 = 2159 \text{ mm}^2$ (provide 7 bars 20ϕ, $A_{st} = 2198 \text{ mm}^2$)

         $A_{sc} = 771 \text{ mm}^2$ (Provide 4 bars of 16$\phi \cdot A_{sc} = 804 \text{ mm}^2$)

Design can be checked by computing actual x_u and actual M_u.

Example 4.16

A doubly reinforced rectangular beam section is used for a simply supported beam of 6 m effective span carrying a factored u.d.l of 60 kN/m including own weight of the beam. The width of the beam is 250 mm and the overall depth is limited to 450 mm. Consider concrete grade of M25 and steel grade of Fe 415 HYSD with effective concrete cover of 50 mm to both compressive and tensile steel. Determine the necessary areas of tensile and compressive steels.

Solution: $b = 250 \text{ mm}, D = 450 \text{ mm}, d' = 50 \text{ mm}, d = 450 - 50 = 400 \text{ mm}$

         $f_{ck} = 25 \text{ N/mm}^2, f_y = 415 \text{ N/mm}^2, w_u = 60 \text{ kN/m}, L = 6 \text{ m}.$

$$M_u \text{ (max)} = \frac{w_u L^2}{8} = \frac{60 \times 6 \times 6}{8} = 270 \text{ kN-m } (270 \times 10^6 \text{ N-mm})$$

$$\frac{M_u}{bd^2} = \frac{270 \times 10^6}{250 \times 400 \times 400} = 6.75$$

A_{st} and A_{sc} can be found from various tables of areas given in SP-16.

For Fe 415, $x_{u \lim} = 0.48d = 0.48 \times 400 = 192 \text{ mm}$

Also $M_{u (\lim)} = 0.138 f_{ck} bd^2 = 0.138 \times 25 \times 250 (400)^2 = 138 \times 10^6 \text{ N-mm}$

         $M_{u2} = M_u - M_{u \lim} = (270 \times 10^6 - 138 \times 10^6) = 132 \times 10^6 \ (132 \text{ kN-m})$

$$A_{st1} \text{ (singly reinforced} = \frac{138 \times 10^6}{0.87 \times 415 (400 - 0.42 \times 192)} = 1197 \text{ mm}^2$$

Max. strain in the extreme concrete fibre = 0.0035

e_{sc} (at d' below) $= 0.0035 \dfrac{(192-50)}{192} = 0.00259$

$\qquad f_{sc} = 0.00259 \times 2 \times 10^5$ but $\not> 0.87f_y = 518 \not> 361$

Adopt $\qquad f_{sc} = 361 \text{ N/mm}^2$

M_{u2} (steel beam couple) $= f_{sc} A_{sc} (d-d')$ or $A_{st2} (0.87f_y) (d-d')$

$$A_{sc} = \frac{132 \times 10^6}{361 \times 350} = 1045 \text{ mm}^2,$$

Total $\qquad A_{st} = 1197 + 1045 = 2242 \text{ mm}^2$

From design tables w.r.t. $\dfrac{M_u}{bd^2}, \dfrac{d'}{d} = \dfrac{50}{400} = 0.125$

$p_t = 2.240\%, p_c = 1.12\%$ by interpolation

$A_{st} = 2240 \text{ mm}^2, A_{sc} = 1120 \text{ mm}^2$ (these values are almost the same as calculated)

4.12 ANALYSIS AND DESIGN OF FLANGED BEAM SECTIONS

4.12.1 Analysis of Flanged Section

Figure 4.8 shows singly reinforced T-and inverted L-beam section (flanged section).

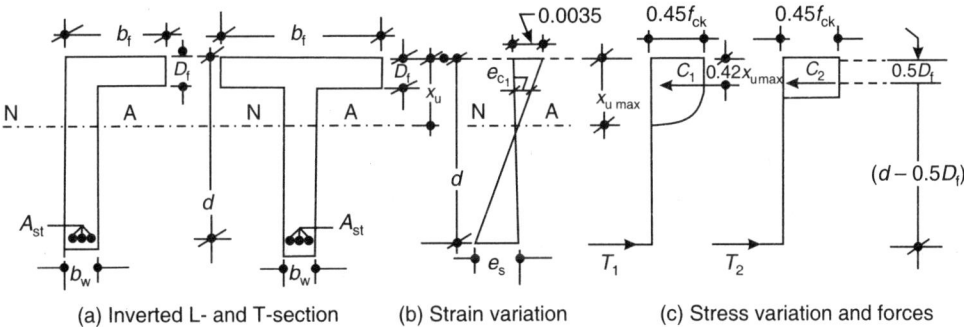

(a) Inverted L- and T-section   (b) Strain variation   (c) Stress variation and forces

Fig. 4.8: Flanged beam section, strain–stress block parameters

The ultimate moment of resistance of the flanged section (L or T) depends on the type of section (balanced, under reinforced or over reinforced) which may be determined by comparing the value of x_u with maximum $x_{u \, lim}$.

$x_u < x_{u \, lim}$: The section shall be under reinforced,

$x_u = x_{u \, lim}$: The section shall be balanced

$x_u > x_{u \, lim}$: The section shall be over reinforced.

Further, the NA may be located within the flange $(x_u < D_f)$, along the bottom of the flange $(x_u = D_f)$ or in the web portion $(x_u > D_f)$.

IS:456-2000 Method of Analysis

IS code has divided the problem of T-beam section into following types according to location of NA and ratio of flange depth (D_f) with effective depth (d), and flange depth (D_f) with the depth of NA (x_u).

Case I: When $x_u \leq D_f, x_u \leq x_{u \, max}$; under reinforced

Case II: When $x_u > D_f, x_u = x_{u \, max}$; balanced

Case III: When $x_u > D_f, x_u < x_{u\,max}$; under reinforced

Case IV: $x_u > x_{u\,max} > D_f$; over reinforced, to be redesigned as balanced or under reinforced.

Case I: When NA is within the flange $(x_u < D_f)$

In this case, the T-or L-beam section acts as rectangle with $b = b_f$ and depth $= d$.

$$\frac{x_u}{d} = \frac{0.87\,f_y \cdot A_{st}}{0.36\,f_{ck} \cdot b_f\,d} = \frac{2.417\,f_y \cdot A_{st}}{f_{ck} \cdot b_f\,d} \qquad \text{... Eq. (4.26)}$$

[*same as* Eq. (4.7)]

The moment of resistance M_u shall be given as

$$M_u = 0.87 f_y\,A_{st}\,(d - 0.416\,x_u) = 0.87 f_y \cdot A_{st}\,d\left(1 - \frac{A_{st} \cdot f_y}{b_f \cdot d\,f_{ck}}\right) \qquad \text{... Eq. (4.27)}$$

[*same as* Eq. (4.9)]

$$\frac{x_{u\,lim}}{d} = \frac{700}{\left(1100 + 0.87\,f_y\right)}, x_{u\,lim} = x_{u\,max} \qquad [\textit{from Eq. (4.6a)}]$$

When $\qquad x_u = x_{u\,lim}$,

$$M_{u\,lim} = 0.36 f_{ck}\,\frac{x_{u\,lim}}{d}\left\{1 - 0.416\frac{x_{u\,lim}}{d}\right\}b_f \cdot d^2 \qquad \text{... Eq. (4.28)}$$

[*same as* Eq. (4.10) with $b = b_f$]

where,

$\qquad b_f$ = Effective width of flange (mm)

$\qquad D_f$ = Depth or thickness of flange (mm)

$\qquad d$ = Effective depth of beam section $(D - \text{cover to } A_{st})$ (mm)

$\qquad f_{ck}$ = Characteristic strength of cement concrete (N/mm^2)

$\qquad f_y$ = Characteristic (yield) strength of steel (N/mm^2)

$\qquad A_{st}$ = Area of tension steel (mm^2)

In case $\frac{x_u}{d} > \frac{x_{u\,lim}}{d}$, the section is redesigned so that $\frac{x_u}{d} \le \frac{x_{u\,lim}}{d}$. This can be done by increasing d or reducing A_{st} [Eq. (4.26)].

Case II: When NA lies outside the flange $(x_u > D_f$ and $x_u = x_{u\,lim})$, i.e. the section is balanced.

In this case, the NA lies outside the flange (i.e. $x_u > D_f$) and $x_u = x_{u\,lim}$. The moment of resistance (M_u) depends on the ratio of D_f and d.

$\qquad$ Consider $\qquad\qquad x_{u\,lim} = \dfrac{7}{3}D_f$

or $\qquad\qquad\qquad \dfrac{x_{u\,lim}}{d} = \dfrac{7}{3}\dfrac{D_f}{d}$ or $\dfrac{D_f}{d} = \dfrac{3}{7}\dfrac{x_{u\,lim}}{d}$

For different steels, $\dfrac{x_{u\,lim}}{d} = \dfrac{700}{\left(1100 + 0.87\,f_y\right)}$ and $\dfrac{D_f}{d} = \dfrac{3}{7}\dfrac{x_{u\,lim}}{d}$ are given in Table 4.6.

Table 4.6: Limiting values of D_f/d ratio for different steels

Grade of steel	$\dfrac{x_{u\,lim}}{d}$	$\dfrac{D_f}{d} = \dfrac{3}{7}\dfrac{x_{u\,lim}}{d}$	Value of D_f/d assumed in IS:456-2000
Fe 250	0.531 (0.53)	0.228	0.20
Fe 415	0.479 (0.48)	0.205	0.20
Fe 500	0.456 (0.46)	0.195	0.20

Thus, important border value of D_f/d will be 0.20 for all grades of steel.

Thus, case II will have two subsections as follows:

II(a): x_u(balanced) $(x_u = x_{u\,lim})$, $x_u > D_f$, and $\dfrac{D_f}{d} \le 0.20$ (or $\ngtr 0.20$)

The moment of resistance is given by the sum of M_r of web portion $(b_w \times d)$ and flange portion $(b_f - b_w) \times D_f$.

$$M_{u\,lim} = 0.36 f_{ck}\,\frac{x_{u\,lim}}{d}\left(1 - 0.416\,\frac{x_{u\,lim}}{d}\right) b_w\, d^2 + 0.446 f_{ck}\,(b_f - b_w)\, D_f\left(d - \frac{D_f}{2}\right)$$

or $\qquad M_{u\,lim} = 0.36 f_{ck}\,\dfrac{x_{u\,lim}}{d}\left(1 - 0.42\,\dfrac{x_{u\,lim}}{d}\right) b_w\, d^2 + 0.45 f_{ck}\,(b_f - b_w)\, D_f\left(d - \dfrac{D_f}{2}\right)$

... Eq. (4.29)

In this case, the IS code has considered the depth of rectangular stress block (y_f) equal to the flange depth (D_f). Width of web b_w and depth $d_w = d$ for the whole web from top to the centre of tensile steel.

II(b): $x_u = x_{u\,lim}$ (balanced), $x_u > D_f$ and $\dfrac{D_f}{d} > 0.20$

The moment of resistance is given by considering depth of stress block (rectangular) up to a depth of y_f and $y_f = (0.15\, x_{u\,lim} + 0.65\, D_f)$ but not more than D_f.

$$M_{u\,lim} = 0.36 f_{ck}\,\frac{x_{u\,lim}}{d}\left(1 - 0.416\,\frac{x_{u\,lim}}{d}\right) b_w\, d^2 + 0.446 f_{ck}\,(b_f - b_w)\, y_f\left(d - \frac{y_f}{2}\right) \quad \text{... Eq. (4.30)}$$

where, $y_f = \{0.15\, x_{u\,lim} + 0.65\, D_f\}$, but not greater than D_f with usual terms and notations.

$$M_{u\,lim} = 0.36 f_{ck}\,\frac{x_{u\,lim}}{d}\left(1 - 0.42\,\frac{x_{u\,lim}}{d}\right) b_w\, d^2 + 0.45 f_{ck}\,(b_f - b_w)\, y_f\left(d - \frac{y_f}{2}\right) \quad \text{... Eq. (4.30a)}$$

Case III: When $x_{u\,lim} > x_u > D_f$, i.e. under reinforced and NA lies outside the flange. In this case also the section will be under reinforced $(x_u < x_{u\,lim})$. The flange falls fully in rectangular portion of the stress block. The moment of resistance (M_u) will depend on whether the value of $\left(\dfrac{D_f}{x_u}\right)$ is equal to or less than 0.43 or $\left(\dfrac{3}{7}\right)$ or the value of $\left(\dfrac{D_f}{x_u}\right)$ is greater than 0.43 $\left(\text{or } \dfrac{3}{7}\right)$. Thus, this case also has two subsections as follows:

III(a): $x_{u\,lim} > x_u > D_f$ and $\dfrac{D_f}{x_u} \le 0.43$ (under reinforced)

When D_f is equal to or less than $0.43x_u\left(\text{or }\dfrac{3x_u}{7}\right)$, the depth of the stress block will be greater than the thickness of the flange (D_f). Thus, the flange thickness will be fully within the rectangular portion of the stress block (Fig. 4.8c). The moment of resistance (M_u) shall be given by:

$$M_u = 0.36 f_{ck} \cdot \frac{x_u}{d}\left(1 - 0.416 \frac{x_u}{d}\right) b_w\, d^2 + 0.446 f_{ck}\,(b_f - b_w)\,D_f\left(d - \frac{D_f}{2}\right)$$

$$\text{... Eq. (4.31)}$$

or $\qquad M_u = 0.36 f_{ck} \cdot \dfrac{x_u}{d}\left(1 - 0.42 \dfrac{x_u}{d}\right) b_w\, d^2 + 0.45 f_{ck}\,(b_f - b_w)\,D_f\left(d - \dfrac{D_f}{2}\right)$

$$\text{... Eq. (4.31a)}$$

$\left(\text{This is similar to Eq. (4.29) except } \dfrac{x_{u\,lim}}{d} \text{ is replaced by } \dfrac{x_u}{d}\right)$

III(b): $x_{u\,lim} > x_u > D_f$ and $\dfrac{D_f}{x_u} > 0.43$ (under reinforced)

When D_f is equal to or less than $0.43x_u\left(\dfrac{3x_u}{7}\right)$, the depth of the stress block (Fig. 4.8(c)) will be less than the depth of the flange (D_f). The distribution of compressive stress in the flange will not be rectangular for the whole thickness (D_f). The moment of resistance is given by:

$$M_u = 0.36 f_{ck}\, \frac{x_u}{d}\left(1 - 0.416 \frac{x_u}{d}\right) b_w \cdot d^2 + 0.446 f_{ck}\,(b_f - b_w)\, y_f\left(d - \frac{y_f}{2}\right) \quad \text{...Eq. (4.32)}$$

where $y_f = (0.15x_u + 0.65D_f)$, but not more than D_f.

or $\qquad M_u = 0.36 f_{ck}\, \dfrac{x_u}{d}\left(1 - 0.42 \dfrac{x_u}{d}\right) b_w \cdot d^2 + 0.45 f_{ck}\,(b_f - b_w)\, y_f\left(d - \dfrac{y_f}{2}\right) \quad \text{... Eq. (4.32a)}$

$(0.416 \cong 0.42,\ 0.446 \cong 0.45$ for simplification$)$

$\left(\text{This is similar to Eq. (4.30a) except } \dfrac{x_{u\,lim}}{d} \text{ is replaced by } \dfrac{x_u}{d}\right)$

Note: x_u is first found by considering NA within flange depth (D_f) but if it works out to be more than D_f (i.e. $x_u > D_f$), then x_u is redetermined by considering compressive area of web ($x_u - D_f$) b_w also for finding the correct value of x_u.

Case IV: When $x_u > x_{u\,lim} > D_f$, i.e. over reinforced and NA lies outside the flange

The code has not specifically dealt this case and recommended redesign of section by increasing the depth d or reducing steel (A_{st}) so that crushing of concrete in compression during collapse is avoided. In this, the strain in steel (e_s) remains less than yield strain (e_{su}). The moment of resistance of such a section is limited to $M_{u\,lim}$ and is given by Eq. (4.29) under Case II(a) when $x_u = x_{max}$ and $x_u > D_f$, and $\dfrac{D_f}{d} \le 0.20$.

$$M_u = M_{u\,lim} = 0.36 f_{ck}\,\frac{x_{u\,lim}}{d}\left(1 - 0.42\,\frac{x_{u\,lim}}{d}\right) b_w \cdot d^2 + 0.45 f_{ck}\,(b_f - b_w)\,D_f \cdot \left(d - \frac{D_f}{2}\right)$$

The procedure will be explained by considering different types of problems of T-beam sections. Generally these problems are of following 4 types:

Type 1: Given, material properties (f_{ck} and f_y), and section properties (e.g. b_f, D_f, b_w, D (or d), A_{st}, and to find moment of resistance (M_u).

Type 2: Given, material properties (f_{ck} and f_y), concrete section properties (b_f, D_f, b_w, D or d) and to find $M_{u\,lim}$ and $A_{st\,lim}$

Type 3: Design problem, given material properties (f_{ck} and f_y) and ultimate design moment $M_{u\,lim}$ and to find area of tension steel (A_{st}).

Type 4: Given, material properties (f_{ck} and f_y) and ultimate design moment $M_{u\,lim}$ and to find A_{st} and A_{sc} in case of doubly reinforced T-beams.

4.12.2 Analysis of Doubly Reinforced Flanged Beam Sections

A given flanged beam section with reinforcement in tensile zone has a limited (maximum) moment of resistance for a given depth. If the applied ultimate moment on the section is more, we will be required to provide steel reinforcement (A_{sc}) in compression zone also and additional tensile steel (A_{st_2}) in tension zone to offer certain additional moment of resistance (M_{u_2}) as moment couple formed by compression steel (A_{sc}) and additional tensile steel (A_{st_2}) as

$$M_{u_2} = A_{sc} \cdot f_{sc} \cdot (d - d') = A_{st_2} \cdot f_{st}\,(d - d') \qquad \text{... Eq. (4.33)}$$

where,

M_{u_2} = Additional ultimate M_r due to compressive steel

A_{sc} = Compressive steel area

f_{sc} = Stress developed in compressive steel due to strain e_{su} at the level of compression steel

d = Effective depth

d' = Concrete cover to compression steel from extreme face.

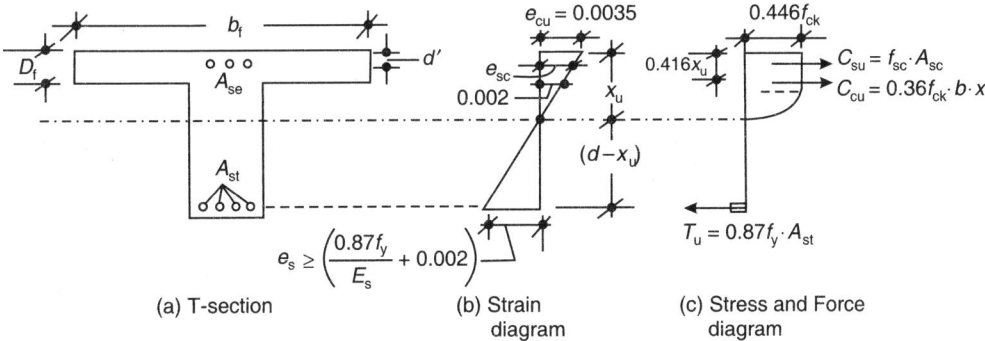

(a) T-section

(b) Strain diagram

(c) Stress and Force diagram

Fig. 4.9: Doubly reinforced T-beam section

Figure 4.9 shows a doubly reinforced T-beam section with compression steel (A_{sc}) at a distance (cover) of d' from the extreme compression fibre. The strain in compressive steel (e_{sc}) is shown Fig. 4.9(b). The compressive force developed in compressive steel will be based on the strain at the level of compressive steel. Thus (f_{sc}) the stress in compressive steel will be:

$$f_{sc} = e_{sc} \times E_s$$

The strain in compressive steel at the time of collapse will be:

$$e_{su} = \frac{0.0035}{x_u} (x_u - d')$$

Thus, compressive force developed by compressive steel (A_{sc}) will be:

$$C_{su} = f_{sc} \cdot A_{sc} = e_{sc} \cdot E_s \cdot A_{sc}$$

or
$$C_{su} = \frac{0.0035 (x_{u\,lim} - d')}{x_{u\,lim}} \cdot E_s \cdot A_{sc}$$

Thus, moment of couple $M_{u_2} = C_{su} \times (d - d')$

or
$$M_{u_2} = \frac{0.0035 (x_{u\,lim} - d')}{x_{u\,lim}} \cdot E_s \cdot A_{sc} (d - d') \qquad \dots \text{Eq. (4.34)}$$

Determination of x_u

When dimensions of the section are given, x_u is found by equating total compressive force and total tensile steel force. To start with, the NA is assumed to lie within the flange as:

$$0.36 f_{ck} \cdot x_u \cdot b_f + f_{sc} \cdot A_{sc} = 0.87 f_y \cdot A_{st} \qquad \dots \text{Eq. (4.35)}$$

(Ignoring the loss of concrete area due to compression steel A_{sc})

If the loss of concrete area is considered, we have

$$0.36 f_{ck} \cdot x_u \cdot b_f + (f_{sc} - 0.446 f_{ck}) A_{sc} = 0.87 f_y A_{st} \qquad \dots \text{Eq. (4.35a)}$$

This is subject to the condition that $d' < \dfrac{3}{7} x_u$, so that the stress in concrete (f_{cc}) at the level of compressive steel is equal to $0.446 f_{ck}$. The stress in steel (f_{sc}) can be found from the stress–strain diagram of steel for the level of strain e_{sc} (or stress = strain × modulus of elasticity).

The stress f_{sc} can also be found from the Table 4.5 giving stress–strain relation for e_{sc} obtained by $e_{sc} = \dfrac{0.0035 (x_u - d')}{x_u}$.

Since x_u depends on f_{sc} and f_{sc} depends on x_u, the value of x_u will be determined by iterative process till x_u converges.

Step (i): Find $x_{u\,lim}$ for a balanced section from strain diagram

$$\text{i.e. } x_{u\,lim} = \frac{0.0035}{\left(0.0055 + \dfrac{0.87 f_y}{E_s}\right)} = \frac{700}{(1100 + 0.87 f_y)} \qquad \textit{[from Eq. (4.6)]}$$

Step (ii): Assume $x_u = x_{u\,lim}$ or $x_u = D_f$ whichever is smaller and find

strain e_{sc} by using $e_{su} = \dfrac{0.0035}{x_u} (x_u - d')$ \qquad \textit{[from Eq. (4.15)]}

Step (iii): Determine f_{sc} from stress–strain curve corresponding to the above value of e_{sc} (alternatively, take f_{sc} slightly less than $0.8 f_y$).

Step (iv): Substitute the value of f_{sc} so obtained in following equation

$$0.36 f_{ck} \cdot b_f \cdot x_u + (f_{sc} - 0.446 f_{ck}) A_{sc} = 0.87 f_y \cdot A_{st} \qquad \textit{[from Eq. (4.14)]}$$

Step (v): Repeat steps (ii) to (iv), till convergence of x_u is achieved. If x_u comes out to be more than D_f, the NA will be outside the flange and a new value of x_u will have to be recalculated by equating the total compressive force to the total tensile force. Total compressive force C_u will be sum of force in web and force in flange, i.e.

$$C_{uw} = 0.36 f_{ck} \cdot x_u \cdot b_w + (f_{sc} - 0.446 f_{ck}) A_{sc} = 0.87 f_y \cdot A_{sw} \qquad \text{[from Eq. (4.21)]}$$

The compressive force C_{uf} can be calculated by considering actual stress block as approximately rectangular for a height y_f

$$y_f = 0.15\, x_u + 0.65\, D_f, \text{ subject to a maximum of } D_f$$

$$C_{uf} = 0.446 f_{ck} (b_f - b_w)\, y_f = 0.87 f_y \cdot A_{sf} \ (A_{sf} = \text{flange equivalent tensile steel})$$

∴ Total force in compression will be equal to the total tensile force, i.e.

$$C_{uw} + C_{uf} = \{0.36 f_{ck} \cdot x_u \cdot b_w + (f_{sc} - 0.446 f_{ck}) A_{sc}\} + 0.446 f_{ck} (b_f - b_w)\, y_f = 0.87 f_y \cdot A_{sw}$$
$$0.87 f_y \cdot A_{sf} \ (A_{sf} = \text{flange equivalent tensile steel}).$$

or $\quad 0.36 f_{ck} \cdot x_u \cdot b_w + (f_{sc} - 0.446 f_{ck}) A_{sc} + 0.446 f_{ck} (b_f - b_w)\, y_f = 0.87 f_y \cdot A_{st} \qquad$ ... Eq. (4.36)

where, $\qquad A_{st} = A_{sw} + A_{sf}$

$\qquad\qquad A_{sw} = $ Web equivalent tensile steel

$\qquad\qquad A_{sf} = $ Flange equivalent tensile steel

Since f_{sc} depends on e_{sc}, which depends on x_u, and x_u depends on f_{sc}, an iterative procedure shall be adopted. Also the value of y_f depends on x_u, and hence the final value of x_u will be valid only if the value of y_f calculated by using the value of x_u ($y_f = 0.15\, x_u + 0.65 D_f$) is less than or equal to D_f. If the value of y_f comes out to be more than D_f, the value of x_u should be recalculated from Eq. (4.36) by taking $y_f = D_f$. This calculated value of x_u should be compared with $x_{u\,lim}$ to find whether the section is under reinforced, balanced or over reinforced.

Moment of Resistance (M_u)

(i) When $x_u < D_f$

In this, beam acts as simple rectangular with $b = b_f$, doubly reinforced. If $x_u < x_{u\,lim}$, the section is under reinforced and the moment of resistance is given as:

$$M_u = 0.36 f_{ck} \cdot x_u \cdot b_f (d - 0.416\, x_u) + (f_{sc} - 0.446 f_{ck}) A_{sc} (d - d') \qquad \text{... Eq. (4.37)}$$

If $x_u = D_f$, we have

$$M_u = 0.36 f_{ck} \cdot b_f \cdot D_f (d - 0.416 D_f) + (f_{sc} - 0.446 f_{ck}) A_{sc} (d - d') \qquad \text{... Eq. (4.37a)}$$

Further, if $x_u = x_{u\,lim}$, we have

$$M_{u\,lim} = 0.36 f_{ck} \cdot x_{u\,lim} \cdot b_f (d - 0.416 x_{u\,lim}) + (f_{sc} - 0.446 f_{ck}) A_{sc} (d - d') \qquad \text{... Eq. (4.38)}$$

If $x_u > x_{u\,lim}$, the section is over reinforced and the M_r will be limited to $M_{u\,lim}$ given by Eq. (4.38). IS:456-2000 recommends redesign of such a section.

(ii) When $x_u > D_f$

The value of x_u obtained from Eq. (4.36) based on equilibrium of forces may be compared with the value of $x_{u\,lim}$ to evaluate whether the section is under reinforced, balanced or over reinforced. The procedure will be same as for singly reinforced except that an additional force in compression steel (A_{sc}) is considered as causing.

$$M_r \text{ of } (f_{sc} - 0.446 f_{ck}) A_{sc} (d - d')$$

(a) For under reinforced section $(x_u < x_{u\,\text{lim}})$, when $\dfrac{D_f}{x_u} \le 0.43\ \left(\text{or } \dfrac{3}{7}\right)$:

$M_{uw}(\text{web}) = 0.36 f_{ck} \cdot \dfrac{x_u}{d} \cdot b_w \left(1 - 0.416 \dfrac{x_u}{d}\right) d^2 + (f_{sc} - 0.446 f_{ck})\, A_{sc}(d - d')$

$M_{uf}(\text{flange}) = 0.446 f_{ck}\,(b_f - b_w)\,D_f\,(d - 0.50 D_f)$

i.e. $M_u = 0.36 f_{ck} \cdot \dfrac{x_u}{d}\left(1 - 0.416 \dfrac{x_u}{d}\right) \cdot b_w \cdot d^2 + (f_{sc} - 0.446 f_{ck})\, A_{sc}(d - d')$

$\qquad + 0.446 f_{ck}\,(b_f - b_w)\,D_f\,(d - 0.5 D_f)$ ... Eq. (4.39)

When $D_f = \dfrac{3}{7} x_u$ (or $x_u = \dfrac{7}{3} D_f$), we have

$M_u = 0.36 f_{ck} \cdot b_w \left(\dfrac{7}{3} D_f\right) + (f_{sc} - 0.446 f_{ck})\, A_{sc}(d - d')$

$\qquad + 0.446 f_{ck}\,(b_f - b_w)\,D_f\,(d - 0.5 D_f)$ ... Eq. (4.39a)

(b) For under reinforced section $(x_u < x_{u\,\text{lim}})$, when $\dfrac{D_f}{x_u} > 0.43$:

$M_{uw}(\text{web}) = 0.36 f_{ck} \cdot \dfrac{x_u}{d}\left(1 - 0.416 \dfrac{x_u}{d}\right) b_w \cdot d^2 + (f_{sc} - 0.446 f_{ck}) A_{sc}(d - d')$

$M_{uf}(\text{flange}) = 0.446 f_{ck}\,(b_f - b_w)\,y_f\,(d - 0.5 y_f)$

$M_u = 0.36 f_{ck} \cdot \dfrac{x_u}{d}\left(1 - 0.416 \dfrac{x_u}{d}\right) \cdot b_w \cdot d^2 + (f_{sc} - 0.446 f_{ck}) A_{sc}(d - d')$

$\qquad + 0.446 f_{ck}\,(b_f - b_w)\,y_f\,(d - 0.5 D_f)$ ... Eq. (4.40)

(iii) When $x_u = x_{u\,\text{lim}} > D_f$

(a) $\dfrac{D_f}{d} \le 0.20$

$M_{u\,\text{lim}} = 0.36 f_{ck} \dfrac{x_{u\,\text{lim}}}{d}\left(1 - 0.416 \dfrac{x_{u\,\text{lim}}}{d}\right) \cdot b_w \cdot d^2 + (f_{sc} - 0.446 f_{ck}) A_{sc}(d - d')$

$\qquad + 0.446 f_{ck}\,(b_f - b_w)\,D_f\,(d - 0.5 D_f)$ ... Eq. (4.41)

f_{sc} is found from the stress–strain curve or from the Table 4.5 for strain of e_{sc} calculated at the level of compression steel. f_{sc} can also be approximately found by $f_{sc} = e_{sc} \times E_s$, when $f_{sc} \le 0.87 f_y$.

e_{sc} is calculated by using the equations $e_{sc} = \dfrac{0.0035\,(x_u - d')}{x_u}$

considering $x_u = x_{u\,\text{lim}}$

(b) $\dfrac{D_f}{d} > 0.20$

$M_{u\,\text{lim}} = 0.36 f_{ck} \cdot \dfrac{x_{u\,\text{lim}}}{d}\left(1 - 0.416 \dfrac{x_{u\,\text{lim}}}{d}\right) \cdot b_w \cdot d^2 + (f_{sc} - 0.446 f_{ck}) A_{sc}(d - d')$

$\qquad + 0.446 f_{ck}\,(b_f - b_w)\,y_f\,(d - 0.5 y_f)$ ... Eq. (4.42)

where, $y_f = (0.15\, x_{u\,\text{lim}} + 0.65 D_f) \le D_f$

Moment of resistance in case of over reinforced beam section will be limited to $M_{u\,lim}$ for:

(a) $x_u > x_{u\,lim}$, and $\dfrac{D_f}{d} \leq 0.20$, given by Eq. (4.41)

(b) $x_u > x_{u\,lim}$, and $\dfrac{D_f}{d} > 0.20$, given by Eq. (4.42).

Code recommends redesign of over reinforced section so as to make it balanced or under reinforced.

Solved examples are given to explain the process of analysis and design of T-beam sections.

Example 4.17

Find the M_r of a T-beam section having the following details:

$b_f = 750$ mm, $d = 400$ mm, $b_w = 250$ mm

$A_{st} = 5$ bars of 20 mm diameter (MS)

$D_f = 100$ mm, M20 grade of CC.

Solution: $A_{st} = 5 \times 314 = 1570$ mm^2 (Fe 250)

Assuming NA to lie in the flange ($x_u < D_f$), we get

$$\frac{x_u}{d} = \frac{0.87\,f_y \cdot A_{st}}{0.36\,f_{ck} \cdot b_f \cdot d} = \frac{0.87 \times 250 \times 1570}{0.36 \times 20 \times 750 \times 400} = 0.158, \qquad x_u = 63.2 \text{ mm}$$

$$\frac{x_{u\,lim}}{d} = \frac{700}{1100 + 0.87 \times 250} = 0.5313 \; (x_u < x_{u\,lim}), \qquad x_{u\,lim} = 212.52 \text{ mm}$$

$x_u < D_f$ (63.2 < 100 mm) and hence NA lies in the flange

$x_u < x_{u\,lim}$, i.e. the section is under reinforced

$$M_r\,(M_u) = 0.87 f_y \cdot A_{st}\, d \left(1 - \frac{f_y}{f_{ck}} \cdot \frac{A_{st}}{b_f \cdot d}\right)$$

$$= 0.87 \times 250 \times 1570 \times 400 \left(1 - \frac{250 \times 1570}{20 \times 750 \times 400}\right) = 127.65 \times 10^6 \text{N-mm} \; (127.65 \text{ kN-m})$$

Example 4.18

Find the M_r of a T-beam section having the following details:

$b_f = 500$ mm, $d = 450$ mm, $b_w = 300$ mm, $A_{st} = 5$ bars of 25 mm (Fe 415)

$D_f = 100$ mm, M30 grade of CC.

Solution: Assuming NA to lie outside the flange $d = 450$ mm, $f_{ck} = 30$ N/mm^2

$f_y = 415$ N/mm^2, $b_f = 500$ mm, $b_w = 300$ mm, $A_{st} = 5 \times 491 = 2455$ mm^2.

$$x_{u\,lim} = \frac{700 \times 450}{1100 + 0.87 \times 415} = 0.4791 \times 450 = 215.6 \text{ mm}$$

Total compressive force = Total tension

$$0.36\,f_{ck} \cdot x_u \cdot b_w + 0.446\,f_{ck}\,(b_f - b_w) \cdot y_f = 0.87 \times 415 \,(5 \times 491)$$

or $0.36 \times 30 x_u \times 300 + 0.446 \times 30 \,(500 - 300)\, y_f = 885700$

or $3240 x_u + 2676\, y_f = 885700,$ $y_f = 0.15 \times x_{u\,lim} + 0.65 \times 100 = 97.34$ mm

or $3240 x_u + 2676 \times 97.34 = 885700$

or $\quad x_u = \dfrac{885700 - 260481}{3240} = 193$ mm, $y_f = 0.15 \times 193 + 0.65 \times 100 = 94$ mm $< D_f$

Substituting $y_f = 94$ mm and revising $x_u = 195.72$ mm,

$$y_f = 0.15 \times 195.72 + 65 = 94.31$$

Substituting y_f and finding x_u:

∴ $3240 x_u + 2676 \times 94.36 = 885700$,   ∴ $x_u = 195.73$ mm

∴ $y_f = 0.15 \times 195.43 + 65 = 94.31$

∴ $x_u = 195.50$ mm, $y_f = 94.4$ mm,   ∴ $\dfrac{D_f}{x_u} = \dfrac{100}{195.5} = 0.512 > 0.43.$

∴ $\dfrac{D_f}{d} = \dfrac{100}{450} = 0.222 > 0.20.$

Thus, M_u will be given by Eq. (4.40)

i.e. $M_u = 0.36 f_{ck} \cdot \dfrac{x_u}{d}\left(1 - 0.416 \dfrac{x_u}{d}\right) b_w \cdot d^2 + 0.446 f_{ck} (b_f - b_w) y_f (d - 0.5 y_f)$

or $M_u = 0.36 \times 30 \times 195.5 (450 - 0.416 \times 195.5) 300$

$\quad + 0.446 \times 30 (200) (94.4) (450 - 47.2)$

$\quad = 233524218 + 101753080 = 335277300$ N-mm $= 335.28$ kN-m

Example 4.19

Find the moment of resistance of a T-beam section shown in Fig. 4.10. Use M20 CC and Fe 415 steel.

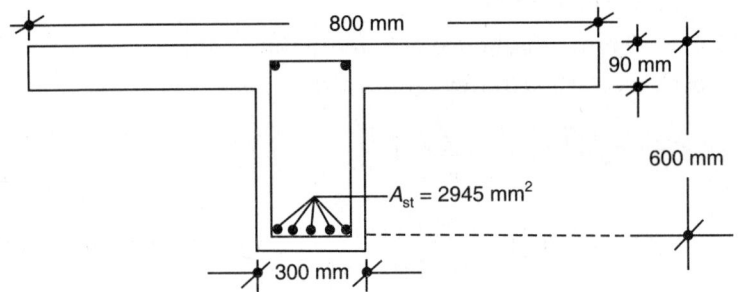

Fig. 4.10: T-beam section dimension

Solution: $b_f = 800$ mm, $D_f = 90$ mm, $d = 600$ mm, $A_{st} = 2945$ mm^2, $f_y = 415$ N/mm^2

$$f_{ck} = 20 \text{ N/mm}^2, \quad \dfrac{D_f}{d} = \dfrac{90}{600} = 0.15 < 0.20$$

Assuming NA to fall in the flange, we have

$$0.36 \times 800 \times 20 \times x_u = 0.87 \times 415 \times 2945$$

$$x_u = \dfrac{0.87 \times 415 \times 2945}{0.36 \times 800 \times 20} = 184.7 \text{ mm} > 90 \text{ mm } (D_f)$$

Since x_u is more than D_f (90 mm), NA lies outside the flange and hence it requires recalculation on this basis.

Let $y_f = 0.15 \times 184.7 + 0.65 \times 90 = 86.2$ mm < 90 mm (D_f)

$$x_{u\,\text{lim}} = \dfrac{700 \times d}{(1100 + 0.87 \times 415)} = 0.48 \times 600 = 288 \text{ mm}$$

Considering internal equilibrium of forces, we have total compressive force C_u = total tensile force T_u

$0.36 f_{ck} \cdot x_u \cdot b_w + 0.446 f_{ck} (b_f - b_w) y_f = 0.87 \times 415 \times 2945$, putting $y_f = 86.2$

$$x_u = \frac{(1063292 - 384452)}{0.36 \times 20 \times 300} = 314.3 \text{ mm} > x_{u\,lim}(288 \text{ mm})$$

Further the value of y_f was based on $x_u = 184.7$ mm. It also requires recalculation with $x_u = 314.3$ mm

$$y_f = 0.15 \times 314.3 + 0.65 \times 90 = 105.65 \text{ mm} > 90 \text{ mm}$$

Thus, recalculating x_u gives

$$0.36 \times 20 \times 300 x_u + 0.446 \times 20 (500) y_f = 1063292$$

or
$$x_u = \frac{(1063292 - 4460 y_f)}{2160} = 274.12 \text{ mm } (< 288 \text{ mm}), \text{ under reinforced.}$$

Recalculate $y_f = 0.15 \times 274.12 + 0.65 \times 90 = 99.62$ mm

Recalculate $x_u = \dfrac{(1063292 - 4460 y_f)}{2160} = 286.57$ mm

Recalculate $y_f = 0.15 \times 286.57 + 0.65 \times 90 = 101.6$ mm (say $y_f = 101$ mm)

Recalculate $x_u = \dfrac{(1063292 - 4460 y_f)}{2160} = 283.7$ mm (say $x_u = 284.00$ mm)

$x_u < x_{u\,lim}$ and hence the section is under reinforced.

$$M_u = 0.36 \frac{x_u}{d}\left(1 - 0.416 \frac{x_u}{d}\right) f_{ck} \cdot b_w \cdot d^2 + 0.446 f_{ck}(b_f - b_w) y_f (d - 0.5 y_f)$$

$= 0.36 \times 20 \times 284 (600 - 0.416 \times 284) \times 300 + 0.446 \times 20(500)101 \times (600 - 50.5)$

$= 295.59 \times 10^6 + 247.53 \times 10^6 = 543.12 \times 10^6$ N-mm

$= 543.12$ kN-m

Example 4.20

A reinforced cement concrete T-beam has the following data:

Flange width (b_f) = 1600 mm, effective depth (d) = 400 mm, thickness of flange (D_f) = 100 mm, web width (b_w) = 300 mm, concrete grade M20, steel grade Fe 500. Determine the limiting M_r and limiting area of steel (A_{st}).

Solution: $b_f = 1600$ mm, $b_w = 300$ mm, $d = 400$ mm, $f_{ck} = 20$ N/mm^2, $f_y = 500$ N/mm^2.

$$x_{u\,lim} = \frac{700 \times d}{(1100 + 0.87 \times 500)} = \frac{(700 \times 400)}{1535} = 182.41 \text{ mm} > D_f (100 \text{ mm})$$

$$\frac{D_f}{d} = \frac{100}{400} = 0.25 > 0.20, \qquad \frac{D_f}{x_u} = \frac{100}{182.41} = 0.548 > 0.43$$

$$y_f = 0.15 \times 182.4 + 0.65 \times 100 = 92.4 \text{ mm} < D_f (100 \text{ mm})$$

The limiting moment of resistance is given by:

$$M_{u\,lim} = 0.36 f_{ck} \times x_{u\,lim} \times b_w (d - 0.416 x_{u\,lim}) + 0.446 f_{ck} (b_f - b_w) y_f \left(d - \frac{y_f}{2}\right)$$

$= 0.36 \times 20 \times 182.41 \times 300 (400 - 0.416 \times 182.41) + 0.446 \times 20 (1600 - 300)$

$92.4 (400 - 46.2)$

$= 127.70 \times 10^6 + 370.09 \times 10^6 = 506.79 \times 10^6$ N-mm (506.79 kN-m)

Considering internal forces of equilibrium, we have

Total compressive force = Total tensile force, i.e.

$0.87 f_y \cdot A_{st} = 0.36 f_{ck} \times b_w \times x_{u\,lim} + 0.446 \times 20 \, (1600 - 300) \, y_f$

$0.87 \times 500 A_{st} = 0.36 \times 20 \times 300 \times 182.41 + 1071470 = 394006 + 1071470$

$$A_{st} = \frac{1465476}{435} = 3369 \text{ mm}^2$$

Example 4.21

A T-beam has the following data:

Width of flange (b_f) = 800 mm, width of web (b_w) = 300 mm

Effective depth (d) = 500 mm, flange thickness (D_f) = 90 mm

Applied moment = 150 kN-m. Design the beam section (tensile) steel using M20 CC and Fe 415 steel.

Solution: b_f = 800 mm, D_f = 90 mm, b_w = 300 mm, d = 500 mm, M_{uD} = 1.5 × 150 = 225 kN-m (225 × 10^6 N-mm)

$$x_{u\,lim} = \frac{700 \times 500}{(1100 + 0.87 \times 415)} = 239.6 \text{ mm}$$

Assume $x_u = D_f$ (say), we get

M_u(flange) = $0.36 f_{ck} \times b_f \times D_f \, (d - 0.416 D_f)$

M_u(flange) = $0.36 \times 20 \times 800 \times 90 \, (500 - 0.416 \times 90) = 239.8 \times 10^6$ N-mm $> M_{uD}$

$\qquad\qquad (225 \times 10^6$ N-mm)

∴ $x_u < D_f$, i.e. NA lies in the flange only.

$$M_{uD} = 225 \times 10^6 = 0.87 f_y \cdot A_{st} \times d \left(1 - \frac{f_y}{f_{ck}} \cdot \frac{A_{st}}{b_f \times d}\right)$$

or $\quad 225 \times 10^6 = 0.87 \times 415 \times 500. A_{st} \left(1 - \frac{415}{20} \times \frac{A_{st}}{800 \times 500}\right)$

or $\quad 225 \times 10^6 = 180525 A_{st} \, (1 - 5.1875 \times 10^{-5} A_{st})$

or $\quad 5.1875 \times 10^{-5} A_{st}^2 - 1 A_{st} + 1246.4 = 0$

$\qquad A_{st}^2 - 0.19277 \times 10^5 A_{st} + 240.27 \times 10^5 = 0, A_{st} = 1340$ mm^2

Alternatively, solution of quadratic equation gives

$$A_{st} = \frac{0.5 f_{ck}}{f_y} \left[1 - \sqrt{1 - \frac{4.6 M_{uD}}{f_{ck} b_f d^2}}\right] b_f \cdot d \qquad\qquad \text{... Eq. (4.43)}$$

$$A_{st} = \frac{0.5 \times 20}{415} \left[1 - \sqrt{1 - \frac{4.6 \times 225 \times 10^6}{20 \times 800 \times 500^2}}\right] (800)(500)$$

$$A_{st} = \frac{4000 \times 10^3}{415} \, (1 - 0.86096) = 1340.16 \text{ mm}^2$$

Example 4.22

Redesign the T-beam reinforcement if the applied moment is 260 kN-m in Example 4.21.

$d = 500$ mm, $b_f = 800$ mm, $b_w = 300$ mm, $D_f = 90$ mm, $f_{ck} = 20$ N/mm^2
$f_y = 415$ N/mm^2

Solution: M_u design $= 1.5 \times 260 = 390$ kN-m $(390 \times 10^6$ N-mm)

M_u (flange) $= 239.8 \times 10^6$ N-mm (from Example 4.21). Also $x_u < D_f$

M_{uD} $(390 \times 10^6$ N-mm) is more than M_u (flange) $(239.8 \times 10^6$ N-mm) and hence the beam section may require compression reinforcement and additional tensile reinforcement or may be singly reinforced with NA outside the flange.

$$x_{u\,lim} = \frac{700 \times d}{(1100 + 0.87\,f_y)} = \frac{700 \times 500}{(1100 + 0.87 \times 415)} = 239.55 \text{ mm} > D_f\,(90 \text{ mm})$$

$$\frac{7}{3}D_f = \frac{7}{3} \times 90 = 210 \text{ mm}, x_{u\,lim} = 239.55 \text{ mm}$$

Assuming $x_u \le \dfrac{7}{3}D_f$, calculate compressive forces in the flange and web

$(x_u = 210$ mm say)

C_{u1}web $= 0.36\,f_{ck} \cdot x_u \cdot b_w = 0.36 \times 20 \times 210 \times 300 = 453600$ N
C_{u2} flange $= 0.446\,f_{ck}\,(b_f - b_w) \cdot D_f = 0.446 \times 20 \times (800 - 300)\,90 = 401400$ N
M_u flange does not depend on x_u and hence

$$M_{uf} = 401400\left(d - \frac{D_f}{2}\right) = 401400\,(500 - 0.50 \times 90) = 182637000 \text{ N-mm}$$

M_u web $= 453600 \times (500 - 0.416 \times 210) = 184315824 = 184315824$ N-mm
Total $M_u = M_{uf} + M_{uw} = (182.64 \times 10^6 + 184.32 \times 10^6) = 366.96 \times 10^6$ N-mm

$$M_{uD}\,(390 \times 10^6) > M_{uf} + M_{uw}\,(366.96 \times 10^6)$$

Since M_{uf} does not depend on x_u, extra moment can still be taken by web since $x_{u\,lim} > 210$ mm.

Hence M_{uw} $(390 \times 10^6 - M_{uf}) = 390 \times 10^6 - 182.64 \times 10^6 = 207.36 \times 10^6$ N-mm

$$M_{uw} = 0.87\,f_y\,A_{sw} \times 500\left(1 - \frac{f_y}{f_{ck}} \cdot \frac{A_{sw}}{b_w \cdot d}\right) \qquad \dots \text{Eq. (4.44)}$$

or $\qquad 207.36 \times 10^6 = 0.87 \times 415 \times 500\,A_{sw}\left(1 - \dfrac{415\,A_{sw}}{20 \times 300 \times 500}\right)$

or $\qquad 1148.65 + (1.383 \times 10^{-4}\,A_{sw} - 1)\,A_{sw} = 0$

$\qquad A^2_{sw} - 7230.66\,A_{sw} + 8.3055 \times 10^6 = 0$

$A_{sw} = 1433$ mm^2, $A_{sf} = \dfrac{C_{uf}}{0.87\,f_y} = \dfrac{401400}{0.87 \times 415} = 1112$ mm^2 (Assuming $f_{sc} = 0.87f_y$)

Total $A_{st} = A_{sw} + A_{sf} = 1433 + 1112 = 2545$ mm^2; check $x_u = \dfrac{C_u\,\text{web}}{0.36\,f_{ck} \times b_w}$

$x_u = \dfrac{0.87 \times 415 \times 1433}{0.36 \times 20 \times 300} = 239.53$ mm $= x_{u\,max}$ (239.55 mm), hence the section is balanced.

Example 4.23

Design a doubly reinforced T-beam section of flange width 1500 mm, flange depth 100 mm, and web width 250 mm and subjected to a bending moment of 267 kN-m. Use M20 grade CC and Fe 415 grade steel. NA may be assumed to lie at bottom of the flange.

Solution: $f_{ck} = 20 \text{ N/mm}^2, f_y = 415 \text{ N/mm}^2, b_f = 1500 \text{ mm}, D_f = 100 \text{ mm}, b_w = 250 \text{ mm}$

$M_u = 1.50 \times 267 = 400.5 \text{ kN-m} = 400.5 \times 10^6 \text{ N-mm}, d = ?, A_{st} = ?, A_{sc} = ?$

Assuming $x_u = D_f = 100$ mm (say)

$$\frac{x_{u\,lim}}{d} = \frac{0.0035}{\left(0.0055 + \dfrac{0.87 f_y}{E_s}\right)} = \frac{700}{(1100 + 361)} = 0.48, \text{ or } x_{u\,lim} = 0.48d = 100 \text{ mm}$$

$$d = \frac{100}{0.48} = 208.8 \text{ mm}$$

Assume effective concrete cover = 35 mm

$D = 208.8 + 35 = 243.8$ mm, provide D (overall depth) = 250 mm (say)

Actual $d = 250 - 35 = 215$ mm. The section may be doubly reinforced, if necessary.

$M_{u\,lim}$ (singly reinforced portion), $x_{u\,lim} = 0.48 \times 215 = 103.2 \text{ mm} > D_f (100 \text{ mm})$

$$M_{u\,lim} = 0.36 f_{ck}\, x_{u\,lim}\, b_w\, (d - 0.416\, x_u) + 0.446 f_{ck}\, (b_f - b_w)\, y_f \left(1 - \frac{y_f}{2}\right)$$

$y_f = 0.15 x_u + 0.65 D_f = 0.15 \times 103.2 + 65 = 80.48 \text{ mm} < D_f (100 \text{ mm})$

∴ $M_{u\,lim}$ (singly reinforced) = $0.36 \times 20 \times 103.2 \times 250$

$(215 - 0.416 \times 103.2) + 0.446 \times 20\ (1250)\ 80.48 \times (215 - 40.24)$

$M_{u\,lim} = \{31.964 \times 10^6 + 156.82 \times 10^6\}$ N-mm

$= 188.79 \times 10^6$ N-mm $< 400.5 \times 10^6$ N-mm

Balance moment $(400.5 - 188.8) \times 10^6 = 211.7 \times 10^6$

as a steel beam theory $= 211.7 \times 10^6$ N-mm.

This is quite large moment and hence depth may be increased.

Let us increase the depth, $d = 270$ mm, $D = 310$ mm (cover 40 mm)

$x_{u\,lim} = 0.48 \times 270 = 129.6 \text{ mm}, d = 270 \text{ mm}, d' = 40 \text{ mm}$

$$M_{u\,lim} = 0.36 f_{ck} \cdot x_{u\,lim} \cdot b_w\, (d - 0.416\, x_{u\,lim}) + 0.446 f_{ck}\, (b_f - b_w)\, y_f \left(d - \frac{y_f}{2}\right)$$

$y_f = 0.15 x_u + 0.65 D_f = 0.15 \times 129.6 + 0.65 \times 100 = 84.44 \text{ mm}$

$M_{u\,lim}$ (singly reinforced) = $0.36 \times 20 \times 129.6 \times 250\ (270 - 0.416 \times 129.6) + 0.446$

$\times 20 \times 1250 \times 84.44 \times 227.78$

$= 50.41 \times 10^6 + 214.46 \times 10^6 = 264.87 \times 10^6$ N-mm

M_{u2} required (steel beam couple) $= M_u - M_{u\,lim}$

$= (400.5 - 264.87)10^6 = 135.63 \times 10^6$ N-mm

Providing $d' = 40$ mm cover to A_{sc}, lever arm of couple = 230 mm

Assume $f_{sc} = 0.87 f_y = 361 \text{ N/mm}^2$ and

$f_{cc} = 0.446 f_{ck}$ (rectangular stress block) = 8.92 N/mm^2

$$A_{st_2} = A_{sc} = \frac{M_{u_2}}{361 \times 230} = \frac{135.63 \times 10^6}{361 \times 230} = 1634 \text{ mm}^2$$

$$A_{st_1} = \frac{M_{u\ lim}}{0.87 f_y \cdot jd}, \quad jd = \text{lever arm of compressive force in concrete}$$

Let CG of compressive forces in concrete be $\tilde{y}$ from the extreme top fibre,

$$\tilde{y} = \frac{0.36 f_{ck} \times b_w \times x_{u\ lim} \times 0.446 x_{u\ lim} + 0.446 f_{ck} \times (b_f - b_w) \, y_f \times \frac{y_f}{2}}{(0.36 f_{ck} \times b_w \times x_{u\ lim}) + 0.446 f_{ck}(b_f - b_w) \, y_f}$$

$$= \frac{0.36 \times 20 \times 250 \times 129.6 \times 0.446 \times 129.6) + (0.446 \times 20 \times 1250 \times 84.44 \times 42.22)}{(0.36 \times 20 \times 250 \times 129.6) + (0.446 \times 20 \times 1250 \times 84.44)}$$

$$= \frac{(12.577 + 39.75) \, 10^6}{(233280 + 941506)} = 44.54 \text{ mm}$$

$$jd = (d - \tilde{y}) = (270 - 44.54) = 225.46 \text{ mm}$$

$$\therefore \quad A_{st_1} = \frac{264.87 \times 10^6}{361 \times 225.46} = 3254.3 \text{ mm}^2$$

Total A_{st} = 3254.3 + 1634 = 4888.3 mm², A_{sc} = 1634 mm²

x_u can be recalculated from known values of total A_{st} (4888.3 mm²) and A_{sc} (1634 mm²). y_f can also be recalculated with new value of x_u and M_r can be checked with applied BM for safety.

Example 4.24

Design a T-beam section of 1500 mm flange width, 100 mm flange depth, 300 mm web width, and 500 mm effective depth to sustain an ultimate BM of (a) 600 kN-m, and (b) 750 kN-m. Use M20 grade CC and Fe 415 grade steel.

Solution: b_f = 1500 mm, D_f = 100 mm, b_w = 300 mm, d = 500 mm, A_{st} = ?, A_{sc} = ?

   (a) $M_u = 600 \times 10^6$ N-mm   (b) $M_u = 750 \times 10^6$ N-mm

   The design may require singly reinforced or doubly reinforced section.

   (a) M_u may be compared with $M_{u\ lim}$ to decide regarding singly reinforced or doubly reinforced section.

$$x_{u\ lim} = \frac{700 \times d}{(1100 + 0.87 f_y)} = 0.48 \times 500 = 240 \text{ mm} > D_f \ (100 \text{ mm})$$

y_f = 0.15 $x_{u\ lim}$ + 0.65 D_f or D_f, whichever is smaller.

   = 0.15 × 240 + 0.65 × 100 = 101 or 100 mm, whichever is smaller = 100 mm

$M_{u\ lim}$ = 0.36 $f_{ck} b_w \cdot x_{u\ lim}$ $(d - 0.416 \, x_{u\ lim})$ + 0.446 f_{ck} $(b_f - b_w) \, y_f$ $(d - 0.5 f_y)$

   = 0.36 × 20 × 300 × 240 (500 − 0.446 × 240) + 0.446 × 20 (1500 − 300)

      100 (500 − 50)

   = 203.71 × 10⁶ + 481.68 × 10⁶ = 685.39 × 10⁶ N-mm > M_u (600 × 10⁶ N-mm)

Since $M_{u\ lim} > M_u$, the beam section will be singly reinforced. The NA may lie within the flange or outside the flange. This may be ascertained by comparing M_u with $M_{u\ lim}$, when $x_u = D_f$.

M_{uf} (when $x_u = D_f = 100$ mm) $= 0.36 f_{ck} \cdot b_f \cdot D_f (d - 0.416 D_f)$

or M_u flange $= 0.36 \times 20 \times 100 \times 1500 (500 - 0.416 \times 100) = 495.0$ kN-m $< M_u (600)$

$M_{u\,lim}$ (685.39×10^6 N-mm) is more than M_u (600×10^6 N-mm). The section is under reinforced. The NA shall lie in the web since M_u flange (495 kN-m) is less than M_u required (600 kN-m).

Compression will have two components F_{cuf} and F_{cuw}.

Let CG of these compressive forces be at $\tilde{y}$ from the extreme compression fibre. Find x_u and y_f by iterative process till convergence in two consecutive values occur.

Alternatively, M_u is found by trial and error method.

(i) $M_u < M_{u\,lim}$ $\therefore$ $x_u < x_{u\,lim}$. Assume $x_u = 200$ mm,

$y_f = 95$ mm ($600 < 685.39$ kN-m)

$$M_u \,(x_u = 200) = 0.36 f_{ck} \times b_w \times x_u (d - 0.42 x_u) + 0.45 f_{ck} (b_f - b_w) y_f \left(d - \frac{y_f}{2}\right)$$

$$= (179.7 + 464.27) \, 10^6 = 643 \times 10^6 \text{ N-mm}$$

(ii) $x_u = 180$ mm, $y_f = 92$ mm, $M_u = 165 \times 10^6 + 451 \times 10^6 = 616 \times 10^6$ N-mm

(iii) $x_u = 170$ mm, $y_f = 90.5$ mm, $M_u = 157.4 \times 10^6 + 444.47 \times 10^6 = 601.87 \times 10^6$ N-mm

Thus, tensile steel may be determined with $x_u = 170$ mm, $y_f = 90.5$ mm.

$$M_u = 0.87 f_y \cdot A_{st} \cdot jd \text{ or } A_{st} = \frac{M_u}{0.87 f_y \times jd} = \frac{601.87 \times 10^6}{361 jd} = \frac{166.7230 \times 10^4}{jd} \text{ mm}^2$$

CG of compressive forces from extreme concrete fibre $= \tilde{y}$

Lever arm $= (d - \tilde{y})$

$$\tilde{y} = \frac{\left\{0.36 f_{ck} \cdot b_w \cdot x_u \times 0.42 x_u + 0.45 f_{ck} \times (b_f - b_w) y_f \left(\frac{y_f}{2}\right)\right\}}{\left\{0.36 f_{ck} \cdot b_w \cdot x_u + 0.45 f_{ck} (b_f - b_w) \cdot y_f\right\}}$$

or

$$\tilde{y} = \frac{\left\{0.36 \times 20 \times 300 \times 170 \times 0.42 \times 170 + 0.45 \times 20 \times 1200 \times 95 \times \frac{95}{2}\right\}}{\{0.36 \times 20 \times 300 \times 170 + 0.45 \times 20 \times 1200 \times 95\}}$$

$$= \frac{26218080 + 48735000}{367200 + 1026000}$$

or

$$\tilde{y} = \frac{74953080}{1393200} = 53.80 \text{ mm}$$

Lever arm $jd = (d - \tilde{y}) = (500 - 53.8) = 446.2$ mm

$\therefore$

$$A_{st} = \frac{166.7230 \times 10^4}{446.2} = 3737 \text{ mm}^2$$

(b) $M_u = 750 \times 10^6$ N-mm $> M_{u\,lim}$ (685.39×10^6) N-mm,

hence section has to be made doubly reinforced (i.e. provide A_{sc} and A_{st})

$M_{u\,lim} = 685.39 \times 10^6$ N-mm.

$M_{u_2} = (750 - 685.39) 10^6$ N-mm $= 64.61 \times 10^6$ N-mm (MR due to steel couple)

Assume d' (cover to A_{sc}) $= 40$ mm, say and maximum stress in compression steel (A_{sc}) $= f_{sc} = 0.87 f_y = 0.87 \times 415 = 361$ N/mm^2

Concrete stress around compression steel $f_{cc} = 0.45 \times 20 = 9.0 \text{ N/mm}^2$

$$A_{st_2} = A_{sc} = \frac{M_{u_2}}{f_{sc} \times (d-d')} = \frac{64.61 \times 10^6}{361 \times (500 - 40)} = 389 \text{ mm}^2$$

$$A_{st_1} = \frac{M_{u \, lim}}{0.87 f_y (d - \tilde{y})} = \frac{685.39 \times 10^6}{361 \times (500 - \tilde{y})}$$

So find $\tilde{y}$ CG of compressive force C_1 (flange) and C_2 (web).

Alternatively, total compressive force $(C_1 + C_2)$ = Total tensile force $(T_1 + T_2)$

i.e. A_{stw} (web) $= 0.36 f_{ck} b_w \times x_u$ or $A_{stw} = \dfrac{0.36 f_{ck} \cdot b_w \cdot x_{u \, lim}}{0.87 \times 415}$

and A_{stf} (flange equivalent) $= \dfrac{0.45 f_{ck} (b_f - b_w) \times y_f}{0.87 \times 415} = \dfrac{0.45 \times 20 \times 1200 \times 100}{361}$

$(y_f = 0.15 \times 240 + 65 = 101$ or 100 mm whichever is smaller)

Total $A_{st_1} = A_{stw} + A_{stf} = \dfrac{0.36 \times 20 \times 300 \times 240}{361} + \dfrac{0.45 \times 20 \times 1200 \times 100}{361}$

$A_{st_1} = 1436 + 2992 = 4428 \text{ mm}^2$ (flange plus web equivalent reinforcement)

$A_{sc} = A_{st_2} = 389 \text{ mm}^2$.

Total tensile steel $A_{st} = A_{st_1} + A_{st_2} = (A_{stw} + A_{stf}) + A_{st_2}$

∴ $A_{st} = 4428 + 389 = 4817 \text{ mm}^2$, $A_{sc} = 389 \text{ mm}^2$

{*Note:* $0.446 \cong 0.45$ and $0.416 \cong 0.42$}

Alternatively: $x_{u \, lim} = 240$ mm, $y_f = 101$ or $100 = 100$ mm

$$\tilde{y} \text{ (CG)} = \frac{0.36 \times 20 \times 300 \times 240 \times 0.42 \times 240 + 0.45 \times 20 \times (1200) \, y_f \times \dfrac{y_f}{2}}{(0.36 \times 20 \times 300 \times 240 + 0.45 \times 20 \times 1200 \times y_f)}$$

$$= \frac{(52254720 + 54000000)}{(518400 + 1080000)} = \frac{106254720}{1598400} = 66.5$$

$jd = d - \tilde{y} = 500 - 66.5 = 433.5$ mm

$$A_{st_1} = \frac{685.39 \times 10^6}{361(500 - 66.5)} = 4380 \text{ mm}^2 \text{ (almost same as 4428 mm}^2)$$

$$A_{st_2} = \frac{(750 - 685.39) \times 10^6}{0.87 \times 415 (500 - 40)} = 389 \text{ mm}^2 = A_{sc} \text{ (with } f_{sc} = 0.87 f_y)$$

Total $A_{st} = A_{st_1} + A_{st_2} = 4380 + 389 = 4769 \text{ mm}^2$

$A_{sc} = 389 \text{ mm}^2$ (in compression zone).

SUMMARY

Limit state design method is based on various limits of failures, viz. collapse under different loads or serviceability conditions (excessive deflections, wide cracks, etc.). It considers both working stress method for serviceability and ultimate load for collapse conditions. The limit state method is now fully incorporated in the design code

IS:456-2000. Limit state design method ensures safe and satisfactory design and acceptable probability during the specified service life of the structure. Limit states of failure are:

- Failure due to flexure, shear, torsion or combination
- Failure due to fatigue
- Failure due to bond and anchorage
- Failure due to elastic instability
- Failure due to impact, earthquake, wind, fire etc.
- Failure due to destructive chemicals, corrosion, etc.

Limit state design is based on characteristic material strength and characteristic load. Different materials have different safety factors (e.g. concrete has $\gamma_m = 1.50$, for steel $\gamma_m = 1.15$, etc.). Loads are based on 95% probability of not being exceeded. Generally partial load factor is considered as 1.5.

Assumptions in limit state of collapse in flexure are:

- Plane sections before bending remain plane after bending
- Maximum strain in concrete at failure will be 0.0035
- Maximum strain in tensile steel at failure will not be less than $\left(\dfrac{f_y}{1.15E_s}+0.002\right)$
- The stresses in reinforcement are derived from stress–strain curves
- The modulus of elasticity (E_s) of steel is assumed same in compression and tension
- The stress-strain relation for concrete is parabolic

Indian standard code IS:456-2000 have adopted various design parameters for the stress block for various grades of concrete and grades of steel. The stress–strain variations across the depth of beam in flexure varies parabolically. Different design parameters are considered based on actual failure pattern.

The stress in extreme concrete fibre is considered as $0.446\,f_{ck}$ (approximately $0.45\,f_{ck}$) and in tensile steel as $0.87\,f_y$ with usual notations. Uniform stress in concrete equivalent to $0.36f_{ck}$ is assumed at failure. The CG of stress block is considered at $0.416x_u\ (\cong 0.42x_u)$ from the extreme compressive concrete fibre. Hence NA lies at x_u and CG of stress block at $0.416x_u$ from the extreme concrete fibre. NA of a balanced section is found when steel (stressed at $0.87f_y$) and concrete (strained at $e_{cu} = 0.0035$) reach their limit states simultaneously.

$$\frac{x_{u\,\text{lim}}}{d} = \frac{0.0035}{\left(0.0055 + \dfrac{0.87\,f_y}{E_s}\right)} = \frac{700}{\left(1100 + 0.87\,f_y\right)}, \text{(putting } E_s = 2 \times 10^5\,\text{N/mm}^2)$$

$\dfrac{x_{u\,\text{lim}}}{d}$ for various steels are: Fe 250 (0.53), Fe 415 (0.48), Fe 500 (0.46)

We have $T_u = C_{cu}$ for equilibrium of internal forces. The section may be under reinforced $(x_u < x_{u\,\text{lim}})$, balanced $(x_u = x_{u\,\text{lim}})$, or over reinforced $(x_u > x_{u\,\text{lim}})$.

$$p_{t\,\text{lim}} = 0.414\frac{f_{ck}}{f_y}\left(\frac{x_{u\,\text{lim}}}{d}\right) \text{ for rectangular beam sections}$$

$$M_u \text{ (under reinforced rectangular sections)} = 0.87 f_y \cdot A_{st} \cdot d \left\{ 1 - \frac{A_{st}}{bd} \cdot \frac{f_y}{f_{ck}} \right\}$$

$$M_{u\,lim} = 0.36 f_{ck} \cdot \frac{x_{u\,lim}}{d} \left(1 - 0.42 \frac{x_{u\,lim}}{d} \right) bd^2 = R_{u\,lim} \times bd^2$$

Also
$$A_{st} = \frac{0.5 f_{ck}}{f_y} \left[1 - \sqrt{1 - \frac{4.6 M_u}{f_{ck} \cdot bd^2}} \right] bd$$

$$x_u = \frac{\{0.87 f_y \cdot A_{st} - (f_{sc} - f_{cc}) A_{sc}\}}{0.36 f_{ck} \cdot b}$$

$$e_{sc} \text{ (strain in compressive steel)} = \frac{0.0035 (x_u - d')}{x_u},$$

$f_{sc} = e_{sc} \times E_s$ or obtained from the stress–strain curves.

Lever arm in doubly reinforced rectangular section will be

$$\alpha = (d - \tilde{y}) = \left[d - \frac{\{0.36 f_{ck} \cdot x_u \times b \times 0.416 x_u + f_{sc} \cdot A_{sc} \cdot d'\}}{\{0.36 f_{ck} \cdot x_u \cdot b + f_{sc} \cdot A_{sc}\}} \right],$$

$$M_u = [0.36 f_{ck} \cdot x_u \cdot b(d - 0.416 x_u) + A_{sc} (f_{sc} - 0.446 f_{ck})(d - d')] \qquad \text{... as in Eq. (4.20)}$$

$$C_u = T_u, \text{ i.e. } 0.87 f_y \cdot A_{st} = \{0.36 f_{ck} \cdot x_u \cdot b + (f_{sc} - 0.446 f_{ck}) \cdot A_{sc}\} \qquad \text{... as in Eq. (4.21)}$$

(Assuming $f_{cc} = 0.446 f_{ck}$)

$$M_{u\,lim} = 0.36 f_{ck} \cdot \frac{x_{u\,lim}}{d} \left(1 - 0.416 \frac{x_{u\,lim}}{d} \right) bd^2 + (f_{sc} - 0.446 f_{ck}) A_{sc} (d - d') \text{ ... as in Eq. (4.22)}$$

$$f_{st} \cdot A_{st} = 0.87 f_y \cdot A_{st} = 0.36 f_{ck} \cdot b \cdot x_u + (f_{sc} - 0.446 f_{ck}) A_{sc} \qquad \text{... as in Eq. (4.23)}$$

$$x_u = \frac{\{0.87 f_y \cdot A_{st} - (f_{sc} - f_{cc}) A_{sc}\}}{0.36 f_{ck} \cdot b} \qquad \text{... as in Eq. (4.24)}$$

$$\frac{x_u}{d} = \frac{0.87 f_y \cdot A_{st}}{0.36 f_{ck} \cdot b_f \cdot d} = \frac{2.417 f_y \cdot A_{st}}{f_{ck} \cdot b_f \cdot d}, \text{ singly reinforced T-section} \qquad \text{... as in Eq. (4.26)}$$

$$M_u = 0.87 f_{ck} \cdot A_{st} (d - 0.42 x_u) = 0.87 f_y \cdot A_{st} \cdot d \left(1 - \frac{A_{st} \cdot f_y}{b_f \cdot d \cdot f_{ck}} \right) \qquad \text{... as in Eq. (4.27)}$$

$$M_{u\,lim} (x_u \text{ within flange}) = 0.36 f_{ck} \cdot \frac{x_{u\,lim}}{d} \left\{ 1 - 0.42 \frac{x_{u\,lim}}{d} \right\} b_f \cdot d^2 \qquad \text{... as in Eq. (4.28)}$$

$$M_{u\,lim} (x_u \text{ outside flange}) = 0.36 f_{ck} \cdot \frac{x_{u\,lim}}{d} \left\{ 1 - 0.42 \frac{x_{u\,lim}}{d} \right\} b_w \cdot d^2$$

$$+ 0.45 f_{ck} (b_f - b_w) D_f \left(d - \frac{D_f}{2} \right) \qquad \text{... as in Eq. (4.29)}$$

$y_f = 0.15 x_{u\,lim} + 0.65 D_f$ but not greater than D_f.

$$M_{u\,lim} = 0.36f_{ck} \cdot \frac{x_{u\,lim}}{d} \left\{ 1 - 0.42\, \frac{x_{u\,lim}}{d} \right\} b_w \cdot d^2 + 0.45f_{ck}\,(b_f - b_w)\,y_f \left(d - \frac{y_f}{2} \right)$$

$$\dots \textit{as in Eq. (4.30)}$$

$$\text{Also } M_u = 0.36f_{ck} \cdot \frac{x_u}{d} \left\{ 1 - 0.42\, \frac{x_u}{d} \right\} b_w \cdot d^2 + 0.45f_{ck}\,(b_f - b_w)\,y_f \left(d - \frac{y_f}{2} \right)$$

$$\dots \textit{as in Eq. (4.32)}$$

$$M_{u2}\,(\text{moment couple}) = \frac{0.0035\,(x_u - d')}{x_u} \cdot E_s \cdot A_{sc}\,(d - d') \qquad \dots \textit{as in Eq. (4.34)}$$

Total compression $= T_u, \; 0.36f_{ck} \cdot x_u \cdot b_f + f_{sc} \cdot A_{sc} = 0.87f_y \cdot A_{st}$

Also $0.87f_y \cdot A_{st} = 0.36f_{ck} \cdot x_u \cdot b_w + (f_{sc} - 0.45f_{ck}) \cdot A_{sc} + 0.45f_{ck}\,(b_f - b_w)\,y_f$

$$\dots \textit{as in Eq. (4.36)}$$

M_u (within flange) $= 0.36\,f_{ck} \cdot x_u \cdot b_f\,(d - 0.42x_u) + (f_{sc} - 0.45f_{ck})A_{sc}\,(d - d')$

$$\dots \textit{as in Eq. (4.37)}$$

$$M_u \text{ (within web)} = 0.36f_{ck} \cdot \frac{x_u}{d} \left\{ 1 - 0.42\, \frac{x_u}{d} \right\} b_w \cdot d^2 + (f_{sc} - 0.45f_{ck})\,A_{sc}\,(d - d')$$

$+\, 0.45f_{ck}\,(b_f - b_w)\,y_f\,(d - 0.5y_f)$ $\qquad\qquad\qquad\qquad\qquad\qquad \dots \textit{as in Eq. (4.40)}$

PRACTICE QUESTIONS

(I) Objective Questions

Q. 4.1 Select the correct response given after each statement to complete it correctly and fill in the response sheet provided.

(i) The limit state method of design is not concerned with ....
   (a) limit state of failure due to fatigue
   (b) limit state of failure due to torsion
   (c) limit state of failure of vibrator during concrete placing
   (d) limit state of failure due to impact

(ii) A RC beam section at the time of failure will have a ....
   (a) strain of 0.002 in extreme concrete fibre
   (b) strain of 0.0035 in extreme concrete fibre
   (c) strain of 0.0035 in extreme tensile steel fibre
   (d) strain of (2×10^{-5}) in extreme tensile steel fibre

(iii) The limit state method of design for beam sections is not based on the assumption that the ....
   (a) concrete shall have high grades only with good characteristic strength
   (b) plane sections before bending remain plane after bending
   (c) modulus of elasticity of materials is the same in compression and tension
   (d) stress–strain curve of concrete is parabolic

(iv) IS:456-2000 code for design and analysis of RCC beams recommends that the ....

    (a) concrete stress at collapse is constant for all grades

    (b) concrete strain at collapse is constant for all grades

    (c) stress strain curve of steel is a straight line up to failure

    (d) neutral axis factor $\left(\dfrac{x_u}{d}\right)$ is independent of grade of steel

(v) The maximum NA depth factor $\left(\dfrac{x_{u\,lim}}{d}\right)$ with usual notations is given by ....

$$\text{(a)}\quad \frac{x_{u\,lim}}{d} = \frac{0.0035}{\left(0.0055 + \dfrac{0.87\,f_{ck}}{E_s}\right)} \qquad\qquad \text{(b)}\quad \frac{x_{u\,lim}}{d} = \frac{700}{\left(1100 + 0.87\,f_{ck}\right)}$$

$$\text{(c)}\quad \frac{x_{u\,lim}}{d} = \frac{700}{\left(1100 + 0.87\,f_y\right)} \qquad\qquad \text{(d)}\quad \frac{x_{u\,lim}}{d} = \frac{0.0035}{\left(0.0020 \times 0.87\,f_y\right)}$$

(vi) According to equilibrium of internal forces in a doubly reinforced T-beam section, the limit state equation will be ....

    (a) total compression in steel (C_{su}) = Total tension in steel (T_{su})

    (b) total compression in concrete (C_{cu}) = Total tension in steel (T_{su})

    (c) total compression in flange (C_{uf}) = Total compression in web (C_{uw})

    (d) total compression in compression zone (C_u) = Total tension in steel (T_{su})

(vii) In a balanced rectangular beam section, the percentage of tensile steel (p_t) is given by ....

$$\text{(a)}\quad p_{t\,lim} = 0.414\,\frac{f_y}{f_{ck}}\left(\frac{x_{u\,lim}}{d}\right)$$

$$\text{(b)}\quad p_{t\,lim} = 0.414\,\frac{f_{ck}}{f_y}\left(\frac{x_{u\,lim}}{d}\right)$$

$$\text{(c)}\quad p_{t\,lim} = 0.36 f_{ck}\,\frac{x_{u\,lim}}{d}\left(1 - 0.42\,\frac{x_{u\,lim}}{d}\right) bd$$

$$\text{(d)}\quad p_{t\,lim} = \frac{0.5 f_{ck}}{f_y}\left[1 - \sqrt{1 - \frac{4.6\,M_u}{f_{ck}\cdot bd^2}}\,\right] bd$$

(viii) In an under reinforced rectangular beam $(b \times d)$, the tensile steel (A_{st}) required to resist safely an ultimate bending moment (M_u) will be given by ....

$$\text{(a)}\quad A_{st} = \frac{0.5 f_{ck}}{f_y}\left[1 - \sqrt{1 - \frac{4.6\,M_u}{f_{ck}\cdot bd^2}}\,\right] bd$$

$$\text{(b)}\quad A_{st} = \frac{0.5 f_y}{f_{ck}}\left[1 - \sqrt{1 - \frac{4.6\,M_u}{f_{ck}\cdot bd^2}}\,\right] bd$$

(c) $A_{st} = \dfrac{0.87 f_y}{0.446 f_{ck}}\left[1-\sqrt{1-\dfrac{M_u}{f_y \cdot bd^2}}\right] bd$

(d) $A_{st} = \dfrac{0.36 f_{ck}}{0.87 f_y}\left[1-\sqrt{1-\dfrac{M_u}{f_{ck}\cdot bd^2}}\right] bd$

(ix) In a doubly reinforced rectangular beam section, the strain (e_{sc}) in compressive steel (A_{sc}) is given by ....

(a) $e_{sc} = \dfrac{0.0035 x_u}{(x_u - d')}$

(b) $e_{sc} = \dfrac{0.0035 d'}{x_u}$

(c) $e_{sc} = \dfrac{0.0035 x_u}{(d - d')}$

(d) $e_{sc} = \dfrac{0.0035 (x_u - d')}{x_u}$

(x) Lever arm (α) in doubly reinforced rectangular beam section, will be ....

(a) $\alpha = (d - \bar{y}) = \left[d - \left\{\dfrac{0.36 f_{ck}\cdot x_u \cdot b \times 0.416 x_u + f_{sc}\cdot A_{sc}\cdot d'}{0.36 f_{ck}\cdot x_u \cdot b + f_{sc}\cdot A_{sc}}\right\}\right]$

(b) $\alpha = (d - \bar{y}) = \left[d - \left\{\dfrac{0.36 f_{ck}\cdot x_u \times b \times 0.33 x_u + f_{sc}\cdot A_{sc}\cdot d}{0.36 f_{ck}\times x_u \times b + f_{sc}\cdot A_{sc}}\right\}\right]$

(c) $\alpha = (d - d') = \left[d - \left\{\dfrac{0.87 f_y \times x_u \times b \times 0.33 x_u + f_{sc}\cdot A_{sc}\cdot d'}{0.87 f_y \times x_u \times b + f_{sc}\cdot A_{sc}}\right\}\right]$

(d) $\alpha = (d - d') = \left[d - \left\{\dfrac{0.87 f_y \cdot x_u \times b \times 0.416 x_u + f_{sc}\cdot A_{sc}\cdot d}{0.87 f_y \cdot x_u \cdot b + 0.87 f_y \cdot A_{st}}\right\}\right]$

(xi) In a singly reinforced T-beam section, the limiting ultimate moment of resistance $(M_{u\,lim})$ is given by ...., when the NA lies outside the flange.

(a) $M_{u\,lim} = 0.36 f_{ck}\cdot \dfrac{x_{u\,lim}}{d}\left\{1 - 0.416\,\dfrac{x_{u\,lim}}{d}\right\} b_w \cdot d^2$

(b) $M_{u\,lim} = 0.36 f_y \cdot \dfrac{x_{u\,lim}}{d}\left\{1 - 0.416\,\dfrac{x_{u\,lim}}{d}\right\} b_f \cdot d^2$

(c) $M_{u\,lim} = 0.36 f_{ck}\cdot \dfrac{x_{u\,lim}}{d}\left\{1 - 0.416\,\dfrac{x_{u\,lim}}{d}\right\} b_w \cdot d^2 + 0.446 f_{ck}\,(b_f - b_w)\, D_f\left(d - \dfrac{D_f}{2}\right)$

(d) $M_{u\,lim} = 0.36 f_{ck}\cdot \dfrac{x_{u\,lim}}{d}\left\{1 - 0.416\,\dfrac{x_{u\,lim}}{d}\right\} (b_f + b_w)d^2$

(xii) In a doubly reinforced T-beam section, the ultimate moment of resistance (M_u) when the NA lies outside the flange will be given by ....

(a) $M_u = 0.36 f_{ck}\cdot x_u \cdot b_w\,(d - 0.42 x_u) + (f_{sc} - 0.45 f_{ck})A_{sc}\,(d - d')$

(b) $M_u = 0.36 f_{ck}\cdot x_u \cdot b_f\,(d - 0.42 x_u) + 0.45 f_{ck}\,(b_f - b_w)\, D_f\left(d - \dfrac{D_f}{2}\right)$

(c) $M_u = 0.36f_{ck} \cdot x_u \cdot b_w (d - 0.42x_u) + 0.45f_{ck} (b_f - b_w) D_f \left(d - \dfrac{D_f}{2} \right)$
$+ (f_{sc} - 0.45f_{ck}) A_{sc} (d - d')$

(d) $M_u = 0.36f_{ck} \cdot x_u \cdot b_w (d - 0.42x_u) + 0.36f_{ck} (b_f - b_w) D_f \left(d - \dfrac{D_f}{2} \right)$
$+ (f_{sc} - f_{ck}) A_{sc} \left(d - \dfrac{D_f}{2} \right)$

(xiii) Ultimate moment of resistance (M_u) in a doubly reinforced T-beam section when the NA lies within the flange will be given by ....

(a) $M_u = 0.36f_{ck} \cdot x_u \cdot b_w (d - 0.42x_u) + 0.45f_{ck} (b_f - b_w) y_f \left(d - \dfrac{y_f}{2} \right)$
$+ (f_{sc} - 0.45f_{ck}) A_{sc} (d - d')$

(b) $M_u = 0.36f_{ck} \cdot x_u (d - 0.42x_u) b_f + (f_{sc} - 0.45f_{ck}) A_{sc} (d - d')$

(c) $M_u = 0.45f_{ck} \cdot x_u (d - 0.36x_u) b_f + (f_{sc} - 0.45f_{ck}) A_{sc} (d - d') + 0.36f_{ck} \cdot x_u \cdot b_w \cdot d^2$

(d) $M_u = 0.45f_{ck} \cdot x_u (d - 0.36x_u) b_w + 0.45f_{ck} \times b_f \cdot D_f \left(d - \dfrac{D_f}{2} \right)$

(xiv) A rectangular beam section of 300 mm width and 500 mm effective depth has tensile reinforcement of 1346 mm². The beam is constructed using M20 grade CC and Fe 415 grade steel. Calculate actual NA (x_u).

(a) $x_u = \dfrac{0.87 \times 415 \times 1346}{0.36 \times 20 \times 300} = 225$ mm

(b) $x_u = \dfrac{0.36 \times 20 \times 300 \times 500}{0.87 \times 415 \times 1346} = 2.22$ mm

(c) $x_u = \dfrac{0.87 \times 415 \times 1346}{0.36 \times 20 \times 500} = 135$ mm

(d) $x_u = 0.48 \times 500 = 240$ mm

(xv) A rectangular beam section of 300 mm width and 500 mm effective depth has tensile reinforcement of 1346 mm². The beam uses M20 CC and Fe 415 steel. Calculate its actual ultimate moment of resistance (M_u) if its actual NA is considered as 225 mm.

(a) $M_u = 0.87f_{ck} \cdot A_{st} \cdot (d - 0.42x_u) = 9.497 \times 10^6$ N-mm

(b) $M_u = 0.87f_y \cdot A_{st} \cdot (d - 0.42x_u) = 197.06 \times 10^6$ N-mm

(c) $M_u = 0.36f_{ck} \cdot b \cdot d (d - 0.42x_u) = 437.94 \times 10^6$ N-mm

(d) $M_u = 0.36f_{ck} \cdot x_u \cdot d (d - 0.42x_u) = 76.545 \times 10^6$ N-mm

(xvi) Calculate the tensile reinforcement for a balanced singly reinforced beam section of 300 mm × 400 mm (effective depth), if the beam uses M30 grade CC and Fe 500 grade steel. Critical NA may be taken as 184 mm. Consider ultimate moment of resistance to be 192 kN-m.

(a) $A_{st} = \dfrac{M_u}{0.87\, f_{ck}\, (d - 0.42x_u)} = 22795$ mm²

(b) $A_{st} = \dfrac{M_u}{0.87\, f_y\, (d - 0.42x_u)} = 1368$ mm²

(c) $A_{st} = \dfrac{0.87 f_y (d - 0.42 x_u) \times bd^2}{M_u} = 35096 \text{ mm}^2$

(d) $A_{st} = \dfrac{0.36 f_{ck} \times b \times x_u \cdot d^2}{M_u} = 497.8 \text{ mm}^2$

(xvii) Calculate the maximum ultimate moment of resistance ($M_{u \, lim}$) for a rectangular beam of 300 mm width and 500 mm effective depth using M20 grade CC and Fe 500 grade steel.

  (a) $M_{u \, lim} = 0.148 f_{ck} \cdot bd^2 = 222 \times 10^6 \text{ N-mm}$

  (b) $M_{u \, lim} = 0.138 f_{ck} \cdot bd^2 = 207 \times 10^6 \text{ N-mm}$

  (c) $M_{u \, lim} = 0.138 f_y \cdot bd^2 = 5175 \times 10^6 \text{ N-mm}$

  (d) $M_{u \, lim} = 0.133 f_{ck} \cdot bd^2 = 199.5 \times 10^6 \text{ N-mm}$

(xviii) A T-beam has 1200 mm flange width and 120 mm flange thickness, effective depth of web 500 mm and web width of 300 mm. It has 1964 mm² tensile reinforcement at an effective cover of 50 mm. Determine the ultimate moment of resistance of the T-beam section. Use M 20 CC and Fe 415 grade steel. The section is under reinforced and NA lies in the flange.

$x_{u \, lim} = 0.48 \times 500 = 240 \text{ mm}.$

  (a) $M_u = 0.87 f_y \cdot A_{st} \cdot d \left(1 - \dfrac{A_{st} f_y}{b_f \cdot d \cdot f_{ck}}\right) = 330.47 \times 10^6 \text{ N-mm}$

  (b) $M_u = 0.36 f_{ck} \cdot x_{u \, lim} \cdot b_f (d - 0.42 \, x_{u \, lim}) = 827.78 \times 10^6 \text{ N-mm}$

  (c) $M_u = 0.36 f_{ck} \cdot x_{u \, lim} \cdot b_w (d - 0.42 \, x_{u \, lim}) = 4294.15 \times 10^6 \text{ N-mm}$

  (d) $M_u = 0.87 f_y \cdot A_{st} \cdot \left(1 - 0.42 \dfrac{x_{u \, lim}}{d}\right) d = 283.307 \times 10^6 \text{ N-mm}$

(xix) Maximum ultimate moment of resistance for a rectangular beam of width (b) and effective depth (d) with using M30 grade of concrete and Fe 500 grade of steel will be approximately given by ....

  (a) $M_{u \, lim} = 0.53 \times 500 \times bd$

  (b) $M_{u \, lim} = 0.148 \times 30 \times bd^2$

  (c) $M_{u \, lim} = 0.138 \times 30 \times bd^3$

  (d) $M_{u \, lim} = 0.133 \times 30 \times bd^2$

(xx) If in a singly reinforced rectangular beam ($b \times d$), the value of actual NA (x_u) comes out to be more than the critical value of NA ($x_{u \, lim}$), the section will be considered as ....

  (a) under reinforced section and failure occurs in compression

  (b) balanced section and failure occurs either in compression or tension

  (c) over reinforced section and failure occurs in compression

  (d) both under reinforced or balanced and failure of neutral zone

Response sheet to Q. 4.1 (i to xx)

Question	(i)	(ii)	(iii)	(iv)	(v)	(vi)	(vii)	(viii)	(ix)	(x)
Response (a/b/c/d)										

Question	(xi)	(xii)	(xiii)	(xiv)	(xv)	(xvi)	(xvii)	(xviii)	(xix)	(xx)
Response (a/b/c/d)										

(II) Descriptive Questions

Q. 4.2   Explain different limit states of failure and differentiate between working stress and limit state methods of design.

Q. 4.3   Explain characteristic and design data for RCC materials.

Q. 4.4   Define characteristic strength of concrete and grade of concrete (meaning of M30 grade CC).

Q. 4.5   State the meaning of partial safety factor for materials and partial safety load factor.

Q. 4.6   State at least 5 important assumptions in limit state of collapse in flexure.

Q. 4.7   Sketch stress–strain curves of concrete in (a) cube, (b) structure and (c) design.

Q. 4.8   Show approximate stress block and strain variations in a concrete beam section.

Q. 4.9   Explain the basic difference in compression failure and tension failure of a RCC beam section.

Q. 4.10   Differentiate between stress–strain characteristic curve and design curve of steel giving sketch and explanation.

Q. 4.11   Explain briefly the meaning of following stress block parameters in RCC flexural members in not more than one line:

$A_{st}, A_{sc}, f_{ck}, f_y, j, Q, n_c, n_a, x_u, x_{u\,lim}, e_{sc}, e_{cu}, E_s, p_t, p_{t\,lim}, M_u, M_{u\,lim}.$

Q. 4.12   Define: under reinforced, balanced, and over reinforced beam sections.

Q. 4.13   Explain advantages of under reinforced sections w.r.t. failure.

Q. 4.14   Determine the ultimate moment of resistance of a singly reinforced rectangular cement concrete beam section of width 200 mm, effective depth 400 mm, and reinforced with 3 bars of 16 mm diameter of Fe 415 grade steel in tension and use M20 grade of concrete.

[**Hint:** $\dfrac{x_u}{d} = 0.378$, $\dfrac{x_{u\,lim}}{d} = 0.48$, under reinforced, MR = 74 kN-m]

Q. 4.15   A reinforced concrete beam of rectangular section 200 mm × 550 mm deep is reinforced with 6 bars of 20 mm diameter at an effective depth of 500 mm. Using M20 grade of CC and Fe 415 HYSD steel bars, calculate the safe moment of resistance of the section.

[**Hint:** $\dfrac{x_u}{d} = 0.945 > 0.48$, over reinforced, $M_u = 138$ kN-m, $M_r = 92$ kN-m]

Q. 4.16 A RC beam section of 300 mm × 490 mm effective depth is reinforced with 7 bars of 16 mm diameter of Fe 415 grade steel. If M20 grade of CC is used, calculate the ultimate moment of resistance of the section.

[**Hint:** $\dfrac{x_u}{d} = 0.48 = \dfrac{x_{u\,lim}}{d}$, balanced section, $M_u = 199$ kN-m]

Q. 4.17 Determine the area of tension reinforcement required for a singly reinforced beam of 300 mm width and 650 mm overall depth with 50 mm effective cover to resist a factored moment of 200 kN-m if concrete of M25 grade and steel of Fe 415 grade are used.

[**Hint:** $A_{st} = 1019$ mm^2 (4 bars of 18 mm diameter or 5 bars of 16 mm)]

Q. 4.18 Design a rectangular RC beam section of 300 mm width to bear a factored moment of 375 kN-m. M20 grade CC and Fe 415 grade steel are used in the beam section. Determine the depth and tension steel.

[**Hint:** Eff. depth $d = 675$ mm, $A_{st} = 2423$ mm^2 (5 bars of 25 mm ϕ)]

Q. 4.19 A simply supported beam of 4 m effective span carries a total factored u.d.l. of 30 kN/m. The width of rectangular beam section is 200 mm. Find the safe depth of the section if M20 grade CC and Fe 415 grade steel are used in the construction. Also find the tensile steel required.

[**Hint:** $M_u = 60$ kN-m, $d = 330$ mm, $A_{st} = 632$ mm^2 (2 bars of 20 mm)]

Q. 4.20 Design a singly reinforced rectangular beam of 8 m simply supported span to carry a uniformly distributed service load of 20 kN/m over the entire span. Beam construction uses M30 grade CC and Fe 500 grade steel. Assume suitable self-weight of beam and b/d ratio of 1/2.

[**Hint:** $w_d = 5$ kN/m (say), $w_u = 37.5$ kN/m, $M_u = 300$ kN-m, $\dfrac{x_{ul}}{d} = 0.46$, $d = 532$ mm,

$b = 266$ mm (270 mm × 540 mm, say), $A_{st} = 1617$ mm^2, 5 bars of 22ϕ (1900 mm^2)]

Q. 4.21 A RCC beam of rectangular section 300 mm wide × 400 mm effective depth is reinforced with 4 bars of 16 mm diameter in tension zone. The beam is constructed with M25 grade CC and Fe 500 grade steel. Find the actual stresses in the material when the beam is subjected to a total factored BM of 120 kN-m.

[**Hint:** $\dfrac{x_u}{d} = 0.3242 < \dfrac{x_{u\,lim}}{d}$ of 0.456, under reinforced, $f_{st} = 431$ N/mm^2 (< 435 N/mm^2)

$f_{cc} = \dfrac{f_{st} \cdot A_{st}}{0.36 \times b \cdot x_u} = \dfrac{431 \times 804}{0.36 \times 0.3242 \times 400 \times 300} = 24.74$ N/mm^2 ($< f_{ck}$ of 25)]

Q. 4.22 Determine the actual stresses in concrete and steel in a rectangular singly reinforced beam section 330 mm wide × 500 mm effective depth if it is reinforced with 6 bars of 25 mm and subjected to an ultimate moment of 385 kN-m. Concrete of M30 grade and steel of Fe 415 grade are used in beam construction.

[**Hint:** $x_u = 298.4$ mm, $x_{u\,lim} = 239.5$ mm, over reinforced since $x_u > x_{u\,lim}$, $e_s = 0.00237$, $f_{st} = 0.00237 \times 2 \times 10^5 = 341.3$ N/mm^2, (actual stress is found from Table 4.5

of stress-strain), M_u (over reinforced) $= 0.36f_{ck} \cdot b \cdot x_u(d - 0.42x_u) = 385 \times 10^6$, $f_{cc} = 13.4 \text{ N/mm}^2 (0.446f_{ck})$]

Q. 4.23 A doubly reinforced rectangular RC beam has 300 mm width and 600 mm effective depth. The beam is reinforced with 5 bars of 25 mm diameter in tension zone and 2 bars of 25 mm in compression zone at an effective cover of 60 mm. Using M20 grade CC and Fe 415 grade steel, determine the ultimate moment of resistance.

[**Hint:** $x_{u \lim} = 288$ mm, e_{sc}(strain) $= 0.0028$, $f_{sc} = 352 \text{ N/mm}^2$, $A_{st_2} = \dfrac{352 \times 2 \times 491}{0.87 \times 415}$

$= 957 \text{ mm}^2$, $A_{st_1} = (5 \times 491 - 957) = 1498 \text{ mm}^2$, actual $x_u = 251 \text{ mm} < x_{u \lim}$ under reinforced, $M_u = M_{u_1} + M_{u_2} = 268 \times 10^6 + 185 \times 10^6 = 453 \times 10^6 \text{ N-mm}$]

Q. 4.24 A doubly reinforced beam of M25 grade CC and Fe 415 grade steel is used for 6 m simply supported span. The beam carries an u.d.l. factored load of 60 kN/m over the entire span, including its self-weight. The beam section is limited to 250 mm × 400 mm effective depth. Concrete cover to compression steel may be taken as 50 mm (effective). Determine the compression and tension steel areas.

[**Hint:** $M_u = 270 \times 10^6$ N-mm, $M_{u_1} = 138$ kN-m, $A_{st} = 2240 \text{ mm}^2$, $A_{sc} = 1040 \text{ mm}^2$]

Q. 4.25 Design a doubly reinforced rectangular beam section subjected to an ultimate moment of 150 kN-m. Use M20 grade CC and Fe 415 grade steel.

[**Hint:** Find b, d, A_{st} and A_{sc}

Assume $d/b = 2$

$M_{u_1} = 0.6$ m, $0.6 \times 150 = 0.36f_{ck} \, b \cdot x_{u \lim} (d - 0.42x_{u \lim})$,

$x_{u \lim} = 0.48d$, $bd^2 = 32.6 \times 10^6 \text{ mm}^3$, $b = 202$ mm, $d = 404$ mm,

adopt 220 mm × 440 mm, $A_{st} (A_{st_1} + A_{st_2}) = 1210 \text{ mm}^2$, $A_{sc} = 330 \text{ mm}^2$]

Q. 4.26 Design a reinforced beam section 300 mm wide × 450 mm effective depth, using M20 CC and Fe 250 MS. The beam is subjected to an ultimate BM of 235 kN-m. Assume effective concrete cover of 50 mm.

[**Hint:** $x_{u \lim} = \dfrac{700d}{1100 + 0.87 \times 250} = 239$ mm, $M_{u \lim} = 0.36 \times 20 \times 239 (450 - 0.42 \times 239)$

$300 = 181 \times 10^6$ N-mm $< M_u$ requires doubly reinforced section.

$M_{u_1} = M_{u \lim}$, $M_{u_2} = (235 - 181)10^6 = 54 \times 10^6$

$M_{u_2} = (54 \times 10^6) = 0.87f_y \times (450 - 50)A_{st_2}$,

$A_{st_2} = \dfrac{54 \times 10^6}{0.87 \times 250 \times 400} = 621 \text{ mm}^2$

$A_{st_1} = \dfrac{M_u}{0.87 f_y (d - 0.42x_{u \lim})} = \dfrac{18 \times 10^6}{0.87 \times 250 (450 - 0.42 \times 239)} = 2380 \text{ mm}^2$

Total $A_{st} = 621 + 2380 = 3001 \text{ mm}^2$, $M_{u_2} = f_{sc} \cdot A_{sc} (d - d')$,

$f_{sc} \cong 0.87f_y$

$A_{sc} = \dfrac{M_{u_2}}{0.87 \times 250 \times (450 - 50)} = 621 \text{ mm}^2$]

Q. 4.27 Determine the ultimate moment of resistance of a rectangular beam 300 mm wide and 500 mm effective depth. Use M20 CC and Fe 415 steel with effective cover of 50 mm both for tensile and compressive steels. Tension face has 4 bars of 25 mm diameter while compression face has 4 bars of 16 mm diameter. Use steel beam theory.

[**Hint:** $A_{st} = 1964$ mm^2, $A_{sc} = 804$ mm^2, $d = 500$ mm, $d' = 50$ mm, $x_{u\,lim} = 240$ mm,

$$f_{sc} = \frac{0.0035}{240} \times 190 \times 2 \times 10^5 = 554 > 0.87 \times 415, \text{ so adopt } 361 \text{ N/mm}^2.$$

$$A_{st_2} = \frac{f_{sc} \cdot A_{sc}}{0.87 f_y} = 804 \text{ mm}^2, A_{st_1} = 1160 \text{ mm}^2, T_1 = C_1, x_u = \frac{361 \times 1160}{0.36 \times 20 \times 300}$$

$$= 194 \text{ mm}$$

$$M_u = 0.87 f_y \cdot A_{st_1} (d - 0.42 x_u) + f_{sc} \cdot A_{sc} (d - d') = (175.6 + 130.6)10^6 = 306.2 \text{ kN-m}]$$

Q. 4.28 Determine the ultimate moment of resistance for a T-section, if
(a) $A_{st} = 1964$ mm^2 (b) $A_{st} = 4925$ mm^2 for given sectional data:
$b_f = 1500$ mm, $D_f = 100$ mm, $b_w = 300$, effective $d = 500$ mm, M20 CC and Fe 415 steel are used in the construction.

[**Hint:** (a) $x_u = 65.6$ mm $< D_f$, $x_{u\,lim} = 240$ mm, under reinforced, NA within flange
$M_u = 335$ kN-m.
(b) $x_u = 328$ mm $> D_f > x_{u\,lim}$, $f_s = 361$ N/mm^2 ($e_s = 0.0038 \times E_s$ or $0.87 f_y$)
x_u by alternative process $= 291.65$ mm, $y_f = 100$ mm,
$M_u = (238.5 + 481.7)10^6 = 720.2$ kN-m]

Q. 4.29 Determine the ultimate moment of resistance of T-beam section having following dimensions:
$b_f = 1500$ mm, D_f (flange thickness) $= 120$ mm, $b_w = 300$ mm, effective $d = 600$ mm, A_{st} (8 bars of 25 mm) $= 3925$ mm^2, M20 CC and Fe 415 steel

[**Hint:** $x_u = 152.33$ mm $> D_f$ (120 mm), $0.87 f_y \cdot A_{st} = 0.36 f_{ck} \cdot b_w \cdot x_u + 0.45 f_{ck} (b_f - b_w) \times$
$(0.15 x_u + 0.65 D_f)$

$$\frac{x_u}{d} = 0.254 < \frac{x_{u\,lim}}{d} = 0.48, \text{ under reinforced.}$$

$$M_u = 0.36 \left(\frac{x_u}{d}\right)\left(1 - 0.42 \frac{x_u}{d}\right) f_{ck} \cdot b_w \cdot d^2 + 0.45 f_{ck}(b_f - b_w) y_f (d - 0.5 y_f)$$

$$= 775 \text{ kN-m}]$$

Q. 4.30 A T-beam section has following dimensions:
$b_f = 3000$ mm, $D_f = 150$ mm, $b_w = 300$ mm, eff. depth $d = 1000$ mm.
Determine the balanced ultimate moment of resistance and corresponding area of tension steel. Use M20 CC and Fe 415 steel.

[**Hint:** $x_{u\,lim} = 0.48d = 480$ mm, $\frac{D_f}{d} = 0.15 < 0.2$, $\frac{D_f}{x_u} = \frac{150}{480} = 0.3125 < 0.43$.

$M_{u\,lim} = 0.138 f_{ck} \cdot b_w \cdot d^2 + 0.45 f_{ck} (b_f - b_w) D_f (d - 0.5 D_f) = 4200$ kN-m.]

Q. 4.31 A RCC T-beam section constructed with M20 CC and Fe 500 steel has following data:
$b_f = 1600$ mm, $b_w = 300$ mm, $d = 400$ mm, $D_f = 100$ mm, $f_{ck} = 20$ N/mm^2,
$f_y = 500$ N/mm^2

[Hint: $x_{u\,lim} = 0.46d = 184$ mm $> D_f$, $\dfrac{D_f}{d} = \dfrac{100}{400} = 0.25 > 0.20$, $\dfrac{D_f}{x_u} = 0.5435 > 0.43$

$M_{u\,lim} = (127.7 + 370.1)10^6 = 506.8 \times 10^6$ N-mm, $C = T$, $A_{st} = 3370$ mm^2]

Q. 4.32 A RCC T-beam has following data:

Flange width $(b_f) = 800$ mm, web width $(b_w) = 300$ mm, effective depth $(d) = 500$ mm, flange thickness $(D_f) = 90$ mm, applied moment $= 160$ kN-m. Design the section's tensile steel A_{st} using M20 CC and Fe 415 steel.

[Hint: $M_u = 240$ kN-m (240×10^6 N-mm), $x_{u\,lim} = 0.48 \times 500 = 240$ mm,

Let $x_u = D_f$, M_u flange $= 0.36f_{ck} \cdot b_f \cdot D_f \cdot (d - 0.42x_u) = 239.8 \times 10^6$ N-mm $= 240 \times 10^6$ N-mm

Hence $x_u = D_f$

$$M_u = 240 \times 10^6 = 0.87 \times 415 \times 500\, A_{st}\left(1 - \dfrac{415}{20} \cdot \dfrac{A_{st}}{800 \times 500}\right)$$

Solving $A_{st} = 1438$ mm^2

Also $A_{st} = \dfrac{0.50 f_{ck}}{f_y}\left[1 - \sqrt{1 - \dfrac{4.6 M_{uD}}{f_{ck} \cdot b_f \cdot d^2}}\,\right] b_f \cdot d = 1438$ mm^2]

Q. 4.33 Determine the ultimate moment of resistance of a T- beam section shown in Fig. 4.11. Use M20 CC and Fe 415 steel.

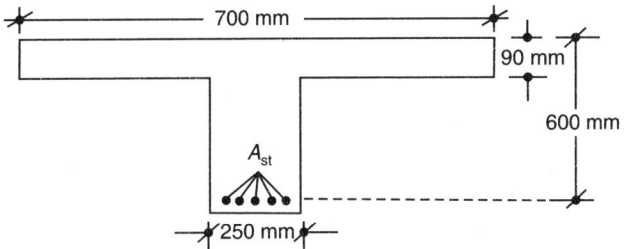

Fig. 4.11

[Hint: $A_{st} = 2455$ mm^2, $x_u = 175.87$ mm > 90 mm,   NA falls outside flange,

$T_u = 886378$ N, $y_f = 0.15\, x_u + 0.65 \times 90 = 84.9$ mm

$C_u = 0.36 \times 20 \times x_u \times 250 + 0.45 \times 20 \times 450 \times 84.9$ ($1800x_u + 343845$)

$T_u = C_u$, $x_u = 301.4$ mm $> x_{u\,lim}$ (288 mm), over reinforced, $y_f = 90$ mm (revised)

$$M_u = M_{u\,lim} = 0.36x_{u\,lim} \cdot f_{ck} \cdot b_w (d - 0.42x_{u\,lim}) + 0.45 \cdot f_{ck} (b_f - b_w)\, 90\left(600 - \dfrac{90}{2}\right)$$

$$= 68.68 \times 10^6 + 202.3 \times 10^6 = 270.98 \times 10^6 \text{ N-mm]}$$

Q. 4.34 Design the T-beam section's tensile steel if the T-beam has the following data:

$M_u = $ (ultimate MR) $= 355$ kN-m, $b_w = 250$ mm, $d = 500$ mm, $b_f = 750$ mm, D_f (flange thickness) $= 90$ mm, M20 CC and Fe 415 steel.

[Hint: $x_u = \dfrac{7}{3} D_f = 210$ mm (say), $C_{uw} = 378000$ N, $C_{uf} = 401400$ N,

$M_{uw} = 156 \times 10^6$ N-mm

$M_{uf} = 182.6 \times 10^6$ N-mm, $M_u = 338.60 \times 10^6$ N-mm $(< 345$ kN-m $\therefore x_u > \dfrac{7}{3} D_f)$

required $M_{uw} = (355 - 182.6)10^6 = 172.4 \times 10^6$ N-mm

$= 0.87 \times 415 \times A_{sw} \times 500 \left(1 - \dfrac{415}{20} \cdot \dfrac{A_{sw}}{250 \times 500}\right)$ or $\dfrac{172.4 \times 10^6}{180525}$

$$= A_{sw}(1 - 166 \times 10^{-6} A_{sw})$$

or $A_{sw} = \dfrac{0.5 \times 20}{415} \times 250 \times 500 \left[1 - \sqrt{1 - \dfrac{4.6 \times 172.4 \times 10^6}{20 \times 250 \, (500)^2}}\right] = 1191$ mm^2

$A_{sw} = 1191$ mm^2, $A_{sf} = \dfrac{C_{uf}}{0.87 f_y} = 1112$ mm^2

Total $A_{st} = A_{sw} + A_{sf} = 2303$ mm^2, actual $x_u = \dfrac{C_{uw}}{0.36 f_{ck} b_w} = 210$ mm

< 240 mm $(x_{u\,lim})]$

Q. 4.35 Design the tensile reinforcement for a T-beam in a beam slab structure having the following data:

Effective span of beam = 8 m (simply supported), spacing of beams c/c = 3.30 m

Thickness of slab = 130 mm, width of web b_w = 300 mm,

Depth (effective) = 330 mm, overall depth = 400 mm,

Live load on the floor $(w_l) = 10$ kN/m^2

Floor finish load = 0.75 kN/m^2

Beam supports a partition wall which transfers 12 kN/m run

Use M20 CC and Fe 415 steel.

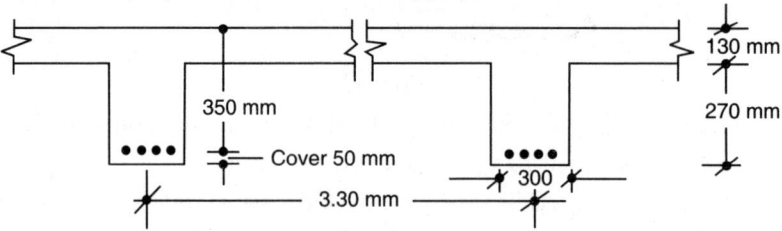

Fig. 4.12

[**Hint:** $b_f = 2413$ mm, $L_0 = 8000$ mm, $w_u = 90.4$ kN/m, $M_{uD} = 723.2$ kN/m, $d = 330$, $x_u = D_f = 130$, say $M_{uf} = 622 \times 10^6$ N-mm $< M_{uD}$ (723.2 kN-m),

Let $x_u = \dfrac{7}{3} D_f = 303.3$, $C_{uw} = 655128$ N, $C_{uf} = 2472210$ N,

$M_u = 787.88$ kN-m > 723.2 kN-m, $y_f = 130.00$ mm $\leq D_f x_u < \dfrac{7}{3} D_f$,

$x_u = 232$ mm (assume) and find M_u (web and flange),

$y_f = 0.15 \times 232 + 0.65 \times 330 = 119.3$ mm

$$M_u = \frac{0.36 \times 20 \times 232}{330}\left(1 - 0.42 \times \frac{232}{330}\right) 300 \times (330)^2 + 0.45 \times 20 \times (2413 - 300)$$

119.3 (330 – 0.5 × 119.3) = (116.54 × 10^6 + 613.35 × 10^6)

= 729.89 × 10^6 N-mm (approximately equal to 723.2 × 10^6 N-mm)

Hence y_f = 119.3 mm, x_u = 232 mm, d = 330 mm, b_w = 300 mm, $(b_f - b_w)$ = 2113 mm

C_{uw} = 0.36 × 232 × 20 × 300 = 501120 N

C_{uf} = 0.45 × 20 × 119.3 (2113) = 2268730 N

C_u = 2769850 N

Total compression = Total tension or 0.87 × 415 × A_{st} = (C_{uw} + C_{uf}) = 2769850 N

$$\therefore A_{st} = \frac{2769850}{0.87 \times 415} = 7672 \text{ mm}^2]$$

Q. 4.36 Determine the ultimate moment of resistance of a T-beam section shown in Fig. 4.13. The beam section is reinforced with 5 bars of 25 mm diameter in tension face and 3 bars of 25 mm diameter in compression face at an effective cover of 40 mm. The beam comprises M15 CC and Fe 415 steel.

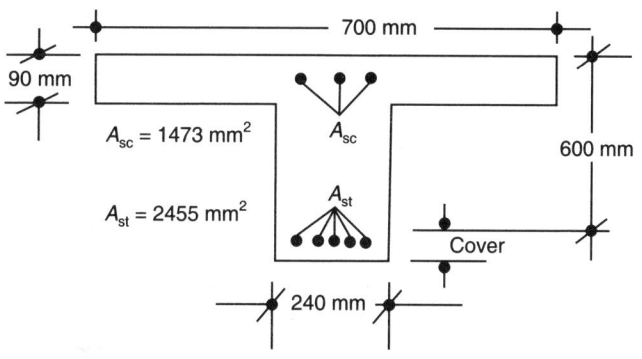

Fig. 4.13

[**Hint:** Let x_u(NA) lies beyond flange, and compressive stress f_{sc} in steel = $0.87f_y$ and in concrete at the compressive steel level = $0.45f_{ck}$ (compressive stress)

Let $x_u > D_f$ and d' (40 mm) $\leq \dfrac{3}{7} x_u$

Total tensile force will be equal to total compressive force ($C_{uf} + C_{uw} = C_{sc}$)

T_u = 0.87 × 415 × 2455 N = 886378 N, y_f = $(0.15x_u + 58.5)$, max D_f = 90mm

C_u = 0.36 × 15x_u × 240 + $(f_{sc} - 0.45 \times 15)$ 1473 + 0.45 × 15 × (700 – 240)

$(0.15x_u + 58.5)$

C_u = T_u, x_u = 407.4 – 0.8378f_{sc}

$$e_{sc} = \frac{0.0035\left(x_u - d'\right)}{x_u} = 0.0021 \ (f_{sc} = 328 \text{ N/mm}^2)$$

Trial values: x_u = 132.65 mm, y_f = 78.4 mm < D_f,

$$x_u = 132.6 \text{ mm, } e_{sc} = 0.0035\left(\frac{132.6 - 40}{132.6}\right) = 0.0024, f_{sc} = 342 \text{ N/mm}^2$$

$x_u (x_u > D_f) = 122$ mm, $e_{sc} = 0.00235$, $f_{sc} = 340.5$ N/mm^2

$x_{u\,lim} = 0.48d = 288$ mm, thus under reinforced $= \dfrac{D_f}{x_u} = \dfrac{90}{122} = 0.74 > 0.43$

$M_u = 0.36 f_{ck} \cdot \dfrac{x_u}{d}\left(1 - 0.42 \dfrac{x_u}{d}\right) b_w \cdot d^2 + (f_{sc} - 0.45 f_{ck})(A_{sc})(d - d') + 0.45 f_{ck}$

$(b_f - b_w) \times y_f (d - 0.5 y_f) = (86.85 + 275 + 132.75)10^6 = 494.6 \times 10^6$ N-mm]

Q. 4.37 Design a T-beam section if the ultimate MR is 450 kN-m. The beam details are shown in Fig. 4.14.

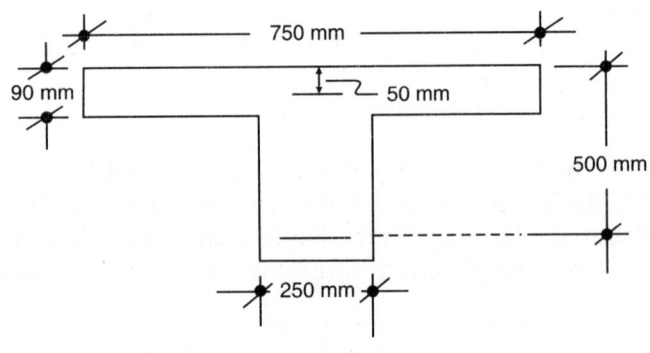

Fig. 4.14

[**Hint:** $M_{uD} = 450$ kN-m, assume $x_u = D_f$

$M_{uf} = 225$ kN-m $< M_{uD}$, NA outside flange.

Let $x_u = \dfrac{7}{3} D_f = 210$ mm, $M_{uf} = 338.6 \times 10^6$ N-mm $< M_{uD}$

$x_{u\,lim} = 0.48 \times 500 = 240$ mm, $M_{u\,lim} = 355$ kN-m $< M_{uD}$
Thus, the section may be doubly reinforced.

$M_{u_2} = M_{uD} - M_{u\,lim}$
$M_{u_2} = 95 \times 10^6$ N-mm. Equate total compressive force and tensile tone
$A_{sw} + A_{sf} = 1194 + 1112 = 2306$ mm^2 (A_{st_1})
$A_{st_2} = A_{so}$ (assuming $f_{sc} = 0.87 f_y$) $= 583$ mm^2, A_{st} (total) $= 2889$ mm^2]

Q. 4.38 Design a cantilever rectangular beam section for bending with the following data:

span of the cantilever = 3.0 m, working load = 15 kN/m, self-load of beam and finishing = 5 kN/m, effective depth = 450 mm, width of web = 300 mm.

[**Hint:** $w_u = 30$ kN/m, $M_{uD} = 135$ kN-m, $M_{u\,lim} = 0.138 f_{ck} \cdot bd^2 = 167$ kN-m $> M_{uD}$, under reinforced, singly reinforced, $A_{st} = 980$ mm^2 (2 bars of 25 mm)]

Q. 4.39 A T-beam is continuous over 10 m span and having flange width of 1200 mm, flange thickness of 100 mm, web width of 300 mm, effective depth of 500 mm. Use M20 grade CC and F$_e$ 415 grade steel. Design the reinforcement if the beam is subjected to an ultimate moment of (a) 600 kN-m (b) 300 kN-m.

[**Hint:** $x_{u\,lim} = 240$ mm, $M_{u\,lim} = 571$ kN-m

(a) $M_{ud} = 600$ kN-m $> M_{ulim}$ (571 kN-m), doubly reinforced section

$M_{uf} = (x_u = D_f) = 395.7$ kN-m $< M_{uD}$, NA lies in web,

$A_{st} = A_{st_1} + A_{st_2} = 3858$ mm^2, $A_{sc} = 185$ mm^2, $f_{sc} = 0.45 \times f_{ck}$

(b) $M_{ud} = 300$ kN-m $< M_{u\ lim}$ (571 kN-m), $M_{uD} < M_u$ flange ($x_u = D_f$) = 395.7 kN-m

NA lies in the flange $x_u \le D_f = 74.04$ mm,

A_{st} (singly reinforced) = 1772 mm^2]

Chapter 5

Limit State Design for Shear, Torsion, Bond and Development Length of Reinforcement in Beams

LEARNING OBJECTIVES

After the study of *limit state design for shear, torsion, bond and development length of reinforcement in beams,* the learner will understand the basic principles and design procedures for RCC beams and will be able to:

- ◉ Define nominal shear, design shear and limit state of shear
- ◉ Calculate nominal shear stress in beams subjected to shear force
- ◉ Calculate design shear strength from IS:456-2000 for different grades of concrete and different percentages of tensile steel
- ◉ Design beam sections and shear reinforcement for a given shear force
- ◉ Calculate shear capacity of a section with given details of reinforcement and size of the section
- ◉ Explain torsion in beam section and simple concept of equivalent shear and equivalent bending moment
- ◉ Design shear and longitudinal reinforcement in a beam with given SF, BM and torsion
- ◉ Explain the importance of bond between concrete and steel and development length beyond the point of requirement
- ◉ Explain the safety checks at the points of inflexion and points of curtailment
- ◉ Explain anchorage bond stress and flexural bond stress
- ◉ Explain the curtailment of longitudinal reinforcement in a beam

5.1 LIMIT STATE DESIGN FOR SHEAR

5.1.1 Introduction

A vast majority of structural elements such as beams and slabs loaded with transverse loads develop shear forces along with bending moments at various sections along the span. We have already studied the design procedure for resisting bending moments safely by working stress and limit state methods of design. During earlier study of strength of materials, we have seen that shear forces cause shear stresses and diagonal tensile and diagonal compressive stresses. To avoid tension cracks, we need to provide steel reinforcement to offer resistance to such diagonal tension. The design recommendations are based on empirical relations derived from practical tests.

The simply supported (SS) beams carrying u.d.l. over the entire span, the maximum BM occurs at the mid span point, while the maximum shear force occurs at the supports. Thus, the mid span point is critical for flexure design while the support section is important for shear design. Thus, the SS beam is provided with maximum longitudinal steel reinforcement near the mid span point to offer resistance to bending in tension zone. Maximum reinforcement in the form of vertical or inclined stirrups is provided near the support in case of SS beams to offer resistance to diagonal tension to avoid cracking.

The cantilever beams develop maximum BM and maximum SF at the supports and hence the support requires maximum longitudinal reinforcement in tension zone as well as maximum shear reinforcement in the form of vertical or inclined stirrups. The reinforcement may be reduced towards free end in relation to BM and SF variations.

We shall study the behaviour of beams under shear and design of stirrups in accordance with the provision of IS:456-2000.

5.1.2 Shear Stress Failure

Shear failure occurs near support in case of SS, cantilever, and continuous beams. The important types of failures are as follows:

- Diagonal tension failure
- Flexure shear failure
- Shear compression failure
- Shear bond failure

Shear failures are influenced by the principal parameter, the effective depth and shear span. The transverse shear force (V) is resisted by the following mechanisms:

(a) Shear resistance of the uncracked portion of concrete (V_c)

(b) Vertical component of interface (aggregate interlock force (V_a)

(c) Dowel force (V_d) developed due to the tension steel

(d) The shear resistance (V_s) developed in shear reinforcement.

Different types of failures may occur based on whether the beam is very deep or short or normal type. In case of I-beams having thin webs, failure is due to web crushing of concrete which can be prevented by suitable design of reinforcement and using high strength concrete.

5.1.3 Nominal Shear Stress

The shear stress distribution in a RC beam section is dependent on the SF acting on the section and the shape of the cross-section in the elastic stage. At the ultimate stage, the concrete below the NA becomes ineffective due to cracking under tension zone. Hence, for simplicity the nominal shear stress across the section is calculated as average shear stress (shown in Fig. 5.1) and is expressed as:

$$\tau_v = \frac{V_u}{bd} \qquad\qquad \text{... Eq. (5.1)}$$

where,
V_u = Ultimate shear force at the section

τ_v = Nominal shear stress

b = Width (b_w in case of flanged beams)

d = Effective depth

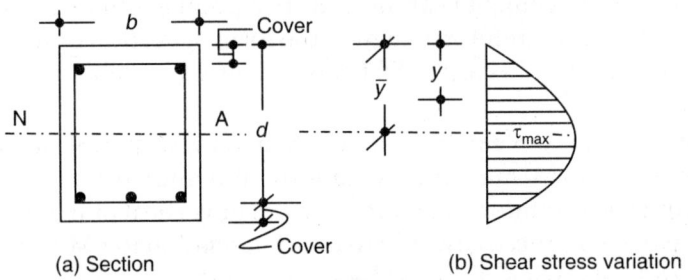

(a) Section   (b) Shear stress variation

Fig. 5.1: Shear stress distribution across the section

Sometimes the beams (specially cantilever beams) are provided with variable depth d to resist variable moment and shear along the span. The nominal shear stress equation for uniform beam section $\left(\tau_v = \dfrac{V_u}{bd}\right)$ shall be modified to Eq. (5.1a) for a beam section with variable depth.

$$\tau_v = \frac{V_u}{bd} \pm \frac{M_u \tan\beta}{bd^2} \qquad\qquad \text{... Eq. (5.1a)}$$

where,
τ_v = Nominal shear stress (N/mm^2)

b = Uniform width of the beam section (mm)

d = Effective depth at the section (mm)

M_u = Ultimate bending moment (N-mm) at the section

V_u = Ultimate shear force at the section (N)

β = Angle between the top and bottom side of the beam

The negative sign in Eq. (5.1a) shall be applicable when the BM (M_u) and effective depth (d) increases in the same direction along the span.

5.1.4 Design Strength of Concrete

Design shear strength in beams without shear reinforcement depends upon the grade of concrete (f_{ck}) and the percentage of tensile reinforcement in the section.

The allowable design shear strength (τ_c) of concrete in beams without shear reinforcement is given in Table 5.1 (Table 19 of IS:456-2000). These values given in Table 5.1 are applicable for beams. In the case of slabs having an overall depth less than 300 mm, the shear strength being higher, the IS code 456-2000 suggests an enhanced shear strength calculated as $k\tau_c$, where k is a multiplying factor depending upon the overall depth of the slab as given in Table 5.2 (Clause 40.2.1.1 of IS:456-2000).

The code also specifies an upper limit for the design shear strength of concrete strengthened by shear reinforcement. Accordingly the maximum shear stress in concrete should not exceed the values of $\tau_{c\,max}$ given in Table 5.3 (Table 20 of IS:456-2000). If the value of the nominal shear stress (τ_v) exceeds the value of $\tau_{c\,max}$, the section should be redesigned with increased cross-sectional dimensions (specially d wherever possible).

Table 5.1: Design shear strength of concrete (τ_c in N/mm²)

(as per Table 19, IS:456-2000)

Percentage of tensile steel 100 (A_{st}/bd)	Grades of concrete					
	M15	M20	M25	M30	M35	M40 and higher grades
$\leq 0.15\%$	0.28	0.28	0.29	0.29	0.29	0.30
0.25%	0.35	0.36	0.36	0.37	0.37	0.38
0.50%	0.46	0.48	0.49	0.50	0.50	0.51
0.75%	0.54	0.56	0.57	0.59	0.59	0.60
1.00%	0.60	0.62	0.64	0.66	0.67	0.68
1.25%	0.64	0.67	0.70	0.71	0.73	0.74
1.50%	0.68	0.72	0.74	0.76	0.78	0.79
1.75%	0.71	0.75	0.78	0.80	0.82	0.84
2.00%	0.71	0.79	0.82	0.84	0.86	0.88
2.25%	0.71	0.81	0.85	0.88	0.90	0.92
2.50%	0.71	0.82	0.88	0.91	0.93	0.95
2.75%	0.71	0.82	0.90	0.94	0.96	0.98
3.00% and above	0.71	0.82	0.92	0.96	0.99	1.01

Table 5.2: Values of k for solid concrete slabs for enhancing τ_c

Overall depth of slab (mm)	300 or more	275	250	225	200	175	150 or less
Value of k	1.0	1.05	1.10	1.15	1.20	1.25	1.30

Table 5.3: Maximum shear stress ($\tau_{c\,max}$ in N/mm²) (Table 20 of IS:456-2000)

Concrete grade	M15	M20	M25	M30	M35	M40 and above
$\tau_{c\,max}$, (N/mm²)	2.50	2.80	3.10	3.50	3.70	4.00

5.1.5 Design of Shear Reinforcement

In case of RC section, if the nominal shear stress (τ_v) exceeds the design shear strength of concrete (τ_c), shear reinforcements are necessary. The shear reinforcement comprises following types:

- Vertical stirrups or inclined stirrups
- Inclined bent up bars near supports

Let $\quad\quad\quad\quad\quad\quad V_u$ = Total design shear force

V_c = Shear resistance of concrete

V_{us} = Shear to be resisted by shear reinforcement in the form of stirrups or bent up bars.

We have,

$$V_{us} = (V_u - V_c) = (\tau_v - \tau_c)\, b.d \quad\quad\quad ... \text{Eq. (5.2)}$$

where, $\quad\quad\quad\quad\tau_v$ = Nominal shear stress $\left(\dfrac{V_u}{bd}\right)$

τ_c = Design shear strength of concrete (Table 5.1)

A_{sv} = Total area of shear reinforcement legs of stirrups

S_v = Spacing of the stirrups

d = Effective depth of section

Then $\quad\quad\quad\quad\quad S_v = \dfrac{0.87\, f_y \cdot A_{sv} \cdot d}{V_{us}} \quad\quad\quad ... \text{Eq. (5.3)}$

The shear resisted by the bent up inclined bars (main tensile) inclined at an angle α to the horizontal is given by

$$V_{us} = 0.87 f_y \cdot A_{sv} \sin \alpha \quad\quad\quad ... \text{Eq. (5.4)}$$

where, $\quad\quad\quad\quad A_{sv}$ = Area of the inclined bent up bars

f_y = Characteristic or proof stress of steel

α = Inclination of bent up bars with horizontal

Example 5.1

A SS reinforced concrete beam section of 300 mm width and 500 mm effective depth has 3 bars of 20 mm diameter on the tension face near the support. The support section is provided with 8 mm diameter two legged stirrups at a spacing of 200 mm centre to centre. Using M20 grade CC and Fe 415 grade steel, calculate the shear resistance of the support section.

Solution: $\quad f_{ck} = 20\,\text{N/mm}^2, f_y = 415\,\text{N/mm}^2, d = 500\,\text{mm}, b = 300\,\text{mm},$

$A_{st} = 942\,\text{mm}^2, A_{sv} = 2 \times 50 = 100\,\text{mm}^2, S_v = 200\,\text{mm c/c},$

$$p_t = \frac{942 \times 100}{300 \times 500} = 0.63\%.$$

For p_t of 0.63% (Table 5.1), $\tau_c = 0.52\,\text{N/mm}^2$ (for M20 grade CC)

Shear resisted by concrete $V_{uc} = 0.52\,(300 \times 500) = 78000\,\text{N} = 78\,\text{kN}$

Shear resisted by shear reinforcement $= \dfrac{0.87\, f_y \cdot A_{sv} \cdot d}{S_v} = 90263\,\text{N} = 90.263\,\text{kN}$

Total resistance for the section $= 78 + 90.263 = 168.263\,\text{kN}$

Example 5.2

A SS CC beam of rectangular section of 300 mm width and 600 mm effective depth is reinforced with 4 bars of 25 mm diameter in tension face. Two of tensile reinforcement bars are bent up at 45° with the horizontal near ends. The beam provides 8 mm two legged vertical stirrups at a spacing of 200 mm near supports. Using M30 grade CC and Fe 415 grade steel, calculate the ultimate shear resistance of the support section as per provision of IS:456-2000 code.

Solution: $b = 300$ mm, $d = 600$ mm, $A_{st} = (4 - 2)$ 491 $= 982$ mm²,

$$A_{sv} = 2 \times 50 = 100 \text{ mm}^2, S_v = 200 \text{ mm}, \alpha = 45°, p_t \text{ (\% tensile steel)} = \frac{982 \times 100}{300 \times 600} = 0.55\%,$$

$$\tau_c \text{ (Table 5.1)} = 0.52 \text{ N/mm}^2 \text{ (for M30 CC)}$$
$$V_{uc} \text{ (concrete)} = 0.52 \times 300 \times 600 = 93600 \text{ N} = 93.6 \text{ kN}$$

$$V_{us} \text{ (vertical stirrups)} = \frac{0.87 f_y \cdot A_{sv} \cdot d}{S_v} = \frac{0.87 \times 415 \times 100 \times 600}{200} = 108315 \text{ N}$$

$$= 108.32 \text{ kN}$$

$$V_{us1} \text{ (inclined steel bars)} = 0.87 f_y \cdot A_{st} \times \sin \alpha = 0.87 \times 415 \times 982 \times 0.7071$$
$$= 250700 \text{ N} = 250.7 \text{ kN}$$

Total shear resistance of support $= 93.6 + 108.32 + 250.7 = 452.62$ kN

Example 5.3

A simply supported rectangular beam section of 300 mm width and 500 mm effective depth is reinforced with 5 bars of 20 mm diameter. The beam is subjected to a service shear force of (a) 50 kN (b) 80 kN (c) 200 kN and (d) 300 kN. Use M20 CC and Fe 415 steel. Determine the shear reinforcement.

Solution:
$$f_{ck} = 20 \text{ N/mm}^2, f_y = 415 \text{ N/mm}^2, b = 300 \text{ mm}, d = 500 \text{ mm},$$
$$A_{st} = 1570 \text{ mm}^2 (p_t = 1.05\%)$$
$$\tau_c \text{ (M20) for } p_t \text{ of } 1.05\% = 0.63 \text{ N/mm}^2$$
$$8\phi \text{ 2 legged } A_{sv} = 2 \times 50 = 100 \text{ mm}^2$$

$$\therefore \qquad V_{uc} = 0.63 \times 300 \times 500 = 94500 \text{ N (94.5 kN)}$$

(a) $V_u = 1.50 \times 50 = 75$ kN $< V_{uc}$ of 94.5 kN and hence provide only minimum shear reinforcement, provide 8 mm 2 legged stirrups.

$$S_v \text{ (max)} = \frac{2.5 A_{sv} \cdot f_y}{b} \not> 0.75d \text{ (375 mm c/c) or} \not> 450 \text{ mm}$$

$$= \frac{2.5 \times 100 \times 415}{300} = 346 \not> 375 \text{ mm or} \not> 450 \text{ mm c/c}$$

Provide 8 mm ϕ 2 legged vertical stirrups at 345 mm c/c (say)

(b) $V_u = 1.5 \times 80 = 120$ kN $> V_{uc}$ of 94.5 kN, provide shear reinforcement
$$V_{uc} \text{(max)} = 94.5 \text{ kN (as in above part)}$$
$$V_{us} \text{ (SF for stirrups)} = (120 - 94.5) = 25.5 \text{ kN}$$
$$V_{uc \text{ max}} = 2.8 \times 300 \times 500 = 420 \times 10^3 \text{ N} = 420 \text{ kN} > V_u \text{ (120 kN)}$$

Design the shear reinforcement. Consider 8 mm ϕ 2 legged stirrups.
$$A_{sv} = 100 \text{ mm}^2, S_v = ?$$

$$S_v = \frac{0.87 \times 415 \times 100 \times d}{V_{us}}, \quad V_{us} = 25.5 \times 10^3 \text{ N}$$

$$= \frac{0.87 \times 415 \times 100 \times 500}{25.5 \times 10^3} = 708 \text{ mm c/c}, \nleq 0.75d, \nleq 450 \text{ mm c/c},$$

$$\nleq 345 \text{ c/c} \left(\frac{2.5 \, A_{sv} \cdot f_y}{b} \right),$$

Thus, provide 8 mm ϕ 2 legged vertical stirrups @ 345 mm c/c

(c) $V_u = 1.5 \times 200 = 300 \text{ kN} > V_{uc}$ of 94.5 kN

Hence provide shear reinforcement and $< V_{uc\,max}$ of 420 kN

$$V_{us} \text{ (stirrups} = V_u - V_{uc} = (300 - 94.5) = 205.5 \text{ kN} = 205.5 \times 10^3 \text{ N}$$

Providing 8 ϕ2 legged vertical stirrups $\quad S_v = \dfrac{0.87 \times 415 \times A_{sv} \times d}{V_u - V_{uc}}$

$$= 87.8 \text{ mm} \nleq 345.8 \text{ mm},$$

or provide 10 ϕ2 legged vertical stirrups $S_v = \dfrac{0.87 \times 415 \times 157 \times 500}{205.5 \times 10^3}$

$$= 137.9 \text{ mm} \nleq 345.8 \text{ mm}$$

Thus, provide 10 mm 2 legged stirrups at 135 mm c/c.

(d) $V_u = 300 \times 1.5 = 450 \text{ kN} > V_{uc\,max} = 420 \text{ kN}$, the design of section needs to be revised by increasing the section so that $V_{uc\,max}$ is more than 450 kN.

$$d \text{ (reqd.)} = \frac{450 \times 1000}{2.80 \times 300} = 535.7 \text{ mm (provide } d = 550 \text{ mm and appropriate shear}$$

stirrups)

Example 5.4

Determine the shear reinforcement required for a flanged beam section shown in Fig. 5.2. The beam is subjected to an ultimate shear force of 300 kN. Consider concrete grade of M20 and steel grade of Fe 415.

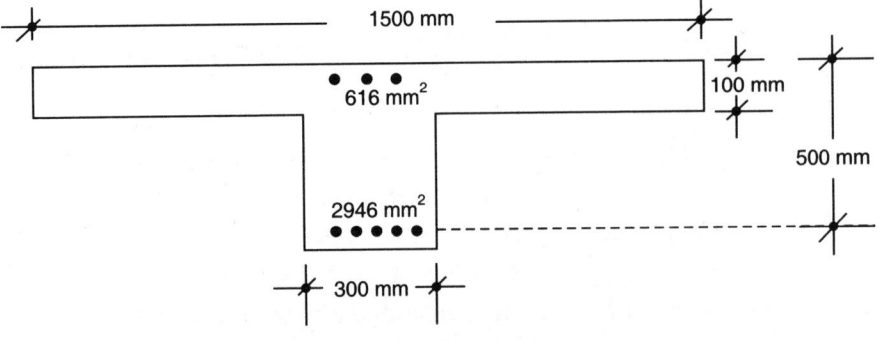

Fig. 5.2: T-beam section

Solution: For flanged beam section mainly web resists the shear.

$$f_{ck} = 20 \text{ N/mm}^2, f_y = 415 \text{ N/mm}^2, d = 500 \text{ mm}, b_w = 300 \text{ mm}, b_f = 1500 \text{ mm},$$

$$D_f = 100 \text{ mm}, A_{st} = 2946 \text{ mm}^2, A_{sc} = 616 \text{ mm}^2$$

$$p_t \text{ (\% tensile steel)} = \frac{2946 \times 100}{300 \times 500} = 1.96\%,$$

$$\tau_{c\,max} = 2.8 \text{ N/mm}^2, V_{uc\,max} = 2.8 \times 300 \times 500$$
$$= 420000 \text{ N (420 kN)} > 300 \text{ kN}$$
$$\tau_c \, (p_t = 1.96) = 0.784 \text{ N/mm}^2 \text{ (from Table 5.1)}$$
$$V_{uc} = 0.784 \times b_w \times d = 0.784 \times 300 \times 500 = 117600 \text{ N (117.6 kN)}$$
$$V_{uc\,max} \text{ (420 kN)} > 300 \text{ kN} > 117.6 \text{ kN}$$

Provide shear reinforcement for $V_{us} = (V_u - V_{uc}) = (300 - 117.6) = 182.4$ kN

Provide 10 mm ϕ 2 legged bars $S_v = \dfrac{0.87 \, f_y \cdot A_{sv} \cdot d}{(V_u - V_{uc})}$

$$= \frac{0.87 \times 415 \times 157 \times 500}{182.4 \times 1000}$$

$$= 155.4 \text{ mm c/c (say 150 mm c/c)}$$

$$S_{v\,max} = \frac{2.5 \, A_{sv} \cdot f_y}{b} = 543 \text{ mm}, 0.75 \times 500 = 375 \text{ mm}$$

Thus, provide 10 mm ϕ 2 legged @ 150 mm c/c.

Example 5.5

A cantilever beam of 3 m span carries a factored u.d.l. of 60 kN/m including its own weight. The beam section varies 250 mm × 400 mm (effective) at the free end to 250 mm × 550 mm (effective) at the fixed end. Use M20 grade CC and Fe 415 grade steel. The beam is reinforced with 5 bars of 25 mm diameter at the fixed end and curtailed to 3 bars of 25 mm diameter at the free end (Fig. 5.3).

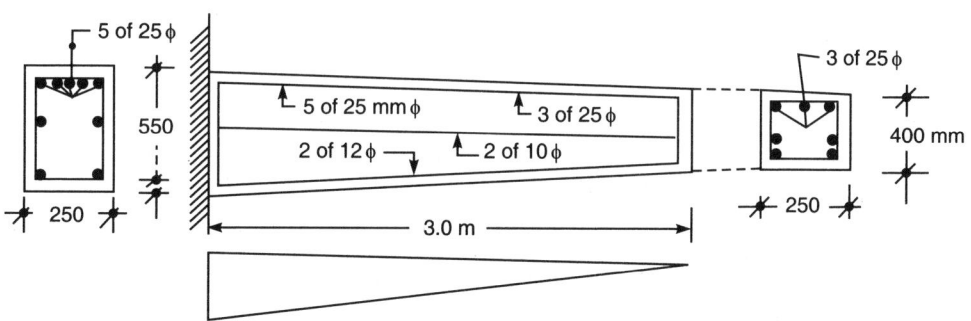

Fig 5.3: Cantilever beam with variable sections

Solution: Let us analyse two sections, viz. one near the fixed end and other near the mid span point.

    (i) SF $V_{u1} = 60 \times 3 = 180$ kN

        SF V_{u2} (near mid span point) $= 60 \text{ kN/m} \times 1.5 \text{ m} = 90$ kN

    (ii) Size of section

        Fixed end $= 250 \text{ mm} \times 550$ mm

Section $(2-2)$, $d = 250 \left[400 + \dfrac{(550-400)1.5}{3.0} \right] = 475$ mm, section (250 mm × 475 mm)

(iii) Slope $\tan\theta = \dfrac{150 \text{ mm}}{3000 \text{ mm}} = 0.05$, thus, $\tan\theta = 0.05$

(iv) BMS at fixed end $(1-1)$ and near the mid span point $(2-2)$

fixed end $(1-1)$, $M_{u1} = \dfrac{60 \times 3 \times 3}{2} = 270$ kN-m

mid span $(2-2)$, $M_{u2} = \dfrac{60 \times 1.5 \times 1.5}{2} = 67.5$ kN-m

(v) Net V_{u1} considering effect of moment

$$V_{u1} = \left(180 - \dfrac{M_u \cdot \tan\theta}{d} \right) = \left(180 - \dfrac{270 \times 1000 \times 0.05}{550} \right) = 155.45 \text{ kN}$$

$$V_{u2} = \left(90 - \dfrac{67.5 \times 1000 \times 0.05}{475} \right) = 82.9 \text{ kN}$$

(vi) Shear reinforcement

Percentage tensile steel, at sections $1-1$ and $2-2$:

$$p_{t1} = \dfrac{5 \times 491 \times 100}{250 \times 550} = 1.79\%$$

$$p_{t2} = \dfrac{3 \times 491 \times 100}{250 \times 475} = 1.24\%$$

(vii) τ_c (Table 5.1)

$$\tau_{c1} = 0.756 \text{ N/mm}^2, \quad \tau_{c2} = 0.67 \text{ N/mm}^2$$

(viii) V_{uc1} (concrete resistance),

$$V_{uc1} = 0.756 \times 250 \times 550 = 104000 \text{ N (104 kN)}$$
$$V_{uc2} = 0.67 \times 250 \times 475 = 79563 \text{ N (79.56 kN)}$$

(ix) Shear force for shear reinforcement design,

$$V_{us1} = (155.45 - 104) \, 10^3 = 51450 \text{ N}$$
$$V_{us2} = (82.9 - 79.56) \, 10^3 = 3340 \text{ N}$$

(x) Minimum shear reinforcement 8 mm ϕ 2 legged stirrups ($A_{sc} = 100$ mm^2)

$$S_v \not> \dfrac{2.5 \times 100 \times 415}{b} = 415 \text{ mm c/c}, \, 0.75d = 412 \text{ c/c}, \, (0.75 \times 475 = 356 \text{ mm c/c})$$

or   450 mm c/c

$$S_{v1} = \dfrac{0.87 \times 415 \times 100 \times 550}{51450} = 386 \text{ mm c/c}$$

$$S_{v2} = \dfrac{0.87 \times 415 \times 100 \times 475}{3340} = 5134 \text{ mm c/c (provide } S_v = 350 \text{ mm c/c)}$$

Thus, provide 8 mm ϕ 2 legged stirrups at 350 mm c/c in both the sections.

5.2 LIMIT STATE DESIGN FOR TORSION

5.2.1 Introduction

Reinforced concrete (RC) structural elements are subjected to torsion in addition to bending moments and shear forces depending on the loading and shape of the structural

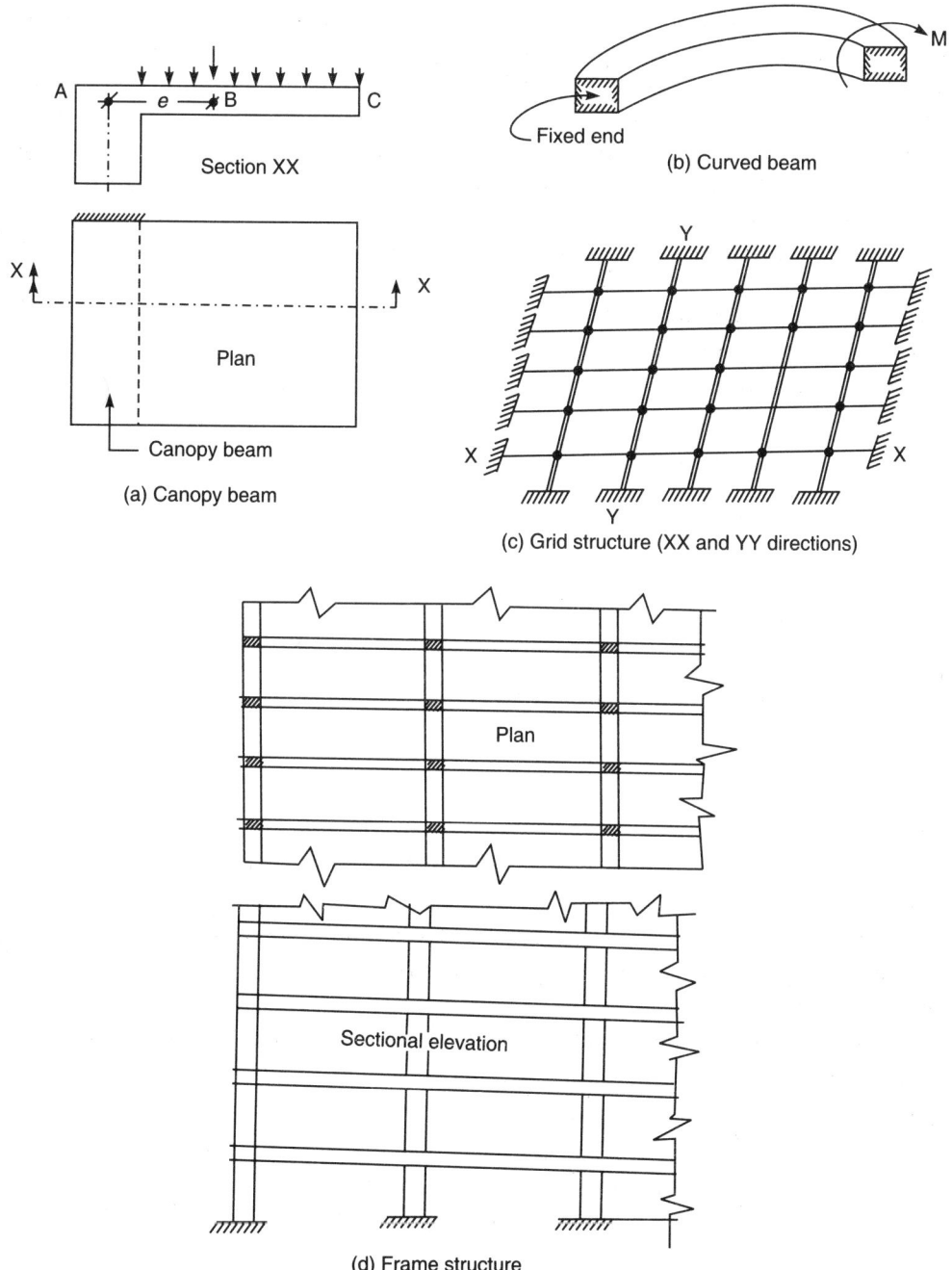

Fig. 5.4: RCC structural elements subjected to torsion

members. Pure torsion in RCC elements is a rare occurrence. Generally the torsion occurs in combination with the shear force and bending moment. Torsion is caused by eccentricity of loading to the axis of the structural element. Some of the elements are shown in Fig. 5.4

5.2.2 Limit State Design of Beam for Torsion

Torsion is caused when the loads are eccentric to the line of reactions. For example, in canopy beam, the load W_o acts at a distance of e from the axis of the beam (reaction line), the beam is subjected to a torsion $T = W_o e$. Equilibrium conditions are sufficient to evaluate the torsional moments in case of determinate structures at critical sections.

The effect of torsion is to induce shear stresses and it causes warping of non-circular sections. The failure of plain concrete members in torsion is due to diagonal tensile cracks since concrete is weak in tension. IS:456-2000 provides a method of designing suitable reinforcements in concrete sections subjected to the combined effects of torsion, flexure and shear by introducing the concept of enhanced equivalent shear (V_e), and bending moment (M_e) due to torsion.

Sections subjected to torsion and shear are to be designed for an equivalent shear force (V_e) computed as:

$$V_e = V_u + 1.6 \left(\frac{T_u}{b} \right) \qquad \text{... Eq. (5.5)}$$

where, V_e = Enhanced equivalent shear due to torsional moment

V_u = Ultimate transverse shear

T_u = Torsional moment

b = Breadth of the beam

The equivalent nominal shear stress is expressed as:

$$\tau_{ve} = \left(\frac{V_u + 1.6\dfrac{T_u}{b}}{bd} \right) = \left(\frac{V_u}{bd} + \frac{1.6\,T_u}{b^2 d} \right) \qquad \text{... Eq. (5.6)}$$

The values of τ_{ve} should lie between the design shear stress τ_c (Table 5.1) and the maximum shear stress $\tau_{c\,max}$ (Table 5.3). In case, where τ_{ve} exceeds $\tau_{c\,max}$, the section needs to be redesigned suitably by increasing the sectional area or selecting higher grade of concrete.

If $\tau_{ve} < \tau_c$, minimum shear reinforcement may be provided according to IS:456-2000.

Longitudinal reinforcements are designed to resist an equivalent bending moment expressed as:

$$M_e = (M_u + M_t) \qquad \text{... Eq. (5.7)}$$

where, M_e = Equivalent bending moment

M_u = Ultimate (design) bending moment

M_t = Bending moment developed due to torsion and expressed as

$$M_t = T_u \left(\frac{1 + \dfrac{D}{b}}{1.7} \right) = \frac{T_u}{1.7} \left(1 + \frac{D}{b} \right) \qquad \qquad \text{... Eq. (5.8)}$$

where, T_u = Torsional moment

D = Overall depth of the member

b = Breadth of the section.

In case where the numerical value of M_t exceeds the numerical value of M_u, the longitudinal reinforcement should be provided on the flexural compression face so that the beam can also withstand an equivalent moment M_e calculated as $M_{e2} = (M_t - M_u)$, the moment M_{e2} being taken as acting in the opposite sense to the moment M_u. Transverse reinforcements comprising two legged closed stirrups or links enclosing the corner longitudinal bars should be provided. The cross-sectional area of stirrups should have

$$A_{sv} = \left[\frac{T_u \cdot S_v}{b_1 d_1 \, (0.87 f_y)} + \frac{V_u \cdot S_v}{2.5 d_1 \, (0.87 f_y)} \right] \qquad \qquad \text{... Eq. (5.9)}$$

However, the total transverse reinforcement should be not less than the value calculated as:

$$A_{sv} \nleftarrow \frac{(\tau_{ve} - \tau_c) \, b \, S_v}{0.87 f_y} \qquad \qquad \text{... Eq. (5.10)}$$

or $$\text{spacing } S_v = \frac{A_{sv} \, (0.87 f_y)}{(\tau_{ve} - \tau_c) \, b} \qquad \qquad \text{... Eq. (5.10a)}$$

where, T_u = Torsional moment

V_u = Transverse shear force

S_v = Spacing of vertical links (stirrups)

b_1 = Centre to centre distance between corner bars in the direction of width

d_1 = Centre to centre distance between corner bars in the direction of depth

f_y = Characteristic strength of stirrups reinforcement

τ_{ve} = Equivalent shear stress $\left(\dfrac{V_u}{bd} + \dfrac{1.6 \, T_u}{b^2 d} \right)$

τ_c = Shear strength of concrete as given in Table 5.1

5.2.3 Longitudinal Reinforcement

Longitudinal reinforcement shall be provided to resist equivalent moment $M_e = (M_u \pm M_t)$. If $M_u \geq M_t$, then the design shall be based on the equivalent moment $M_e = M_u + M_t$, only as $(M_u - M_t)$ will be smaller than $(M_u + M_t)$ and act in the same direction. If $M_u < M_t$, the design shall also be checked for $M'_e - (M_u - M_t)$, which acts in opposite direction to M_e, i.e. the longitudinal steel shall be provided on the compression face also.

5.2.4 Transverse Reinforcement (Stirrups)

The transverse reinforcement shall consist of vertical stirrups enclosing the longitudinal reinforcement. The design of stirrups requires determination of spacing for its chosen bar diameters as follows:

$$S_v = \text{Smaller of} \quad \frac{0.87\, f_y\, A_{sv}\, b_1\, d_1}{(T_u + 0.4\, V_u\, b_1)} \quad \text{or} \quad \frac{0.87\, f_y\, A_{sv}\, d_1}{(V_{ue} - V_{uc})}, \quad S_{v\,\text{max}} \qquad \dots \text{Eq. (5.11)}$$

where,

b_1 = Centre to centre distance between the corner bars in the direction of width

d_1 = Centre to centre distance between the corner bars in the direction of the depth

f_y = The tensile strength of shear reinforcement (stirrups) $\not>$ 415 N/mm^2

$$S_{v\,\text{max}} \leq \text{shorter dimension of stirrups } (b_1 \text{ or } d_1)$$

$$\leq \frac{1}{4}\,(\text{shorter} + \text{longer dimension of stirrups})$$

$$\leq 0.75d$$

$$\leq 300 \text{ mm}$$

Example 5.6

A reinforced concrete beam of rectangular section having a width of 300 mm and overall depth of 600 mm is reinforced with 4 bars of 25 mm diameter placed one bar on each corner at an effective cover of 50 mm in the direction of depth and side covers of 25 mm in the direction of width. Two legged stirrups of 8 mm ϕ are provided at 120 mm centre to centre. Determine the torsional strength of the section adopting Fe 415 grade steel for the following cases:

(a) $V_u = 0$,   (b) $V_u = 160$ kN.

Solution: $b = 300$ mm, $D = 600$ mm, $b_1 = 300 - 25 \times 2 = 250$ mm

$d_1 = 600 - 2 \times 50 = 500$ mm, $d = 600 - 50 = 550$ mm

$A_{st} = 2 \times 491 = 982$ mm^2, $A_{sc} = 982$ mm^2, $A_{sv} = 2 \times 50 = 100$ mm^2

$S_v = 120$ mm c/c.

(a) Torsional strength (when $V_u = 0$)

$$T_u = \frac{0.87\, f_y\, A_{sv}\, b_1\, d_1}{S_v} = \frac{0.87 \times 415 \times 100 \times 250 \times 500}{120} = 37.61 \times 10^6 \text{ N-mm}$$

$$= 37.61 \text{ kN-m}$$

(b) Torsional strength (when $V_u = 160$ kN)

$$A_{sv} = \frac{T_u\, S_v}{b_1 d_1\,(0.87\, f_y)} + \frac{V_u\, S_v}{2.5\, d_1\,(0.87\, f_y)}$$

$$100 = \frac{T_u\, 120}{250 \times 500 \times (0.87 \times 415)} + \frac{160 \times 10^3 \times 120}{2.5 \times 500 \times (0.87 \times 415)} = \frac{T_u}{376094} + 42.54$$

or $T_u = 376094\,(100 - 42.54) = 21610346$ N-mm $= 21.6104$ kN-m

Thus, design ultimate torsional strength is the smaller of the two values (i.e. 21.61 kN-m is the torsional strength of the beam).

Example 5.7

A RC beam of rectangular section with width of 300 mm, effective depth of 600 mm is reinforced with 4 bars of 16 mm diameter in tension zone. The section is subjected to a factored BM of 150 kN-m. If M25 grade CC and Fe 415 grade steel are used, find the maximum ultimate torsional resistance that can be allowed.

Solution: $b = 300$ mm, $d = 600$ mm, $f_{ck} = 25$ N/mm², $f_y = 415$ N/mm²,

$$M_u = 150 \times 10^6 \text{ N-mm}$$
$$A_{st} = 804 \text{ mm}^2$$
$$x_{u \lim} = 0.48 \times 600 = 288 \text{ mm}$$

$$x_u = \frac{0.87 f_y A_{st}}{0.36 f_{ck} b} = \frac{0.87 \times 415 \times 804}{0.36 \times 25 \times 300}$$

$$= 107.513 \text{ mm} < x_{u \lim}(288 \text{ mm}) \text{ under reinforced.}$$

$$M_e = 0.87 f_y A_{st} d \left[1 - \frac{A_{st} f_y}{bd f_{ck}}\right]$$

$$= 0.87 \times 415 \times 804 \times 600 \ (1 - 0.07415) = 161.26 \times 10^6 \text{ N-mm}$$

Also $[0.87 f_y A_{st} \ (d - 0.42 x_u)] = M_e$

Thus, $$M_e = M_u + M_t = M_u + T_u \left(\frac{1 + \dfrac{D}{b}}{1.7}\right) = 150 \times 10^6 + T_u \left(\frac{1 + \dfrac{650}{300}}{1.7}\right)$$

$$(161.26 \times 10^6 - 150 \times 10^6) = T_u \left[\frac{1 + 2.167}{1.7}\right] = T_u \ (1.863)$$

$$T_u = 6.045 \times 10^6 \text{ N-mm (6.045 kN-m).}$$

Example 5.8

Design a rectangular beam section 300 mm width and 500 mm effective depth subjected to ultimate BM of 180 kN-m, ultimate shear force of 30 kN and ultimate torsional moment of 10 kN-m. Consider M20 grade CC and Fe 415 grade steel. Assume 50 mm effective cover along verticals and 25 mm effective cover on sides.

Solution: $b = 300$ mm, $d = 500$ mm, $D = 550$ mm, $b_1 = 250$ mm, $d_1 = 450$ mm

$$V_u = 30000 \text{ N}, M_u = 180 \times 10^6 \text{ N-mm}, T_u = 10 \times 10^6 \text{ N-mm}$$
$$f_{ck} = 20 \text{ N/mm}^2, f_y = 415 \text{ N/mm}^2$$

$$V_{ue} = 30 + \frac{1.6 \times 10}{0.3} = 83.3 \text{ kN } (83.3 \times 10^3 \text{ N})$$

$V_{ue} = \tau_c bd$, τ_c (from Table 5.1) depends on p_t(% of tensile steel. The design of longitudinal steel shall be based on BM only if $V_{ue} \leq V_{uc}$ or on equivalent M_{ue} if $V_{ue} > V_{uc}$.

Consider that $V_{ue} \leq V_{uc}$, then the longitudinal reinforcement shall be provided for BM only.

The design of section for moment may result in singly or doubly reinforced which may be ascertained by comparing the given BM (M_u) with the $M_{u\,lim}$ of singly reinforced balanced section.

$$M_{u\,lim} = 0.36 f_{ck}\, bx_{u\,lim}\,(d - 0.42 x_{u\,lim}),$$

$$x_{u\,lim} = 0.48d = 0.48 \times 500 = 240 \text{ mm}$$

or $\qquad M_{u\,lim} = 0.36 \times 20 \times 300 \times 240\,(500 - 240 \times 0.42)$

$$= 206.94 \times 10^6 \text{ N-mm} > M_u\,(180 \times 10^6), \text{ hence singly reinforced.}$$

Now $\qquad A_{st} = \dfrac{M_u}{0.87 f_y\,(d - 0.42 x_u)} \qquad$ and $\qquad x_u = \dfrac{0.87\,A_{st}\,f_y}{0.36\,f_{ck}\,b}$

From above two expressions, we get $A_{st} = \dfrac{0.5 f_{ck}}{f_y}\left[1 - \sqrt{1 - \dfrac{4.6 \times M_u}{f_{ck} \times bd^2}}\right] bd$

$$A_{st} = \dfrac{10 \times 300 \times 500}{415}\left[1 - \sqrt{1 - \dfrac{4.6 \times 180 \times 10^6}{20 \times 300 \times 500^2}}\right] = 1195 \text{ mm}^2$$

Provide tension reinforcement 4 bars 20 mm ϕ ($A_{st} = 1256 \text{ mm}^2$)
Provide 2 bars 12 mm in compression face ($A_{sc} = 226 \text{ mm}^2$)

$$\therefore \qquad\qquad x_u = \dfrac{0.87 \times 1195 \times 415}{0.36 \times 20 \times 300} = 199.8 \text{ mm},$$

$$p_t\,(\% \text{ tensile steel})\; p_t = \dfrac{1256 \times 100}{300 \times 500} = 0.84$$

τ_c (Table 5.1) for 0.84% p_t and M20 CC = 0.582 N/mm^2

$$V_{uc} = 0.582 \times 300 \times 500$$

$$= 87.3 \times 10^3 \text{ N} \,(87.3 \text{ kN}) > V_{ue}\,(83.3 \text{ kN}), \text{ OK.}$$

Therefore, the design of longitudinal reinforcement for BM alone is correct. Since V_{uc} (87.3 kN) > V_{ue} (83.3 kN), provide minimum shear reinforcement. Use 8 mm ϕ 2 legged vertical stirrups ($A_{sv} = 100 \text{ mm}^2$). Spacing S_v is given by

$$S_{v\,max} = \dfrac{2.5\,A_{sv}\,f_y}{b} \not> S_{v\,max} = \dfrac{2.5 \times 100 \times 415}{300} = 348.8 \text{ mm}$$

$S_{v\,max}$ = minimum of $\quad$ (i) $\;0.75d = 375$ mm

$$\text{(ii)} \;\left(\dfrac{b_1 + d_1}{4}\right) = \dfrac{(450 + 250)}{4} = 175 \text{ mm}$$

(iii) $\;b_1$ or $d_1 = 250$ mm

(iv) $\;300 = 300$ mm

Thus, provide 8 mm ϕ 2 legged vertical stirrups at a spacing of 175 mm c/c as shown in Fig. 5.5.

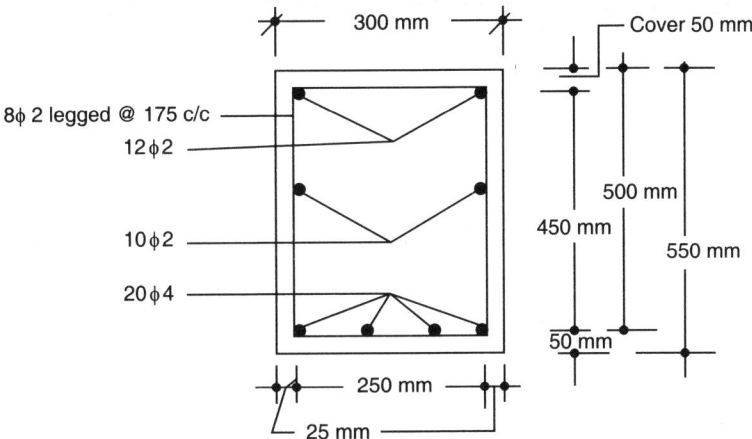

300 mm

Cover 50 mm

8φ 2 legged @ 175 c/c

12φ2

10φ2

20φ4

500 mm

450 mm

550 mm

50 mm

250 mm

25 mm

Fig. 5.5: Details of beam section and reinforcement

Since $d > 450$ mm, provide an additional reinforcement (A_s) of 0.1% which shall be equally distributed along the vertical faces at spacing not exceeding 300 mm or width of the beam whichever is small.

$$A_s = \frac{0.1 \times 300 \times 500}{100} = 150 \text{ mm}^2 \text{ (provide 2 bars of 10 mm diameter } A_s = 157 \text{ mm}^2).$$

Example 5.9

Design a rectangular RC beam section of 300 mm width and 500 mm effective depth which is subjected to an ultimate moment of 50 kN-m, ultimate shear force of 55 kN and ultimate torsional moment of 42 kN-m. Use M20 grade CC and Fe 415 grade steel reinforcement. Consider uniform effective concrete cover of 35 mm on all faces (Fig. 5.6).

Solution: $b = 300$ mm, $d = 500$ mm, $b_1 = 230$ mm, $d_1 = 465$ mm, $f_{ck} = 20$ N/mm^2,

$f_y = 415$ N/mm^2, $V_u = 55$ kN, $M_u = 50$ kN-m, $T_u = 42$ kN-m, $D = 535$ mm

The design of beam is subjected to combined effect of flexure (bending), design shear force and torsion. The design shall depend on whether

$$V_{ue} \le V_{uc}, V_{ue} > V_{uc}, \text{ where } V_{ue} = \left(V_u + \frac{1.6\,T_u}{b}\right) = 55 + \frac{1.6 \times 42}{0.3} = 279 \text{ kN}$$

$V_{uc} = \tau_c.bd$, τ_c depends on % steel in tension (p_t).

The design for longitudinal reinforcement shall be made for ultimate BM only if $V_{ue} \le V_{uc}$ or for equivalent moment M_{ue} if $V_{ue} > V_{uc}$. Consider $V_{ue} > V_{uc}$, then the longitudinal reinforcement shall be provided for equivalent moment M_{ue}.

$$M_{ue} = M_u \pm M_{ut}, \text{ where } M_{ut} = T_u \frac{\left(1 + \dfrac{D}{b}\right)}{1.7} = \frac{42}{1.7}\left(1 + \frac{535}{300}\right) = 68.77 \text{ kN-m}$$

M_u (50 kN-m) $< M_{ut}$ (68.77 kN-m)

$M_{ue1} = 50 + 68.77 = 118.77$ kN-m

$M_{ue2} = 50 - 68.77 = -18.77$ kN-m

The design of section may result in singly reinforced or doubly reinforced. This may be ascertained by comparing the equivalent moment (M_{ue1}) with the ultimate moment of resistance of singly reinforced critical or balanced section ($M_{u\,lim}$).

$$M_{u\,lim} = 0.36 f_{ck}\, b\, x_{u\,lim}\, (d - 0.42 x_{u\,lim}),\ x_{u\,lim} = 0.48 \times 500 = 240\ \text{mm}$$
$$M_{u\,lim} = 0.36 \times 20 \times 300 \times 240\ (500 - 0.42 \times 240)$$
$$= 206.95 \times 10^6\,\text{N-mm} > M_u\ (50 \times 10^6)\ \text{and also} > M_{ue1}\ (118.77\ \text{kN-m})$$

Hence, the section will be singly under reinforced.

Now $\quad A_{st} = \dfrac{M_{ue1}}{0.87\, f_y (d - 0.42 x_u)}$ and $\quad x_u = \dfrac{0.87\, f_y\, A_{st}}{0.36\, f_{ck}\, b}\quad (C_u = T_u)$

From the above two expressions, $A_{st} = \dfrac{0.5 f_{ck}\,(bd)}{f_y}\left[1 - \sqrt{1 - \dfrac{4.6\, M_{ue1}}{f_{ck} \cdot b \cdot d^2}}\right] = 733\ \text{mm}^2$

(3 bars @ 18 ϕ, $A_{st} = 763\ \text{mm}^2$)

$\therefore \qquad x_u = \dfrac{0.87 \times 415 \times 763}{0.36 \times 20 \times 300} = 127.54\ \text{mm}$

Design for M_{ue2} (−18.77 kN-m), $A_s = \dfrac{18.77 \times 10^6}{0.87 \times 415 \times (d - 0.42 \times 127.54)} = 116.5\ \text{mm}^2$

Provide 2 bars of 12 mm ϕ ($A_s = 226\ \text{mm}^2$).

By steel beam theory, $A_s = \dfrac{18.77 \times 10^6}{0.87\, f_y\,(d - d_1)} = \dfrac{18.77 \times 10^6}{0.87 \times 415\,(500 - 35)} = 112\ \text{mm}^2$

($d_1 = 35$ mm given)

Thus, provide 2 bars of 12 mm ϕ ($A_s = 226\ \text{mm}^2$) at top face.

$$p_t = \left(\dfrac{76300}{300 \times 500}\right) = 0.5087\%\ \text{(from Table 5.1), } \tau_c = 0.4828\ \text{N/mm}^2,$$

$$V_{uc} = 0.4828 \times 300 \times 500 = 72.42 \times 10^3\,\text{N}$$

$\tau_{c\,max}$ (from Table 5.3) = 2.80 N/mm^2, $V_{uc\,(max)} = 420000\ \text{N} > 272000\ \text{N}$

$V_{uc} < V_{ue} < V_{uc\,max}$, i.e. (72.42 kN) < (279 kN) < (420 kN)

Shear reinforcement is necessary ($\because V_{ue} > V_{uc}$) and shear reinforcement will be provided for the equivalent shear force (V_{ue}). The longitudinal bars will be provided for the equivalent moment (M_{ue}). Consider shear reinforcement of 10 mm ϕ stirrups ($A_{sv} = 157\ \text{mm}^2$) and spacing S_v given as follows:

$$S_v = \text{smaller of}\ \dfrac{0.87\, f_y \cdot A_{sv} \cdot b_1 \cdot d_1}{(T_u + 0.4\, V_u\, b_1)}\ \text{or}\ \dfrac{0.87\, f_y \cdot A_{sv} \cdot d}{(V_{ue} - V_{uc})} \not> S_{v\,max}$$

where, $\qquad b_1 = 300 - 35 - 35 = 230\ \text{mm}, d_1 = 500 - 35 = 465\ \text{mm}$

$S_{v\,max} = \text{smaller of}\qquad$ (i) $x_1 = 300 - 2 \times 35 + 18 = 248\ \text{mm}\ (x_1)$

$\qquad\qquad$ (ii) $\left(\dfrac{x_1 + y_1}{4}\right) = \dfrac{248 + (465 + 18)}{4} = 237\ \text{mm}$

$\qquad\qquad$ (iii) $0.75d = 0.75 \times 500 = 375\ \text{mm}$

$\qquad\qquad$ (iv) $b = 300\ \text{mm}$

$$\text{Smaller} = 237 \text{ mm}$$

$$\text{Smaller of } \frac{0.87 \times 415 \times 157 \times 230 \times 465}{(42 \times 10^6 + 0.4 \times 55 \times 1000 \times 248)} \text{ or } \frac{0.87 \times 415 \times 157 \times 500}{(279 - 72.42) 10^3}$$

i.e. $S_{v\,max} = 128$ or 137.2, i.e. 128 mm

Thus, provide 10 mm ϕ 2 legged vertical stirrups at 125 mm c/c.

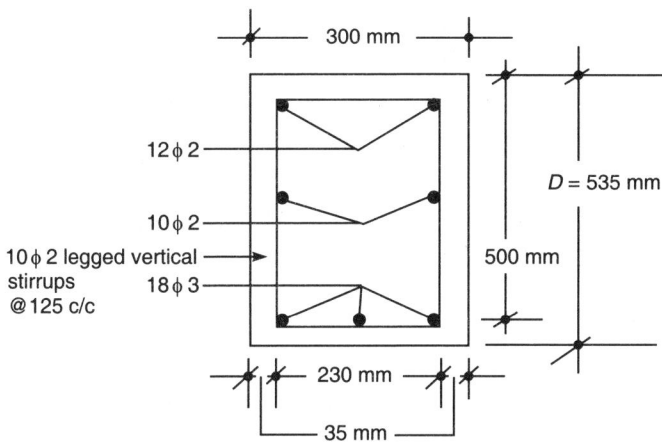

Fig. 5.6: Details of cross-section of beam

5.3 LIMIT STATE DESIGN FOR BOND

5.3.1 Introduction

The concept of reinforcing concrete with steel or other materials was developed on the basis of bonding between steel and cement concrete. When the reinforcing steel bars are embedded in the cement concrete, the concrete adheres to its surface and resist any force that tries to cause slippage of the bar relative to its surrounding concrete. This is achieved by the development of shear resistance at the interface of the bar and concrete. This resistance to slippage at the interface of the bar and surrounding concrete is known as *bond stress*. Bond is the force per unit of nominal surface area of the reinforcing bar acting on the interface between the bar and the surrounding concrete.

5.3.2 Mechanism of Bond Failure

The bond stress comprises 3 components: (i) pure adhesion, (ii) frictional resistance, and (iii) mechanical resistance. The bond resistance of a plain bar is due to the adhesion and friction between concrete and the steel bar. The adhesion between concrete and steel breaks even at low tensile stress causing the slippage of steel bars. With the occurrence of the slip, further bond is developed by friction between the steel bar and the concrete around the bar surface. Bond failure occurs when the bar is pulled overcoming the frictional force and leaving a round hole in concrete. To prevent this slippage, end anchorage in the form of hooks are provided. Modern reinforcing steel bars are manufactured as deformed bars with surface deformations and surface projections (protrusions) causing roughness with higher surface area and increase in mechanical resistance in addition to adhesion and frictional resistance. The interacting force between the deformed bar and the surrounding concrete is shown in Fig. 5.7.

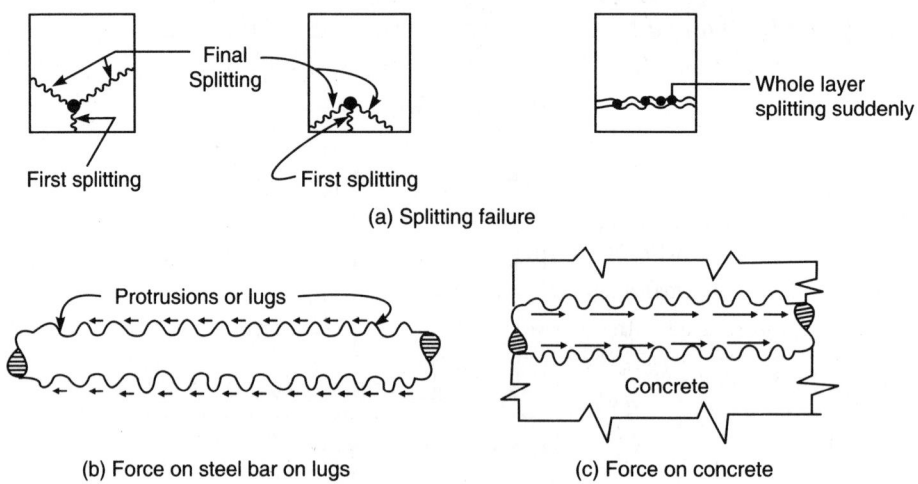

Fig. 5.7: Bond failure and bond force

Concrete beam sections split by first initiating vertical splits and then diagonal splits as shown in Fig. 5.7(a) before final bond failure by pullout of bar. When the area of concrete around the reinforcement is less, failure occurs by splitting as shown in Fig. 5.7(b). When the bar diameters are small and the concrete cover is adequate, the concrete around the lugs get crushed and results in pull out failure Fig. 5.7(c).

5.3.3 Flexural Bond Stress

The shear stress developed at the interface of concrete and reinforcing bar is called *flexural bond stress*. Bond stress is the force per unit of nominal surface area of the reinforcing bar acting on the interface between the bar and surrounding concrete. It develops along the length of the steel bar around the surface so that both steel and concrete act monolithically. This is called flexural bond. The stress developed in anchorage zone at the ends or near the cut off point of bar is known as *anchorage bond* (Fig. 5.8).

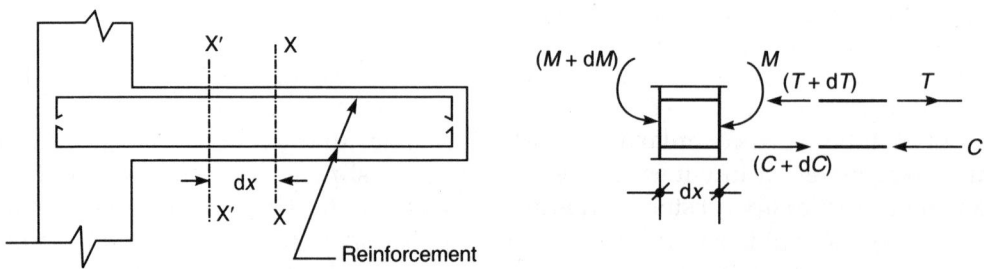

Fig. 5.8: Flexural bond stress

Consider a section XX in a cantilever beam subjected to BM is M. At a distance dx from XX, let the BM = $(M + dM)$. The reinforcing bar is subjected to a tensile force of

$T = \dfrac{M}{jd}$, and at X′X′ $(T + dT) = \left(\dfrac{M + dM}{jd} \right)$. The bar is subjected to unbalanced force

$$(T + dT) - T = \left(\dfrac{M + dM}{jd} \right) - \dfrac{M}{jd}$$

or $$dT = \dfrac{M}{jd}$$

This unbalanced force dT in rod is in internal equilibrium with the bond stress (μ_b) in interface of concrete and reinforcing bar, i.e.

$$\mu_b\,(\pi\phi)\,dx = dT$$

where, μ_b = Bond stress

$(\pi \cdot \phi)$ = Circumference of the bar

dx = Length of bar of strip XX–X′X′

dT = Unbalanced tensile force in bar

$$\mu_b\,\pi\phi\,dx = dT = \dfrac{dM}{jd}$$

or $\mu_b = \dfrac{dM}{jd(\pi\phi) \cdot dx} = \dfrac{dM}{dx(jd\,\pi \cdot \phi)} = \dfrac{V}{jd(\pi \cdot \phi)}$ $\left(\text{since } \dfrac{dM}{dx} = V \right)$, i.e.

$$\mu_b\,(\text{bond stress}) = \dfrac{V}{jd(\pi \cdot \phi)} \qquad \text{... Eq. (5.12)}$$

where, μ_b = Bond stress

$$V = \text{Shear force} \left(\dfrac{dM}{dx} \right)$$

jd = Lever arm of the section

ϕ = Diameter of the bar

$(\pi \cdot \phi)$ = Circumference of the bar

In case beam section provides n number of bars, the total circumference is $\Sigma O = (n\pi\phi)$ and the bond stress μ_b shall be

$$\mu_b = \dfrac{V}{jd\,n\pi \cdot \phi} = \dfrac{V}{jd\,\Sigma O} \qquad \text{... Eq. (5.13)}$$

where, $\Sigma O = n\pi\phi$ (total circumference of reinforcing bars)

Bond stress μ_b can similarly be calculated for bars in compression zone. If the compression reinforcement shares a fraction of moment (α times) by a single bar of diameter ϕ, then the bond stress is given as

$$\mu_b = \dfrac{\alpha V}{\pi\phi\,jd} \qquad \text{... Eq. (5.14)}$$

For a beam provided with n bars of the same diameter in compression then

$$\mu_b = \frac{\alpha V}{jd \, \Sigma O} \qquad\qquad \text{... Eq. (5.14a)}$$

where, $\quad\quad \Sigma O = n\,\pi\phi$ (sum of circumference of all the bars in compression)

α = Fraction of moment shared by compression steel

jd = Lever arm of RC section

V = SF at the section.

5.3.4 Anchorage Bond and Development Length

The bond developed in the anchorage area at the ends of a beam or at the cut off points of a reinforcing bar within the span of a beam causes slippage of the bar at the ends. This type of bond resisting slippage of reinforcing bar is called *anchorage bond*. The load that causes the limiting value of slip from serviceability requirement gives the measure of bond resistance between the concrete and steel. This bond resistance is considered uniformly distributed over a length (L_d) required to transfer the axial force of the bar to the surrounding concrete safely. This length of the bar is known as the *development length*.

The values of the anchorage bond stress for plain and deformed bars in tension and compression are given in Table 5.4.

Table 5.4: Anchorage bond stress for plain and HYSD bars (N/mm²)

Types of bond stress	Nature of stress	Type of reinforcement	Grade of concrete				
			M20	M25	M30	M35	M40
Permissible	Tensile	Plain	0.80	0.90	1.00	1.10	1.20
value		Deformed	1.12	1.26	1.40	1.54	1.68
(N/mm²)	Compressive	Plain	1.00	1.125	1.25	1.375	1.50
	(1.25 times)	Deformed	1.40	1.575	1.75	1.925	2.10
Ultimate	Tensile	Plain	1.20	1.40	1.50	1.70	1.90
value		Deformed	1.92	2.24	2.40	2.72	3.04
(N/mm²)	Compressive	Plain	1.50	1.75	1.875	2.125	2.375
	(1.25 times tensile)	Deformed	2.40	2.80	3.00	3.40	3.80

The reinforcement bar must extend in concrete adequately so that it can develop the required stress in bar or the bar force is transferred to concrete safely without failure (crushing). The extended length of the bar for developing the required stress is known as development length (L_d) as shown in Fig. 5.9.

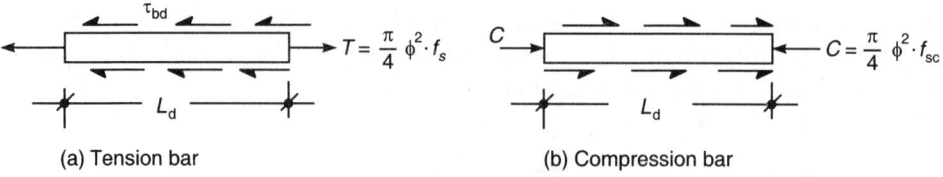

(a) Tension bar $\qquad\qquad\qquad\qquad$ (b) Compression bar

Fig 5.9: Development length (L_d)

It is determined as follows:

$$\text{Total axial force } T = \frac{\pi}{4} \times \phi^2 \times f_{st}$$

Total resistance in length $(L_d) = \tau_{bd} \cdot \pi\phi \cdot L_d$

By equilibrium: $\dfrac{\pi}{4} \phi^2 \cdot f_{st} = \tau_{bd} \cdot \pi\phi \cdot L_d$

or $$L_d = \frac{\pi}{4} \cdot \frac{f_{st}}{\tau_{bd}} = 0.25\phi \cdot f_{st}/\tau_{bd} \qquad \text{... Eq. (5.15)}$$

Similarly for axial compressive force

$$L_d = 0.25\phi f_{sc}/\tau_{bd}, \text{ i.e.} \qquad \text{... Eq. (5.15a)}$$
$$L_d = 0.25\phi f_s/\tau_{bd} \qquad \text{... Eq. (5.15b)}$$

where, L_d = Development length
f_s = Axial tensile or compressive stress
τ_{bd} = Resisting anchorage bond stress
ϕ = Diameter of the reinforcing bar

The development lengths of tension and compression bars based on limit state design method are given in Table 5.5.

$$L_d = \frac{0.25 \, \phi \, f_s}{\tau_{bd}}$$

Table 5.5: Development length of tension and compression bars based on limit state method

Types of bars	Characteristic strength of steel (f_y) N/mm^2	L_d/ϕ							
		Tension bar			Compression bar				
		Concrete grade			Concrete grade				
		M20	M25	M30	M20	M25	M30	M35	M40
Plain MS bar	250	46	39	37	37	32	29	25	23
HYS deformed bar	250	29	25	23	23	20	19	16	15
	415	47	41	38	38	33	31	27	24
	500	57	49	46	46	39	37	32	29

When the available length of the bar is less than the development length required for anchorage, the bar ends can be bent to form hooks. The anchorage value of the hook may be taken as 4 times the diameter for each 45° bend subject to a maximum of 16 times the diameter of the bar. As good construction practice, following specifications must be adopted for reinforcement:

(i) The ends of the stirrups should be bent through an angle of at least 90° around a bar of at least its own diameter and continued beyond the end of the curve for a length of at least 8 times the bar diameter (Fig. 5.10(a)).

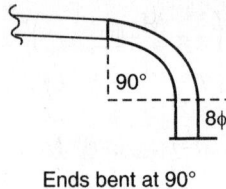

Ends bent at 90°
(continued 8φ)

Fig. 5.10(a): 90° end bend

(ii) The ends of the bar should be bent through an angle of 135° around a bar of at least its own diameter and continued beyond the end of the curve for a length of at least 6 times the diameter of the bar.

(iii) The ends of the bar should be bent through an angle of 180° around a bar of at least its own diameter and continued beyond the end of the curve for a length of at least 4 times the bar diameter.

The development length can be expressed in terms of the flexural bond by limiting the anchorage and flexural bond stresses to their maximum value and equating them, i.e. $\tau_{bd} = \mu$.

Substituting the value of τ_{bd} from Eq. (5.15b) and the value of μ_b from Eq. (5.13) in terms of shear V_u, we get

$$\frac{0.25\,\phi\,f_s}{ld} = \frac{V_u}{jd\,\Sigma O} \qquad\qquad \text{or } l_d = \frac{0.25\phi\cdot f_s\cdot jd\,\Sigma O}{V_u}$$

Substituting $0.25\phi\,\Sigma O = A_{st}$, we have $l_d = \dfrac{A_{st}\cdot f_s\cdot jd}{V_u}$

Substituting $A_{st}\cdot f_s\cdot jd = M_{u\,lim}$, we have $L_d = \dfrac{M_{u\,lim}}{V_u}$ $\qquad$ ... Eq. (5.16)

The ratio of $\dfrac{M_{u\,lim}}{V_u}$ should be larger than the required development length L_d for bond stress τ_{bd} which should not exceed its design value specified. IS code recommends an additional anchorage length equal to ld to ensure the anchorage bond stress to be within its maximum limit.

$$L_d \le \left(\frac{M_{u\,lim}}{V_u} + l_d \right) \qquad\qquad \text{... Eq. (5.17)}$$

where, $\qquad\qquad$ l_d = Embedment length beyond the centre of support
+ equivalent embedment length of any hook or mechanical anchorage at simply supported end
= larger of d or 12ϕ at the point of inflexion ($M_u = 0$)

IS code has also recommended to increase the value of $\left(\dfrac{M_u}{V_u} \right)$ by 30% when the ends of the reinforcement are contained by compressive reaction such as simply supported end.

$\therefore$ $$L_d \le \left(1.3\frac{M_{u\,\text{lim}}}{V_u} + l_d\right)$$ ... Eq. (5.18)

In all designs of reinforcement, it should be ensured that the necessary development length is available at every section. However, critical sections for l_d are the sections of maximum moment, point of curtailment and point of inflexion shown in Fig. 5.10(b).

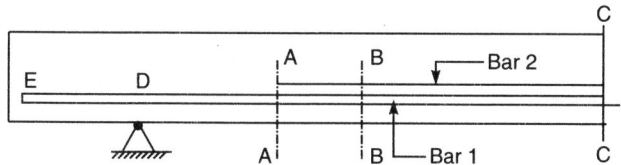

Bar 2: Theoretical curtailment BB and actual curtailment AA.
CA ≥ Ld, BA ≥ d or 12ϕ (whichever is more)

Fig. 5.10(b): Curtailment and development length

5.3.5 Curtailment of Reinforcement

The bending moment in a beam varies along the span, and hence the requirement of tensile steel reinforcement also varies along the span. The design of reinforcement is always based on the maximum moment. Since steel is very costly material, there is no point in providing large quantity of steel wherever it is no more required to bear various loads. It is, therefore, efficient and economical to curtail steel wherever it is not required for any force or other design consideration.

Generally simply supported beam have maximum BM at the centre of span when the beam is subjected to u.d.l. or point load at the centre and hence there is maximum requirement of tensile steel at the centre. At supports of simply supported beam, the bending moment is zero and theoretically requires no tensile steel but practically some reinforcement always continues up to end supports also.

Important principle in reinforcement curtailment is to continue reinforcement beyond theoretical cut off point of any bar. Each bar is continued beyond its cut off point by a curtain distance d or 12ϕ (whichever is more). The reinforcement detailing must be carried out according to the codal provisions to have adequate safety against failures in shear, bond or flexure, etc.

Shear reinforcement must also be adequate to ensure safety against shear, i.e.

$$V_{uc} + V_{us} \ge 1.5V_u$$ ... Eq. (5.19)

where, V_{uc} = Shear capacity of concrete ($\tau_c \cdot bd$)

V_{us} = Shear capacity of shear stirrups (reinforcement)

V_u = Applied shear force (ultimate)

Excess flexural reinforcement together with excess shear capacity is provided at the point of curtailment of main flexural reinforcement.

The following conditions should be satisfied for the curtailment of reinforcement for the positive moment:

(i) At least one third the positive moment reinforcement in simply supported beam and one fourth the positive moment reinforcement in the continuous beam shall extend along the same face of the member into the support to a length equal to $\dfrac{L_d}{3}$.

(ii) At the simply supported end and at the point of inflexion in continuous span, the following conditions should be satisfied:

At simply supported end, $L_d \leq 1.3 \dfrac{M_{u\,lim}}{V_u} + l_d$

At point of inflexion, $L_d \leq \left(\dfrac{M_{u\,lim}}{V_u} + l_d \right)$

For negative moment reinforcement, at least one third of the total reinforcement provided at the support shall extend beyond the point of inflexion for a length not less than the effective depth of the member (d) or 12ϕ or one sixteenth of the clear span $\left(\dfrac{L}{16} \right)$ whichever is greater.

The lap length for flexural tensile reinforcement bar shall be L_d or 30ϕ, whichever is greater. Lap length for bar subjected to direct tension shall be $2L_d$ or 30ϕ, whichever is greater. The straight length of lap shall not be less than 15ϕ or 200 mm, whichever is greater.

Various details of reinforcement can be studied and provided for in actual designs. Some of the basic requirements can be understood by examples.

Example 5.10

A simply supported rectangular beam section of 300 mm width and 500 mm effective depth is reinforced with 3 bars of 20 mmϕ in the mid span section subjected to maximum BM. The beam is subjected to an ultimate shear force of 300 kN at the centre of the end simple support. Considering CC of M20 grade and steel of Fe 415 grade, determine the anchorage length of bars at the simply supported end of the RC beam (Fig. 5.11).

Solution: $f_{ck} = 20\,N/mm^2, f_y = 415\,N/mm^2, b = 300\,mm, d = 500\,mm, R = 550\,mm,$
$V_u = 300\,kN, A_{st} = 3 \times 314 = 942\,mm^2$

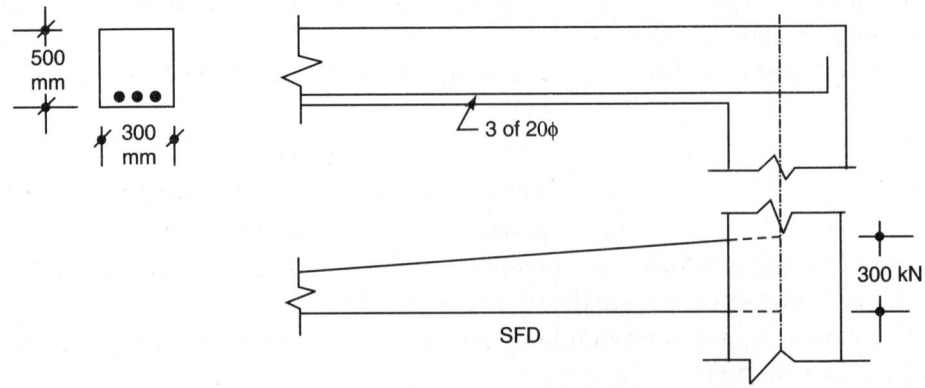

Fig. 5.11

Anchorage length at support,

$$L_d \le \frac{1.3\, M_{u\,\text{lim}}}{V_u} + l_d$$

or $\quad l_d = \left(L_d - \dfrac{1.3\, M_{u\,\text{lim}}}{V_u} \right), \quad L_d = \dfrac{0.87\, f_y\, \phi}{4\, \tau_{bd}} = \dfrac{0.87 \times 415 \times 20}{4 \times 1.92} = 940.2 \text{ mm}$

$\tau_{bd} = 1.92 \text{ N/mm}^2$ (from Table 5.4) for deformed bars

$M_{u\,\text{lim}} = f_s \cdot A_{st}\,(d - 0.42 x_u)$,

$\quad f_s = 0.87 f_y$, if balanced or under reinforced

$\hspace{5cm} < 0.87 f_y$, if the section is over reinforced

$$x_u = \frac{f_s A_{st}}{0.36\, f_{ck} b} \text{ by equating } C \text{ and } T.$$

$$x_u = \frac{0.87 \times 415 \times 942}{0.36 \times 20 \times 300} = 157.46 \text{ mm}$$

$$x_{u\,\text{lim}} = \frac{700 \times d}{1100 + 0.87 \times 415} = 0.48 \times 500 = 240 \text{ mm}$$

$x_u \le x_{u\,\text{lim}}$, hence the section is under reinforced.

$M_{u\,\text{lim}} = 0.87 \times 415 \times 942 \times (500 - 0.42 \times 157.46)$

$\hspace{2cm} = 147.56 \times 10^6 \text{ N-mm (147.56 kN-m)}$

Hence, $l_d \ge \left(L_d - \dfrac{1.3 \times 147.56 \times 10^6}{300 \times 1000} \right) \ge (942.0 - 639.43) = 300.17 \text{ mm}$

The anchorage length of bar shall be not less than the minimum length of bar to be extended beyond the centre of support. The minimum length of bar to be extended beyond the centre of support $= L_d/3 -$ width of support$/2$

$= \dfrac{940.2}{3} - \dfrac{300}{2} = 163.4 \text{ mm} < l_d$ (300.77). Provide anchorage length of 301 mm.

Example 5.11

A continuous beam of 300 mm width and 500 mm effective depth has 3 bars of 20 mm diameter in tension face. The beam has a point of inflexion at certain distance from the support. The SF diagram is shown in Fig. 5.12, which has SF of 300 kN at the point of inflexion. Use M20 grade CC and Fe 415 grade steel (Fig. 5.12).

Solution: $f_{ck} = 20 \text{ N/mm}^2, f_y = 415 \text{ N/mm}^2, b = 300 \text{ mm}, d = 500 \text{ mm}, \phi = 20 \text{ mm}$,

$\quad A_{st} = 3 \times 314 = 942 \text{ mm}^2, \tau_{bd} = 1.92 \text{ N/mm}^2$ (from Table 5.4) for deformed bars and M20 CC.

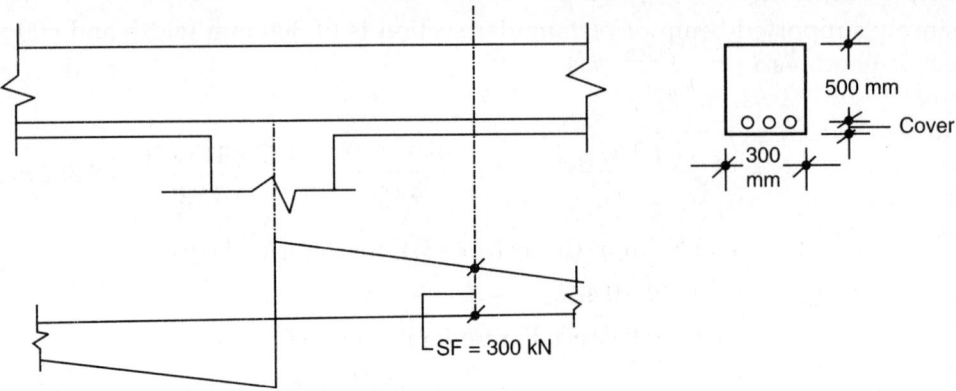

Fig. 5.12: Ultimate SF diagram

$$L_d = \left(\frac{M_{u\,lim}}{V_u} + l_d \right)$$

where,

$$L_d = \frac{0.87\,f_y\phi}{4\,\tau_{bd}}$$

or

$$L_d = \frac{0.87 \times 415 \times 20}{4 \times 1.92} = 940.2 \text{ mm}$$

$$x_u = \frac{f_s \cdot A_{st}}{0.36\,f_{ck} \cdot b} = \frac{0.87 \times 415 \times 942}{0.36 \times 20 \times 300} = 157.46 \text{ mm (equating } C = T)$$

Assuming the sections balanced or under reinforced, $f_s = 0.87 f_y$

$$x_{u\,lim} = \frac{700 \times d}{\left(1100 + 0.87\,f_y\right)} = 0.48 \times 500 = 240 \text{ mm}$$

$x_u < x_{u\,lim}$, the section is under reinforced and $x_u = 157.46$ mm

$M_{u\,lim} = 0.36 f_{ck} \cdot b \cdot x_u (d - 0.42 x_u) = 0.36 \times 20 \times 300 \; x_u(500 - 0.42 x_u)$

$M_{u\,lim} = 2160 \; x_u(500 - 0.42 x_u)$

∴ $M_{u\,lim} = 2160 \times 157.46 \,(500 - 0.42 \times 157.46) = 147.56 \times 10^6$ N-mm

$l_d = d$ or 12ϕ, whichever is greater (i.e. 500 mm or 240 mm)

∴ $l_d = 500$ mm

∴ $\dfrac{M_{u\,lim}}{V_u} + l_d = \dfrac{147.56 \times 10^6}{300 \times 1000} + 500 = (491.9 + 500)$

$$= 991.9 \text{ mm} > L_d \; (= 940.2 \text{ mm})$$

Thus, the bond stress τ_{bd} will be within the safe value of 1.92 N/mm^2

τ_{bd} (actual) $= 1.92 \times 940.2/991.9 = 1.82$ N/mm^2

Example 5.12

A simply supported beam of rectangular section is of 300 mm width and 600 mm effective depth. Effective cover of 40 mm is provided on all sides. The beam uses M20 grade CC and Fe 415 grade steel. The beam carries an ultimate u.d.l. of 80 kN/m including self-weight over the entire span of 8 m c/c of simple supports of 400 mm width. The SF diagram is shown in Fig. 5.13. The beam is reinforced with 7 bars of 28ϕ in tension zone at the mid span point. The beam is also reinforced with 3 bars of 28 mm diameter in the compression zone throughout. The tension zone reinforcement is curtailed and only 4 bars of 28ϕ continue in the support. The beam is provided with 8 mm ϕ 2 legged stirrups at 125 mm c/c spacing in end quarter spans and 8 mm ϕ 2 legged stirrups at 250 mm c/c in the middle half span. Check for shear at section BB, 1.50 m from the support, where 3 tensile bars are curtailed.

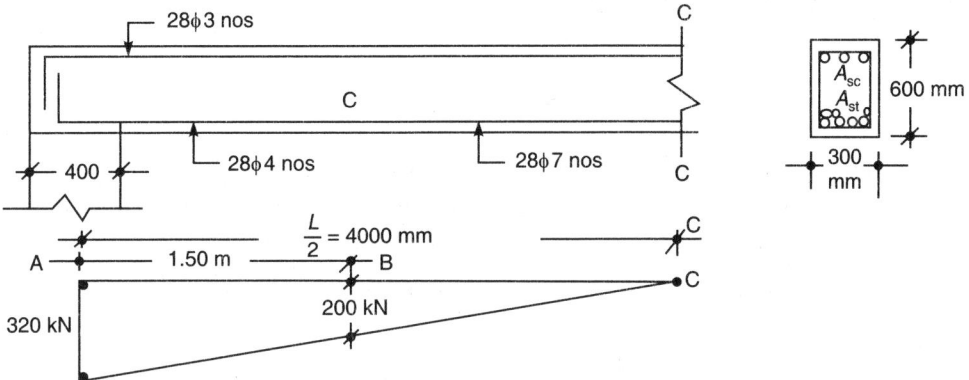

Fig. 5.13: SF diagram

Solution: $f_{ck} = 20 \text{ N/mm}^2, f_y = 415 \text{ N/mm}^2, b = 300 \text{ mm}, d = 600 \text{ mm}, d' = 40 \text{ mm},$
$A_{st1} = 7 \times 615.75 = 4310 \text{ mm}^2, A_{sc} = 3 \times 615.75 = 1847 \text{ mm}^2,$
$A_{st2} = 4 \times 615.75 = 2463 \text{ mm}^2, A_{sc} = 1847 \text{ mm}^2$

V_u (at 1.50 m from the support) = 200 kN

For safety check of shear we have to satisfy the condition that

(i) $(V_{us} + V_{uc}) \geq 1.5 V_u$

(ii) Area of continuing reinforcement ≥ 2 times the reinforcement required to resist the applied moment at 1.50 m.

$$M_u \text{ (max)} = \frac{80 \times 8^2}{8} = 640 \text{ kN-m}$$

$$M_u \text{ (BB at 1.50 m from the support AA)} = 320 \times 1.5 - \frac{80 \times 1.5}{2} = 390 \text{ kN-m}$$

$$p_t = \frac{100 \times 2463}{300 \times 600} = 1.37\%, \tau_{uc} = 0.695 \text{ N/mm}^2 \text{ (Table 5.1)}$$

$$x_{u\,lim} = 0.48d = 0.48 \times 600 = 288 \text{ mm},$$
$$M_{u\,lim} = 0.36 f_{ck} \cdot b \cdot x_{u\,lim} (d - 0.42 x_{u\,lim}) = 298.0 \text{ kN-m} < 640 \text{ kN-m}$$

$$x_u = \frac{0.87\, A_{st} \cdot f_y}{0.36\, f_{ck} \cdot b} \quad \text{(assuming under reinforced singly reinforced)}$$

For doubly reinforced,

$$0.87 f_y \cdot A_{st} = (0.36 f_{ck} \cdot b \cdot x_u + f_{sc} \cdot A_{sc}), \text{ neglecting the loss of concrete due to } A_{sc}$$

$$x_u = \frac{(0.87\, f_y \cdot A_{st} - f_{sc} \cdot A_{sc})}{0.36\, f_{ck} \cdot b} = \frac{0.87\, f_y (A_{st} - A_{sc})}{0.36\, f_{ck} \cdot b}$$

$$= \frac{0.87 \times 415\,(2463 - 1847)}{0.36 \times 20 \times 300} = 103 \text{ mm} < 288 \text{ mm}$$

($f_{sc} = 0.87 f_y$ maximum, exact f_{sc} depends on x_u and can be found by trial, hence the section is under reinforced)

V_{uc} (shear capacity of concrete) $= \tau_{uc} \cdot b \cdot d = 0.695 \times 300 \times 600 = 125.1 \times 10^3 \text{ N}$

$$V_{us} \text{ (stirrups capacity)} = 0.87 \times 415 \times A_{sv} \times \frac{600}{125} = 173.34 \times 10^3 \text{ N}$$

Total capacity $V_{uc} + V_{us} = (125.1 + 173.34) \text{ kN} = 298.44 \text{ kN}$

$1.5 V_u = 1.5\,(200) = 300 \text{ kN} > 298.44 \text{ kN}$, thus just unsafe.

Since $M_{u\,lim}$ (298 kN-m) is less than $M_{u\,max}$, the section is doubly reinforced.

Steel A_{st2} required at 1.5 m from the support section, we have total

$$A_{st2}\ (M_u \text{ at } 1.5 \text{ m} = 390 \text{ kN-m}) = \frac{M_{u\,lim}}{0.87\, f_y (d - 0.42 x_{u\,lim})} + \frac{M_u - M_{u\,lim}}{0.87\, f_y (d - d')}$$

$$= \frac{298 \times 10^6}{0.87 \times 415\,(600 - 0.42 \times 288)} + \frac{(390 - 298)\,10^6}{0.87 \times 415\,(600 - 40)}$$

$$= (1723 + 455) \text{ mm}^2 = 2178 \text{ mm}^2$$

Area of continuing bars 2463 mm^2 < 2 × 2178 (4356 mm^2).

Thus, this condition is not satisfied and there is need to provide extra shear stirrups in 0.75 × 600 = 450 mm distance from the cut off point.

$$\therefore \text{ Spacing of extra stirrups} = \frac{2.5\, f_y}{b} \cdot A_{sv} \ngtr 0.125 d / \beta_b$$

$$\beta_b = \frac{\text{Area of longitudinal bars curtailed}}{\text{Total area of bars at the section}} = \frac{3 \times 612.75}{7 \times 612.75} = 0.43$$

$$\therefore \qquad S_v = 2.5 \times 415 \times \frac{100}{300} \ngtr 0.125 \times \frac{600}{0.43}$$

or $\qquad S_v = 346 \ngtr 174.5$ mm, thus continue extra 8 mm ϕ 2 legged vertical stirrups @ 125 c/c in 0.75 × 600 = 450 mm length from the cut off point.

SUMMARY

Flexural member designs need to be checked for shear forces in addition to design for bending moments. Vertical or inclined shear reinforcements in the form of stirrups may be required. Shear failures may occur due to diagonal tension, diagonal compression, flexural shear and shear bond failure.

Nominal shear stress $\quad \tau_v = \dfrac{V_u}{bd}$ (with usual notations)

Design shear strength of concrete (τ_c) depends on the percentage of tensile steel and the grade of CC. This can be evaluated from Table 5.1 (Table 19 of IS:456-2000). Design shear strength of concrete in slabs is taken as $k\tau_c$, where k can be found from Table 5.2 (Clause 40.2.1.1 of IS:456-2000). The maximum shear stress in concrete ($\tau_{c\,max}$) should not exceed the values given in Table 5.3 (Table 20 of IS:456-2000) and if τ_v exceeds $\tau_{c\,max}$, the section needs to be redesigned with higher cross-section ($b \times d$).

Shear reinforcement is designed when the shear capacity of concrete (V_{uc}) is less than V_u (total design SF). Shear resistance required of shear reinforcement (V_{us}) = ($V_u - V_{uc}$)

Spacing of vertical shear stirrups, $S_v = \dfrac{0.87\, f_y \cdot A_{sv}\, d}{V_{us}}$

For inclined bent up bars, shear resistance

$$V_{us} = 0.87 f_y A_{sv} \cdot \sin \alpha \text{ (with usual notations)}$$

Sometimes RCC elements are subjected to torsion in addition to BM and SF depending on the loading and shape of the structure (member). Eccentric loads may also cause torsion. Members subjected to torsion and shear are to be designed for an equivalent shear force (V_{ue}), which may be found by the following relation:

$$V_{ue} = V_u + 1.6 \left(\frac{T_u}{b} \right)$$

Thus, equivalent nominal shear stress is expressed as:

$$\tau_{ve} = \left(\frac{V_u + 1.6\dfrac{T_u}{b}}{bd} \right) = \left(\frac{V_u}{bd} + \frac{1.6\, T_u}{b^2 d} \right)$$

It should be ensured that τ_{ve} lies between τ_c and $\tau_{c\,max}$.

The design is similar to nominal shear stress except τ_{ve} replaces τ_v.

In case of BM (M_u) acting with torsional moment (M_t), equivalent bending moment (M_e) is calculated as

$$M_e = M_u \pm M_t$$

where, $\quad M_t - \dfrac{T_u}{1.7} \left(1 + \dfrac{D}{b} \right)$

Longitudinal reinforcement is designed to resist equivalent moment ($M_e = M_u \pm M_t$).

If M_t is more than M_u, then the equivalent moments are
$M_{e1} = M_u + M_t$ and $M_{e2} = (M_u - M_t)$, of opposite nature.

Two legged vertical stirrups area (A_{sv}) should be calculated as

$$A_{sv} = \frac{T_u \, S_v}{b_1 \, d_1 \, (0.87 f_y)} + \frac{V_u \, S_v}{2.5 \, d_1 \, (0.87 f_y)}$$

Total transverse reinforcement should not be less than

$$A_{sv} = \frac{(\tau_{ve} - \tau_c) \, b S_v}{0.87 f_y} , \text{ or } S_v = \frac{0.87 \, f_y \, A_{sv}}{(\tau_{ve} - \tau_c) b}$$

Transverse reinforcement shall consist of vertical stirrups with spacing S_v given as smaller of

$$S_v = \frac{0.87 \, f_y \, A_{sv} \, b_1 \, d_1}{(T_u + 0.4 V_u \, b_1)} \quad \text{or} \quad \frac{0.87 \, f_y \, A_{sv} \, d_1}{(V_{ue} - V_{uc})} \not> S_{v\,max}$$

where, $S_{v\,max} \le$ shorter dimension of stirrups $(b_1$ or $d_1)$

$$\le \frac{1}{4} \left(\text{shorter} + \text{longer dimension of stirrup, i.e. } \frac{b_1 + d_1}{4} \right)$$

$$\le 0.75d$$

$$\le 300 \text{ mm}$$

Bond of concrete with steel bar is important to develop axial force in the reinforcement without slippage. Bond of concrete with steel comprises three components:

(i) pure adhesion (ii) frictional resistance and (iii) mechanical resistance

Flexural bond stress develops along the beam span while anchorage bond stress develops near the supports.

Bond stress in beam $\mu_b = \dfrac{V}{jd(\pi\phi)n} = \dfrac{V}{jd \, \Sigma O}$

In compression zone (with usual notations) $\mu_b = \dfrac{\alpha \cdot V}{jd \, \Sigma O}$

Permissible and ultimate bond stresses are given in the Table 5.4.

Development length,

$$L_d = \frac{\phi}{4} \cdot \frac{f_{st}}{\tau_{bd}}, \ (\phi = \text{diameter of the bar})$$

Development length can be found for tension or compression from the Table 5.5 for different grades of CC.

$$L_d = \frac{M_{u\,lim}}{V_u}$$

$$L_d \le \left(\frac{M_{u\,lim}}{V_u} + l_d \right)$$

where, $\qquad$ l_d = Embedment length beyond the support centre.

Also in compression $L_d \leq 1.3 \dfrac{M_{u\,lim}}{V_u} + l_d$

At the point of curtailment of reinforcement, we have to ensure that
$$V_{uc} + V_{us} \geq 1.5V_u$$
Also ensure that at the point of curtailment, the area of continuing reinforcement is greater than 2 × area of reinforcement required to resist the applied moment at the point.

PRACTICE QUESTIONS

(I) Objective Questions

Q. 5.1 Select the correct response given after each statement to complete it correctly and fill in the response sheet provided.

(i) A simply supported rectangular RC beam section is designed for a maximum ultimate BM at the centre. The section also needs to be checked for design shear stress at ....

    (a) middle point of span only

    (b) quarter span points only

    (c) both middle and quarter span points

    (d) near simple support points.

(ii) Shear stress capacity of SS beam section cannot be enhanced by ....

    (a) simply providing higher grade of CC in the beam section

    (b) simply providing higher grade of steel in straight longitudinal bars in compression face.

    (c) enhancing the cross-sectional area of the beam section

    (d) enhancing the cross-sectional depth of the beam section.

(iii) Which type of failure does not belong to the shear failure?

    (a) bond failure

    (b) diagonal compressive stress

    (c) diagonal tensile stress

    (d) crushing under direct load

(iv) Nominal shear stress in a rectangular beam section is an average shear stress and is given as ....

    (a) $\dfrac{\text{Ultimate shear force}}{\text{Surface area of reinforcement}}$

    (b) $\dfrac{\text{Ultimate BM}}{\text{Cross-sectional area}}$

    (c) $\dfrac{\text{Ultimate shear force}}{\text{Cross-sectional area}}$

    (d) $\dfrac{\text{Ultimate BM} + \text{ultimate SF} \times d}{\text{Moment of inertia}}$

(v) In a beam having rectangular cross-section, the nominal shear stress at a point is given as .... with usual notations.

(a) $\tau_v = \dfrac{V_u}{bd}$

(b) $\tau_v = \dfrac{V_u}{bd^2}$

(c) $\tau_v = \dfrac{\text{Maximum ultimate SF } (V_u)}{\dfrac{\pi}{4}(d^2)}$

(d) $\tau_v = \dfrac{\text{Maximum ultimate SF}}{b\pi\, d^2}$

(vi) Design shear strength in a RC beam section increases with ....
   (a) increase in spacing of stirrups
   (b) decrease in grade of concrete
   (c) decrease in percentage of tensile steel
   (d) increase in percentage of tensile steel

(vii) The section of a beam is redesigned for shear when the nominal shear stress ....
   (a) exceeds the design shear stress for the given section and grade of CC
   (b) exceeds the value of $\tau_{c\,max}$ for the given grade of CC
   (c) exceeds the value of characteristic strength of CC (f_{ck})
   (d) is less than the yield strength of steel and more than the f_{ck} of CC

(viii) Vertical shear stirrups (area A_{sv}) are suitably spaced at S_v given by ....

(a) $S_v = \dfrac{0.87 f_{ck} \cdot A_{sv} \cdot d}{V_{us}}$, where $V_{us} = (V_u - V_{uc})$

(b) $S_v = \dfrac{0.87 f_y \cdot A_{sv} \cdot d}{V_{uc}}$, where $V_{us} = \tau_c \cdot bd$

(c) $S_v = \dfrac{0.87 f_y \cdot A_{sv} \cdot d}{V_{us}}$, where $V_{us} = (V_u - V_{uc})$

(d) $S_v = \dfrac{0.36 f_{ck} \cdot b \cdot d}{0.87 f_y \phi}$, where ϕ = diameter of stirrups bar

(ix) The shear capacity (V_{us}) of bent up inclined bars (area A_{sv}) with usual notations is given by ....
   (a) $(V_{us} - V_{uc}) = 0.87 f_y \cdot d \cdot \sin \alpha$

   (b) $V_{us} = \dfrac{0.36 f_{ck} \, d \, A_{sv}}{\sin \alpha}$

(c) $V_{us} = 0.87 f_y \cdot A_{sv} \cdot \sin \alpha$

(d) $(V_{uc} - V_{us}) = \dfrac{\left(0.87 \, f_y - 0.36 f_{ck}\right) A_{sv}}{\sin \alpha}$

(x) Limit state design for torsion (T_u) in beams is checked for safety by considering equivalent shear (V_e) and equivalent BM (M_e) given by
....

(a) $V_e = \left(V_u + 1.6\dfrac{T_u}{d}\right)$ and $M_e = \left[M_u + 1.6\, T_u\left(1+\dfrac{D}{b}\right)\right]$

(b) $V_e = \left(1.6\, V_u + \dfrac{T_u}{d}\right)$ and $M_e = \left[1.6\, M_u + T_u\left(1+\dfrac{D}{b}\right)\right]$

(c) $V_e = \left(V_u + \dfrac{T_u}{d}\right)$ and $M_e = \left[M_u + \dfrac{T_u}{1.7}\left(1+\dfrac{D}{b}\right)\right]$

(d) $V_e = \left(V_u + 1.6\dfrac{T_u}{b}\right)$ and $M_e = \left[M_u + \dfrac{T_u}{1.7}\left(1+\dfrac{D}{b}\right)\right]$

(xi) For safety against shear we have to ensure that shear stirrups are adequate and satisfy the condition that ....

(a) $(V_{uc} - V_{us}) \geq 1.5 V_u$

(b) $(V_{uc} + V_{us}) \geq 1.5 V_u$

(c) $(V_{uc} + V_{us}) \geq \dfrac{V_u}{1.5}$

(d) $(V_{uc} - V_{us}) \leq \dfrac{V_u}{1.5}$

(xii) Maximum spacing of vertical stirrups $(S_{v\,max})$ is given by ....
Identify the wrong statement/conditions.

(a) $S_{v\,max} \leq$ shorter dimension of stirrups $(b_1$ or $d_1)$

(b) $S_{v\,max} \leq \dfrac{1}{4}$ (Shorter + longer dimension of stirrups)

(c) $S_{v\,max} \leq 0.75d$

(d) $S_{v\,max} \leq 0.50d$

(xiii) If 8 mm diameter 2 legged $(A_{sv} = 100 \text{ mm}^2)$ vertical stirrups of 250 mm × 500 mm size are provided at (S_v) 150 mm c/c to resist torsion (T_u) by using Fe 415 steel, determine T_u when ultimate SF is zero.

(a) $T_u = \dfrac{A_{sv} \cdot b_1 \cdot d_1 \cdot (0.87 f_y)}{S_v} = 30.88 \times 10^6 \text{ N-mm}$

(b) $T_u = \dfrac{d_1 \cdot b_1^3 \cdot (0.87 f_y)}{A_{sv}\, S_v} = 752.188 \times 10^6 \text{ N-mm}$

(c) $T_u = \dfrac{b_1^3 \cdot (0.87 f_y)\, S_v}{A_{sv}} = 135.39 \times 10^6 \text{ N-mm}$

(d) $T_u = \dfrac{\left(0.87 f_y\right) A_{sv}\, S_v}{b_1 d_1} = 43.33 \times 10^6 \text{ N-mm}$

(xiv) Bond resistance in reinforcing bars in concrete is developed due to
....

(a) pure adhesion alone

(b) pure frictional resistance alone

(c) mechanical resistance and compression

(d) adhesion, friction and mechanical resistances

(xv) Bond resistance of HYSD (deformed bars) in compression zone shall be approximately .... the plain bars of the same diameter in tension zone.

(a) equal to   (b) 40% higher than

(c) 75% higher than   (d) 25% lower than

(xvi) Total development length (L_d) of a reinforcing bar (diameter ϕ) to transfer stress f_{st} at any point in a stressed beam to the surrounding concrete (characteristic strength f_{ck}) shall be given by ....

(a) $L_d = \dfrac{0.25\phi f_{st}}{f_{ck}}$

(b) $L_d = \dfrac{0.25 f_{st}\phi}{\tau_{bd}}$

(c) $L_d = \dfrac{4 f_{st}\phi}{f_{ck}}$

(d) $L_d = \dfrac{4 f_{st}\phi}{\tau_{bd}}$

(xvii) Development length (L_d) of a reinforcement bar (ϕ diameter) .... with higher grade of concrete.

(a) decreases   (b) increases

(c) remains the same   (d) none of the above

(xviii) For safety against slippage, each reinforcing bar (diameter ϕ) in a beam of effective dimensions ($b \times d$) should be .... the theoretical cut off point.

(a) cut at a distance of 0.75b beyond

(b) cut at a distance of 5 ϕ beyond

(c) cut at a distance of 'd' or 12 ϕ beyond

(d) cut at a distance of (d + 12 ϕ) before

(xix) With usual notations, the nominal shear stress = ...

(a) $\tau_v = \dfrac{V_{us}}{bd}$, where V_{us} = shear capacity of the stirrups at the section

(b) $\tau_v = \dfrac{V_{uc}}{bD}$, where V_{uc} = shear capacity of concrete at the section

(c) $\tau_v = \dfrac{V_u}{bd}$, where V_u = ultimate shear force at the section

(d) $\tau_v = \dfrac{(V_{uc} + V_{us})}{bD}$, where (V_{uc} + V_{us}) is total shear capacity at the section

(xx) The design of vertical shear stirrups, considering spacing (S_v) is such that $S_v = $ ....

(a) $S_v = \dfrac{f_y A_{sv} d}{V_u}$

(b) $S_v = \dfrac{f_y A_{sv} d}{V_{us}}$

(c) $S_v = \dfrac{0.87 f_y A_{sv} d}{V_{us}}$

(d) $S_v = \dfrac{0.87 f_y A_{sv} b}{V_u}$

Response sheet to Q. 5.1 (i to xx)

Question	(i)	(ii)	(iii)	(iv)	(v)	(vi)	(vii)	(viii)	(ix)	(x)
Response (a/b/c/d)										

Question	(xi)	(xii)	(xiii)	(xiv)	(xv)	(xvi)	(xvii)	(xviii)	(xix)	(xx)
Response (a/b/d/c)										

(II) Numerical Questions

Q. 5.2 A cantilever beam of rectangular section 300 mm width and 600 mm effective depth has 5 bars of 20 mm diameter in top tensile face near the fixed support. The beam section has vertical stirrups of 8 ϕ 2 legged at a spacing of 150 mm c/c near the fixed support. Using M30 grade of CC and Fe 415 grade steel, calculate the ultimate shear resistance of the beam section near the support.

[**Hint:** $\tau_c = 0.62$ N/mm^2 (for M30), $V_u = V_{uc} + V_{us} = (112.3 + 144.4) = 256.7$ kN]

Q. 5.3 A SS RCC rectangular beam section of 300 mm × 500 mm (effective) has 5 bars of 25 mm ϕ in the tensile face at the middle span point. Two bars are bent up near the support at 45° with the horizontal, 8 mm ϕ 2 legged vertical stirrups are provided at 120 mm c/c near the support. Use M20 grade CC and Fe 415 grade steel. Determine the total shear resistance.

[**Hint:** $\tau_c = 0.636$ N/mm^2, $V_u = V_{uc} + V_{us} + V_{ur} = (95.25 + 301.24 + 250.74) = 647.23$ kN]

Q. 5.4 A simply supported rectangular beam section of 300 mm × 550 mm effective depth is reinforced with 5 bars of 22 mm ϕ. The beam is subjected to a service shear force of (a) 60 kN (b) 90 kN (c) 200 kN. Design the shear reinforcement for the beam section if concrete of M20 grade and steel of Fe 415 grade are used.

[**Hint:** $\tau_c = 0.65$ N/mm^2, $V_{uc} = 107.25$ kN, $V_u = 1.5 \times 60 = 90$ kN ($\ngtr V_{uc}$, provide minimum A_{sv})]

(a) $S_v = \dfrac{2.5\,A_{sc}f_y}{b}$, $0.75d$, 450, 8 mm ϕ 2 legged @ 345 c/c.

(b) $V_u = 1.5 \times 90 = 135$ kN > 107.25 kN, $S_v = 715$ mm, 412.5 mm, 345 mm, 450 mm, 8ϕ 2 legged @ 345 mm c/c

(c) $V_u = 1.5 \times 200 = 300$ kN $< V_{uc\,max}$ (462 kN) and > 107.25 kN

$V_{us} = (300 - 107.25)$ kN,

$$S_v\,(10\phi\ 2\ \text{legged}) = \frac{0.87 \times 415 \times 157 \times 550}{(300 - 107.25) \times 10^3} = 161.7\ \text{mm c/c}$$

10ϕ 2 legged @ 161.7 mm, 450 mm, 412.5 mm, 345 mm c/c.

Thus, provide 10ϕ 2 legged @ 160 c/c.]

Q. 5.5 A T-beam has flange width $b_f = 1600$ mm, flange thickness $D_f = 100$ mm, effective depth $d = 500$ mm, web width $(b_w) = 300$ mm, CC M20 grade, and steel Fe 415 grade. The beam is reinforced with 6 bars of 25 mm in tension zone and 2 bars of 25 mm ϕ in compression zone. Design the shear reinforcement if the beam is subjected to a service SF of 210 kN.

[Hint: $V_u = 1.50 \times 210 = 315$ kN, $p_t = \dfrac{6 \times 491 \times 100}{300 \times 500} = 1.964\%$, $\tau_c = 0.784$ N/mm^2,

$\tau_{c\,max} = 2.8$ N/mm^2, $V_{uc} = 0.784 \times 300 \times 500 = 117.6 \times 10^3$ N, $V_{uc\,max} = 2.8 \times 300 \times 500 = 420 \times 10^3$ N $> 315 \times 10^3$ N $> 117.6 \times 10^3$ N,

$\tau_v = \dfrac{315 \times 10^3}{300 \times 500} = 2.1$ N/mm$^2 > 0.784$ N/mm^2,

provide 10ϕ 2 legged @ $S_v = 140$ mm c/c, $S_{v\,max} = \dfrac{2.5\,f_y\,A_{sc}}{b}$, $0.75d$, 450 c/c]

Q. 5.6 A simply supported beam of 8 m effective span c/c carries a service u.d.l. of 30 kN/m including self-loads. The beam section is rectangular 300 mm × 500 mm overall with 6 bars 22ϕ in tension zone at an effective cover of 40 mm. The beam has 2 bars of 22ϕ in compression zone at an effective cover of 40 mm. Use M30 grade CC and Fe 415 grade steel. Design the shear reinforcement for the maximum shear force if 2 of tensile bars are bent up an angle of 45° with the horizontal near the support.

[Hint: $V_u = 180$ kN, p_t (remaining) $= 0.905\%$, $\tau_c = 0.6334$ N/mm^2 (for M30 CC)

$V_{uc} = 106.4$ kN, V_{uR} (2 bars at 45°) $= 0.87 \times 415 \times (2 \times 380) \sin 45° = 194$ kN, total shear resistance $= 300.4$ kN $> V_u$ (180 kN). Provide minimum 8ϕ 2 legged at 300 mm c/c (S_v max $= 345$, 450, 420 mm c/c)]

Q. 5.7 A RC beam of rectangular section has 300 mm width and overall depth of 650 mm reinforced with one 28 mm diameter ($A_{st} = 615.4$ mm^2) in each corner at an effective cover of 50 mm in vertical direction and side effective cover of 25 mm in the direction of width. Two legged stirrups of 8 mm ϕ are provided at a spacing (S_v) of 150 mm c/c. Determine the torsional strength of the section using Fe 415 grade steel when (a) $V_u = 0$, (b) $V_u = 200$ kN.

[Hint: (a) $V_u = 0$, $T_u = 33.096$ kN-m (b) $V_u = 200$ kN, $T_u = 13.096$ kN-m]

Q. 5.8 A RC rectangular beam section 300 mm × 650 mm overall size is reinforced with 4 bars of 20 mm in tension zone. The section is subjected to a factored BM of 160 kN-m. If M20 grade CC and Fe 415 grade steel are used, find the maximum ultimate torsional resistance that can be allowed. Effective side cover is 25 mm and vertical cover is 50 mm.

[Hint: $x_{u\,lim} = 240$ mm, $x_u = \dfrac{0.87 f_y A_{st}}{0.36 f_{ck} \cdot bd} = 210$ mm $< x_{u\,lim}$, under reinforced

$$M_e = 0.87 f_y \cdot A_{st} \cdot d \left(1 - \dfrac{A_{st} f_y}{bd\, f_{ck}}\right) = 232.7 \times 10^6\,\text{N-mm}$$

$$M_e = M_u + M_t = M_u + \dfrac{T_u}{1.7}\left(1 + \dfrac{D}{b}\right),$$

$$232.7 \times 10^6 = 160 \times 10^6 + \dfrac{T_u}{1.7}\left(1 + \dfrac{650}{300}\right) = 160 \times 10^6 + 1.863 T_u$$

$T_u = 39.023 \times 10^6\,\text{N-mm}]$

Q. 5.9 Design a rectangular beam section 300 mm width and 600 mm overall depth subjected to ultimate BM of 200 kN-m, ultimate SF of 35 kN and ultimate torsional moment of 12 kN-m. Use M20 grade CC and Fe 415 grade steel. Assume 50 mm effective cover in the direction of depth and 30 mm effective cover in the direction of width.

[Hint: $V_u = 35000$ N, $M_u = 200 \times 10^6$ N-mm, $T_u = 12 \times 10^6$ N-mm, $f_{ck} = 20$ N/mm², $f_y = 415$ N/mm², $V_{ue} = 99000$ N, $V_{uc} > V_{ue}$ (assumed), $x_{u\,lim} = 0.48d = 264$ mm, $M_{u\,lim} = 0.36 f_{ck} \cdot b \cdot x_{u\,lim}\,(d - 0.42 x_{u\,lim}) = 250.4 \times 10^6 > M_u$ (200×10^6), hence under reinforced

$$A_{st} = \dfrac{0.5 f_{ck}}{f_y}\left[1 - \sqrt{1 - \dfrac{4.6\,M_u}{f_{ck} \cdot bd^2}}\right] bd = 1184\,\text{mm}^2\,\text{(provide 4 bars 20}\phi\text{)}$$

$A_{st} = 1256$ mm², $p_t = \dfrac{125600}{300 \times 550} = 0.761$, τ_c (from Table 5.1) $= 0.56$ N/mm²,

$$V_{uc} = 92400\,\text{N}$$

$x_u = \dfrac{0.87 f_y \cdot A_{st}}{0.36 f_{ck} \cdot b} = 210$ mm < 264, under reinforced (assumption is OK).

Longitudinal bars based on BM.

Since V_{ue} (99000 N) $> V_{uc}$ (92400), provide vertical stirrups 8ϕ 2 legged @ 180 mm c/c (minimum of 345 mm, 412 mm, 185 mm, 240 mm, 300 mm).

Since $d > 450$, provide additional bars, and $A_s = \dfrac{0.1}{100} \times 300 \times 550 = 165$ mm²

Provide 2 bars 12ϕ in the middle ($A_o = 226$ mm²) on sides.]

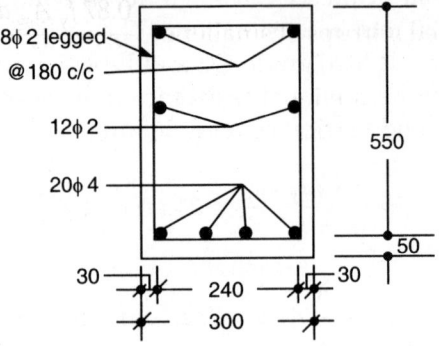

Fig. 5.14

Q. 5.10 Design a rectangular RC beam section of 300 mm width and 500 mm effective depth which is subjected to an ultimate moment of 55 kN-m, ultimate shear force of 60 kN and ultimate torsional moment of 40 kN-m. Use M20 grade CC and Fe 415 grade steel reinforcement. Consider uniform effective concrete cover of 40 mm.

[Hint: $\qquad V_u = 60000$ N, $M_u = 55 \times 10^6$ N-mm, $T_u = 40 \times 10^6$ N-mm,

$$V_{ue} = V_u + \frac{1.6\,T_u}{b} = 273.3 \times 10^3\,\text{N}, \; M_{ue} = M_u \pm M_t$$

$$= 55 \times 10^6 \pm \frac{40 \times 10^6}{1.7}\left(1 + \frac{D}{b}\right) = (55 \pm 66)10^6$$

$M_{e1} = 121 \times 10^6$ N-mm, $M_{e2} = -11 \times 10^6$ N-mm, $x_{u\,lim} = 240$ mm
$M_{u\,lim} = 0.36 f_{ck} \cdot b \cdot x_{u\,lim}\,(d - 0.42 x_{u\,lim}) = 207 \times 10^6 > M_u$, under reinforced

$$A_{st1} = \frac{M_{e1}}{0.87\,f_y\,(d - 0.42\,x_u)}, \; x_u = \frac{0.87\,f_y \cdot A_{st1}}{0.36\,f_{ck} \cdot b},$$

$$A_{st} = \frac{0.5 f_{ck} \cdot bd}{f_y}\left[1 - \sqrt{1 - \frac{4.6\,M_{e1}}{f_{ck} \cdot bd^2}}\right]$$

$\qquad = 748$ mm^2 (4 bars 16ϕ = 804 mm^2).

$$x_u = \frac{0.87 \times 415 \times 804}{0.36 \times 20 \times 300} = 134\;\text{mm} < x_{u\,lim}, \text{under reinforced}$$

$$M_{e2} = \frac{-11 \times 10^6}{0.87 \times 415 \times (500 - 0.42 \times 134)} = 69\;\text{mm}^2 \;(\text{provide } 10\phi\,2,\, A_s = 157\;\text{mm}^2)$$

$$p_{t1} = \frac{804 \times 100}{300 \times 500} = 0.54,\; \therefore\; \tau_c\,(\text{from Table 5.1}) = 0.493\;\text{N/mm}^2,\; V_{uc} = 74000\;\text{N}$$

$\qquad V_{uc\,max} = 2.8 \times 300 \times 500 = 420000$ N, thus $V_{ue} > V_{uc} < V_{uc\,max}$ (420 kN)

Provide 10ϕ 2 legged stirrups at smaller of $\dfrac{0.87\,f_y\,A_{sv}\,d}{(V_{ue}-V_{uc})}$ and $\dfrac{0.87\,f_y\,A_{sv}\,b_1\,d_1}{(T_u+0.4\,V_u\cdot b_1)}$

or $S_{v\,max}\,x_1,\ \dfrac{x_1+y_1}{4}$, $0.75d$, 300, i.e. 236, 178, 375, 300, 141, 127.

Thus, provide 10ϕ 2 legged @ 125 mm c/c and 2 bars 12ϕ near NA zone.]

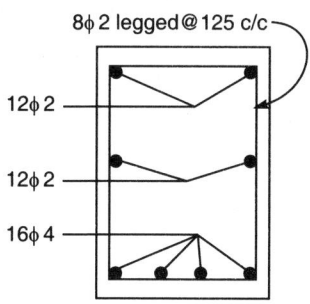

Fig. 5.15

Chapter 6

Limit States of Serviceability for Bending Elements

LEARNING OBJECTIVES

After the study of *limit states of serviceability*, the learner will understand the basic principles of serviceability in bending elements and will be able to:

- State various limits of serviceability for bending elements
- Explain limit state of deflection in bending elements
- Explain design of bending elements for deflection limits
- Explain the deflection control in bending elements
- Explain the serviceability limit state for cracks
- Explain the serviceability limit state for lateral stability

6.1 INTRODUCTION

In the limit state design method, a RCC beam element has to satisfy a number of serviceability limit states at working loads apart from satisfying the limit state of collapse. We have already discussed the important limit state of collapse under different loads and shall consider limit state of deflection at service loads. The important serviceability limits of any bending element (beam and slab) are those of deflection and cracks. Excessive deflection can result in crushing of partition walls, droppings of overhang balconies, sagging of floors, etc. Excessive cracking may result in corrosion of steel reinforcement leading to bond failure and spoiling the appearance of the structure. The two important serviceability limit states are:

(a) Limit state of deflection in bending elements

(b) Limit state of cracking in structural bending element under service loads. The width of cracks should remain within certain limits prescribed by IS codes

(c) Other limit states of serviceability include durability, vibrations and stability, etc.

The Indian standard code IS:456-2000 has prescribed partial safety factors for various combinations of loads for checking deflection and cracking limits. The largest combination should be used for design of structural elements.

The use of high grade steel (like Fe 415, Fe 500) and high strength concrete (like M30, M40, M50) generally results in slender or thin structural elements. In such a case, the limit state of serviceability becomes more critical.

6.2 LIMIT STATES FOR DEFLECTION

The limit states for deflection are specified empirically or theoretically.

6.2.1 Empirical Approach

In empirical approach, span/depth ratios are specified as a measure of deflection control for different structural elements. IS:456-2000, in its Clause 23.2(page 37–38) specifies span/depth ratios for different structural elements which are reproduced here. According to this clause, the deflection limits are satisfied when:

(a) The final deflection including the effects of all loads, temperature, creep and shrinkage of horizontal bending elements shall not exceed the value of $\dfrac{\text{span}}{250}$ (measured from the as cast level of the supports).

(b) The deflection including the effects of temperature, creep and shrinkage occurring after erection of partitions, and the application of finishes (dead loads) shall not exceed the value of $\dfrac{\text{span}}{350}$ or 20 mm.

According to IS:456-2000, the basic span/depth ratios for controlling deflection limit state in structural concrete bending members, are specified in Table 6.1.

Table 6.1: Basic span to effective depth ratios for beam and slabs
(Clause 23.2, IS:456-2000)

Type of support/beam	Rectangular section	Flanged sections
Cantilever	07	Multiply values for rectangular sections by factor K_f shown in Fig. 6.1 for various ratios of b_w and b_f
Simply supported	20	
Continuous	26	

The basic span to effective depth ratios are modified by multiplying with modifying constants (K) dependent on %age of tension steel (K_t) and on %age of compression steel (K_c) as determined from Figs 6.2 and 6.3 respectively. For flanged beam, the %age of tension and compression reinforcement is based on area of section ($b_f d$).

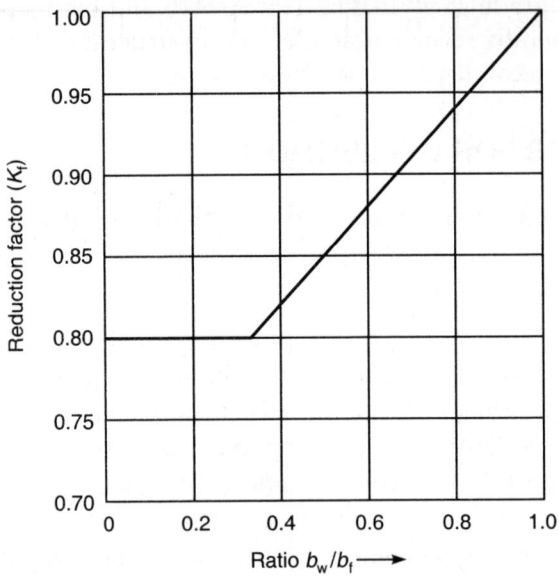

Fig 6.1: Reduction factor for (span/effective) depth ratios K_f

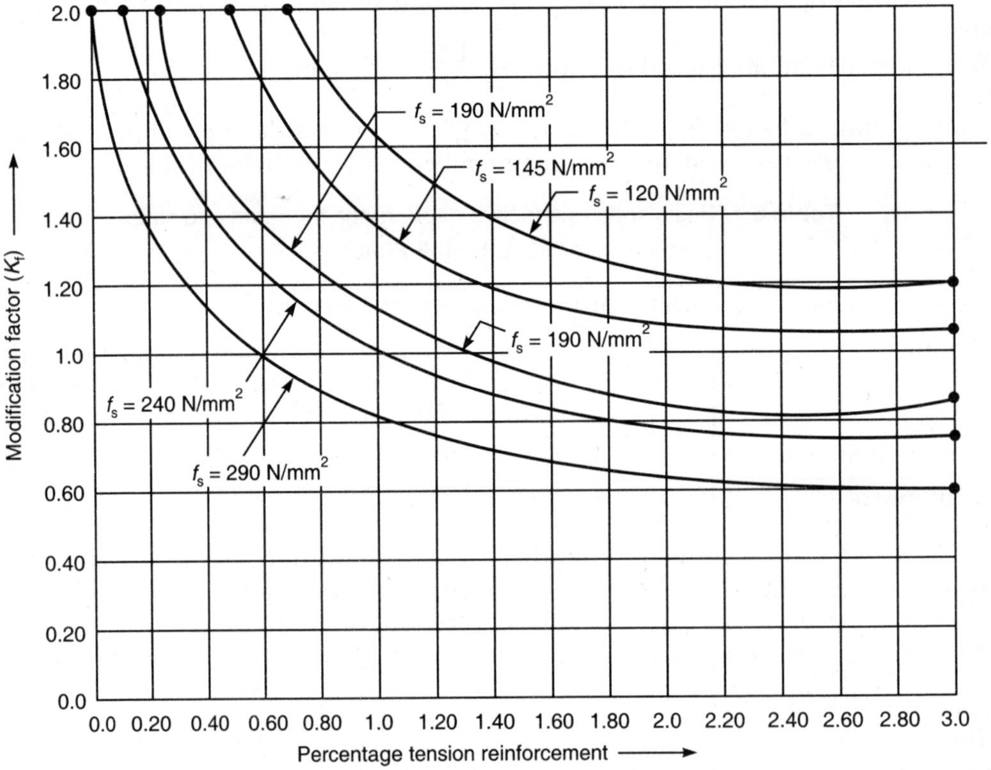

$$f_s = 0.58 f_y \times \frac{\text{area of cross-section of steel required}}{\text{area of cross-section of steel provided}}$$

Fig 6.2: Modification factor (K_t) for %age tension reinforcement

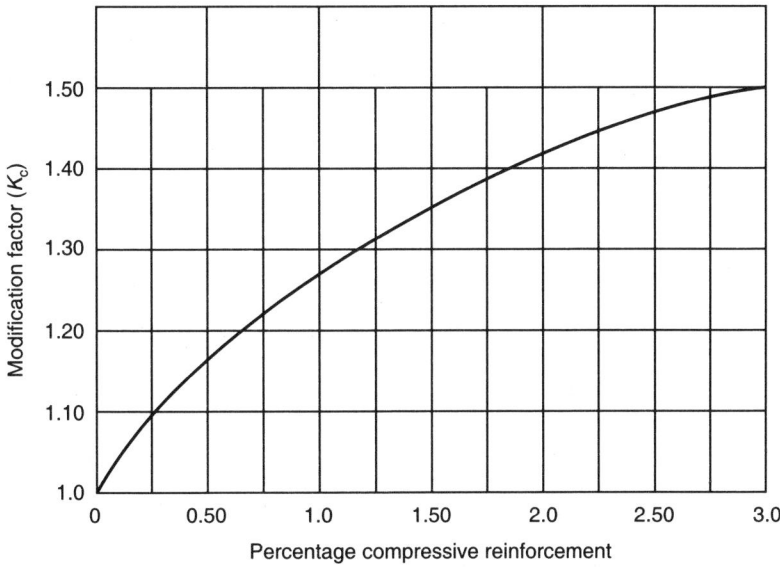

Fig 6.3: Modification factor (K_c) for %age compressive steel

The (span/effective) depth ratios are modified according to modification factors obtained from Fig. 6.1(K_f), Fig. 6.2 (K_t), and Fig. 6.3 (K_c). For rectangular beam section $K_f = 1$, and for flanged beam the values of K_t and K_c are obtained on the basis of %age tensile and %age compressive steels provided.

The modified expression for the (span/effective depth) ratio is specified as

$$\left(\frac{L}{d}\right)_{modified} = \left(\frac{L}{d}\right)_{basic} \times K_t \times K_c \times K_f$$

where, $\left(\dfrac{L}{d}\right)_{basic}$ values are given in Table 6.1

K_t = Modification factor for %age of tension steel (Fig. 6.2)

K_c = Modification factor for %age of compression steel (Fig. 6.3)

K_f = Modification factor or reduction factor for flanged section (Fig. 6.1)

Percentage of tension or compression steel is based on cross-section area ($b_f d$) in case of flanged beams.

For preliminary proportioning or design of the slab using Fe 415 steel, it is preferable to assume 0.40% tension steel (p_t) for which the value of $K_t = 1.25$, giving span/depth ratio = $1.25 \times 20 = 25$, for SS beams or solid slabs.

When the beams are supporting heavy loads, the span/depth ratio of 10 to 12 is recommended from practical considerations. For slabs supported in two perpendicular directions and spans less than 4 m carrying imposed loads less than 3 kN/m², the code recommends span/overall depth (D) ratios as given in Table 6.2 (as per Clause 24.1 of IS:456-2000, page 39).

Table 6.2: Span/overall depth ratios for two-way slabs
(IS:456-2000, Clause 24.1)

Support conditions	Span/overall depth (D) ratio	
	Fe 250 grade steel bars	Fe 415 HYSD bars
Simply supported slabs	35	28
Continuous slabs	40	32

Notes: D is the overall depth including cover.

The deflection limits are applicable for loading class up to 3 kN/m².

These span/depth ratios are generally applicable for shorter spans up to 3.5 m.

The span/depth ratio is found with respect to shorter span.

Deflection limit is checked so that span/effective depth ratios are satisfied as modified from values given in Tables 6.1 and 6.2 according to the type of structure. Modification constants K_f, K_t and K_c are obtained from Figs 6.1, 6.2 and 6.3 respectively.

6.2.2 Theoretical Approach

Theoretical method of deflection calculation is quite involved and requires number of constants and may be adopted when very accurate details are required and design is done with the help of a computer. As per IS:456-2000, deflections (short term and long term) are given as follows:

(i) *Short term deflection*

The short term deflection is calculated using short term modulus of elasticity of concrete (E_c) and the effective second moment of area of the cracked concrete section (I_{eff}) expressed as:

$$I_{eff} = \cfrac{I_{cr}}{\left[1.2 - \left(\dfrac{M_r}{M}\right)\left(\dfrac{Z}{d}\right)\left\{1 - \left(\dfrac{x}{d}\right)\right\}\left(\dfrac{b_w}{b}\right)\right]}, I_{cr} \leq I_{eff} \leq I_{gr}$$

where,

I_{gr} = MI of the gross cross-section about centroidal axis neglecting the

reinforcement $\left(I_{gr} = \dfrac{bD^3}{12}\right)$

I_{cr} = MI of the cracked section about NA accounting for reinforcements

$(I_{cr} = \dfrac{bx^3}{3} + (1.5m - 1) A_{sc} (x - d')^2 + mA_{st} (d - x)^2$

M_r = Cracking moment ($= f_{cr}I_{gr}/y_t$)

f_{cr} = Modulus of rupture (flexural strength of concrete) $= 0.7 \sqrt{f_{ck}}$

y_t = Distance of extreme tension fibre from centroid of gross cross-section neglecting reinforcement

m = Modular ratio (E_s/E_c)

M = Maximum service load BM

z = Lever arm

x = Depth of NA

d = Effective depth of rectangular section

D = Overall depth of rectangular section

b_w = Breadth of web

d' = Effective cover of compression steel

b = Breadth of compression face (or compression flange b_f)

A_{sc} = Area of compression reinforcement (if any)

A_{st} = Area of tension reinforcement

Short term (instantaneous) deflection is found by:

$$\delta_{i\,(perm)} = K_w \left(\frac{WL^3}{E_c I_{eff}} \right)$$

where,

K_w = Constant depending on the type of load (u.d.l. or point load) and support conditions (cantilever or SS)

$E_c = 5000 \sqrt{f_{ck}}$ (modulus of elasticity of concrete) in N/mm²

W = Total load on the beam (as point load or u.d.l., etc.)

L = Span of the beam

(ii) Deflection due to shrinkage (δ_{cs})

$$\delta_{cs} = K_3 \cdot \Psi_{cs} L^2$$

where,

K_3 = A constant depending upon the support conditions and values are:

= 0.5 for cantilevers

= 0.125 for simply supported members

= 0.086 for members continuous at one end

= 0.063 for continuous on both supports

Ψ_{cs} = Shrinkage curvature = $K_4 (e_{cs}/D)$

e_{cs} = Total shrinkage strain of concrete

(Clause 6.2.4, page 16 of IS:456-2000)

= 0.0003 (approximately)

D = Overall depth of section

L = Span of the member

$$K_4 = 0.72 \left[\frac{p_t - p_c}{\sqrt{p_t}} \right] \le 1.0 \text{ for } 0.25 \le (p_t - p_c) < 1.0$$

$$= 0.65 \left[\frac{p_t - p_c}{\sqrt{p_t}} \right] \le 1.0 \text{ for } (p_t - p_c) \ge 1.0$$

where,

$$p_t = \frac{100A_{st}}{bd} \quad \text{and} \quad p_c = \frac{100A_{sc}}{bd}$$

(A_{st}, A_{sc} = Area of tension and compression steel respectively)

b, d = Breadth and effective depth of rectangular section respectively.

(iii) *Deflection due to creep* (δ_{cp})

The creep deflection caused by permanent loading is given by

$$\delta_{cp\,(perm)} = [\delta_{icp\,(perm)} - \delta_{i\,(perm)}]$$

δ_{icp} = Initial plus creep deflection due to permanent loads obtained by elastic analysis with an effective modulus of elasticity E_{ce}

$[(E_{ce} = E_c/(1 + \theta_{cp})]$, where θ_{cp} = creep coefficient

$\delta_{i\,(perm)}$ = Short term deflection due to permanent loads using E_c.

(iv) *Long term deflection*

The long term deflection (δ_{ld}) due to load, shrinkage and creep shall be evaluated by the following equation:

$$\delta_{ld} = [\text{short term deflection}] + [\text{shrinkage deflection}] + [\text{creep deflection}]$$

$$\delta_{ld} = \delta_{i(perm)} + \delta_{cs} + \delta_{cp\,(perm)}$$

The detailed calculations may be done for actual designs, whereas IS codal provisions for approximate designs shall be quite adequate.

6.3 CRACKING IN CONCRETE ELEMENTS

6.3.1 Causes of Concrete Cracks

Concrete members develop cracks due to flexural tensile stress, diagonal tensile stress due to shear, differential shrinkage between different parts, creep, thermal stresses, aggressive environmental conditions, bond and anchorage failures, etc. Indian standard code IS:456-2000, Clause 35.3.2, page 67 specifies crack widths for serviceability limits.

6.3.2 Permissible Crack Width

- Surface width in general where members are not exposed to harmful environment, permissible crack width = 0.30 mm.
- Width of cracks in tension or exposed to moisture and environment harmful to the member = 0.20 mm (maximum).
- Width of cracks in aggressive environment (severe type) not to exceed 0.10 mm.

  Flexural cracks must also be checked.

6.3.3 Empirical Method to Check Cracking

The reinforcement size and spacing results in check on cracking within limits. Clause 26.3.3 of IS:456-2000 specifies maximum spacing of reinforcement bars. Table 6.3 (Table 15 of IS:456-2000, page 46) specifies minimum clear distances between reinforcing bars. Table 15 considers the percentage redistribution as a function of the clear distance between bars of different grades of steel used in concrete elements.

Table 6.3: Clear distance between bars (as per IS:456-2000)

f_y (N/mm²)	Percentage redistribution from section considered				
	−30	−15	0	+15	+30
	Clear distance between bars (mm)				
250	215	260	300	300	300
415	125	155	180	210	235
500	105	130	150	175	195

> *Note:* These spacings given in Table 6.3 are not applicable when the members are subjected to aggressive environment unless in the calculation of the moment of resistance, f_y has been limited to 300 N/mm² in limit state design method and σ_{st} limited to 165 N/mm² in working stress method.

In case of slabs, the horizontal distance between main parallel bars should not exceed three times the effective depth or 300 mm whichever is lesser. For distribution or secondary reinforcement, the distance between the bars is limited to 5 times the effective depth or 450 mm whichever is lesser.

For beams of overall depth more than 750 mm, side face reinforcement of 0.10% of web area should be placed equally on both the side faces with a vertical spacing not exceeding 300 mm or web width whichever is smaller.

The minimum tension reinforcement in beams to prevent cracking failure in tension zone shall be specified as:

$$A_{st} = \left(\frac{0.85bd}{f_y} \right), \text{ where } f_y \text{ is characteristic strength of reinforcement in N/mm}^2.$$

For slabs, the minimum reinforcement is specified as 0.15% for Fe 250 grade steel (MS) while 0.12% for Fe 415 HYSD and Fe 500 HYSD on the basis of shrinkage and temperature effects.

6.3.4 Crack Width

In case of aggressive environments, crack widths are calculated and compared to permissible crack widths. IS:456-2000 has specified crack width formula as:

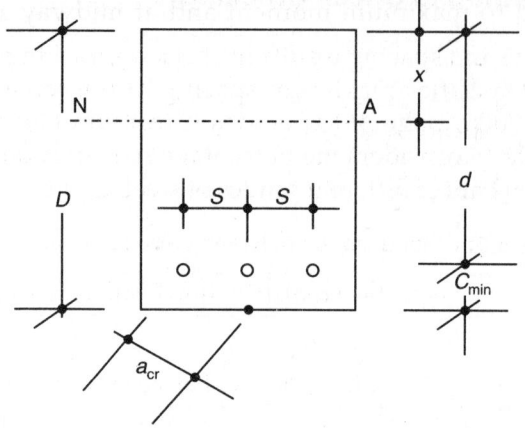

Fig 6.4: Crack details

$$W_{cr} = \left[\frac{3a_{cr} \cdot e_m}{1 + 2\left\{ \dfrac{a_{cr} - C_{min}}{D - x} \right\}} \right]$$

where,

W_{cr} = Width of crack

C_{min} = Minimum cover to the longitudinal bars

a_{cr} = distance from the point considered to the surface of the nearest bar

$[a_{cr} = \{ (0.5S)^2 + C^2_{min} \}^{1/2}]$, S = spacing of bars and C_{min} = minimum cover

x = Depth of neutral axis from the compression face

D = Overall depth of the member

e_m = Average steel strain at the level considered and obtained as

$$e_m = e_1 - \left\{ \frac{b(D - x)(a - x)}{3E_s A_s (d - x)} \right\}$$

e_1 = Strain at the level considered, calculated ignoring the stiffening of concrete in tension zone

a = Distance from the compression face to the point at which the crack width is being calculated

For crack at the beam soffit, $a = D_1$

$$e_1 = \left(\frac{f_s}{E_s} \right) \left(\frac{D - x}{d - x} \right), \text{ and}$$

$$e_m = \frac{1}{E_s} \left(\frac{D - x}{d - x} \right) \left[f_s - \frac{b(D - x)}{3A_s} \right]$$

where, f_s = Stress at the centroid of tension reinforcement (N/mm²)

Generally the maximum width of cracks are observed at the soffit of the beams and at sections subjected to maximum moment and at midway at soffit between the reinforcing bars and at corners as shown in Fig. 6.4.

Example 6.1

A simply supported beam of 8 m span and singly reinforced rectangular section has an effective depth of 600 mm and tension reinforcement of 1% using Fe 415 grade of steel. Check for the deflection control by empirical method. Use M20 grade CC.

Solution: $f_y = 415 \text{ N/mm}^2, f_{ck} = 20 \text{ N/mm}^2, L = 8000 \text{ mm}, d = 600 \text{ mm}, p_t = 1\%.$

$$\text{Actual span/depth ratio} = \frac{8000}{600} = 13.33$$

$$\text{Allowable span/ depth ratio} \left(\frac{L}{d}\right) = \left(\frac{L}{d}\right)_{basic} \times K_t \cdot K_c \cdot K_f$$

$$\left(\frac{L}{d}\right)_{basic} = 20 \text{ (from Table 6.1)}$$

From Fig. 6.1,

$$p_t = 1\% \text{ for Fe 415}, K_t = 1.0, K_c = 1.0,$$
$$K_f = 1.0 \text{ for rectangular singly reinforced}$$

$$\left(\frac{L}{d}\right)_{max} \text{permissible} = \left(\frac{L}{d}\right)_{basic} \times K_t \cdot K_c \cdot K_f = 20 \times 1 \times 1 = 20$$

Thus, actual $\left(\frac{L}{d}\right)$ ratio (13.33) is less than allowable ratio (20), hence the beam section satisfies the serviceability limit of deflection control.

Example 6.2

A simply supported beam of 8 m span and 300 mm × 600 mm (effective depth) is reinforced with 6 bars of 20 mm diameter in tension face and 2 bars of 20 mm in compression face at an effective cover of 40 mm. Use M20 grade CC and Fe 415 grade steel and check limit state of deflection by using empirical method.

Solution: $L = 8000 \text{ mm}, b = 300 \text{ mm}, d = 600 \text{ mm}, d' = 40 \text{ mm}$

$$p_t = \frac{6 \times 314 \times 100}{300 \times 600} = 1.05\%, p_c = \frac{2 \times 314 \times 100}{300 \times 600} = 0.35\%$$

$$\text{Actual } \frac{\text{span}}{\text{depth}} = \frac{8000}{600} = 13.3, \text{ basic} \left(\frac{L}{d}\right) \text{ allowed} = 20 \text{ (from Table 6.1)}$$

K_t (1.05% p_t and Fe 415) = 0.99, K_c (0.35%) = 1.10, K_f = 1.0 (for rectangular)

$$\text{Permissible} \left(\frac{L}{d}\right)_{max} = 20 \times 0.99 \times 1.1 = 21.78 > 13.3,$$

OK (limit state of deflection is satisfied).

Example 6.3

A simply supported beam of 8 m span has T-beam cross-section with flange width of 1000 mm, flange thickness of 120 mm, web width of 300 mm and overall depth of 450 mm with 50 mm effective cover. Beam is reinforced with 5 bars of 20 mm in tension face and 2 bars of 20 mm in compression face. Use Fe 415 steel and M30 grade CC.

Solution: $d = 450 - 50 = 400$ mm, $D_f = 120$ mm, $L = 8000$ mm, $b_f = 1000$ mm, $b_w = 300$ mm, $f_y = 415$ N/mm², $(\sigma_{st} = 240$ N/mm²$)$

$$p_t = \frac{5 \times 314 \times 100}{1000 \times 400} = 0.39\%, \ p_c = \frac{2 \times 314 \times 100}{1000 \times 400} = 0.160\%, \ \frac{b_w}{b_f} = \frac{300}{1000} = 0.30$$

$$\text{Actual} \left(\frac{L}{d}\right) = \frac{8000}{400} = 20, f_s = 415 \times 0.58 \times 1 = 240 \ \text{N/mm}^2$$

From graph (IS:456-2000, Figs 4.0, 5.0 and 6.0)

$$K_t = 1.38, \ K_c = 1.05, \ K_f = 0.80$$

Thus, permissible $\left(\dfrac{L}{d}\right) = \left(\dfrac{L}{d}\right)_{\text{basic}} \times K_t \times K_c \times K_f$

$$= 20 \times 1.38 \times 1.05 \times 0.80 = 23.18$$

$\text{Actual} \left(\dfrac{L}{d}\right) = 20 <$ permissible 23.18, hence the beam section satisfies the limit state of deflection.

Example 6.4

Determine the short term deflection due to dead and live load and long term shrinkage and creep deflection due to permanent load of a cantilever of 4 m span carrying a service dead load of 12 kN/m and service live load of 16 kN/m over the entire span. The beam has rectangular section of 300 mm × 640 mm overall size with concrete cover of 40 mm. The beam is reinforced with 5 bars of 25 mm in tension face and 2 bars of 25 mm in compression face at an effective concrete cover of 40 mm. Concrete of M20 grade and steel of Fe 415 grade are used in the beam. Also $E_s = 2 \times 10^5$ N/mm².

Solution: $L = 4000$ mm, $A_{st} = 2455$ mm², $(p = 1.36\%)$, $A_c = 982$ mm², $(p' = 0.55\%)$

$b = 300$ mm, $d = 640 - 40 = 600$ mm, $D = 640$ mm, $d' = 40$ mm

$w_d = 12$ kN/m, $w_l = 16$ kN/m

$$E_c = 5000 \sqrt{f_{ck}} = 5000 \sqrt{20} = 22.36 \times 10^3 \ \text{N/mm}^2$$

$$E_s = 2 \times 10^5 \ \text{N/mm}^2, \ m = \frac{E_s}{E_c} = \frac{2 \times 10^5}{22.36 \times 10^3} = 8.95 \cong 9.0$$

$$I_{cr} = \frac{bx^3}{3} + (1.5m - 1) A_{sc} (x - d')^2 + m A_{st} (d - x)^2,$$

about the NA of cracked section.

$$x = [\{m^2 (p + 1.5p')^2 + 2m (p + 1.5p)\}^{1/2} - m (p + 1.5p')]d$$

$$x = \left[\left\{81\left(\frac{1.36 + 0.82}{100}\right)^2 + 18\left(\frac{1.36 + 0.82 \times \dfrac{40}{600}}{100}\right)\right\}^{1/2} - 9\left(\frac{1.36 + 0.82}{100}\right)\right] 600$$

$x = 600 \, [\{ 0.03867 + 0.25464\}^{1/2} - 0.1962]$

$\quad = 600 \, [0.54158 - 0.1962] = 207.2 \text{ mm}$

$I_{cr} = \dfrac{300(207.2)^3}{3} + (1.5 \times 9 - 1)982 \, (207.2 - 40)^2 + 9 \times 2455 \, (600 - 207.2)^2$

$\quad = 889.55 \times 10^6 + 343.16 \times 10^6 + 3409.08 \times 10^6 = 4641.788 \times 10^6 \text{ mm}^4$

$M_{cr} = \dfrac{f_{cr} I_{gr}}{y_t}, f_{cr} = 0.7\sqrt{20} = 3.13 \text{ N/mm}^2, y_t = \dfrac{640}{2} = 320 \text{ mm}$

$I_{gr} = \dfrac{bD^3}{12} = \dfrac{300 \times 640^3}{12} = 6553.6 \times 10^6 \text{ mm}^4$

$M_{cr} = \dfrac{3.13 \times 6553.6 \times 10^6}{320} = 64102400 \text{ N-mm} = 64.1024 \text{ kN-m}$

$$\text{Max M (cantilever)} = \frac{(12 + 16)4^2}{2} = 224 \text{ kN-m}$$

$$\text{Lever arm } z = \left(d - \frac{x}{3}\right) = \left(600 - \frac{207.2}{3}\right) = 530.93 \text{ mm}$$

$$I_e = \frac{I_{cr}}{\left[1.2 - \dfrac{M_{cr}}{M} \cdot \dfrac{z}{d}\left(1 - \dfrac{x}{d}\right)\dfrac{b_w}{b_f}\right]} = \frac{4641.788 \times 10^6}{\left[1.2 - \dfrac{64.1024}{224}\left(\dfrac{530.93}{600}\right)\left(1 - \dfrac{207.2}{600}\right)\right]}$$

$$I_e = \frac{4641.788 \times 10^6}{\left[1.2 - 0.16578\right]} = 4488.2 \times 10^6 \text{ mm}^4$$

$I_{cr} = 4641.788 \times 10^6, I_e = 4488.2 \times 10^6, I_{gr} = 6553.6 \times 10^6 \text{ mm}^4$

$\therefore \quad I_e = I_{cr} = 4641.788 \times 10^6 \text{ mm}^4, \{I_{cr} \le I_e \le I_{gr}\}$

$$\delta_s(\text{short term}) = \frac{1}{8}\frac{WL^3}{E_c I_e} = \frac{1}{8} \times \frac{28 \times 4000 \times (4000)^3}{(22.36 \times 10^3) \times \left(4641.788 \times 10^6\right)} = 8.633 \text{ mm}$$

Long term deflection due to shrinkage and creep

$$\delta_e = \delta_{sh} + \delta_{cp}, \, \delta_{sh} = 0.5 \, \phi_{sh} \cdot l^2, \quad \phi_{sh} = \frac{0.0003\beta}{D} \quad \text{(as per IS:456-2000)}$$

$$\beta = \frac{0.72(p_t - p_c)}{\sqrt{p_t}} \leq 1.0 \text{ for } 0.25 \leq (p_t - p_c) \leq 1.0$$

$$= \frac{0.65(p_t - p_c)}{\sqrt{p_t}} \leq 1.0 \text{ for } (p_t - p_c) \geq 1.0$$

$$\frac{(p_t - p_c)}{\sqrt{p_t}} = \frac{(1.36 - 0.55)}{\sqrt{1.36}} = 0.695, \beta = 0.72 \times 0.695 = 0.500$$

$$\therefore \quad \delta_{sh} = \frac{0.5 \times 0.0003}{640} \times 0.50 \times (4000)^2 = 1.875 \text{ mm}$$

$$\delta_{c \text{ (perm)}} = \delta_{sc \text{ (perm)}} - \delta_{s \text{ (perm)}}$$

$$\delta_{sc \text{ (perm)}} = \frac{1}{8} \times \frac{W_d \cdot l^3}{E_{ce} \cdot I_e}, \; E_{ce} = \frac{E_c}{(1 + C_c)} = \frac{22.36 \times 10^3}{(1 + 1.6)} = 8.6 \times 10^3 \text{ N/mm}^2$$

$$\delta_{sc \text{ (perm)}} = \frac{1}{8} \times \frac{(12 \times 4 \times 1000)(4000)^3}{8.6 \times 10^3 \times 4641.788 \times 10^6} = 9.62 \text{ mm}$$

$$\delta_{s \text{ (perm)}} = \frac{1}{8} \times \frac{(12 \times 4000)(4000)^3}{22.36 \times 10^3 \times 4641.788 \times 10^6} = 3.700 \text{ mm}$$

$$\delta_{c \text{ (perm)}} = (9.62 - 3.70) = 5.92 \text{ mm}$$

$$\delta_e = \delta_{sh} + \delta_{c \text{ (perm)}} = 1.875 + 5.92 = 7.795 \text{ mm}$$

Combined short term and long term deflection,

$$\delta = (\delta_s + \delta_e) = 8.633 + 7.795 = 16.428 \cong 16.43 \text{ mm.}$$

Example 6.5

A simply supported beam of rectangular section 300 mm × 500 mm has overall depth and effective span of 4 m. The beam is reinforced with 4 bars of 20 mm diameter (Fe 415 HYSD) at an effective depth of 460 mm in tension face. Two bars of 12 mm diameter are provided in compression face to support stirrups. The beam has self-weight of 5 kN/m including dead load. A service live load of 10 kN/m also acts on the entire span. Use M20 CC. Calculate (i) the short term deflection, (ii) long term deflection according to the provisions of IS:456-2000.

Solution: $L = 4$ m, $b = 300$ mm, $D = 500$ mm, $d = 460$ mm, $f_{ck} = 20$ N/mm^2

$$f_y = 415 \text{ N/mm}^2, f_{cr} = 0.7 \sqrt{f_{ck}} = 3.13 \text{ N/mm}^2, w_d = 5 \text{ kN/m}$$

$$(W_d = 4 \times 5 = 20 \text{ kN}), w_1 = 10 \text{ kN/m} \; (W_L = 40 \text{ kN})$$

$$E_c = 5000 \sqrt{f_{ck}} = 22.36 \times 10^3 \text{ N/mm}^2$$

$$A_{st} = 4 \times 314 = 1256 \text{ mm}^2, A_c = 2 \times 113 = 226 \text{ mm}^2, m = 13$$

Let $\quad$ NA $= x$,

$$\frac{b}{2} x^2 = mA_{st} (d - x), \text{ neglecting the effect of hanger bars.}$$

$$\frac{300}{2} x^2 = 13 \times 1256\,(460 - x), \text{ or } 150x^2 + 16328x - 7510880 = 0$$

$$x = 175.87 \text{ mm}, \quad \frac{x}{d} = 0.3823$$

$$(d - x) = 460 - 175.87 = 284.13 \text{ mm}$$

$$I_{cr} \text{ (cracked)} = \frac{300}{3} \times 175.87^3 + mA_{st}\,(d - x)^2 \text{ (neglecting } A_{sc})$$

$$= 543.97 \times 10^6 + 1318.16 \times 10^6$$

$$I_{cr} = 1862.13 \times 10^6 \text{ mm}^4$$

$$I_{gr} = \frac{1}{12}\,300\,(500)^3 = 3125 \times 10^6 \text{ mm}^4$$

Service load moment $\quad M = \dfrac{(5 + 10)\,4^2}{8} = 30 \text{ kN-m}$

Cracking moment $\quad M_{cr} = \dfrac{f_{cr} \cdot I_{gr}}{y_t}$

$$= \frac{3.13 \times 3125 \times 10^6}{250} = 39125000 \text{ N-mm (39.125 kN-m)}$$

Lever arm $jd = \left(d - \dfrac{x}{3}\right) = \left(460 - \dfrac{175.87}{3}\right) = 401.38 \text{ mm}$

$$I_e = \frac{I_{cr}}{\left[1.2 - \dfrac{M_r}{M}\cdot\left(\dfrac{jd}{d}\right)\left(1 - \dfrac{x}{d}\right)\dfrac{b_w}{b}\right]} = \frac{1862.13 \times 10^6}{\left\{1.2 - \left(\dfrac{39.125}{30}\right)\left(\dfrac{401.38}{460}\right)(0.6177)1\right\}}$$

$$= 3746.17 \times 10^6 \text{ mm}^4 > I_{gr}$$

$\therefore \qquad I_e = I_{cr} = 1862.13 \times 10^6$

Maximum short term deflection $= \dfrac{5}{384}\left\{\dfrac{(W_d + W_l)L^4}{E_c \cdot I_e}\right\}$

or $\quad \delta_{i\,\text{(perm)}} = \dfrac{5 \times (5 + 10) \times 4 \times 1000 \times (4000)^3}{384 \times 22.36 \times 10^3 \times 1862.13 \times 10^6} = 1.201 \text{ mm}$

Long term deflection: $\qquad \delta_L = [\delta_{i\,\text{(perm)}} + \delta_{cs} + \delta_{cp\,\text{(perm)}}]$

Shrinkage deflection: $\qquad \delta_{cs} = K_3 \cdot \psi_{cs} L^2$,

where, $\qquad\qquad\quad K_3 = 0.125$ for SS $\qquad\qquad\qquad$ (constant)

$$\psi_{cs} - \text{Shrinkage curvature} = \frac{K_4 e_{cs}}{D}$$

$$e_{cs} = \text{Ultimate shrinkage strain} = 0.0003$$

$$K_4 = \left\{ \frac{0.72(p_t - p_c)}{\sqrt{p_t}} \right\} \leq 1.0 \text{ for } 0.25 \leq (p_t - p_c) < 1.0$$

$$p_t = \frac{1256 \times 100}{300 \times 460} = 0.91, p_c = \frac{226 \times 100}{300 \times 460} = 0.164$$

$$K_4 = \left\{ \frac{0.72(0.91 - 0.164)}{\sqrt{0.91}} \right\} = 0.563 < 1.0$$

$$\psi_{cs} = K_4 \frac{e_{cs}}{D} = \frac{0.563 \times 0.0003}{500} = 3.378 \times 10^{-7}$$

$$\delta_{cs} = K_3 \psi_{cs} L^2 = 0.125 \times 3.378 \times 10^{-7} (4000)^2 = 0.6756 \text{ mm} = 0.68 \text{ mm}$$

Creep deflection:

$$\delta_{icp \text{ (perm)}} = \text{Initial} + \text{creep deflection due to permanent load}$$

$$E_{ce} = \left[\frac{E_c}{1 + \theta} \right], \text{ where } \theta = \text{creep coefficient} = 1.6 \text{ for 28 days}$$

$$= \frac{E_c}{2.6}$$

$$\delta_{icp \text{ (perm)}} = \frac{5}{384} \left\{ \frac{(W_d + W_l)L^4}{E_{cs} \cdot I_{eff}} \right\} = \frac{5}{384} \times \frac{15 \times 1000 \times 4 \times (4000)^3}{\dfrac{E_c}{2.6} \times 1862.13 \times 10^6} = 3.122 \text{ mm}$$

$$\delta_{cp \text{ (perm)}} = \delta_{icp \text{ (perm)}} - \delta_{i \text{ (perm)}} = 3.122 - 1.201 = 1.9212 \text{ mm}$$

Total long term deflection $\delta_L = 1.201 + 0.68 + 1.9212 = 3.802$ mm

$$(\delta_{i \text{ (perm)}} + \delta_{cs} + \delta_{cp \text{ (perm)}})$$

Maximum permissible long term deflection (IS:456-2000) $= \dfrac{\text{span}}{250}$

$$\delta_{\text{(perm)}} = \frac{4000}{250} = 16.0 \text{ mm} > 3.802 \text{ mm (actual), (OK)}$$

Example 6.6

A SS beam of rectangular section spanning 8.0 m has a section of 300 mm width and 650 mm effective depth. The beam is reinforced with 5 bars of 25 mm diameter on the tension face at an effective cover of 50 mm. The bars are placed at 50 mm centre to centre horizontal space. Using M20 grade of CC and Fe 415 grade of steel, check the beam for the limit state of crack width according to IS:456-2000 code method. The beam is subjected to a maximum service BM = 200 kN-m.

Solution: $b = 300$ mm, $d = 650$ mm, $D = 700$ mm, $L = 8$ m (8000 mm), $f_{ck} = 20$ N/mm^2

$f_y = 415$ N/mm^2, $m = 9.0$, $A_{st} = 5 \times 491 = 2455$ mm^2, $(p = 1.26\%)$

$M = 200$ kN-m $(200 \times 10^6$ N-mm$)$, $S = 50$ mm, $E_s = 2 \times 10^5$ N/mm^2

$E_c = 5000 \sqrt{20} = 22.36 \times 10^3$ N/mm^2, $\dfrac{E_s}{m} = 22.22 \times 10^3$ N/mm^2

NA depth $= x$, $\dfrac{1}{2} b . x^2 = m A_{st} (d - x)$, or $150 x^2 + 22095 x - 14361750 = 0$

or $\qquad x^2 + 147.3 x - 95745 = 0$, $x = 244.42$ mm

MI (I_{cr}) of cracked section:

$$I_{cr} = \frac{b \cdot x^3}{3} + m \cdot A_{st}(d-x)^2 = \frac{300}{3}(244.42)^3 + 9 \times 2455 (650 - 244.42)^2$$

$$I_{cr} = 10^6 (1460.20) + 10^6 (3634.52) = 10^6 (5094.72) \text{ mm}^4$$

Clear cover, $\qquad C_{min} = 50 - \dfrac{25}{2} = 37.5 \text{ mm}$

$\dfrac{1}{2}$ (spacing between bars) $= \dfrac{50}{2} = 25 \text{ mm}$

$$a_{cr} = [(0.5S)^2 + (C_{min})^2]^{1/2} = [625 + 1406.25]^{1/2} = 45.07 \text{ mm}$$

Crack width shall be maximum at the soffit between 2 bars.

CG of steel bars from NA (y) = (650 – 244.42) = 405.58 mm

$$e_1 = \frac{f_s}{E_s}\left[\frac{D-x}{d-x}\right], f_s = \frac{M \cdot y \cdot m}{I_{cr}} = \frac{200 \times 10^6 \times (650 - 244.42) \times 9}{5094.72 \times 10^6} = 143.3 \text{ N/mm}^2$$

$$e_1 = \frac{143.3}{2 \times 10^5}\frac{[700 - 244.42]}{(650 - 244.42)} = 8.048 \times 10^{-4}$$

$$e_m = e_1 - \left[\frac{b(D-x)(700-x)}{3E_s \cdot A_s(d-x)}\right], \qquad a' = D$$

$$e_m = (8.048 \times 10^{-4}) - \left[\frac{300(700-244.42)(700-244.42)}{3 \times 2 \times 10^5 \times 2455(650-244.42)}\right] = (8.048 - 1.042) \times 10^{-4}$$

$$e_m = 7.006 \times 10^{-4}$$

$$W_{cr} = \left[\frac{3a_{cr}e_m}{1+2\left\{\dfrac{a_{cr}-C_{min}}{D-x}\right\}}\right] = \frac{3 \times 45.07 \times 7.006 \times 10^{-4}}{1+2\left\{\dfrac{45.07 - 37.5}{700 - 112.1}\right\}} = 0.0918 \text{ mm}$$

Permissible W_{cr} (as per IS:456-2000)

$\qquad W_{cr} > 0.3$ mm, under normal conditions

$\qquad 0.0895 \not> 0.004 \, C_{min}$, i.e. $\not> 0.004 \times 37.5$ mm = 0.15 mm

$\qquad 0.0895 \not> 0.15$ mm, thus serviceability limit state of crack width is satisfied.

Example 6.7

A cantilever beam of 4 m span carries a service dead load of 12 kN/m and a service live load of 16 kN/m over the entire span. The beam has rectangular section of 300 mm width and 640 mm overall depth with effective concrete cover of 40 mm. The beam is reinforced with 5 bars of 25 mm diameter in tension face with horizontal spacing of 50 mm and 2 bars of 25 mm in compression face at an effective cover of 40 mm. Concrete of M20 grade and steel of Fe 415 grade are used in the beam. Also $E_s = 2 \times 10^5 \text{ N/mm}^2$.

Solution: Clear cover on tension face $= 40 - \dfrac{25}{2} = 27.5$ mm $\cong 28$ mm (a_{cr})

$$E_s = 2 \times 10^5 \, \text{N/mm}^2, \; E_c = 5000 \sqrt{20} = 22.36 \times 10^3 \, \text{N/mm}^2, \; m = 9.0,$$

$$D = 640 \text{ mm}, \; d = 600 \text{ mm}, \; b = 300 \text{ mm}, \; A_{st} = 2455 \text{ mm}^2 \; (p = 1.364\%)$$

$$A_c = 982 \text{ mm}^2 \; (p' = 0.55\%), \; C_{min} = 40 - \dfrac{25}{2} = 27.5 \text{ mm} \cong 28 \text{ mm}, \; S = 50 \text{ mm}$$

$$M_{max} = \dfrac{(12+16)4^2}{2} = 224 \text{ kN-m} \; (224 \times 10^6 \, \text{N-mm})$$

Neutral axis depth $= x$

$$\dfrac{b \cdot x^2}{2} + (1.5m - 1) \, A_{sc} \, (x - d') = mA_{st} \, (d - x)$$

or $\quad 150x^2 + 12.5 \times 982 \, (x - 40) = 9 \times 2455 \, (600 - x)$

$$x^2 + 229.13x - 91653.33 = 0, \qquad\qquad\qquad x = 209.13 \text{ mm}$$

$$a_{cr} = \sqrt{25^2 + 27.5^2} = 37.17 \text{ mm} \cong 37 \text{ mm}$$

$$I_{cr} = \dfrac{300}{3} (209.13)^3 + 12.5 \times 982 \times 169.13^2 + 9 \times 2455 \, (600 - 209.13)^2$$

$$= 4641.43 \times 10^6 \text{ mm}^4$$

Surface crack width W_{cr}

$$W_{cr} = \dfrac{3a_{cr}e_m}{\left[1 + 2\left\{\dfrac{a_{cr} - C_{min}}{D - x}\right\}\right]}, \quad e_m = e_1 - \dfrac{0.75bD\left(a' - x\right)}{A_{st} \cdot f_s \left(D - x\right)} \times 10^{-3}$$

$$e_1 = \left(\dfrac{D - x}{d - x}\right)\dfrac{f_s}{E_s} = \left(\dfrac{430.87}{390.87}\right) \times \dfrac{169.8}{2 \times 10^5} = 9.36 \times 10^{-4}$$

$$f_s = \dfrac{m \cdot M(d - x)}{I_{cr}} = \dfrac{9 \times 224 \times 10^6 \times 390.87}{4641.43 \times 10^6} = 169.8 \text{ N/mm}^2$$

$$e_m = \left[9.36 \times 10^{-4} - \dfrac{0.75 \times 300 \times 640 \, (640 - 209.13)}{2455 \times 169.8 \, (640 - 209.13)} \times 10^{-3}\right]$$

$$= (9.36 \times 10^{-4} - 3.45 \times 10^{-4}) = 5.91 \times 10^{-4}$$

$$C_{min} = 28 \text{ mm}$$

$$W_{cr} = \dfrac{3 \times 37 \times 5.91 \times 10^{-4}}{\left[1 + 2\left\{\dfrac{37 - 28}{640 - 209.13}\right\}\right]} = \dfrac{0.0656}{\left[1 + 0.0418\right]}$$

$= 0.0630$ mm < 0.3 mm (normal exposure) < 0.15 mm (aggressive exposure)

Thus, width of crack is less than the permissible values, and hence the serviceability of crack width is satisfied.

SUMMARY

Apart from satisfying limit state of collapse, a bending element must also satisfy the limit state of serviceability (deflection, crack width, etc.). Excessive deflections result in crushing of partitions, sagging of floors, balconies, etc. Excessive cracking may cause dampness and corrosion of reinforcement and loss of bond.

Limit states of serviceability become more critical when high strength concretes and steels are used which results in thin sections. As per IS:456-2000, final deflections should not exceed (span/250) and deflections including effects of temperature, creep, shrinkage, etc. after erection of partitions, etc. should not exceed (span/350) or 20 mm. Normally empirical method is used to satisfy the limit states of deflections by limiting (span/depth) ratio as:

	Rectangular sections	T-beam sections	
Cantilever	7	$7K_f$	where, K_f is constant
Simply supported	20	$20K_f$	depending on b_w and b_f
Continuous	26	$26K_f$	ratios

The permissible ratio of span to effective depth is also dependant on percentage (p_t) of tensile steel and percentage of steel (p_c') in compression face. The constants K_t, K_c and K_f are found from graphs provided in Figs 4, 5 and 6 in IS:456-2000.

For preliminary proportioning, depth is assumed as:

For heavy loads: $d \geq$ (span/10 to 12)

For medium loads: $d \geq$ (span/12 to 15)

For light loads: $d \geq$ (span/15 to 20)

For two way simply supported slabs, span/depth ratio = 35 and for two way continuous slabs, span/depth ratio = 45, when HYSD bars are used this ratio is multiplied by 0.80. These ratios are generally applicable for spans up to 3.50 m.

In theoretical method actual deflection is computed and checked with the limit state of deflections allowed. For actual computation of deflection, moment of inertia (I) is found for gross-section (I_{gr}), effective section (I_e) and cracked section (I_{cr}). Formula given in IS:456-2000 should be used for computation. Deflections corresponding to shrinkage and creep are also calculated by taking appropriate values of constants.

Instantaneous short term deflection $\delta_s = K_w \dfrac{WL^3}{E_c \cdot I_e}$, where K_w is a constant and depends on type of load and type of beam.

$\delta_{sh} = K_3 \Psi_{sh} \cdot L^2$, where K_3 depends on type of the beam (0.50 for cantilever, 0.125 for SS, 0.086 for continuous at one end and 0.063 for continuous on both ends).

$e_{sh} = 0.0003$ (generally), $\Psi_{sh} = K_4 \dfrac{e_{sh}}{D}$, $K_4 = \dfrac{0.72(p_t - p_c)}{\sqrt{p_t}}$ or $\dfrac{0.65(p_t - p_c)}{\sqrt{p_t}}$

$$\delta_{cp\ (perm)} = \delta_{icp} - \delta_{i\ (perm)}$$
$$\delta_{ld} = \delta_{i\ (perm)} + \delta_{sh} + \delta_{cp}$$

Limit state of crack width is also checked with reference to permissible values as:

- Normal environment crack width   not more than 0.30 mm
- Exposed to moisture harmful to the members   not more than 0.20 mm
- Exposed to aggressive (severe) environment   not more than 0.10 mm

Various formulae specified in IS:456-2000 are used to compute crack width according to depth, distance of surface from the tensile steel bars and spacing of bars, etc. Limit state of crack widths are satisfied by maintaining spacing between the bars as specified in IS:456-2000.

In slabs, spacing of main bars ≯ 3 × eff. depth or 300 mm, whichever is less.

Spacing of secondary bars ≯ 5 × eff. depth or 450 mm, whichever is less. When the depth of beam is more than 750 mm, additional side face reinforcement of 0.10% of web area shall be provided so that the maximum vertical spacing of the additional horizontal bars does not exceed 300 mm or web width (b_w) whichever is smaller.

PRACTICE QUESTIONS

(I) Objective Questions

Q. 6.1 Select the correct response given after each statement to complete it correctly and fill in the response sheet provided.

  (i) Serviceability limit state does not include ....
- (a) collapse load limit on beam
- (b) crack width limit of beam section
- (c) deflection limit of beam
- (d) lateral stability limit of beam

  (ii) According to the empirical method specified in IS:456-2000 code, the limit state of deflection is ensured by ....
- (a) computing collapse load causing deflections
- (b) span/effective depth ratio of beam
- (c) collapse load/elastic load ratio of beam
- (d) modulus of elasticity of steel/modulus of elasticity of concrete

(iii) According to IS:456-2000, the maximum deflection limit in one way supported continuous slab is assumed to be satisfied when the effective depth of a rectangular slab section is ....
- (a) less than or equal to span/20
- (b) more than or equal to 20 mm
- (c) less than span/26
- (d) more than or equal to span/26

(iv) Final deflection of RCC beam members as measured from the as cast level of the support including all loads, effect of temperature, creep and shrinkage, etc. shall not exceed the value of ....
- (a) span/400 mm   (b) span/350 mm
- (c) span/250 mm   (d) span/20 mm

(v) As per IS:456-2000, when the beam section has flanged width b_w/b_f ratio of 0.25, the value of span/depth ratio is modified by multiplying a factor K_f for deflection control check. The value of K_f will be ....

    (a) 1.0   (b) 0.80

    (c) 0.25   (d) $0.80 \times 0.25 = 0.20$

(vi) For preliminary design of a simply supported slab provided with 0.40% tensile steel (p_t) of Fe 415 grade for which the modification factor K_t is 1.25, what will be the permissible span/depth ratio to satisfy the deflection limit?

    (a) 35   (b) 26

    (c) 25   (d) 20

(vii) A two way simply supported solid slab of 3.50 m × 4.50 m size and carrying a total u.d.l. of 3 kN/m on the entire span, has steel of Fe 415 grade and CC of M30 grade. Suggest minimum suitable overall thickness of slab so that the deflection limits are satisfied.

    (a) 100 mm   (b) 112.5 mm

    (c) 125.0 mm   (d) 160.7 mm

(viii) A continuous two way slab has effective size of 3.50 m × 4.0 m and carries an u.d.l. of 2.5 kN/m². The slab uses M25 grade CC and Fe 415 grade steel. Find the minimum overall thickness of the slab required to control the deflection limit state as per IS:456-2000.

    (a) 88 mm   (b) 100 mm

    (c) 110 mm   (d) 125 mm

(ix) Total maximum deflection due to uniformly distributed dead $\left(\dfrac{w_d}{m}\right)$ and live loads $\left(\dfrac{w_1}{m}\right)$ in case of SS beam of span L shall be given by δ = ...

    (a) $\dfrac{(w_d + w_1)L^3}{48EI_e}$ (b) $\dfrac{5(w_d + w_1)L^4}{384EI_e}$

    (c) $\dfrac{(w_d + w_1)L^3}{384EI_{gr}}$ (d) $\dfrac{(w_d + w_1)L^4}{48EI_{gr}}$

(x) Maximum total deflection in case of a cantilever beam of span L mm and carrying total u.d.l. of W newton will be ....

    (a) $\dfrac{WL^3}{8EI_e}$ (b) $\dfrac{WL^2(1000)}{8EI_{gr}}$

    (c) $\dfrac{WL^3}{384EI_{gr}}$ (d) $\dfrac{WL^2(1000)}{24EI_e}$

Response sheet to Q. 6.1 (i to x)

Question	(i)	(ii)	(iii)	(iv)	(v)	(vi)	(vii)	(viii)	(ix)	(x)
Response (a/b/c/d)										

(II) Numerical Questions

Q. 6.2 A simply supported beam of 6 m span and doubly reinforced rectangular section 300 mm × 500 mm (overall depth) has tension reinforcement of 1% and compression reinforcement of 1.0% with steel of Fe 415 (f_s = 190 N/mm²) and concrete cover of 40 mm. Check if the section satisfies the deflection limits by empirical method.

[**Hint:** Actual $\dfrac{\text{span}}{\text{depth}}$ ratio = 13.04 < permissible 30, satisfies the limits.]

Q. 6.3 A simply supported beam of 10 m effective span has T-beam cross-section with 1500 mm flange width and 300 mm web width. The beam section has 1% tensile steel and 1% compressive steel. The section has 400 mm effective depth. The steel used has permissible stress f_s = 230 N/mm². Check by empirical method if the beam section satisfies the deflection limit state.

[**Hint:** $\dfrac{L}{d}$ (actual) = 25 > permissible (20 × 1.04 × 1.25 × 0.80 = 20.8), does not satisfy the deflection limit.]

Q. 6.4 A rectangular beam section of 250 mm × 540 mm overall size is used for a cantilever beam of span 3 m. The beam section is reinforced with 4 bars of 20 mm diameter in tension zone at an effective cover of 40 mm. Two bars of 20 mm diameter are provided in compression zone at an effective cover of 40 mm. Concrete of M20 grade and steel of Fe 415 grade are used in beam. Also E_s = 2 × 10⁵ N/mm². The beam is subjected to a u.d.l. of w_d = 10 kN/m and w_l = 10 kN/m. Compute the total long term deflection.

[**Hint:** M = 90 kN-m, m = 9.00, E_c = 22.36 × 10³, E_{ce} = 8.6 × 10³ N/mm², x = 155 mm, I_{cr} = 1759.58 × 10⁶ mm⁴, I_g = 3280.5 × 10⁶, I_e = 1877.25 × 10⁶, f_{cr} = 3.13 N/mm², M_{cr} = 38.03 kN-m.

$$\delta_s = \frac{(10+10)3000\times(3000)^3}{8E_c\times1877.25\times10^6} = 4.824 \text{ mm}$$

$$\delta_{sh} = \frac{0.5(0.0003)\beta\cdot l^2}{D} = 0.900 \text{ mm},\ \beta = 0.36$$

$$\delta_{sc\,(perm)} = \frac{W_d\cdot L^3}{8E_{ce}I_e} = \frac{30000\times(3000)^3}{8\times8.6\times10^3\times1877.25\times10^6} = 6.272 \text{ mm}$$

$$\delta_{s\,(perm)} = \frac{30000\times(3000)^3}{8E_cI_e} = 2.412 \text{ mm}$$

$$\delta_{cp} = (6.272-2.412) = 3.86 \text{ mm},\ \delta_{long\,term} = 9.59 \text{ mm}.]$$

Q. 6.5 Determine the short term deflection due to dead and live loads of 10 kN/m and 10 kN/m respectively on a cantilever beam of 5 m span. Also compute long term deflections due to shrinkage and creep. The beam section has 400 mm width and 750 mm overall depth with clear cover of 25 mm on tension and compression faces. The beam is reinforced with 6 bars of 25 mm in tension and 2 bars of 25 mm in compression zone. Concrete of M20 grade and steel of Fe 415 grade are used. Take $E_c = 22.4 \times 10^3$ N/mm² and $f_{cr} =$

3.13 N/mm³ for concrete and $m = 8.93$, $jd = \left(d - \dfrac{x}{3}\right)$. Take $e_{sh} = 0.0003$ and

$\beta = 0.4881$, creep coeff. $C_c = 1.6$.

[Hint: $I_{gr} = 14062.5 \times 10^6$ mm⁴, $I_{cr} = 8038.32 \times 10^6$ mm⁴, $I_e = I_{cr}$, $\delta_s = 8.68$ mm,

$\delta_{sh} = 2.44$ mm, $\delta_{sc\ (perm)} = 11.28$ mm, $\delta_{s\ (perm)} = 4.34$ mm, $\delta_{c\ (perm)} = 6.94$ mm

Total (short term and long term) $\delta = \delta_s + \delta_e = 8.68 + 9.38 = 18.06$ mm

Q. 6.6 Determine the surface crack width directly under a bar on the tension face at a section of maximum moment in case of a cantilever beam of 5 m span and carrying u.d.l. of 15 kN/m (dead load) and 15 kN/m (live load). The cross-section of beam has 400 mm width and 750 mm overall depth with clear covers of 25 mm both on tension and compression zones. The beam is reinforced with 6 bars of 25 mm diameter in tension zone and 2 bars of 25 mm in compression zone. The beam is constructed with M20 grade CC and Fe 415 grade steel.

[Hint: $W_{cr} = \dfrac{3a_{cr} \cdot e_m}{1 + 2\left(\dfrac{a_{cr} - C_{min}}{D - x}\right)}$, $D = 750$ mm, $x = 226.8$ mm, $a_{cr} = 25$ mm,

$C_{min} = 25$ mm, $f_s = 202.4$ N/mm², $e_m = 0.0007376$, W_{cr} (width) = 0.0553 mm]

Q. 6.7 A simply supported beam of rectangular section has 300 mm width and 600 mm overall depth with an effective cover of 50 mm. The beam is reinforced with 4 bars of 25 mm diameter in tension zone. The beam is subjected to a working load moment of 160 kN-m at the mid span point. Using M25 grade CC and Fe 415 HYSD bars. Check the serviceability limit state of cracking in accordance with IS:456-2000 code method.

[Hint: $x = 220$ mm, $I_r = 34.1 \times 10^8$ mm⁴, $a_{cr} = 45$ mm, $f_s = 170$ N/mm², $e_1 = 9.78 \times 10^{-4}$,

$$e_m = e_1 - \left[\dfrac{b_t(D-x)(a'-x)}{3E_s A_s(d-x)}\right] = 8.67 \times 10^{-4}, (a'-D)$$

$$W_{cr} = \left[\dfrac{3a_{cr} \cdot e_m}{1 + 2\left(\dfrac{a_{cr} - C_{min}}{D - x}\right)}\right] = 0.113 \text{ mm}$$

< max 0.30 mm crack width allowed under normal conditions.
< 0.15 mm under severe conditions OK]

Chapter 7

Limit State Design of Slabs

LEARNING OBJECTIVES

After the study of *limit state design of slabs*, the learner will be able to apply the principles of limit state of collapse and serviceability for the design of different type of slabs and will be able to:

- ⊙ Explain the meaning of one-way, two-way and flat slabs
- ⊙ Design simply supported one-way slab for the given service loads and other service conditions
- ⊙ Design continuous slabs spanning in one-way for the given service loads and other service conditions
- ⊙ Design one-way continuous beam–slab roof for the given service loads and other service conditions
- ⊙ Design a simply supported two-way slab for the given service loads and other service conditions
- ⊙ Design a two-way continuous beam–slab roof for the given service loads and other service conditions
- ⊙ Explain the advantages of flat slab construction
- ⊙ Design the flat roof slab for the given service loads and other service conditions for the given area.

7.1 LIMIT STATE DESIGN OF ONE-WAY SIMPLY SUPPORTED SLABS

7.1.1 Introduction

Reinforced concrete slabs constitute the most common elements to construct floors and roofs of buildings. Slabs may be supported on walls or beams or directly on columns. The slabs supported on walls or beams are called *edge supported slabs*. The slabs directly supported on the columns are known as *flat slabs*. The edge supported slabs are called *one-way slabs* when the slabs are supported on the two parallel supports (walls or beams) (Fig. 7.1(a)).When the edge supported slabs are supported on all the four edges (Fig. 7.1(c)) and the ratio of length and breadth is not more than 2, it may behave as *two-way slab* and the loads are shared in two perpendicular directions. Bending also

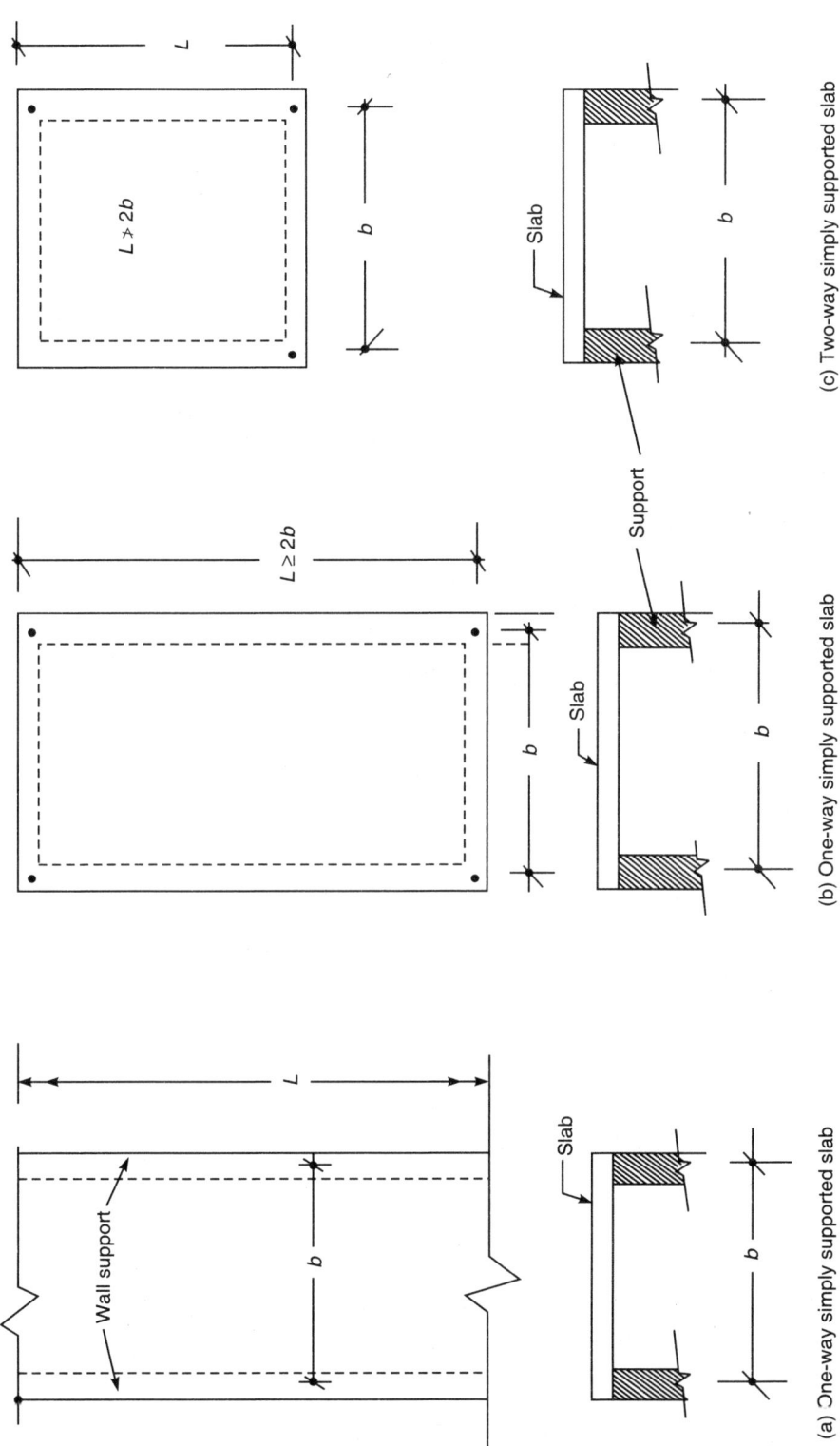

(a) One-way simply supported slab

(b) One-way simply supported slab

(c) Two-way simply supported slab

Fig. 7.1: One-way and two-way simply supported slabs

occurs in two perpendicular directions. If the ratio of length and breadth is more than 2 (Fig. 7.1(b)), the slab bends mainly in one direction only and the slab is known as *one-way slab*. The behavior and design of one-way simply supported (Fig. 7.1(a) and (b)) and one-way continuous slabs (Fig. 7.2(a)) will be explained.

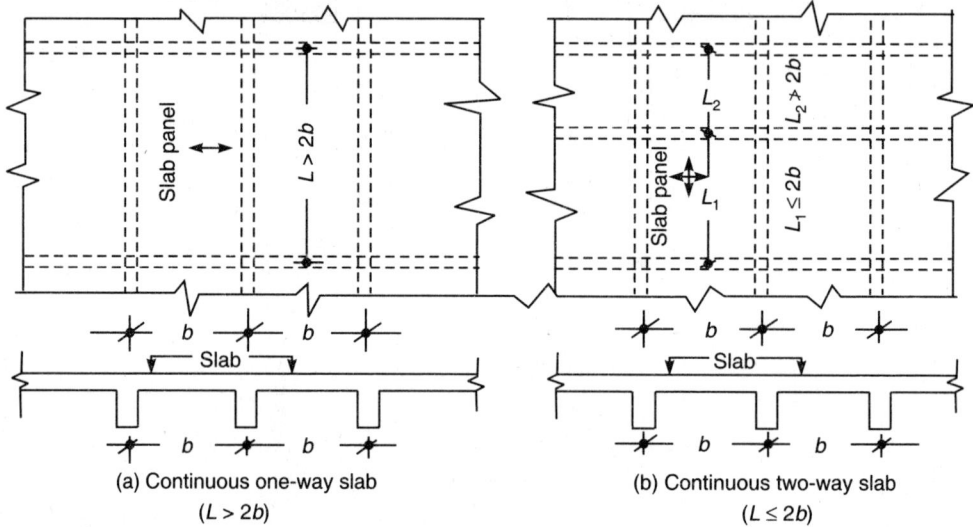

(a) Continuous one-way slab
(L > 2b)

(b) Continuous two-way slab
(L ≤ 2b)

Fig 7.2: Continuous slabs

7.1.2 Behaviour of Edge Supported Slabs

Edge supported slabs are of two types — one-way slabs and two-way slabs. Edge supported slabs supported on 4 side supports can behave as one-way if the length (L) is more than or equal to twice the breadth ($L \geq 2b$) as shown in Fig. 7.1(c). One-way

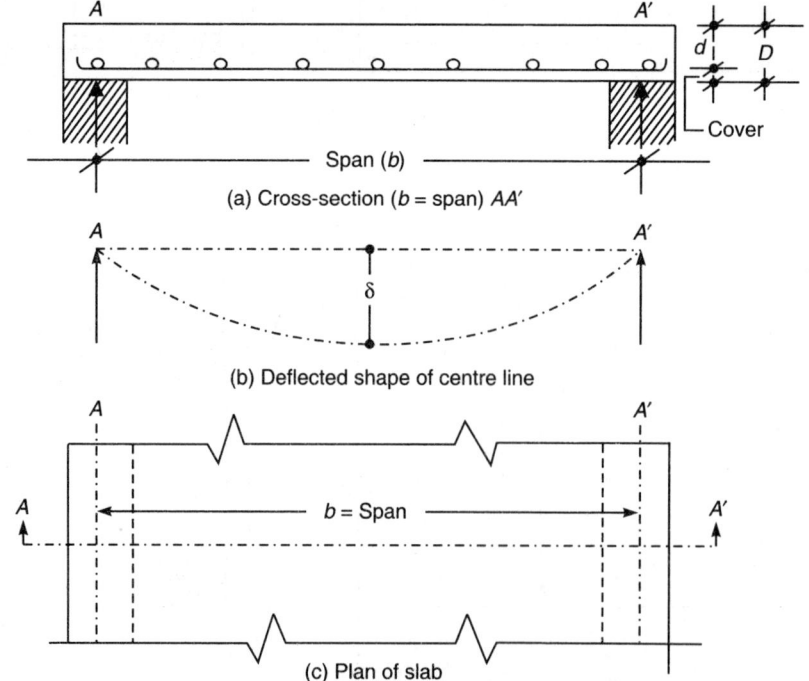

(a) Cross-section (b = span) AA′

(b) Deflected shape of centre line

(c) Plan of slab

Fig 7.3: Simply supported (one-way slab)

slabs mainly bend (deflects) in one direction and it is assumed to share full load by bending in one direction only (Fig. 7.3). These slabs are designed as beams considering width of beam = length of slab, and depth of beam as depth of the slab.

7.1.3 Design of One-Way Slabs

Design of one-way slab is done in a similar manner as beams except that the width of the slab is considered 1 m and depth is considered equal to thickness of the slab. The effective depth is taken as thickness minus effective cover $(D - C_e = d)$. Effective cover is equal to clear cover plus half the diameter of main bar $(C_e = C_c + \dfrac{\phi}{2})$. Clear cover (C_c) will be considered as per IS:456-2000 depending on the surrounding environment (mild, moderate, severe and extreme, etc.). A minimum nominal cover of 20 mm shall be adopted for mild exposure. Adoption of minimum concrete cover is essential to get adequate durability (life of the element) under the given exposure conditions and water tightness of the concrete. Thus, adoption of appropriate concrete cover to steel forms one of the most important criteria of design for durability.

Another important factor in design is considered by choosing suitable loading which depends on the type of structure (residential, commercial, educational, industrial, etc.). Maximum bending moments, shear forces and deflections are computed according to the type of slab (simply supported, cantilever and continuous, etc.) and type of loading (uniformly distributed, concentrated load, triangular distributed, etc.). For the selected section and given data NA depth (x) maybe calculated and limit states of moment (M_r) and shear (V_r), etc. calculated for limit state of stress and equated to the corresponding limit states of bending moments and shear forces developed due to ultimate factored loads. From this size of the section depth d, NA depth (x), tensile steel (A_{st}), etc. may be obtained for the design of the section. The percent of tension steel in slabs may be 0.3–0.5% and hence the constant (k_t) for tension reinforcement may result in the span/depth ratio of 25 to 30 for one-way slabs. For slabs, no shear reinforcement is provided.

Generally from practical considerations, the depth of slab is provided more than the minimum depth required for the balanced section. Thus, mostly the slab sections are under reinforced or balanced. Generally, the design equations shall be based on balanced or under reinforced sections.

These details are explained through following solved examples.

Example 7.1

Design a one-way slab with a clear span of 3.0 m, simply supported on 200 mm thick masonry walls to support a live load of 5 kN/m² in addition to its self-weight. Use M30 grade CC and Fe 415 grade steel bars.

Solution: Clear span = 3 m (3000 mm), L_e = 3000 + 200 or (3000 + d) = 3180 mm

(Assume d = 180 mm), f_{ck} = 30 N/mm², f_y = 415 N/mm²

$$m = \frac{280}{f_{ck}} = 9.33 \cong 9.0, \text{say, Assume } d = \frac{\text{span}}{25} = \frac{3180}{25} = 127.2 \text{ mm}$$

say d = 135 mm, D = 160 mm

Consider *Im strip, b = 1 m = 1000 mm*

$$x_u = 0.48 \times 135 = 65 \text{ mm (critical NA)}$$

Self-weight $= 0.16 \times 25 = 4.0 \text{ kN/m}^2$, plus weight of finish (say) $= 1 \text{ kN/m}^2$

Total $w_d = 5 \text{ kN/m}^2$, w_l (live load) $= 5 \text{ kN/m}^2$

Total u.d.l. $w = 5 + 5 = 10 \text{ kN/m}^2$, ultimate load $w_u = 1.5 \times 10 = 15 \text{ kN/m}^2$

Max. BM, $M_u = \dfrac{15 \times 3.18^2}{8} = 18.96 \text{ kN-m} \ (18.96 \times 10^6 \text{ N-mm})$

Max. SF, $V_u = \dfrac{15 \times 3.18}{2} = 23.85 \text{ kN} \ (23850 \text{ N})$

$M_{u \text{ lim}}$ (for 1 *m* wide strip) $= 0.138 f_{ck} bd^2 = 0.138 \times 30 \times 1000 \ (135)^2$

$M_{u \text{ lim}} = 75.45 \times 10^6 \text{ N-mm}$

$M_u = 18.96 \times 10^6 \text{ N-mm} < M_{u \text{ lim}}$,

hence the section will be under reinforced.

Main steel in tension

$$18.96 \times 10^6 = 0.87 f_y A_{st} d \left[1 - \frac{A_{st} f_y}{bd f_{ck}} \right]$$

$$= 0.87 \times 415 A_{st} \times 135 \left[1 - \frac{415 A_{st}}{1000 \times 135 \times 30} \right]$$

or $48742 A_{st} [1 - 1.0247 \times 10^{-4} A_{st}] = 18.96 \times 10^6$

or $A_{st}^2 - 9758.95 A_{st} + 379.6 \times 10^4 = 0$

$$A_{st} = 406 \text{ mm}^2 \ (p_t = 0.30\%)$$

Also by direct formula:

$$A_{st} = \frac{0.5 f_{ck} \cdot b \cdot d}{f_y} \left[1 - \sqrt{1 - \frac{4.6 M_u}{f_{ck} \cdot b \, (d)^2}} \right]$$

$$= \frac{0.5 \times 30 \times 1000 \times 135}{415} \left[1 - \sqrt{1 - \frac{4.6 \times 18.96 \times 10^6}{30 \times 10^3 \ (135^2)}} \right]$$

$A_{st} = 406.1 \text{ mm}^2$ (provide 10ϕ bars @ 190 c/c, $A_{st} = 413 \text{ mm}^2/\text{m}$)

Using 10 mm diameter bars, spacing of bars $= \dfrac{1000 \times 78.5}{406} = 193 \text{ mm c/c, i.e.}$

provide 10 mm diameter @ spacing of 190 mm c/c ($A_{st} = 413 \text{ mm}^2$)

(alternate bars are bent up at ends in 0.15L)

Distribution bars: $A_{st2} = 0.12\%$ of gross cross-sectional area

$$= \frac{0.12}{100} \times 1000 \times 160 = 192 \text{ mm}^2/\text{m}$$

(provide 8 mm diameter bars @ 250 c/c, ($A_{st2} = 200 \text{ mm}^2/\text{m}$)

Check for shear: $\qquad \tau_v = \dfrac{V_u}{bd} = \dfrac{23850}{1000 \times 135} = 0.18 \text{ N/mm}^2$

$$p_t = \frac{413 \times 100}{1000 \times 135} = 0.306\%,$$

τ_c (Table 19 of IS:456-2000, page 73) = 0.395 N/mm²

Permissible shear stress $K\tau_c$ = 1.27 × 0.395 = 0.502 N/mm²

(K for D = 160 mm is 1.27, from clause 40.2.1.1, page 72 of IS:456-2000).

Thus, the slab is safe in shear (0.18 << 0.502 N/mm²).

Check for deflection (empirical approach):

$(L/d)_{max} = (L/d)_{basic} \times K_t \times K_c \times K_f, \qquad\qquad K_c = 1, K_f = 1,$

$\qquad p_t$ = 0.306%, K_t = 1.45 (Fig. 7.4 of IS:456-2000 for f_s = 240 N/mm², Fe 415)

$(L/d)_{max} = 20 \times 1.45 = 29,$ actual $(L/d) = \dfrac{3180}{135} = 23.55 < (L/d)$ permissible

Reinforcement details (Fig. 7.4):

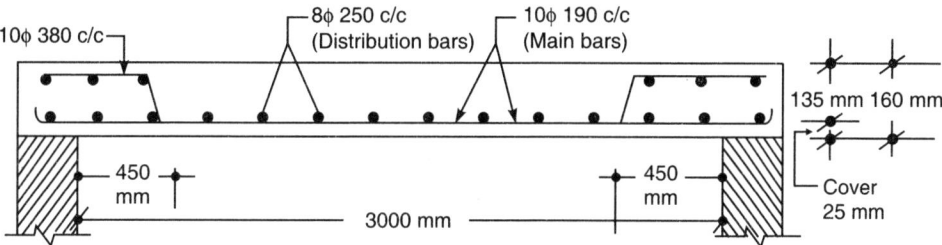

Fig. 7.4: Reinforcement details of slab

Example 7.2

Design a simply supported one-way slab supported on 230 mm walls with clear span of 4 m between the walls and located in severe environment. The slab is subjected to a live load of 4 kN/m² in addition to self-load and finishing load. Use M20 concrete and Fe 415 steel.

Solution: Effective span L_e = 4000 + 230 or 400 + d, whichever is smaller

$$\text{Assume } d = \frac{(4000 + 230)}{20K}, K = 1.25 \text{ (say)}$$

$$= 169.2 \text{ mm,}$$

Assume d = 170 mm, D = 200 mm

(under severe environmental conditions) Cover = 30 mm

$$L_e = \text{(effective span)} = 4000 + 170 = 4170 \text{ mm (4.17 m)}$$

Self-weight = 25 × 0.20 × 1 = 5 kN/m², let finishing weight = 1 kN/m²

Total load w = 5 + 1 + 4 = 10 kN/m², factored load w_u=1.5 × 10= 15 kN/m²

$$\text{Max. BM, } M_u = \frac{15 \times 4.17^2}{8} = 32.6 \text{ kN-m (32.6} \times 10^6 \text{ N-mm)}$$

Max. SF, $V_u = \dfrac{15 \times 4.17}{2} = 31.28$ kN (31280 N)

Design for moment ($b = 1$ m $= 1000$ mm)

$\quad\quad x$ (balanced section) $= 0.48d = 0.48 \times 170 = 81.6$ mm

$\quad\quad\quad\quad\quad\quad M_{u\,lim} = 0.138f_{ck}\cdot bd^2 = 79.76 \times 10^6$ N-mm $> 32.6 \times 10^6$ N-mm

Hence, under reinforced section.

$$M_u = 0.36 f_{ck}\,\dfrac{x_u}{d}\left(1 - 0.42\,\dfrac{x_u}{d}\right)bd^2 = 0.87 f_y \cdot A_{st}(d - 0.42\,x_u)$$

$$M_u = 0.87 f_y\,A_{st}d\left[1 - \dfrac{A_{st}\,f_y}{b \cdot d \cdot f_{ck}}\right]$$

or $\quad\quad 32.6 \times 10^6 = 0.87 \times 415\,A_{st}170\left[1 - \dfrac{415\,A_{st}}{1000 \times 170 \times 20}\right]$

or $531.13 - A_{st} + 0.000122\,A_{st}^2 = 0,\ A_{st}^2 - 8193\,A_{st} + 4351426 = 0$

$\quad\quad\quad\quad A_{st} = 571$ mm^2 (10ϕ @ 137 c/c, say 10ϕ @ 135 c/c,

$\quad\quad\quad\quad A_{st} = 581$ mm^2)

($< 3d$ or 300 mm c/c whichever is less)

$\quad\quad\quad\quad p_t = 0.342\%,\ \tau_c$(Table 19 of IS:456-2000, page 73) $= 0.403$ N/mm^2

$\quad$ Distribution bars $= \dfrac{0.12}{100} \times 200 \times 1000$

$\quad\quad\quad\quad = 240$ mm^2 (8ϕ @ 200 c/c, $A_{st} = 250$ mm^2)

Check for shear:

$$\tau_v = \dfrac{31280}{1000 \times 170} = 0.184 < 0.403 \times 1.2\ (< \text{design } \tau_c = 0.403\text{ N/mm}^2)$$

$$(K = 1.2 \text{ for } D = 200 \text{ mm})$$

or $\quad\quad\quad\quad \tau_v = 0.184$ N/mm$^2 < 0.484$, OK

Thus, the slab is safe in shear stress.

Check for deflection:

$$\left(\dfrac{L_e}{d}\right) = 20,\ K_t \text{ for } (p_t = 0.342\%) = 1.4,\ K_f = 1,\ K_c = 1$$

$$\left(\dfrac{L_e}{d}\right)_{\text{permissible}} = 20 \times 1.4 \times 1 \times 1 = 28$$

$$\left(\dfrac{L_e}{d}\right)_{\text{actual}} = \dfrac{4170}{170} = 24.5 < 28, \text{ OK by empirical approach.}$$

Curtailment or bent up bars (Fig. 7.5): From ends at a distance of 0.15L, alternate bars must be bent up for taking care of any fixity moment developed due to fixity in the walls.

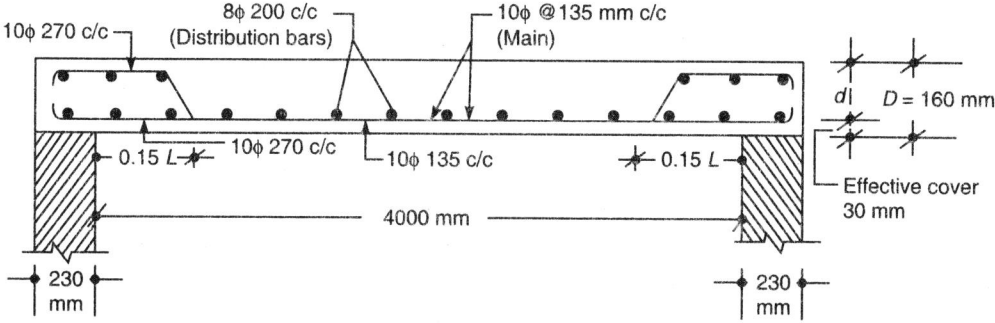

Fig 7.5: Details of reinforcement in slab

Example 7.3

A reinforced concrete slab is supported on two parallel walls 300 mm thick, spaced at a clear distance of 5.0 m. The slab carries a superimposed load of 6.0 kN/m². Design the slab using M20 grade of CC and HYSD bars of Fe 415 steel.

Solution: Clear span $L = 5$ m

Effective span $L_e = (5000 + 300)$ or $(5000 + d)$ whichever is less.

$$\text{Assume } \frac{L_e}{d} = 25, d = \frac{500 + 200}{25} = 210 \text{ mm (say)}$$

$$D = 210 + 20 + \frac{10}{2} = 235 \text{ mm}$$

Provide $d = 210$ mm, $\therefore L_e = 5.21$ m

Loads: $w_d = 0.235 \times 25 + \text{finishes} = 5.875 + 1.125 = 7.0 \text{ kN/m}^2$

Total load $w = 7.0 + 6.0 = 13.0 \text{ kN/m}^2$

Factored load $w_u = 1.5 \times 13 = 19.5 \text{ kN/m}^2$

Assume 1 m strip, $b = 1000$ mm, $d = 210$ mm

(SS) Max. BM (M_u) $= \dfrac{19.5 \times 5.21^2}{8} = 66.16 \text{ kN-m } (66.16 \times 10^6 \text{ N-mm})$

(SS) Max. SF (V_u) $= \dfrac{19.5 \times 5.21}{2} = 50.8 \text{ kN } (50.8 \times 10^3 \text{ N})$

Design of section: $b = 1000$ mm, $d = 210$ mm, $A_{st} = ?, f_{ck} = 20,$

$$f_y = 415 \text{ N/mm}^2 \ (M_u = 66.16 \times 10^6 \text{ N-mm}),$$

$$M_{u(lim)} = R_u bd^2 = 2.76 \times 1000 \times 210^2 = 121.716 \times 10^6 \text{ N-mm} > M_u$$

The section will be under reinforced since $M_{u\,lim} > M_u = 66.16 \times 10^6$ N-mm

Tensile steel A_{st} will be given by

$$A_{st} = \frac{0.5 f_{ck}}{f_y}\left[1 - \sqrt{1 - \frac{4.6 \, M_u}{f_{ck} \cdot b(d)^2}}\right] bd$$

or

$$A_{st} = \frac{0.5 \times 20}{415}\left[1 - \sqrt{1 - \frac{4.6 \times 66.16 \times 10^6}{20 \times 1000 \times 210^2}}\right] 1000 \times 210$$

or $\qquad A_{st} = \dfrac{21 \times 10^5}{415}\left[1 - \sqrt{1 - 0.345}\right] = \dfrac{21 \times 10^5}{415} \times (0.190712) = 965 \text{ mm}^2$

Provide 12 mmϕ bars at 115 mm c/c ($A_{st} = 983 \text{ mm}^2$, OK)

Percent steel $\qquad\qquad\qquad p_t = 0.47\%$

Check for shear:

$$V_u = 50.8 \times 10^3 \text{ N, tensile steel } p_t = 0.47\%$$

τ_c(design shear stress) from Table 19 of IS:456-2000 will be 0.47 N/mm^2
(K for d of 210 mm = 1.18).

$$\tau_v \text{ (nominal shear stress)} = \frac{V_u}{bd} = \frac{50800}{1000 \times 210} = 0.242$$

Permissible shear stress = $K \cdot \tau_c = 1.18 \times 0.47 = 0.56 \text{ N/mm}^2 >> 0.242$, OK

Distribution bars (temperature reinforcement):

$$A_{st2} = \frac{0.12}{100} \times 1000 \times 235 = 282 \text{ mm}^2 \text{ (Use 8 mm}\phi \text{ @ 175 mm c/c, } A_{st2} = 286 \text{ mm}^2)$$

Deflection check:

$$\frac{L}{d} = \frac{5210}{210} = 24.8 < 25, (1.25 \times 20 = 25) \text{ OK}$$

Details of reinforcement (Fig. 7.6):

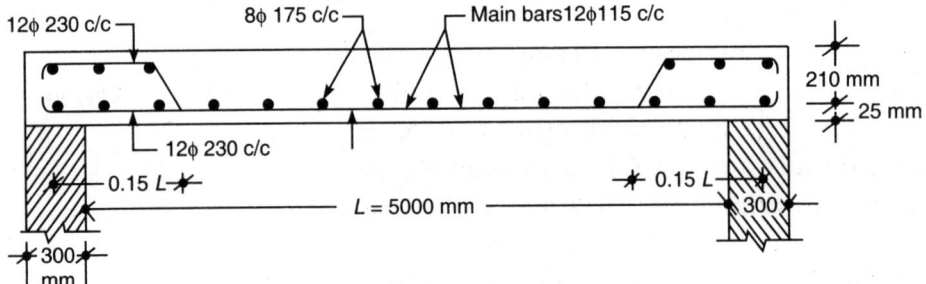

12ϕ 230 c/c 8ϕ 175 c/c Main bars12ϕ115 c/c
12ϕ 230 c/c
0.15 L 0.15 L
210 mm
25 mm
L = 5000 mm 300
300 mm

Fig 7.6: Details of reinforcement in slab

7.2 LIMIT STATE DESIGN OF ONE-WAY CONTINUOUS SLAB

Design of one-way continuous slab is similar to one-way simply supported slab except that the bending moment due to service loads exists at the supports also (instead of zero in simply supported). The hogging moments develop at the continuous supports while the sagging moments develop at the mid span sections. Generally the hogging moments at continuous support sections shall be more than the sagging moments at mid span sections. The design of the slab shall be based on hogging as well as sagging moments. The depth d is computed with respect to greater value of hogging or sagging moments and the same depth is provided for the whole slab. Tensile steel is computed separately for hogging (A_{st1}) and sagging (A_{st2}) moments. Tensile reinforcement for mid span section for sagging moment is A_{st2}. Half of the sagging moment reinforcement (A_{st2}) will be bent up for hogging moment near the supports (0.15L distance from the centre of the support). The remaining hogging moment reinforcement $(A_{st1} - \dfrac{1}{2} A_{st2})$ will be provided by extra bars. Normally support section shall be designed as the balanced section and the depth d shall be kept the same throughout. Hence, the mid span section will be under reinforcement when the support section is balanced. The procedure will be explained by following solved examples.

Example 7.4

Design a RCC continuous slab for a hall of 15 m × 6 m. The slab is supported on beams 250 m wide and monolithic with the slab. The ends of slab are supported on 300 mm thick walls. The slab carries a load of 4 kN/m² in addition to self-load and load of finishing. Use M20 CC and Fe 415 steel.

Solution: Arrangement of the beams and slab are shown in Fig. 7.7.

Span of beams = 6.30 m, assuming d of beams > 0.3 m. Span of each panel of slab = (3.1 + 0.25) or (3.1 + 0.15) intermediate continuous, L_{e1} (slab) = 3.25 m (lesser), and

$$\text{end span} = \left(2.85 + \frac{0.25}{2} + \frac{0.30}{2} \right) \text{ or } (2.85 + 0.15) \text{ whichever is less, } L_{e2} = 3.0 \text{ m.}$$

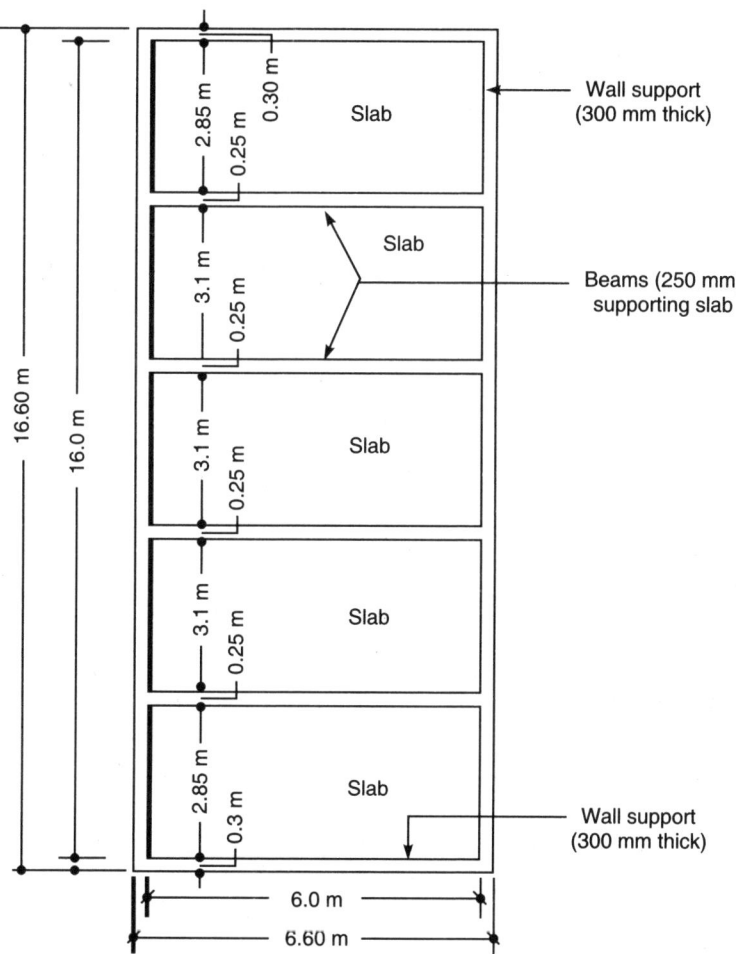

Fig. 7.7: Arrangement of beam and slab

L_{e1} and L_{e2}, the difference is less than 15% and may be considered equal for the purpose of moment coefficients given in IS:456-2000 (Table 12, page 36 and Table 13, page 36) for shear force coefficients.

Fixing of d:

For one-way $\dfrac{\text{span}}{d} = 20$, or $d = \dfrac{\text{span}}{26}$ for continuous spans and for end spans,

$$d = \dfrac{\text{span}}{\left(\dfrac{26+20}{2}\right)} = \dfrac{L}{23}.$$

Assuming 0.30% steel in tension zone, and $f_s = 240$ N/mm², modification factor $K = 1.46$ (IS:456-2000 code, Fig. 4, page 38). Thus, $d = \dfrac{3250}{(26 \times 1.46)}$ or $\dfrac{3000}{23 \times 1.46}$, whichever is greater. Thus, $d = 85.62$ mm or 89.34 mm. Adopt $d = 100$ mm, $D = 125$ mm. Clear cover = 20 mm, 10 ϕ bars, thus effective $d = 125 - 20 - \dfrac{10}{2} = 100$ mm, OK

Design constants in limit state:

$$\dfrac{x_{u\,\text{lim}}}{d} = \dfrac{700}{1100 + 0.87 \times 415} = 0.48,$$

$$R_u = 0.36 f_{ck} \times \dfrac{x_{u\,\text{lim}}}{d}\left[1 - 0.42\dfrac{x_{u\,\text{lim}}}{d}\right] = 0.36 \times 20 \times 0.48\,[1 - 0.2016] = 2.76$$

Loads, BM and SF computations:

Self-load $\quad(w_d) = 0.125 \times 25 + \text{finishing } (1.125) = 3.125 + 1.125 = 4.25 \text{ kN/m}^2$

Live load $\quad(w_l) = 4 \text{ kN/m}^2 \text{ (given)}$

Factored load $\quad w_{ud} = 4.25 \times 1.5 = 6.38 \text{ kN/m}^2$

$\qquad\qquad\qquad w_{ul} = 4 \times 1.5 = 6.00 \text{ kN/m}^2$

$\qquad\qquad\qquad L_{e2} = 3.0 \text{ m}, L_{e1} = 3.25 \text{ m}$

For end spans (IS:456-2000, Tables 12 and 13, page 36)

Mid span points $\qquad M_2 = \dfrac{+W_{ud} \cdot (L_{e2}^2)}{12} + \dfrac{W_{ul} \cdot (L_{e2}^2)}{10} = \dfrac{6.38 \times 3.0^2}{12} + \dfrac{6 \times 3.0^2}{10}$

$\qquad\qquad\qquad\qquad = 4.78 + 5.4 = + 10.18 \text{ kN-m}$

Support moment $M'_2 = -\left(\dfrac{6.38 \times 3.0^2}{10} + \dfrac{6 \times 3.0^2}{9}\right) = -(5.74 + 6) = -11.74 \text{ kN-m}$

For continuous intermediate spans:

Mid span sections: $\qquad M_1 = \dfrac{+W_{ud} \cdot L_{e1}^2}{16} + \dfrac{W_{ul} \cdot L_{e1}^2}{12} = \dfrac{6.38 \times 3.25^2}{16} + \dfrac{6 \times 3.25^2}{12}$

$\qquad\qquad\qquad\qquad = 4.21 + 5.28 = + 9.49 \text{ kN-m}$

Support moment $\qquad M'_1 = -\left[\dfrac{W_{ud} \cdot 3.25^2}{12} + \dfrac{W_{ul} \cdot 3.25^2}{9}\right]$

$$= -(5.62 + 7.04) = -12.66 \text{ kN-m}$$

Positive moments $= +10.18$ or $9.49 = +10.18 \text{ kN/m}$

Negative (hogging) moments $= -11.74$ or $-12.66 = -12.66 \text{ kN-mm}$

Thus, the depth shall be found for $M_u = 12.66 \times 10^6 \text{ N-mm}$

Design of depth d for M_u:

$$d = \sqrt{\dfrac{M_u}{R_u \cdot b}} = \sqrt{\dfrac{12.66 \times 10^6}{2.76 \times 1000}} = \sqrt{4587} = 67.73 \text{ mm}$$

Available d from serviceability criteria (deflection) $= 100 \text{ mm}$, hence the section will be under reinforced ($D = 125 \text{ mm}$).

Computation of reinforcement:

At mid span, M_2 (end span) : $A_{st2} = \dfrac{0.5 f_{ck} \cdot b \cdot d}{f_y}\left[1 - \sqrt{1 - \dfrac{4.6\, M_u}{f_{ck} \cdot bd^2}}\right] = 301 \text{ mm}^2$

$A_{st2} = 301 \text{ mm}^2/\text{m}$ (8ϕ 166 c/c, 300, 3 × 100), thus provide 8ϕ @ 160 mm c/c

$$A_{st} = 313 \text{ mm}^2, \text{ OK}$$

$$p_t = \dfrac{313 \times 100}{1000 \times 100} = 0.313\%, \text{ (bend half bars up at support)}$$

Distribution bars $= \dfrac{0.12}{100} \times 1000 \times 100 = 120 \text{ mm}^2$ (8ϕ 416 mm c/c, max. spacing 500, 450 mm. Thus provide 8 mm ϕ 400 c/c).

At middle of intermediate spans, provide the same reinforcement, i.e. 8ϕ 160 mm c/c and bend up half the reinforcement at a distance of $0.15L$ from the centre of the support and continue half the bars into the support in bottom face of the slab.

Reinforcement in the support next to the end support:

$$M'_u = -12.66 \text{ kN-m/m}$$

$$A_{st1} = \dfrac{0.5 f_{ck} \cdot b \cdot d}{f_y}\left[1 - \sqrt{1 - \dfrac{4.6\, M_u}{f_{ck} \cdot bd^2}}\right]$$

$$= \dfrac{10 \times 1000 \times 100}{415}\left[1 - \sqrt{1 - \dfrac{4.6 \times 12.66 \times 10^6}{20 \times 1000 \times (100)^2}}\right]$$

$$A_{st1} = \dfrac{10^6}{415}\left[1 - \sqrt{1 - 0.2919}\right] = \dfrac{10^6}{415}[1 - 0.8419] = 381 \text{ mm}^2/\text{m}$$

Steel already available from bent up bars from both adjacent spans $= 313 \text{ mm}^2$ (8ϕ 160 c/c)

Additional steel required $= (381 - 313) = 68 \text{ mm}^2$

Thus, provide additional pieces of 8 mm ϕ @ 735 mm c/c but max, spacing of these bars = 5 × 100 = 500 mm or 450, whichever is lesser. Thus, provide multiple of earlier bar spacing's (i.e. 8ϕ @ 160 mm c/c).

Provide additional bars 8ϕ 320 mm c/c (multiple of 160 mm c/c for uniform distribution) for a length of 0.3L = 980 mm (say)

Total A_{st} (for hogging moment) = 313 + 156 = 469 mm² > 381 mm² (reqd)

Check for shear:
$$p_{t2} = 0.469\% \text{ and } p_{t1} = 0.313\%, K = 1.30 \text{ for 100 mm}$$
τ_c (Table 19, page 73 of IS:456-2000) = 0.465 N/mm², at supports
$$= 0.39 \text{ N/mm}^2 \text{ near mid span points.}$$

Distribution (temperature) reinforcement = $\dfrac{0.12}{100}$ × 1000 × 125 = 150 mm²

Provide 8ϕ 300 mm c/c (< 450 mm, 5 × 100 mm), A_{st} = 167 mm² OK

Max V_u (Table 13 of IS:456-2000) = 0.6 × 3.25 × (6.38 + 6) = 24.57 kN

$$\tau_v = \frac{24.570 \times 10^3}{1000 \times 100} = 0.246 \text{ N/mm}^2, < 0.605 \text{ N/mm}^2 \text{ (permitted)}$$

(τ_c permitted at supports = 1.3 × 0.465 = 0.605 N/mm²), hence OK

Check for deflection:

Mid span (p_t = 0.313%), K_t (Fig. 4, page 38 of IS:456-2000) = 1.4

$$\left(\frac{l}{d}\right)_{basic} = \left(\frac{23+26}{2}\right)_{for\ end}, \left(\frac{1}{d}\right)_{permitted} = 1.4 \times 24.5 = 34.3 \text{ and } 1.4 \times 26 = 36.4$$

Actual ratio: $\quad \dfrac{L}{d}$ (at ends) = $\dfrac{3000}{100}$ = 30 < 34.3 OK

$\qquad\qquad \dfrac{L}{d}$ (intermediate) = $\dfrac{3250}{100}$ = 32.5 < 36.4, OK

Reinforcement details:

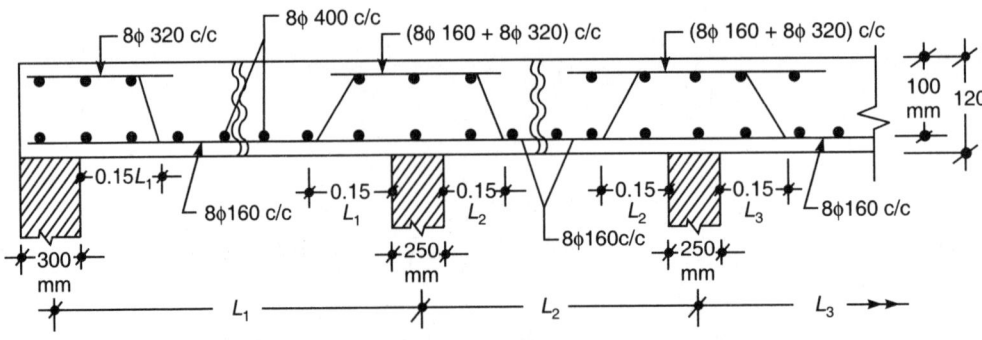

Fig. 7.8: Details of reinforcement in the continuous slab

Example 7.5

Design intermediate beams from Example 7.4.

Solution: Effective span $L_e = 6.30$ m, Assumed $d = \dfrac{L_e}{20} = 315$, say 320 mm, $D = 360$ mm

$b_w = 250$ mm, b_f (flanged beam) $= 3250$ or $\left(\dfrac{6300}{6} + 250 + 6 \times 125\right)$ (lesser of the two),

$b_f = 2050$ mm, $D_f = 125$ mm

Factored load on beam (slab + beam) $= (6.38 + 6)\,3.25 \times 1 + \left(25 \times 0.25 \times \dfrac{360 \times 125}{1000}\right)$

$\times\ 1.5$

$$w_u = 42.44 \text{ kN/m}$$

$$\frac{b_w}{b_f} = \frac{250}{2050}\ 0.122$$

SF and BM developed:

$$\text{Max. } V_u = \frac{42.44 \times 6.3}{2} = 133.7 \text{ kN (133700 N)}$$

$$\text{Max. BM } M_u = \frac{42.44 \times 6.3^2}{8} = 210.6 \text{ kN-m } (210.6 \times 10^6 \text{ N-mm})$$

Critical NA depth $x_c = 0.48 \times 320 = 153.6$ mm

Assume actual NA depth $x_u \le D_f$ ($b_f = 2050$ mm, $D_f = 125$ mm, $d = 320$ mm)

$210.6 \times 10^6 = 0.36 \times 2050 \times 20 \times x_u(320 - 0.42_u)$, $x_u^2 - 761.9\,x_u + 33972 = 0$

$x_u = 47.6$ mm ($<x_c = 153.6$ mm), under reinforced and NA is within flange.

Section considered as rectangular and $b = b_f$, $f_{ck} = 20$, $f_y = 415$ N/mm^2

Reinforcement A_{st}:

$$A_{st} = \frac{0.5 f_{ck} \cdot b_f \cdot d}{f_y}\left[1 - \sqrt{1 - \frac{4.6\,M_u}{f_{ck} \cdot b_f d^2}}\right]$$

$$= \frac{10 \times 2050 \times 320}{415}\left[1 - \sqrt{1 - \frac{4.6 \times 210.6 \times 10^6}{20 \times 2050 \times (320)^2}}\right] = 1944 \text{ mm}^2$$

Provide 4 bars 25 mmϕ ($A_{st} = 1964$ mm^2)

$$p_t\,(\% \text{ tensile steel}) = \frac{1964 \times 100}{250 \times 320} = 2.455\%,\ K_t = 0.80$$

$K_f = 0.80$ (for $\dfrac{b_w}{b_f} = 0.122$), Fig. 6 of IS:456-200 and $K_c = 1$, Fig. 5 of IS:456-2000

$$\text{Span/depth} = \left(\frac{\text{Span}}{\text{depth}}\right) \times K_t \times K_c \times K_f = 20 \times 0.80 \times 0.80 = 12.8$$

Check for deflection:

Actual ratio $\left(\dfrac{\text{Span}}{\text{depth}}\right) = \dfrac{6.30 \times 1000}{320} = 19.69 > 12.8$, unsafe for deflection, thus increase the depth or provide comp. reinforcement for $K_c > 1$.

(D = 400 mm, A_{sc}= 628 mm^2)

Recheck deflection with revised dimension.

Check for shear:

V_u = 133700 N, τ_v (nominal shear stress) = $\dfrac{133700}{250 \times 320}$ = 1.67 N/mm^2, $\tau_v > \tau_c < \tau_{c\,max}$

Provide shear reinforcement

τ_c (design stress) = 0.82 N/mm^2, V_{us} = $(V_u - 0.82 \times 250 \times 320)$

$$= (133700 - 65600) = 68100 \text{ N}$$

Provide 8ϕ 2 legged at spacing S_v = $\dfrac{0.87\ f_y \cdot A_{sv} \cdot d}{V_{us}}$

$$= \dfrac{0.87 \times 415 \times 100 \times 320}{68100} = 169.7 \text{ mm c/c}$$

Thus, provide 8ϕ 2 legged vertical stirrups at 160 mm c/c, 4 bars 25ϕ in tension face and 2 bars 20ϕ in compression face. Provide an overall depth 400 mm instead of 360 mm.

Reinforcement details (Fig. 7.9):

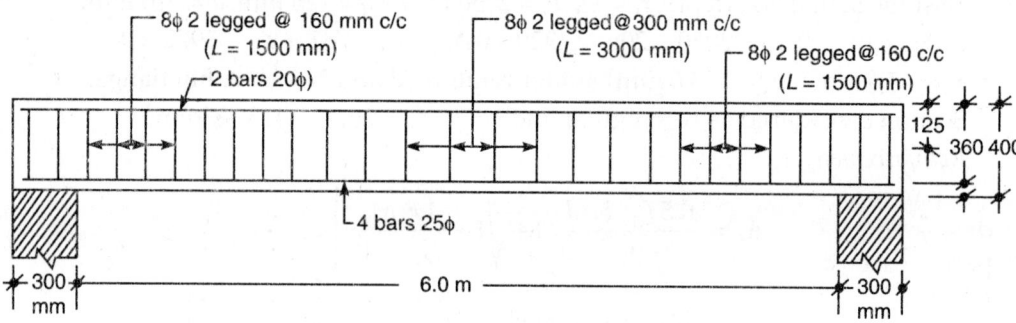

(a) Details of reinforcement of beam

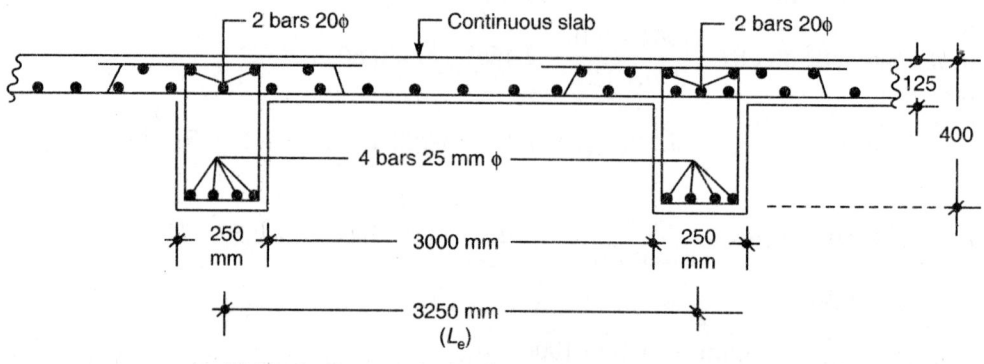

(b) Details of cross-section and beam reinforcement

Fig 7.9: Reinforcement of beam and continuous slab

7.3 LIMIT STATE DESIGN OF TWO-WAY SIMPLY SUPPORTED SLABS

7.3.1 Introduction

When a slab is supported on all the four edges, and the ratio of long span (l_y) to shorter span (l_x) is less than 2, the load is shared in both the transverse directions. Deflection and bending occurs along both the spans. Sharing of loads in both spans is found by considering consistency of deflections at any point from both the directions. A slab supported on all the four sides and the ratio of spans $\left(\dfrac{l_y}{l_x}\right)$ is less than 2, the slab is called *two-way slab* or the *slab spanning in two directions*. The maximum bending moments and maximum deflections in two-way slabs are much less than that of a one-way slab and hence designs result in thinner slabs in case of two-way slabs. In two-way slabs, special torsional reinforcement at the corners is necessary to check the cracking of slab at corners. The moments developed in two directions of a two-way slab is influenced by the ratio (r) of $\left(\dfrac{l_y}{l_x}\right)$, type of supporting edge (free, fixed or continuous), and the type and magnitude of load (u.d.l., concentrated, etc.). These slabs are classified into following 3 types:

   (i) Slabs simply supported on the four edges, with corners not held down and carrying uniformly distributed load
   (ii) Slabs simply supported on the four edges, with corners held down and carrying uniformly distributed load
   (iii) Slabs with edges fixed or continuous and carrying uniformly distributed load.

   The magnitude of reinforcement in the direction of short and long spans are designed to resist the maximum design moments developed in the mutually perpendicular directions. The analysis of a two-way slab shall be explained by IS code method.

7.3.2 Design of Simply Supported Slabs

The simply supported slabs with free edges, loading on the slab develops sagging moments towards the centre of the span and lifting of the slab at corners due to non-uniform variation of load transferred to the supports. Thus, for simply supported slabs which do not have corners held down, the maximum moments are computed as per IS:456-2000 code (Eq. 7.1).

$$M_x = \alpha_x w l_x^2, \quad M_y = \alpha_y w l_x^2 \qquad \text{... Eq. (7.1)}$$

   where,
   M_x and M_y are the maximum design moments in the x and y directions
   $$w = \text{Uniformly distributed service load on the slab}$$
   l_x and l_y are the short and long spans respectively.
   α_x and α_y are the moment coefficients whose values are given in Table 7.1.

Table 7.1: Bending moment coefficients for different ratios (r)

(According to Table 27 of IS:456-2000)

l_y/l_x	1.00	1.10	1.20	1.30	1.40	1.50	1.75	2.00	2.50	3.00
α_x	0.062	0.074	0.084	0.093	0.099	0.104	0.113	0.118	0.122	0.124
α_y	0.062	0.061	0.059	0.055	0.051	0.046	0.037	0.029	0.020	0.014

These moment coefficients are based on Rankine–Groshoff theory in which moments are computed by using the *principle of compatibility of deflection* (consistency) at a mid span point by considering strips in two orthogonal directions. According to IS:456-2000, 50% of tensile reinforcement required at mid span point is extended to the supports. Remaining 50% reinforcement is curtailed at $0.1l_x$ or $0.1l_y$ distance from the support.

7.3.3 Design of Two-Way Restrained Slabs with Corners Held Down

When the slabs are supported in such a way that the corners are prevented from lifting, such a slab is called *restrained slab*. These slabs may be discontinued over edges or may be continuous. The edges of the two-way slab are considered as rigidly supported against vertical translation. The design moments in restrained slabs are calculated by using moment coefficients given in Table 7.2 (Table 26 of IS:456-2000 code).

Moment per metre width $M_x = \alpha_x wL_x^2$, $M_y = \alpha_y wL_x^2$ *same as* ... Eq. (7.1)

For the design of two-way continuous slab, following assumptions are made in provision of reinforcement:

(i)   The positive moment reinforcement is uniformly distributed over the middle strip spreading over 75% of the span in either direction.

(ii) Edge strips comprise a width equal to $(l_x/8$ and $l_y/8)$.

(iii) Minimum reinforcement as specified in IS:456-2000 code for slabs should be provided in edge strips.

(iv) At the corners of simply supported slabs, torsion reinforcement comprising 75% of the area required for maximum middle span moment in the slab is provided in each of the four layers in the form of a mesh extending to a length of one-fifth of the shorter span (see Fig. 7.10).

7.3.4 Span/Depth Ratio in Two-Way Slabs

In the case of two-way slabs, the bending occurs in two mutually perpendicular directions. The magnitude of bending moments in either direction will be smaller than the bending moments developed in one-way slabs of similar span and similar loadings. This results in lesser main reinforcement. Thus in case of two-way slabs, the percent of tensile reinforcement (p_t) gets reduced and hence $\dfrac{\text{span}}{\text{depth}}$ ratio modification factor (K_t) for deflection gets increased from basic values of 20 for simply supported and 26 for continuous to 28 for SS and 32 for continuous slabs respectively for a loading class up to 3 kN/m².

Table 7.2: Bending moment coefficients for rectangular panels supported on four sides with provision for torsion at corners

(Table 26, IS:456-2000, page 91)

Case No.	Type of panel and moments considered	Short span coefficients α_x (value of l_y/l_x)								Long span coefficients α_y for all value of l_y/l_x
		1.0	1.1	1.2	1.3	1.4	1.5	1.75	2.0	
1.	Interior panels:									
	Negative moment at continuous edge	0.032	0.037	0.043	0.047	0.051	0.053	0.06	0.065	0.032
	Positive moment at mid-span	0.024	0.028	0.032	0.036	0.039	0.041	0.045	0.049	0.024
2.	One short edge discontinuous:									
	Negative moment at continuous edge	0.037	0.043	0.048	0.051	0.055	0.057	0.064	0.068	0.037
	Positive moment at mid-span	0.028	0.032	0.036	0.039	0.041	0.044	0.048	0.052	0.028
3.	One long edge discontinuous:									
	Negative moment at continuous edge	0.037	0.044	0.052	0.057	0.063	0.067	0.077	0.085	0.037
	Positive moment at mid-span	0.028	0.033	0.039	0.044	0.047	0.051	0.059	0.065	0.028
4.	Two adjacent edges discontinuous:									
	Negative moment at continuous edge	0.047	0.053	0.060	0.065	0.071	0.075	0.084	0.091	0.047
	Positive moment at mid-span	0.035	0.04	0.045	0.049	0.053	0.056	0.063	0.069	0.035
5.	Two short edge discontinuous:									
	Negative moment at continuous edge	0.045	0.049	0.052	0.056	0.059	0.060	0.065	0.069	–
	Positive moment at mid-span	0.035	0.037	0.040	0.043	0.044	0.045	0.049	0.052	0.035
6.	Two long edges discontinuous:									
	Negative moment at continuous edge	–	–	–	–	–	–	–	–	0.045
	Positive moment at mid-span	0.035	0.043	0.051	0.057	0.063	0.068	0.080	0.088	0.035
7	Three edges discontinuous: (One long edge continuous)									
	Negative moment at continuous edge	0.057	0.064	0.071	0.076	0.08	0.084	0.091	0.097	–
	Positive moment at mid-span	0.043	0.048	0.053	0.057	0.06	0.064	0.069	0.073	0.043
8.	Three edges discontinuous: (One short edge continuous)									
	Negative moment at continuous edge	–	–	–	–	–	–	–	–	0.057
	Positive moment at mid-span	0.043	0.051	0.059	0.065	0.071	0.076	0.087	0.096	0.043
9.	Four edges discontinuous:									
	Positive moment at mid-span	0.056	0.064	0.072	0.079	0.085	0.089	0.100	0.107	0.056

Similarly, the deflections are also smaller in case of two-way slabs than one-way slabs due to higher flexural rigidity in mutually perpendicular directions. The $\dfrac{span}{depth}$

ratio is directly influenced by the deflection modification factor K_t which is influenced by percent of tensile steel. Empirical method provided in IS:456-2000 will be generally adopted.

Example 7.6

Design a two-way slab for an office room of size 3.0 m × 5.0 m with discontinuous and simply supported edges on all the sides and with corners restrained from lifting. The slab supports a service load of 4 kN/m². Use M20 grade CC and Fe 415 HYSD grade bars.

Solution: $f_{ck} = 20$ N/mm², $f_y = 415$ N/mm², live load = 4 kN/m², $l_x = 3$ m clear

$l_y = 5$ m clear, $\dfrac{l_y}{l_x} = \dfrac{5}{3} = 1.67 < 2$, slab behaves as two-way with corners

held down by torsional reinforcement in corners.

Effective depth:

$$d_x \text{ (effective)} = \frac{3.0 \times 1000}{25} \text{ (since loading of 4 kN/m² is higher than 3 kN/m²)}$$

$$d_x = 125 \text{ mm, overall depth } (D) = 125 + 20 + \frac{10}{2} = 150 \text{ mm}$$

∴ d_x provided = $150 - 20 - \dfrac{10}{2} = 125$ mm, $d_y = 125 - 10 = 115$ mm

Effective span:

$$l_x = \text{clear span} + \text{effective depth} = 3 + 0.125 = 3.125 \text{ m}$$

Load:

Self weight = (25 × 0.15) + finishing = (3.75 + 1.25) = 5.0 kN/m²

Live load = 4.0 kN/m²

Total service load w = 5 + 4 = 9 kN/m²

Ultimate design load = 1.5 × 9 = 13.5 kN/m² (w_u)

Design moments and shear forces:

r = 1.67 from Table 27 of IS:456-2000, page 91.

$\alpha_x = 0.110$, $\alpha_y = 0.040$

∴ $M_x = \alpha_x w_u l^2_x = 0.110 \times 13.5 \times 3.125^2$

= 14.506 kN-m (14.51 × 10⁶ N-mm)

$M_y = \alpha_y w_u l^2_x = 0.040 \times 13.5 \times 3.125^2 = 5.28$ kN-m (5.28 × 10⁶ N-mm)

$V_{ux} = 0.5\, w_u l_x = 0.5 \times 13.5 \times 3.125 = 21.09$ kN (21090 N)

Design for depth:

$$M_{max} = 0.138 f_{ck} \cdot b \cdot d^2, d \text{ (critical)} = \sqrt{\frac{M_{max}}{0.138 \times f_{ck} \cdot b}} = \sqrt{\frac{14.51 \times 10^6}{0.138 \times 20 \times 1000}}$$

= 72.38 mm < d = 125 mm

Hence, selected $d_x = 125$ mm is OK, $d_y = 125 - \dfrac{10}{2} - \dfrac{10}{2} = 115$ mm

Reinforcement (for under reinforced):

$$A_{stx} = \frac{0.5\,f_{ck} \cdot b \cdot d_x}{f_y}\left[1 - \sqrt{1 - \frac{4.6\,M_u}{bd_x^{\,2}\,f_{ck}}}\right] = 341 \text{ mm}^2$$

(10ϕ 200 c/c, $A_{st} = 393$ mm^2 in x-direction)

$$A_{sty} = \frac{0.5\,f_{ck} \cdot b \cdot d_x}{f_y}\left[1 - \sqrt{1 - \frac{4.6\,M_u}{bd_y^{\,2}\,f_{ck}}}\right] = 131 \text{ mm}^2$$

(8ϕ 300 c/c, $A_{st} = 167$ mm^2 in y-direction), OK

Check for shear (short direction):

$$\tau_c \text{ (nominal shear stress)} = \frac{V_u}{bd_x} = \frac{21090}{1000 \times 125} = 0.17 \text{ N/mm}^2$$

$$p_{tx} = \frac{39300}{1000 \times 125} = 0.315\%$$

Table 19 of IS:456-2000 gives basic $\tau_c = 0.39$ N/mm^2

Design shear stress $\tau_c = 1.3 \times 0.39 = 0.507$ N/mm$^2 > 0.17$ N/mm^2, OK.

Check for deflection:

$$K_t \text{ (for } p_t = 0.315\text{), } K_t = 1.41,\ K_c = 1,\ K_f = 1,\ \left(\frac{L}{d}\right)_{basic} = 20$$

$$\left(\frac{L}{d}\right)_{max} \text{ allowed} = 20 \times 1.41 = 28.2$$

Actual $\left(\dfrac{L}{d}\right) = \dfrac{3125}{125} = 25 < 28.2$ (allowed value) and hence safe in deflection.

Torsion reinforcement:

At each corner of the slab, provide torsional reinforcement in 4 layers = 0.75×393

= 295 mm^2. Torsion reinforcement is provided in $\dfrac{l_x}{5} = \dfrac{3125}{5} = 625$ mm length ($A_{st} = 295$

$\times\, 0.625 = 185$ mm^2/m), provide 6 mm ϕ bars at 150 mm c/c at 4 corners in 4 layers as shown in Fig. 7.10.

Reinforcement in edge strips (Fig. 7.10):

$$\text{Minimum steel} = \frac{0.12}{100} \times 1000 \times 150 = 180 \text{ mm}^2/\text{m in edge strips } (8\phi\ @\ 270 \text{ c/c,}$$

$$A_{st} = 185 \text{ mm}^2)$$

Check for crack control:

$$\text{Minimum steel} = \frac{0.12}{100} \times 1000 \times 150 = 180 \text{ mm}^2$$

Spacing of main reinforcement $\ngtr 3 \times 125 = 375$ c/c (10ϕ @ 200 c/c, OK)

Diameter of reinforcing bars $< \dfrac{D}{8} = \dfrac{150}{8} = 18.75$ mm, hence the cracks will be within limits.

Provide 8 ϕ @ 270 c/c ($A_{st} = 185$ mm^2/m) in edge strips.

Details of reinforcement are shown in Fig. 7.10.

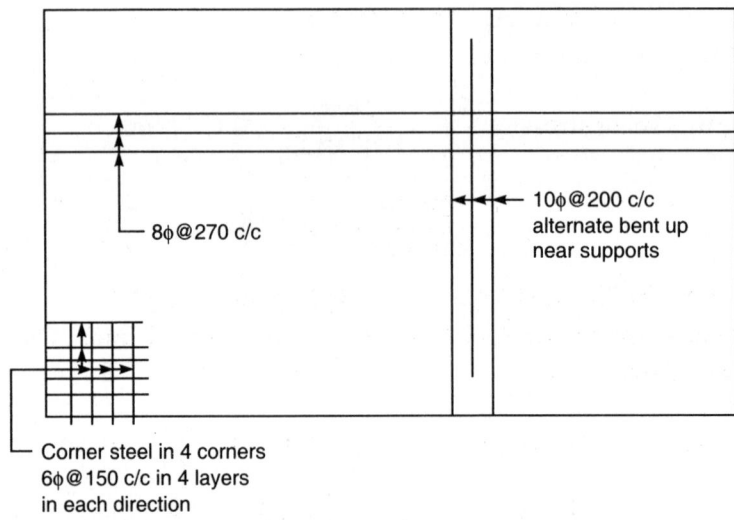

8ϕ@270 c/c

10ϕ@200 c/c
alternate bent up
near supports

Corner steel in 4 corners
6ϕ@150 c/c in 4 layers
in each direction

Fig. 7.10: Two-way slab with reinforcement details

Example 7.7

Design a beam–slab roof for a hall of 10 m × 12 m clear dimensions and supported on 300 mm thick walls on all the four edges. The beam–slab structure uses M20 CC and Fe 415 steel. The slab carries a live load of 4 kN/m^2. Assume dead load of slab and its finishing as 5 kN/m^2 area. Assume width of both secondary and main beams as 300 mm. Additional dead loads of beams may be taken as 2 kN/m run for secondary beams and 3 kN/m for main beams. The slab is cast monolithically with beams and is also continuous over beams. The slab and beams are simply supported on end walls. Secondary beams are continuous over 10 m long main beams. The beam–slab structure is so adjusted that the internal panel is 4.20 m c/c of beams while the end panels are 4.05 m centre of the end beam to the centre of the wall. Assume any other suitable data required.

Solution: $f_{ck} = 20$ N/mm^2, $f_y = 415$ N/mm^2, w_d (slab) $= 5$ kN/m^2, w_l (slab) $= 4$ kN/m^2

w_u (slab) $= 1.5 \times (5 + 4) = 13.5$ kN/m^2. Consider 1m wide strip in slab. End panel ($S_1 = 4.85 \times 3.75$ m clear), effective span $= 3.88$ m × 4.98 m $= 3.875$ m × 4.975 m

Assuming $d = 0.125$ m, $\dfrac{l_y}{l_x} = \dfrac{4.980}{3.88} = 1.284 < 2$. The slab will behave as two-way

slab continuous on two adjacent sides (Fig. 7.11(a)).

From Table 26 of IS:456-2000, page 91.

$$\alpha_x = -0.064, +0.0482, \alpha_y = -0.047, +0.035$$
$$M_{x\,(\text{max})} = -0.064 \times 13.5 \times 3.88^2 = -13.01 \text{ kN-m}$$
$$M_{x\,(\text{max})} = +0.0482 \times 13.5 \times 3.88^2 = +9.80 \text{ kN-m}$$
$$M_{y\,(\text{max})} = -0.047 \times 13.5 \times 3.88^2 = -9.6 \text{ kN-m}$$
$$M_{y\,(\text{max})} = +0.035 \times 13.5 \times 3.88^2 = +7.15 \text{ kN-m}$$

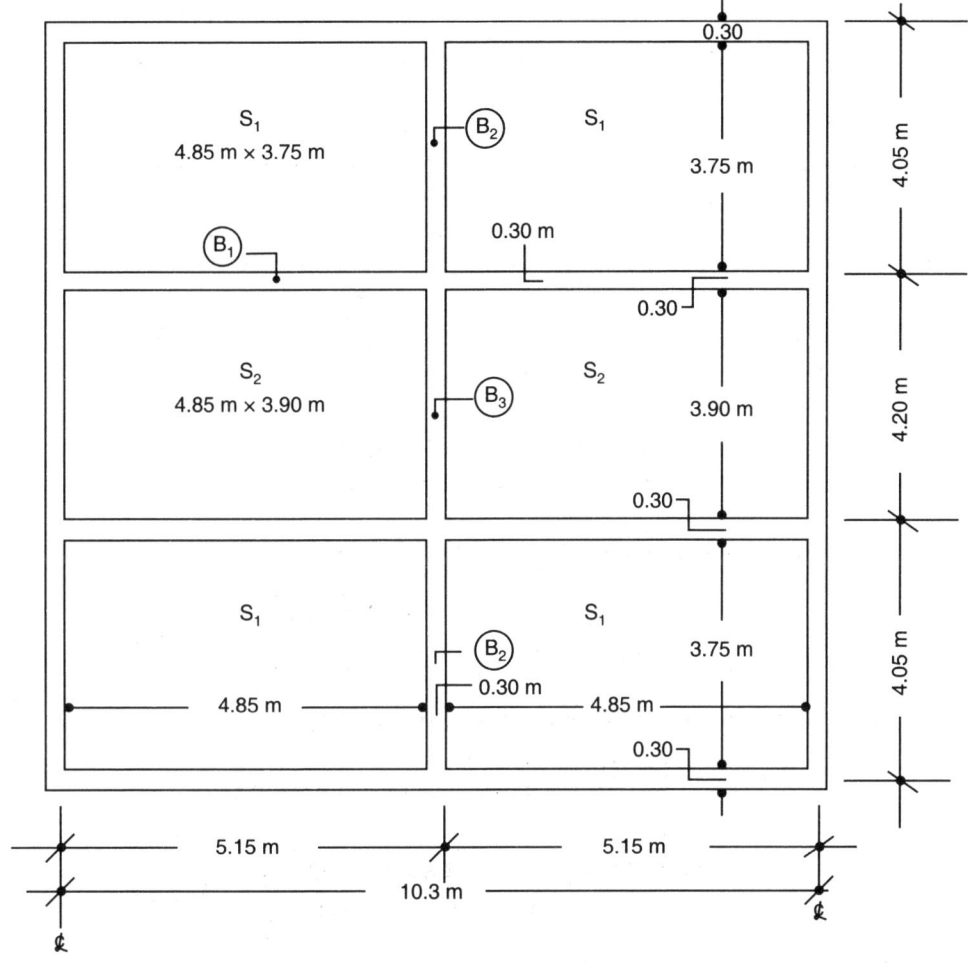

Fig 7.11(a): Arrangement of beam–slab (10 m × 12 m)

Internal panel (S_2):

Effective spans = $4.975 \times 4.025 \cong (4.98 \times 4.03)$

$$r = \dfrac{4.98}{4.03} = 1.24 < 2, \text{ one short edge discontinuous (two-way)}$$

From Table 26 of IS:456-2000:

$\alpha_x = -0.0492, +0.0372, M_{max}(M_x) = -0.0492 \times 13.5 \times 4.03^2 = -10.8$ kN-m

$= +0.0372 \times 13.5 \times 4.03^2 = +8.16$ kN-m

$\alpha_y = -0.037, +0.028,$ $\quad M_y(max) = -0.037 \times 13.5 \times 4.03^2 = -8.11$ kN-m

$= +0.028 \times 13.5 \times 4.03^2 = +6.14$ kN-m

It is seen that the slab (S_1) edge is most critical and the design shall be based on data of slab S_1. M_x (max) $= 13.01 \times 10^6$ N-mm

Consider $b = 1000$ mm

Balanced section $\qquad d = \sqrt{\dfrac{M_{max}}{0.138 \cdot b \cdot f_{ck}}} = \sqrt{\dfrac{13.01 \times 10^6}{0.138 \times 1000 \times 20}} = 68.7$ mm

$$d = \frac{L}{35} = \frac{4030}{35} = 115.1 \text{ mm}$$

Adopt 125 mm (effective) and $D = 150$ mm.

Since d (available) 125 mm is more than 68.7 mm for moment resistance, the beam section will be under reinforced. Reinforcements shall be computed for maximum negative and positive bending moments. For end panel (S_1), the steel reinforcements both for −ve BM (supports) and +ve BM (mid spans) shall be computed.

$$A_{stx} = \frac{0.5 f_{ck} \cdot b \cdot d_x}{f_y} \left[1 - \sqrt{1 - \frac{4.6 M_u}{f_{ck} \cdot b(d_x)^2}} \right]$$

$$= \frac{10 \times 1000 \times 125}{415} \left[1 - \sqrt{1 - \frac{4.6 \times 13.01 \times 10^6}{20 \times 1000 \times 125^2}} \right]$$

$A_{stx}(-ve) = 304$ mm^2 (10ϕ 250 c/c, $A_{st} = 314$ mm^2)

$$A_{stx}(+ve) = \frac{10 \times 1000 \times (125)}{415} \left[1 - \sqrt{1 - \frac{4.6 \times 9.8 \times 10^6}{20 \times 1000 \times 125^2}} \right] = 226 \text{ mm}^2$$

(10ϕ 250 c/c, 314 mm^2 alternate bent up near beam supports)

$$d_y = 125 - \frac{(10 + 10)}{2} = 115 \text{ mm (using } 10\phi \text{ bars in both directions)}$$

$$A_{sty}(-ve) = \frac{10 \times 1000 \times 115}{415} \left[1 - \sqrt{1 - \frac{4.6 \times 9.6 \times 10^6}{20 \times 1000 \times (115)^2}} \right] = 242 \text{ mm}^2$$

(10ϕ 300 c/c, $A_{st} = 262$ mm$^2 > 242$)

$$A_{sty}(+ve) = \frac{10 \times 1000 \times 115}{415} \left[1 - \sqrt{1 - \frac{4.6 \times 7.15 \times 10^6}{20 \times 1000 \times (115)^2}} \right] = 178 \text{ mm}^2 \text{ (maximum}$$

spacing 300 c/c) (hence, 10ϕ 300 c/c, alternate bent up in top face near beam supports at a distance of 0.15L).

Thus, provide main reinforcement in bottom face 10ϕ 250 c/c and bend up alternate bars at a distance of 0.15L (600 mm approximately) and extend these bars in top face (650 mm) beyond the support. Provide in transverse direction (y-direction) 10ϕ 300 c/c in bottom face at mid span section and bend alternate bars in top face near supports (0.15L = 750 mm).

Main bars (x-direction): 10 ϕ 250 c/c in bottom face at mid span section and alternate bars are bent up in top face near support (600 mm before centre of support) and extended 650 mm beyond centre of support.

Transverse bars (y-direction): 10 ϕ 300 c/c in bottom face (above main bars) at mid span section and alternate bars are bent up in top face near support (0.15L) 750 mm before centre of support and extended 800 mm beyond centre of support.

It may be noted that at intermediate supports the bent up bars must be provided from both the spans so that the main steel in top face over the beam support will be 10 ϕ 250 c/c and also at mid span the main steel will be 10 ϕ 250 (bottom face). Similarly, in transverse direction the distribution bars shall be 10 ϕ 300 c/c in bottom face and alternate bars are bent up in top face near the support beam at 750 mm from the support and extended 800 mm beyond the centre of the support.

Design of beams:

Secondary beams (3 spans continuous), b_w = 300 m.

L_1 = 4.05 m (end span), L_2 = 4.20 m (middle span), L_3 = 4.05 m (end span)

Total load from slab on span 1,

$$W_{u1} = \frac{4.05}{2} \times \frac{4.05}{2} \times 2 \times 13.5 + \text{web } (4.05 \times 2 \times 1.5) \qquad \text{(since given 2 kN/m)}$$

W_{u1} = (110.72 + 12.15) = 122.90 kN (122900 N)

Using BM and SF coefficients from IS:456-2000 (Table 12, Table 13, page 36)

$$\text{Max support BM} = -\frac{1}{9}(122.9 \times 4.05) = -55.3 \text{ kN-m } (55.3 \times 10^6 \text{ N-mm})$$

$$\text{Max mid span BM} = +\frac{1}{10}(122.9 \times 4.05) = +49.77 \text{ kN-m } (49.77 \times 10^6 \text{ N-mm})$$

Max SF (near 1st inner support) = 0.6 × 122.9 = 73.7 kN (73700 N)

For mid span BM, the beam section behaves as T-beam while for –ve support BM, the section behaves as rectangular. Let us first find d for rectangular section for M_u = 55.3 × 10⁶ N-mm

Assume the member to be balanced and b = 300 mm

$$d = \sqrt{\frac{M_u}{0.138\, f_{ck} \cdot b}} = \sqrt{\frac{55.3 \times 10^6}{0.138 \times 20 \times 300}} = 258.4 \text{ mm, provide } d = 300 \text{ mm}, D = 350 \text{ mm}$$

$$A_{st1} \text{ (support)} = \frac{0.5\, f_{ck} \cdot b \cdot d}{f_y}\left[1 - \sqrt{1 - \frac{4.6\, M_u}{f_{ck} \cdot bd^2}}\right]$$

$$= \frac{10 \times 300 \times 300}{415} \left[1 - \sqrt{1 - \frac{4.6 \times 55.3 \times 10^6}{20 \times 300 \times (300)^2}} \right] = 592 \text{ mm}^2$$

A_{st1} = 592 mm^2 (top face over supports), provide 12ϕ 4 bars in top (i.e. 2 bars bent up from either side spans) and provide 1 bar 14ϕ (154 mm^2),

$$A_{st1} = 606 \text{ mm}^2 \text{ (total = 452 + 154 = 606 mm}^2\text{)}.$$

$$p_{t1} = \frac{60600}{300 \times 300} = 0.67\%, \text{ design shear stress } \tau_c = 0.53 \text{ N/mm}^2$$

$\tau_{c(max)}$ = 2.80 N/mm^2, b_f (for T-beam) = $\dfrac{4050}{6}$ + 6 × 150 + 300 = 1875 mm

A_{st2} (for +ve BM) for T-section, assuming NA within flange and under reinforced, $b = b_f$

$$A_{st2} = \frac{0.5 \, f_{ck} \cdot b \cdot d}{f_y} \left[1 - \sqrt{1 - \frac{4.6 \, M_u}{f_{ck} \cdot bd^2}} \right]$$

$$= \frac{10 \times 1875 \times 300}{415} \left[1 - \sqrt{1 - \frac{4.6 \times 49.77 \times 10^6}{20 \times 1875 \times (300)^2}} \right] = 452 \text{ mm}^2$$

(4 bars of 12 ϕ in bottom face, A_{st} = 4 × 113 = 452 mm^2. Bend 2 bars in top face near main beam support from both adjacent spans).

Check for shear over support:

Nominal shear stress $\tau_v = \dfrac{V_u}{b_w \cdot d} = \dfrac{73700}{300 \times 300} = 0.82 \text{ N/mm}^2 > 0.53$ and $< 2.8 \text{ N/mm}^2$

$$V_{us} = (73700 - 0.53 \times 300 \times 300) = 73700 - 47700 = 26000 \text{ N}$$

Provide 8 ϕ 2 legged vertical stirrups spacing $S_v = \dfrac{0.87 \, f_y \cdot A_{sv} \cdot d}{V_{us}}$

$$S_v = \frac{0.87 \times 415 \times (2 \times 50) \times 300}{26000} = 417 \text{ mm c/c}$$

(max S_v = 300 or 300 = 300, whichever is less)

Thus provide 8ϕ 2 legged vertical stirrups at 300 mm c/c

Check for deflection at mid span point:

$$p_t = \frac{45200}{300 \times 300} = 0.50\%, \, K_t = 1.2 \text{ (Fig. 4, IS:456-2000)}, \, p_c = 0.25\%, \, K_c = 1.08,$$

$$K_f = \left(\text{for } \frac{b_w}{b_f} = 0.16 \right) = 0.80$$

$$\text{permissible} \left(\frac{\text{span}}{d} \right) = (20)\ 1.2 \times 1.08 \times 0.80 = 20.74$$

Actual $\dfrac{\text{span}}{d} = \dfrac{42000}{300}$ or $\dfrac{4050}{300} = 14 < 20.74$, OK.

Design of main beam:

Span = 10 m clear, effective = $(10 + d)$ or centre to centre of wall, whichever is less.

$L_e = 10.30$ m, simply supported, loading is heavy, assume $\dfrac{L_e}{d} = 15$

$\therefore \qquad d = \dfrac{10300}{15} = 686$, adopt $d = 690$ mm, $D = 740$ mm

Load on the beam from secondary beams and directly from the slab plus load of projected portion of web below slab (given as 3×1.5 kN/m).

W_{u1} (point load from secondary beams), max = $(73.7 + 73.7) = 147.4$ kN

$$W_{u2} \text{ (from slab plus web)} = 2 \left[\frac{4.05}{2} \times \frac{4.05}{2} \times 2 + 4.05 \times 1.10 \right] 13.5 + 3 \times 1.5 \times 10.3$$

$W_{u2} = 341.72 + 46.35 = 388.07$ kN (total u.d.l.)

Simply supported span = 10.3 m

$$\text{Max. SF} = \frac{147.4}{2} + \frac{388.07}{2} = 267.74 \text{ kN } (267740 \text{ N})$$

$$\text{Max. BM} = \frac{147.4 \times 10.3}{4} + \frac{388.07 \times 10.3}{8} = 379.555 + 499.64 = 879.2 \text{ kN-m}$$

$$= (879.2 \times 10^6 \text{ N-mm})$$

The beam acts as T-beam for +ve (sagging) moment at mid span point.

$\qquad b_w = 300$ mm, $D_f = 150$ mm, $L_e = 10.3$ m.

$$b_f = \frac{10.3 \times 10^3}{6} + 6 \times 150 + 300 = 2917 \text{ mm}, d \text{ (assumed)} = 690 \text{ mm}$$

Assuming NA to lie in the flange and under reinforced, $b = b_f$,

$$A_{st} \text{ (reqd.)} = \frac{0.5\ f_{ck} \cdot b \cdot d}{f_y} \left[1 - \sqrt{1 - \frac{4.6\ M_u}{f_{ck} \cdot bd^2}} \right]$$

$$A_{st} = \frac{10 \times 2917 \times 690}{415} \left[1 - \sqrt{1 - \frac{4.6 \times 879.2 \times 10^6}{20 \times 2917 \times 690^2}} \right] = 3670 \text{ mm}^2$$

(provide 6 bars 28 ϕ in 2 layers $(4 + 2)$, $A_{st} = 3693$ mm^2)

Also provide 2 bars 28 ϕ in top face to hold vertical stirrups.

$$p_t = \frac{369300}{300 \times 690} = 1.78\%, K_t \text{ (for deflection)} = 0.84, p_c = 0.59\%, K_c = 1.16,$$

$$K_f\left(\frac{b_w}{b_f} = 0.103\right) = 0.80, \text{ permissible } \frac{\text{Span}}{\text{depth}} = 20 \times 0.84 \times 1.16 \times 0.80 = 15.6$$

Actual $\dfrac{\text{Span}}{\text{depth}} = \dfrac{10300}{690} = 14.93 < 15.6$, permitted and hence OK.

Hence, provide 6 bars 28 ϕ in bottom face at mid span point. Near supports bend 2 bars 28 ϕ in top face at a distance of $0.20 \times 10.30 = 2.06$ m from the centre of the wall support.

Check for shear:

Max. SF $V_u = 267740$ N

Nominal shear stress $= \dfrac{267740}{300 \times 690} = 1.293 \text{ N/mm}^2 < 2.80 \text{ N/mm}^2 \text{ (max)} > 0.66 \text{ N/mm}^2,$

thus, provide shear reinforcement

$$p_t \text{ (at supports)} = \frac{4 \times 615.4 \times 100}{300 \times 690} = 1.19\% \ \tau_c \text{ (design shear)} = 0.66 \text{ N/mm}^2$$

Design of stirrups (Fig. 7.11(b):

$V_{us} = 267740$ N $- 0.66 \times 300 \times 690 - 2 \times 615.4 \times 0.87 f_y \sin 45°$ (2 bars 28 ϕ bent at 45° near support)

$V_{us} = 267740 - 136620 - 314177 = -183057$ N

Thus, provide nominal shear reinforcement 8 ϕ 2 legged at spacing S_v (maximum) @ 300 c/c.

Also provide 2 bars 16 mm ϕ in the middle of height to take care of torsion since depth of beam is more than 450 mm.

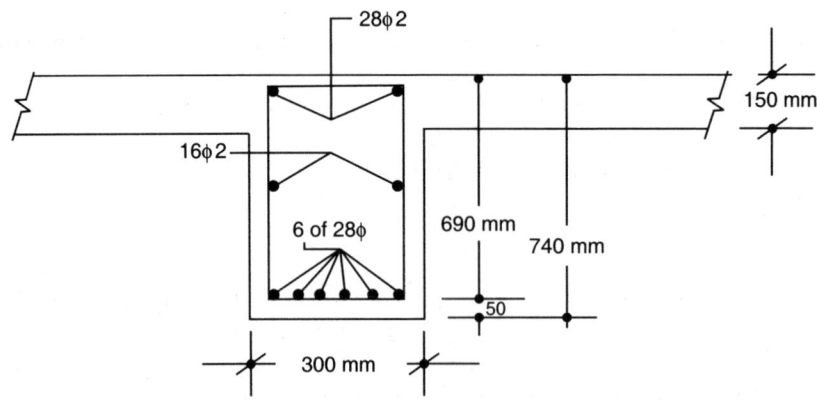

Fig 7.11(b): Section of main beam (middle span point)

7.4 FLAT SLAB

7.4.1 Introduction

Slabs supported directly on the columns without beams and cast monolithically with the columns, are known as *flat slabs.* Such flat slabs are generally reinforced in two perpendicular directions. Sometimes flat slabs are reinforced in more than two perpendiculars directions. The loads from the flat slabs are directly transferred to the supporting columns. Flat slabs provide plain ceiling free from the beam projections giving much better appearance. The flat slab requires cheaper form work. The maximum area of flat slab lies in compression zone, and hence using concrete most efficiently and making the total cost considerably less due to lower quantity of reinforcement. In flat slabs, large bending moments and shear forces develop near the columns and hence the column heads are flared at the top called column heads or column capitals. The slabs are thickened surrounding the column capitals (called drops) for reducing stresses due to moments and shears near columns. The flat slab is shown in Fig. 7.12.

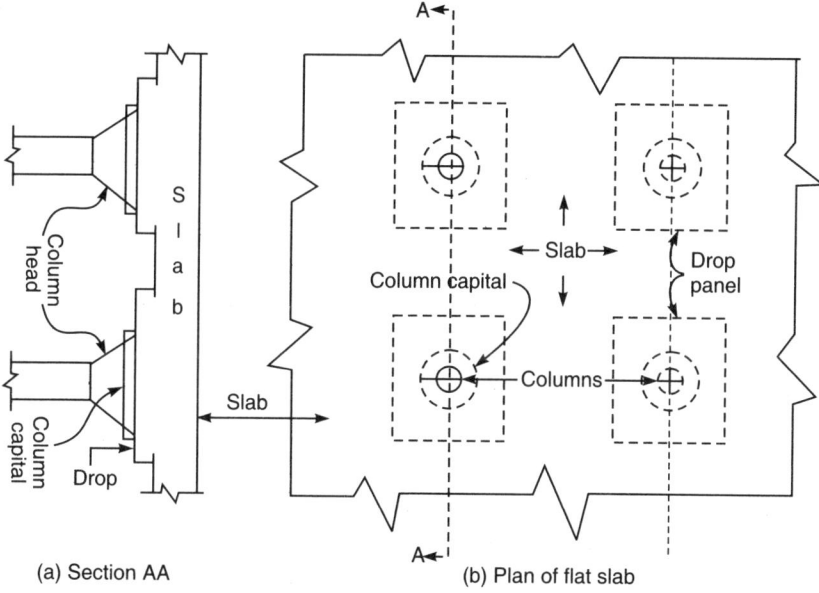

(a) Section AA   (b) Plan of flat slab

Fig 7.12: Flat slab

Various components of flat slab are shown in Fig. 7.12. We shall now understand limit state design procedure by empirical approach specified in IS:456-2000. The flat slab bends as two-way beam–slab construction. Whole slab is divided in column strips (just like beams between columns) and middle strip between two adjacent column strips. Column strips of certain breadth acts as beam supported by the columns and it acts as continuous beam. The deflections are maximum in middle of column strips. While the deflections are minimum near the supports of column strips, the deflections of central panel will be in both directions and the panel becomes like a saucer after deflection. Middle strips are supported by the column strips and thus the loads from the middle strips are transferred to column strips. Maximum shear occurs in column strips near the column in both direction. The failure may occur around the periphery of column by punching shear.

The design of flat slab shall be considered by considering panels. A panel of a flat slab is the area enclosed between the centre lines connecting adjacent columns in two directions and the outline of the column heads are shown in Figs 7.13 and 7.14.

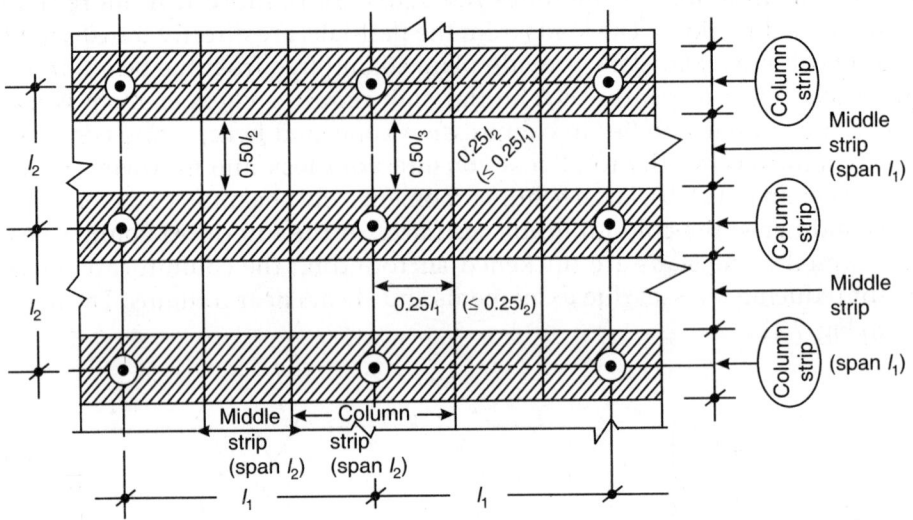

Fig. 7.13: Column and middle strips in flat slab panels

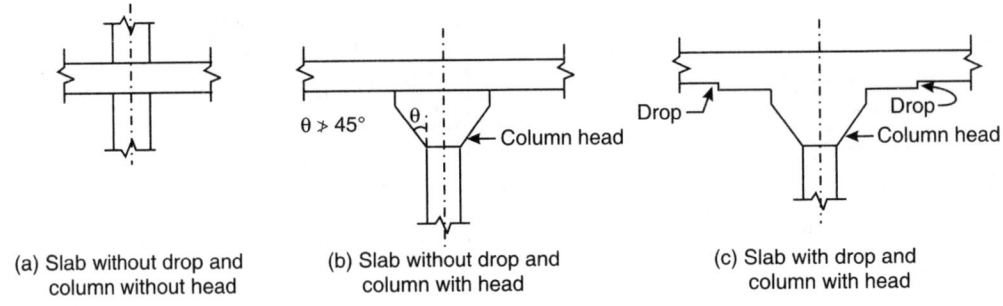

(a) Slab without drop and column without head

(b) Slab without drop and column with head

(c) Slab with drop and column with head

Fig. 7.14: Different types of flat slabs and its support

7.4.2 Design Recommendations According to IS:456-2000

Flat slabs are specified in terms of panels ($l_1 \times l_2$) and strips in two perpendicular directions. In each direction the panels are divided into *column strips* and *middle strips*.

Panel is the area of flat slab enclosed by centre lines joining adjacent columns in two perpendicular directions and the outline of the column heads. *Column strip* means a design strip having a width of $0.25l_2$ but not greater than $0.25l_1$ on each side of the column centre line, where l_1 is the span in the direction in which moments are being computed, measured centre to centre of supports and l_2 is the span transverse to l_1 measured centre to centre of supports.

Middle strip means a design strip bounded on each of its opposite sides by the column strips.

Proportioning

(a) Thickness of the flat slab shall be generally controlled by considering the effective span to effective depth ratio to control deflection as specified in IS:456-2000, e.g. simply supported slab:

$$d = \frac{\text{effective span}}{20K_t}, \text{ for two-way } d = \frac{\text{span}}{35K_t}$$

Continuous slab: $d = \dfrac{\text{effective span}}{26K_t}$, for two-way continuous slab $d = \dfrac{\text{span}}{40K_t}$

These values may be modified according to the percentage of tensile steel provided in the slab by modifying constant K_t obtained from the Fig. 4, of IS:456-2000, page 38.

(b) For slabs with drops rectangular in plan, and have a length in each direction not less than one-third of the panel length in that direction, in such a case the depth obtained for deflection control shall be directly adopted. In case this condition is not satisfied, the deflection control ratio of $\left(\dfrac{\text{span}}{d}\right)$ is multiplied by 0.90. For this, longer span must be considered.

The minimum thickness (D) of the flat slab shall be 125 mm.

Slabs with drops: $\dfrac{\text{span}}{\text{depth}} = 40$ approximately

Slab without drops: $\dfrac{\text{span}}{\text{depth}} = 36$ approximately

7.4.3 Design Moments in Flat Slab

Bending moments in flat slab can be determined in two-ways:
(a) The *direct method* as specified in IS:456-2000 (Section 31.4)
(b) The *equivalent frame* method as specified in IS:456-2000 (Section 31.5)

(a) *Direct Method*

The direct method has following limitations:

(i) The direct method is applicable only when there are minimum 3 spans in each direction.

(ii) The panels shall be rectangular and the ratio of longer span to the smaller span shall not exceed 2 $\left(\text{i.e. } \dfrac{l_x}{l_y} \not> 2\right)$.

(iii) The successive span lengths in each direction shall not differ by more than one-third of the longer span. The end spans shall be shorter but not greater than the interior spans in each directions.

(iv) The columns may be offset by a maximum of 10% of the span in the direction of offset from either axis between centre line of successive columns, not with-standing the provisions in (ii) above.

(v) The design live loads shall not exceed three times the design dead loads.

Design moments:

The absolute sum of the positive and negative bending moments in each direction is given as

$$M_o = \frac{W_u \times l_n}{8} \qquad \qquad \text{... Eq. (7.2)}$$

where,

M_o = Total moment

W_u = Total design load on the area ($l_2 \times l_n$)

l_n = Clear span extending from face to face of columns, capitals, brackets or walls but not less than $0.65 l_1$

l_1 = Length of span in the direction of M_o (l_1 = c/c of columns)

l_2 = Length of transverse to L_1 (i.e. c/c of columns in transverse direction)

IS:456-2000 recommends proportion of total moment in different strips (in column strip and middle strip) in case of interior panels and exterior panels.

In an interior span, the total design moment M_o shall be distributed in the following proportions:

- Negative design moment = $0.65 M_o$
- Positive design moment = $0.35 M_o$

when the transverse span of the panels on either side of the centre line of supports varies, l_2 shall be taken as the average of the transverse spans.

Negative and positive moments at an interior support are shared by column and middle strips as:

	Negative	Positive
Column strip –	75% of total negative (65% M_o)	60% of total positive moment (35% M_o)
Middle strip –	25% of total negative (65% M_o)	40% of total positive moment (35% M_o)

Moments in Interior Panel

Type of moments	Bending moment distribution % of M_o	
	Column strip	Middle strip
Negative	$0.65 \times 0.75 M_o$ (49% M_o)	$0.65 \times 0.25 M_o \simeq 16\% M_o \simeq 15\% M_o$
Positive	$0.35 \times 0.60 M_o$ (21% M_o)	$0.35 \times 0.40 M_o \simeq 14\% M_o \simeq 15\% M_o$
Total	70% M_o	30% M_o

Moments in Exterior Panel

The moments in the exterior panel are affected by the flexural stiffness of columns and slab. The total design moment M_o is distributed in the following proportions:

Interior support negative design moment = $\left[0.75 - \left(\dfrac{0.10}{1 + \dfrac{1}{\alpha_c}} \right) \right] M_o$ ... Eq. (7.3)

Exterior support negative design moment = $\left[\dfrac{0.65}{1 + \dfrac{1}{\alpha_c}} \right] M_o$ ... Eq. (7.4)

Positive design moment (mid span point) = $\left[0.63 - \left(\dfrac{0.28}{1 + \dfrac{1}{\alpha_c}} \right) \right] M_o$ ... Eq. (7.5)

where, α_c = Ratio of flexural stiffness of exterior columns to the flexural stiffness of the slab at a joint taken in the direction of moments

and $\alpha_c = \dfrac{\Sigma K_c}{K_s}$

where, ΣK_c = Sum of the flexural stiffness of the columns meeting at the joint

            K_s = flexural stiffness of the slab, expressed as moment per unit rotation

At the exterior support, the column strip shall be designed to resist the total negative moment in the panel at that support.

(b) Equivalent Frame Method

The structure is analysed as a continuous frame with the following assumptions:

(i) The structure is considered to be made up of equivalent frames longitudinally and transversely consisting of row of columns and strip of slab with a breadth equal to the distance between the centre lines of the panel on each side of the row of columns.

(ii) Each frame is analysed by moment distribution or any other established method. Each strip of floor and roof slab shall be analysed as a separate frame with columns above and below considered as fixed at their ends.

(iii) The relative stiffness is calculated by taking gross cross-section of the concrete alone in the calculations of the moment of inertia.

(iv) The provision of drops in the slab along the axis causes variation of moment of inertia and shall be considered in the analysis. In case of recessed slab which is made solid around the columns, the stiffening effect may be neglected if the solid part of the slab does not extend more than $0.15l_{eff}$ in to the span measured from the centre line of the columns. The stiffening effect of flared column heads may be ignored.

We shall discuss the empirical method specified in IS:456-2000 with examples. Equivalent frame method shall not be discussed in this volume dealing with fundamentals of structural concrete design only.

7.4.4 Design Shear in Flat Slab

The critical section for shear stress design lies at a distance of $\dfrac{d}{2}$ from the surface of the column/capital/drop panel, perpendicular to the plane of the slab, where d is the effective depth of the slab section. The shape in plan is geometrically similar to the support immediately below the roof or floor slab.

The nominal shear stress in flat slabs is calculated as $\left(\dfrac{V}{b_o d}\right)$, where V is the shear force due to design loads and b_o is the width of the most critical section and d is the effective depth at the critical section.

When the shear reinforcement is not provided, the calculated shear stress at the critical section shall be compared with $K_s \tau_c$,

where $\qquad K_s = (0.5 + \beta_c)$, but not greater than 1

$\qquad \beta_c$ = Ratio of short side to long side of the column/capital

$$\tau_c = 0.25\sqrt{f_{ck}} \text{ in limit state design method} \qquad \ldots \text{Eq. (7.6)}$$

$$= 0.16\sqrt{f_{ck}} \text{ in working stress design method} \qquad \ldots \text{Eq. (7.7)}$$

When the shear stress exceeds the above value, a suitable reinforcement system may be provided according to the codal provisions. In flat slab, the shear stress is controlled by the thickness of the slab and hence if nominal shear exceeds, the thickness of the flat slab is increased to make nominal shear stress within design shear stress or nominal shear stress within permissible stress in elastic or working stress method.

The design of flat slab shall be explained through following examples.

Example 7.8

Design the interior panel of a flat slab with drops for an office floor to suit the following data:

Size of office floor: 30 m × 30 m, size of panel: 6 m × 6 m, live loading: 4 kN/m², materials: concrete grade M20, steel grade Fe 415 HYSD bars.

Solution: As per IS:456-2000, for two-way continuous slab, $d = \dfrac{\text{span}}{40} = \dfrac{6000}{40} = 150$ mm

Adopt overall thickness $D = 150 + 20 + \dfrac{10 + 10}{2} = 180$ mm, (> 125 mm minimum)

Let the slab thickness at drops = 180 mm + 50 = 230 mm

Column head diameter $D \ngtr 0.25l = 0.25 \times 6000 = 1500$ mm

Length of drop $\ngtr \dfrac{L}{3}$ in either direction $\geq \dfrac{6000}{3} = 2000$ mm

Adopt drop width = width of column strip = 3000 mm (say)

∴ Middle strip = 6000 − 3000 = 3000 mm

Span of flat slab panel = 6000 mm × 6000 mm

Loads:

Self-weight of slab = 25 × 0.175 = 4.375 kN/m² (in middle strip)

Self-weight of slab (column strip) = (4.375) + (0.05 × 25) = 5.625 kN/m²

Live load $(w_l) = 4$ kN/m^2, finishes (assumed) $= 1.375$ kN/m^2

Total working load $w = (5.625 + 1.375 + 4) = 11$ kN/m^2

Total ultimate load (or factored load) $w_u = 1.5 \times 11 = 16.5$ kN/m^2

$L_1 = L_2 = 6000$ mm, $L_n = (6000 - 1500) = 4500$ mm ($> 0.65 \times L_1 = 3900$ mm)

Bending moment:

Ultimate BM:
$$M_o = \frac{W_u \cdot L_n}{8}, \quad W_u = 16.5 \times 6.0 \times 4.50 = 445.50 \text{ kN}$$

$$M_o = \frac{445.5 \times 4.5}{8} = 250.6 \text{ kN-m}$$

Interior panel (6 m $\times$ 6 m)

(i) Column stripss:

Negative moment $= \dfrac{49}{100} \times M_o = 0.49 \times 250.6 = 122.8$ kN-m (–ve)

Positive moment $= \dfrac{21}{100} \times M_o = 0.21 \times 250.6 = 52.6$ kN-m (+ve)

(ii) Middle strip:

Negative moment $= \dfrac{15}{100} \times M_o = 0.15 \times 250.6 = 37.6$ kN-m (–ve)

Positive moment $= \dfrac{15}{100} \times M_o = 0.15 \times 250.6 = 37.6$ kN-m (+ve)

Check for thickness of slab:

Thickness required at drops for –ve BM of column strip ($b = 3000$ mm)

$$d = \sqrt{\frac{M_u}{0.138\, f_{ck}\, b}} = \sqrt{\frac{122.8 \times 10^6}{0.138 \times 20 \times 3000}} = 121.8 \text{ mm } (< 230 \text{ mm provided, OK)}$$

$D = 122 + 30 = 152$ mm

d provided $(230 - 30) = 200$ mm

Thickness required in the middle strip

$$d = \sqrt{\frac{M_u}{0.138\, f_{ck}\, b}} = \sqrt{\frac{37.6 \times 10^6}{0.138 \times 20 \times 3000}} = 67.4 \text{ mm } (<180 \text{ mm provided, OK)}$$

$D = 97.4$ mm

Check for shear:

Shear stress is checked near the column head at a section $(D + d)$, i.e.

D (column head diameter) $= 1500$ mm, $d = (230 - 30) = 200$ mm

Total load on the circular area $(D + d)$ diameter,

$$W_1 = \frac{\pi}{4} (1.50 + 0.2)^2 \times 16.5 \text{ kN} = 37.45 \text{ kN (37450 N)}$$

Shear force along the periphery $= [(\text{total load}) - (\text{load on the circular area})]$

$V_s = (16.5 \times 6 \times 6 - 37.45) = (594 - 37.45) = 556.55$ kN (556.55×10^3 N)

$$V_u = \text{shear force per } m \text{ perimeter} = \frac{556.55}{\pi\,(1.70)} = 104.21 \text{ kN/m}$$

Nominal shear stress $= \dfrac{104.21 \times 10^3}{1000 \times 200} = 0.52\ \text{N/mm}^2$

According to IS:456-2000: permissible shear stress $= K_s \tau_c$

$K_s = (0.5 + \beta_c),\ \beta_c = \dfrac{L_1}{L_2} = \dfrac{6}{6} = 1,\ K_s = 1.5 \not> 1.0,\ \text{thus}\ K_s = 1$

$\tau_c = 0.25\ \sqrt{f_{ck}} = 0.25\ \sqrt{20} = 1.12\ \text{N/mm}^2$

$K_s \tau_c = 1 \times 1.12 = 1.12\ \text{N/mm}^2 > 0.52\ \text{N/mm}^2$ (actual), hence safe in shear.

Reinforcement in column and middle strips:

Column strip: $M_u = 122.8 \times 10^6\ \text{N-mm}\ (-ve),\ d = 200\ \text{mm},\ b = 3000\ \text{mm}$

$$A_{st1} = \dfrac{0.5 f_{ck}\ (b \cdot d)}{f_y}\left[1 - \sqrt{1 - \dfrac{4.6\ M_u}{f_{ck} \cdot bd^2}}\right]$$

$$= \dfrac{10 \times 3000 \times 200}{415}\left[1 - \sqrt{1 - \dfrac{4.6 \times 122.8 \times 10^6}{20 \times 3000\ (200)^2}}\right] = 1815\ \text{mm}^2$$

$$\dfrac{A_{st}}{m} = \dfrac{1815}{3.0} = 605\ \text{mm}^2/\text{m}\ (12\ \phi\ 180\ \text{mm c/c},\ A_{st} = 628\ \text{mm}^2/\text{m})$$

A_{st} for +ve moment $(M_u = 52.6\ \text{kN-m}),\ d = 150\ \text{mm}$

$$A_{st}\ (+ve) = \dfrac{0.5 f_{ck} \cdot b \cdot d}{f_y}\left[1 - \sqrt{1 - \dfrac{4.6\ M_u}{f_{ck} \cdot bd^2}}\right]$$

$$= \dfrac{10 \times 3000 \times 150}{415}\left[1 - \sqrt{1 - \dfrac{4.6 \times 52.6 \times 10^6}{20 \times 3000\ (150)^2}}\right]$$

$$= 1020\ \text{mm}^2\ \left(\dfrac{1020}{3} = 340\ \text{mm}^2/\text{m}\right)$$

$\dfrac{A_{st}}{m}\ (+ve) = 340\ \text{mm}^2/\text{m}\ (10\ \text{mm}\ \phi\ @\ 230\ \text{c/c},\ A_{st} = 341\ \text{mm}^2/\text{m})$

Middle strip: $(b = 3000\ \text{mm}),\ M_u = +37.6\ \text{kN-m and} -37.6\ \text{kN-m},\ d = 150\ \text{mm}$

$$A_{st}\ (\pm BM) = \dfrac{0.5 f_{ck} \cdot b \cdot d}{f_y}\left[1 - \sqrt{1 - \dfrac{4.6\ M_u}{f_{ck} \cdot bd^2}}\right]$$

$$= \dfrac{10 \times 3000 \times 150}{415}\left[1 - \sqrt{1 - \dfrac{4.6\ M_u}{f_{ck} \cdot b(d)^2}}\right] = 719\ \text{mm}^2$$

$\dfrac{A_{st}}{m} = 239.5\ \text{mm}^2/\text{m}\ (10\phi\ 300\ \text{c/c},\ A_{st} = 262\ \text{mm}^2/\text{m}\ (> 239.5\ \text{mm}^2/\text{m}),\ \text{OK}.$

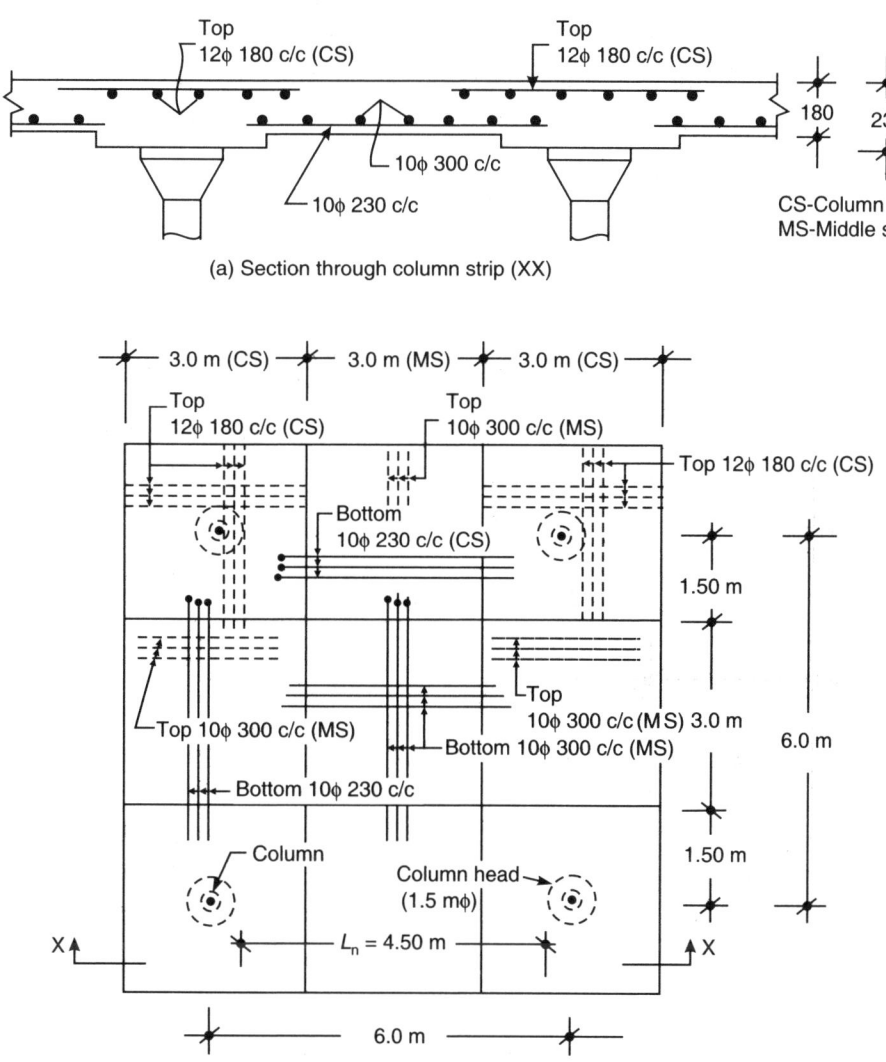

Top
12φ 180 c/c (CS)

Top
12φ 180 c/c (CS)

180   230

10φ 300 c/c

10φ 230 c/c

CS-Column strip
MS-Middle strip

(a) Section through column strip (XX)

3.0 m (CS)   3.0 m (MS)   3.0 m (CS)

Top
12φ 180 c/c (CS)

Top
10φ 300 c/c (MS)

Top 12φ 180 c/c (CS)

Bottom
10φ 230 c/c (CS)

1.50 m

Top
10φ 300 c/c (MS) 3.0 m

Top 10φ 300 c/c (MS)

Bottom 10φ 300 c/c (MS)

6.0 m

Bottom 10φ 230 c/c

Column

1.50 m

Column head
(1.5 mφ)

X L_n = 4.50 m   X

6.0 m

(b) Plan of flat slab

Fig. 7.15: Sample details of reinforcement in flat slab

Example 7.9

Design a flat slab for moment for an office room of 32 m × 24 m with columns of 450 mm × 600 mm placed at 6 m c/c in one direction and 8 m c/c in the other direction. Drops of 3 m × 3 m may be provided over the columns. Consider live load of 4 kN/m² and weight of floor finish 1 kN/m² in addition to self-weight of slab. Use M20 CC and Fe 415 grade steel.

Solution: Room is divided into panels of 6 m × 8 m. There are four types of flat slab panels S_1, S_2, S_3 and S_4 as shown in Fig. 7.16.

Thickness of slab:

For serviceability: $d = \dfrac{\text{span}}{K \, (20 \text{ or } 26)} = \dfrac{8000}{1.0(23)\,K} = \dfrac{8000}{23 \times 1.6} = 217.4$

$$D \text{ for slab} = 217.4 + \text{eff. cover} = 240 \text{ mm (say)}$$

Thickness of drop $= 1.25 \times 240 = 300$ mm

Coloumn size: 450 mm × 600 mm. Drop size 3000 mm × 3000 mm

Coloumn Head: 1500 mm × 1500 mm

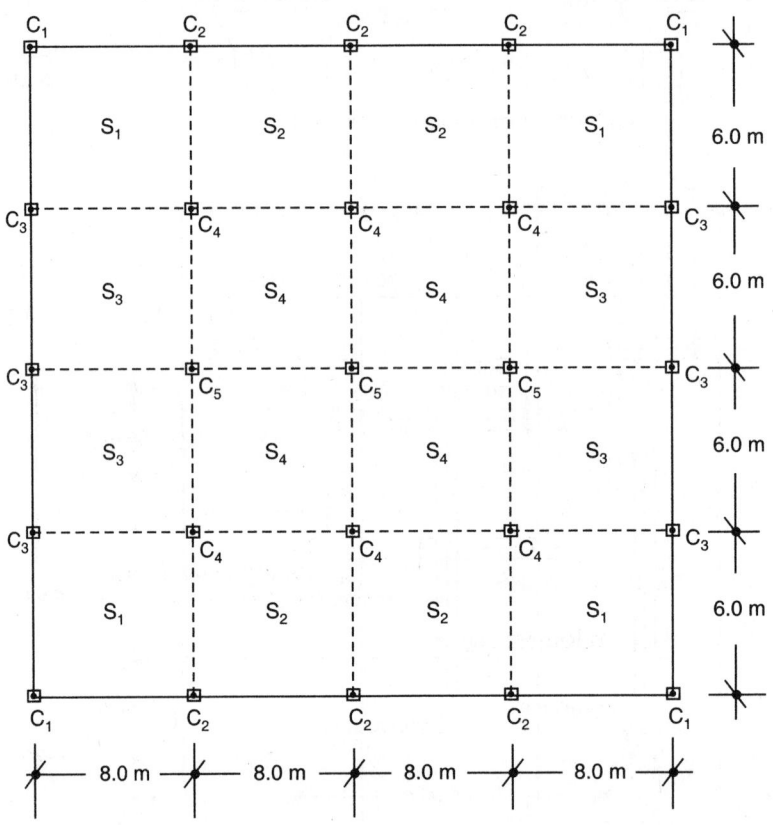

Fig. 7.16: Arrangement of flat slab panels
Slab panels: S_1, S_2, S_3 and S_4
Columns: C_1, C_2, C_3 and C_4

Loads:

Dead load of slab $= 25 \times 1 \times 1 \times 0.24 + \dfrac{25 \times 3 \times 3 \times (0.30 - 0.24)}{8 \times 6}$

$+ 1 \text{ kN/m}^2 \text{ (finishing)}$

$= (6 + 0.28125 + 1) = 7.28125 \text{ kN/m}^2$

Total u.d.l. $w = (4 + 7.28125) = 11.28125 \text{ kN/m}^2$

Ultimate design load $w_u = 1.5 (11.28125) = 16.922 \text{ N} \cong 17 \text{ kN/m}^2$

Span: Long span $l_1 = 8$ m, shorter span $l_2 = 6$ m

Clear span longer side $= (8 - 1.5) \nleftarrow 0.65 \times 8 = 6.5 \text{ m } (l_{2n})$

Clear span shorter side $= (6 - 1.5) \nleftarrow 0.65 \times 6 = 4.5 \text{ m } (l_{1n})$

(a) Moment in longer direction: $M_{uol} = \dfrac{w_u l_2 l_{1n}^2}{8} = \dfrac{17 \times 6 \times (6.5)^2}{8} = 538.7 \text{ kN-m}$

(b) Moment in Shorter direction: $M_{uos} = \dfrac{w_u l_1 l_{2n}^2}{8} = \dfrac{17 \times 8 \times 4.5^2}{8} = 344.25$ kN-m

(a) *Longer Direction Moment:*
Moments in exterior and interior panels:

Total exterior negative moment $= \dfrac{0.65 M_{uol}}{\left(1 + \dfrac{1}{\alpha_c}\right)}$

Height of column = 3.50 m (assumed)

$\alpha_c = \dfrac{\text{Summation of flexural stiffness of columns meeting at the joint (upper and lower columns)}}{\text{flexural stiffness of slab}}$

$\alpha_c = \dfrac{\Sigma K_c}{K_s} = \dfrac{2\left(\dfrac{450 \times 600^3}{12}\right)}{\dfrac{3500}{\left(\dfrac{6000 \times 240^3}{12}\right)}} = \dfrac{8000 \times 2}{3500}\left(\dfrac{450 \times 600^3}{6000 \times 240^3}\right) = 5.357$

Hence –ve moment in longer direction exterior panel

Total M_{une} (exterior) $= \dfrac{0.65 \times 538.7}{\left(1 + \dfrac{1}{5.357}\right)} = \dfrac{0.65 \times 538.7 \times 5.357}{6.357} = 295.08$ kN-m

The entire moment M_{une} goes to column strip only.
Column strip (per metre): Negative moment (exterior)

$= \dfrac{1 \times 295.08}{3} = 98.36$ kN-m/m (–ve)

Total interior end negative moment in exterior panel

Total M_{unl} (interior) $= \left\{0.75 - \dfrac{1}{1 + \dfrac{1}{\alpha_c}}\right\} M_{uol}$

$= \left\{0.75 - \dfrac{0.10 \times 5.357}{6.357}\right\} 538.7 = 358.63$ kN-m

Column strip:

Negative moment in the interior column strip/m $= \dfrac{0.75 \times 358.63}{3}$

$= 89.66$ kN-m/m (–ve)

Middle strip:

Negative moment in the interior middle strip/m $= \dfrac{0.25 \times 358.63}{3}$

$$= 29.89 \text{ kN-m/m (–ve)}$$

Total positive moment (M_{up}):

$$M_{up} = +\left(0.63 - \dfrac{0.28}{1 + \dfrac{1}{\alpha_c}}\right) M_{uol} = +\left(0.63 - \dfrac{0.28 \times 5.357}{6.357}\right) 538.7 = +212.27 \text{ kN-m}$$

Positive moment:

(Column strip)/m width $= \dfrac{0.6 \times 212.27}{3} = +42.45 \text{ kN-m/m}$

(Middle strip)/m width $= \dfrac{0.4 \times 212.27}{3} = +28.3 \text{ kN-m/m}$

Interior panel:

Total negative moment (M_{un}) $= -0.65 M_{oul} = -0.65 \times 538.7 = 350.16$ kN-m

Column strip (–ve moment) $= 0.75 M_{un} = 0.75 \times 350.16$

$$= 262.62 \text{ kN-m, } 87.54 \text{ kN-m/m}$$

Middle strip (–ve moment) $= 0.25 M_{un} = 0.25 \times 350.16$

$$= 87.54 \text{ kN-m, } 29.18 \text{ kN-m/m}$$

Total positive moment (M_{up}) $= +0.35 M_{oul} = 0.35 \times 538.7 = +188.55$ kN-m

Column strip (+ve moment) $= 0.6 \times (M_{up}) = 0.6 \times 188.55$

$$= +113.13 \text{ kN-m, } 37.71 \text{ kN-m/m}$$

Middle strip (+ve moment) $= 0.4 \times (M_{up}) = 0.4 \times 188.55$

$$= +75.42 \text{ kN-m, } 25.14 \text{ kN-m/m}$$

(b) *Shorter direction moment*

(Distribution in the exterior and interior panels)

Exterior panel:

Total exterior negative moment $M_{une} = \dfrac{0.65 \, M_{uos}}{1 + \dfrac{1}{\alpha_c}}$

$$\alpha_c = \dfrac{\Sigma K_c}{K_s} = \dfrac{\text{Summation of flexural stiffness of columns}}{\text{flexural stiffness of slab}}$$
(upper and lower columns)

$$\alpha = \dfrac{2\left(\dfrac{600 \times 450^3}{12}\right)}{3500} = \dfrac{2 \times 600 \times 450^3}{8000 \times 240^3} \times \dfrac{6000}{3500} = 1.695 \cong 1.70$$

$$M_{une} \text{ (total negative moment)} = \frac{0.65\,M_{uos}}{\left(1+\dfrac{1}{1.70}\right)} = \frac{0.65 \times 344.25 \times 1.7}{\left(1+1.7\right)}$$

$$= 140.89 \text{ kN-m } (-ve)$$

Column strip (exterior end):

$$\text{Negative moment} = 1 \times 140.89/3 = 47.0 \text{ kN-m/m}$$

Total interior negative moment:

$$M_{uni} = \left\{0.75 - \frac{0.1}{1+\dfrac{1}{\alpha_c}}\right\}, M_{uos} = \left\{0.75 - \frac{0.10}{1+\dfrac{1}{1.70}}\right\} 344.25 = 236.5 \text{ kN-m } (-ve)$$

Column strip (interior end):

$$\text{Negative moment} = 0.75 \times M_{uni}/\text{ width} = \frac{0.75 \times 236.5}{3} \text{ kN-m/m}$$

$$= 59.13 \text{ kN-m/m } (-ve)$$

$$\text{Middle strip negative (interior)} = \frac{0.25\,M_{uni}}{\text{width middle strip}} = \frac{0.25 \times 236.5}{5}$$

$$= 11.83 \text{ kN-m/m}(-ve)$$

$$\text{Total positive moment } (M_{up}) = \left(0.63 - \frac{0.28}{1+\dfrac{1}{\alpha_c}}\right) M_{uos}$$

$$M_{up} \text{ (+ve)} = \left(0.63 - \frac{0.28 \times 1.7}{2.7}\right) 344.25 = +156.2 \text{ kN-m}$$

$$\text{Column strip (positive) (width = 3 m)} = \frac{0.6\,M_{up}}{\text{width}} = \frac{0.6 \times 156.2}{3}$$

$$= 31.25 \text{ kN-m/m (+ve)}$$

$$\text{Middle strip (positive) (width = 5 m)} = \frac{0.4\,M_{up}}{\text{width}} = \frac{0.4 \times 156.2}{5} = +12.5 \text{ kN-m/m}$$

Interior panel:

Total negative moment $(M_{uni}) = 0.65 M_{uos} = 0.65 \times 344.25 = 223.76$ kN-m $(-ve)$

Column strip $(b = 3$ m),

$$\text{Negative moment} = \frac{0.75 \times 223.76}{3} = 55.94 \text{ kN-m/m } (-ve)$$

$$\text{Middle strip (negative moment)}(b = 5 \text{ m}) = \frac{0.25 \times 223.76}{5} = 11.2 \text{ kN-m } (-ve)$$

Total positive moment $(M_{upi}) = +0.35M_{uos} = 0.35 \times 344.25 = +120.5$ kN-m

Column strip ($b = 3$ m) positive moment/m $= \dfrac{0.6\,M_{upi}}{b} = \dfrac{0.6 \times 120.15}{3}$

$$= +24.1 \text{ kN-m/m}$$

Middle strip ($b = 5$ m) positive moment/m $= \dfrac{0.6\,M_{upi}}{b} = \dfrac{0.4 \times 120.15}{5}$

$$= +9.6 \text{ kN-m/m}$$

The negative moments at the junction of slab panels S_1, S_2, S_3 and S_4 are different and hence maximum negative value shall be considered for design. Maximum moment (negative in column strip) shall be greater of: (55.94, 59.13) or (87.54, 89.66), M_{max} (–ve) = 89.66 kN-m/m (internal end of external panel). Exterior end of exterior panel (–ve) maximum $M_{om} = -98.36$ kN-m/m, Max. positive moment = 42.45 kN-m/m, 37.71 kN-m/m, 31.24, 24.1 = +42.45 kN-m/m.

Design of depth is based on higher moment.

$\qquad$ Maximum –ve moment = –98.36 kN-m/m

$\qquad$ Maximum +ve moment = +42.45 kN-m/m

$$d = \sqrt{\dfrac{98.36 \times 10^6}{0.138 \times 1000 \times 20}} = 188.8 \text{ mm, say } 190 + 15 + 12 + \dfrac{12}{2} = 223 \text{ mm}$$

Provided $\quad D = 300 > 223$ mm, OK.

$$d \text{ (slab)} = \sqrt{\dfrac{42.45 \times 10^6}{0.138 \times 20 \times 1000}} = 124 \text{ mm}, D = 124 + 15 + 12 + \dfrac{12}{2} = 157 \text{ mm}$$

Provided $\qquad\qquad D = 240 > 157$ mm, OK.

Steel reinforcement shall be computed for the respective moment.

Effective depth in longer (x)-and shorter (y)-direction:

($D = 240$ mm, clear cover = 15 mm)

(Using 16ϕ bars in column strip and 12 ϕ bars in middle strip)

$$d_x = 240 - \left(15 + \dfrac{12}{2}\right) = 219 \text{ mm}, \quad d_y = 240 - \left(15 + 12 + \dfrac{12}{2}\right) = 207 \text{ mm}$$

At the junction of column strips in two directions:

$$d_x' = 300 - \left(15 + \dfrac{16}{2}\right) = 277 \text{ mm}, \quad d_y' = 300 - \left(15 + 16 + \dfrac{16}{2}\right) = 261 \text{ mm}$$

At the inter section of column strip and middle strip

$$d_x \text{ (column strip) and (middle strip)} = 240 - \left(15 + \dfrac{16}{2}\right) = 217 \text{ mm (as only 1 layer)}$$

$$d_x' \text{ (middle strip)} = 240 - \left(15 + \dfrac{12}{2}\right) = 219 \text{ mm}$$

d'_y (intersection of middle strips) $= 240 - \left(15 + 12 + \dfrac{12}{2}\right) = 207$ mm

Calculation of some reinforcements:

$$A_{stx} \text{ (–ve) column strip} = \frac{0.5 f_{ck} \cdot b \cdot d}{f_y}\left[1 - \sqrt{1 - \frac{4.6\, M_u}{f_{ck} \cdot bd^2}}\right]$$

$$= \frac{10 \times 1000 \times 277}{415}\left[1 - \sqrt{1 - \frac{4.6 \times 98.36 \times 10^6}{20 \times 1000\,(277)^2}}\right] = 1070 \text{ mm}^2$$

$(16\,\phi\ 185$ c/c, $A_{st} = 1086$ mm$^2)$
A_{sty} (–ve) column strip $(M_u = 59.13$ kN-m/m$)$

$$= \frac{0.5 f_{ck} \cdot b \cdot d}{f_y}\left[1 - \sqrt{1 - \frac{4.6 \times 59.13}{20 \times 1000\,(261)^2}}\right]$$

$$= 663 \text{ mm}^2 \ (16\,\phi\ 300 \text{ c/c}, A_{st} = 670 \text{ mm}^2)$$

$$A_{stx} \text{ (+ve) column strip} = \frac{0.5 f_{ck} \cdot b \cdot d_x}{f_y}\left[1 - \sqrt{1 - \frac{4.6\, M_u}{f_{ck} \cdot b \times d_x^2}}\right]$$

$$= \frac{10000 \times 219}{415}\left[1 - \sqrt{1 - \frac{4.6 \times 42.45 \times 10^6}{f_{ck} \times 1000\,(219)^2}}\right] = 568 \text{ mm}^2$$

$(12\,\phi\ 185$ c/c, $A_{st} = 611$ mm$^2)$

$$A_{stx} \text{ (+ve) middle strip} = \frac{0.5 f_{ck} \cdot b \cdot d_x}{f_y}\left[1 - \sqrt{1 - \frac{4.6 \times 28.3 \times 10^6}{f_{ck} \cdot b \cdot d_x^2}}\right]$$

$$= \frac{10000 \times 219}{415}\left[1 - \sqrt{1 - \frac{4.6 \times 28.3 \times 10^6}{20 \times 1000\,(219)^2}}\right] = 371 \text{ mm}^2$$

$(12\,\phi\ 300$ c/c, $A_{st} = 377$ mm$^2)$

$$A_{sty} \text{ (+ve) middle strip} = \frac{0.5 f_{ck} \cdot b \cdot d_y}{f_y}\left[1 - \sqrt{1 - \frac{4.6 \times 28.3 \times 10^6}{f_{ck} \cdot b (d_y)^2}}\right]$$

$$= \frac{10000 \times 207}{415}\left[1 - \sqrt{1 - \frac{4.6 \times 12.5 \times 10^6}{20 \times 1000\,(207)^2}}\right] = 170 \text{ mm}^2$$

$10\phi\ 300$ c/c, $A_{st} = 262$ mm^2(OK) in bottom face above transverse steel of $12\phi\ 300$ c/c in longer direction.

Thus, reinforcement details are shown in Fig. 7.17.

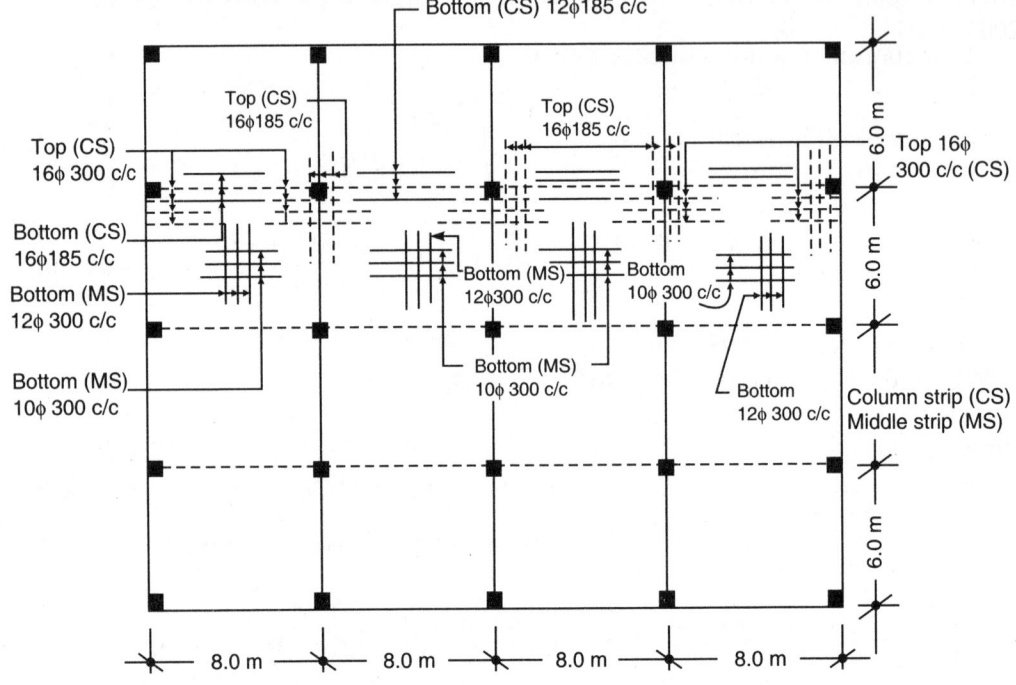

Fig. 7.17: Sample details of reinforcement (column and middle strips)

Each panel (summary of reinforcement):

Column strip – Top face near column support:	Long direction 16φ 185 c/c
	Short direction 16φ 300 c/c
– Bottom face (mid span):	Long direction 12φ 185 c/c
Middle strip – Bottom face (mid span):	Long direction 12φ 300 c/c
	Short direction 10φ 300 c/c
– Top face (near column strip):	Long direction 12φ 300 c/c
	Short direction 10φ 300 c/c

SUMMARY

Reinforced cement concrete slab forms the most common element of any building. The RCC slab may be edge supported on walls and beams or columns directly. The flat slabs may be analysed by empirical method specified in IS:456-2000. Edge supported slabs may be simply supported or continuous. The RCC slabs may be one-way or two-way. The slabs supported on all the four edges may behave as two-way supported if length *l* is less than or equal to 2 × breadth.

Effective span is considered as centre to centre of supports or clear span plus effective *d* of slab, whichever is lesser. The bending moment is computed according to the support conditions and loading pattern. The ultimate loads on slab are calculated from service loads by multiplying with suitable load factor (generally load factor = 1.50).

Effective depth of slab is assumed to control deflection by using different ratios of effective span to effective depth, for specific support conditions (as specified in IS:456-2000, page 37 and 39).

	Effective span/effective depth
• Cantilever	7
• Simply supported	20
• Continuous	26
• Two-way simply supported	35
• Two-way continuous	40

These values are modified by multiplying with a factor K_t based on percent of tension steel (p_t) as found from IS:456-2000.

The section of slab is designed by finding shear forces and bending moments for the given support conditions, loads and spans using usual limit state procedures.

$$M_{u\,lim} = R_u f_{ck} \cdot bd^2, \text{ with usual notations}$$

$$d = \sqrt{\frac{M_{u\,lim}}{R_u \cdot f_{ck} \cdot b}} \; (\text{in slabs usually } b = 1\,\text{m (1000 mm)}),$$

Steel reinforcement $\quad A_{st} = \dfrac{0.5\, f_{ck} \cdot b \cdot d}{f_y}\left[1 - \sqrt{1 - \dfrac{4.6\, M_u}{f_{ck} \cdot b(d)^2}}\right]$

By solving quadratic equation, $M_u = 0.87 f_y \cdot A_{st} \cdot d \left(1 - \dfrac{A_{st} f_y}{bd\, f_{ck}}\right)$

The check for shear stress is done by computing nominal shear stress and comparing with design shear stress specified in Table 19 in IS:456-2000, page 73.

Nominal shear stress $\tau_v = \dfrac{V_u}{b \cdot d}$.

Deflection may also be checked by computing maximum (span/depth) ratio.

For continuous slabs, depth is computed for greater of negative and positive moments while steel reinforcements are computed for both negative and positive moments separately. Usual checks for shear stress and deflection are conducted. Reinforcement is provided in appropriate tension faces according to the nature of bending moments.

Whenever the slab is supported on all the four sides and ratio of longer side to smaller side of the slab is less than or equal to 2, it is called two-way supported slab $\left(\dfrac{l_y}{l_x} \le 2\right)$. Values of BM developed in each direction depends on this ratio and can be computed by using various coefficients as specified (in Tables 26 and 27 of IS:456-2000, page 91) based on its support conditions.

$$M_x = \alpha_x \cdot w\, l_x^2, \quad M_y = \alpha_y \cdot w l_x^2$$

where, α_x, α_y are the values of coefficients obtained on the basis of ratio $\left(\dfrac{l_y}{l_x}\right)$ and support conditions (Tables 26 and 27 of IS:456-2000).

w_u = Ultimate loads (total u.d.l.)/m

l_x, l_y = Lengths of shorter and longer spans respectively.

Rest of the design procedure remains the same as per flexural members having width $b = 1000$ mm (1 m). Corners or simply supported two-way slab are reinforced for torsion.

Whenever the slab is directly supported on column, the slab is known as flat slab. The slab may be of uniform thickness or may be thickened with drop slab around the column to suit the higher moments and shears around the column. The supporting columns may also be provided with column heads (or capitals) to increase the support area. Design of flat slab is made by empirical method as specified in IS:456-2000. Flat slab load is shared in both the transverse directions. Minimum thickness of flat slab is taken as 125 mm.

Flat slab is divided in column strips and middle strips for the purpose of design. Column strips are generally taken with width = $0.50l_1$ and $0.5l_2$ in each direction. Alternatively width of column strip is taken equal to the width of drop slab. Remaining portion of slab in each direction may be taken as width of the middle strip.Initial depth of the slab is assumed by using deflection control modified ratios of span to effective depth. Design is based on bending moments, shear forces and deflection limits and further the adopted section is checked for appropriate shear stress and deflections.

Steel reinforcement is computed corresponding to the (bending moments in external and internal panels) share of BM in column strip and middle strip in both the directions. Bending moment in each panel and share of BM in column strips and middle strips is given according to IS:456-2000 as follows:

Sum of positive and negative moments in each panel

$$M_o = \frac{w_u l_n}{8} = \frac{w_u l_2 l_n^2}{8}$$

where, M_o = Absolute sum of –ve and +ve BMs

W_u = Total u.d.l. on each panel clear area ($W_u = w_u \cdot l_2 l_n$)

l_n = Clear span extending from the face of columns, capitals, or brackets to the respective faces of the span but not less than $0.65l_1$

l_1 = Length of the span in the direction of M_o (l_1 = c/c of columns)

l_2 = Length of the transverse span perpendicular to l_1.

In interior panels:

	Negative moments	Positive moments
Column strips (70% M_o)	0.65 × 0.75 M_o (49% M_o)	0.35 × 0.60 M_o (21% M_o)
Middle strips (30% M_o)	15% M_o	15% M_o

In exterior panels:

Exterior panel moments are affected by the flexural stiffness of columns and the slab. If $\alpha_c = \dfrac{\Sigma K_c}{K_s}$, the share of moments in column and middle strips shall be:

$$\text{Interior support negative design moment} = \left[0.75 - \frac{0.10}{1 + \dfrac{1}{\alpha_c}} \right] M_o$$

Exterior support negative design moment $= \left[\dfrac{0.65}{1 + \dfrac{1}{\alpha_c}} \right] M_o$

Positive design moment $= \left[0.63 - \left(\dfrac{0.28}{1 + \dfrac{1}{\alpha_c}} \right) \right] M_o$

Column strip at the exterior support shall take up the total negative moment

$\left[\dfrac{0.65\, M_o}{1 + \dfrac{1}{\alpha_c}} \right]$ at the external support.

The flat slab depth is checked for shear stress. The actual (nominal) shear stress

$= \dfrac{V_u}{b_o d}$

where, V_u is the maximum shear force at the critical section

b_o is the width of critical section

d is the effective depth of the slab at the critical section.

This shear stress at the critical section should be less than the design shear stress permitted.

τ_c: design shear stress in flat slab $= 0.25 \sqrt{f_{ck}}$ in limit state method

τ_c: permissible stress in flat slab $= 0.16 \sqrt{f_{ck}}$ in working stress method

Nominal shear stress τ_v should be less than $K_s \tau_c$, where K_s is a coefficient, calculated as $K_s = (0.5 + \beta_c)$, but not greater than 1.0, where β_c = Ratio of short side to long side of the column or capital. In case of excess shear, the depth of the slab may be increased.

PRACTICE QUESTIONS

(I) Objective Questions

Q. 7.1 Select the correct response given after each statement to complete it correctly and fill in the response sheet provided.

    (i) A rectangular two-way slab is supported on ....

      (a) two opposite sides   (b) any two sides

      (c) three sides   (d) four sides

    (ii) A roof slab, for a rectangular room ($L \times B$) is simply supported on all the four sides, will be considered as one-way slab if ....

      (a) $L > 2B$ (b) $L < 2\,B$

      (c) $L = B$ (d) $L \leq 2\,B$

(iii) Effective span of simply supported slab will be equal to ....
   (a) clear distance between the supporting walls
   (b) clear distance between the supporting walls plus half thickness of wall
   (c) centre to centre distance between the supports or clear distance between supports plus effective depth of the slab whichever is smaller
   (d) centre to centre distance between the supports or clear distance between supports plus effective depth of the slab whichever is larger

(iv) Deflection limit is checked by evaluating span/effective depth ratio modified by coefficient K_t, where K_t is based on ....
   (a) percentage of total steel in the slab $\left(\dfrac{A_{st1} + A_{st2}}{b \cdot d} \right) 100$

   (b) percentage of tensile steel $\dfrac{A_{st}}{b \cdot d} \times 100$

   (c) ratio of length to width of the slab
   (d) ratio of steel for negative BM to steel for positive BM.

(v) In limit state design approach, a 160 mm thick slab will have design dead load for 1 m strip $w/m =$ ....
   (a) $0.16 \times 25 = 4 \text{ kN/m}$
   (b) $1.50 \times 0.160 \times 25 = 6 \text{ kN/m}$
   (c) $160 \times 25 = 4000 \text{ kg/m}$
   (d) $160 \times 25 \times 1.5 = 6000 \text{ kg/m}$

(vi) Shear check for slabs is done by comparing nominal shear stress $\tau_v = \dfrac{V_u}{b \cdot d}$ with modified shear stress $K\tau_c$, where K depends on ....
   (a) percent steel in compression zone
   (b) percent steel in tension zone
   (c) thickness of the solid slab
   (d) span/effective depth ratio

(vii) In continuous slabs, for uniformity of maximum BM in intermediate and end spans, the end spans are generally kept .... in comparison to intermediate spans.
   (a) more
   (b) less
   (c) equal
   (d) independent

(viii) In two-way simply supported slab, the maximum moments depend on the ratio of longer span (l_y) to smaller span (l_x), with usual notations, the moments in two directions are computed by formulae (where, l_x = smaller side, l_y = longer side, w = total u.d.l./m on the slab).
   (a) $M_x = \alpha_x w l_x^2$ and $M_y = \alpha_y w l_y^2$
   (b) $M_x = \alpha_y w l_x^2$ and $M_y = \alpha_x w l_y^2$
   (c) $M_x = \alpha_x w l_x^2$ and $M_y = \alpha_y w l_x^2$
   (d) $M_x = \alpha_x w l_y^2$ and $M_y = \alpha_y w l_x^2$

(ix) In two-way simply supported slab, the value of longer side moment coefficient (α_y) for (M_y) ....
  (a) remains constant
  (b) increases with the ratio (l_y/l_x)
  (c) decreases with the ratio (l_y/l_x)
  (d) is independent of the ratio

 (x) The maximum total negative moment in the direction considered in an intermediate panel of a flat slab shall be ....(with usual notations).

$$\left(\text{where } M_o = \frac{W_u\, l_2\, l_n^{\,2}}{8}\right)$$

  (a) 65% of M_o (b) 49% of M_o
  (c) 21% of M_o (d) 15% of M_o

(xi) The maximum positive moment in the direction considered in column strip in an intermediate panel of a flat slab according to IS:456-2000 shall be .... with usual notations.
  (a) 65% of M_o (b) 49% of M_o
  (c) 21% of M_o (d) 15% of M_o

(xii) Which one of the following is not an assumption in the direct method of design of flat slab?
  (a) The successive span length in each direction shall not differ by more than one-third of the longer span
  (b) The panels shall be rectangular and the ratio of longer span to the smaller span shall not exceed 2
  (c) The direct method is applicable only when there are minimum 3 spans
  (d) The design live loads shall exceed three times the design dead loads

Response sheet to Q. 7.1 (i to xii)

Question	(i)	(ii)	(iii)	(iv)	(v)	(vi)	(vii)	(viii)	(ix)	(x)	(xi)	(xii)
Response (a/b/c/d)												

(II) Numerical Questions

Q. 7.2 Design a simply supported slab resting on beams of 300 mm width and spaced at 4 m clear distance between the supports. The slab carries an imposed load of 5 kN/m² in addition to a load of 2.75 kN/m² of flooring. Use M 30 grade CC and Fe 415 grade.

[**Hint:** $d = 145$ mm, $L_e = 4.145$ m, $w_d = 7.0$ kN/m², $w_1 = 5$ kN/m², $w_u = 18.0$ kN/m²
  $M_u = 38.8 \times 10^6$ N-mm, $V_u = 36000$ N, $M_{u\,lim} > M_u$ (UR)
  A_{st} (12 ϕ 140 c/c, $A_{st} = 885$ mm²), $p_t = 0.61\%$ ($K_t = 1.19$), $l/d = 23.8$ (allowed)
  Increase $d = 180$ mm, $D = 205$ mm, A_{st} (revised): 12 ϕ 150 c/c, $A_{st} = 753$ mm²
  $\tau_v = 0.226$ N/mm² (OK), distribution steel 8ϕ 300 c/c.

Q. 7.3 Design a two-way slab for an office floor size 4 m × 5 m with simply supported edges on all the 4 sides with corners held down from lifting. The slab supports a service load of 4 kN/m². Use M20 grade CC and Fe 415 HYSD steel bars.

[**Hint:** $r = \dfrac{5.17}{4.17} = 1.24 < 2$ Þ two-way, $L_x = 4.17$ m, $L_y = 5.17$ m, $d = \dfrac{4170}{25} = 170$ mm

$D = 200$ mm, $\alpha_x = 0.0880$, $\alpha_y = 0.0570$, $M_x = 23.06$ kN-m, $M_y = 15.0$ kN-m

$V_{ux} = 31.28$ kN-m, under reinforcement

$A_{stx} = 395$ mm², (10 ϕ 180 c/c $A_{st} = 436$ mm²)

$A_{sty} = 269$ mm², (10 ϕ 250 c/c $A_{st} = 314$ mm²)

$\tau_v = 0.184$ N/mm² $< 1.26 \times 0.36$ N/mm², OK.

Torsion steel corners: 4 layers of 6 mm @ 280 c/c

l/d (for deflection) $= 24.53 < 1.6 \times 20$, OK.

Q. 7.4 Design a simply supported one-way slab having clear distance of 4 m between the supports. The slab carries live load of 5 kN/m² and assume floor finish weight of 1.75 kN/m². Use M20 grade CC and Fe 415 HYSD steel. Check for shear and deflection.

[**Hint:** $l_e = 4.18$ m, $d = 180$ mm, A_{st} (main) $= 12\phi$ 170 c/c ($A_{st} = 665$ mm²), $p_t = 0.37\%$

$w_u = 18$ kN/m, $K_s = 1.13$, $\tau_c = 1.13 \times 0.42 = 0.475$ N/mm² $> \tau_v$, OK

$K_t = 1.35$, $\dfrac{L_e}{d} = 1.35 \times 20 = 27$, OK $>$ actual 23.2.

Q. 7.5 Design a two-way slab for an office floor of size 3.50 m × 4.50 m with simply supported edges on all the four sides and corners prevented from lifting. Use M20 grade CC and Fe 415 grade steel. A service load of 4 kN/m² acts on the floor. Assume necessary data.

[**Hint:** $w_u \cong 13.5$ kN/m², $d = 140$ mm, $l_{ex} = 3.64$ m, $l_{ey} = 4.64$ m, $r = 1.27$, $\alpha_x = 0.0903$,

$\alpha_y = 0.0562$, $\tau_v = 0.176$ N/mm², $\tau_c \cdot K = 1.26 \times 0.36 = 0.454$ N/mm², OK.

$A_{stx} = 10\phi$ 230 c/c, $A_{st} = 341$ mm² > 336 mm² (reqd.), $A_{sty} = 8\phi$ 220 c/c,

$A_{st} = 227$ mm². Provide 4 layer reinforcement at corners.]

Q. 7.6 An office room has clear size of 10 m × 6 m and the roof slab rests on 400 mm thick walls. Roof slab comprises 3 bays with 2 beams each of 300 mm width. The central bay has clear distance of 3.40 m between the beams and the end bays are of 3 m clear width. Design the slab and beam of the roof if the roof carries a superimposed load of 2 kN/m² in addition to self loads. Use M20 grade CC and Fe 415 HYSD grade steel. Assume any suitable data not given.

[**Hint:** End span $l_{e1} = 3.05$ m clear, intermediate span $l_{e2} = 3.40$ m clear,

L_e(beam) $= 6.40$ m, w_u (slabs) $= 10.5$ kN/m², $d_s = 100$ mm, ($D_s = 130$ mm).

M_u (slab) $= (-13.5, -10.85, +10.12, +9.77$ kN-m/m),

A_{st}(−ve) top face over supports 10α @ 190 c/c, $A_{st} = 413$ mm²/m

($p_t = 0.41\%$, $K_t = 1.30$)

A_{st}(+ve) bottom face (mid span) 10α @ 250 c/c, $A_{st} = 314$ mm²/m

($p_t = 0.31\%$, $K_t = 1.45$)

$$\frac{span}{d} = 34 \ngtr 26 \times 1.45, OK.$$

Nominal shear $\tau_v = 0.214 \, N/mm^2 < 1.3 \times 0.437 \, N/mm^2$, OK.

Beam: Total u.d.l. $W_u = 179 \, kN$, V_u (SF) $= 89.5 \, kN$, $M_u = \dfrac{W_u \cdot l_e}{8} = 143.2 \, kN\text{-}m$

$d = 400 \, mm$, $D = 450 \, mm$, τ_v (nominal) $= 0.75 \, N/mm^2$, provide vertical stirrups. A_{st} (main bottom face at mid span), 4 bars 20ϕ ($A_{st} = 1256 \, mm^2$), $p_t = 1.05\%$, $\tau_c = 0.63 \, N/mm^2$

Vertical stirrups: $8\phi \, 300 \, c/c$ (minimum)

Top face anchor bars: 2 bars of 16ϕ. At ends bend 2 bars 20 mmϕ up]

Q. 7.7 A typical floor plan is shown in Fig. 7.18. Design the interior panel of flat slab. The flat slab is subjected to a live load of 4 kN/m^2 and floor finish load of 2 kN/m^2 in addition to an average self weight of 5 kN/m^2. The flat slab is supported on square columns (400 mm × 400 mm) placed at 7000 mm c/c in one direction and 6000 mm c/c in transverse direction. Use M20 grade CC and Fe 415 grade of steel.

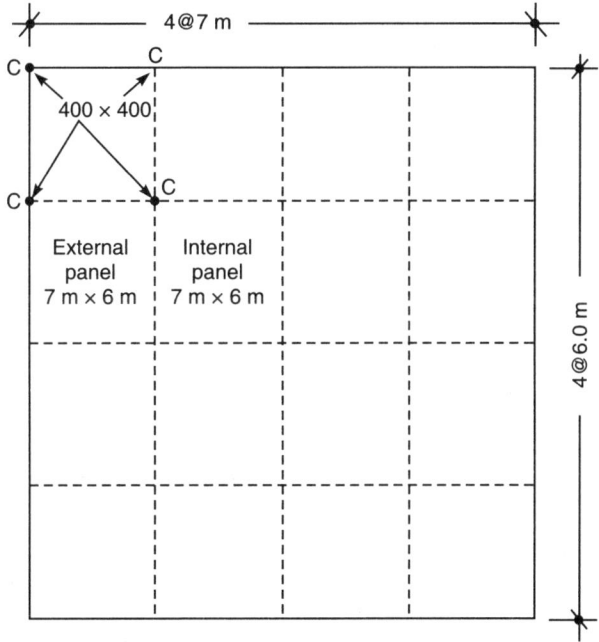

Fig. 7.18: Floor layout plan of a hall 28 m × 24 m

[**Hint:** Column 400 mm × 400 mm, column head 1.50 m × 1.50 m, drop slab = 3 m × 3 m

$d_{slab} = 190 \, mm$, drop slab = 265 mm, ($d_d = 245 \, mm$), panel = 7 m × 6 m, $l_{n1} = 5.5 \, m$, $l_2 = 6 \, m$, $w_u = 16.5 \, kN/m^2$, $W_{uol} = 545 \, kN$, $M_{ol} = 375 \, kN\text{-}m$, $M_{os} = 292.4 \, kN\text{-}m$

Top face column strip (long) = $12\phi \, 140 \, c/c$ ($A_{st} = -807 \, mm^2$)

(Short) = 12ϕ 180 c/c (A_{st} = –628 mm^2)

Bottom face middle strip (long) = +10ϕ 295 c/c (A_{st} = 266 mm^2)

(Short) = 8 ϕ 300 c/c (A_{st} = +167 mm^2)]

Q. 7.8 Design the exterior panel of the flat slab given in question 7.7 (panel size 7 m × 6 m), w = 16.5 kN/m^2. Assume column height = 3.50 m.

[**Hint:** Column 400 mm × 400 mm, column head 1.50 m × 1.50 m, drop slab = 3 m × 3 m

Assume d_s = 190 mm, d_{ds} = 244 mm, panel 7 m × 6 m, l_{n1} = 5.50 m, l_2 = 6 m,

Long side: $M_{uol} \cong$ 375 kN-m, Short side M_{uos} = 292.4 kN-m

$$\text{Long side: } \alpha_c = \frac{\dfrac{2}{3500}\left(\dfrac{400}{12} \times 400^3\right)}{6000 \times \dfrac{210^3}{12}} = 1.843, \left(1 + \frac{1}{\alpha_c}\right) = 1.5426$$
$$\frac{}{7000}$$

Exterior panel:

Total exterior –ve moment: $M_{une} = \dfrac{0.65\, M_{uol}}{\left(1 + \dfrac{1}{\alpha_c}\right)} = -158$ kN-m

Carried totally by exterior column strip (3 m) = $\dfrac{1 \times 158}{3} = -52.70$ kN-m/m

(12ϕ 170 c/c, A_{st} – 665 mm^2)

Total interior (full panel) negative moment = $\left(0.75 - \dfrac{0.1}{\left(1 + \dfrac{1}{\alpha_c}\right)}\right)$

$$M_{uol} = 257 \text{ kN-m}$$

Column strip (3 m) negative moment = $\dfrac{0.75 \times 257}{3} = -64.24$ kN-m/m

(A_{st} = 12 ϕ 140 c/c, A_{st} = 807 mm^2/m)

Middle strip (3 m) negative moment = $\dfrac{0.25 \times 257}{3} = -21.42$ kN-m/m

(10ϕ 240 c/c, A_{st} = 327 mm^2/m)

Total positive moment (long side):

$$M_{up} = \left(0.63 - \dfrac{0.28}{\left(1 + \dfrac{1}{\alpha_c}\right)}\right) M_{uol} = +168.2 \text{ kN-m}$$

Column strip ($b = 3$ m) positive $= \dfrac{0.6 \times 168.2}{3} = + 33.64$ kN-m/m

(10ϕ 150 c/c, $A_{st} = 523$ mm^2/m)

Middle strip ($b = 3$ m) positive $= \dfrac{0.4 \times 168.2}{3} = + 22.42$ kN-m/m

(10ϕ 230 c/c, $A_{st} = 341$ mm^2/m)

Short direction:

$$\alpha_c = 1.354, \left(1 + \dfrac{1}{\alpha_c}\right) = 1.74$$

M_{une} (total) $= \dfrac{0.65 \times 292.4}{1.74} = - 109.32$ kN-m

Column strip ($b = 3$ m, exterior) $= -36.44$ kN-m/m (12ϕ 250 c/c, $A_{st} = 452$ mm^2/m) (top)

Total negative on interior support $= \left(0.75 - \dfrac{0.1}{1.74}\right) M_{uol} = - 202.5$ kN-m

Column strip (–ve) $= \dfrac{202.5}{3} \times 0.75 = - 50.6$ kN-m/m (12ϕ 170c/c,

$A_{st} = 665$ mm^2/m)

Middle strip (–ve) $= \dfrac{0.25 \times 202.5}{4} = - 12.7$ kN-m/m (10ϕ 300c/c,

$A_{st} = 262$ mm^2/m)

Total positive $M_{up} = \left(0.63 - \dfrac{0.28}{1 + \dfrac{1}{\alpha_c}}\right) M_{uos} = 137.1$ kN-m

Column strip $= \dfrac{+ 0.60 \times 137.1}{3} = + 27.4$ kN-m/m (column strip 10ϕ 180 c/c,

$A_{st} = 436$ mm^2/m)

Middle strip $= \dfrac{+ 0.4 \times 137.1}{4} = + 13.71$ kN-m/m

(Middle strip, 10ϕ 300 c/c, $A_{st} = 262$ mm^2/m).

UNIT – IV Design of Compression Members (Columns)

Chapter 8

Limit State Design of Columns

LEARNING OBJECTIVES

After the study of *limit state design of columns*, the learner will understand the principles and procedure of limit state design of different types of columns and will be able to:

- ◉ Differentiate between long and short columns
- ◉ State effect of support conditions of columns on its design
- ◉ Compute load carrying capacity of short columns of square, rectangular, and circular sections
- ◉ Compute load carrying capacity of long columns of different sections
- ◉ Compute load carrying capacity of columns subjected to eccentric loads
- ◉ Design column sections for the given loads and other conditions

8.1 INTRODUCTION

8.1.1 General

Columns are defined as structural members used primarily to support direct compressive loads. A *column* or *strut* is a compression member, the effective length of which exceeds three times the least lateral dimension. When the member is supporting vertical axial load, it is called *column*. Inclined or horizontal member carrying axial compressive load is termed *strut*. Based on structural or architectural requirements, columns may be of different shapes, such as square, rectangular, circular or hexagonal. Based on the type of lateral bracings to the longitudinal bars and composition of the cross-section, the columns are classified as follows:

(i) *Tied columns*: The longitudinal bars of column are tied by closed loops called ties at close intervals to avoid buckling of longitudinal bars.

(ii) *Spiral columns*: These columns are circular and the longitudinal bars along with concrete core is warped with a closely spaced helix of transverse steel instead of simple ties.

(iii) *Composite columns*: These columns consist of structural steel or cast iron sections encased in concrete reinforced with both longitudinal bars and transverse ties.

(iv) *Concrete filled steel tubes*: These columns are circular, square or rectangular hollow sections filled with concrete without any additional reinforcement.

The sketches of these columns are shown in Fig. 8.1.

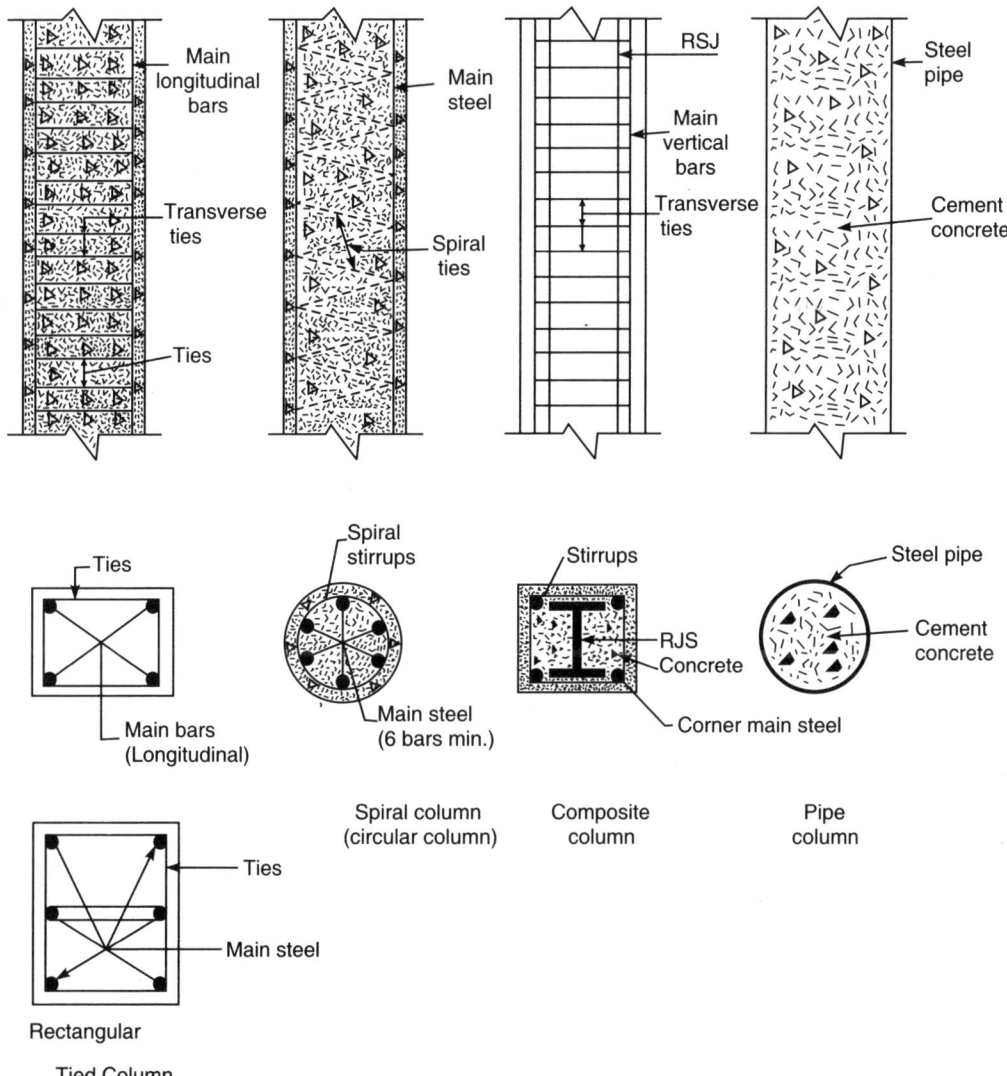

Fig. 8.1: Different types of columns

8.1.2 Type of Columns

Concentrically loaded columns are subjected to axial stresses but such columns rarely exist in practical life. Generally columns in actual structure are subjected to moments along with axial loads. If the moment acts about one axis only, the columns are classified as uniaxially eccentrically loaded columns. If moments act about both the axes, the columns are classified as biaxially eccentrically loaded columns. Columns are further classified as pedestal, short and long (slender) columns depending on its effective length and lateral dimensions ratio. The effective length is the distance between the points of inflexion of the column. Very short columns having effective lengths less than 3 times the least lateral dimension are called *pedestal columns*. Columns having effective lengths equal to or less than 12 times the least lateral dimensions are called *short columns*. Columns having effective lengths greater than 12 times the least lateral dimensions are called *long or slender columns*. The behaviour and design of short and slender columns are based on different principles.

8.1.3 Behaviour of Columns Under Load

A column is subjected to gradually increasing axial load, stresses develop both in concrete and steel. Concrete and steel stresses developed remain in proportion to their elastic moduli up to certain load. With passage of time, the concrete develops shrinkage and creep strains and at this stage additional load is shared by steel bars till steel attains its yield strength. Further increase of load is shared by concrete till concrete attains its maximum strength. During this stage spirally reinforced columns undergo large deformations before failure (ductile failure). Tied columns undergo brittle failure after yield point of steel. In this the concrete fails suddenly without undergoing much deformations (Fig. 8.2).

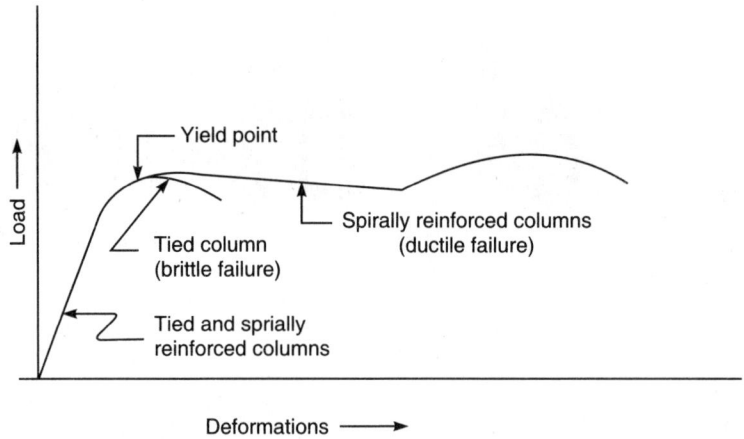

Fig. 8.2: Load versus deformations curve for RCC columns

8.1.4 Design Approach

The design of columns is quite complex specially when it is subjected to eccentric loads in addition to axial loading. The design of column requires determination of cross-sectional dimensions, the area of main longitudinal reinforcement and transverse ties or spiral reinforcement to carry safely the given loads. The design is governed by the maximum axial load and the maximum moments. Lateral ties are provided to safeguard buckling of longitudinal bars.

The column design is carried out either by limit state method or by working stress method. Limit state design is adopted by most of the countries and details are provided in their country's standard code. Indian standard code IS:456-2000 provides the details of this method. Limit state design method shall be explained for different cases of column design.

The limit state method of design of column is based on the behavior of structural element under compressive loads at the collapse and ensures adequate margin of safety. The serviceability condition is assumed to be satisfied as the deflections and deformations under compression are very small. The working stress method of design is based on the behavior of the structural element at working load and ensures the adequate safety margin in the stresses of concrete and steel.

8.1.5 Effective Length of Column

Unsupported length (l) of a compression member shall be taken as the clear distance between end restraints. It is the clear distance between the floor and lateral restraint such as bracket, lateral beam or strut, etc.

The effective length (l_e) of column or compressive member in a framed structure may be taken as the distance between the two consecutive points of inflexion created by bending or buckling along the length of the column. The effective length of a column influences its load carrying capacity. The effective length depends on the end support restraints of the column. Table 28 of IS:456-2000 provides guidelines to assess the effective length of compression member according to the support conditions and is reproduced in Table 8.1.

Table 8.1: Effective length of compression members
(As per Table 28 of IS:456-2000)

Degree of end restraint of compression members	Symbol	Theoretical value of effective length	Recommended value of effective length
(1)	(2)	(3)	(4)
Effectively held in position and restrained against rotation in both ends		0.50 l	0.65 l
Effectively held in position at both ends, restrained against rotation at one end		0.70 l	0.80 l
Effectively held in position at both ends, but not restrained against rotation		1.00 l	1.00 l

(Contd.)

Table 8.1: Effective length of compression members (contd.)
(As per Table 28 of IS:456-2000)

Degree of end restraint of compression members	Symbol	Theoretical value of effective length	Recommended value of effective length
(1)	(2)	(3)	(4)
Effectively held in position and restrained against rotation at one end, and at the other restrained against rotation but not held in position		1.00 l	1.20 l
Effectively held in position and restrained against rotation in one end, and at the other partially restrained against rotation but not held in position		$\dfrac{1}{N}$	1.50 l
Effectively held in position at one end but not restrained against rotation, and at the other end restrained against rotation but not held in position		2.00 l	2.00 l
Effectively held in position and restrained against rotation at one end but not held in position not restrained against rotation at the other end		2.00 l	2.00 l

Note: l is the unsupported length of compression member.

8.2 LIMIT STATE DESIGN OF COLUMNS

8.2.1 Assumptions in Design

Limit state design of columns is based on the following assumptions:

(i) The plane section normal to the axis of the column before deformation remains plane after deformation, i.e. the strain at any point is proportional to its distance from the neutral axis.

(ii) The tensile strength of concrete is ignored.

(iii) The short term static modulus of elasticity of concrete $E_c = 5000\sqrt{f_{ck}}$ N/mm^2

(iv) The failure of concrete is governed by the maximum strain criteria. For concentric load, the ultimate strain in concrete is taken as 0.002. The ultimate strain in concrete in bending is taken as 0.0035.

(v) The compressive strength of concrete is $\dfrac{0.67\,f_{ck}}{1.50} = 0.446f_{ck}$. The partial safety factor for concrete $\gamma_m = 1.50$, and hence design strength of concrete $= 0.446f_{ck}$.

(vi) The design strength of steel $= \dfrac{f_y}{1.15} = 0.87f_y$,

(assuming γ_m for steel $= 1.15$).

8.2.2 Short and Long Columns

Columns are divided into two categories: (i) short and (ii) long depending on the ratio of effective length to least lateral dimension. When this ratio of effective length to the least lateral dimension is less than 12, the columns are called *short*. When the ratio of effective length (l_e) to the least lateral dimension is not less than 12, the columns are called *long (or slender)*.

Load on the long column is reduced from the load on short column by multiplying with a reduction factor C_r and is given by

$$C_r = \left(1.25 - \frac{l_e}{48b}\right)$$

where, C_r = Reduction factor
l_e = Effective length of column
b = Least lateral dimension of column

When the column is circular and has spiral reinforcement, the least lateral dimension is equal to the diameter of the core. *Slenderness ratio* is defined as ratio of effective length to least radius of gyration (k). Radius of gyration (k) is defined as square root of ratio of MI and area, i.e. $k = \sqrt{\dfrac{I}{A}}$.

If the slenderness ratio is less than 50, the column is called short. For more accurate computation of load on long columns, a reduction factor (C_r) based on minimum slenderness ratio is used which is given by

$$C_r = \left(1.25 - \frac{l_e}{160k_{min}}\right)$$

where, C_r = Reduction factor
l_e – Effective length

k_{min} = Minimum radius of gyration $\left(k = \sqrt{\dfrac{I_{min}}{A}}\right)$

8.2.3 Functions of Reinforcement

Reinforcement in columns serve the following functions:

(a) Longitudinal reinforcement:
   (i) To share the vertical compressive loads and reduce the size of the column
   (ii) To resist tensile stresses caused by eccentricity, moment or lateral loading
   (iii) To prevent sudden brittle failure of the column
   (iv) To impart ductility to the column
   (v) To reduce the effect of creep and shrinkage

(b) Transverse reinforcement:
   (i) To prevent buckling of longitudinal reinforcement
   (ii) To resist diagonal tension caused by transverse shear and transverse loads
   (iii) To hold the longitudinal bars in position during concreting
   (iv) To prevent longitudinal splitting by confining concrete
   (v) To prevent sudden brittle failure of the columns
   (vi) To impart ductility to the columns
      Suitable reinforcement is provided to serve above functions.

8.2.4 Design Equations for Load Carrying Capacity

(a) Short column with ties:

The ultimate concentric load capacity is determined as follows:

$P_u = 0.446 f_{ck} A_c + 0.87 f_y A_{sc}$ (for mild steel)   ... Eq. (8.1)

$P_u = 0.446 f_{ck} A_c + 0.75 f_y A_{sc}$ (for high strength deformed bars)   ... Eq. (8.2)

where,

P_u = Ultimate concentric axial load (N)

f_{ck} = Characteristic strength of concrete (N/mm²)

f_y = Characteristic (yield or proof) strength of steel (N/mm²)

A_c = Area of concrete

A_{sc} = Area of compression steel

The ultimate load capacity with lateral ties is attained when it develops a limiting strain of 0.002, resulting in the uniform column stress of $0.446 f_{ck}$ in concrete. The corresponding stress induced in MS is $0.87 f_y$ and in Fe 415 and Fe 500 steels is $0.75 f_y$. It is very rare that the column is subjected to purely concentric load. Therefore, all compression members are designed for a minimum eccentricity of load in two principal directions given as minimum eccentricity (e_{min})

$$e_{min} = \left[\left(\frac{\text{unsupported length of column}}{500} \right) + \left(\frac{\text{lateral dimension}}{30} \right) \right] \text{ or } \not< 20 \text{ mm}$$

When the minimum eccentricity (e_{min}) is less than 0.05 times the lateral dimension, the design ultimate load (P_u) on the column is given by the following equations (about 11% less):

$P_u = 0.4 f_{ck} A_c + 0.77 f_y A_{sc} = 0.40 f_{ck} A_g + (0.77 f_y - 0.4 f_{ck}) A_{sc}$ ... Eq. (8.3)

(for mild steel)

$P_u = 0.4f_{ck} A_c + 0.67f_y A_{sc} = 0.40f_{ck} A_g + (0.67f_y - 0.4f_{ck}) A_{sc}$ ... Eq. (8.4)

(for high strength steels)

where, P_u = Ultimate axial load on the column (N)

   f_{ck} = Characteristic compressive strength of concrete (N/mm²)

   f_y = Characteristic compressive (yield or proof) strength of steel

   A_c = Area of concrete = $(A_g - A_{sc})$ in mm²

   A_{sc} = Area of longitudinal reinforcement (mm²) in the column

   A_g = Gross area of column ($A_c + A_{sc}$) in mm².

(b) *Short columns with spiral (helical) reinforcement:*

When the short columns are laterally tied with spiral reinforcement (helical ties) and the minimum eccentricity does not exceed 0.05 times the lateral dimension, the ultimate design load (P_u) on the column is given by the following equations (5% more than tied columns):

   $P_u = 1.05 [0.4f_{ck}A_c + 0.77f_y A_{sc}]$ (for mild steel) ... Eq. (8.5)

   $P_u = 1.05 [0.4f_{ck}A_c + 0.67f_y A_{sc}]$ (for high strength steels) ... Eq. (8.6)

IS:456-2000 specifies that above Eqs. (8.5) and (8.6) shall be applicable only if the

minimum volume of helical reinforcement is not less than $\left[0.36 \left(\dfrac{A_g}{A_c} - 1 \right) \dfrac{f_{ck}}{f_y} \right]$ times

the volume of concrete core

where, A_g = Gross area of column section

   A_c = Area of concrete

   f_{ck} = Characteristic strength of concrete

   f_y = Characteristic (yield or proof) strength of helical reinforcement but not exceeding 415 N/mm².

(c) *Short columns subjected to combined axial load and uniaxial bending:*

Columns subjected to combined axial load and uniaxial bending are designed by assuming dimensions and verifying adequacy of safety by trial and error approach. This involves lengthy calculations by trial and error and is quite cumbersome. Indian standard code SP-16 (design aids) provides various charts of interaction diagrams in non-dimensional parameters $\dfrac{P_u}{f_{ck}BD}, \dfrac{M_u}{f_{ck} \cdot BD^2}$,

$\dfrac{d'}{D}$, etc. to determine $\dfrac{p}{f_{ck}}$ (% steel) for various values of f_y.

Set of curves are drawn for various values of $\dfrac{P}{f_{ck}}$ for non-dimensional

quantities $\dfrac{P_u}{f_{ck}BD}$ and $\dfrac{M_u}{f_{ck}BD^2}$ and for different values of $\dfrac{d'}{D}$ and f_y.

These sample curves are shown in Fig. 8.3 (a-c). Complete set of curves are available in SP-16 (design aids) and should be used for design when the columns are subjected to combined axial load and uniaxial bending. Various quantities are as under:

P_u = Axial ultimate load

M_u = Limiting (ultimate) moment in combination with P_u

f_{ck} = Characteristic compressive strength of concrete

f_y = Characteristic (yield) strength of steel reinforcement

B = Width of column

D = Overall depth of column in the direction of moment (M_u)

d' = Effective concrete cover to steel in the direction of D

p = Percent of longitudinal steel $\left(\dfrac{A_{sc}}{BD} \times 100 \right)$

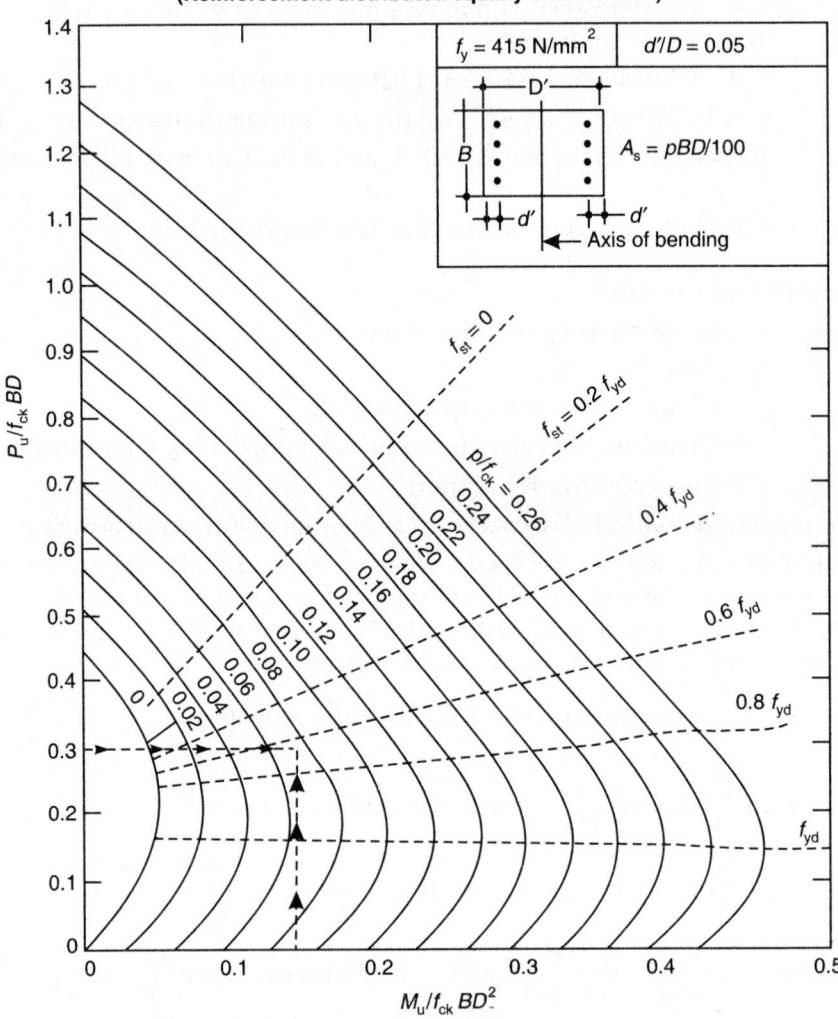

Compression with bending : Rectangular section
(Reinforcement distributed equally on two sides)

Fig 8.3 (a) Sample interaction chart for columns with axial load and bending moment

for different $\dfrac{d'}{D}$ ratios

(Ref. *Design Aids for Reinforced Concrete*, SP-16)

The percent of steel p is found from these charts from assumed values of $(B \times D)$ and computed values of $\dfrac{P_u}{f_{ck}BD}$ and $\dfrac{M_u}{f_{ck}BD^2}$ by getting the point S as inter-

section of vertical line from x $\left(\dfrac{M_u}{f_{ck}BD^2}\right)$ and horizontal line from y $\left(\dfrac{P_u}{f_{ck}BD}\right)$.

The point S represents the value of $\dfrac{p}{f_{ck}}$, which gives area of steel A_{sc} (equal to

$p \times BD \times 100 f_{ck}$). The method shall be explained through Example 8.6.

Compression with bending : Rectangular section
(Reinforcement distributed equally on two sides)

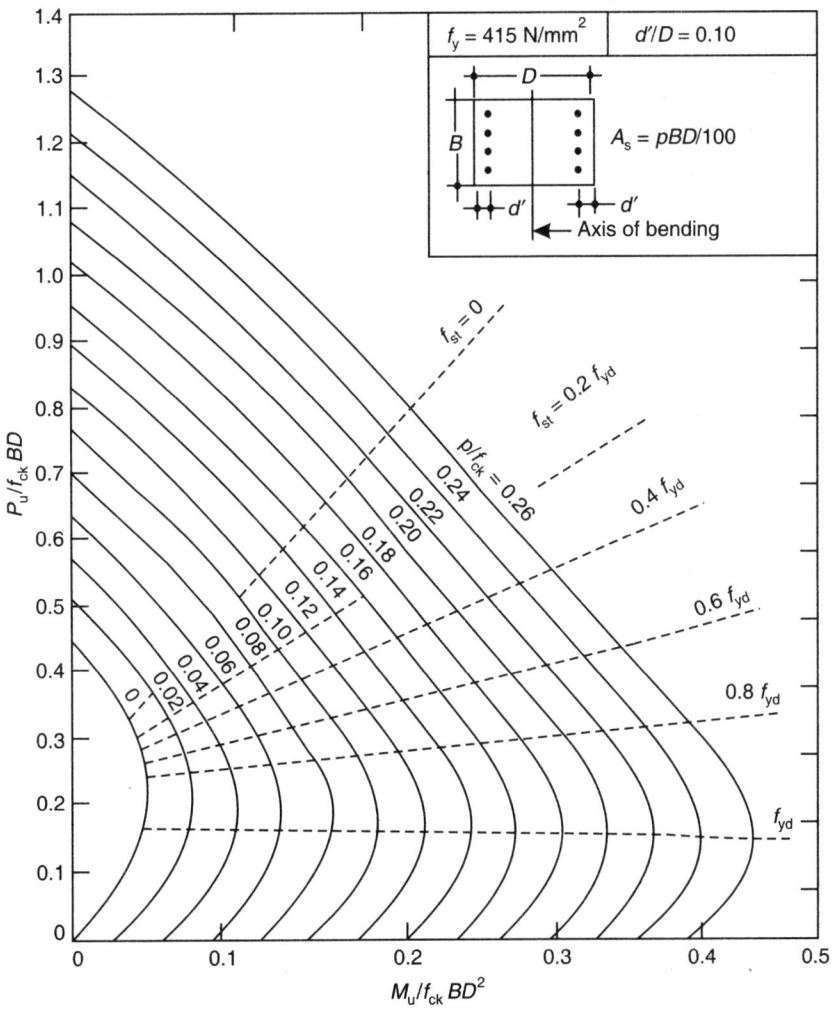

Fig. 8.3 (b) Sample interaction chart for columns with axial load and

bending moment for different $\dfrac{d'}{D}$ ratios

(Ref. *Design Aids for Reinforieng Concrete*, SP-16)

Compression with bending : Circular section

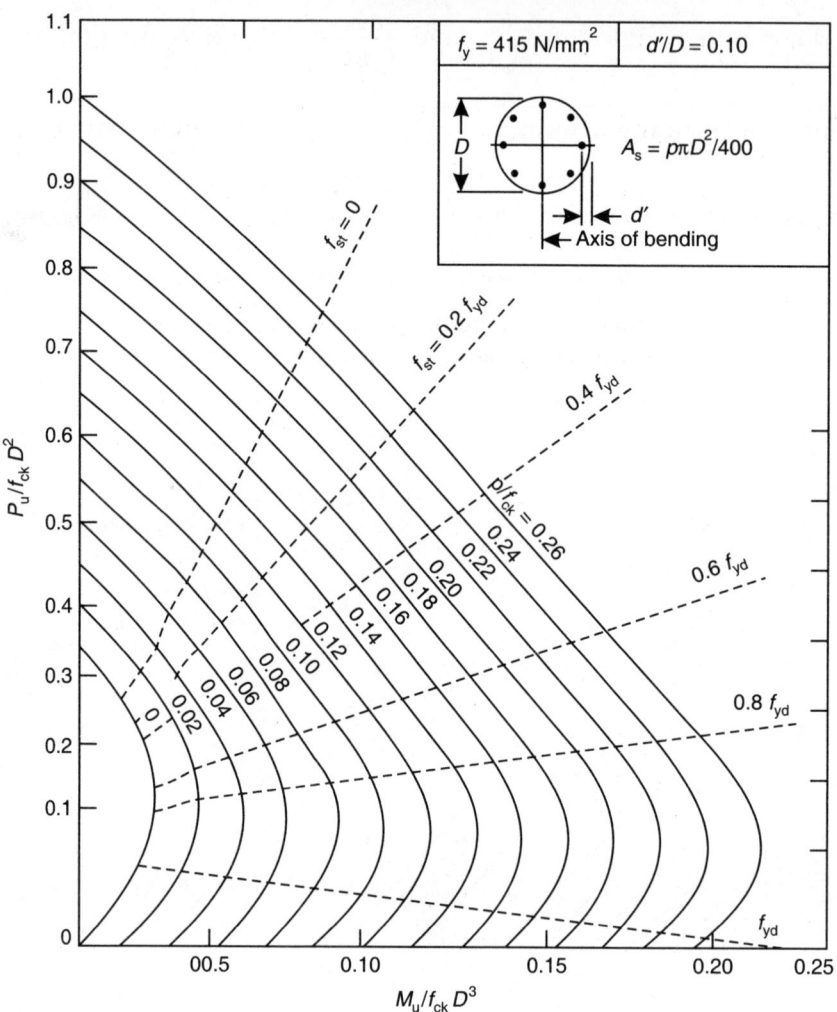

Fig. 8.3(c) Sample interaction chart for columns with axial load and bending moment for different $\dfrac{d'}{D}$ ratios

(Ref. *Design Aids for Reinforcing Concrete*, SP-16)

(d) *Short Columns Carrying Axial load with Biaxial Bending:*

The columns located at corners in multistoreyed building with rigidly connected beams at right angles and are subjected to compressive load along with biaxial moments (M_{ux} and M_{uy}). Analysis and design of such columns require lengthy calculations by trial and error method. Indian standard code recommends a simple approach based on empirical formulation in terms of design parameters. The design interaction equation is expressed as:

$$\left(\frac{M_{ux}}{M_{uxl}}\right)^{\alpha_n} + \left(\frac{M_{uy}}{M_{uyl}}\right)^{\alpha_n} < 1.0 \qquad \qquad \text{... Eq. (8.7)}$$

where,

α_n is an exponent whose value depends on the ratio of $\left(\dfrac{P_u}{P_{uz}} \right)$

P_u = Axial load when $M = 0$

$P_{uz} = [0.45 f_{ck} A_c + 0.75 f_y A_{sc}]$ ...Eq. (8.8)

α_n varies linearly from 1 to 2 with ratio $\left(\dfrac{P_u}{P_{uz}} \right)$ from 0.2 to 0.80

$\alpha_n = 1$, for value of $\left(\dfrac{P_u}{P_{uz}} \right) \leq 0.20$

$\alpha_n = 2$, for value of $\left(\dfrac{P_u}{P_{uz}} \right) \geq 0.80$

M_{ux} = BM about XX and M_{uy} = BM about YY axes due to design loads
M_{uxl} and M_{uyl} are the maximum uniaxial moment capacities for an axial load of P_u bending about XX and YY axes respectively (M_{uxl} limit).

A simpler approach for the selection of reinforcement in the column section has been suggested by Devdas Menon based on the equivalent uniaxial moment computed as:

$$M_{ue} = 1.15 \sqrt{(M_{ux}^2 + M_{uy}^2)} \qquad \ldots \text{Eq. (8.9)}$$

The equivalent moment is assumed to act along with axial compressive load P_u. Using non-dimensional design parameters $\dfrac{P_u}{f_{ck}BD}$ and $\dfrac{M_{ue}}{f_{ck}BD^2}$ for percent of steel $\left(\dfrac{p}{f_{ck}} \right)$ is found from the charts provided in SP-16 (Design Aids) in previous case with axial load and uniaxial bending. The design can be checked by interaction diagrams $\left(\dfrac{M_{ux}}{M_{uxl}} \right), \left(\dfrac{M_{uy}}{M_{uyl}} \right)$ and $\left(\dfrac{P_u}{P_{uz}} \right)$ and provided in SP-16 (Design Aids). The method shall be explained through examples.

(e) *Long (Slender) Column:*

In long columns, the failure is caused by buckling and additional moments are generated. Long columns are designed for the axial load and moments including additional moments as computed below:

$$M_{ax} = \frac{P_u D}{2000} \left\{ \frac{l_{ex}}{D} \right\}^2, \qquad M_{ay} = \frac{P_u B}{2000} \left\{ \frac{l_{ey}}{B} \right\}^2$$

where,

M_{ax} = Additional moments in *x-x* direction

M_{ay} = Additional moments in y-y direction

P_u = Axial load on the member

l_{ex} = Effective length in respect of the major axis depending on support restrain

l_{ey} = Effective length in respect of the minor axis depending on support restrain

D = Depth of the cross-section at right angles to the major axis

B = Width of the cross-section

The design of the section is carried out in a similar manner as done for short column subjected to axial load with biaxial or uniaxial moments. In case of long columns, the stresses in concrete and steel shall not exceed maximum permissible stresses multiplied by a reduction factor C_r given by

$$C_r = \left(1.25 - \frac{l_{ef}}{48b}\right) \text{ or } C_r = \left(1.25 - \frac{l_{ef}}{160\, k_{min}}\right)$$

where, l_e = Effective span

b = Least lateral dimension

$$k_{min} = \text{Least radius of gyration} = \sqrt{\frac{I_{min}}{A}}$$

Column sections are checked for safety by using interaction equation,

$$\left(\frac{M_{ux}}{M_{uxl}}\right)^{\alpha_n} + \left(\frac{M_{uy}}{M_{uyl}}\right)^{\alpha_n} \leq 1.0$$

where, M_{ux} is the ultimate moment about XX

M_{uy} is the ultimate moment about YY

M_{uxl}, M_{uyl} are the maximum uniaxial moment capacities of the section for axial load of P_u, bending about X and Y axes respectively.

$$\alpha_n = 1, \text{ for } \frac{P_u}{P_{uz}} \leq 0.2$$

$$\alpha_n = 2, \text{ for } \frac{P_u}{P_{uz}} \geq 0.80$$

$$\alpha_n = 1 \text{ to } 2, \text{ varying linearly with } \frac{P_u}{P_{uz}} \text{ value from } 0.20 \text{ to } 0.80$$

$$P_{uz} = 0.45\, f_{ck}.\, A_c + 0.75\, f_y\, A_{sc}.$$

Example 8.1

Determine the ultimate load carrying capacity of square column 250 mm x 250 mm, reinforced with 4 bars of 20 mm diameter, if the effective length of the column is 4.50 m and M25 CC and Fe 415 steel are used.

Solution: $A_g = 250 \times 250 = 62500$ mm^2, $A_{sc} = 4 \times \dfrac{\pi}{4}\,(20)^2 = 1256$ mm^2

$$L_e = 4500 \text{ mm}, \frac{L_e}{b} = \frac{4500}{250} = 18 > 12, \text{ hence column is long}$$

$$C_r = \left(1.25 - \frac{4500}{48 \times 250}\right) = 0.875$$

$P_u \text{ (long)} = C_r [0.4 f_{ck} A_g + (0.67 f_y - 0.40 f_{ck}) A_{sc}]$
$\qquad = 0.875 [625000 + 268.05 \times 1256\} = 841462 \text{ N} = 841.46 \text{ kN}.$

Example 8.2

A RCC rectangular section of 300 mm × 600 mm carries an axial factored load of 2000 kN. Design the suitable reinforcement if M20 grade CC and Fe 415 grade HYSD steel bars are used in the column. The effective length of the column is 3.0 m.

Solution: $P_u = 2000 \text{ kN}, f_{ck} = 20 \text{ N/mm}^2, f_y = 415 \text{ N/mm}^2, b = 300 \text{ mm}, D = 600 \text{ mm}$

$$\frac{l}{b} = \frac{3000}{300} = 10, \text{ the column is short.}$$

$$P_u = 0.40 f_{ck} A_g + (0.67 f_y - 0.4 f_{ck}) A_{sc}$$
$$2000 \times 10^3 = 0.40 \times 20 \times (300 \times 600) + (0.67 \times 415 - 0.40 \times 20) A_{sc}$$

$$A_{sc} = \left(\frac{2 \times 10^6 - 1.44 \times 10^6}{270.05}\right) = \frac{0.56 \times 10^6}{270.05} = 2074.0 \text{ mm}^2$$

(6 bars 22ϕ, A_{sc} = 2281 mm²)

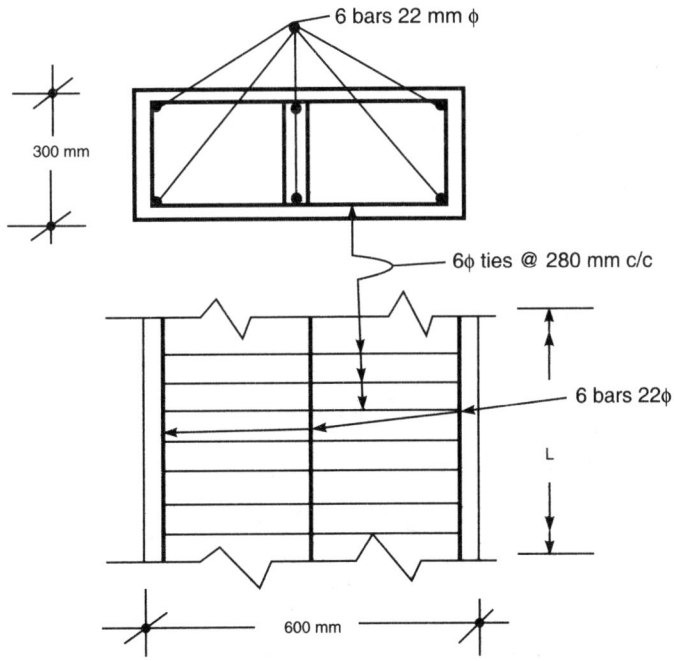

Fig. 8.4: Rectangular column

Design of ties:

Diameter of ties $\not< \dfrac{1}{4} \times 22 = 5.5$ mm, provide 6 mmϕ ties

Spacing: Least of 300 mm, or $48 \times 6 = 288$ mm, or $16 \times 22 = 352$ mm

Provide 6 mm ties @ 288 mm c/c (say @ 280 mm c/c)

Example 8.3

An internal column has 3 m length between the floors. The column carries a total load of 20 m² of floor area. The floor is subjected to a factored load of 40 kN/m² including the self-load of the slab. Use M25 grade CC and Fe 415 steel. Design the columns.

Solution: l_e (effective length) = 3.0 m, least lateral dimension $B = \dfrac{l_e}{12} = \dfrac{3000}{12} = 250$ mm

(since column is short)

Factored load = $40 \times 20 = 800$ kN, $f_{ck} = 25$ N/mm²

$$\text{Assume 1\% steel} = \dfrac{A_g}{100}$$

$$\text{Load } 800000 \text{ N} = \left[0.40 \times 25 A_g + \left(0.67 f_y - 0.4 \times f_{ck}\right)\dfrac{A_g}{100}\right]$$

$$800000 = \left[10 A_g + \left(0.67 \times 415 - 10\right)\dfrac{A_g}{100}\right] = 12.68\, A_g$$

$A_g = \dfrac{800000}{12.68} = 63092$ mm², Assuming $B = 250$, $D = \dfrac{63092}{250} = 252.4$ mm

Thus, provide 250 mm × 250 mm size. Find A_{sc} (exact) to have capacity of 800 kN load.

$800000 = [0.4 \times 25 \times 250 \times 250 + (0.67 \times 415 - 0.4 \times 25)\, A_{sc}]$

$A_{sc} = \dfrac{(800000 - 625000)}{268.05} = 653$ mm² (16ϕ 4 bars, $A_{sc} = 804$ mm²) or 1.286%, OK.

Lateral ties:

Min diameter = $\dfrac{1}{4}$ (16) = 4 mm, adopt 6 mm diameter

Spacing least of: 250 mm, $48 \times 6 = 288$ mm, $16 \times 16 = 256$ mm,

Least = 250 mm c/c (i.e. 6ϕ @ 250 mm c/c lateral ties).

Example 8.4

Design the reinforcement in a circular column of diameter 250 mm to support a service load of 600 kN. The column has unsupported length of 3 m and is restrained from side sway. The column is reinforced with helical ties. Use M20 grade CC and Fe 415 grade steel.

Solution: $L = 3000$ mm, diameter $D = 250$ mm, $f_{ck} = 20$ N/mm², $f_y = 415$ N/mm²

$P_u = 600 \times 1000 \times 1.5 = 9 \times 10^5\,\text{N}, \dfrac{L}{D} = \dfrac{3000}{250} = 12$, column is short.

Minimum eccentricity = 20 mm,

or $\quad \left(\dfrac{L}{500} + \dfrac{D}{30}\right) = \dfrac{3000}{500} = \dfrac{250}{30} = 6 + 8.33 = 14.33$ mm

$$A_g = (250)^2 = 49087.4\,\text{mm}^2$$

Main Reinforcement (Fig 8.5):

$$P_u = 1.05\,[0.4f_{ck}\,A_g + (0.67f_y - 0.4f_{ck})A_{sc}],\ \text{for helical reinforcement.}$$

$\dfrac{9 \times 10^5}{1.05} = [0.4 \times 20 \times 49087.4 + 270.05\,A_{sc}]$ or $A_{sc} = \dfrac{464442}{270.05} = 1720\,\text{mm}^2$

Minimum steel $A_{sc} = \dfrac{0.80}{100} \times \dfrac{\pi}{4} \times (250)^2 = 393\,\text{mm}^2$

Maximum steel $A_{sc} = \dfrac{6}{100} \times \dfrac{\pi}{4}\,(250)^2 = 2945\,\text{mm}^2$

Thus, 1720 mm² is OK.

Provide 6 bars of 22 mmφ ($A_{sc} = 2281\,\text{mm}^2 < 2945\,\text{mm}^2$ and $> 393\,\text{mm}^2$

Provide clear cover of 40 mm (minimum)

Diameter of helix = $250 - 2 \times 40 = 170$ mm

Area of core $A_c = \dfrac{\pi}{4}\,(170)^2 = 22698\,\text{mm}^2,\ A_g = \dfrac{\pi}{4}\,(250)^2 = 49087.4\,\text{mm}^2$

Volume of core/m = $22698 \times 1000\,\text{mm}^3/\text{m}$

Let pitch of helical ties = p, no of turns = $\dfrac{1000}{p}$ per metre

Volume of helical ties = $\pi\,(170 - 6) \times \dfrac{\pi}{4}\,(6)^2 \times \left(\dfrac{1000}{p}\right) = \dfrac{(14567.5 \times 1000)}{p}$

Ratio of volume of helical ties to the volume of core = $0.36\left(\dfrac{A_g}{A_c} - 1\right)\dfrac{f_{ck}}{f_y}$

$$= 0.36\left(\dfrac{49087.4}{22698} - 1\right)\dfrac{20}{415} = 0.0202$$

Volume of helical reinforcement = $0.0202 \times (49087.4 \times 1000) = 991557\,\text{mm}^3/\text{m}$

Thus, $\pi(170 - 6)\,\dfrac{\pi}{4}\,(6)^2 \times \dfrac{1000}{p} = 991557$

$p = \pi^2 \times 164 \times \dfrac{36}{4} \times \dfrac{1000}{991557} = 14.69$ mm

p (less than 75 mm, $\dfrac{170}{6} = 28.3$ mm and more than 25 mm, $3 \times 6 = 18$ mm)

Thus, provide pitch p of 28 mm c/c (more than 25 mm and less than 75 mm).

Details:

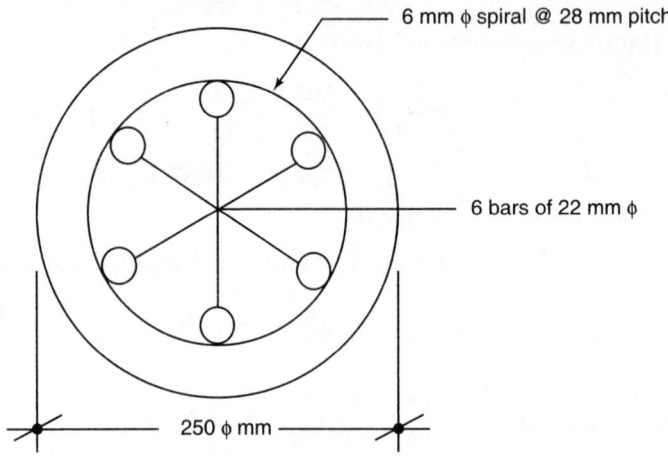

6 mm φ spiral @ 28 mm pitch

6 bars of 22 mm φ

250 φ mm

Fig. 8.5

Example 8.5

Design reinforcement in a circular column of 400 mm diameter to support a factored load of 800 kN together with a factored moment of 80 kNm. Use M20 CC and Fe 415 HYSD bars with 50 mm cover.

Solution: $D = 400$ mm, $d' = 50$ mm, D_o (core) $= (400 - 2 \times 50) = 300$ mm

$$A_g = \frac{\pi}{4}(400)^2 = 125664 \text{ mm}^2, \ P_u = 800000 \text{ N}, \ \frac{P_u}{f_{ck} \cdot BD} = \frac{8 \times 10^5}{20(400)^2} = 0.25$$

$$M_u = 80 \times 10^6 \text{ N-mm}, \ \frac{M_u}{f_{ck} \cdot BD^2} = \frac{8 \times 10^6}{20(400)^3} = 0.0625$$

$$\frac{d'}{D} = \frac{50}{400} = 0.125 \text{ (say 0.10), from the chart of SP-16 for } f_y = 415 \text{ N/mm}^2$$

$$\frac{P_u}{f_{ck} \cdot BD} = 0.25, \text{ and } \frac{M_u}{f_{ck} \cdot BD^2} = 0.0625, \text{ we have } \frac{p}{f_{ck}} = 0.062$$

$$p \text{ (\% steel)} = 0.062 \times 20 = 1.25\%, \ A_{sc} = \frac{1.25}{100} \times \frac{\pi}{4} \times (400)^2 = 1571 \text{ mm}^2$$

(provide 20φ 6 bars, $A_{sc} = 1884$ mm²). Minimum 6 bars in circular section.

Lateral ties:

Diameter of ties not less than: 6 mm, $\frac{20}{4}$ mm, provide 6 mm ties

Spacing not more than: 300 mm, $16 \times 20 = 320$ mm, or $48 \times 6 = 288$ mm
Provide spacing 280 mm c/c.

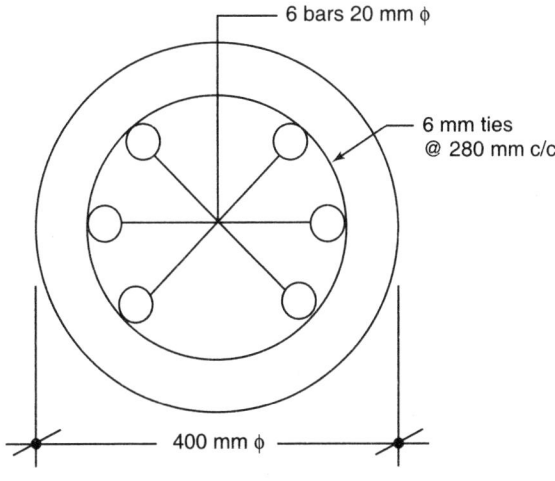

6 bars 20 mm φ

6 mm ties
@ 280 mm c/c

400 mm φ

Fig 8.6

Example 8.6
Design a square and circular column section subjected to an ultimate axial load of 1600 kN. Consider concrete of M25 grade and steel of Fe 415 grade.

Solution: $f_{ck} = 25 \text{ N/mm}^2, f_y = 415 \text{ N/mm}^2, P_u = 1600000 \text{ N}$

Design for $e_{min} \ngtr 0.05 \times$ lateral dimension (or 20 mm)

$$P_u = 0.4 f_{ck} A_g + (0.67 f_y - 0.4 f_{ck}) A_{sc}, \text{ assume } A_{sc} = 2\% \text{ of } A_g = \frac{2 A_g}{100}$$

(a) *Lateral ties of square section:*

$$1600000 = 10 A_g + (0.67 \times 415 - 10) \frac{2 A_g}{100} = (10 A_g + 5.36 A_g) = 15.36 A_g$$

$$A_g = \frac{1600000}{15.361} = 104160, B = D = \sqrt{104160} = 322.74 \text{ mm}$$

Provide 350 mm × 350 mm size. Recalculate the reinforcement required.

$$1600000 = 10(350 \times 350) + (268.05) A_{sc} = 1225000 + 268.05 A_{sc}$$

$$A_{sc} = \frac{(1600000 - 1225000)}{268.05} = 1399 \text{ mm}^2$$

(8 bars of 16φ, $A_{sc} = 1608 \text{ mm}^2 > 1399$)

Lateral ties: Provide ties of 6 mmφ spacing: 48 × 6 or 16 × 16 or 350 mm c/c = 250 mm c/c

(b) *Circular section:*

   (i) *Lateral ties:*

$$A_g = 104160 \text{ mm}^2 \text{ as in (a) above} = \frac{\pi}{4} (D)^2, D = \sqrt{\frac{104160 \times 4}{\pi}} = 364.17 \text{ mm}$$

Provide circular section of 360 mm diameter.

A_{sc} required to be calculated.

$$10 \left(\frac{\pi}{4} \times 360^2 \right) + (268.05) A_{sc} = 1600000$$

$$A_{sc} = \frac{(1600000 - 1014876)}{268.05} = 2172 \text{ mm}^2$$

(Provide 6 bars 22 mmϕ, $A_{sc} = 2281$ mm$^2 = 2.14\%$, OK)

Provide lateral ties of 6 mm at 280 mm c/c.

(ii) *Helical ties*:

Provide same diameter of 360 mm, $A_g = 101787.6$ mm^2,

$$A_{sc} = \left(\frac{\dfrac{1600000}{1.05} - 1017876}{268.05} \right) = 1885 \text{ mm}^2$$

$$\text{Diameter of concrete core} = \left\{ 360 - 2 \times 50 + 2 \left(\frac{20}{2} + \frac{6}{2} \right) \right\} = 286 \text{ mm}$$

$$\text{Area of concrete core} = \frac{\pi}{4} (286)^2 = 64242 \text{ mm}^2 \ (A_c)$$

Let the pitch of ties $= p$

$$\text{Volume of helical ties} = 0.36 \left(\frac{A_g}{A_c} - 1 \right) \frac{f_{ck}}{f_y} \times \text{volume of concrete core}$$

$$= 0.36 \left(\frac{101789}{64242} - 1 \right) \frac{25}{415} \times 64242 \times 1000$$

or

$$\pi (286) \times \frac{\pi}{4} (6)^2 \times \frac{1000}{p} < 0.36 (1.5845 - 1) \frac{25}{415} \times 64242 \times 1000$$

or

$$p > \frac{\pi^2 (286) 9000 \times 415}{0.36 \times 0.5845 \times 25 \times 64242000} > 31.2 \text{ mm c/c}$$

$$\text{Maximum } p = 75 \text{ mm}, \ \frac{286}{6} = 47.7 \text{ mm}$$

Minimum $p = 25$ mm, $3 \times 6 = 18$ mm

Pitch of spiral ties $= 35$ mm

Provide 6 bars of 20ϕ ($A_{sc} = 1885$ mm^2) and 6 mm spiral lateral dimension at a pitch of 35 mm c/c.

Example 8.7

Determine the ultimate load capacity of column 300 mm × 500 mm if the eccentricity of load is 100 mm about one of the axis. 6 bars of 28 mm are used as reinforcement. Appropriate lateral ties are used. M25 CC and Fe 415 HYSD bars are used (Fig. 8.7). Clear cover = 40 mm.

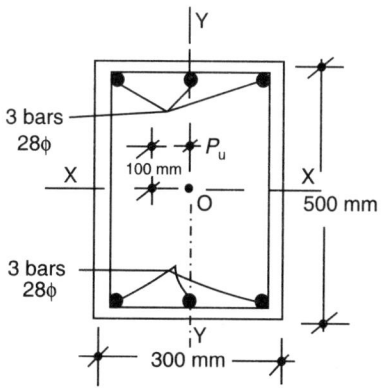

Fig 8.7

Solution: $B = 300$ mm, $D = 500$ mm, clear cover = 40 mm

$f_{ck} = 25$ N/mm^2, $f_y = 415$ N/mm^2, $e = 100$ mm

$$\frac{M_u}{f_{ck} \cdot BD^2} = \frac{P_u \cdot e}{25 \times 300 \times 500^2} \quad \text{and} \quad \frac{P_u}{f_{ck} \cdot BD} = \frac{P_u}{25 \times 300 \times 500}$$

$$A_{sc} = 6 \times 615.75 = 3694 \text{ mm}^2$$

$$\frac{\dfrac{M_u}{f_{ck} \cdot BD^2}}{\dfrac{P_u}{f_{ck} \cdot BD}} = \frac{\dfrac{P_u \cdot e}{D}}{\dfrac{P_u}{1}} = \frac{100}{500} = \frac{1}{5}$$

Using 28ϕ bars effective cover $= 40 + \dfrac{28}{2} = 54$ mm

$$\frac{p}{f_{ck}} = \frac{100 A_{sc}}{BD \cdot f_{ck}} = \frac{100 \times 3694}{300 \times 500 \times 25} = 0.0985$$

$$\frac{d'}{D} = \frac{54}{500} = 0.108$$

From interaction curves select the curves with $\dfrac{d'}{D} = 0.10$, $f_y = 415$ N/mm^2

Since $$\frac{\dfrac{M_u}{f_{ck} \cdot BD^2}}{\dfrac{P_u}{f_{ck} \cdot BD}} = \frac{1}{5} = \frac{e_y}{D}$$

Draw a line from the origin with a slope of $\dfrac{1}{5}$ with the Y-axis $\left(\dfrac{P_u}{f_{ck} \cdot BD}\right)$.

This straight line crosses $\dfrac{P}{f_{ck}} = 0.0985$, the intersection point S represents

intersection of vertical and horizontal projections on $\dfrac{M_u}{f_{ck} \cdot BD^2}$ on (X-X) and $\dfrac{P_u}{f_{ck} \cdot BD}$

on (Y-Y) axes. Thus, from these curves we get,

$\dfrac{P_u}{f_{ck} \cdot BD} = 0.580$, thus $P_u = 0.580 \times 25 \times 300 \times 500 = 2175 \times 10^3 \, N = 2175 \, kN$

Example 8.8

Design reinforcement for a eccentrically loaded rectangular column of 300 mm × 500 mm size subjected to ultimate load of 1500 kN with an eccentricity of $e_y = 150$ mm. Use M25 grade CC and Fe 415 grade steel. Clear cover may be taken as 40 mm.

Solution: $f_{ck} = 25 \, N/mm^2, f_y = 415 \, N/mm^2, e_y = 150$ mm, $B = 300$ mm, $D = 500$ mm,

$$d' = 40 + \dfrac{20}{2} = 50 \text{ mm (assuming longitudinal bars of 20 mm)},$$

$$P_u = 1500 \text{ kN}, \dfrac{d'}{D} = \dfrac{50}{500} = 0.10,$$

$e_y = 150$ mm, assume the column to be short

$$\dfrac{P_u}{f_{ck} \cdot BD} = \dfrac{1500 \times 1000}{25 \times 300 \times 500} = 0.40, \quad \dfrac{M_u}{f_{ck} \cdot BD^2} = \dfrac{15 \times 10^5 \times 150}{25 \times 300 \times (500)^2} = 0.12$$

Consider SP-16 design curves for $\dfrac{d'}{D} = 0.10, f_y = 415 \, N/mm^2$

Take horizontal line at $\dfrac{P_u}{f_{ck} \cdot BD} = 0.4$, take vertical line at $\dfrac{M_u}{f_{ck} \cdot BD^2} = 0.12.$

Point of intersection S of vertical and horizontal lines give

$$\dfrac{p}{f_{ck}} = 0.082, p \text{ (\%age steel)} = 0.082 \times 25$$

$$A_{sc} = \dfrac{0.082}{100} \times 25 \times 300 \times 500 = 3075 \text{ mm}^2$$

(provide 10 bars of 20 mm ϕ) $A_{sc} = 3140$ mm², OK.

Provide 6 mm ties

Spacing: 300 mm, 48 × 6 = 288 mm, 16 × 20 = 320 mm, i.e. 280 mm c/c

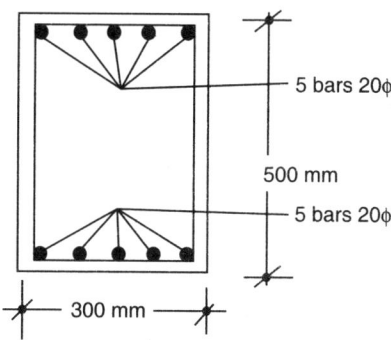

Fig 8.8

Example 8.9

Design the reinforcement in a circular column of 300 mm diameter to support a service axial load of 800 KN. The column has unsupported length of 3 m and is braced adequately against sway. Column uses M30 grade of CC and Fe 415 grade of steel. Column is reinforced with helical spiral ties.

Solution: $L = 3000$ mm, $D = 300$ mm, $f_{ck} = 30$ N/mm^2, $f_y = 415$ N/mm^2
$P_u = 1.5 \times 800 = 1200$ kN. Assume clear cover $= 40$ mm

Slenderness ratio $= \dfrac{3000}{300} = 10 < 12$. The column is short.

Minimum eccentricity $e_{min} = \left(\dfrac{L}{500} + \dfrac{D}{30} \right) = \left(\dfrac{3000}{500} + \dfrac{300}{30} \right) = 16 < 20$ mm

Also $\qquad\qquad 0.05D = 0.05 \times 300 = 15$ mm < 20, OK

$$A_g = \frac{\pi}{4} (300)^2 = 2.25 \, \pi \times 10^4 \, \text{mm}^2$$

According to IS:456-2000

$$P_u = 1.05 \, [0.4 \, f_{ck} \cdot A_g + (0.67 f_y - 0.4 f_{ck}) \, A_{sc}]$$

$$\frac{1200000}{1.05} = [0.4 \times 30 \times 2.25 \, \pi \times 10^4 + (0.67 \times 415 - 0.4 \times 30) \, A_{sc}]$$

$$A_{sc} = \frac{\left[\dfrac{12 \times 10^5}{1.05} - 848230 \right]}{266.05} = 1322 \, \text{mm}^2$$

(provide 6 bars 18ϕ, $A_{sc} = 1527$ mm$^2 > 1322$)

Minimum $A_{sc} = \dfrac{0.80}{100} \dfrac{\pi}{4} \times (300)^2 = 565$ mm^2

Maximum $A_{sc} = \dfrac{6}{100} \dfrac{\pi}{4} \times (300)^2 = 4241$ mm^2

1527 mm^2 provided > 565 mm^2 and < 4241 mm^2 thus, OK.

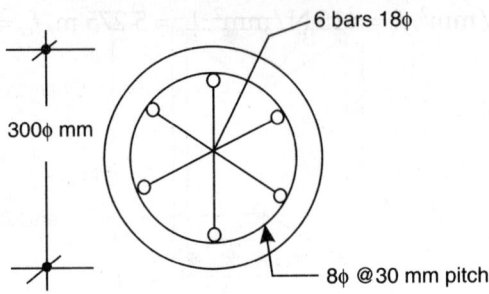

Fig 8.9

Helical reinforcement:

$$\text{Effective cover} = 40 + \frac{18}{2} = 49 \cong 50 \text{ mm, say}$$

$$\text{Diameter of concrete core} = (300 - 2 \times 50) = 200 \text{ mm}$$

$$A_g \text{ (gross area)} = \frac{\pi}{4}(300)^2 = 70686 \text{ mm}^2$$

$$A_c \text{ (concrete core)} = \left[\frac{\pi}{4}(200)^2 - 1527\right] = 29890 \text{ mm}^2$$

$$\text{Volume of core} = 29890 \times 10^3 \text{ mm}^3$$

$$\left(\frac{A_g}{A_c} - 1\right) = \left(\frac{70686}{29890} - 1\right) = 1.3649$$

Let the pitch of 8 mm diameter spiral ties $= p$

$$\text{Volume of ties per meter} = \pi\,(200 - 8)\,\frac{1000}{p} \times \frac{\pi}{4} \times (8)^2 = \frac{3016 \times 10^4}{p} \text{ mm}^3/\text{m}$$

$$\text{Volume of ties} < 0.36 \left(\frac{A_g}{A_c} - 1\right)\frac{f_{ck}}{f_y} \text{ (volume of core)} = \frac{0.36 \times 1.3649 \times 30}{415} \times 29890 \times 10^3$$

$$\frac{3016 \times 10^4}{p} < 106.17 \times 10^4,\ p > \frac{3016}{106.17} = 28.4 \text{ mm}$$

$$< 75 \text{ mm},\ \frac{200}{6} = 33.33 \text{ mm},$$

$$> 25 \text{ mm},\ 3 \times 8 = 24 \text{ m}$$

Provide 8ϕ spiral ties at a pitch of 30 mm.

Example 8.10

Design a long tied column carrying 700 kN load. Effective height of the column is 5.275 m. Use M20 grade CC and Fe 415 HYSD bars.

Solution: $f_{ck} = 20$ N/mm², $f_y = 415$ N/mm², $L_e = 5.275$ m, $f_{cc} = 0.4f_{ck} = 8$ N/mm²

$$f_{sc} = 0.67f_y = 278 \text{ N/mm}^2, \ \alpha = \frac{f_{sc}}{f_{cc}} = \frac{278}{8} = 34.75$$

Design load $(P_u) = 1.5 (700 + \text{self-load}) = 1.5 (700 + 7) = 1061$ kN (Fig. 8.10)

Assume 1% longitudinal reinforcement (0.8 to 6%) for the short column,

i.e. $A_{sc} = \dfrac{A_g}{100}$

$$P_u = 1061 \times 10^3 = [0.4f_{ck} \cdot A_g + (0.67f_y - 0.4f_{ck}) A_{sc}]$$

$$= 0.4f_{ck} \cdot A_g \left[1 + p \left(\frac{0.67 f_y}{0.4 f_{ck}} - 1 \right) \right], \ 0.4 \times 20 = 8$$

or $1061 \times 10^3 = 8A_g \left[1 + 0.01 \left(\dfrac{278}{8} - 1 \right) \right] = 8A_g [1 + 0.3375] = 10.70A_g$

$$A_g = \frac{1061 \times 10^3}{10.70} = 99159 \text{ mm}^2, \text{ square section } B = D = \sqrt{99159} = 314.8 \text{ mm}$$

Thus, provide 330 mm × 330 mm, $\dfrac{l}{b} = \dfrac{5275}{330} = 15.99 > 12$, column is long

Reduction factor $C_r = \left(1.25 - \dfrac{L_e}{48b} \right) = \left(1.25 - \dfrac{5275}{48 \times 330} \right) = 0.917$

Equivalent short column load,

$$P_o = 0.40 \times 20 \times A_g + (0.67 \times 415 - 0.4 \times 20) \frac{A_g}{100}$$

$$= 8 \times 330 \times 330 + 270 \times \frac{330 \times 330}{100} = 8.712 \times 10^5 + 2.9403 \times 10^5$$

$$= 11.6523 \times 10^5 \text{ N} = 1165.23 \text{ kN}$$

Load on long column = $C_r P_o = 0.917 \times 1165.23 = 1069$ kN > 1061 kN, safe

1% longitudinal reinforcement = $\dfrac{330 \times 330}{100} = 1089$ mm²

Provide 4 bars of 18ϕ, $A_{sc} = 1018$ mm² < 1089 mm² (required)

Provide 4 bars of 20ϕ, $A_{sc} = 1256$ mm² > 1089 mm², OK.

Design of lateral ties:

Diameter 6 mm or $\dfrac{1}{4} \times 20 = 5$ mm, provide 6 mmϕ lateral ties.

Spacing: 330 mm, 6 × 48 mm = 288 mm, 16 × 20 = 320 mm

Provide 6 mm @ 280 mm c/c

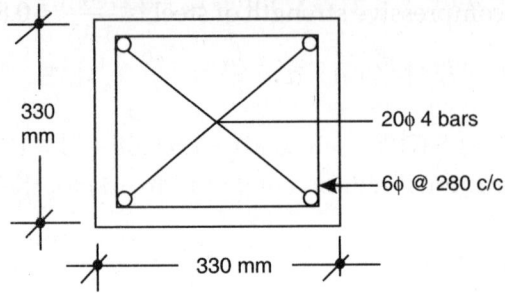

330
mm

330 mm

20φ 4 bars

6φ @ 280 c/c

Fig 8.10

SUMMARY

Columns are structural members primarily supporting compressive loads. The column (or strut) is a compression member whose effective length exceeds 3 times the least lateral dimension. Columns can be *tied, spiral, composite or concrete filled tubes*. Columns are reinforced longitudinally and tied by lateral ties or helical spirals. Columns are loaded either concentrically or eccentrically.

The *slenderness ratio* of column is defined as the **ratio** of effective length to its least lateral dimension. The effective length (l_e) is the distance between two consecutive points of inflexion created by buckling. Effective length is influenced by the end support conditions and is specified in Table 28 of IS:456-2000. Columns are classified as *pedestals* $(l_e < 3$ times least lateral dimension), *short* $(l_e \le 12$ times least lateral dimension) and long $(l_e > 12$ times least lateral dimension). Load carrying capacity of column depends on its slenderness ratio. Pedestals fail by direct crushing under compressive loading while long columns primarily fail due to buckling and short columns fail by combination of crushing and buckling.

Stresses developed in concrete and steel remain in proportion to their elastic moduli up to certain load. With passage of time, the concrete develops shrinkage and creep strains. Spirally reinforced columns can undergo large deformations before failure (ductile failure) while tied column undergo brittle failure.

Design of column based on IS:456-2000 ensures adequate margin of safety on compressive loads at collapse in limit state method while it ensures adequate margin of safety on concrete and steel stresses at working loads.

Assumptions in limit state design of columns are:

- Plane section normal to the axis before deformation remains plane after deformation, i.e. the strain at any point is proportional to its distance from the NA.
- Tensile strength of concrete is ignored.
- Modulus of elasticity of concrete $E_c = 5000 \sqrt{f_{ck}}$ N/mm^2
- Failure of concrete is governed by maximum strain of 0.002 under concentric axial load while 0.0035 under eccentric load or bending.
- The design compressive strength of concrete is $\dfrac{0.67 f_{ck}}{1.5} = 0.446 f_{ck}$.

- The design compressive strength of steel is $\dfrac{f_y}{1.15} = 0.87f_y$.

  The loads on long column shall be reduced by a factor $C_r = \left(1.25 - \dfrac{l_e}{48b}\right)$

  or $C_r = \left(1.25 - \dfrac{l_e}{160\,k_{min}}\right)$,   where $k_{min} = \sqrt{\dfrac{I_{min}}{A}}$

Functions of the reinforcement in column:
- Shares the load and reduces the size.
- Resists any tensile stress.
- Imparts ductility to avoid sudden brittle failure.
- Reduces the effect of creep and shrinkage.
- Lateral ties prevent buckling.
- Lateral ties resist diagonal tension.
- Lateral ties hold longitudinal bars during construction.
- Lateral ties improve ductility and prevent brittle failure.

Load carrying capacities:

   (*a*) **Short columns with ties carrying concentric loads:**
      (i) Mild steel: $P_u = [0.446f_{ck} \cdot A_c + 0.87f_y \cdot A_{sc}]$
      (ii) HYSD steel bars: $P_u = [0.446f_{ck} \cdot A_c + 0.75f_y \cdot A_{sc}]$

   (*b*) **Short columns with ties with eccentric loads:**

$$e_{min} \leq 20 \text{ mm or } \left(\frac{l_e}{500} + \frac{D}{30}\right) \text{ or } 0.05D.$$

      (i) Mild steel: $P_u = [0.40f_{ck} \cdot A_c + 0.77f_y \cdot A_{sc}]$
      (ii) HYSD steel bars: $P_u = [0.40f_{ck} \cdot A_c + 0.67f_y \cdot A_{sc}]$

   (*c*) **Short columns with spiral ties:**
      $P_u = 1.05\,[0.4f_{ck} \cdot A_c + 0.77f_y \cdot A_{sc}]$ with MS bars
      $P_u = 1.05\,[0.4f_{ck} \cdot A_c + 0.67f_y \cdot A_{sc}]$ with HYSD steel

   (*d*) **Short columns subjected to combined axial load and uniaxial bending:**
      Design is based on assumed section and safety checked by trials. SP-16 provides various charts in non-dimensional parameters $\dfrac{P_u}{f_{ck} \cdot BD}$, $\dfrac{M_u}{f_{ck} \cdot BD^2}$, $\dfrac{d'}{D}$, etc. to determine $\dfrac{p}{f_{ck}}$ (% steel) for different values of f_y.

   (*e*) **Short columns carrying axial load with biaxial bending:**
      Design interaction expressed as:

$$\left(\frac{M_{ux}}{M_{uxl}}\right)^{\alpha_n} + \left(\frac{M_{uy}}{M_{uyl}}\right)^{\alpha_n} \leq 1.0 \text{ (with usual notations)}$$

      where,

      α_n based on $\dfrac{P_u}{P_{uz}}$, P_u = axial load when $M = 0$, and

$$P_{uz} = [0.45f_{ck} \cdot A_c + 0.75f_y A_{sc}]$$

Biaxial moment can be simplified into uniaxial by

$$M_{ue} = 1.15 \sqrt{M_{ux}^2 + M_{uy}^2}$$

SP-16 charts are used to design columns carrying axial load and equivalent uniaxial bending M_{ue}.

(f) Long (slender) columns:

Long columns develop additional moments as calculated below:

$$M_{ax} = \frac{P_u \cdot D}{2000} \left(\frac{l_{ex}}{D}\right)^2, \quad M_{ay} = \frac{P_u \cdot B}{2000} \left(\frac{l_{ex}}{B}\right)^2$$

Safety is checked by interaction equation

$$\left(\frac{M_{ux}}{M_{uxl}}\right)^{\alpha_n} + \left(\frac{M_{uy}}{M_{uyl}}\right)^{\alpha_n} \le 1.0, \text{ with usual notations, } \alpha_n \text{ based on } \frac{P_u}{P_{vz}}$$

$$P_u \text{ reduced by reduction factor } C_r = \left(1.25 - \frac{l_e}{48b}\right) \text{ or } \left(1.25 - \frac{l_e}{160\, k_{min}}\right)$$

PRACTICE QUESTIONS

(I) Objective Questions

Q. 8.1 Select the correct response given after each statement to complete it correctly and fill in the response sheet provided.

(i) The difference in tied circular and spiral circular column of the same size does not relate to ....

    (a) their load carrying capacity

    (b) their nature of failure

    (c) their effective lengths

    (d) their slenderness ratios

(ii) A column of rectangular section ($B \times D$) and length L with both ends fixed $\left(l_e = \frac{L}{2}, \text{ say}\right)$ shall be categorized as long if the slenderness ratio ....

    (a) $\frac{l_e}{B} \ge 3,$ (b) $\frac{l_e}{B} > 12$ (c) $\frac{l_e}{B} \le 12$ (d) $\frac{l_e}{B} = 4.8$

(iii) Failure of columns under compressive loads shall be ductile in case of ....

    (a) tied columns   (b) composite columns

    (c) concrete filled pipe columns   (d) spiral columns

(iv) Effective length (l_e) of the column depends on ....
    (a) slenderness ratio   (b) least lateral dimensions
    (c) larger lateral dimension   (d) support restrains at ends

(v) In case of long column, the axial load carrying capacity is obtained by multiplying with a factor (C_r) with the load carrying capacity of a short column of the same size and the value of C_r with usual notations will be found as C_r = ...

(a) $C_r = \left(1.25 - \dfrac{l_e}{48B}\right)$ (b) $C_r = \left(1.25 - \dfrac{l_e}{48k_{min}}\right)$

(c) $C_r = \left(1.50 - \dfrac{l_e}{160k_{min}}\right)$ (d) $C_r = \left(1.5 - \dfrac{l_e}{160B}\right)$

(vi) Provision of reinforcement in columns does not serve the function of ....

    (a) sharing vertical loads
    (b) imparting ductility of the columns
    (c) increasing the effect of creep and shrinkage
    (d) preventing buckling of longitudinal bars.

(vii) In the limit state design method, the ultimate axial load P_u on a short tied column reinforcement with HYSD steel bars when the eccentricity of load does not exceed 0.05D or 20 mm or $\left(\dfrac{L_e}{500} + \dfrac{D}{30}\right)$ with usual notations shall be P_u = ...

    (a) $P_u = \{0.40 f_{ck} \cdot A_c + 0.67 f_y A_{sc}\}$
    (b) $P_u = \{0.4 f_{ck} \cdot A_c + 0.77 f_y A_{sc}\}$
    (c) $P_u = \{0.446 f_{ck} \cdot A_c + 0.75 f_y A_{sc}\}$
    (d) $P_u = \{0.446 f_{ck} \cdot A_c + 0.87 f_y A_{sc}\}$

(viii) In the limit state design, the ultimate axial load P_u on a short column with helical (spiral) reinforcement having HYSD steel bars with e_{min} not exceeding 0.05D (20 mm) with usual notations when the volume of helical reinforcement is not less than $\left[0.36\left(\dfrac{A_g}{A_c} - 1\right)\dfrac{f_{ck}}{f_y}\right]$ volume of concrete core shall be P_u = ...

    (a) $P_u = [0.4 f_{ck} \cdot A_c + 0.67 f_y A_{sc}] \dfrac{1}{1.05}$
    (b) $P_u = 1.05 [0.4 f_{ck} \cdot A_c + 0.67 f_y A_{sc}]$
    (c) $P_u = 1.05 [0.446 f_{ck} \cdot A_c + 0.87 f_y A_{sc}]$
    (d) $P_u = \dfrac{1}{1.05} [0.446 f_{ck} \cdot A_c + 0.77 f_y A_{sc}]$

(ix) Short column subjected to combined axial load and uniaxial bending are designed by assuming dimensions of the column and then checking the design by trial and error which require lengthy calculations but SP-16 has developed interaction charts for finding %age p of steel reinforcement. These charts are drawn between .... for various values of $\dfrac{d'}{D}$ and f_y.

(a) $\dfrac{P_u}{f_{ck}}$ and $\dfrac{M_u}{f_{ck}BD}$

(b) $\dfrac{P_u}{f_yBD}$ and $\dfrac{M_u}{f_yBD^2}$

(c) $\dfrac{P_u}{f_yBD^2}$ and $\dfrac{M_u}{f_yBD^3}$

(d) $\dfrac{P_u}{f_{ck}BD}$ and $\dfrac{M_u}{f_{ck}BD^2}$

(x) In case of short columns carrying axial load with biaxial bending moments, the safety is checked by interaction equation with usual notations can be expressed as ....

(a) $\left(\dfrac{M_{ux}}{M_{uxl}}\right)^{\alpha_n} + \left(\dfrac{M_{uy}}{M_{uyl}}\right)^{\alpha_n} \leq 1.0$

(b) $\left(\dfrac{M_{uxl}}{M_{ux}}\right)^{0.8} + \left(\dfrac{M_{uyl}}{M_{uy}}\right)^{0.8} \geq 1.0$

(c) $\left(\dfrac{M_{ux}}{M_{uxl}}\right)^{2} + \left(\dfrac{M_{uy}}{M_{uyl}}\right)^{2} \geq 1.0$

(d) $\left(\dfrac{M_{ux}}{P_{uxl}}\right)^{\alpha_n} + \left(\dfrac{M_{uy}}{P_{uyl}}\right)^{\alpha_n} \leq 1.0$

Response sheet to Q. 8.1:

Question	(i)	(ii)	(iii)	(iv)	(v)	(vi)	(vii)	(viii)	(ix)	(x)
Response (a/b/c/d)										

(II) Numerical Questions

Q. 8.2 A short square column of 450 mm × 450 mm size is reinforced with 4 bars of 20 mm diameter. Use M20 grade CC and Fe 415 grade steel. Determine the service load capacity of the column.

[Hint: $P_u = 0.40 f_{ck} A_c + 0.67 f_y \cdot A_{sc} = 1960$ kN, service load = 1306 kN]

Q. 8.3 Design the reinforcement for a short axially loaded column of 500 mm × 500 mm with a service load of 2000 kN. Use M30 grade CC and Fe 415 grade steel.

[Hint: $P_u = 3000$ kN, $0.05D = 25$ mm $> e_{min}$, $P_u = 0.4 \times 30 \,(A_g - A_{sc}) + 0.67 \times 415\, A_{sc}$

Reqd. $A_{sc} = 0$, provide minimum steel $A_{sc} = \dfrac{0.80}{100} \times 500 \times 500 = 2000$ mm^2

(4 bars of 28 mm, $A_{sc} = 2463$ mm^2). Lateral ties 8 mmϕ @ 350 c/c]

Q. 8.4 Design the reinforcement for a short axially loaded column of 500 mm × 500 mm size carrying an axial ultimate load of 3000 kN. Use M20 grade CC and Fe 415 grade steel.

[**Hint:** $P_u = 0.4 \times 20\ (500 \times 500 - A_{sc}) + 0.67 \times 415\ A_{sc}$, $A_{sc} = 3703\ mm^2$ (8 bars 25ϕ), $A_{sc} = 3928\ mm^2$, 6ϕ lateral ties @ 280 c/c]

Q. 8.5 Design a short circular column to carry an axial service load of 1000 kN. Use M20 grade CC and Fe 415 grade steel.

[**Hint:** $P_u = 1500\ kN$, $P_u = 0.4 \times 20\ (A_g - A_{sc}) + 0.67 \times 415\ A_{sc}$ (assuming

$e_{min} < 0.05D$), $A_{sc} = \dfrac{A_g}{100}$, $A_g = 140180\ mm^2$, $D = 422.5\ mm$ (425 mm), $A_{sc} =$

$1401.8\ mm^2$ (7 bars, 16ϕ, $A_{sc} = 1407\ mm^2$), 6ϕ ties @ 250 c/c]

Q. 8.6 Design the short circular spiral column to carry an ultimate axial load of 1500 kN. Use M20 grade CC and Fe 415 grade steel.

[**Hint:** $\dfrac{P_u}{1.05} = 1429000 = 0.4 \times 20 \left(A_g - \dfrac{A_g}{100} \right) + 0.67 \times 415 \left(\dfrac{A_g}{100} \right)$, Assuming 1% A_{sc}.

$A_g = 133505\ mm^2$, $D = 412.3\ mm$ (assume 420 mm diameter), $A_{sc} = 1335\ mm^2$
e_{min} (20 mm < 0.05 × 420 = 210 mm), use 7 bars 16ϕ bars, clear cover 40 mm

$(A_{sc} = 1407\ mm^2)$, $A_k = \dfrac{\pi}{4}\ (340)^2 - 1407 = 89385\ mm^2$, $A_g = 138544\ mm^2$,

$\dfrac{0.36 f_{ck}}{f_y} \left(\dfrac{A_g}{A_k} - 1 \right) = 0.00954$. Use 8ϕ spiral reinforcement, volume of

$\dfrac{\pi 332}{p} \left(\dfrac{\pi}{4} 8^2 \right) \times 0.00954 \times 89385$, $p = 61.5\ mm > 75\ mm$, $\dfrac{340}{6} = 56.6\ mm$

$< 25\ mm$, $3 \times 8 = 24\ mm$.

Thus, provide 8ϕ spiral reinforcement with a pitch of 50 mm c/c]

Q. 8.7 Design a rectangular column of 4.5 m length with both ends restrained in position and direction to carry an axial ultimate load of 1800 kN. Use M20 grade CC and Fe 415 grade steel.

[**Hint:** $L_e = 0.65 \times 4500 = 2925\ mm$, 1800×1000

$= \left[0.4 \times 20 \left(A_g - \dfrac{A_g}{100} \right) + 0.67 \times 415 \times \dfrac{A_g}{100} \right]$, $A_g = 168216\ mm^2$, Assume $b =$

400 mm ($0.05b = 20\ mm = e_{min}$), $D = 425\ mm$, $A_{sc} = 8$ bars 16ϕ, $(A_{sc} = 1608\ mm^2)$, 6 mmϕ ties at 250 mm c/c.]

Q. 8.8 A short RCC column 300 mm × 500 mm is reinforced with 6 bars 20ϕ (3 bars each of shorter side). Determine the max. BM, M_u about the axis bisecting the depth, if the column carries an ultimate axial load of 800 kN also. Use M20 grade CC and Fe 415 grade steel. Effective cover to steel = 50 mm.

[**Hint:** $\dfrac{P_u}{f_{ck}BD} = 0.267$, $p = 1.257\%$, $\dfrac{p}{f_{ck}} = 0.06285 = 0.063$, from the chart SP-16 with

$\dfrac{P_u}{f_{ck}BD}$ and $\dfrac{p_t}{f_{ck}}$, $\dfrac{M_u}{f_{ck}BD^2}$, $= 0.13$, $M_u = 195 \times 10^6\ kN\text{-}mm$.]

Q. 8.9 A short column 300 mm × 500 mm is reinforced with 6 bars of 20φ (3 bars each on short edge) with effective concrete cover of 50 mm. The column is subjected to an ultimate moment M_u = 100 kN-m. Determine axial load (P_u) at the time of failure. Use M20 and Fe 415.

[Hint: From SP-16 charts for $\dfrac{M_u}{f_{ck}BD^2}$ = 0.067 and $\dfrac{p}{f_{ck}}$ = 0.063, $\dfrac{P_u}{f_{ck}BD}$ = 0.49,

P_u = 1470 kN.]

Q. 8.10 A column of 300 mm × 500 mm is subjected to an eccentric loading with an eccentricity of 100 mm to the major axis. The column is reinforced with 3 bars 20 mm on each short edge at an effective cover of 50 mm.

[Hint: $\dfrac{e}{D} = \dfrac{100}{500}$ = 0.20, assume $\dfrac{P_u}{f_{ck}BD}$ = 1, $\dfrac{M_u}{f_{ck}BD^2} = \dfrac{P_u}{f_{ck}BD}\dfrac{e}{D}$ = 0.20, $\dfrac{p}{f_{ck}}$

= 0.06283 from chart of SP-16, point of intersection of $\dfrac{P_u}{f_{ck}BD}$ and $\dfrac{M_u}{f_{ck}BD^2}$

= 1.0 and = 0.20, w.r.t. = 0.063, gives $\dfrac{P_u}{f_{ck}BD}$ = 0.437, P_u = 1310 kN.]

Q. 8.11 Design the reinforcement in a circular column of 400 mm diameter to support a factored load of 800 kN along with a factored moment of 80 kN-m. Adopt M20 grade concrete and Fe 415 grade steel. Adopt effective concrete cover of 40 mm. Use (a) ties and (b) helical bars.

[Hint: (a) $\dfrac{P_u}{f_{ck}BD}$ = 0.25, $\dfrac{M_u}{f_{ck}BD^2}$ = 0.0625 = 0.063, from chart (SP-16), p = 1.25%,

$\dfrac{d'}{D}$ = 0.10, A_{sc} = 1570 mm² (provide 6 bars 20 mm, A_{sc} = 1884 mm²), 6 mm ties @ 280 c/c

(b) $\dfrac{762000}{f_{ck}BD}$ = 0.24, $\dfrac{M_u}{f_{ck}BD^2}$ = 0.063, $\dfrac{d'}{D}$ = 0.10, from charts SP-16 (56 P 141),

p = 1.2%, A_{sc} = 1508 mm². Provide 6 bars 20 mmφ, spiral ties 6 mm @ 30 mm c/c pitch (> 25, 18, < 75, 69, OK)]

Q. 8.12 Design a short column 400 mm × 400 mm at the corner subjected to an axial service load of 1000 kN, along with biaxial service moments of 33.3 kN-m acting in perpendicular planes. Use M20 grade CC and Fe 415 grade HYSD bars.

[Hint: $\dfrac{P_u}{f_{ck}BD} = \dfrac{1500000}{20 \times 400 \times 400}$ = 0.47, $M_{ue} = 1.15\sqrt{\left(50^2 + 50^2\right)}$ = 81.32 kN-m,

$\dfrac{M_{ue}}{f_{ck}BD^2} = \dfrac{81.32 \times 10^6}{20 \times 400 \times 400^2}$ = 0.06353

Assumed: $\dfrac{d'}{D} = 0.10$, from chart (SP-16), $\dfrac{p}{f_{ck}} = 0.058$, $p = 1.16\%$, $A_{sc} =$

1856 mm², (8 of 18ϕ, $A_{sc} = 2035$ mm²)

Alternatively: 4 of 20ϕ plus 4 of 16ϕ ($A_{sc} = 2060$ mm²)

Check: $\left(\dfrac{p}{f_{ck}}\right) = \dfrac{2035 \times 100}{20 \times 400 \times 400} = 0.064$, from chart 44, SP-16, $\dfrac{M_{uxl}}{f_{ck}BD^2} =$

0.06352, $M_{ux1} = M_{uy1} = 81.31$ kN-m, $P_{uz} = [0.45 f_{ck} \cdot A_c + 0.75 f_y A_{sc}] = 2055$ kN,

$\dfrac{P_u}{P_{uz}} = \dfrac{1500}{2055} = 0.73$, $\alpha_n = 1.8$,

$$\left[\left(\dfrac{M_{ux}}{M_{uxl}}\right)^{1.80} + \left(\dfrac{M_{uy}}{M_{uyl}}\right)^{1.80}\right] \leq , 2(0.615)^{1.8} = 0.834 \leq 1.0, \text{OK.}]$$

Q. 8.13 Design the reinforcement for a square column size 400 mm × 400 mm × 6000 mm long using M30 grade concrete and Fe 415 steel to carry factored axial load of 1500 kN along with factored moment of 40 kN-m on two mutually perpendicular axes. Use M20 grade CC and Fe 415 grade steel and 40 mm cover.

[Hint: Slenderness ratio $\dfrac{l_e}{B} = \dfrac{6000}{400} = 15 > 12$, long

For $\dfrac{l_e}{B}$ of 15: $\dfrac{e_x}{D} = \dfrac{e_y}{B} = 0.113$ (TI SP-16), $M_{ax} = M_{ay} = P_u$ (0.113 × 400) = 67.8 kN-m,

$P_{uz} = 3600$ kN (SP-16, $\dfrac{P_{vz}}{A_g} = 22.5$ N/mm² for 3% steel)

$P_{bx} = P_{by} = 1151$ kN (T 60 of SP 16), $K_x = K_y = 0.85$

$M_{ax} = M_{ay} = 57.63$ kN-m, $e_x = e_y = 27.3$ mm (> 20 mm), $M_{ux} = M_{uy} = 41$

Total $M_{ux} = M_{uy} = 41 + 57.63 = 98.63$ kN-m

$\alpha_n = 1.35$, $\left(\dfrac{M_{ux}}{M_{uxl}}\right)^{1.35} + \left(\dfrac{M_{uy}}{M_{uyl}}\right)^{1.35} = 0.50 < 1.0$, OK.

Chapter 9

Limit State Design of Footings

LEARNING OBJECTIVES

After the study of *limit state design of footings*, the learner will understand the principles of design of foundations and will be able to:

- ⊙ Define foundations
- ⊙ State the purpose of foundations
- ⊙ Sketch different foundations
- ⊙ Differentiate between various type of foundations
- ⊙ Explain safe bearing capacity (SBC) of soil
- ⊙ Design foundation for a wall
- ⊙ Design an isolated foundation of a column
- ⊙ Design a combined foundation for two columns
- ⊙ Design a raft (mat) foundation for set of columns
- ⊙ Design a pile foundation for a structure

9.1 INTRODUCTION

9.1.1 Function

All structures are required to carry and transfer loads (forces) to their connecting structural elements (media) by undergoing strains (deformations) and offering different type of resistances. For example, *slabs* transfer load to supporting beams (or walls) by bending, similarly beams transfer loads to columns or wall support, *columns* transfer loads to foundations (footings), and *foundations* ultimately transfer the loads to the ground (earth in contact with foundation). These structures are divided into *superstructure* (above plinth) and *substructure* (below plinth). Substructure comprises foundation which interfaces with superstructure on one side and with the ground on the other side. The purpose of foundation is to transfer all loads from superstructure to the ground safely and provide a stable base for the superstructure. Foundations distribute the loads over a larger area so that the pressure on the soil does not exceed its allowable capacity and settlement and differential settlements remain within permissible limits.

There are different types of foundations for transfer of superstructure loads to the ground. The type of foundation depends on the type and magnitude of the load and the bearing capacity of soil. These foundations are described in following sub-sections.

9.1.2 Types of Foundations

Foundations are classified as follows:

(i) Shallow Foundation

Shallow foundation has limited depth not more than width of the foundation and the load of superstructure spreads on a large area so that the pressure intensity is reduced to a safe limit of soil capacity. Shallow foundations are further divided into *isolated* and *combined footings.*

An *isolated footing* supports one wall or one column as shown in Fig. 9.1. A wall footing is a continuous strip, either flat or stepped transferring the wall load to the soil. An isolated footing supports a single column when the loads on the column are small and columns are not located closely. These footings are square, rectangular or circular.

Combined footing supports two or more columns. Combined footings are specially suited for comparatively heavy loads and when columns are located near property line. Combined footings may be rectangular, trapezoidal or T-shaped in plan. These combined footings may be formed by connecting two isolated footings with a strap beam.

When the bearing capacity of soil is very low and the isolated footings are located very close to each other, continuous strip footings supporting more than two columns in a row are provided. Such strip footings provided in both directions are known as *grid foundation* (Fig. 9.1). For very low bearing capacity of soil, the strip or isolated footings merge with each other and in such a case mat (raft) foundation is provided. Such a mat foundation supports the entire structure. Differential settlement in case of mat foundation is also much less.

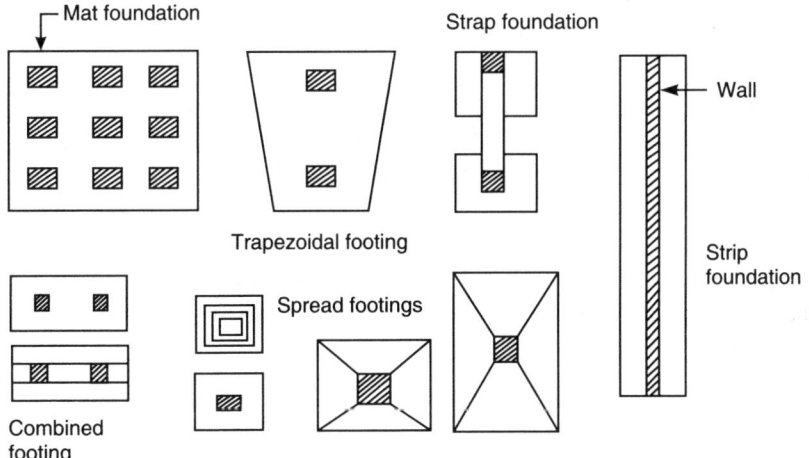

Fig. 9.1: Different type of shallow foundations

(ii) Deep Foundations

When the top layer of soil is very weak to support the structure on a shallow foundation, the depth of foundation is increased till more suitable soil or hard strata is found to support the structure. *Pile foundation* and *well foundation* are important deep foundations (Fig. 9.2). Bearing piles transfer the loads by direct bearing on hard strata while friction piles bear loads by (surface) skin friction along the length of the pile. Batter piles are used to resist lateral horizontal loads. These piles are driven along the inclined plane. Well foundations are used to transfer heavy loads of piers and abutments of bridge.

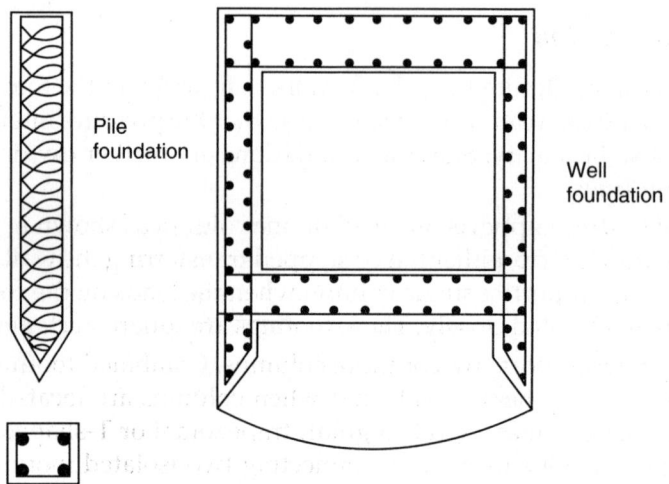

Fig. 9.2: Different types of deep foundations

9.1.3 Bearing Capacity of Soil

Bearing capacity of soil is the maximum intensity of load or pressure developed under the foundation without causing failure of soil and excessive settlement of the structure resting on the foundation. Safe bearing capacity (SBC) of the soil is specified with adequate factor of safety on ultimate bearing capacity of soil and sufficient margin on excessive settlement. The ultimate load capacity of soil corresponds to load beyond which settlement increases rapidly.

An idealised soil pressure on the entire base is assumed as uniform for simplicity and this provides a conservative design as the bending moments and shear forces calculated by this method are more than those developed actually. Bearing pressure developed shall be in equilibrium with the external load supported by the footing. The point of application of the load supported by the foundation shall coincide with the centre of gravity of the soil pressure of the base. Due to eccentricity of the load, the CG of the soil pressure shall also shift and the pressure intensity shall also be variable from maximum at one end to minimum at the other end.

9.2 DESIGN OF SHALLOW FOUNDATIONS

9.2.1 Wall Footing

The shallow foundations are wall footing, isolated column footing, combined footing and mat (or raft) foundation. Consider a simple wall footing of total width B with equal projections $\{\frac{1}{2}(B-b)\}$ on either side of the wall (width b). Width B of foundation depends on the bearing capacity of the soil (p). If the service load on the wall is w kN/m length, the ultimate load $w_u = 1.5w$ kN/m, and the required width B shall be

$$B = \frac{\text{ultimate load }(w_u)/m}{\text{ultimate bearing capacity }(p_u)} = \frac{1.5\,w}{1.5\,\text{SBC}} = \frac{w}{\text{SBC}\,(p)} \qquad \text{... Eq. (9.1)}$$

Projection of footing on either side $= \frac{1}{2}(B-b)$

Assume load of footing = 10% of imposed load $(0.1w)$

Total load on foundation/m length $= \left(w_u + \frac{w_u}{10}\right) = 1.5\,(1.1)w$

Pressure intensity $p = \frac{1.65\,w}{B \times 1} = 1.65\,\frac{w}{B}$

Net soil pressure (upward) on the projected slab $(p_u) = \dfrac{1.5\,w}{B}$ $\qquad$ ... Eq. (9.2)

Max. SF on the projected slab at a distance of d from the face

$$V_u\,(\text{max}) = p_u\left\{\frac{1}{2}(B-b) - d\right\} \qquad \text{... Eq. (9.3)}$$

Max. BM on the face of the wall $M_u = p_u \cdot \dfrac{1}{2}(B-b)\left(\dfrac{B-b}{4}\right) = \dfrac{p_u}{8}(B-b)^2$, i.e.

$$M_u = \frac{p_u(B-b)^2}{8} \qquad \text{... Eq. (9.4)}$$

Depth from BM criteria $\quad d = \sqrt{\dfrac{M_u}{1000\left[0.36 f_{ck} \cdot \dfrac{x_u}{d}\left(1-0.416\,\dfrac{x_u}{d}\right)\right]}}$

$d_1 = \sqrt{\dfrac{M_u}{1000 R_u \cdot f_{ck}}}$, where $R_u = \left[0.36 f_{ck} \cdot \dfrac{x_u}{d}\left(1-0.416\,\dfrac{x_u}{d}\right)\right]$ $\qquad$ ... Eq. (9.5)

If τ_c is the design shear stress, the depth required for shear will be

$$d_2 = \frac{V_{u\,\text{max}}}{\tau_c \cdot 1000} \qquad \text{... Eq. (9.6)}$$

τ_c is obtained from IS code 456-2000. Actual d adopted will be greater of d_1, d_2 or 150 mm. Minimum concrete cover of 50 mm must be provided in foundations.

9.2.2 Isolated Column Footing

Consider a column footing of $(L \times B)$ size for a rectangular column section $(l \times b)$ as shown in Fig. 9.3. The projection of footing slab on shorter and longer sides are:

$$\text{Short edge} = \frac{1}{2}(L-l), \qquad \text{longer edge} = \frac{1}{2}(B-b)$$

The net pressure on soil $(p_u) = \dfrac{W_u}{L \times B} \quad$ or $\quad W_u = p_u \times L \times B$

where, W_u = ultimate load on the column

The maximum BM occurs at the face of column

$$M_{u\,max1}\ (\text{short edge}) = p_u \frac{B}{2}(L-l)\frac{(L-l)}{4} = \frac{p_u \cdot B}{8}(L-l)^2 = \frac{W_u(L-l)^2}{8L} \qquad \text{... Eq. (9.7)}$$

$$M_{u\,max2}\ (\text{long edge}) = \frac{p_u L\,(B-b)^2}{8} = \frac{W_u(B-b)^2}{8B} \qquad \text{... Eq. (9.7a)}$$

Max. SF is checked at a distance of d from the face of the column.

$$V_{us}\ (\text{short edge}) = p_u\,B\left\{\frac{1}{2}(L-l)-d\right\} = \frac{W_u}{L}\left\{\frac{1}{2}(L-l)-d\right\}$$

$$V_{us}\ (\text{long edge}) = p_u\,L\left\{\frac{1}{2}(B-b)-d\right\} = \frac{W_u}{B}\left\{\frac{1}{2}(B-b)-d\right\}$$

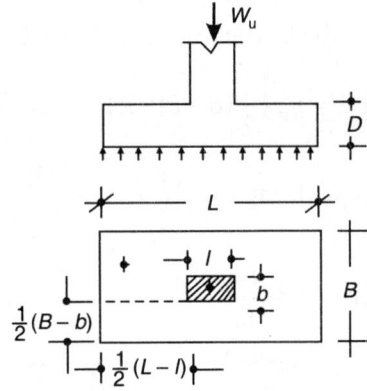

Fig. 9.3: Isolated column footing

Shear Stress

$$\tau_{vs} = \frac{V_{us}}{Bd} = \frac{W_u\{0.5(L-l)-d\}}{L \cdot B \cdot d} = \frac{p_u}{d}\{0.5\,(L-l)-d\}, \text{ since } \frac{W_u}{L \cdot B} = p_u \qquad \text{... (9.8)}$$

$$\tau_{vl} = \frac{V_{ul}}{L \cdot d} = \frac{W_u}{L \cdot d \cdot B}\{0.5\,(B-b)-d\} = \frac{p_u}{d}\{0.5\,(B-b)-d\} \qquad \text{... (9.8a)}$$

Greater of τ_{vs} or τ_{vl} is compared with τ_c corresponding to the grade of concrete and assumed %age of tensile steel (generally assumed 0.2%-0.3% initially to start check on design depth). τ_c is obtained from Table 19 of IS:456-2000. If the actual shear stress τ_{vs} or τ_{vl} is greater than τ_c, the design depth d is increased so as to bring τ_{vs} or τ_{vl} within the safe value of τ_c. Tensile steel is recalculated with the adopted value of d and external moment (M_{u1} or M_{u2}).

$$A_{st} = \frac{0.5 f_{ck}}{f_y} bd \left[1 - \sqrt{\frac{4.6 M_u}{f_{ck} \cdot b \cdot d^2}} \right] \text{ for a balanced section or minimum } A_{st} = 0.12\%.$$

Example 9.1

Design a RCC footing for RCC wall of 250 mm thickness and supporting a service load of 200 kN/m length including self-weight of wall. The safe bearing capacity of soil is 150 kN/m² and unit weight of soil, $w_e = 20$ kN/m³, the angle of repose $\phi = 30°$. Concrete of M20 grade and steel of Fe 415 grade shall be used.

Solution: $f_{ck} = 20$ N/mm², $f_y = 415$ N/mm², $\phi = 30°$, $w_e = 20$ kN/m³, $q_o = 150$ kN/m²

Depth of foundation (Rankine's formula) $h = \dfrac{q_o}{w_e} \dfrac{(1 - \sin \phi)^2}{(1 + \sin \phi)^2}$

$h = \dfrac{150}{20} \left(\dfrac{1}{3} \right) = 0.83$ m, provide depth of 1.0 m, so that foundation is not affected by variations in environment.

W_u (ultimate load on foundation) $= 1.5 \left(200 + \dfrac{10}{100} \times 200 \right) = 330$ kN

Area of foundation required $= \dfrac{330}{1.5 \times 150} = 1.47$ m (provide say 1.50 m wide and 1 m deep)

Projection from the wall face $= \left(\dfrac{1.50 - 0.25}{2} \right) = 0.625$ m

Net ultimate soil reaction under the footing $= \dfrac{200 \times 1.5}{1.5 \times 1} = 200$ kN/m²

$$(< 1.5 \times 150 = 225 \text{ kN/m}^2)$$

M_u (max) $= 200 \times 1 \times 0.625 \times \dfrac{0.625}{2} = 39.063$ kN-m/m

For Fe 415 and $f_{ck} = 20$ N/mm², $\dfrac{x_{ul}}{d} = 0.48$, $R_u = 0.1383$

$$d = \sqrt{\frac{M_u}{R_u f_{ck} b}} = \sqrt{\frac{39.063 \times 10^6}{0.1383 \times 20 \times 1000}} = \sqrt{14123} = 118.84 \text{ mm}$$

$$(d = 120 \text{ mm}, D = 180 \text{ mm})$$

SF at a distance of 0.12 m from the wall face
$$V_u = 1 \times \{0.625 - 0.12\} \times 200 = 101 \text{ kN}$$
Assuming 0.25% steel, $\tau_c = 0.36$ N/mm² (Table 19 of IS:456-2000, Page 73)
K for D of 180 mm $= 1.26$

Thus, $(0.36 \times 1.26) \geq \dfrac{101 \times 1000}{1000\,d}$ or $d \geq \dfrac{101}{1.26 \times 0.36}$, $d \geq 223$ mm

Provide $d = 250$ mm, $D = 300$ mm

$$A_{st} = \frac{0.5 \times 20 \times (1000 \times 250)}{415}\left[1-\sqrt{1-\frac{4.6 \times 39.063 \times 10^6}{20 \times 1000 \times (250)^2}}\right] = 450 \text{ mm}^2$$

Provide 12ϕ 250 c/c (0.15%)

τ_c (0.18% steel) $= 0.28 \times K = 0.28$ N/mm^2, $K = 1$ say.

Actual $\tau_v = \dfrac{101 \times 1000}{1000 \times 250} = 0.404$ N/mm$^2 > 0.28$ N/mm^2, revise the design

Increase the depth $d = \dfrac{0.404}{0.28} \times 250 = 361$ mm, provide $d = 400$ mm, $D = 450$ mm

$$A_{st} = \frac{10 \times 1000 \times 400}{415}\left[1-\sqrt{1-\frac{4.6 \times 39.063 \times 10^6}{20 \times 1000 \times (400)^2}}\right] = 275 \text{ mm}^2 \ (0.12\% = 480 \text{ mm}^2)$$

Provide minimum $A_{st} = 480$ mm^2

Actual $\tau_v = \dfrac{101 \times 1000}{1000 \times 400} = 0.2525$ N/mm$^2 \leq \tau_c = 1 \times 0.28$ N/mm^2, OK

Thus, provide 12ϕ bars @ 230 c/c and thickness $D = 450$ mm ($d = 400$ mm) Taper to 200 mm at edges.

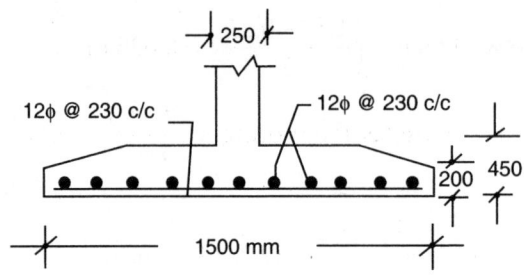

Fig. 9.4: Reinforcement details

Example 9.2

Design a RCC footing for rectangular column of 300 mm × 500 mm supporting a service load of 1000 kN. The safe bearing capacity of soil at site is 180 kN/m^2. Use M20 grade CC and Fe 415 grade steel.

Solution: $f_{ck} = 20$ N/m^2, $f_y = 415$ N/mm^2, $b = 300$ mm, $D = 500$ mm, $q_o = 180$ kN/m^2

$W_u = 1.5$, $W = 1500$ kN, self-weight of footing (10%) = 150 kN

Total load = 1500 + 150 = 1650 kN

Area of footing reqd. $= \dfrac{1650}{(1.5 \times 180)} = 6.11$ m^2

Keeping the ratio of $\dfrac{L}{B} = \dfrac{D}{b} = \dfrac{5}{3}$, we have $3x \times 5x = 6.11$ or $x^2 = \dfrac{6.11}{15} = 0.407$

or $x = 0.638$, $L = 5 \times 0.638 = 3.19$ m, $B = 3 \times 0.638 = 1.91$, Assume 3.10 m $\times$ 2.0 m

Net soil pressure $p_u = \dfrac{1500}{3.1 \times 2} = 242$ kN/m² ($< 1.5 \times 180$ kN/m²), OK

Cantilever projections:

From short edge $\dfrac{1}{2}(3.1 - 0.5) = 1.30$ m, from long edge $= \dfrac{1}{2}(2.0 - 0.3) = 0.85$ m

M_{u1} (BM) about short edge/m $= \dfrac{242}{2}(1.3)^2 = 204.5$ kN-m/m

M_{u2} (BM) about long edge/m $= \dfrac{242}{2}(0.85)^2 = 87.4$ kN-m/m

Depth footing w.r.t BM (204.5 kN-m/m)

$$d = \sqrt{\dfrac{M_u}{0.138\, f_{ck} \cdot b}} = \sqrt{\dfrac{204.5 \times 10^6}{0.138 \times 20 \times 1000}} = 272.2 \text{ mm}$$

(provide $d = 300$ mm, $D = 350$ mm)

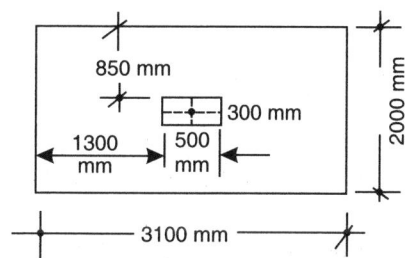

Fig. 9.5: Plan layout of footing

V_{u1} (ultimate SF) at a distance d from the face of the column
$$= 204.5\,(1.30 - d) \times 1 \text{ kN} = (265.85 - 204.5d)$$
where, d is in meters.

Assuming initial %age of tensile reinforcement = 0.25%
$$\tau_c = 0.36 \text{ N/mm}^2 \text{ (Table 19, IS:456-2000)}$$

$$\tau_v = \dfrac{V_u}{b \cdot d} = \dfrac{204.5 \times 1000\,(1.3 - d)}{1000 \times (d \times 1000)} \leq \tau_c$$

or $d \geq \dfrac{204.5 \times 1000\,(1.3 - d)}{1000 \times 1000 \times 0.36} = 0.568\,(1.3 - d)$, or $(1 + 0.568)\,d \geq 0.7385$

or $d \geq 0.471$ m (adopt $d = 480$ mm, $D = 530$ mm)
Thus, V_{u1} ($d = 0.48$ m) $= 204.5\,(1.3 - 0.48) = 167.7$ kN/m

$$\tau_v = \dfrac{167.7 \times 1000}{1000 \times 480} = 0.35 \text{ N/mm}^2 \ (< 0.36\text{N/mm}^2) \text{ OK}$$

Reinforcement for tension:

$$A_{st} = \frac{0.5 f_{ck} \cdot b \cdot d}{f_y} \left[1 - \sqrt{1 - \frac{4.6 M_u}{f_{ck} \cdot b(d^2)}} \right]$$

$$= \frac{10 \times 1000 \times 480}{415} \left[1 - \sqrt{1 - \frac{4.6 \times 204.5 \times 10^6}{20 \times 1000 \times (480)^2}} \right] = 1248 \text{ mm}^2$$

(16ϕ@ 150mm c/c, A_{st}=1340 mm², OK, %age p = 0.28%, τ_c> 0.36 N/mm² OK)

Reinforcement in other direction:

$$A_{st2} = \frac{0.5 \times 20 \times 1000 (480 - 16)}{415} \left[1 - \sqrt{1 - \frac{4.6 \times 87.4 \times 10^6}{20 \times 1000 \times (464)^2}} \right] = 534.8 \text{ mm}^2/\text{m}$$

Provide 12ϕ 200 c/c, A_{st} = 565 mm²/m

Ratio of long side to short side $\beta = \dfrac{3.10}{2.0} = 1.55$

Reinforcement in central band of 2 m $= \dfrac{2 \times 2 A_{st}}{(\beta + 1)} = \dfrac{4 \times 535}{(1.55 + 1)} = 840 \text{ mm}^2$

Minimum reinforcement $= \dfrac{0.12}{100} \times 2000 \times 530 = 1272 \text{ mm}^2 > 840 \text{ mm}^2$

Hence, provide 12ϕ @ 170 mm c/c in 2 m
$\quad$ (A_{st} = 665 mm²/m or total 1330 mm² in 2 m band)

Check for shear stress in other direction $\left(d_2 = 480 - \dfrac{16 + 12}{2} = 466 \text{ mm} \right)$

$$V_{u2} = 204.5 \times 1 \times (0.85 - 0.466) = 78.53 \text{ kN}$$

$$\tau_{v2} = \frac{78.53 \times 1000}{1000 \times 466} = 0.17 \text{ N/mm}^2 \ (< 0.28 \text{ N/mm}^2), \text{ OK}$$

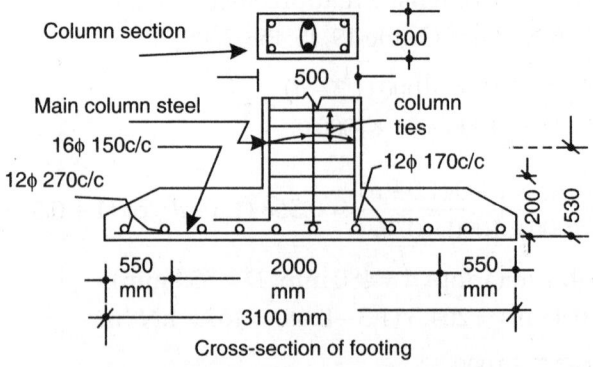

Fig. 9.6: Reinforcement details of a rectangular isolated footing

Example 9.3

Design a RCC circular footing for a circular column of 300 mm diameter and supporting a service load of 500 kN. The safe bearing capacity of the soil is 200 kN/m². Use
concrete grade M20 and steel grade Fe 415 HYSD.

Solution: $f_{ck} = 20 \text{ N/mm}^2, f_y = 415 \text{ N/mm}^2, q_o = 200 \text{ kN/m}^2, p_u = 1.5 \times 500 = 750 \text{ kN},$
$D = 300 \text{ mm}, q_u = 1.5 \times 200 = 300 \text{ kN/m}^2$

Size of footing:

Total load (W_u) of column plus footing $= 750 + 75 = 825$ kN

If the diameter of footing is D_f, area $= \dfrac{\pi}{4} (D_f)^2$

$$\dfrac{\pi}{4} (D_f)^2 = \dfrac{825}{300} = 2.75 \text{ m}^2, D_f^2 = \dfrac{2.75 \times 4}{\pi}, D_f = 1.871 \text{ m}$$

Adopt diameter of footing $= 2000$ mm (2.0 m)

Net upward pressure of soil $p_u = \dfrac{750 \times 4}{\pi(2)^2} = 238.7 \text{ kN/m}^2 < 300 \text{ kN/m}^2$, OK.

Consider one quadrant of footing, centre of gravity of quadrant of footing
OBB'C'CO will be $\bar{x}$

$$\bar{x} = \dfrac{0.6 \, (R^2 + \gamma^2 + R \cdot \gamma)}{R + \gamma} = \dfrac{0.6 \, (1000^2 + 150^2 + 150 \times 1000)}{(1000 + 150)} = 612 \text{ mm (611.74 mm)}$$

Upward load on area BB'C'C (quadrant),

$$238.7 \times \dfrac{\pi}{4} (1^2 - 0.15^2) = 183.26 \text{ kN}$$

BM (max) on the column face,

$$M_u = 183.26 \, (0.6117 - 0.15) = 84.16 \text{ kN-m about column face}$$

Breadth of footing at column face $\dfrac{\pi}{4} (300) = 235.6$ mm

Depth of footing $d = \sqrt{\dfrac{M_u}{0.138 f_{ck} \cdot b}} = \sqrt{\dfrac{84.61 \times 10^6}{0.138 \times 20 \times 235.6}}$ or $d = 360.72$ mm

Generally d required for shear is about 1.5 times that required for moment.
Thus, adopt $d = 540$ mm, $D = 540 + 60 = 600$ mm

Tensile steel A_{st}:

$$A_{st} = \dfrac{0.5 f_{ck} \cdot b \cdot d}{f_y} \left[1 - \sqrt{1 - \dfrac{4.6 M_u}{f_{ck} \cdot bd^2}} \right]$$

$$= \dfrac{0.5 \times 20 \times 235.6 \times 540}{415} \left[1 - \sqrt{1 - \dfrac{4.6 \times 84.6 \times 10^6}{20 \times 235.6 \times 540^2}} \right] = 471 \text{ mm}^2 \text{ in } 235.6 \text{ mm}$$

Minimum steel = 0.0012 × 235.6 × 600 = 169.6 mm²

Steel in 2 quadrants of 2 × 235.6 width = 2 × 471 = 942 mm²/471.2 mm

Per meter width steel = 942 × $\dfrac{1000}{417.2}$ = 1999 mm²/m. Provide mesh 20 mm @

150 c/c (A_{st} = 2093 mm²) in mutually perpendicular directions.

Check for shear:

Ultimate shear force at distance of d (540 mm) from the face of column,

$$V_u = 238.7\ (2^2 - 1.38^2)\ \frac{\pi}{4} = 393\ \text{kN}$$

Shear per meter of perimeter = $\dfrac{393}{\pi(1.38)}$ = 90.62 kN/m

$$\tau_v = \frac{V_u}{b \cdot d} = \frac{90.62 \times 1000}{1000 \times 540} = 0.17\ \text{N/mm}^2$$

% tensile steel = $\dfrac{2093 \times 100}{1000 \times 540}$ = 0.39%, τ_c (Table 19 of IS:456-2000) = 0.427 N/mm²

τ_v = 0.17 N/mm² < $K\tau_c$ (1 × 0.427 N/mm²), shear stress is within limits.

Reinforcement details:

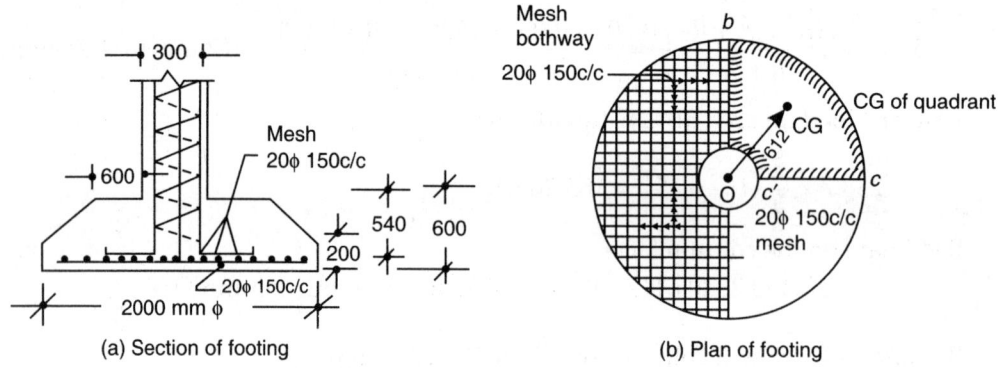

(a) Section of footing   (b) Plan of footing

Fig. 9.7: Reinforcement in circular footing

Example 9.4

Design a footing for 300 mm thick RCC wall supporting a service load of 200 kN/m and service moment of 20 kN-m/m. Consider unit weight of soil w = 20 kN/m³, angle of repose for soil, ϕ = 30°, allowable bearing capacity of soil, q_o = 200 kN/m², concrete grade M20 and steel grade of Fe 415.

Solution: f_{ck} = 20 N/mm², f_y = 415 N/mm², w_e = 20 kN/m³, ϕ = 30°, SBC(q_o) =

200 kN/m², Load W = 200 kN/m, M = 20 kN-m/m, self weight of foundation = $\dfrac{200}{10}$

= 20 kN/m

Total ultimate load W_u = 1.5 (200 + 20) 330 kN/m

Bearing area A (L × B) = $\dfrac{330}{1.5\,(200)}$ = 1.1 m²

Width of footing $= \dfrac{1.1}{1.0} = 1.10$ m

Depth of foundation $h = \dfrac{q_o}{w} \dfrac{(1-\sin\theta)^2}{(1+\sin\theta)^2} = \dfrac{150}{20}\left(\dfrac{1^2}{3^2}\right) = 0.83$ m $\cong 1.0$ m

Moment causes eccentricity $e = \dfrac{20\times10^6}{200\times1000} = 100$ mm (0.10 m)

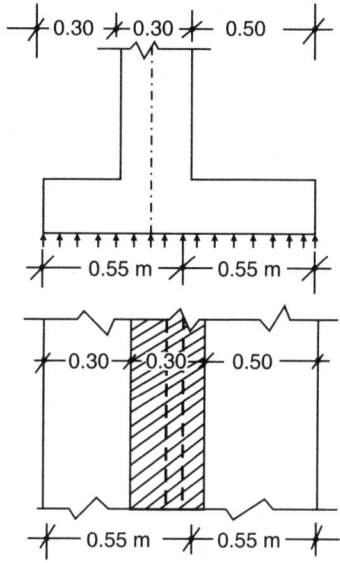

Fig. 9.8: Rectangular footing with eccentric loading

For uniform soil pressure, the resultant of wall loads must coincide with the CG of the base. Resultant of the load and moment acts at 0.10 m from the centre of the wall, i.e. $0.15 - 0.10 = 0.05$ m from the wall face. The centre of the base must be located along this point. Thus, projection of base $= 0.55 - 0.05 = 0.50$ m on one side.

Other side projection $= 0.55 - 0.15 - 0.10 = 0.30$ m

(Check) Total 0.30 m + 0.30 + 0.50 = 1.10 m

Net ultimate pressure on soil p_u is

$$p_u = \dfrac{300}{1\times1.1} = 273 \text{ kN/m}^2 \; (< 1.5 \times 200 \text{ kN/m}^2) \text{ OK.}$$

Max. moment M_u at the face of RCC wall,

$$M_u = 273 \,(0.50 \times 1) \times \dfrac{0.50}{2} = 34.10 \text{ kN-m}$$

$\dfrac{x_{ul}}{d} = 0.48$ (for steel Fe 415), $d = \sqrt{\dfrac{M_u}{R_u \, f_{ck} \, b}} = \sqrt{\dfrac{34.10\times10^6}{0.1383\times20\times1000}} = 111.0$ mm,

Assume $d = 150$ mm

$$A_{st} = \dfrac{0.5 f_{ck} \cdot b \cdot d}{f_y}\left[1 - \sqrt{1 - \dfrac{4.6 M_u}{f_{ck} \cdot bd^2}}\right]$$

$$= \frac{10 \times 1000 \times 150}{415} \left[1 - \sqrt{1 - \frac{4.6 \times 34.1 \times 10^6}{20 \times 1000 \times 150^2}} \right] = 697 \text{ mm}^2 \ (0.465\%)$$

Max. SF $V_u = 273 \ (0.50 \times 1) = 137$ kN, V_{u1} at d from face $= 273 \ (0.50 - d)$

$\tau_c = 0.46$ N/mm^2 (assuming $p = 0.45\%$), $\tau_v = \dfrac{273 \ (0.50 - d) \times 1000}{1000 \ (d \times 1000)}$ N/mm^2

$$\frac{273 \ (0.50 - d)}{1000 \, d} \leq 0.46 \text{ or } \frac{273 \ (0.50 - d)}{0.46 \times 1000} \leq d$$

or $(0.297 - 0.5935d) \leq d$ or $0.297 \leq 1.5935d$

$d \geq 0.186$ m, adopt $d = 190$ mm, $D = 250$ mm (cover 60 mm)

Thus, adopt $d = 190$ mm, $D = 250$ mm

$$A_{st} = \frac{10 \times 1000 \times 190}{415} \left[1 - \sqrt{1 - \frac{4.6 \times 34.1 \times 10^6}{20 \times 1000 \times 190^2}} \right] = 528 \text{ mm}^2$$

(12ϕ 200 c/c, $A_{st} = 565$ mm^2)

$$A_{st \, min} = \frac{0.85 \times 1000 \times 190}{415} = 390 \text{ mm}^2$$

Thus, 12ϕ 200 c/c is provided up to end.

Temp. steel $= \dfrac{0.12}{100} \times 1000 \times 250 = 300$ mm^2

(10ϕ @ 250 c/c, $A_{st} = 314$ mm^2) $< 5d$ or 450 c/c.

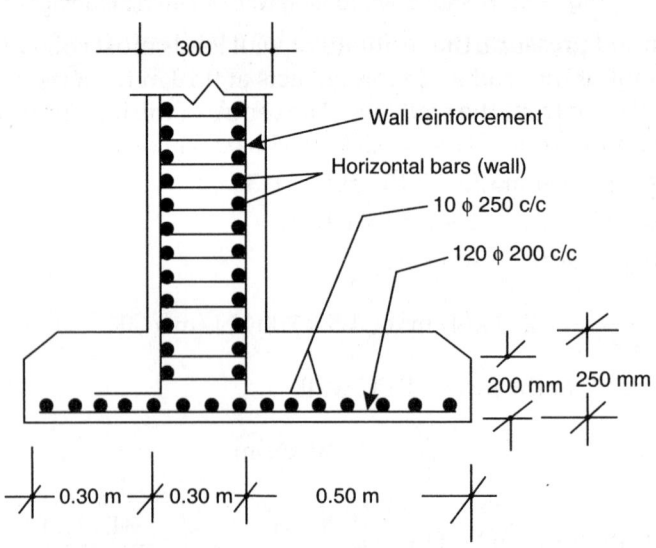

Fig. 9.9: Details of reinforcement in section of footing of wall

Example 9.5

Design a square RCC footing for a 400 mm × 400 mm column carrying a service load of 1000 kN. Weight of soil is 20 kN/m^3, bearing capacity of soil is 150 kN/m^2, and angle of repose ϕ is 30°. Use M20 grade CC and Fe 415 grade steel.

Solution: $f_{ck} = 20\,\text{N/mm}^2, f_y = 415\,\text{N/mm}^2, W = 1000\,\text{kN}, q_o = 150\,\text{kN/m}^2, w_e = 20\,\text{kN/m}^3,$
$\phi = 30°$

Depth of footing:

$$h = \frac{q_o}{w_e}\frac{(1-\sin\phi)^2}{(1+\sin\phi)^2} = \frac{150}{20}\left(\frac{1}{9}\right) = 0.83\,\text{m} \cong 1.0\,\text{m}$$

Footing size:

$W_u = 1.5 \times 1000 = 1500\,\text{kN}$, weight of footing = 10% of $W_u = 150\,\text{kN}$

Total $W_u = 1500 + 150 = 1650\,\text{kN}$

$$\text{Area of footing } A = \frac{1650}{1.5 \times 150} = 7.333\,\text{m}^2$$

$$\text{size} = \sqrt{7.3333} = 2.71\,\text{m} \times 2.71\,\text{m}\ [\text{adopt } (2.75\,\text{m} \times 2.75\,\text{m})]$$

Net upward soil pressure on footing,

$$p_u = \frac{(1.5 \times 1000)}{2.75 \times 2.75} = 198.35\,\text{kN/m}^2\ (< 1.5 \times 150\,\text{kN/m}^2),\ \text{OK.}$$

$$M_{u\,max} = 198.35\frac{(2.75-0.40)^2}{2 \times 4} = 137.0\,\text{kN-m/m}$$

$\dfrac{x_{ul}}{d}$ (for Fe 415) $= 0.48d, R_u = 0.138$

$$d = \sqrt{\frac{M_u}{b \cdot f_{ck}\,(0.138)}} = \sqrt{\frac{137 \times 10^6}{0.1383 \times 1000 \times 20}} = 223\,\text{mm}$$

Max. SF (one way) at a distance d (m),

$$V_u = 198.35 \times 2.75\left(\frac{2.35}{2} - d\right) = (233.06 - 198.35d)\,2.75\,\text{kN},\ b_o = 2.75\,\text{m}$$

(Assume $\tau_c = 0.36\,\text{N/mm}^2$, for M20, $p = 0.25\%$)

$$\tau_v = \frac{(233.06 - 198.35\,d)\,2750}{2.75 \times 1000 \times (d \times 1000)} \leq 0.36\,\text{N/mm}^2$$

or

$$d \geq \frac{(233.06 - 198.35\,d)\,2750}{0.36 \times 2.75 \times 1000 \times 1000} = (0.6474 - 0.551d)$$

$$d \geq \frac{0.6474}{1.55} = 0.4180\,\text{m (assume 420 mm)}$$

Assume d greater of 223 (for BM) and 418 mm (for shear stress) = 420 mm

$D = 420 + 40$ mm cover + bar diameter (16) = 476 mm (say 480 mm), $d = 424$ mm

Area of steel for BM,

$$A_{st} = \frac{0.5 f_{ck} \cdot b \cdot d}{f_y}\left[1 - \sqrt{1 - \frac{4.6M_u}{f_{ck} \cdot bd^2}}\right]$$

$$= \frac{10 \times 2750 \times 424}{415} \left[1 - \sqrt{1 - \frac{4.6 \times 137 \times 10^6}{20 \times 2750 \times (424)^2}}\right]$$

$$= 911 \text{ mm}^2 < \left(A_{st\,min} = \frac{0.12}{100} \times 2750 \times 480 = 1584 \text{ mm}^2\right)$$

Provide 1584 mm² in 2750 mm width (576 mm²/m)

(12ϕ 190 c/c, A_{st} = 635 mm² in 2750 mm width)

(190 < 3 × 424 = 1272 mm, < 450 mm, OK).

$$L_d = \frac{0.87\,f_y\,\phi}{4\,\tau_{bd}} = \frac{0.87 \times 415 \times 12}{4 \times 1.92} = 564 \text{ mm}$$

$$\left(\frac{2.75 - 0.40}{2}\right) - 25 = (1.15 \text{ m}) > 0.564 \text{ mm, OK.}$$

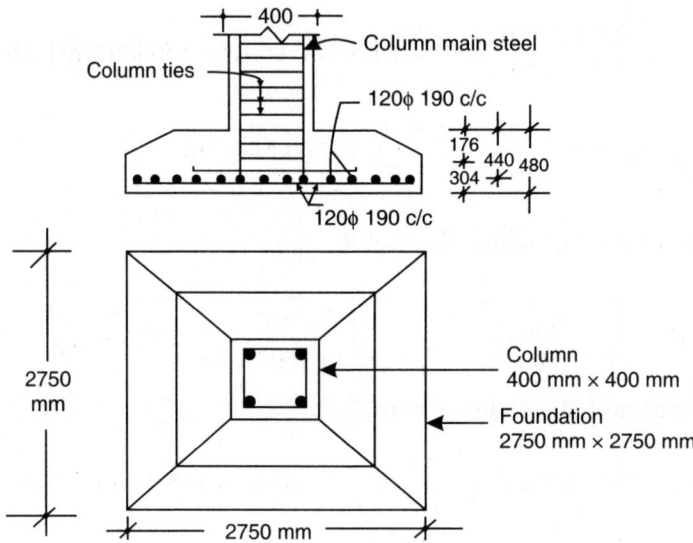

Fig. 9.10: Isolated square column footing with reinforcement details

Example 9.6

Design a combined footing with a strap beam of 400 mm width for two columns (size 400 mm × 400 mm) spaced at 4 m c/c. Each column supports a factored axial load of 600 kN. The ultimate bearing capacity of soil is 200 kN/m². Use M20 grade CC and Fe 415 grade steel.

Solution: Column 400 mm × 400 mm, spacing 4 m c/c, factored load each column W_u = 600 kN, f_{ck} = 20 N/mm², f_y = 415 N/mm², ultimate bearing capacity q_u = 200 kN/m²

Total load = 2 × 600 kN, total load + footing weight = 1200 + 120 = 1320 kN

$$\text{Area of footing required} = \frac{1320}{200} = 6.6 \text{ m}^2 \text{ (adopt 1.5 m × 5.0 m)}$$

Strap beam b = 400 mm

$$\text{Net upward } (p_u) \text{ soil pressure} = \frac{1200}{1.5 \times 5.0} = 160 \text{ kN/m}^2 \text{ (< 200 kN/m}^2\text{, OK)}$$

Cantilever projection of footing beyond strap beam face $= \left(\dfrac{1.5-0.40}{2}\right) = 0.55$ m

Ultimate max. moment at the face of beam,

$$M_u = 160 \times (1 \times 0.55)\left(\dfrac{0.55}{2}\right) = 24.2 \text{ kN-m } (24.2 \times 10^6 \text{ N-mm})$$

V_u (SF at a distance d from the beam face) $= 160 \times 1 \times (0.55 - d)$
$= (88 - 160d)$ kN (where d is in meters).

Depth d required for $M_u = \sqrt{\dfrac{24.2 \times 10^6}{0.138 \times 1000 \times 20}} = 93.64$ mm

say $d = 100$ mm, $D = 150$ mm

Minimum temp. steel $= \dfrac{0.12}{100} \times 1000 \times 150 = 180 \text{ mm}^2/\text{m}$

τ_c (Table 19 of IS:456-2000) $= 0.28 \text{ N/mm}^2$ for $p \le 0.15$ and M20

$$\tau_v = \dfrac{V_u}{1000 \times d} \le \tau_c \text{ or } \dfrac{(88-160d)\times 1000}{1000 \times d \times 1000} \le 0.28$$

or $(0.3143 - 0.57143d) \le d$ or $1.57143d \ge 0.3143$

or $d \ge \dfrac{0.3143}{1.57143} = 0.20001$ m (200.01 mm), provide $d = 250$ mm, $D = 300$ mm

$$\text{Reinforcement } A_{st} = \dfrac{0.5 f_{ck} \cdot 1000 \times 250}{415}\left[1 - \sqrt{1 - \dfrac{4.6 \times 24.2 \times 10^6}{20 \times 1000 \times (250)^2}}\right] = 275 \text{ mm}^2/\text{m}$$

Minmum temp. steel $= \dfrac{0.12}{100} \times 1000 \times 300 = 360 \text{ mm}^2/\text{m}$

(Thus, provide 10ϕ 200 c/c, $A_{st} = 393 \text{ mm}^2/\text{m}$),

$$p = \dfrac{393 \times 100}{1000 \times 250} = 0.157\%, \tau_c = 0.285 \text{ N/mm}^2, K = 1 \text{ for } D = 300 \text{ mm}$$

Check for shear:

$$\tau_v = \dfrac{(88 - 160 \times 0.25)1000}{1000 \times 0.25 \times 1000} = 0.192 \text{ N/mm}^2 < 0.285 \text{ N/mm}^2, \text{ OK.}$$

Thus, provide 10ϕ bars @ 200 c/c in both directions.
Design of strap beam (400 mm wide):

c/c span $= 4.0$ m, projection on both side $\dfrac{(5-4-0.40)}{2} = 0.30$ m beyond column face.

Consider 4 m SS span approximately: $M_u = \dfrac{W_u \cdot L^2}{8}$

w_u (factored u.d.l.) = 160 kN/m upward, $M_u = \dfrac{160 \times 4^2}{8}$ = 320 kN-m (as SS beam)

M_u (with overhangs 0.50 m) = $\dfrac{160 \times 2.5^2}{2} - 400 \times 2 = 300$ kN-m

$V_u = 0.5 \times 160 \times 4 = 320$ kN (320000 N),

Depth d for M_u:

$$d = \sqrt{\dfrac{320 \times 10^6}{0.138 \times 400 \times 20}} = 538.4 \text{ mm}$$

Depth d for V_u:

$$d = \dfrac{320000}{400 \times 0.75} = 1070 \text{ mm. (provide } D = 1150 \text{ mm and } d = 1100 \text{ mm)}$$

(Assuming $\tau_c = 0.75$ N/mm^2)

A_{st} (as rectangular singly reinforced) = $\dfrac{0.5 f_{ck} \cdot b \cdot d}{f_y} \left[1 - \sqrt{1 - \dfrac{4.6 \times 320 \times 10^6}{f_{ck} \cdot (bd^2)}} \right]$

$$= \dfrac{10 \times 400 \times 1100}{415} \left[1 - \sqrt{1 - \dfrac{4.6 \times 320 \times 10^6}{20 \times 400 \times 1100^2}} \right]$$

$$= 840 \text{ mm}^2$$

(provide 4 bars of 20ϕ, $A_{st} = 1256$ mm^2, $p = 0.29\%$, $\tau_c = 0.38$ N/mm^2

Minimum temp. steel = $\dfrac{0.12}{100} \times 400 \times 1150 = 552$ mm^2.

$\tau_v = \dfrac{320000 \text{ N}}{400 \times 1100} = 0.7272$ N/mm^2 > 0.38 N/mm^2 (τ_c), provide vertical stirrups in beam.

$V_{us} = (V_u - \tau_c \cdot b \cdot d) = (32 \times 10^4 - 0.38 \times 400 \times 1100) = 152800$ N

Using 8ϕ 2 legged vertical stirrups, the spacing S_v shall be

$$S_v = \dfrac{0.87 \times 415 \times (2 \times 50) 1100}{152800} = 259.9 \text{ mm c/c}$$

(provide 8ϕ 2 legged vertical stirrups at 250 c/c in strap beam)

Since beam is more than 450 mm deep, provide 0.1% reinforcement in side faces

of strap beam $\left(\dfrac{0.1}{100} \times 400 \times 1150 = 460 \text{ mm}^2 \text{, i.e. } 230 \text{ mm}^2 \text{ on each face} \right)$. Provide 5 bars

8ϕ on each face.

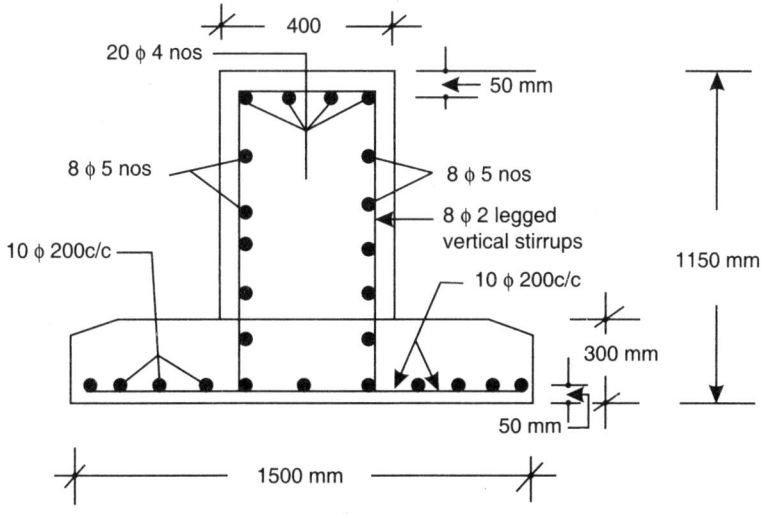

Fig. 9.11: Strap beam footing

Example 9.7

Design a combined rectangular footing for two columns spaced 6.0 m c/c. One of the column of size 400 mm × 400 mm and carrying a service load of 1620 kN is connected to other column of size 500 mm × 500 mm and carrying a service load of 2700 kN. Face of first column of 400 mm × 400 mm size is located at the property line. Safe bearing capacity of soil is 162 kN/m² and angle of repose of soil ϕ is 30° and unit weight of soil is 21.6 kN/m³. Use M20 grade CC and Fe 415 grade steel.

Solution: Depth of foundation $h = \dfrac{q_o}{W_e} \dfrac{(1-\sin\phi)^2}{(1+\sin\phi)^2} = \dfrac{162}{21.6}\left(\dfrac{1}{3}\right)^2 = 0.83 \text{ m} \cong 1.0 \text{ m}$

Size of footing $L \times B$ (10% weight of footing) $= \dfrac{(P_1 + P_2)\,1.1 \times 1.5}{q_o \times 1.5} = 29.33 \text{ m}^2$

For uniform soil pressure, the CG of footing must coincide with the resultant point of the loads. Let x_1 be the distance of resultant point from the property column C_1 centre:

$$x_1 = \frac{P_2 \times 6.0}{(P_1 + P_2)} = \frac{2700 \times 6.0}{(2700 + 1620)} = 3.75 \text{ m}$$

Thus, length of the footing $= \left(\dfrac{0.40}{2} + 3.75\right) \times 2 = 7.90 \text{ m}$

$B \times 7.9 = 29.33 \text{ m}^2$, $B = \dfrac{29.33}{7.9} = 3.713 \text{ m} \cong 3.75 \text{ m}$, say

Footing size $= 7.90 \text{ m} \times 3.75 \text{ m}$

Net soil pressure (factored) $= \dfrac{1.5\,(2700 + 1620)}{7.9 \times 3.75} = 218.8 \text{ kN/m}^2 \cong 219 \text{ kN/m}^2$

Consider footing as beam with width $b = 3.75 \text{ m}$ and $L = 7.9 \text{ m}$

Net pressure (ultimate) per meter length on the whole width = 218.8 × 3.75
$$= 820.5 \text{ kN/m}$$

Beam sketch is shown in Fig. 9.12.

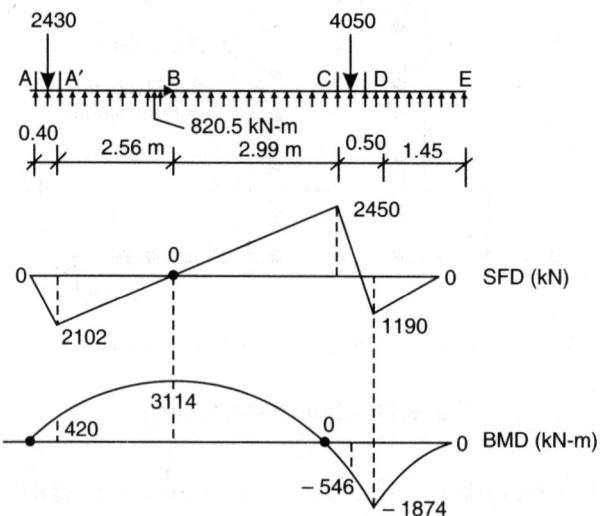

Fig. 9.12: Footing as beam is 7.90 m long ($b = 3.75$ m)

With upward soil pressure of 820.5 kN/m and column reactions (loads) of 2430 kN and 4050 kN downward at their respective centers, the SF and BM is found as follows:

SF at A' (inner face) = 820.5 × 0.4 – 2430 = – 2102 kN

SF at $B = 0$ at a distance of 2.56 m, SF at $C = + 2450$ kN, SF at $D = -1190$ kN

BM at $A' = + 420$ kN-m, BM at B (max) = 3114 kN-m, BM at $C = -546$ kN-m

BM at $D = -1874$ kN-m,

Maximum design SF = +2450 kN, –2102 kN

Maximum design BM = +3114 kN-m, –1874 kN-m,

Depth required for M_u (+3144 kN-m)

$$3750 \, d^2 = \frac{3114 \times 10^6}{0.138 \times 20} \text{ or } d = \sqrt{\frac{3114 \times 10^6}{0.138 \times 20 \times 3750}} = 549 \text{ mm (550 mm)}$$

Consider transverse moment under column C_1 and column C_2:

$$p_{u1} = 1620 \times 1.5/3.75 = 648 \text{ kN/m}, \quad M_{u1} = 648 \frac{(3.75 - 0.4)^2}{2^2} \times \frac{1}{2} = 909 \text{ kN-m}$$

($b = 400 + 1140 = 1540$ mm)

$$p_{u2} = \frac{2700 \times 1.5}{3.75} = 1080 \text{ kN/m}, \quad M_{u2} = 1080 \left(\frac{(3.75 - 0.50)^2}{2^2} \right) \times \frac{1}{2} = 1426 \text{ kN-m}$$

($b = 500 + 1140 = 1640$ mm)

Width for transverse moment = (400 + d) mm under column C_1

Width for transverse moment = (500 + d) mm under column C_2

(i) $(400 + d)d^2 = \dfrac{909 \times 10^6}{0.138 \times 20} = 329347826$, or $400d^2 + d^3 = 329347826$,

$d = 580$ mm by trial

(ii) $(500 + d)d^2 = \dfrac{1426 \times 10^6}{0.138 \times 20} = 516666667$, or $500d^2 + d^3 = 516666667$,

$d = 665$ mm by trial

Thickness based on shear:

$V_{u\,max} = \tau_{uc} \cdot b_o \cdot d$ or $d = \dfrac{V_u}{\tau_{uc} \cdot b_o}$, τ_{uc} (minimum) $= 0.28$, $K = 1$, $K\tau_c = 0.28$ N/mm^2

V_u (at a distance d from the inner face of column C_2) $= (2450 - 820.5 \times d)1000$

Thus, $1000d = \dfrac{(2450 - 820.5d)1000}{0.28 \times 3750}$ or $d = 2.3333 - 0.7814d$, $d = 1.31$ m

Thus, assume $d = 1140$ mm and $D = 1200$ mm with a cover of 60 mm
Check for punching shear around the column C_2.

Punching SF at $\dfrac{1}{2}d$ from the face $= [4050 - 820.5 \times (0.5 + 1.14)(1.64)] = 1843.2$ kN

maximum design punching shear stress $= 0.25\sqrt{f_{ck}} = 0.25\sqrt{20} = 1.12$ N/mm^2

$\tau_u = \dfrac{1843.2 \times 1000}{\{1.64 \times 1000 \times 4 \times 1140\}} = 0.2465$ N/mm^2 (< 1.12 N/mm^2), OK

Longitudinal steel:

$A_{st} = \dfrac{10 \times 3750 \times 1140}{415}\left[1 - \sqrt{1 - \dfrac{4.6 \times 3114 \times 10^6}{20 \times 3750 \times (1140)^2}}\right] = 7870$ mm^2

(3750 mm width), (2099 mm^2/m) in top face
Provide 20 mmϕ bars in top face at 140 c/c ($A_{st} = 2243$ mm^2/m)

Minimum steel for temperature: $\dfrac{0.12}{100} \times 1200 \times 1000 = 1440$ mm^2/m

(< 2099 mm^2/m)

Provide transverse reinforcement in two layers 16ϕ 250 c/c ($A_{st} = 2 \times 804$

$= 1608$ mm^2/m)

A_{st} for transverse moment (1426 kN-m)

$A_{sty} = \dfrac{10 \times 1640 \times 1120}{415}\left[1 - \sqrt{1 - \dfrac{4.6 \times 1426 \times 10^6}{20 \times 1640 \times (1120)^2}}\right]$

$= 3682$ mm^2/1640 mm $= 2245$ mm^2/m (20ϕ 140c/c)
$M_{u1} = 909$ kN-m ($b = 1540$ mm)

$$A_{sty1} = \frac{0.5 f_{ck} \cdot b \cdot d}{f_y} \left[1 - \sqrt{1 - \frac{4.6 M_u}{f_{ck} \cdot bd^2}} \right]$$

$= 2270 \text{ mm}^2$ (1474 mm^2/m, 20 mm ϕ @ 200 c/c, $A_{st} = 1570$ mm^2/m)

$M_u = -1874$ kN-m in longitudinal direction ($b = 3750$ mm)

$$A_{st} = \frac{0.5 f_{ck} \cdot b \cdot d}{f_y} \left[1 - \sqrt{1 - \frac{4.6 \times 1874 \times 10^6}{20 \times 3750 \times (1140)^2}} \right] = 4661 \text{ mm}^2 \text{ (1243 mm}^2\text{/m)}$$

Provide 20ϕ @ 250 mm c/c ($A_{st} = 1256$ mm^2/m) in bottom face.

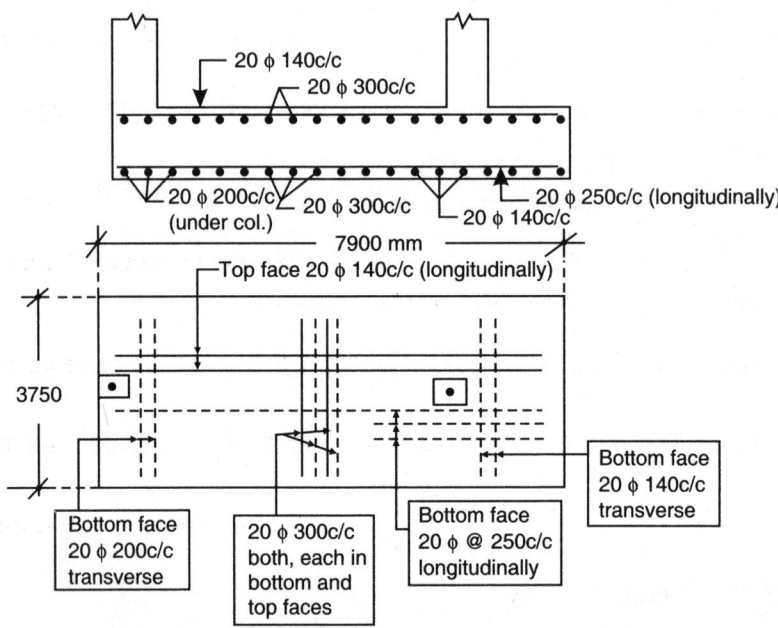

Fig. 9.13: Details of reinforcement in combined footing

Example 9.8

Design a raft foundation for columns of 400 mm × 400 mm size and subjected to service loads of 900 kN on corner columns and 1200 kN in interior columns as shown in Fig. 9.14. Soil has unit weight $w_e = 20$ kN/m^3, angle of repose $\phi = 30°$, and safe bearing capacity $q_o = 120$ kN/m^2. The columns are spaced at 5.0 m c/c longitudinally and 4.0 m c/c in transverse direction. Use M20 grade CC and Fe 415 grade steel.

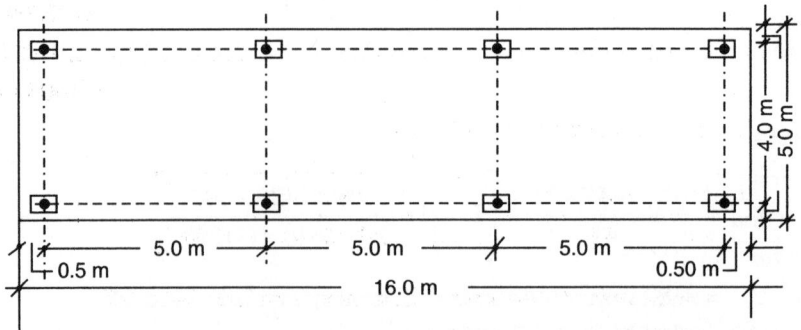

Fig. 9.14: Arrangement of raft slab and columns

Solution: $f_{ck} = 20$ kN/mm^2, $f_y = 415$ N/mm^2, $q_o = 120$ kN/m^2, $w_e = 20$ kN/m^3,
$W_1 = 900$ kN, $W_2 = 1200$ kN, weight of foundation = 10% of total load.

Depth of foundation required is $h = \dfrac{q_o}{w_e} \dfrac{(1 - \sin\phi)^2}{(1 + \sin\phi)^2} = \dfrac{120}{20} \left(\dfrac{1}{3}\right)^2 = 0.667$ m

Provide $h = 1.0$ m

Size of footing $L \times B = \dfrac{\left\{(900 \times 4 + 1200 \times 4) 1.5\right\} 1.1}{1.5 \times 120} = 77$ m^2 (provide 16 m × 5 m)

For uniform soil pressure, the CG of the footing (raft slab) shall coincide with the CG of the loads (Fig. 9.14).

Net soil pressure (upward) $p_u = \dfrac{(3600 + 4800) 1.5}{16 \times 5} = 157.5$ kN/m^2 (ultimate soil pressure)

Mat slab is divided into strips of widths in the ratio of loads.

Thus, widths of strips carrying loads of 900 × 1.5 kN (ultimate) and 1200 × 1.5 kN

ultimate will have strips of width $b_e = \dfrac{(900 \times 2 \times 1.5) \times 16}{(900 \times 4 \times 1.5 + 1200 \times 4 \times 1.5)} = 3.43$ m

$$b_i = \dfrac{1200 \times 2 \times 1.5 \times 16}{(8400 \times 1.5)} = 4.57 \text{ m}$$

Transverse direction each strip shall be of 2.5 m.

Consider longitudinal strip of 2.50 m width and 16 m length as shown in Fig. 9.15

Loading = 157.5 kN/m^2 (width $b = 2.50$ m)

Loading on the whole strip of 2.5 m = 157.5 × 2.5 = 394 kN/m.

SF at A = 394 × 0.5 = 197 kN and (197 − 900 × 1.5) = − 1153 kN = SF at D.

SF at E = 0, (p_t of max. BM) = SF at G

SF at B = + 816 kN, − 984 kN

SF at F = 0, (point of max. BM)

SF at C = + 984 kN, − 816 kN

Bending moment ($b = 2.50$ m):

BM at A = − 49 kN-m (width)

BM at E = + 1639 kN-m (on strip width) = BM at G

BM at B = − 795 kN-m (on strip width) = BM at C

BM at F = + 2025 kN-m (on strip width)

Max. SF = 1153 kN ($b = 2.5$ m)

Max. BM = + 2025 kN-m (on $b = 2.50$ m), − 795 kN-m (on $b = 2.50$ m)

Transverse strip ($b = 3.43$ m)

Ultimate loading = 3.43 × 157.5 = 540 kN/m

SF diagram ($b = 3.43$ m):

SF at A = 270 kN, − 1080 kN

SF at C = − 270 kN, + 1080 kN

SF at B = 0

Max. transverse SF = 1080 kN,

BM diagram (b = 3.43 m):

BM at A = – 68 kN-m = BM at C.

BM at B = + 1013 kN-m

Max. transverse BM = 1013 kN-m

Depth required for strip width = 2500 mm

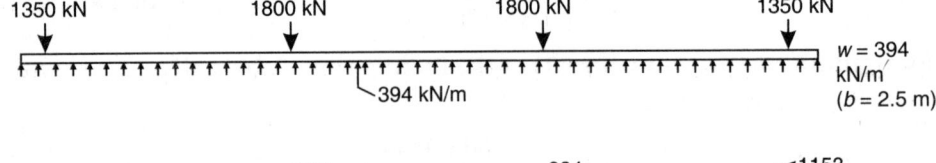

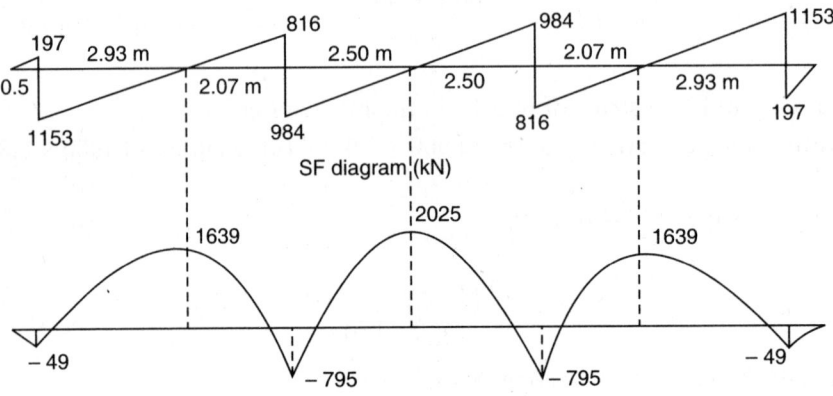

Fig. 9.15: Loading, SF and BM diagrams longitudinal strip (width b = 2.50 m)

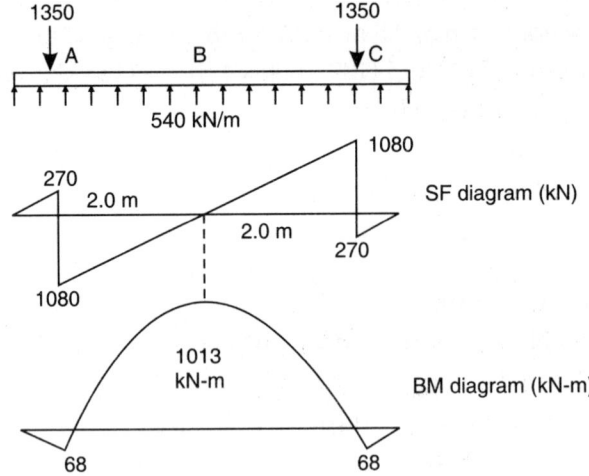

Fig. 9.16: Loading, SF and BM diagrams (transverse strip), width b = 3.43 m, loading $w_u = 3.43 \times 157.5 = 540$, kN/m

For BM:

$$d = \sqrt{\frac{M_u}{0.138 \times 20 \times b}} = \sqrt{\frac{2025 \times 10^6}{0.138 \times 2500 \times 20}} = 542 \text{ mm}$$

For SF at critical section at a distance of *d* m from the face of the column,

$V_u = 1350 - 394 (0.5 + 0.2 + d) = (1054 - 394\,d)$ kN $= (1054 - 394d)1000$ N. Assuming design stress $\tau_c = 0.36$ N/mm^2

(assuming tensile reinforcement $p = 0.25\%$),

$$d \text{ (based on SF)} = \frac{(1054 - 394\,d)\,1000}{0.36 \times 2500 \times 1000} = (1.1711 - 0.4378d) \text{ or}$$

$$d = \frac{1.1711}{1.4378} = 0.815 \text{ m.}$$

Thus, adopt $D = 870$ mm, $d = 820$ mm with clear cover of 40 mm,

$$\text{minimum temp. steel} = \frac{0.12}{100} \times 2500 \times 870 = 2610 \text{ mm}^2 \text{ in 2.5 m width}$$

$$(1044 \text{ mm}^2/\text{m strip})$$

provide on both faces $\dfrac{1044}{2} = 522$ mm^2/m each

(16ϕ 300 c/c, $A_{st} = 670$ mm^2/m on each face).

Tension steel (top face):

$$A_{st} = \frac{0.5 f_{ck} \cdot b \cdot d}{f_y}\left[1 - \sqrt{1 - \frac{4.6 M_u}{f_{ck} \cdot bd^2}}\right]$$

$$= \frac{10 \times 2500 \times 820}{415}\left[1 - \sqrt{1 - \frac{4.6 \times 2025 \times 10^6}{20 \times 2500 \times (820)^2}}\right] = 7397 \text{ mm}^2 \text{ (full width 2500 mm)}$$

$= 2959$ mm^2/m (provide 20ϕ @ 100 c/c, $A_{st} = 3140$ mm^2/m) > minimum, OK.
Consider transverse strip $b = 3430$ mm

$$\text{Max. BM at B} = 1013 \text{ kN-m}, d = \sqrt{\frac{1013 \times 10^6}{0.138 \times 20 \times 3430}} = 327.12 \text{ mm}$$

$$< \text{adopted depth is OK.}$$

SF at d from the column face:
$V_u = 1350 - (0.5 + 0.2 + d)540 = (972 - 540d)$
Assuming design shear stress $\tau_c = 0.36$ N/mm^2,

$$d = \frac{(972 - 540d)}{(0.36 \times 3430)} = (0.7872 - 0.43732d) \quad \text{or} \quad d = \frac{0.7872}{1.43732} = 0.548 \text{ m}$$

Thus, adopted $D = 870$ mm, and $d = 820$ mm, d' (transverse) $= 820 - 10 - \dfrac{16}{2}$

$$= 802 \text{ mm}$$

$$\text{Transverse steel } A'_{st} \text{ } (b = 3430 \text{ mm}) = \frac{0.5 f_{ck} \cdot b \cdot d'}{f_y}\left[1 - \sqrt{1 - \frac{4.6 M_u}{f_{ck} \cdot bd'^2}}\right]$$

$$A'_{st} = \frac{10 \times 3430 \times 802}{415}\left[1 - \sqrt{1 - \frac{4.6 \times 1013 \times 10^6}{20 \times 3430 \times (802)^2}}\right] = 3598 \text{ mm}^2 \text{ (1049 mm}^2/\text{m)}$$

16ϕ 190 c/c in top face (A_{st}= 1058 mm²/m) OK.

Bottom face minimum temp. steel (on each face) 16ϕ 300 c/c (A_{st} = 670 mm²/m)

Check for two way shear (punching shear):

τ_c (permissible punching) = $0.25\sqrt{f_{ck}}$ = $0.25\sqrt{20}$ = 1.12 N/mm²

Consider internal columns (load = 1800 kN), side = $\left(400 + \dfrac{820}{2}\right)$ = 810 mm

SF at a distance $\dfrac{1}{2}d$ = 1800 – 157.5 (0.81 × 0.810)

$$= (1800 - 103.34) = 1696.66 \text{ kN}$$

$$\text{Punching stress} = \frac{1696.666 \times 1000}{(0.810 \times d \times 4)\,1000} = \frac{1696666}{810 \times 820 \times 4}$$

$$= 0.64 \text{ N/mm}^2 < 1.12 \text{ N/mm}^2,\ OK.$$

Separate steel in exterior and interior spans will result in lot of extra lengths as development length, and hence the same reinforcement may be provided throughout. Thus, provide throughout the spans the same reinforcement unless too much different requirements are there.

Longitudinal bars 20ϕ 100 c/c in top face throughout.

Longitudinal bars 16ϕ 300 c/c in bottom face throughout (as temp. steel).

Transverse bars:

Top face throughout 16ϕ @ 190 c/c

Bottom face throughout 16ϕ @ 300 c/c

Details of reinforcement:

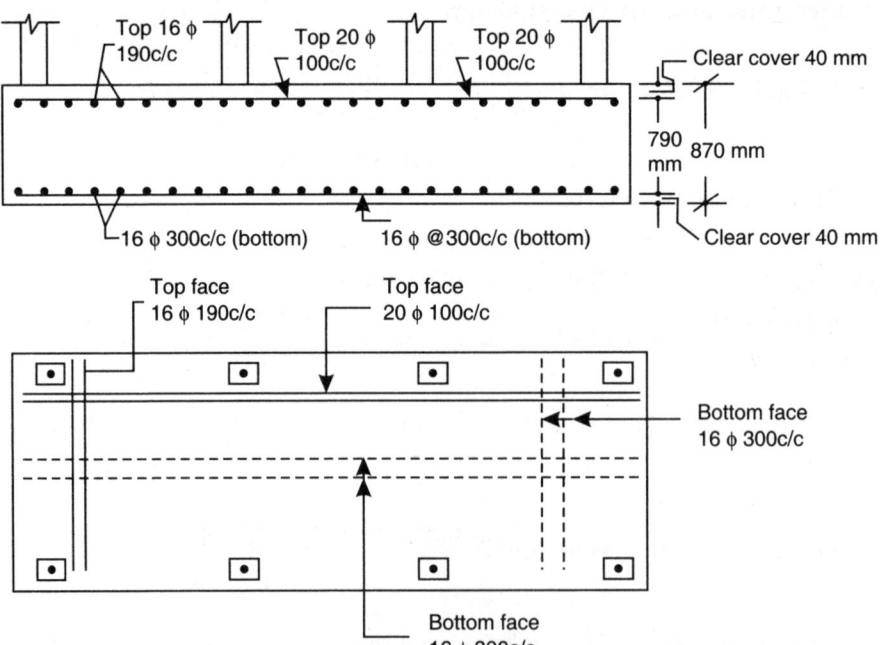

Fig. 9.17: Details of reinforcement in raft foundation

9.3 PILE FOUNDATION

9.3.1 Introduction

When soil under the foundation is weak to support the load of the superstructure on a shallow foundation, piles are provided in the foundation to transmit loads to the rigid hard rock at a greater depth and through surface friction to the surrounding soil. When the load is transferred through soft soil to a suitable bearing stratum by means of bearing of the piles alone, the piles are known as end bearing or point bearing piles. When the load is transferred by means of surface friction of soil along the length of the piles, such piles are called *friction piles*. Generally the piles are used to support loads partly by skin friction and partly by point bearing.

Compaction piles are used for compacting the loose granular loose soil deposits.

Piles may also be classified according to the material used in its construction such as *timber, RCC, steel or composite piles*. Concrete piles may further be classified as precast or cast in place (Fig. 9.18).

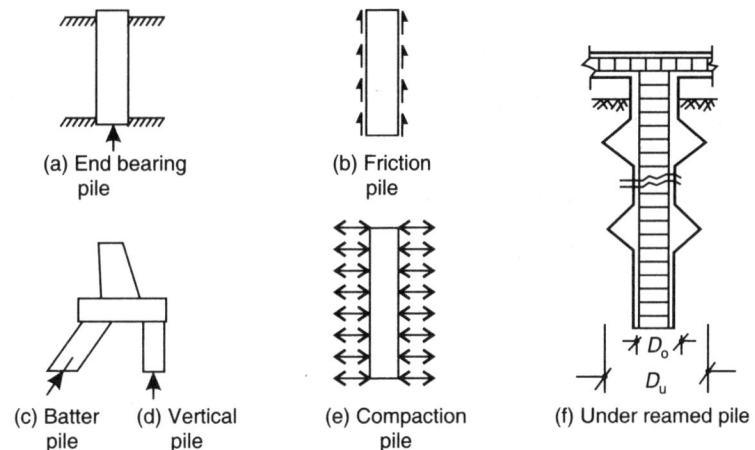

| (a) End bearing pile | (b) Friction pile |

| (c) Batter pile | (d) Vertical pile | (e) Compaction pile | (f) Under reamed pile |

Fig. 9.18: Different types of piles

9.3.2 Pile Capacity

Pile capacity depends on the soil surrounding the pile to support loads on the pile. It can be determined by the load test. The ultimate load bearing capacity Q_u of a pile in a homogeneous soil may be expressed as the sum of point resistance Q_p and skin resistance Q_s, i.e.

$$Q_u = Q_p + Q_s = (A_p \cdot q_p + \Sigma A_s \cdot f_s) \qquad \dots \text{Eq. (9.9)}$$

where,

A_p = Area of pile point

q_p = Unit bearing capacity = $C N_c \xi_c + \gamma D N_q \xi_q$

C = Unit cohesion

N_c, N_q = Bearing capacity factors

ξ_c, ξ_q = Shape factors

γ = Unit weight of soils

D = Depth of foundation

A_s= Effective pile surface area on which f_s acts and computed as perimeter x incremental embedment length to allow for variations in pile shaft and stratification of soil.

f_s = Unit skin friction

The allowable pile capacity $Q_a = \left(\dfrac{Q_p}{F_p} + \dfrac{Q_s}{F_s} \right)$ ... Eq. (9.10)

where, F_p, F_s = factors of safety which vary in the range 2.5 – 4 or more depending on their uncertainties.

Pile Group

Generally more than two piles are provided in a foundation to avoid problems of alignment and inadvertent eccentricities. The provision of several piles in a group lead to overlapping of soil pressure envelopes developed in the soil for resisting the pile load. This may cause failure of soil in shear or excessive settlement of pile group. To avoid this, spacing of piles could be increased. However, large spacings are impractical because a pile cap is constructed over a group of piles for column base and transmit applied loads to all piles.

The efficiency of the group of friction piles is defined as the ratio of the actual group capacity to the sum of the individual pile capacities depending on the spacing of piles.

Let there be m rows and n columns of piles, the total number of piles = $m \times n$, the perimeter of group is $p = [2 \{(m-1)S + (n-1)S\} + 4D]$

or $p = [2(m+n-2)S + 4D)]$

For 100% efficiency, we have $\dfrac{[2(m+n-2)S+4D]f_s\,L_f}{mn\,(\pi D)f_s\,L_f} = \dfrac{100}{100}$

or $[2(m+n-2)S + 4D] = mn\,(\pi D)$

or $S = \dfrac{(mn\pi D - 4D)}{2(m+n-2)} = \dfrac{(1.57D)mn - 2D}{(m+n-2)}$ ... Eq. (9.11)

Pile Cap

Pile cap is provided over a group of piles to distribute the load from the superstructure to piles in the group. These caps act as rigid footing which is subjected to concentrated reactions from the piles. The resultant reaction of the piles is equal to the sum of loads on the cap, self-weight of cap, and weight of the soil above cap, if any.

Reaction of the pile $P_p = \dfrac{Q}{n}$ ... Eq. (9.12)

where, Q = Concentric load on the cap

n = Number of piles

When the pile cap is eccentrically loaded and the eccentricity causes moment in two perpendicular directions as M_x and M_y, then the reaction in pile is determined as:

$$P_p = \frac{Q}{n} \pm \frac{M_y \cdot x}{\sum x^2} \pm \frac{M_x \cdot y}{\sum y^2} \qquad \text{... Eq. (9.13)}$$

where, M_x, M_y = Moments w.r.t X and Y axes respectively

x, y = Distances from X and Y axes to the piles.

Example 9.9

Design a pile cap for supporting a column of size 400 mm × 400 mm and carrying a service load of 800 kN. The pile cap has a group of 4 friction piles of 250 mm diameter to transfer the column load to soil through piles. Use M20 grade CC and Fe 415 grade steel.

Solution: D = 250 mm, column 400 mm × 400 mm. Generally pile cap is projected beyond piles to overhang with minimum of 150 mm. Spacing of piles S shall not be less than minimum $3D$ or 900 mm. For 100% efficiency, $S = \dfrac{\pi D(mn - 4D)}{2(m + n - 2)}$,

where $(m = n = 2)$

$$S = \frac{\pi \times 250 \times 2 \times 2 - 4 \times 250}{2(2 + 2 - 2)} \quad \text{or} \quad S = 535.4 \nless 900 \text{ or } 3 \times 250, S = 900 \text{ mm}$$

$L = S + D + 2 \times 150 = 1450$ mm, $\qquad (B = 1450$ mm$)$

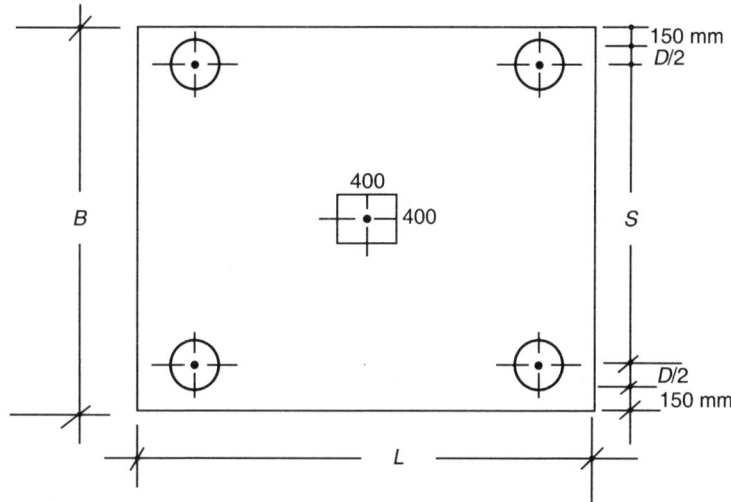

Fig. 9.19: Pile cap plan layout

Depth of pile cap:

$$M_{u\,max} = \left[P_{up} \times \left(\frac{S - l}{2} \right) \right] 2$$

where, P_{up} = ultimate reaction of each pile considering concentric loading

$$P_{up} \text{ (each pile)} = \{1.5 (800 + 80)\} \frac{1}{4} = 330 \text{ kN}$$

M_u (max.) at the face of column (due to concentrated pile loads)

$$= 2 \times 330 \times \frac{(0.9 - 0.4)}{2} = 165 \text{ kN-m}$$

$$d_1 = \sqrt{\frac{M_u}{0.138 \, f_{ck} \cdot b}} = \sqrt{\frac{165 \times 10^6}{0.138 \times 1450 \times 20}} = 203 \text{ mm}$$

Depth based on shear:

Maximum SF $V_{u \, max}$ (one way at critical section at a distance of d)

$$V_{u \, max} = 2P_u = 2 \times 330 = 660 \text{ kN}$$

τ_{uc} (Table 19 of IS:456-2000), assuming 0.25% tensile steel,

$$\tau_{uc} = 0.36\% \text{ N/mm}^2$$

Thus, $d = \dfrac{V_{u \, max}}{B\tau_{uc}} = \dfrac{660 \times 1000}{0.36 \times 1450} = 1265$ mm

For two way shear: τ_{uc} (punching) $= 0.25 \sqrt{20} = 1.118 \text{ N/mm}^2$

$$B = (400 + d') \times 4 \text{ for punching at } \frac{d}{2} \text{ from the column face}$$

$$d' = \frac{(800 \times 1.5) \, 1000}{(400 + d') \, 4 \times 1.118} \text{ or } (d'^2) + (400d') - 268336 = 0, d' = 355.3 \text{ mm}$$

Punching of piles (load P_{up} = 330 kN)

$$d'' = \frac{330 \times 1000}{\pi (250 + d'') 1.118} \text{ or } (d''^2) + 250d'' - 93955.5 = 0, d'' = 206 \text{ mm}$$

Thus, maximum depth required for one way shear is equal to 1265 mm and may be adopted without providing any shear reinforcement.

Alternatively, for one way shear, stirrups may be provided if the depth provided is less than 1265 mm. Let us provide the greatest of depth required 203 mm (for BM), 355.3 mm (for punching of column.) or 206 mm (punching of pile), hence adopt d =

$$360 \text{ mm}, \ D = 360 + 40 + \frac{20}{2} = 410 \text{ mm}$$

Minimum temperature steel $= \dfrac{0.12}{100} \times 1450 \times 410 = 714 \text{ mm}^2$ in full width.

Steel required for bottom tension zone for BM:

$$A_{st} = \frac{0.5 f_{ck} \cdot b \cdot d}{f_y} \left[1 - \sqrt{1 - \frac{4.6 \, M_u}{f_{ck} \cdot b \cdot d^2}} \right]$$

$$= \frac{10 \times 1450 \times 360}{415} \left[1 - \sqrt{1 - \frac{4.6 \times 165 \times 10^6}{20 \times 1450 \times 360^2}} \right]$$

$$= 1342 \text{ mm}^2 \left[\geq \min. \frac{0.85 \, B \cdot d}{415} \text{ or } \frac{0.12}{100} \times B \times D \text{ i.e. } 1070 \text{ or } 714 \text{ mm}^2 \right]$$

Thus, in bottom face provide 16ϕ 8 bars in each direction (A_{st} provided = 1608 mm²). In perpendicular direction in bottom face also provide 16ϕ 8 nos. (i.e. 16ϕ @ 180 c/c).

At the end of reinforcement provide adequate development length by bending in vertical direction.

$$p_t \left(\% \text{ tensile steel} = \frac{1608 \times 100}{1450 \times 360} = 0.31\% \right).$$

Also provide 8ϕ 8 legged ($A_{sv} = 8 \times 50 = 400$ mm² vertical stirrups).

8 legged stirrups also take care of temperature steel in top face.

Spacing of 8ϕ 8 legged stirrups (S_v),

$$\text{Spacing } (S_v) = \frac{0.87 f_y A_{sv}}{(\tau_{uv} - \tau_{uc}) \cdot B}$$

$$= \frac{660 \times 1000}{1450 \times 360} = 1.264 \text{ N/mm}^2 > \tau_{uc}, \text{ provide stirrups.}$$

τ_{uc} (for 0.31% tensile steel) = 0.388 N/mm² (Table 19 of IS:456-2000)

$$S_v = \frac{0.87 \times 415 \times 400}{(1.264 - 0.388)} = 113.7 \text{ mm c/c } (S_{vmax}) \text{ OK.}$$

$$\left\{ S_{vmax} = 0.75d \ (= 270 \text{ mm}), \ \frac{2.5 A_{sv} f_y}{B} = 286 \text{ mm} \right\}$$

Thus, provide 8ϕ 8 legged vertical stirrups at 100 mm c/c.

Details of reinforcement in pile cap:

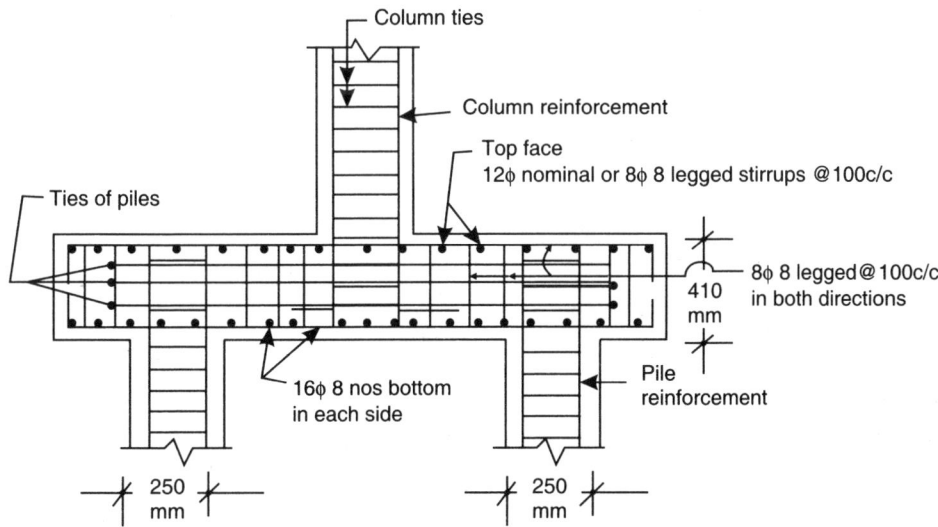

Fig 9.20: Reinforcement details in pile cap

Example 9.10

Design a pile cap for supporting a column of section 500 mm × 500 mm carrying an axial service load of 2100 kN and a uniaxial service moment of 300 kN-m. The pile cap contains a group of 9(3 × 3) friction piles, each of 250 mm diameter to transfer column load to soil. Use M20 grade CC ad Fe 415 grade steel.

Solution: $f_{ck} = 20 \text{ N/mm}^2, f_y = 415 \text{ N/mm}^2, P = 2100 \text{ kN}, m = n = 3, D = 250 \text{ mm}$,
$M = 300 \text{ kN-m} (300 \times 10^6 \text{ N-mm})$

Assuming 100% efficiency of group of piles, we have

$$\frac{100}{100} = \frac{2(m+n-2)S+4D}{\pi Dmn} \quad \text{or} \quad S = \frac{\pi D(mn-4D)}{2(m+n-2)} \nless S_{min}$$

$(S_{min} = 3D$, or 900 mm whichever is more)

or

$$S = \frac{\pi \cdot (250)3 \times 3 - 4 \times 250}{2(4)} = 758.6 \text{ mm (should not be} < 900 \text{ mm)}$$

∴ $S = 900 \text{ mm}$

$L = B = 2 \times 900 + 250 + 2 \times 50 \text{ (overhang)} = 2350 \text{ mm}$

Depth of pile cap by moment or shear force.

Total ultimate load on the pile $= 1.5 \{2100 + 210 (10\% \text{ weight of cap})\} = 1.5 \times$
2310 kN

Thus, ultimate load reactions from piles,

$$P_{uc1}, P_{uc2}, P_{uc3} = \frac{1.5 \times 2310}{9} + \frac{\sum (1.5 \times 300 \times 1000) \, y_c}{\sum y_c^2}$$

$$= 385 \text{ kN} + \frac{450 \times 900 \times 1000}{6(900)^2} = 468.33 \text{ kN}$$

$$M_{ax} \cdot M_{uy} = 3 \times 468.33 \left(\frac{900}{1000} - \frac{0.50}{2} \right) = 913.25 \text{ kN-m} (913.25 \times 10^6 \text{ N-mm})$$

$$\text{Depth of cap (BM) } d_1 = \sqrt{\frac{913.25 \times 10^6}{0.138 \times 20 \times 2350}} = 375.24 \text{ mm}$$

Depth for one way shear:

$V_{u \, max}$ at the critical section $= 468.33 \times 3 = 1405 \text{ kN (uniform)}$

τ_{uc} (Table 19 of IS:456-2000) for assumed tensile steel of about 0.25% $= 0.36 \text{ N/mm}^2$

$$\text{Thus, } d_2 = \frac{V_{u \, max}}{0.36 \times B} = \frac{1405 \times 1000}{0.36 \times 2350} = 1660 \text{ mm}$$

Two way punching shear:

Punching load of column $= 1.5 (2100 + 210) = 3465 \text{ kN}$

Punching shear of each pile $P_{uc1} = 468.33 \text{ kN}$

Ultimate two way punching design stress $\tau_{uc} = 0.25 \sqrt{f_{ck}} = 1.118 \text{ N/mm}^2$

B for column punching (at a distance d) $= (500 + d)4$ (square)

b for pile punching (at a distance d) $= (250 + d)\pi$ (circular)

$$\text{Thus, for column punching, } d = \frac{3465 \times 1000}{1.118 \, (500+d)4} \quad \text{or } d^2 + 500d - 774821 = 0,$$

$d = 665 \text{ mm}$

For pile punching $d = \dfrac{468.33 \times 1000}{\pi (250 + d) 1.118}$ or $d^2 + 250d - 133340 = 0$, $d = 261$ mm

Thus, one way shear governs the depth of pile cap. This depth is too much and is uneconomical. Alternatively, we can provide depth on the basis of two way shear or bending moment (greater of the two). Adopt d (= 665, 261 or 375.24 mm) = 700 mm, $D = 750$ mm

$$A_{st} \text{ (tensile steel for BM)} = \dfrac{0.5 f_{ck} \cdot b \cdot d}{f_y} \left[1 - \sqrt{1 - \dfrac{4.6 M_u}{f_{ck} \cdot b(d)^2}} \right]$$

$$b = 2350$$

or $A_{st} = \dfrac{10 \times 2350 \times 700}{415} \left[1 - \sqrt{1 - \dfrac{4.6 \times 913.25 \times 10^6}{20 \times 2350 \times (700)^2}} \right]$

$= 3797$ mm^2 (1616 mm^2/m), 16ϕ 120 c/c, $A_{st} = 1675$ mm^2)

$p_t = 0.24\%$, $\tau_{uc} = 0.35$ N/mm^2, or 20ϕ @ 190 c/c, $A_{st} = 1653$ mm^2

Check:

$$\tau_{uv} \text{ (one way)} = \dfrac{1405000}{2350 \times 700} = 0.854 \text{ N/mm}^2 \, (> 0.35 \text{ N/mm}^2)$$

Provide 8 legged 8ϕ vertical stirrups for shear reinforcement

$$S_v = \dfrac{0.87 f_y \cdot A_{sv}}{(\tau_{uv} - \tau_{uc}) \cdot B} \ngtr S_{max}$$

$$S_v = \dfrac{0.87 \times 415 \times (8 \times 50)}{(0.504) 1000} = 286 \text{ c/c}, \; S_{v \, max} = \dfrac{2.5 A_{sv} f_y}{B}, 0.75d, = 525, 415$$

Thus, provide 8ϕ 8 legged vertical stirrups @ 280 c/c.

$$M_{ux \, max} = (P_{ua1} + P_{ub1} + P_{uc1}) \left(s - \dfrac{b}{2} \right) \text{ about column face}$$

Pile reactions ($P_{ua1} + P_{ub1} + P_{uc1}$):

$$P_{ua1} = \dfrac{1.5 (2100 + 210)}{9} - 300 \times 1000 \times \dfrac{900}{6 \times 900^2} = 385 - 55.55 = 329.45 \text{ kN}$$

$$P_{ub1} = \dfrac{1.5(2310)}{9} - \dfrac{300 \times 1000 \times 0}{6 \times 900^2} = 385 - 0 = 385 \text{ kN}$$

$$P_{uc1} = \dfrac{1.5 (2310)}{9} + 300 \times 1000 \times \dfrac{900}{6 \times 900^2} = 385 + 55.55 = 440.55 \text{ kN}$$

$\therefore$ Moment $M_{ux \, max} = (329.45 + 385 + 440.55) \left(s - \dfrac{500}{2} \right)$

$$= 1155 \text{ kN} \times \dfrac{(900 - 250)}{1000} = 750.75 \text{ kN-m} = 751 \text{ kN-m}$$

$$A_{st} = \frac{0.5 f_{ck} \cdot b \cdot d}{f_y} \left[1 - \sqrt{1 - \frac{4.6 M_u}{f_{ck} \cdot b(d)^2}} \right]$$

$$= \frac{10 \times 2350 \times 700}{415} \left[1 - \sqrt{1 - \frac{4.6 \times 751 \times 10^6}{20 \times 2350 \times (700)^2}} \right]$$

$= 3094$ mm^2 (in 2350 mm), i.e. 1316 mm^2/m strip

Use 20 mm ϕ @ 230 c/c ($A_{st} = 1365$ mm^2) > 1316 mm^2 (required), OK.

Other direction 20ϕ @ 190 c/c in bottom face.

$$\text{Minumum temp. steel} = \frac{0.12}{100} \times 1000 \times 750 = 900 \text{ mm}^2/\text{m}$$

(12ϕ @ 120 c/c, $A_{st} = 941$ mm^2/m)

Provide $\dfrac{1}{2}$ temp. steel in top face to act as anchor bars. Temperature steel is not required in bottom face since main steel is available. Thus, provide steel as below:

Depth $D = 750$ mm, $d = 700$ mm, $d' = 680$ mm

Bottom face: 20ϕ bars @ 190 mm c/c in one direction and

20ϕ bars @ 230 mm c/c in perpendicular direction

Top face: 12ϕ 240 c/c in both the directions.

Shear reinforcement: Provide 8ϕ 8 legged vertical stirrups @ 280 c/c.

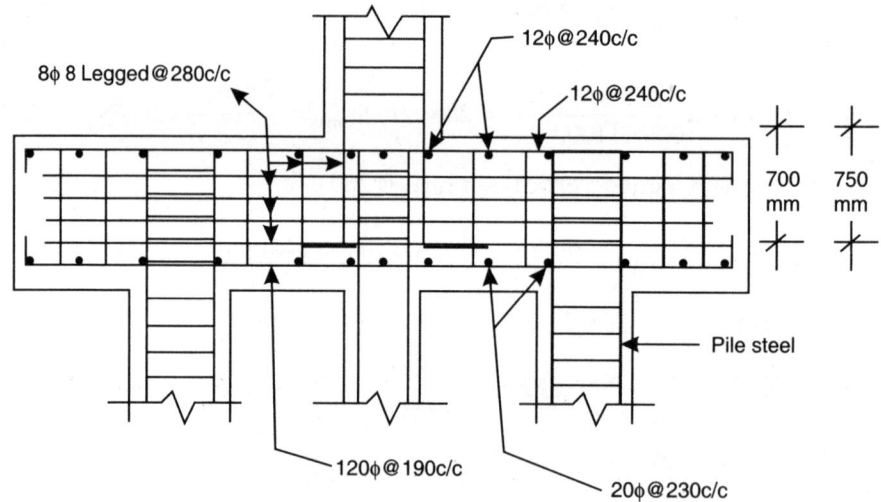

Fig 9.21: Details of reinforcement

SUMMARY

Foundations are provided to transfer the loads of superstructure to the ground and soil in contact with it. The foundation transfer the loads by getting different types of strains and stresses which shall be within safe limits. The foundations are provided at certain minimum depth to avoid the effects of environmental variations.

The foundations are classified as *shallow* or *deep*. *Shallow foundations* are provided for light structures when located on soils with good bearing capacity. *Deep foundations* are required for heavy structures specially when located on soils with poor bearing capacity. The size and design of foundations depend on bearing capacity of soil and on loading quantity and pattern.

Foundations for column can be isolated, combined or mat (raft) slab type depending on the loads and type of soil. Combined footing for columns become necessary specially when one of the column is located on the property line. Mat (raft) slab foundation becomes necessary when the loads are eccentric and there may be excessive differential settlement and the bearing capacity of soil is also small. One very important consideration in design is that all external loads of structure shall always remain in equilibrium, i.e. external load W_u will be balanced by internal resistance of foundation structure.

$$\text{Width of footing } (b) = \frac{\text{ultimate load } (W_u)}{\text{ultimate bearing capacity}}$$

Upward soil reaction $P_u = 1.5 \times 1.1W = 1.65W$

$$\text{Net ultimate upward soil pressure } (p_u) = \frac{1.65\,W}{B \times 1} = \frac{1.65\,W}{B}$$

$$M_u = \frac{p_u}{8}\,(B - b)^2$$

Depth d for moment $\qquad d_1 = \sqrt{\dfrac{M_u}{R_u\, b\, f_{ck}}}$

For shear, $\qquad d_2 = \dfrac{V_u \text{ max at critical section at a distance } d}{\tau_{uc}\, b}$

Greater depth (d_1 or d_2) is adopted for the footing.

$$\text{Steel } A_{st} \text{ (for moment } M_u) = \frac{0.5 f_{ck}\, bd}{f_y}\left[1 - \sqrt{1 - \frac{4.6\, M_u}{f_{ck}\, b(d)^2}}\right] \text{ for adopted } d.$$

For isolated footing:

$$\text{Net } p_u = \frac{W_u}{(L \times B)}, \text{ for rectangular}$$

$$M_{u1} \text{ (short edge)} = \frac{W_u (L - l)^2}{8L},$$

$$M_{u2} \text{ (long edge)} = \frac{W_u (B - b)^2}{8B}$$

Shear Force (at critical section):

$$V_{us} \text{ (short edge)} = \frac{W_u}{L} \left\{ \frac{1}{2}(L-l) - d \right\}$$

$$V_{us} \text{ (long edge)} = \frac{W_u}{B} \left\{ \frac{1}{2}(B-b) - d \right\}$$

In case of raft for uniform soil pressure, the CG of resultant of all column loads on the raft must coincide with the centroid of the mat slab. Depth of the mat slab is greater of the depth from moment or shear stress criterion at critical section.

When the soil below the foundation is very soft, piles are provided to support the load of structure by skin friction. Pile cap is designed on the basis of the maximum moment developed at the face of the column due to upward pile reactions. Similarly, the design depth is calculated at a critical shear section at a distance d from the face of the column for one way shear.

Similarly, two way shear (punching) is checked or depth is computed at a distance of $(d/2)$ from the face of the column. Ultimate permissible punching shear

$$= 0.25 \sqrt{f_{ck}} \text{ N/mm}^2$$

Pile Capacity:

$$Q_u = Q_p \text{ (point bearing)} + Q_s \text{ (skin bearing)}$$
$$Q_u = (A_p q_p + \Sigma A_s f_s)$$

Allowable pile capacity $Q_a = \left(\dfrac{Q_p}{F_p} + \dfrac{Q_s}{F_s} \right)$, where F_p, F_s are safety factors.

Spacing S between piles in pile cap is determined by considering the efficiency of pile group $(m \times n)$ as 100% of individual pile, i.e.

$$\frac{\left[2(m+n-2)S + 4D \right] f_s L_f}{mn(\pi D) f_s L_f} = \frac{100}{100}, \text{ i.e. } S = \frac{(mn\pi D - 4D)}{2(m+n-2)}$$

Reaction of each pile $P_p = \dfrac{Q \text{ (total load)}}{n \text{ (number of piles)}}$

In case of eccentricity or moment,

$$P_p = \frac{Q}{n} \pm \frac{M_y \cdot x}{\Sigma x^2} \pm \frac{M_x \cdot y}{\Sigma y^2}$$

Pile cap is designed for maximum moment on the face of the column and critical one way shear at a distance of d from the face of column. τ_{uc} (design shear stress) is taken from Table 19 of IS:456-2000. Two-way punching shear is checked at a distance of $d/2$ from the column face with reference to allowable ultimate punching stress of $\tau_{uc} = 0.25 \sqrt{f_{ck}}$.

PRACTICE QUESTIONS

(I) Objective Questions

Q. 9.1 Select the correct response given after each statement to complete it correctly and fill in the response sheet provided.

(i) The foundations are defined as structural part which ....

    (a) interface with roof on one side and column on the other side

    (b) interface with superstructure on one side and plinth on the other side

    (c) interface with superstructure on one side and earth on the other side

    (d) is shorter and transfers load to the beam.

(ii) The function of the foundation is to ....

    (a) hold loads of superstructure

    (b) add load on the superstructure

    (c) make plinth uniformly loaded

    (d) transfer the superstructure load to the soil.

(iii) Deep foundation becomes necessary when the bearing capacity of soil is ....

    (a) very low

    (b) very high

    (c) less than the load/m^2 from the superstructure

    (d) equal to the load/m^2 from the superstructure.

(iv) Spacing (S) of piles in a group of piles ($m \times n$) in a pile cap is determined so that the efficiency of pile group is the same as the individual piles and the value of $S = $ .... (if diameter of pile is D).

    (a) $\dfrac{2(m+n-2)}{mn\pi(D-4D)}$
        (b) $\dfrac{(mn\pi D - 4D)}{2(m+n-2)}$

    (c) $900 + D$
        (d) diameter of pile (D)

(v) In case of column carrying a service load P and located on the ground having soil of safe bearing capacity q_o, the area of footing (A) is obtained as $A = $ ...., if the load factor is 1.5 and self-weight of footing is 10% of the column load (P).

    (a) $\dfrac{1.5P}{q_o}$
    (b) $\dfrac{P}{1.5q_o}$
    (c) $\dfrac{1.1P}{q_o}$
    (d) $\dfrac{P}{q_o}$

(vi) The ultimate moment (M_u) in footing of a wall of thickness b and carrying a load of W/m length shall be .... (if load factor = 1.5)

    (a) $M_u = \dfrac{3(B-b)^2 W}{16B}$
        (b) $M_u = \dfrac{W(B-b)^2}{8B}$

    (c) $M_u = \dfrac{W(B-b)^3}{4Bb}$
        (d) $M_u = \dfrac{3}{16}\dfrac{W(B)^2}{(B-b)}$

(vii) The safe effective depth d (mm) required on the basis of shear at a critical section at a distance d from the wall face for the footing of width B for a wall of thickness b and carrying an axial ultimate load W_u/m length shall be found by ...., if the design shear stress is τ_{uc} in (N/mm^2) and W_u in N.

(a) $d = \dfrac{W_u}{B\tau_{uc}} \{B - b - d\}$
(b) $d = \dfrac{W_u}{b\tau_{uc}} \left\{ \dfrac{B - b - d)}{2} \right\}$

(c) $d = \dfrac{W_u \cdot \tau_{uc}}{1.5B} \{0.5\,(B - b) - d\}$
(d) $d = \dfrac{W_u}{1000\,B\tau_{uc}} \{0.5\,(B - b) - d\}$

(viii) When a column is located on the property line, the best way of providing foundation will be .... type of foundation.

(a) isolated footing
(b) well
(c) combined footing
(d) pile

(ix) In case of group of piles with pile cap subjected to bending moment in addition to concentric column load Q, the reaction in each pile (P_p) is found by using formula $P_p = $ .... with usual notations.

(a) $\dfrac{M_y \cdot x}{\sum x^2} \pm \dfrac{M_x \cdot y}{\sum y^2}$
(b) $\dfrac{Q}{n} \pm \dfrac{M_y \cdot x}{\sum x^2} \pm \dfrac{M_x \cdot y}{\sum y^2}$

(c) $\dfrac{Q}{n} \pm \dfrac{M_x \cdot x}{\sum x^2} \pm \dfrac{M_y \cdot y}{\sum y^2}$
(d) $\pm \dfrac{M_y \cdot y}{\sum y^2} \pm \dfrac{M_x \cdot x}{\sum x^2}$

(x) The piles generally support loads of superstructures by ....
(a) point bearing and surface skin resistance
(b) point bearing alone
(c) surface skin resistance alone
(d) uniform shear resistance alone

Response sheet to Q. 9.1 (i to x)

Question	(i)	(ii)	(iii)	(iv)	(v)	(vi)	(vii)	(viii)	(ix)	(x)
Response (a/b/c/d)										

(II) Numerical Questions

Note: Design answers may vary from person to person based on assumptions and the answers given here are just for guidance.

Q. 9.2 Design a footing for 250 mm thick masonry wall which supports a service load of 200 kN/m length. The safe bearing capacity of soil is 150 kN/m^2. Angle of repose of soil $\phi = 30°$, unit weight of soil $w_e = 20$ kN/m^3. Use M20 grade CC and Fe 415 grade steel.

[**Hint:** Width of footing $B = 1.50$ m, p_u(net pressure intensity) $= 200$ kN/m^2, d(BM) $= 131$ mm, d(shear) $= 199$ mm (τ_{uc} assumed $= 0.43$ N/mm^2), $D = 250$ mm, A_{st} (for M_u) $= 12\phi$ @ 150 c/c, 10ϕ @ 250 c/c (temp. steel).]

Q. 9.3 Design a RCC footing for a column of section 400 mm × 400 mm to carry an ultimate axial load of 1800 kN. The safe bearing capacity of soil q_o = 150 kN/m², angle of repose ϕ = 30°, and unit weight of soil w_e = 20 kN/m³. Use M20 grade CC and Fe 415 grade steel.

[Hint: h = 1 m, L = B = 3 m, d(BM) = 238 mm, d_2 (shear at d) = 450 mm, d_3 (punching) = 405 mm, assumed D = 500 mm, d = 450 mm, A_{st}(BM) = 1046 mm² (16ϕ 180 c/c in bottom in both directions, spacing < 450 mm or 3d).]

Q. 9.4 Design a rectangular RCC footing for a rectangular column 250 mm × 500 mm carrying an axial service load of 1500 kN. Safe bearing capacity of soil q_o is 150 kN/m², angle of repose ϕ = 30°, and unit weight of soil w_e = 20 kN/m³.

[Hint: Foundation depth h = 1 m, $L \times B$ = 4.0 m × 2.75 m, net p_u = 204.5 kN/m², M_{uy} = 295.6 kN-m/m, M_{ux} = 147.3 kN-m/m, d(BM) = 327.2 mm, d(shear) = 616 mm, one-way, d(punching) at $\dfrac{d}{2}$ from the column face = 482 mm, provide D = 670 mm, d = 620 mm, $A_{st}(M_{uy})$ = 1368 mm² (16ϕ @ 140 c/c), A_{st} = 1436 mm²/m > 804 mm² (minimum steel), $A_{st}(M_{ux})$ = 702 mm² < $A_{st\ min}$ = $\dfrac{0.12}{100} \times 1000 \times 670$ = 804 mm² (12ϕ @ 140 c/c), spacing < 3d or 450 c/c, A_{sty} (central zone of 2.75 m) 12ϕ 130 c/c.]

Q. 9.5 Design a RCC circular footing for a circular column of 300 mm diameter and supporting an ultimate load of 750 kN. The ultimate bearing capacity of soil may be assumed as 300 kN/m². Use M20 grade CC and Fe 415 grade steel.

[Hint: Diameter of footing = 2.0 m, CG of quadrant $\cong$ 612 mm, net upward pressure p_u = 238.7 kN/m², M_u = 84.6 kN-m quadrant, d(BM) = 361 mm, assume d = 540 mm, A_{st} = 471 mm², (942 mm² in 2 quadrants), 12ϕ 120 c/c both-ways, check shear.]

Q. 9.6 Design a RCC wall footing for supporting an ultimate load of 300 kN/m length and a factored moment of 30 kN-m/m length. An ultimate bearing capacity of soil q_u = 300 kN/m². Concrete grade M20 and steel grade Fe 415 may be used.

[Hint: Width of footing = 1.10 m, e = $\dfrac{30 \times 10^6}{300 \times 1000}$ = 100 mm (0.10 m), CG of footing to be at the resultant point. Cantilever projections 0.5 m and 0.30 m. p_u soil pressure = 273 kN/m², $M_{u\ max}$ = 34.1 kN-m/m, d = 111 mm, V_u = 137 kN, d(shear) = 190 mm, D = 250 mm, A_{st}(BM) = 12ϕ 200 c/c (A_{st} = 565 mm²), temp. steel min 300 mm² (10ϕ 250 c/c).]

Q. 9.7 Design a combined footing with a strap beam of 400 mm width for two columns of 400 mm × 400 mm spaced 4 m c/c. Each column carries a service load of 400 kN. The safe bearing capacity of soil q_o = 130 kN/m². Use M20 grade CC and Fe 415 grade steel.

[Hint: Size of footing = 5 m × 1.5 m, p_u = 160 kN/m², projection = 0.55 m, M_u = 24.2 kN-m, d (M_u) = 100 mm, D = 150 mm, d (shear at d from col.) = 200.1 mm, adopt d = 250 mm, D = 300 mm, Minimum temp. steel = 360 mm² (10ϕ 200 c/c, A_{st} = 393 mm²/m) in both directions. $A_{st}(M_u)$ = 275 mm²/m, thus provide minimum τ_{uv} = 0.192 N/mm² < 0.285 N/mm², OK.]

Q. 9.8 Design a combined rectangular footing for two columns spaced at 6.0 m c/c. One column with section 400 mm × 400 mm and carrying an ultimate load of 2430 kN is located on property line with its face coinciding with the property line. The other column of section 500 mm × 500 mm is carrying an ultimate load of 4050 kN. Safe bearing capacity of soil q_o = 162 kN/m^2 and angle of repose ϕ = 30° and unit weight of soil = 21.6 kN/m^3. Use M20 grade CC and Fe 415 grade steel.

[**Hint:** h = 1 m, size 7.90 m × 3.75 m, net soil pressure p_u = 219 kN/m^2, $M_{u\,(max)}$ = + 3114 kN-m, – 1874 kN-m, maximum. V_u = 2450 kN (– 2102 kN), d(BM) = 550 mm, d(SF) = 1310 mm, adopt D = 1200 mm, d = 1140 mm.

Bottom face: longitudinal bars 20ϕ @ 250 c/c, transverse 20ϕ @ 300 c/c and transverse 20ϕ @ 140 under column bottom.

Top face: longitudinal bars 20ϕ @ 140 c/c, transverse 20ϕ @ 300 c/c.]

Q. 9.9 Design a combined trapezoidal footing for two columns spaced at 5 m c/c. Size of both exterior and interior columns are 400 mm × 400 mm and are carrying service loads of 1250 kN and 1000 kN respectively. Face of exterior column falls on the property line. The safe bearing capacity of soil q_o = 150 kN/m^2, angle of repose of soil ϕ = 30°, and unit weight of soil w_e = 20 kN/m^3, use M20 grade CC and Fe 415 grade steel.

[**Hint:** h (depth of foundation) = 1 m, size of footing (assuming 10% weight) = 33 m^2, centroid of footing to coincide with the resultant of loads W_1 and W_2 so that $\dfrac{B_1}{B_2} = \dfrac{2l - 3(m+n)}{3(m+n) - L}$, where m = 200 mm and assume L = 6.0 m, p_u = 201 kN/m^2, B_1 = 4400 mm, B_2 = 1200 mm, $M_{u\,max}$ = 1830 kN-m, b(width at point of maximum BM or zero shear) = 3.067 m, d(maximum BM) = 465 mm maximum. transverse BM. M_{u1} = 782.5 kN-m, M_{u2} = 146.4 KN-m, d(trans-verse BM) = 257.3 mm, d(one-way shear) assuming (τ_{uc} = 0.36 N/mm^2) = 0.85 m = 850 mm, D = 900 mm, minimum temp. steel = 1080 mm^2/m.

A_{st} (long M_u = 1830 kN-m), 6274 mm^2 in full width (in 3.067 m) minimum temp = 2046 mm^2/m, provide 20ϕ @ 150 c/c.

A_{st} (long. $M_{u\,min}$ = 41 kN-m, b = 1.50 m) = 1332 mm^2, 20ϕ 250 c/c, A_{st} = 1257 mm^2/m.

(Transverse under column. C_1) M_u = 782.5 kN-m (b = 1300 mm), A_{st} = 2684 mm^2, minimum. 2065 mm^2/m, 20ϕ @ 150 c/c

(Transverse under column. C_2) M_u = 146.4 kN-m (b = 1900 mm), A_{st} = 488 m^2, minimum. 1080 mm^2/m, 20ϕ @ 100 c/c (A_{st} = 1131 mm^2/m)

Check for punching: $\dfrac{V_{uc1}}{b_o \cdot d} = \dfrac{1637.7}{3575 \times 850} = 0.539 \le 1.118$ N/mm^2, OK.]

Q. 9.10 Design a raft slab for 8 columns arranged in two rows. The spacing between the two rows is 4.0 m c/c, while columns in each row are 5 m c/c. Section of each column is 400 m × 400 mm and each corner column carries a service load of 750 kN, while interior column carries a service load of 1000 kN each. Soil properties are as follows:

Safe bearing capacity $q_o = 100 \text{ kN/m}^2$,

Angle of repose $\phi = 30°$

Unit weight of soil $w_e = 20 \text{ kN/m}^3$.

[**Hint:** Area required for raft slab = 16 m × 5m, net $p_u = \dfrac{1.5\,(750 \times 4 + 1000 \times 4)}{16 \times 5}$

$= 131.25 \text{ kN/m}^2$

Consider longitudinal strip of 2.5 m width, upward load = 328.125 kN/m and column load reactions 1125, 1500, 1500, 1125 respectively.

M_u (longitudinal) = 326.6 kN-m/m, 253.3 kN-m/m, $d = 750$ mm, A_{st}-16ϕ 200 c/c (1005 mm²/m), bottom face 16ϕ 150 c/c.

M_u (transverse) = 246.1 kN-m/m, $d = 730$ mm, 16ϕ @ 200 c/c top face, bottom face 16ϕ 200 c/c.

Check for shear 0.592 N/mm², 0.5294 N/mm² < 1.118 N/mm² (punching), OK.]

Q. 9.11 Design a pile cap for supporting a column of size 400 mm × 400 mm carrying a service load of 1000 kN. The pile cap contains a group of 4 piles each of 250 mm diameter to transfer the column load by skin friction to soil. Use concrete grade M20 and steel Fe 415 grade.

[**Hint:** For 100% efficiency of the group, $S = \dfrac{[\pi\,250(2 \times 2) - 4 \times 250]}{2\,(2 + 2 - 2)}$ should not be

less than 900 or $3D = 900$ mm.

Assuming 150 mm projection, $L = B = 1450$ mm

$P_{up} = 413$ kN, $M_{u\,max} = 206$ kN-m, $d = 227$ mm

$V_{u\,max} = 825$ kN, $\tau_{uc} = 0.36 \text{ N/mm}^2$ ($p_t = 0.25\%$ say), $d = 1581$ mm,

$V_{u\,max}$ (punching two-way) = 413 kN, (punching of pile by column), $d' = 413$ mm

Assume $D = 500$ mm, $d = 500 - 40 - \dfrac{20}{2} = 450$ mm, provide vertical stirrups

for one-way shear

A_{st} ($M_u = 206$ kN-m) $= \dfrac{0.5 \times 20 \times 1450\,(450)}{415}\left[1 - \sqrt{1 - \dfrac{4.6 \times 206 \times 10^6}{20 \times 1450 \times (450)^2}}\right]$

$= 1324 \text{ mm}^2$ (bottom face) 20ϕ 230 c/c (1365 mm²)

Vertical stirrups 8ϕ 8 legged @ S_v.

$S_v = \dfrac{0.87\,f_y \cdot A_{sv}}{(\tau_{uv} - \tau_{uc})B} = \dfrac{0.87 \times 415 \times (400)}{(1.33 - 0.36)\,1450} = 103$ mm c/c

Minimum temp. steel $= 0.12 \times \dfrac{1000}{100} \times 500 = 600 \text{ mm}^2$ in each direction, top

face 12ϕ 300 c/c ($A_{st} = 376$ mm) OK.

Check for punching shear safety.

Reinforcement details:

20ϕ @ 230 c/c in bottom face in both directions, 12ϕ @ 300 c/c in top face in each direction, 8ϕ 8 legged vertical stirrups @ 100 c/c.]

Q. 9.12 Design a pile cap for supporting a column of section 500 mm × 500 mm and supporting an axial service load of 2000 kN and a uniaxial service moment of 300 kN-m. The pile cap contains 9 friction piles in 3 rows and 3 columns (3 × 3) and each pile has section of 250 mm diameter.

[Hint: $L = B = 2350$ mm, $M_{u\,max} = 878$ kN-m, $d = \sqrt{\dfrac{M_u}{0.138 \times 2350 \times 20}}$, $d = 368$ mm

$V_{u\,max}$ (one-way) $= 1350$ kN, d(shear) $= \dfrac{V_u}{B \cdot \tau_{uc}} = 1596$ mm, two-way shear

(punching) 3000 kN for punching of column.

Bottom face 14 bars of 20ϕ in each direction one above the other in full width 2350 mm also provide 8ϕ 8 legged @ 110 c/c for $V_{u\,max} = 1350$ kN, adopt $D = 680$ mm, $d = 630$ mm.

Check for punching shear, safe within 1.118 N/mm^2.

Top face provide 12ϕ nominal bars for anchor and temp. bars 12ϕ @ 180 c/c ($A_{st} = 627$ mm^2/m). Bottom face main steel is available and we do not require separately temp. steel.]

Q. 9.13 Design the pile cap and sketch the reinforcement for the pile cap which consists of 4 piles each of 300 mm × 300 mm sections. The centers of piles are 600 mm from the centre of the column of 500 mm × 500 mm size. The piles and columns are placed vertically with their faces parallel. Using M20 grade CC and Fe 415 grade steel, design the details of reinforcement. The column is subjected to a service load of 2000 kN.

[Hint: Size of pile cap = 600 + 600 + 2 × 300 = 1800 mm

Total depth D of pile cap = 2 × 300 + 100 = 700 mm

Effective depth $d = 700 - 80 = 620$ mm (say).

Shear span = 600 − 250 = 350 mm from column. face, $P_u = 1.5 \times 2000 = 3000$ kN

$$P_{up} = \frac{1}{4} \times 3000 \times 1.1 = 750 \text{ kN} \times 1.1 = 825 \text{ kN}$$

Tension T in main steel $= \dfrac{P_{up}\left(600 - \dfrac{500}{4}\right)}{d} = \dfrac{825\,(600 - 125)}{620} = T = 632$ kN.

$A_{st} = \dfrac{632.0 \times 1000}{0.87 \times 415} = 1751$ mm^2. Provide 6 bars 20 ϕ under the column width

($A_{st} = 1884$ mm^2) > 1751 mm^2, OK.

From bending moment point of view:

$M_u = 825000 \times 2 \times (600 - 250) = 577.5 \times 10^6$ N-mm.

$b = 1800$ mm, $d = 620$ mm,

$$A_{st} = \frac{10 \times 1800 \times 620}{415}\left[1 - \sqrt{1 - \frac{4.6\,M_u}{20 \times 1800 \times (620)^2}}\right] = 2719 \text{ mm}^2 \text{ (total width}$$

1800)

20ϕ 200 c/c ($A_{st} = 2826$ mm^2 in 1800 mm), OK.

20ϕ 200 c/c in full width in bottom face in both directions or provide 6 bars 20ϕ under the column. width in bottom in both directions. Check for shears. Also check for one way shear, $\tau_{uc} = 0.36 \, N/mm^2$, $d = 252$ mm ($<$ provided 620 mm), OK.

Check for column punching and pile punching also.]

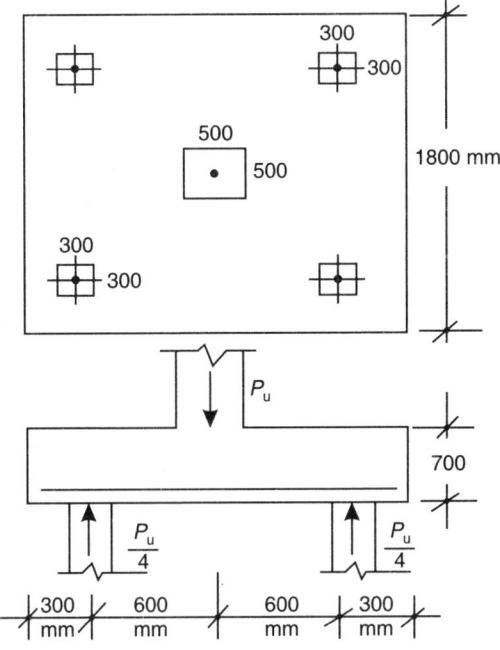

Fig. 9.22: Pile cap layout and section

Q. 9.14 Design a suitable continuous raft slab foundation connecting all the columns provided for a building 12 m × 12 m along the border with columns of 300 mm × 300 mm size, spaced at 4.0 m c/c and carrying a service load of 600 kN. Safe bearing capacity of soil $q_o = 120 \, kN/m^2$. Use M20 grade CC and Fe 415 grade steel.

[**Hint:** No of column. = 2 × 4 + 2 × 2 = 12, total column loads = 12 × 600 = 7200 kN.

$$\text{Area of foundation} = \frac{7200 \times 1.1 \times 1.5}{1.5 \times 120} = 66 \, m^2, \text{ (centre line total length} =$$

48 m)

$$\text{Adopt width } B = \frac{66}{48} = 1.375 \, m = 1.50 \, m \text{ (say), cantilever projection} = 0.60 \, m$$

$$\text{Net ultimate soil pressure } p_u = \frac{7200 \times 1.5}{48 \times 1.5} = 150 \, kN/m^2 \text{ (upward)}$$

$$\text{Raft slab design: } M_u = \frac{150 \times 0.6^2}{2} = 27 \, kN\text{-m/m}, V_u = 150 \times 0.6 = 90 \, kN/m$$

$$d(BM) = \sqrt{\frac{27 \times 10^6}{0.138 \times 20 \times 1000}} = 98.9 \, mm, \text{ assume } d = 180 \, mm \ (D = 230 \, mm)$$

SF at critical section = 150 × 1 (0.6 – 0.18) = 63 kN

Shear stress $\tau_{uv} = \dfrac{63000}{1000 \times 180} = 0.35 \text{ N/mm}^2$

$$A_{st} \; (M_u = 27 \text{ kN-m}) = \frac{10 \times 1000 \,(180)}{415} \left[1 - \sqrt{1 - \frac{4.6 \times 27 \times 10^6}{20 \times 1000 \times (180)^2}} \right] = 438 \text{ mm}^2$$

(12ϕ 250 c/c, 452 mm^2/m, $p_t = 0.25\%$)

$\tau_{uc} \,(0.25\% \; p_t) = 0.36 \; (K = 1.26)$, $\tau_{uc} = 1.26 \times 0.36 = 0.454 \text{ N/mm}^2$

$(> 0.35 \text{ N/mm}^2)$, OK.

Design of continuous beam connecting columns:

$b = 300$ mm, $w_u = 150 \times 1.5$ m $= 225$ kN/m (upward)

$M_u = \pm \dfrac{1}{10} \, w_u \cdot L^2 = 360$ kN-m, $V_u = 0.6$, $w_{ul} = 540$ kN

$$d \text{ (BM)} = \sqrt{\frac{360 \times 10^6}{0.138 \times 20 \times 300}} = 660 \text{ mm, adopt } d = 700 \text{ mm, } D = 750 \text{ mm}$$

$$A_{st} = \frac{10 \times 300 \times 700}{415} \left[1 - \sqrt{1 - \frac{4.6 \times 360 \times 10^6}{20 \times 300 \times (700)^2}} \right] = 1716 \text{ mm}^2$$

(4 bars 25 mm, $A_{st} = 1964$ mm^2, $p_t = 0.935\%$)

τ_{uc} (design) $= 0.576 \text{ N/mm}^2$,

$V_{us} = (\tau_{uv} - \tau_{uc}) \, 300 \times 700 = (540000 - 120960) = 419040$ N

Shear reinforcement: 10ϕ 4 legged, $S_v = \dfrac{0.87 \times 415 \,(4 \times 78.5)\, 700}{419040} = 189$ c/c

Provide 10ϕ 4 legged @ 180 c/c. ($A_{st} = 1744$ mm^2/m)

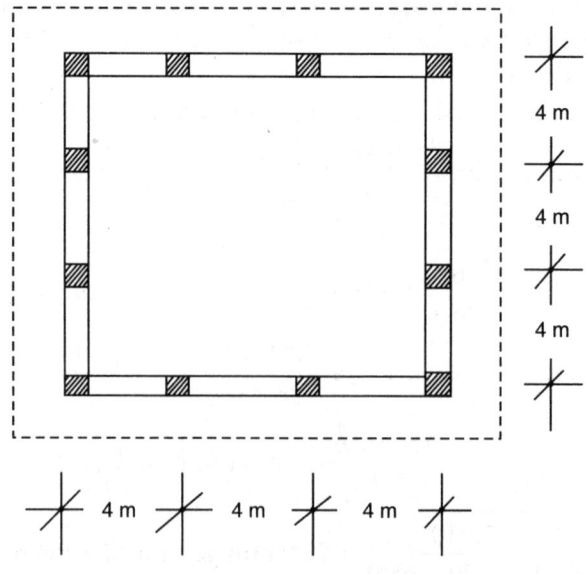

Fig. 9.23: Plan layout of raft slab

Chapter 10

Limit State Design of Retaining Walls

LEARNING OBJECTIVES

After the study of *limit state design of retaining walls*, the learner will understand the principles of design of retaining walls and will be able to:

- Explain the functions of retaining walls
- List different types of retaining walls
- Differentiate between active and passive earth pressures on retaining wall
- Explain different types of forces acting on retaining walls
- Design different components of a cantilever retaining wall
- Design different components of a counterfort retaining wall

10.1 INTRODUCTION

10.1.1 General Function

Retaining walls are provided to retain earth or any other material having tendency to slide. Retaining walls resist lateral thrust exerted by the earth fill or other material. Retaining walls are essential in buildings with basements, buildings located in sloping ground, in bridge abutments and wing walls, etc. If the back fill is above the horizontal level of the top of the retaining walls, it is known as *surcharge*. Surcharge may be sloping or horizontal. In the design of retaining walls, lateral earth pressure is calculated on the basis of various theories or by experimental tests. Generally, Rankine's theory of earth pressure is used for computing lateral pressures on the retaining walls.

10.1.2 Types of Retaining Walls

Retaining walls may be classified according to their shape, component elements and backfill materials, etc. The common types of retaining walls are:

- Gravity walls
- Cantilever retaining walls (T-shaped and L-shaped)
- Counterfort retaining walls
- Buttressed walls.

In case of *gravity walls* (Fig. 10.1(a)), the lateral earth pressure exerted by backfill is balanced by dead weight of the walls. Gravity walls are generally constructed by masonry or plain concrete. These walls are so proportioned that the resultant of weight and lateral earth pressure remains within the middle third of the base so that no tension develops anywhere in the base and stability of the structure is maintained.

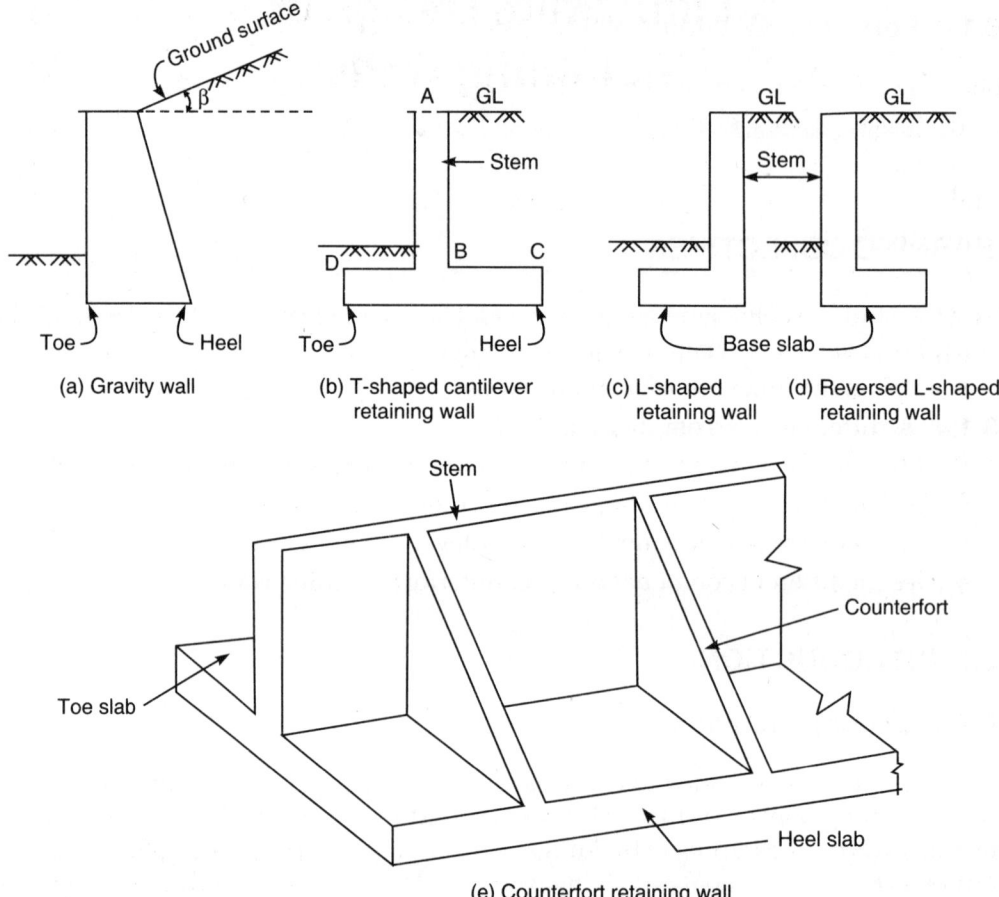

(a) Gravity wall   (b) T-shaped cantilever retaining wall   (c) L-shaped retaining wall   (d) Reversed L-shaped retaining wall

(e) Counterfort retaining wall

Fig. 10.1: Different types of retaining walls

The *cantilever retaining* wall resists the horizontal earth pressure and other vertical pressures by way of bending of various components as cantilevers. Most common type of cantilever retaining wall is T-shaped wall. The wall consists of stem AB, heel slab BC and toe slab DB(Fig. 10.1(b)). Each of these slabs bend as cantilevers about B. These elements are reinforced on tension faces. Cantilever retaining walls can also be L-shaped (Fig. 10.1(c) and (d)) which also resist earth pressure by bending. Cantilever retaining walls are generally used for wall heights between 3 m to 8 m.

The *counterfort retaining wall* (Fig. 10.1(c)) also comprises vertical stem and heel slab, which are strengthened by providing counterforts at some suitable intervals. The provision of counterforts make the vertical stem slab as continuous over counterforts and similarly the heel slab also acts as continuous slab over counterforts. Toe slab acts as cantilever, bending upwards under soil pressure. This type of retaining wall is provided when the backfill of greater heights are to be retained. Counterfort retaining walls are economical for wall heights more than 8 m and the spacing of counterforts is generally kept one third of the height.

A *buttressed* wall is a modification of the counterfort retaining wall in which the counterforts, called buttresses, are provided to the other side of the backfill. Buttresses reduce the front side clearance and hence, generally these are not preferred.

10.1.3 Forces on Retaining Walls

Following forces act on retaining walls:

(i) Self-weight of the retaining wall elements

(ii) Weight of the soil on the base

(iii) Lateral soil pressure on the wall (active and passive earth pressures)

(iv) Surcharge — forces due to loads on the earth surface

(v) Soil reaction (upward soil pressure) on the base slab (footing)

(vi) Frictional force on the contact surface of base slab and soil.

Consider unit length (1 m) and compute various forces.

10.1.4 Active Earth Pressure (Rankine's Theory)

Rankine's theory of lateral earth pressure is applied considering soil to be of uniform and cohesionless properties. The Rankine's theory is applied in all cases to compute lateral earth pressures making following assumptions:

(i) The soil mass is homogeneous, dry and cohesionless

(ii) The ground surface is plane which may be horizontal or inclined

(iii) The back of the wall is vertical and smooth. There is no friction and Shear stress between wall surface and soil.

(iv) The wall bends or yields about the base and satisfies the deformation consistency even in plastic stage.

However, the wall surface is not perfectly smooth and causes some frictional force but practically this does not affect the results too much and accepted for practical designs. Following cases of backfill soil shall be considered for computation of lateral soil pressures:

(i) Dry or moist level backfill with no surcharge

(ii) Submerged level backfill

(iii) Backfill with uniform surcharge

(iv) Backfill with sloping surcharge.

Let the soil has following characteristics:

Unit weight of soil $= w \text{ kN/m}^3$

Angle of repose (friction)$- \phi$

Coefficient of active earth pressure, $k_a = \dfrac{1 - \sin\phi}{1 + \sin\phi}$

Height of soil above base = h_a

Coefficient of passive earth pressure, $k_p = \dfrac{1 + \sin\phi}{1 - \sin\phi}$

(i) Level dry backfill (Fig. 10.2):

$p_y = k_a wy$ (active lateral pressure at y below top)   ... Eq. (10.1)

Total active lateral pressure of soil

$$P_a = \frac{1}{2}\,(k_a wh_a{}^2) = 0.5 k_a wh_a{}^2 \qquad\qquad \text{... Eq. (10.2)}$$

where, $k_a = \dfrac{1 - \sin\phi}{1 + \sin\phi}$ ... Eq. (10.3)

On toe side passive earth pressure at h_b below ground level:

$p_p = k_p wh_b$ ... Eq. (10.4)

where, $k_p = \dfrac{1 + \sin\phi}{1 - \sin\phi}$ ... Eq. (10.5)

Total passive earth pressure $P_p = 0.5 k_p wh_b{}^2$ ... Eq. (10.6)

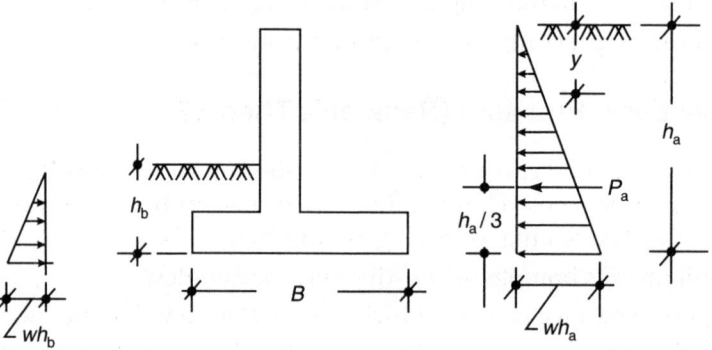

Fig. 10.2: Lateral earth pressure on wall

(ii) Submerged backfill (Fig. 10.3):

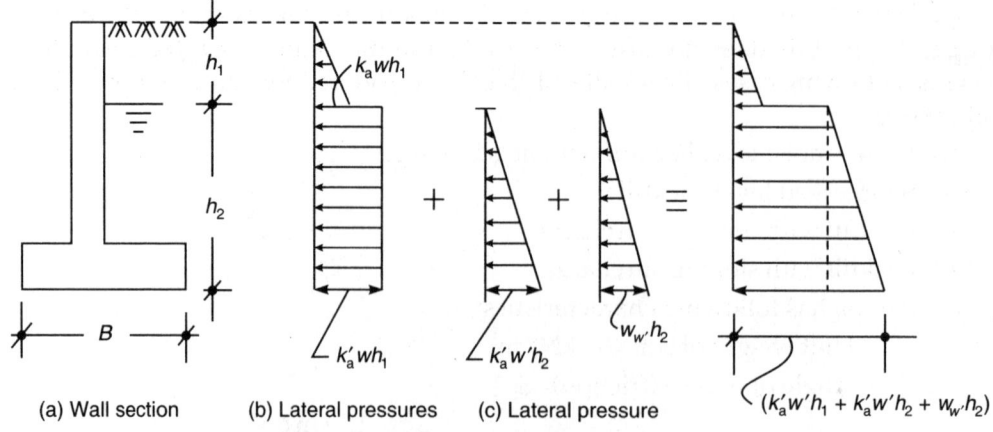

(a) Wall section   (b) Lateral pressures   (c) Lateral pressure $(k_a' w' h_1 + k_a' w' h_2 + w_w \cdot h_2)$

Fig. 10.3: Submerged backfill pressures on wall

Sometimes the backfill gets submerged due to rising water table. Let the backfill gets submerged up to a level h_1, below ground level (GL) and remaining depth h_2 gets submerged. Pressure intensity at any point comprises 3 components, *viz.* due to dry earth (w), due to wet earth (w'submerged weight) and due to water.

Pressure intensity at h_1 due to dry earth below GL:

$$p_1 = k_a w h_1 \qquad \text{... Eq. (10.7)}$$

Pressure intensity at h_1 below GL in wet earth:

$$p_2 = k_a' \, wh_1 \qquad \text{... Eq. (10.8)}$$

Pressure intensity at the base (h_2 below WL) due to wet earth only

$$p_3 = k_a' \, w'h_2 \qquad \text{... Eq. (10.9)}$$

Pressure intensity at the base (h_2 below WL) due to water only

$$p_4 = w_w h_2 \qquad \text{... Eq. (10.10)}$$

Thus, the total pressure intensity due to dry earth, wet earth and water at the bottom of the base will be

$$p = k_a'wh_1 + k_a'w'h_2 + w_w h_2 \qquad \text{... Eq. (10.11)}$$

Total forces and moments are calculated by considering each component of force appropriately.

(iii) Backfill subjected to uniform surcharge W_s (Fig. 10.4):

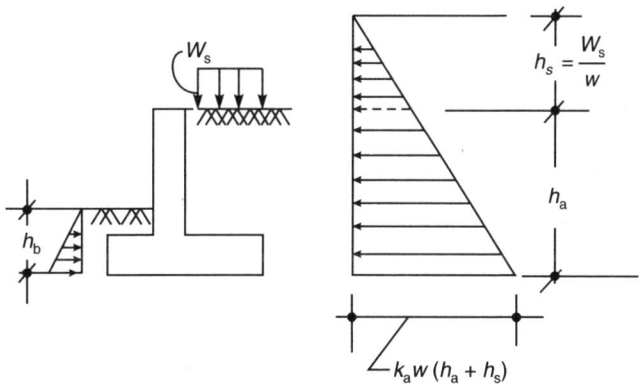

Fig. 10.4: Uniform surcharge

Sometimes, certain load acts uniformly over level earth surface and causes additional lateral pressure on the wall. Let this surcharge load be W_s/m^2. Equivalent height of earth which can cause similar uniform surcharge be $h_s = \dfrac{W_s}{w}$.

Lateral pressure at any height at a level y below the GL,

$$p = k_a w \, (h_s + y) = k_a w \left(\frac{W_s}{w} + y \right) \qquad \text{... Eq. (10.12)}$$

Thus, the pressure intensity at the base h_a below top,

$$p_a = k_a w \left(h_a + \frac{W_s}{w} \right) \qquad \text{... Eq. (10.12a)}$$

where, $\quad k_a = \dfrac{1 - \sin\phi}{1 + \sin\phi}$, W_s = uniform surcharge, w = unit weight of earth

Similarly, on toe side the passive intensity p_b at a depth h_b below GL will be

$$p_b = k_p\, wh_b \qquad\qquad \text{... Eq. (10.13)}$$

where, $\quad p_b$ = Passive earth pressure

$$k_p = \text{Passive earth pressure factor} = \dfrac{1 + \sin\phi}{1 - \sin\phi},$$

ϕ = Angle of repose

w = Unit weight of earth

h_b = Depth of earth below GL

(iv) Sloping backfill (Fig. 10.5):

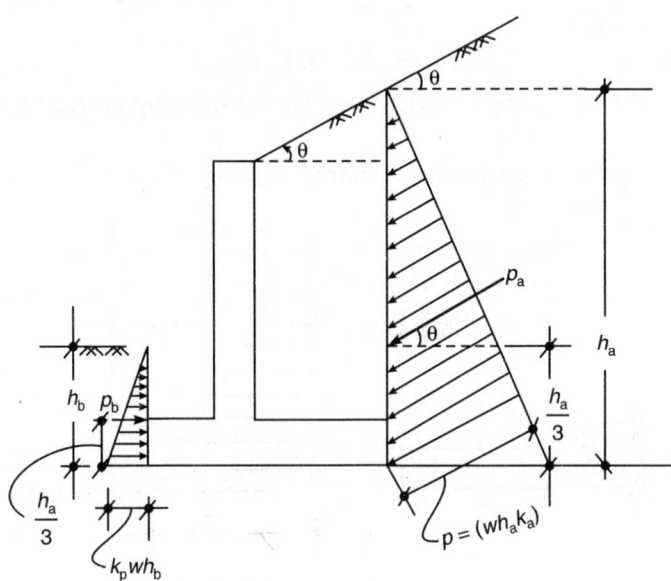

Fig. 10.5: Sloping backfill

Let the backfill be sloping at an angle θ with the horizontal (θ being less than ϕ). The active pressure coefficient (k_a) shall be given by:

$$k_a = \cos\theta \left[\frac{\cos\theta - (\cos^2\theta - \cos^2\phi)^{0.50}}{\cos\theta + (\cos^2\theta - \cos^2\phi)^{0.50}} \right] \qquad \text{... Eq. (10.14)}$$

Similarly, coefficient of passive earth pressure (k_p) shall be given by

$$k_p = \cos\phi \left[\frac{\cos\theta + (\cos^2\theta - \cos^2\phi)^{0.50}}{\cos\theta - (\cos^2\theta - \cos^2\phi)^{0.50}} \right] \qquad \text{... Eq. (10.15)}$$

Total resultant pressure p_a or p_b shall act at one-third the height from the base.

Bending of cantilever retaining walls and provision of tension steel (Fig. 10.6).

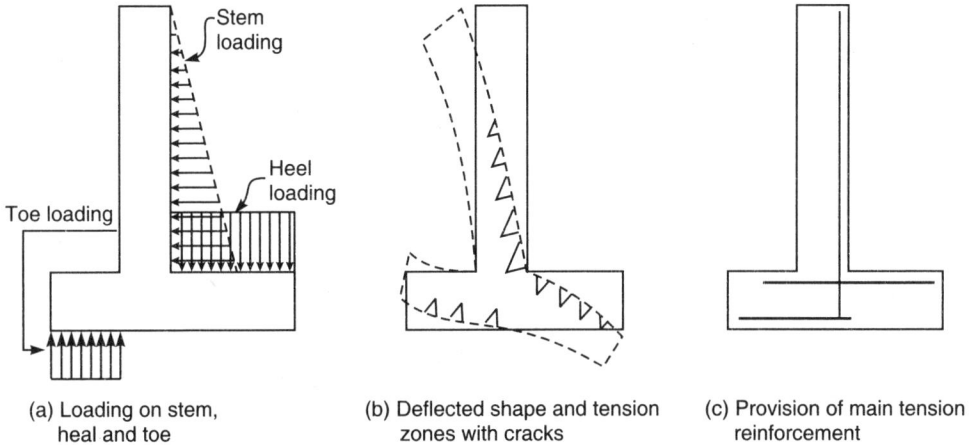

(a) Loading on stem, heal and toe

(b) Deflected shape and tension zones with cracks

(c) Provision of main tension reinforcement

Fig. 10.6: Bending of components of cantilever retaining wall, tension faces and tension reinforcement

10.2 DESIGN OF CANTILEVER RETAINING WALLS

10.2.1 General Features

Cantilever retaining wall consists of stem (vertical wall), heel slab and toe slab. Sometimes a key is also provided in the base slab to enhance the frictional resistance against sliding (Fig. 10.7). Generally cantilever retaining walls are used up to 5.0 m height.

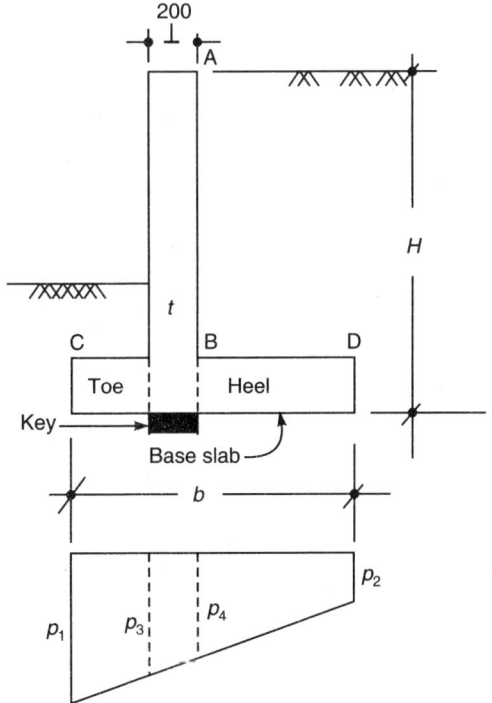

Fig. 10.7: Cantilever retaining wall components

The top thickness of stem shall be kept = 200 mm

The bottom width (thickness) shall be computed from the maximum BM caused by earth pressure at the junction B

Width of the base slab (toe + heel + wall thickness) = b

Height of retained material = H above bottom of base slab

Generally $\qquad b = 0.5H$ to $0.6H$ for walls without surcharge

$\qquad\qquad\qquad\qquad = 0.7H$ for walls with surcharge

$$\text{Toe projection} = \frac{b}{3}$$

The thickness of the base slab = thickness of the stem t.

10.2.2 Design Principles

(i) Heel slab is subjected to upward earth pressure from bottom and downward load of earth and self-weight of heel slab. Design is based on net maximum BM and shear force.

(ii) Toe slab is subjected to upward earth pressure from bottom and weight of earth on the toe slab from top. Design of toe slab is based on the net maximum BM and SF at the junction.

(iii) Check for horizontal sliding of the retaining wall is done by computing net horizontal thrust (P) from earth pressure and comparing with total frictional resistance μW, where μ is the coefficient of friction between soil and base slab, W is the total weight on the base.

$$\text{Factor of safety against sliding} = \frac{\mu W}{P} > 1.5$$

10.2.3 Design of Cantilever Retaining Walls

Design is explained through solved examples using above design principles.

Example 10.1

Design a cantilever retaining wall to retain earth embankment 4 m high above ground level. The unit weight of earth is 20 kN/m³ and angle of repose may be taken as 30°. The embankment is horizontal at its top. The safe bearing capacity of the soil may be taken as 200 kN/m² and the coefficient of friction between soil and concrete is 0.55. Use M20 grade CC and Fe 415 HYSD grade steel bars.

Solution: Unit weight w = 20 kN/m³, height of embankment above GL = 4 m

(Fig. 10.8) $\qquad \phi = 30°$, SBC $q_o = 200$ kN/m², $\mu = 0.55$

$\qquad\qquad f_{ck} = 20$ N/mm², $f_y = 415$ N/mm²

$$\text{Minimum depth of foundation} = \left(\frac{q_o}{w}\right)\frac{(1 - \sin\phi)^2}{(1 + \sin\phi)^2} = \frac{200}{20}\left(\frac{1}{9}\right) = 1.11 = 1.20 \text{ m}$$

Overall depth of wall H = (4 + 1.20) = 5.20 m (up to bottom of base slab).

$$\text{Assume thickness of base slab} \cong \frac{H}{12} = \frac{5200}{12} = 433 \text{ mm, say 450 mm}$$

Assume thickness of stem = 400 mm

Height of stem = 5.20 – 0.45 = 4.75 m

Width of base b = 0.5H to 0.6H = 2.6 m to 3.12 m, adopt b = 3.0 m, say

Dimensions of the cantilever retaining wall are as shown in Fig. 10.8.

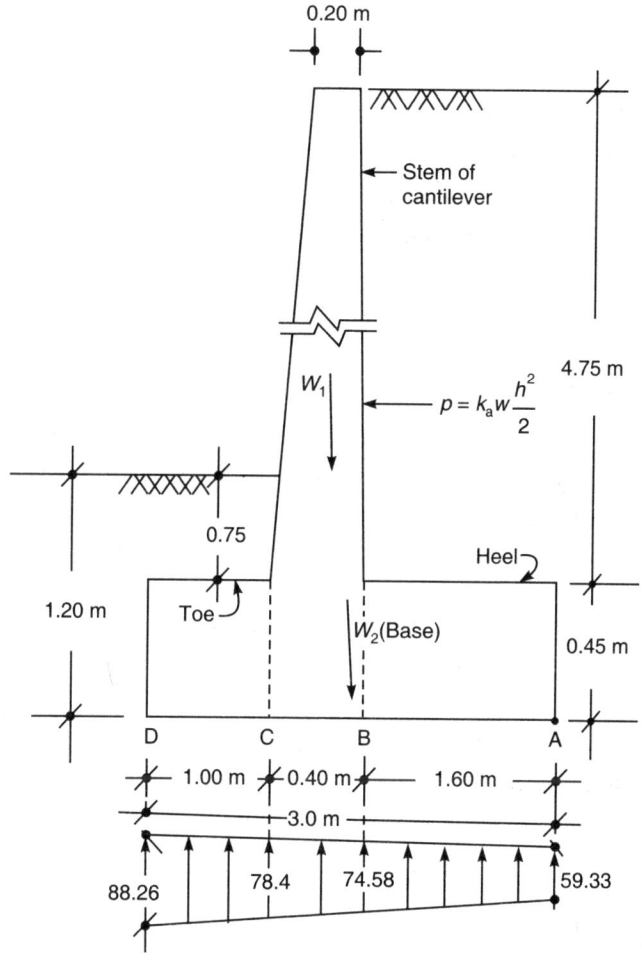

Fig. 10.8: Dimensions and soil pressures of retaining wall

Design of stem:

$$\text{Total earth pressure } P_a = \frac{1}{2}wh^2 k_a, \ k_a = \frac{1 - \sin 30}{1 + \sin 30} = \frac{1}{3}.$$

or

$$P_a = \frac{1}{2} \times \frac{1}{3} \times (4.75)^2 = 75.21 \text{ kN},$$

$$M = 75.21 \times \frac{4.75}{3} = 119.08 \text{ kN-m}$$

Factored moment M_u – 1.50 × 119.08 – 178.62 kN-m/m length

V_u = 1.5 × 75.21 = 112.82 kN/m

$$d(\text{BM}) = \sqrt{\frac{M_u}{0.138 f_{ck} \cdot b}} = \sqrt{\frac{178.62 \times 10^6}{0.138 \times 20 \times 1000}} = 254.4 \text{ mm}$$

(at bottom: adopt $D = 400$ mm, $d = 350$ mm, At top $D = 200$ mm)

$$A_{st} = \left(\frac{0.5 f_{ck} b \cdot d}{f_y}\right)\left[1 - \sqrt{1 - \frac{4.6 M_u}{f_{ck} b d^2}}\right] = \frac{10 \times 1000 \times 350}{415}\left[1 - \sqrt{1 - \frac{4.6 \times 178.62 \times 10^6}{20 \times 1000 (350 \times 350)}}\right]$$

$= 1560$ mm^2 (16ϕ@ 120 c/c vertically, $A_{st} = 1675$ mm^2, $p_t = 0.48\%$

$\tau_{uc} = 0.47$ N/mm^2 (IS:456-2000, Table 19)

Check for shear $\tau_{uv} = \dfrac{112.82 \times 1000}{1000 \times 350} = 0.322$ N/mm$^2 < \tau_{uc} = 0.47$ N/mm^2, OK.

Minimum steel $= \dfrac{0.12}{100} \times 1000 \times 400 = 480$ mm^2 (provide 10ϕ 300 c/c horizontally

on each face, $A_{st} = 2 \times 78.5 \times \dfrac{1000}{300} = 523$ mm^2/m, OK.)

Base slab:

Toe $= 1.0$ m, stem thickness $= 0.40$ m, heel $= 3.0 - 1.0 - 0.40 = 1.60$ m

Consider 1.0 m length of the wall

Check stability:

Weight of stem $W_1 = 0.20 \times 4.75 \times 1.0 \times 25 + 0.2 \times 0.5 \times 4.75 \times 25$

$= 23.75$ kN $+ 11.875$ kN (at 1.7 m, 1.867 m from A),

$M_1 = 40.38 + 22.09 = 62.47$ kN-m

Weight of base slab $W_2 = 3.0 \times 0.45 \times 1 \times 25 = 33.75$ kN at 1.50m from A,

$M_2 = 50.63$ kN-m

Weight of earth over heel

$W_3 = 1.6 \times 4.75 \times 1 \times 20 = 152$ kN at 0.80 m from A,

$M_3 = 121.60$ kN-m

Moment due to earth pressure $= \dfrac{1}{3} \times \dfrac{20 \times 4.75^3}{6} = 119.08$ kN-m about A,

$M_4 = 119.08$ kN-m

Resultant $\bar{x} = \dfrac{\sum M}{\sum W} = \dfrac{62.47 + 50.63 + 121.6 + 119.08}{23.75 + 11.875 + 33.75 + 152} = \dfrac{353.78}{221.38}$

$= 1.598$ m from A, $e = 0.098$ m

Maximum and minimum soil pressures on the base, (Fig. 10.9)

$$p_{max} \text{ or } p_{min} = \frac{\sum W}{b}\left(1 \pm \frac{6e}{b}\right) = \frac{221.38}{3}\left(1 \pm \frac{6 \times 0.098}{3}\right) = 88.26(p_D), 59.33(p_A) \text{ kN/mm}^2$$

$$p_B = 59.33 + \frac{1.6}{3}(28.6) = 15.25 + 59.33 = 74.58 \text{ kN/m}^2$$

$$p_c = 59.33 + \frac{28.6}{3}(1.6 + 0.4) = 78.4 \text{ kN/m}^2$$

Design of heel slab:

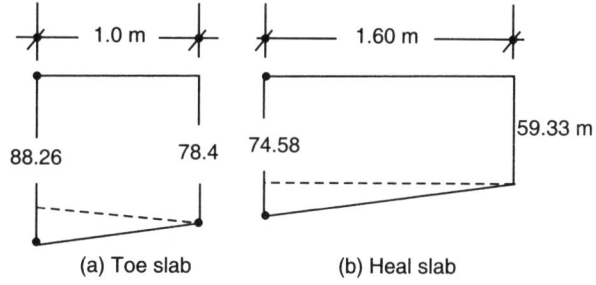

Fig. 10.9: Soil pressure leading on base slab (toe and heel)

Moment at B:

$$M_B = \left[1.6 \times 4.75 \times 1 \times 20 \times \frac{1.6}{2} + 0.45 \times 1.6 \times 1 \times 25 \times \frac{1.6}{2} - 59.33 \times 1.6 \times 1 \right.$$

$$\left. \times \frac{1.6}{2} - \frac{15.25 \times 1.6}{2} \times \frac{1.6}{3} \right]$$

$$= 121.6 + 14.4 - 75.94 - 6.51 = 53.55 \text{ kN-m}$$

Ultimate moment $M_{uB} = 1.5 \times 53.55 = 80.33$ kN-m/m

SF at B: $V_B = 1.6 \times 4.75 \times 1 \times 20 + 0.45 \times 1.6 \times 25 - 59.33 \times 1.6 - \dfrac{15.25 \times 1.6}{2}$

$$= 152 + 18 - 94.93 - 12.2 = 62.87$$

$V_B = 62.87$ kN, Ultimate SF $V_{uB} = 1.5 \times 62.87 = 94.31$ kN

Using $D = 0.450$ m, $d = 0.40$ m (400 mm), $A_{st} = \dfrac{0.5 f_{ck} bd}{f_y} \left[1 - \sqrt{1 - \dfrac{4.6 \times M_u}{f_{ck} b (d)^2}} \right]$

$$A_{st} = \frac{10 \times 1000 \times 400}{415} \left[1 - \sqrt{1 - \frac{4.6 \times 80.33 \times 10^6}{20 \times 1000 (400)^2}} \right] = 574 \text{ mm}^2$$

(provide 12ϕ @ 190 c/c, A_{st} = 595 mm^2 in top face)

Distribution bars = $\dfrac{0.12}{100} \times 1000 \times 450 = 540$ mm^2/m

(provide 12ϕ @ 200 c/c, A_{st} = 565 mm^2/m in top face below main bars).
Provide in bottom face 10ϕ @ 280 c/c in each direction.

Shear stress at the junction:

$$\tau_{uv} = \frac{94.31 \times 1000}{1000 \times 400} = 0.236 \text{ N/mm}^2 \le 0.36 \text{ N/mm}^2, \text{OK.}$$

Design of toe slab:
Maximum BM at C,

$$M_c = 78.4 \times 1 \times 0.5 + \times \frac{9.86 \times 1}{2} \times \frac{2}{3} \times 1 - 0.45 \times 1 \times 25 \times 0.5 - 0.75 \times 1 \times 20 \times 0.5$$

$$= 39.2 + 3.29 - 5.63 - 7.5 = 29.36 \text{ kN-m}$$

M_{uc} (ultimate) $= 1.5 \times 29.36 = 44.04$ kN-m ($b = 1000$ mm), d (reqd.) $= 127$ mm

$$A_{st} = \frac{10 \times 1000 \times (400)}{415}\left[1 - \sqrt{1 - \frac{4.6 \times 44.04 \times 10^6}{20 \times 1000(400)^2}}\right]$$

$$= 310 \text{ mm}^2 \text{ (minimum } A_{st} = 540 \text{ mm}^2)$$

Provide 12ϕ @ 200 c/c ($A_{st} = 565$ mm^2/m), (or provide on each face bottom and top 10ϕ @ 280 c/c, $A_{st} = 2 \times 280$ mm^2), OK.

Check for sliding:

Total horizontal pressure $P = k_a = \left(\dfrac{wH^2}{2}\right) = \dfrac{1}{3} \times \dfrac{20}{2} \times 5.2^2 = 90.13$ kN

$$\text{Total } W = 0.45 \times 3.0 \times 25 + \left(\frac{0.4 + 0.2}{2}\right) \times 4.75 \times 25 + (1.6 \times 4.75 \times 20)$$

$$= 33.75 + 35.63 + 152 = 221.38 \text{ kN (neglecting earth on toe)}$$

Maximum friction resistance $= \mu W = 0.55 \times 221.38 = 121.76$ kN

FS against sliding $= \dfrac{121.76}{90.13} = 1.35 < 1.5$, design a shear key.

Design of key (Figs 10.10a and b):

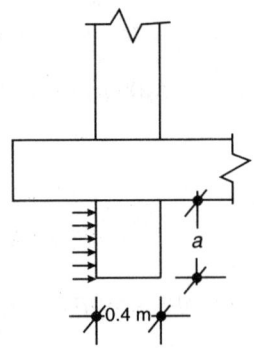

Fig. 10.10(a): Design of key

Provide key just below the stem.

p_c (soil pressure in front of the key) $= 78.4$ kN/m^2

$$\text{Passive lateral pressure} = k_p\, p_c, \quad k_p = \frac{1 + \sin\phi}{1 - \sin\phi} = 3$$

Passive pressure on key surface $= 3 \times p_c = 3 \times 78.4 = 235.2$ kN/m^2.

If a is the depth of key below base slab, then

total passive resistance created $= 235.2 \times (a \times 1) = 235.2a$ kN.

Total resistance $= (\mu W + p_p) = 121.76 + 235.2a$ kN $\geq 1.5 \times 90.13$

or $121.76 + 235.2a \geq 135.20$, $a \geq \dfrac{(135.2 - 121.76)}{235.2} = 0.0571$ m

Thus, provide a key with 0.40 m × 0.3 m height

$$FS = \frac{(235.2 \times 0.3 + 121.76)}{90.13} = 2.14 > 1.5, \text{OK}$$

Reinforcement of stem is extended to key.

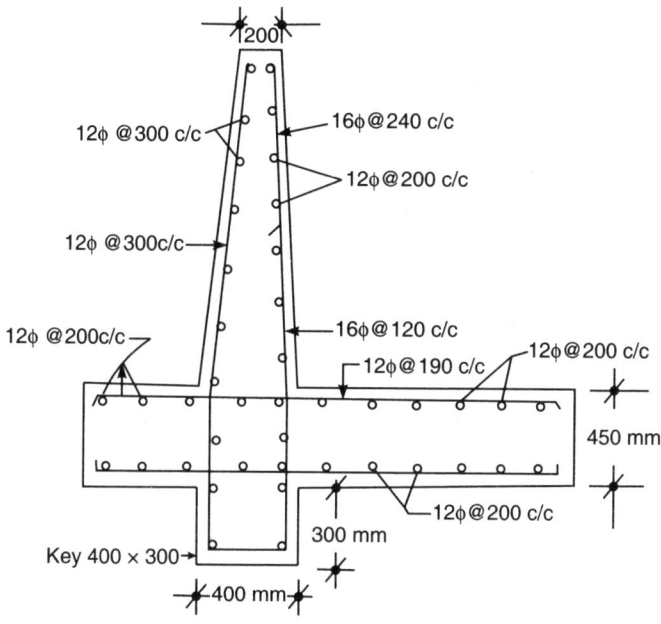

Fig. 10.10(b): Reinforcement details of cantilever retaining wall

Example 10.2

Design a cantilever retaining wall for a road embankment subjected to a traffic load (surcharge) of 18 kN/m². The height of road backfill is 5.0 m above GL. Also provide 1 m height of the parapet wall above stem wall top. Earth/soil has the following characteristics:

Unit weight of earth $w = 18$ kN/m³, angle of repose $\phi = 30°$,

SBC $q_o = 180$ kN/m², coefficient of friction between ground and concrete base $\mu = 0.4$. Use M20 CC and Fe 415 steel bars.

Solution: Minimum depth of foundation = 1.20 m, total height above base = 5 + 1.2 = 6.2 m

$$\text{Thickens of slab} = \frac{H}{12} = 500 \text{ mm (say)}, \ D \text{ (stem)} = 500 \text{ mm}$$

b (base width) – 0.7 II – 4.5 m,

$$\text{Toe projection} \cong \frac{b}{3} = 1.60 \text{ m (say)}, \text{ heel projection} = 4.50 - 0.5 - 1.6 = 2.40 \text{ m}$$

p_a(earth pressure) $= \dfrac{k_a}{2} wh^2 = \dfrac{1}{3} \times \dfrac{1}{2} \times 18 \times 6.2^2 = 115.32$ kN at $\dfrac{6.2}{3} = 2.067$ m above base. Moment due to earth pressure at base $= 238.33$ kN-m. Height of stem $= 5.70$ m ($H = 6.20$ m above base bottom).

Weight, CG and moments about toe end:

Weight (kN)	CG from toe end (m)	Moment about toe end (kN-m)
W_1 (weight of stem) $= 5.7 \times 0.5 \times 25 = 71.25$	1.85	131.813
W_2 (weight of parapet) $= 1 \times 0.2 \times 25 = 5.0$	1.70	8.50
W_3 (weight of base slab) $= 4.5 \times 0.5 \times 25 = 56.25$	2.25	126.56
W_4 (weight of earth) $= 2.4 \times 5.7 \times 18 = 246.24$	3.30	812.59
W_5 (weight of surcharge) $= 2.4 \times 18 = 43.20$	3.30	142.56
$\Sigma W = 421.94$		$\Sigma M = 1222.02$

Lateral pressure intensity due to surcharge $= k_a wh = 6.2 \times \dfrac{18}{3} = 37.20$

Moment $= 115.3$ kN-m (about base)

Neglecting earth above toe slab which may not exist all the time.

Resultant at base from toe end $\bar{x} = \dfrac{(1222.02 - 238.33 - 115.32)}{\sum 421.94} = \dfrac{868.37}{421.94} = 2.06$ m

FSO (over turning) $= \dfrac{1222.02}{(238.33 + 115.32)} = 3.46 > 1.50$, OK.

FS (sliding) $= \dfrac{\mu \sum W}{\text{Lateral pressure}} = \dfrac{0.4 \times 421.94}{(37.20 + 115.32)} = 1.107 < 1.5$, provide key

Resultant force on base, $\bar{x}$ from toe end $= \dfrac{\sum M}{\sum W} = \dfrac{(1222.02 - 238.33 - 115.32)}{421.94}$

$\qquad\qquad\qquad = \dfrac{868.37}{421.94} = 2.058$ m

$e = \dfrac{4.50}{2} - 2.058 = 0.192$ m

Soil pressure $(p_{max}, p_{min}) = \dfrac{\sum W}{b}\left(1 \pm \dfrac{6 \times e}{b}\right)$

$\qquad\qquad = \dfrac{421.94}{4.5}(1 \pm 0.256) = +118$ kN/m^2, $+70$ kN/m$^2 \uparrow$

At junctions of toe and wall, heel and wall

$p_2 = 118 - \dfrac{1.6}{4.5}(118 - 70) = 118 - 17 = 101$ kN/m$^2 \uparrow$

$p_3 = 118 - \dfrac{2.1}{4.5} \times 48 = 118 - 22.4 = 95.6$ kN/m$^2 \uparrow$

Net pressures:

Toe 1.60 m: Downward weight 12.5 kN/m²

(neglecting earth above toe slab)

$p_{1ultimate} = 1.5 (118 - 12.5) = 158.00\uparrow, p_{2u} = 1.5 (101 - 12.5) = 133$ kN/m²↑ (Fig. 10.11)

Heel 2.40 m: Downward weight = 12.5 + 102.6 + 18 = 133.1 kN/m²

$$p_{3\,ultimate} = 1.5 (-95.6 + 133.1) = 56.01 \text{ kN/m}^2$$
$$p_{4u} = 1.5 (133.1 - 70) = 95.0 \text{ kN/m}^2$$

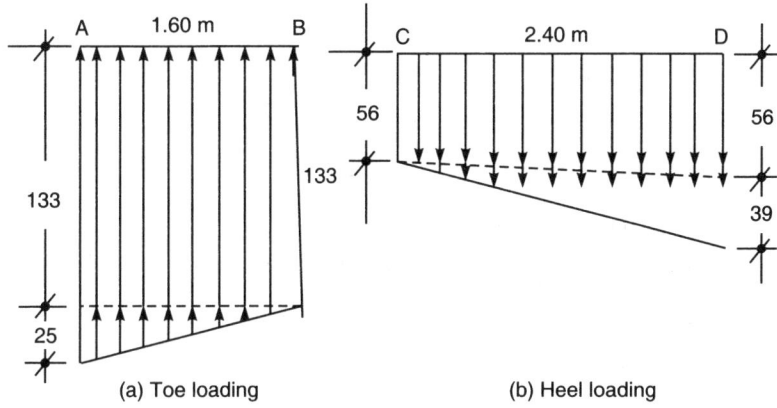

(a) Toe loading

(b) Heel loading

Fig. 10.11: Base slab (toe and heel slabs) loading

$$M_B = 133 \times 1.6 \times 0.8 + \frac{25 \times 1.6}{2} \times \frac{2}{3} \times 1.6 = 192 \text{ kN-m}$$

$$M_c = 56 \times 2.4 \times 1.2 + \frac{39}{3} \times 2.4 \times 2.4 \times \frac{1}{3} = 236 \text{ kN-m}$$

$$\text{SF at B} = 133 \times 1.6 + 25 \times 1.6 \times 0.5 = 233 \text{ kN}$$
$$\text{SF at C} = 56 \times 2.4 + 39 \times 2.4 + 0.5 = 181 \text{ kN}$$

Design of stem:

$$\text{Max. SF} = k_a w \frac{h^2}{2} + k_a w_s h = \frac{1}{3} [18 \times 5.7^2 \times 0.5 + 18 \times 5.7] = 132 \text{ kN}$$

$$\text{Max. BM} = k_a w \frac{h^3}{6} + k_a w_s \frac{h^2}{2} = \frac{1}{3} \times \frac{18 \times 5.7 \times 5.7}{2} \left[\frac{5.7}{3} + 1 \right] = 283 \text{ kN-m}$$

$$d \text{ (BM)} = \sqrt{\frac{283 \times 10^6}{0.138 \times 1000 \times 20}} = 320.2$$

Assumed value $D = 500$ mm, $d = (500 - 50) = 450$ mm

$$d \text{ (SF)} = \frac{132000}{1000 (0.36)} = 367 \text{ mm, by assuming } \tau_{uc} = 0.36 \text{ N/mm}^2$$

Thus, adopted $d = 450$ mm is safe and OK

$$A_{st} \text{ (reqd.)} = \frac{0.5 f_{ck} b \cdot d}{f_y}\left[1 - \sqrt{1 - \frac{4.6 \times M_u}{f_{ck} \cdot b(d)^2}}\right] = 1912 \text{ mm}^2 \text{ on earth face}$$

(20ϕ 150 c/c, A_{st} = 2093 mm², p_t = 0.465%)

τ_{uc} (0.465% steel) = 0.46 N/mm²

$$\tau_{uv} = \frac{132000}{1000 \times 450} = 0.293 \text{ N/mm}^2 < 0.46 \text{ N/mm}^2, \text{ OK.}$$

Minimum temperature steel = $\dfrac{0.12}{100} \times 1000 \times 450 = 540$ mm²

(vertical bars 10ϕ 280 c/c on outer face, A_{st} = 280 mm²/m)

Also provide 10ϕ 280 c/c horizontally on each face (A_{st} = 2 × 280 = 560 mm²/m total).

Design of shear key (Fig. 10.13):

Let the shear key be 'x' metre below base slab (neglecting earth above toe)

$$\text{Passive pressure } p_p = k_p \frac{w(x+0.5)^2}{2} = \frac{3 \times 18}{2}(x+0.5)^2 = 27\,(x+0.5)^2$$

For FS of 1.50, we have (μw = 0.4 × 421.94), $\dfrac{0.4 \times 421.94 + 27(x+0.5)^2}{(115.32 + 37.20)} \geq 1.5$

or 27 $(x+0.5)^2 \geq 1.5\,(152.52) - 168.8 = 60.004$, or $(x+0.5) \geq \sqrt{\dfrac{60.004}{27}} = 1.49$

or $x \geq 1.49 - 0.5 = 0.99$ m, provide shear key 1 m below and of width 500 mm below base.

Design of toe (Figs 10.12 and 10.13):

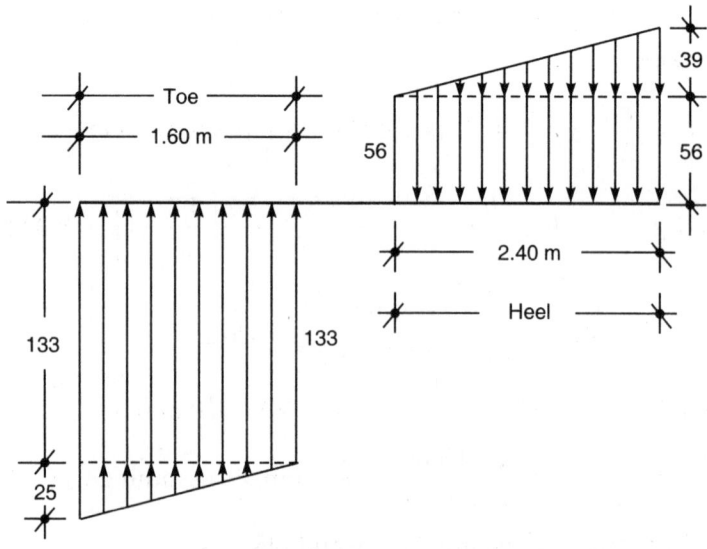

Fig. 10.12: Net ultimate pressure loading on base slab

Max. SF, $V_{uB} = 233$ kN, SF at $d = 0.45$ m from B = $(1.6 - 0.45)\ 133 + \dfrac{25}{1.6} \times \dfrac{0.45^2}{2}$

$$V_u \text{ (critical)} = 155 \text{ kN}$$
$$\text{Max. BM, } M_{uB} = 192 \text{ kN-m}$$

$$d \text{ (BM)} = \sqrt{\dfrac{192 \times 10^6}{0.138 \times 1000 \times 20}} = 264 \text{ mm}$$

Provided 450 mm (to be tapered to 200 mm at toe end)

$$A_{st} = \dfrac{0.5 f_{ck} b \cdot d}{f_y}\left[1 - \sqrt{1 - \dfrac{4.6 \times M_u}{f_{ck} \cdot b (d)^2}}\right]$$

or

$$A_{st} = \dfrac{10000 \times 450}{415}\left[1 - \sqrt{1 - \dfrac{4.6 \times 192 \times 10^6}{20 \times 1000 (450)^2}}\right] = 1255 \text{ mm}^2$$

Provide 16ϕ 140 c/c (bottom face)
($A_{st} = 1436$ mm^2/m, $p_t = 0.30\%$, $\tau_{uc} = 0.384$ N/mm^2, Table 19, IS:456-2000).
Check for shear stress:

$$\tau_{uv}(\text{at B}) = \dfrac{233000}{1000 \times 450} = 0.52 \text{ N/mm}^2 > \tau_{uc} \text{ of } 0.384 \text{ N/mm}^2 \text{ (permitted)}$$

$$\tau_{uv}(\text{at critical point}) = \dfrac{155000}{1000 \times 450} = 0.344 \text{ N/mm}^2 < 0.384, \text{ OK.}$$

Increase the depth to 500 mm ($D = 550$ mm)
Base slab may also be tapered from 550 mm at the junction with stem to 200 mm at the free end.
Min. temp steel = $0.12 \times 5500 = 660$ mm^2
(330 mm^2 on each face: 10ϕ @ 230 c/c, $A_{st} = 393 \times 2 = 786$ mm^2/m)
Design of heel (2.4 m) (Figs 10.12 and 10.13):
Max. SF at C = 181000 N

$$\text{SF at critical point at d from C} = \left(\dfrac{39 + 8.125}{2}\right) \times 1.9 + 56 \times 1.9 = 151.2 \text{ kN}$$

Max. BM at C (M_{uc}) = 236 kN-m

$$d \text{ (BM)} = \sqrt{\dfrac{236 \times 10^6}{0.138 \times 1000 \times 20}} = 293 \text{ mm}$$

Provide $d = -500$ mm, tapered to 200 mm at end.

$$A_{st} = \dfrac{10000 \times 500}{415}\left[1 - \sqrt{1 - \dfrac{4.6 \times 236 \times 10^6}{20 \times 1000 (500)^2}}\right] = 1388 \text{ mm}^2$$

[16φ 140 c/c (in top earth face)]

$(A_{st} = 1436 \text{ mm}^2/\text{m}, p_t = 0.287\%, \tau_{uc} = 0.380 \text{ N}/\text{mm}^2).$

$$\tau_{uv} = \frac{151200}{1000 \times 500} = 0.302 \text{ N}/\text{mm}^2 < 0.380 \text{ N}/\text{mm}^2, \text{ OK.}$$

Thus, the retaining wall is safe in all respects. Reinforcement provided is shown in Fig. 10.13.

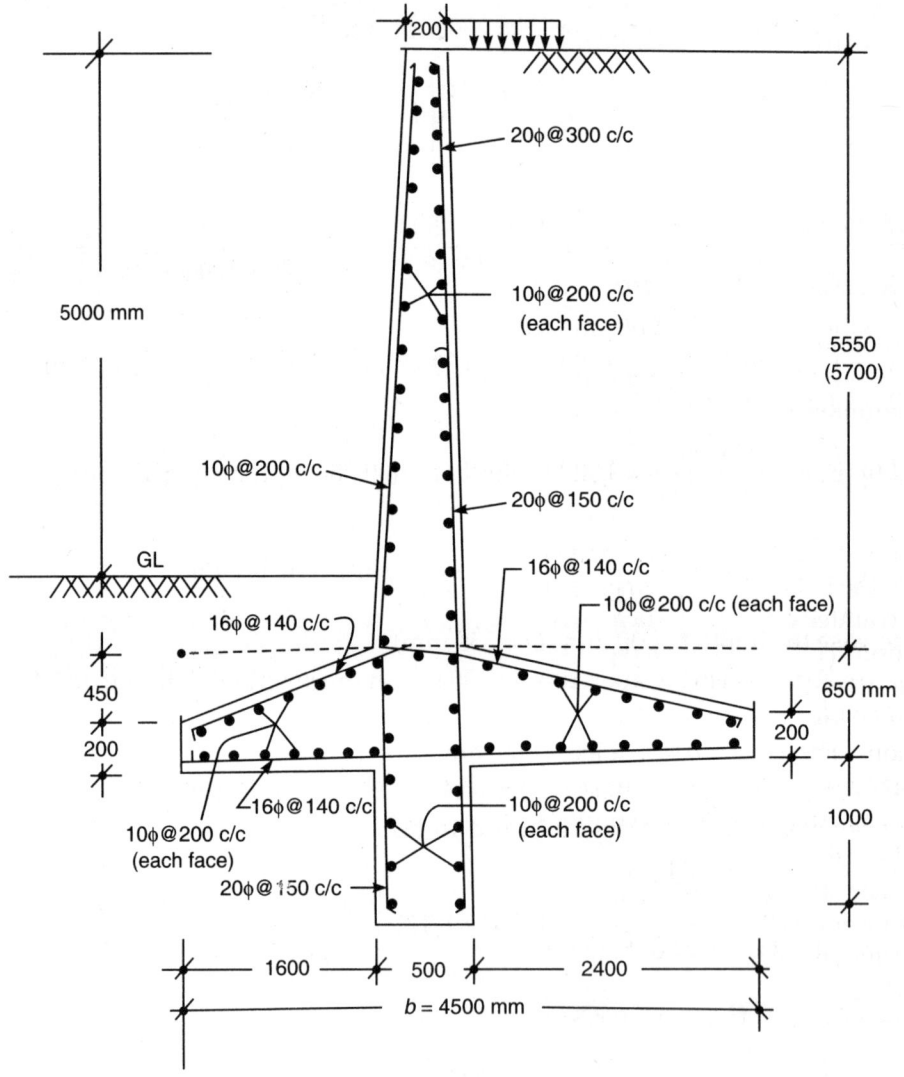

Fig. 10.13: Details of reinforcement and dimensions of cantilever (mm)

- *Wall (stem):* Vertical on earth face 20φ 150 mm c/c and on outer face 10φ 280 mm c/c.

  Horizontal bars on each face (earth and outer) 10φ 280 mm c/c. Curtail main bars as per curtailment rules. Wall thickness is 500 mm at the bottom and tapered to 200 mm at the top.

- *Toe slab*: Provide 550 mm thickness at the junction with vertical wall and tapper to 200 mm at the free end (Fig. 10.13).

    Provide main bars in bottom face 16ϕ @ 140 mm c/c and 10ϕ 200 c/c in top face. Also provide longitudinal distribution (or temperature) bars 10ϕ 200 c/c on each face longitudinally ($A_{st} = 2 \times 393 = 786$ mm^2/m).

- *Heel slab*: Provide heel slab thickness of 550 mm at the junction with stem and tapper it to 200 mm at end (Fig. 10.13).

    Provide 16ϕ @ 140 mm c/c in top (earth fill) face and 10ϕ @ 200 mm c/c in bottom face. Also provide 10ϕ 200 mm c/c longitudinally on each face as distribution (temperature) steel.

    In actual design, the bars may be curtailed as per development length requirement and bending moment requirement. After preliminary calculations, the design may be revised with exact calculations.

10.3 COUNTERFORT RETAINING WALLS

10.3.1 General Feature

The counterfort retaining walls are provided when the height of retaining wall exceeds 6 m. Counterfort retaining walls are constructed by joining vertical wall with heel slab by counterforts at a spacing of about one-third to half of the height of wall (spacing $\frac{1}{3}H$ to $\frac{1}{2}H$). Counterforts provide support to both the vertical wall and the heel slab (Fig. 10.14). The vertical walls act as a continuous slab supported by counterforts and also by the base slab. The vertical wall also exhibits cantilever action at the junction with the base slab. Similarly, the heel slab behaves as a continuous slab supported by the counterforts and vertical wall. The counterforts behave as T-beam of variable cross-section. The toe slab behaves as a cantilever fixed with the vertical slab and heel slab. This may necessitate large thickness of toe slab. In such cases, sometimes front counterforts are also provided to convert these cantilever actions into continuous slab actions. The toe slab shall be subjected to upward soil pressure while the heel slab shall be generally subjected to net downward loading due to retained earth and self-weight, etc. The buttress acts as a compression member due to pressure from vertical wall slab and the toe slab.

The structural behaviour of counterfort retaining wall is shown in Fig.10.14 along with deflection profile and likely location of tension cracks. The reinforcement is located on tension faces. The vertical wall is subjected to lateral earth pressure while the heel slab is subjected to the net downward loads with a tendency of separation from the counterforts. The reinforcement of counterforts should be properly embedded into the vertical wall and the heel slab.

The analysis and design of the counterfort retaining wall can be done after the dimension of its different components have been decided. The probable dimensions of its different components may be determined by following considerations.

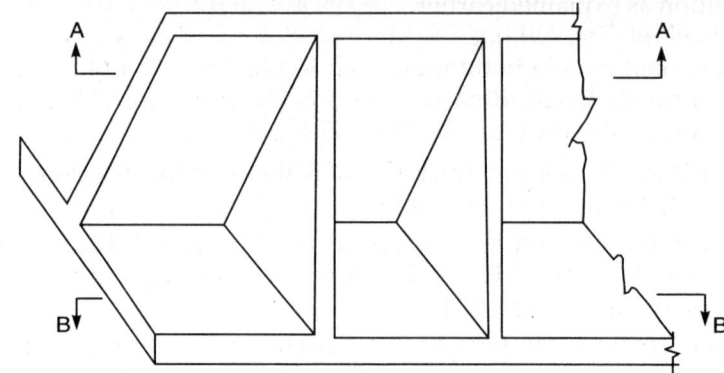

(a) Counterfort retaining wall

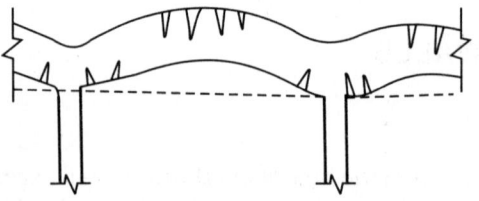

(i) Deflection profile and tension cracks

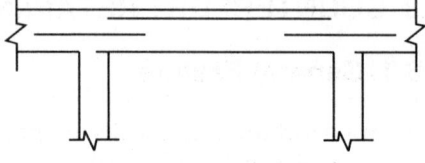

(ii) Tensile reinforcement location

(b) Section AA of vertical wall

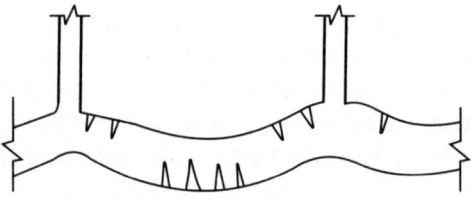

(i) Deflection profile and tension cracks

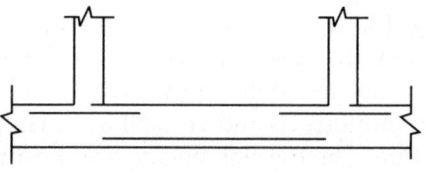

(ii) Tensile reinforcement location

(c) Section BB of heel slab
(d) Cracks at the interface of counterfort, vertical wall and heel slab

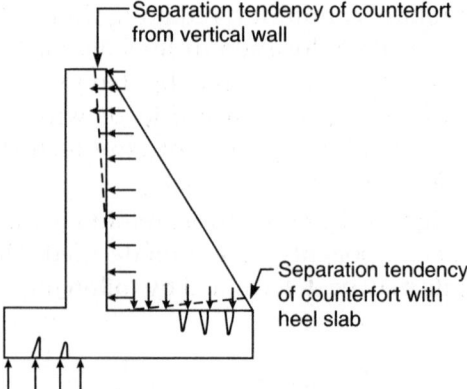

Fig. 10.14: Behaviour of counterfort retaining wall

(i) *Depth of foundation* by Rankine's formula to reduce environmental effect on foundation as explained earlier.

(ii) *Base width (b)* is determined in similar manner as done in case of cantilever retaining walls ($b = 0.5H$ to $0.7H$ depending on surcharge) and check of stability and sliding safety.

(iii) *Spacing of counterforts (L)* is based on the condition that maximum bending moment depends directly on height h, while BM depends on the square of span (i.e. spacing L between the counterforts). It should be so spaced that spacing L is less than height h. It is determined by empirical relations.

$$L = 0.8 \sqrt{h} \text{ to } 1.2 \sqrt{h} \qquad \text{... Eq. (10.16)}$$

The spacing of buttress, whenever provided, shall be kept same as that of counterforts.

The thickness of vertical wall, heel-toe (base) slab and counterforts may be fixed by following empirical equations:

$$\text{Thickness of vertical wall} = \frac{h_a}{40} \qquad \text{... Eq. (10.17)}$$

$$\text{Thickness of base slab} = \frac{h_a}{30} \qquad \text{... Eq. (10.18)}$$

$$\text{Thickness of counterforts} = \frac{L}{10} \qquad \text{... Eq. (10.19)}$$

Vertical wall slabs and heel slabs are fixed on three sides and subjected to uniformly distributed loads or triangular distributed loads, the maximum moments are calculated by using coefficients to be obtained from the Tables 10.1 and 10.2 for the given ratio of for given type of loading by formula

$$M = \alpha w L^2 \qquad \text{... Eq. (10.20)}$$

where, α = Moment coefficient for the given critical location (1, 2, 3, etc. Tables 10.1 and 10.2)

w = Uniformly distributed load or intensity at the base of triangularly distributed load

L = Clear span between the fixed edges.

Table 10.1: Bending moment coefficients for slab with 3 edges fixed and one edge free with uniformly distributed load

Location of critical moments	h/L	Moment coefficients			
		1	2	3	4
	0.60	−0.055	−0.036	0.017	−0.074
	0.70	−0.054	−0.044	0.021	−0.078
	0.80	−0.053	−0.051	0.025	−0.081
	0.90	−0.052	−0.056	0.029	−0.083
	1.00	−0.051	−0.061	0.032	−0.083
	1.25	−0.047	−0.071	0.037	−0.083
	1.50	−0.042	−0.075	0.040	−0.083
	2.0 and more	−0.040	−0.083	0.041	−0.083

Table 10.2: Bending moment coefficients for slab with 3 edges fixed and one edge free carrying triangular loading

Location of critical moments	h/L	Moment coefficients			
		1	2	3	4
	0.60	−0.024	−0.013	0.006	0.0
	0.70	−0.026	−0.017	0.008	0.0
	0.80	−0.028	−0.021	0.010	0.0
	0.90	−0.029	−0.024	0.012	0.0
	1.00	−0.030	−0.027	0.013	0.0
	1.25	−0.031	−0.033	0.017	0.0
	1.50	−0.029	−0.034	0.019	0.0
	2.0 and more	−0.029	−0.040	0.021	0.0

Example 10.3

Design a counterfort retaining wall for retaining 7.3 m high earth above ground level. The properties of earth are as follows:

Weight of soil $w = 15$ kN/m^3, angle of repose $\phi = 30°$, coefficient of friction $\mu = 0.5$, safe bearing capacity $q_o = 150$ kN/m^2, grade of CC is M20 and steel is Fe 415 grade.

Solution: Depth of foundation $h = \dfrac{q_o}{w} \dfrac{(1 - \sin \phi)^2}{(1 + \sin \phi)^2} = \dfrac{150}{15}\left(\dfrac{1}{9}\right) = 1.11$m (say 1.20 m)

Total height of backfill $H = 7.3 + 1.2 = 8.50$ m above bottom of base (Fig. 10.15).

For stability, dimensions of the counterfort are found by successive trials. Approximate dimensions are fixed by empirical formula:

Base width $b = 0.5H$ to $0.6H = 0.5 \times 8.5$ or $0.6 \times 8.5 = 4.25$ m to 5.1 m

Consider $b = 5.0$ m (say), width of toe $= \dfrac{b}{3} = 1.70$ m (say 2.0 m)

Spacing of counterforts $= 0.8 \sqrt{H}$ to $1.2 \sqrt{H}$, i.e. 2.33 m to 3.50 m
Spacing of counterforts $(L) = 3.0$ m (say)

Thickness of counterfort $\cong \dfrac{L}{10} = \dfrac{3000}{10} = 300$ mm

Thickness of base slab $\cong \dfrac{H}{30} = \dfrac{8500}{30} = 283 \cong 300$ mm

Thickness of vertical wall $\cong \dfrac{H}{40} = \dfrac{8500}{40} = 213 \cong 300$ mm

Heel projection $= (5.0 - 0.30 - 2.0) = 2.70$ m

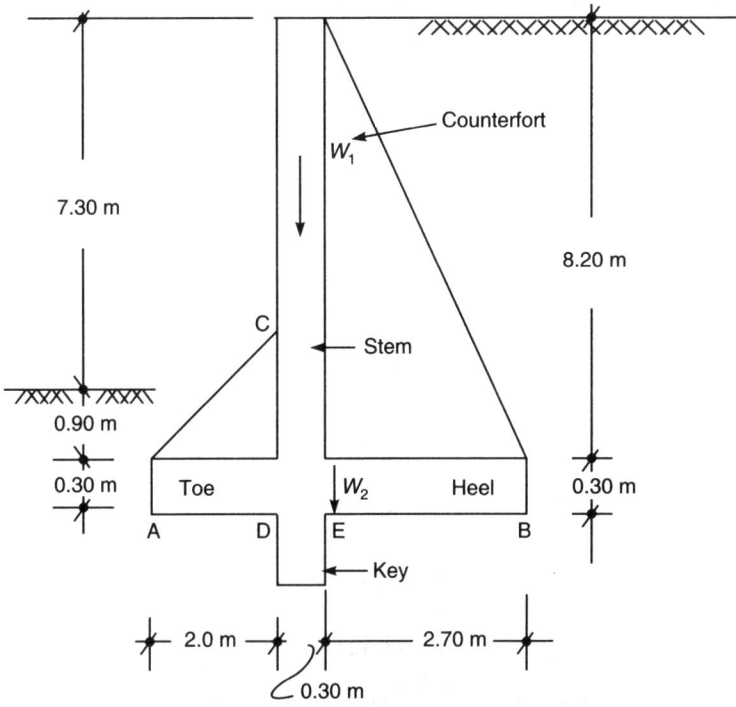

Fig. 10.15: Section of counterfort retaining wall

Factor of safety against overturning (FSO), stabilizing moment M_{wa}/overturning moment M_{pa}

$$P_a = k_a \frac{wH^2}{2} = \frac{1}{3} \times \frac{15 \times 8.5^2}{2} = 180.63 \text{ kN}, M_{pa} = 180.63 \times \frac{8.5}{3} = 812 \text{ kN-m}$$

Calculation of vertical loads and moments about toe end A:

Loads (kN)	Distance of CG from A	Moment about A (kN-m)
Wall $W_1 = 0.3 \times 8.2 \times 25 = 61.5$	$x_{a1} = 2.15$ m	132.23
Base $W_2 = 0.3 \times 5 \times 25 = 37.5$	$x_{a2} = 2.50$ m	93.75
Earth $W_3 = 2.7 \times 8.2 \times 15 = 332.1$	$x_{a3} = 3.65$ m	1212.17
$\Sigma W = 431.1$		$\Sigma M = 1438.15$

$$\overline{x}_a = \frac{\Sigma(1438.15 - 512)}{\Sigma 431.1} = 2.15 \text{ m from A (toe end)}$$

$$e = \left(\frac{b}{2} - 2.15\right) = (2.5 - 2.15) = 0.35 \text{ m}$$

$$\text{FSO (overturning)} = \frac{(1438.15)}{512} = 2.81 > 1.55, \text{ (hence safe against overturning)}$$

$$\text{FS (sliding)} = \frac{\mu W}{P_a} = \frac{0.5 \times 431.1}{180.63} = 1.19 < 1.55$$

Hence, inadequate safety. Provide shear key in base.

Design of shear key (neglect the earth above toe slab, since it may not exist all the time):
Let *a* be the depth of key below base.

$$\text{Passive resistance} = k_p \frac{(0.3+a)^2 w}{2} = \frac{3 \times 15}{2} = (a + 0.3)^2$$

Net sliding force = {180.63 − 22.5 $(a + 0.3)^2$},
resistance to sliding = 0.5 (431.1)

$$\text{FS (sliding)} = \frac{0.5(431.1)}{\{180.63 - 22.5(a+0.3)^2\}} \geq 1.55$$

or $\qquad$ 215.6 ≥ 1.55 $[180.63 - 22.5 (a + 0.3)^2] = 280 - 34.88 (a + 0.3)^2$

or $\qquad (a + 0.3)^2 \geq \dfrac{(280 - 215.6)}{34.88} = 1.847$, $(a + 0.30) \geq 1.36$, $a \geq 1.06$ m

Provide *a* = 1.10 m

Soil pressures on foundation (*e* = 0.35 m), Σ*W* = 431.1 kN (Fig. 10.16)

$$p_{max} \text{ (at toe end } A) = \frac{\Sigma w}{b} \times \left(1 \pm \frac{6e}{b}\right) = \frac{431.1}{5.0}\left(1 \pm \frac{6 \times 0.35}{5}\right)$$

$$= 86.22(1.42) = 122.5 \text{ kN/m}^2 \uparrow$$

$$p_{min} \text{ (at heel end } B) = \frac{431.1}{5.0}(1 - 0.42) = 50 \text{ kN/m}^2 \uparrow$$

$$p_D = 122.5 - \frac{(122.5 - 50)}{5.0} \times 2 = 94.0 \text{ kN/m}^2,$$

$$p_E = 50 + \frac{2.7}{5} \times 72.5 = 89 \text{ kN/m}^2$$

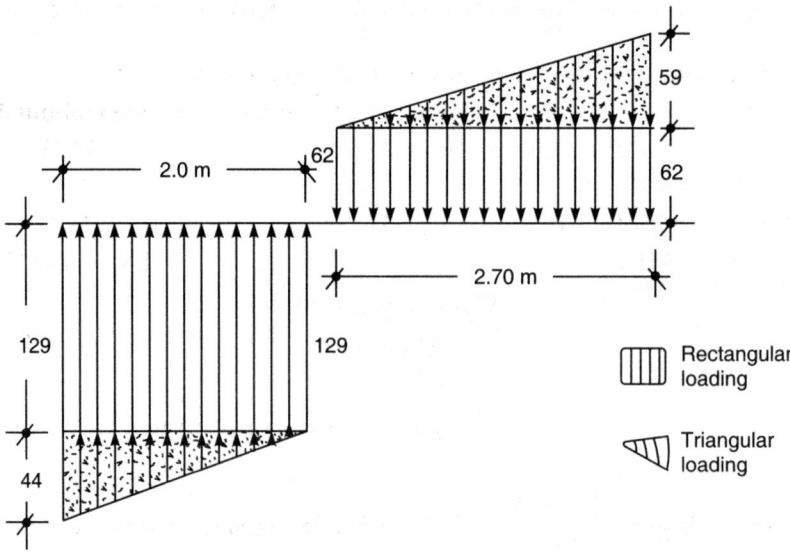

Fig. 10.16: Ultimate soil pressure loading on the base slab

Net loading on base slab downward weight (7.5 kN/m²),
$\quad$ Weight of earth = 8.2 × 15 = 123 kN/m

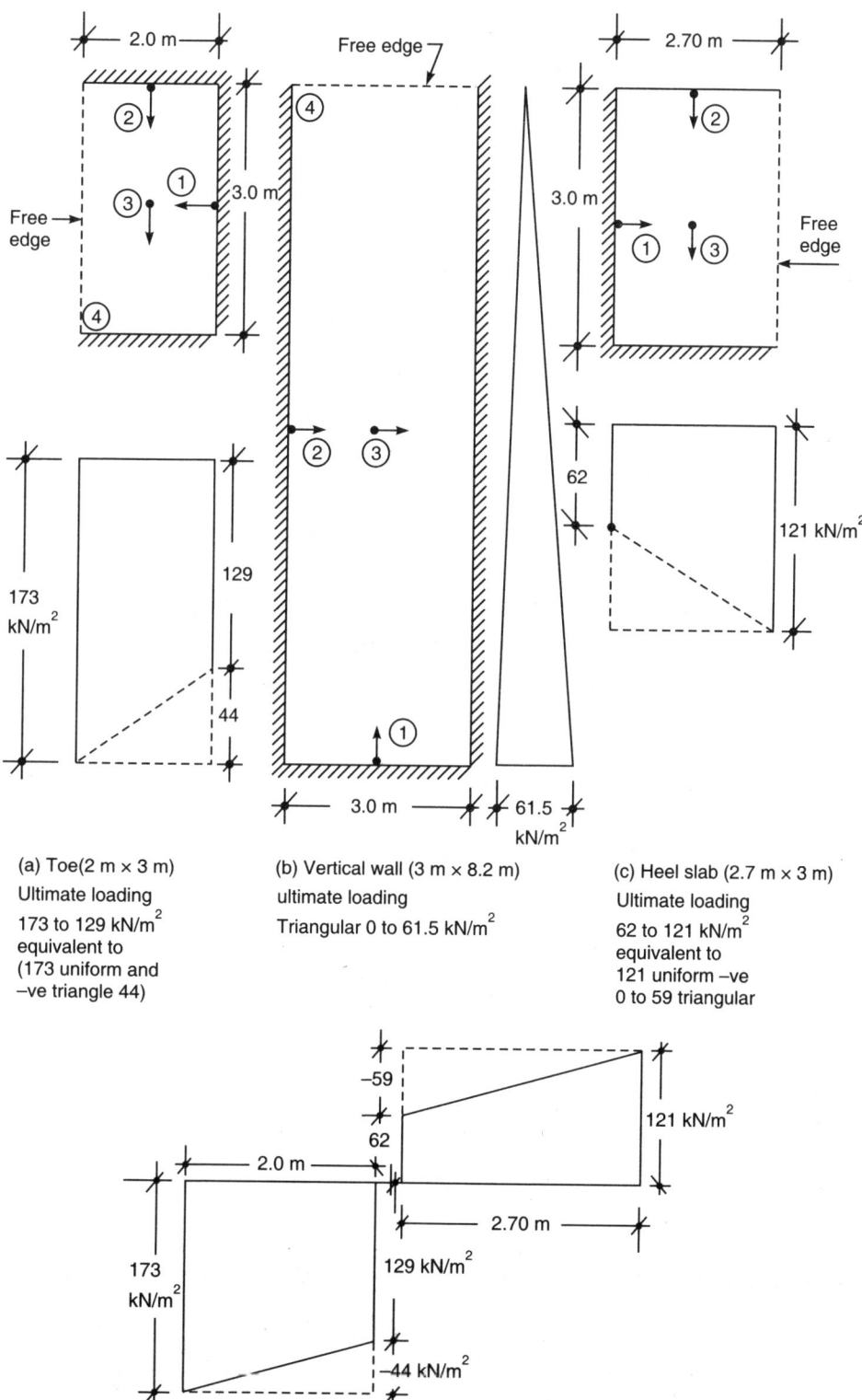

(a) Toe(2 m × 3 m)
Ultimate loading
173 to 129 kN/m^2
equivalent to
(173 uniform and
−ve triangle 44)

(b) Vertical wall (3 m × 8.2 m)
ultimate loading
Triangular 0 to 61.5 kN/m^2

(c) Heel slab (2.7 m × 3 m)
Ultimate loading
62 to 121 kN/m^2
equivalent to
121 uniform −ve
0 to 59 triangular

Fig. 10.17: Ultimate loading on various panels and support conditions
(Toe = 2 m × 3 m, heel = 2.7 m × 3 m, and stem = 3 m × 8.20 m)

Ultimate net loading (Figs 10.16 and 10.17)

$$p_{uA} = 1.5 \, (122.5 - 7.5) = 172.5 \cong 173 \text{ kN/m}^2\uparrow \text{ upward}$$

$$p_{uD} = 1.5 \, (94 - 7.5) = 129 \text{ kN/m}^2\uparrow \text{ upward, toe slab } AD = 2 \text{ m}$$

$$p_{uB} = 1.5 \, (50 - 123 - 7.5) = 121 \text{ kN/m}^2\downarrow \text{ (downward)}$$

$$p_{uE} = 1.5 \, (89 - 123 - 7.5) = 62 \text{ kN/m}^2\downarrow \text{ (downward), heel slab } BE = 2.70 \text{ m}$$

Ultimate lateral pressure (horizontal) on vertical wall panel (3 m × 8.2 m) shall be given as $p_{a\,max}$ at the junction with base slab $= 1.5 \, k_a wh = 1.5 \times \dfrac{1}{3} \times 15 \times 8.2 = 61.5 \text{ kN/m}^2$.

BM coefficients for the type of loading is determined from Tables 10.1 and 10.2 for different critical locations by considering uniform and triangular components of loading.

(i) Toe slab (2 m × 3 m), $\dfrac{B}{L} = \dfrac{2}{3} = 0.67$ (assuming front buttresses also)

Max. BM at location (1) $M_1 = - \, [0.0543 \times 173 \times 3^2 - 0.0254 \times 44 \times 3^2]$
$$= - \, 75 \text{ kN-m/m}$$

Max. BM at location (2) $M_2 = - \, [0.0416 \times 173 \times 3^2 - 0.0158 \times 44 \times 3^2]$
$$= - \, 59 \text{ kN-m/m}$$

Max. BM at location (3) $M_3 = + \, [0.0198 \times 173 \times 3^2 - 0.0074 \times 44 \times 3^2]$
$$= + \, 28 \text{ kN-m/m}$$

Max. BM at location (4) $M_4 = - \, [0.0768 \times 173 \times 3^2 - 0] = - \, 120 \text{ kN-m/m}$

Max. BM between the fixed counterfort supports $= - \, 120 \text{ kN-m/m} \, (-59)$ and $+28 \text{ kN-m/m}$

Perpendicular direction as cantilever $= - \, 75 \text{ kN-m/m}$

(ii) Heel slab (2.70 m × 3 m), $\dfrac{B}{L} = \dfrac{2.7}{3} = 0.90$

Max. BM at location (1) (cantilever) $M_1 = - \, [0.052 \times 121 \times 3^2 - 0.029 \times 59 \times 3^2]$
$$= - \, 41.23 \text{ kN-m/m}$$

Max. BM at location (2), $M_2 = - \, [0.056 \times 121 \times 3^2 - 0.029 \times 59 \times 3^2]$
$$= - \, 45.6 \cong - \, 46 \text{ kN-m/m}$$

Max. BM at location (3), $M_3 = + \, [0.029 \times 121 \times 3^2 - 0.012 \times 59 \times 3^2]$
$$= + \, 25.21 = + \, 25.21 \text{ kN-m/m}$$

Max. BM at location (4), $M_4 = - \, [0.083 \times 121 \times 3^2 - 0]$
$$= - \, 90.4 \text{ kN-m/m}$$

Max. BM $= -90.4 \text{ kN-m/m}$ between the fixed counterfort supports (-46) and $+ 25.2$.

(iii) Vertical stem wall (3 m × 8.20 m), $\dfrac{B}{L} = \dfrac{8.2}{3} = 2.73$ (2 or more)

Max. BM at location (1) (cantilever) $M_1 = - \, [0.029 \times 61.5 \times 3^2] = - \, 16.05 \text{ kN-m/m}$

Max. BM at location (2), $M_2 = - \, [0.040 \times 61.5 \times 3^2] = - \, 22.14 \text{ kN-m/m}$

Max. BM at location (3), $M_3 = + \, [0.021 \times 61.5 \times 3^2] = + \, 11.62 \text{ kN-m/m}$

Max. BM at location (4), $M_4 = - \, [0.0]$

Max. BM = –22.14 kN-m/m and perpendicular BM = –16.05 kN-m/m (vertical cantilevering).

Effective depth in main direction = $300 - 25 - \dfrac{\phi}{2}$ = 269 mm, perpendicular direction $d' = 257$ mm

Reinforcements ($f_{ck} = 20$ N/mm^2, $f_y = 415$ N/mm^2),

$$A_{st} = \frac{0.5 f_{ck} b \cdot d}{f_y}\left[1 - \sqrt{1 - \frac{4.6 \times M_u}{f_{ck} b d^2}}\right]$$

considering different values we get:

(i) Toe slab (–120, +28, perpendicular cantilever –75 kN-m)

Min. temp. steel = $\dfrac{0.12}{100} \times 300 \times 1000 = 360$ mm^2 on both faces

$A_{stx} = 1384$ mm^2 (16ϕ 140 c/c in bottom near counterforts or 20ϕ 200 c/c).

A_{sty} (top face) = 911 mm^2(16ϕ 200 c/c in top face between counterforts)

Min. distribution and temp.steel 10ϕ 300 c/c on faces where main steel is not there.

(ii) Heel slab (–90.4 kN-m/m between the counterforts and –46 kN-m/m in perpendicular direction)

$A_{stx} = -1010$ mm^2 (top face below counterforts between counterforts, i.e. 20ϕ 300 c/c.

A_{sty} = Bottom face between counterforts (+542 mm^2) 12ϕ 200 c/c bottom face up to free end.

Minimum temperature steel 10ϕ 300 c/c.

(iii) Stem (vertical slab panel): $d = 269$ mm, $d' = 257$ mm.

A_{stx} (between counterforts) on earth face = 233 mm^2 (less than temperature steel) thus, provide $A_{st} = 360$ mm^2 (10ϕ 200 c/c, on earth face under counterfort supports).

Also provide in perpendicular direction on earth face 10ϕ 200 c/c ($A_{st} = 393$ mm^2).

For temperature (distribution) steel 10ϕ 300 c/c on each face in each direction.

Shear key to resist sliding (300 mm wide × 1100 mm deep), $d = 270$ mm

Height of earth for design = 1.2 + 1.1 = 2.3 m and 1.20 m or if earth above toe is neglected then the heights will be 1.40 m and 0.30 m. Maximum earth pressure for design of key will be $k_p w$ (0.3) × 1.5 = 20.25 kN/m^2 and $k_p \cdot w$ (1.4) 1.5 = 94.5 kN/m^2

Max. SF = $20.25 \times 1.1 + (94.5 - 20.25)\,\dfrac{1.1}{2}$ = 22.28 + 40.84 = 63.11 kN

Max. BM = $22.28 \times \dfrac{1.1}{2} + 40.84 \times \dfrac{2}{3} \times 1.1$ = 12.25 + 29.95 = 42.20 kN-m

$$A_{st} = \frac{10 \times 1000 \times 270}{415}\left[1 - \sqrt{1 - \frac{4.6 \times 42.2 \times 10^6}{20 \times 1000 (270 \times 270)}}\right] = 449 \text{ mm}^2$$

(12ϕ 200 c/c, A_{st} = 665 mm^2/m)

Alternate bars from heel top shall be bent in shear key on toe face (i.e. 20ϕ 600 c/c) plus provide 12ϕ 200 c/c between 20ϕ 600 c/c on toe side face.

On heel side face continue 10ϕ 200 c/c as temperature steel.

Temperature steel longitudinally on each face 10ϕ 300 c/c. All bars shall be extended beyond the required point by development length.

Design of counterforts (Fig. 10.17):

Let the front counterfort be of the same height as the length of the toe slab, i.e. 2 m above the top of the toe slab. Thus, the most critical point C on the counterfort will be $(8.20 - 2.0) = 6.20$ m below the top surface of the earth back fill.

Angle θ of slope of counterfort surface will be $\tan^{-1}\left(\dfrac{2.7}{8.2}\right) = 18.23°$

Depth of counterfort critical point at C: $D_c = h_c \sin\phi = 6.2 \sin 18.23° = 1.939$ m

Effective depth d_c = 1939 – clear cover 40 mm – $\dfrac{20\text{ mm}}{2}$ (bar dia.) – $\dfrac{\text{spacing}}{\text{in layers}}$

or $\qquad d_c = 1939 - 40 - \dfrac{20}{2} - 20 = 1869 \cong 1870$ mm (assuming 20ϕ)

Max. ultimate BM at C = $1.5\, k_a \dfrac{wL}{6} h^3{}_c = \dfrac{1.5}{3} \times \dfrac{15}{6} \times 3 \times 6.20^3 = 893.73$ kN-m

Since the depth is already fixed, let us find the steel required,

$$A_{st} = \dfrac{0.5 f_{ck} bd}{f_y}\left[1 - \sqrt{1 - \dfrac{4.6 \times M_u}{f_{ck} b (d)^2}}\right]$$

Assuming b of counterfort = 300 mm

or $\qquad A_{st} = \dfrac{10 \times 300 \times 1870}{415}\left[1 - \sqrt{1 - \dfrac{4.6 \times 893.73 \times 10^6}{20 \times 300(1870 \times 1870)}}\right] = 1397$ mm^2

Provide 5 bars of 20 mm (3 + 2), A_{st} = 1570 mm^2 > 1397 mm^2, OK.

Two bars of inner layer may be curtailed after the point where these may not be required. These bars must be extended beyond this point by at least development length.

% steel $p_t = \dfrac{1570 \times 100}{300 \times 1870} = 0.28\%$, $\tau_{uc} = 0.374$ N/mm^2 (Table 19 of IS:456-2000)

Check for shear:

$$V_u\,(SF) = \left[1.5\, k_a \dfrac{wL}{2} h^2{}_c - \dfrac{M_u \sin\phi}{d_c}\right] = \left[\dfrac{1.5}{3} \times \dfrac{15 \times 3}{2} \times (6.20)^2 - \dfrac{893.73 \sin 18.23°}{1.87}\right]$$

or $\qquad V_u = 432.45 - 149.51 = 282.94 \cong 283$ kN

$$d = 6200 \tan 18.23° = \left(40 + \frac{20}{2}\right) = 2042 - 50 = 1992 \text{ mm (outer layer)}$$

$$\tau_{uv} = \frac{283000}{300 \times 1992} = 0.474 \text{ N/mm}^2 > 0.374 \text{ N/mm}^2 \ (\tau_{uc}).$$

Provide shear reinforcement.

$$8\phi \ 2 \text{ legged stirrups at } S_v = \frac{0.87 \times 415 \times 100}{(0.474 - 0.374)b} = 1203 \not> S_{v \max}$$

$$S_{v \max} = \frac{f_y A_{sv}}{0.4b} = \frac{415 \times 100}{0.4 \times 300} = 345 \text{ mm, also max.} = 300 \text{ mm}$$

Thus, provide 8ϕ 2 legged stirrups at 300 c/c.

Design for interface tension between counterfort and vertical wall:

T_u (ultimate tension at critical section/m) = 1.5 $(k_a w h_c L)$

or
$$T_u = 1.5 \times \frac{1}{3} \times 15 \times 6.2 \times 3 = 139.5 \text{ kN}$$

Thus,
$$0.87 f_y A_{sv} \times \frac{1000}{S_v} = 139.5 \times 1000 \text{ N}$$

$$S_v = \frac{0.87 \times 415 \times (100)1000}{139.5 \times 1000} = 258.8 \text{ mm c/c}$$

Thus, provide 8ϕ 2 legged stirrups @ 250 c/c in upper portion of counterfort and spacing may be increased to 345 mm c/c in the bottom portion. Provide stirrups both vertically and horizontally connecting stem wall and heel slab.

Design for tension at interface of counterfort with heel slab:

$T_u = 1.5 \times$ net pressure at the end of heel slab (where net pressure is maximum) $\times L$

or
$$T_u = 1.5 \times L \times 121 = 544.5 \text{ kN}$$

$$S_v = \frac{0.87 \times 415 \times 157 \times 1000}{544500} = 104 \text{ mm c/c}$$

Thus, provide 10ϕ 2 legged at 100 c/c plus extend counterfort main steel in the heel slab.

Design of front counterfort:

It acts as horizontal cantilever beam of varying depth (cross-section) with varying upward soil pressure from the toe slab. Spacing of counterforts = clear + thickness = 3.30 m c/c.

Max. ultimate moment at the support with the vertical wall,

$$M_u = \frac{(129 + 173)}{2} \times 2.0 \times 3.3 \times \left(\frac{2 \times 173 + 129}{173 + 129}\right)\frac{2}{3} = 1045 \text{ kN-m}$$

Total depth $\quad D = 2000 + 300 = 2300 \text{ mm}, d = 2300 - (40 +) = 2250 \text{ mm}$

Reinforcement:

$$A_{st} = \frac{10 \times 300 \times 2250}{415}\left[1 - \sqrt{1 - \frac{4.6 \times 1045 \times 10^6}{20 \times 300(2250 \times 2250)}}\right] = 1343 \text{ mm}^2$$

Min. temp. steel $= \dfrac{0.12}{100} \times 300 \times 2250 = 810 \text{ mm}^2$ (5 bars 20ϕ, $A_{st} = 1570 \text{ mm}^2$)

% tensile steel $= \dfrac{1570 \times 100}{300 \times 2250} = 0.233$, $\tau_{uc} = 0.340 \text{ N/mm}^2$ from IS:456-2000.

Max. SF $V_u = \left(\dfrac{129 + 173}{2}\right) \times 2 \times 3.3 - \dfrac{m \sin 45° \times 1000}{2250} = 669 \text{ kN}$

$\tau_{uv} = \dfrac{669000}{300 \times 2250} = 0.99 \text{ N/mm}^2 > 0.34 \text{ N/mm}^2$ (provide shear reinforcement)

$S_{v\,max} = \dfrac{f_y A_{sv}}{0.4b}$, for 8$\phi$ 2 legged, $S_{v\,max} = \dfrac{415 \times 100}{0.4 \times 300} = 345 \text{ mm c/c}$

$S_v = \dfrac{0.87 f_y A_{sv}}{b(\tau_{uv} - \tau_{uc})} = \dfrac{0.87 \times 415 \times 100}{300(0.99 - 0.34)} = 185 \text{ mm c/c}$ (provide 8ϕ 2 legged stirrups

@ 150 c/c).

Thus, provide 8ϕ 2 legged stirrups vertically connecting front counterfort and toe slab @ 150 c/c. Also provide 8ϕ 2 legged stirrups horizontally to connect vertical stem and front counterforts @ 150 c/c. Also the rear counterfort reinforcement should be extended in the front counterfort.

SUMMARY

The purpose of retaining walls is to retain loose material and resist lateral pressure exerted by this loose material. The retaining wall components are first designed for stability and then each component is designed for resisting shear and moments. Service loads are first converted into ultimate loads to determine ultimate shear forces and ultimate moments. All components are designed for these ultimate shear forces and ultimate bending moments.

Types of retaining walls are gravity walls, cantilever, counterfort and buttressed walls. The forces on retaining walls are self-weight, weight of the soil retained, lateral soil pressure, surcharge, upward soil reaction, frictional force between the soil and wall base and passive earth pressures on toe side. Pressures on base are found by

$p_{1,2} = \dfrac{\Sigma W}{b}\left(1 \pm \dfrac{6e}{b}\right)$, where $e = \left(\dfrac{b}{2} - \bar{x}_r\right)$.

Active earth pressure of soil is $p_a = k_a wh$, $e = \left(\bar{x} - \dfrac{b}{2}\right)$, $\bar{x}_r = \left\{\dfrac{\Sigma M_w - M_{pa}}{\Sigma W}\right\}$

where, k_a (coefficient) $= \dfrac{1-\sin\phi}{1+\sin\phi}$, $w =$ unit weight of soil, $h =$ height of retained soil.

Similarly, passive pressure $p_p = k_p \cdot wh$, where $k_p = \dfrac{1+\sin\phi}{1-\sin\phi}$.

Total active earth pressure force $P_a = k_a \cdot \dfrac{wh^2}{2}$, $P_p = k_p \cdot \dfrac{wh'^2}{2}$

Lateral active earth pressure due to uniform surcharge W_s at a depth y is given by:

$$p = k_a w \left(\dfrac{W_s}{w} + y \right),$$ equivalent height h_s of surcharge $h_s = \dfrac{W_s}{w}$

In cantilever walls, the vertical stem bends as cantilever beam under the active lateral earth pressure, heel slab also bends under net loading due to weight of soil and upward reaction of soil, and the toe slab also bends as cantilever under upward soil reaction (pressure). Generally for stability, the base width b of cantilever walls are taken as

$$b = 0.5H \text{ to } 0.6H$$
$$b = 0.7H \text{ for walls with surcharge}$$

The toe projection may be kept approximately $\dfrac{b}{3}$.

Factor of safety against sliding, $\dfrac{\mu W}{P} \geq 1.50$.

Resultant forces should act within middle one-third of the base, i.e.

eccentricity (e) shall be $\leq \dfrac{b}{6}$.

In counterfort retaining walls, dimensions are decided to satisfy stability requirements. Based on various dimensions, there can be numerous designs and the designs given here are just for guidance. In counterfort retaining walls, the heel slab panels, toe slab panels and vertical slab panels are generally fixed along 3 sides with one side free. For design moment coefficients, α is given in Tables 10.1 and 10.2 for various values of $\dfrac{B}{L}$ ranging from 0.6 to 2.0 and more. From the values of α, L and B, the maximum moments are calculated as $M = \alpha w L^2$.

Spacing (clear) between counterforts is adopted as $0.8\sqrt{h}$ to $1.2\sqrt{h}$

where, h is height of backfill. Thickness of various components are given as under:

Thickness of vertical wall $= \dfrac{h_a}{40}$,

Thickness of base slab $= \dfrac{h_a}{30}$,

Thickness of counterforts $= \dfrac{L}{10}$

These thicknesses are rounded to certain pragmatic values.

Reinforcements are provided on the tension faces in various components of the retaining walls. Reinforcements are designed for maximum ultimate moments and part of this steel is curtailed where it may not be required but such curtailed steel is extended by development length.

Reinforcement A_{st} is computed from the solution of following quadratic equation with usual notations as:

$$A_{st} = \frac{0.5 f_{ck} b \cdot d}{f_y} \left[1 - \sqrt{1 - \frac{4.6 \times M_u}{f_{ck} b (d)^2}} \right].$$

PRACTICE QUESTIONS

(I) Objective Questions

Q. 10.1 Select the correct response given after each statement to complete it correctly and fill in the response sheet provided.

  (i) Which of the following is not a retaining wall type?
    (a) simply supported wall   (b) gravity wall
    (c) cantilever wall   (d) buttress wall

  (ii) What will be the lateral force intensity on the retaining wall subjected to a uniform surcharge w_s (kN/m²) on the top horizontal surface if usual notations are used?
    (a) lateral pressure of the same intensity w_s kN/m²
    (b) variable lateral pressure with 0 at top and maximum w_s kN/m² at bottom
    (c) lateral pressure intensity $p = \dfrac{w_s}{2} k_p$
    (d) lateral pressure of uniform intensity $p = k_a w_s$

  (iii) What will be the total lateral force acting on a retaining wall with horizontal backfill of soil with unit weight w and h metre height above base with usual notations?
    (a) $F = \dfrac{wh}{2}$ (b) $F = wh^2$
    (c) $F = \dfrac{k_a wh^2}{2}$ (d) $F = \dfrac{k_a wh^3}{6}$

  (iv) When a retainingd wall retains earth at a slope α with the horizontal, with the soil properties of unit weight w (kN/m³) and angle of repose ϕ, what will be intensity of active pressure (p_a) at a depth of y metres below top?
    (a) $wy\alpha \left[\dfrac{1 - \sin\phi}{1 + \sin\phi} \right]$

    (b) $wy \cos\alpha \left[\dfrac{\cos\alpha - (\cos^2\alpha - \cos^2\phi)^{0.50}}{\cos\alpha + (\cos^2\alpha - \cos^2\phi)^{0.50}} \right]$

(c) $wy \cos \alpha \left[\dfrac{\sin \alpha - (\sin^2 \alpha - \sin^2 \phi)^{0.50}}{\sin \alpha + (\sin^2 \alpha - \sin^2 \phi)^{0.50}} \right]$

(d) $wy \sin \alpha \left[\dfrac{\cos \alpha + (\cos^2 \alpha - \cos^2 \phi)^{0.50}}{\cos \alpha - (\cos^2 \alpha - \cos^2 \phi)^{0.50}} \right]$

(v) If a cantilever wall has total gravitational load $= \Sigma W$ on the base and the coefficient of friction between soil and concrete base is μ, the factor of safety (FS) against sliding failure shall be ...., if the depth of shear key is h_k below top of toe slab and soil has angle of repose ϕ and unit weight w, with height of backfill h_a above bottom of base slab.

(a) $F_s = \dfrac{2\mu \Sigma W}{\left(k_a w h_a^2 - w h_k^2 k_p\right)}$

(b) $F_s = \dfrac{w(k_a h_a^2 - k_p h_k^2)}{2(\mu \Sigma W)}$

(c) $F_s = \dfrac{2\left(\mu \Sigma W + \dfrac{w}{2} h_k^2 k_p\right)}{(k_a w h_a^2)}$

(d) $F_s = \dfrac{2\mu \Sigma W}{k_a w h_a^2}$

(vi) The clear spacing of the counterforts may generally be taken as $L = $ .... where the height of back fill above bottom of base is h_a.

(a) $L = \sqrt{0.8 h_a}$ to $\sqrt{1.2 h_a}$

(b) $L = 0.8\sqrt{h_a}$ to $1.2\sqrt{h_a}$

(c) $L = 0.8 h_a$ to $1.2 h_a$

(d) $L = h_a \sqrt{0.8}$ to $h_a \sqrt{1.2}$

(vii) In cantilever retaining walls, the vertical stem develops tension on ...., while toe slab develops tension on ....

(a) backfill side, top side

(b) front side, bottom side

(c) backfill side, bottom side

(d) front side, top side

(viii) Passive earth pressure on retaining walls due to earth height (h) m, unit weight w (kN/m³) and angle of repose ϕ shall be $p_p = $ ....

(a) $\left(\dfrac{1 - \sin \phi}{1 + \sin \phi} \right) wh$

(b) $\left(\dfrac{1 + \cos \phi}{1 - \cos \phi} \right) wh$

(c) $\left(\dfrac{1 + \sin \phi}{1 - \sin \phi} \right) wh$

(d) $\left(\dfrac{\cos \phi - \cos^2 \phi}{\cos \phi + \sin^2 \phi} \right) wh$

(ix) In buttress type of counterfort retaining walls, bending moments at critical locations of $L \times B$ panels in case of toe slab panels, heel slab panels and wall slab panels are computed with the help of coefficient α given for uniformly distributed load w (kN/m²) for different ratios of $\dfrac{B}{L}$ by formula $M = $ ....

(a) $\alpha \dfrac{wL^2}{B}$ (b) $\dfrac{\alpha \cdot wB^2}{L}$

(c) αwLB (d) $M = \alpha wL^2$

(x) Counterfort retaining walls are most suitable for backfill heights of

....

(a) 1 to 4 metres   (b) 4 to 5 metres

(c) 6 to 10 metres   (d) More than 20 metres

Response sheet to Q. 10.1 (i to x)

Question	(i)	(ii)	(iii)	(iv)	(v)	(vi)	(vii)	(viii)	(ix)	(x)
Response (a/b/c/d)										

(II) Numerical Questions

Q. 10.2 Design a T-shaped cantilever retaining wall for retaining 5 m high earth above the ground level. Properties of earth are:

Unit weight of earth w = 15 kN/m³, Angle of repose ϕ = 30°, coefficient of friction μ = 0.50, SBC of soil q_o = 150 kN/m². Use M20 grade CC and Fe 415 grade steel.

[Hint: Depth of foundation = 1.20 m, Total height of backfill = 6.20 m above base. Assume base width b = 3.25 m, toe slab projection = 0.90 m, heel slab = 1.9 m,

stem thickness = 0.45 m, $k_a = \dfrac{1}{3}$, k_p = 3, stem $A_{st}(M_u)$ = 20ϕ @ 150 c/c (A_{st} =

2093 mm²/m), stem thickness 450 mm to 200 mm, heel thickness 450 mm to 200 mm, A_{st} = 20 @ 150 c/c, distribution bars 8 @ 300 mm c/c,

toe slab: 16ϕ 200 c/c, D = 450 mm to 200 mm, shear key 450 mm × 450 mm.]

Q. 10.3 Design a T-shaped cantilever retaining wall to retain earth embankment 3 m high above GL. The unit weight of earth is w = 18 kN/m³ and its angle of repose is 30°. The embankment is horizontal at its top. The safe bearing capacity of soil q_o = 100 kN/m² and the coefficient of friction between soil and concrete base slab μ = 0.50. Use M20 mix concrete and Fe 415 grade steel bars.

[Hint: Minimum depth of foundation = 1.0 m, Height of backfill above base = 4 m,

$k_a = \dfrac{1}{3}$, base width b = 0.4 to 0.6 H = 2.40 m, toe width = 0.9 m , thickness of

base = $\dfrac{H}{12}$ = 0.3 m, stem thickness = 0.30 (top thickness = 0.20 m), safe against

overturning (F_S = 2.97 > 2), unsafe in sliding, provide shear key.

e = 0.16 m, p_{1max} = 70.6 kN/m² (<100 kN/m²), p_2 (min) = 30.26 kN/m²,

p_3 (junction toe) = 55.5 kN/m², p_4 (junction heel) = 50.45 kN/m².

Toe design (0.90 m): D_1 = 260 mm, D_2 = 200 mm, A_{st} = 566 mm², 12ϕ @ 180 c/c (A_{st} = 628 mm²/m)

Distribution steel: Average 276 mm² (8ϕ 180 c/c)

Heel Design (1.20 m): D = 260 mm (edge 200 mm), A_{st} = 627 mm² (12ϕ @ 180 c/c, A_{st} = 625 mm²/m)

Distribution bars 276 mm² average (8ϕ 180 c/c, A_{st} = 278 mm²/m), τ_v = 0.20 N/mm² < τ_c.

Stem design: D = 310 mm to 200 mm, A_{st} = 1007 mm²/m (12ϕ 90 c/c, A_{st} = 1256 mm²/m) alternate bars continued in toe bottom 12ϕ 180 c/c.

Average distribution bars = 306 mm²/m (8ϕ 150 c/c and 8ϕ 300 c/c in outer face)

Shear key: 300 mm × 300 mm.]

Q. 10.4 Design a cantilever retaining wall to retain earth embankment 4 m high above GL. The unit weight of earth if 18 kN/m³, its angle of repose ϕ = 30°, SBC of soil q_o = 200 kN/m² and coefficient of friction μ between soil and concrete base = 0.45. The embankment is horizontal at the top. Use M20 and Fe 415.

[**Hint:** Depth of foundation = 1.20 m, depth of earthfill above base h = 5.20 m

Assume stem D = 450 mm, d = 400 mm, height above base slab = 4.75 m, base slab = 450 mm, M_u (stem max) = 160.8 kNm/m, $d_{required}$ = 241.4 mm, adopt d = 400 mm (OK).

Minimum temp steel = $\dfrac{0.12}{100}$ × 450 × 1000 = 540 mm²/m in each direction.

$$A_{st} = \frac{10000 \times 400}{415} \left[1 - \sqrt{1 - \frac{4.6 \times 160.8 \times 10^6}{20 \times 1000(400)^2}} \right] = 1187 \text{ mm}^2/\text{m}$$

(Provide vertical bars on earth side 16ϕ @150 c/c, A_{st} = 1340 mm²)

(Curtail to 16ϕ 300 c/c at top)

Provide longitudinally on each face 8ϕ 180 c/c. Also provide on outer face 10ϕ 200 c/c vertical bars.

Design base slab: b = 3.0 m, toe projection 1.0 m, heel projection 1.55 m

Resultant pressure at 1.23 m from toe end A. eccentricity e = 0.27 m.

Net ultimate pressure on base slab: p_{max} (toe) = 152 kN/m²↑ and p_u junction = 114 kN/m²↑, M_u(toe) = 82.3 kN-m/m, A_{st} = 588 mm²/m (12ϕ 190 c/c in bottom face and 10ϕ 250 c/c in top face)

Also provide longitudinally 10ϕ 250 c/c on each face.

Heel slab: Ultimate net loading downward and M_{umax} = 88.5 kN-m/m

A_{st} (Heel top face) = 634 mm²/m (provide 12ϕ 170 c/c in top face, A_{st} = 665 mm²/m, provide 10ϕ 250 c/c on bottom face)

Also, provide longitudinally 10ϕ 250 c/c on each face.

$$\text{FOS in sliding} = \frac{0.45 \times 219.7}{81.12} = 1.22 < 1.55, \text{ provide shear key below base slab.}$$

Shear key 450 mm wide × 750 mm deep. Continue top steel of heel in to key.

Also continue on the other face, stem main steel alternate bars i.e. 16ϕ 300 c/c and continue 16ϕ 300 c/c from stem to toe bottom face.]

Q. 10.5 Design a counterfort type of retaining wall to retain 6 m earth above GL. Soil properties are w (unit weight) = 16 kN/m³, ϕ (angle of repose) = 30°, q_o(SBC) = 160 kN/m². Spacings of counterforts may be kept 3.0 m clear. Use M20 grade CC and Fe 415 grade HYSD steel bars.

[Hint: Total height of earth above base = 7.2 m, thickness D of base = 450 mm, D (stem) = 220 mm, base slab width b = 4.50 m, toe projection = 1.0 m heel projection = 3.28 m.

Design of stem: Height = 6.75 m above base top, ultimate max. moment M_u (as continuous slab) loading as pressure = $K_a.wh$ = 36 kN/m²

$$M_u = 1.5 \, (K_a.wh) \, \frac{L^2}{12} = \frac{36 \times 3^2 \times 1.5}{12} = 40.5 \text{ kN-m/m}, d = 175 \text{ mm}, D = 220 \text{ mm},$$

A_{st} = 700 mm² (provide 12ϕ @ 150 mm c/c, A_{st} = 754 mm²)

$$\text{Temp. steel} = 0.120 \times 220 \times \frac{1000}{100} = 264 \text{ mm}^2/\text{m} \ (8\phi \ 180 \text{ c/c}, A_{st} = 278 \text{ mm}^2/\text{m})$$

Base slab:

ΣWt = 442 kN, ΣM_A − 331.78 = 836.47 kN-m/m, $\bar{x}_a$ = 1.8925 m

e = (2.25 −1.8925) = 0.3575M, p_{max} = 145 kN/mm²< 160, OK

p_{min} = 51.4 kN/m², p_c = 124.2 kN/m², p_D = 119.6 kN/m².

Net ultimate loading pressures:

Toe 1.0 m, upward 201 to 169 kN/m², heel 3.28 m: downward 0 to 102 kN/m²

Spacing of counterforts 3 m (clear): M_u heel (as continuous) = $\dfrac{102(L^2)}{12}$

Considering one direction bending: $M_{u \, (max)}$ = 77 kN/m

$$M_{u \, max} \text{ (toe)} = \pm \frac{201 \times 3^2}{12} = \pm 151 \text{ kN/m}$$

$$A_{st} \text{ (toe)} = \frac{10000 \times 400}{415} \left[1 - \sqrt{1 - \frac{4.6 \times 151 \times 10^6}{20 \times 1000(400)^2}} \right] = 1110 \text{ mm}^2$$

(16ϕ @ 180c/c, A_{st} = 1117 mm²/m)

$$A_{st} \text{ (heel)} = \frac{10000 \times 400}{415} \left[1 - \sqrt{1 - \frac{4.6 \times 77 \times 10^6}{20 \times 1000(400)^2}} \right] = 550 \text{ mm}^2$$

(12ϕ @ 200c/c, A_{st} = 565 mm²/m)

$$\text{Minimum temp. steel} = \frac{0.12}{100} \times 1000 \times 450 = 540 \text{ mm}^2$$

(provide 8ϕ @ 180 mm c/c on each face, where main steel is not exiting)

Design of counterforts: (b = 450 mm), d at bottom = (3280 + 220 − 50) = 3450 mm

$$M_u \text{ due to earth pressure} = 1.5 \left\{ K_u \frac{wh^3}{6} L \right\} = \frac{1.5 \times 16 \times 6.75^3 \times 3.45}{3 \times 6} = 1415 \text{ kN-m}$$

$$A_{st} = \frac{10000 \times (3450)}{415} \left[1 - \sqrt{1 - \frac{4.6 \times 1415 \times 10^6}{20 \times 1000(3450)^2}} \right] = 1145 \text{ mm}^2$$

Min. temp steel = $\frac{0.12}{100} \times 3450 \times 450 = 1863 \text{ mm}^2$

(provide 6 bars of 20 mm, $A_{st} = 1884 \text{ mm}^2$)

Also, provide vertical stirrups connecting heel slab with counterfort and horizontal stirrups to connect counterfort with the stem slab. Front counterfort if provided may be designed suitably.]

Q. 10.6 Design a counterfort retaining wall for retaining 7.0 m high earth above ground level. Properties of earthfill:

w (unit wt.) = 18 kN/m^3, ϕ (angle of repose) = 30°, μ (coefficient of friction) = 0.50, q_o(SBC) = 200 kN/m^2, use M20 grade CC and Fe 415 grade HYSD bars.

[**Hint:** For guidance: Depth of earth above bottom of base slab = 8.3 m, b = 5.4 m (toe = 1.50 m, heel = 3.30 m, D_{stem} = 0.60 m)]

Q. 10.7 L (spacing) = 3 m, w = 18 kN/m^3, q_o = 200 kN/m^2, ϕ = 30°, μ = 0.50, $k_a = \frac{1}{3}$,

k_p = 3, depth of foundation = 1.3 m, H = 8.30 m, Assume b = 5.0 m

stem thickness = 400 mm, toe = heel = 2.3 m, check stability ΣW = 456 kN

moment about toe end = 1010 kN-m, $\bar{x}_a$ = 2.2144 m, e = 0.2856 m,

p_{toe} (max.) = 123 kN/m^2, net ultimate loading A = 169 kN/m$^2\uparrow$ at C junction = 125 kN/m$^2 \uparrow$

$p_{heel(min)}$ = 60 kN/m^2, net ultimate loading at B = 138 kN/m$^2\downarrow$, at junction 95 kN/m$^2\downarrow$

These loadings can be shown in to u.d.l. and triangular along the breadth of toe and heel panels.

For stem panel the net ultimate loading is triangular with zero value at the top.

[**Hint:** Max. lateral active earth pressure 1.5 $(k_a wh) = \frac{18}{3} \times 7.9 \times 1.5 = 71.1$ kN/m^2.

The net ultimate loading is shown in Fig. 10.18.]

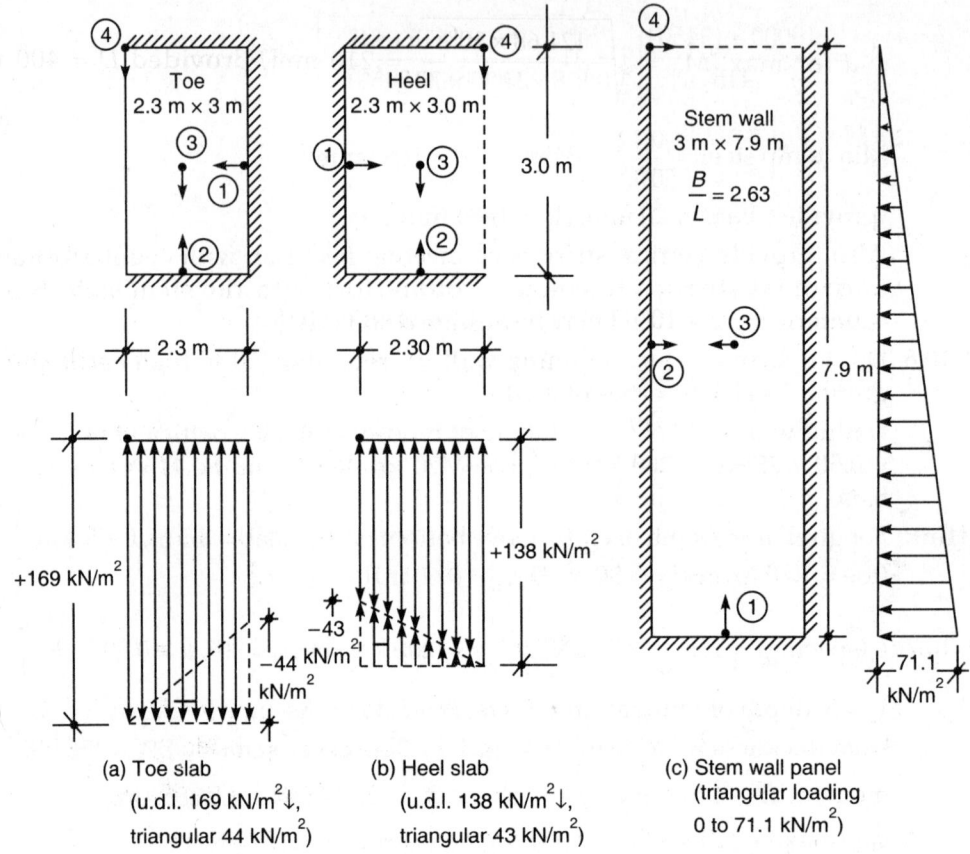

Fig. 10.18: Toe, heel and stem wall panels with loading

Elements		Coefficient moment = $\alpha w L^2$		Moment M
		α_1	α_2	
1. Toe panel points				
	1	$(-0.053, +0.0274)$	$-80.61, +10.85$	-69.76 kN-m
u.d.l.: $169 \times 9 = 1521$	2	$(-0.049, +0.020)$	$-74.53, +7.92$	-66.61 kN-m
Triangular: $44 \times 9 = 396$	3	$(+0.024, -0.0094)$	$+36.50, -3.72$	$+32.78$ kN-m
	4	$(-0.080, +0.0)$	$-121.68, +0$	-121.68 kN-m (maximum)
2. Heel panel points				
	1	$(-0.053, +0.0274)$	$-65.82, +10.60$	-55.22 kN-m
u.d.l.: $138 \times 9 = 1242$	2	$(-0.049, +0.020)$	$-60.86, +7.74$	-53.12 kN-m
Triangular: $43 \times 9 = 387$	3	$(+0.024, -0.0094)$	$+29.81, -3.64$	$+26.17$ kN-m
	4	$(-0.080, +0.0)$	$-99.36, -0.0$	-99.36 kN-m
3. Stem wall panel points				
	1	$(-0.029, +0)$	$-0.029 \times 640, +0$	-18.56 kN-m
	2	$(-0.040, +0)$	$-0.040 \times 640, +0$	-25.60 kN-m
Triangular: $71.1 \times 9 = 640$				
	3	$(+0.021, 0)$	$+0.021 \times 640, +0$	$+13.44$ kN-m
	4	$(-0.0, 0)$	$0.0 + 0.0$	0.0 kN-m

d for max. $M_u = \sqrt{\dfrac{121.68 \times 10^6}{0.138 \times 1000 \times 20}} = 210$ mm, provided $D = 400$ mm,

$d = 350$ mm OK.

$A_{st} = \dfrac{10000 \times 350}{415}\left[1 - \sqrt{1 - \dfrac{4.6 \times 121.68 \times 10^6}{20 \times 350^2 \times 1000}}\right] = 1026$ mm^2, (16ϕ 190 c/c,

$A_{st} = 1058$ mm^2)

Min. $= \dfrac{0.12}{100} \times 1000 \times 400 = 480$ (240 mm^2 each face)

$A_{st} = +264$ mm^2/m (10ϕ 290 c/c each face)

A_{st} (for $M_u = 69.76$) $= 597$ mm^2 (12ϕ 180 c/c, 628 mm^2/m)

Minimum temp. steel in both directions will suffice (10ϕ 280 c/c, 280 mm^2)

Also check for shear stress in usual manner.]

Chapter 11

Water Retaining Structures (Tanks)

LEARNING OBJECTIVES

After the study of *water retaining structures*, the learner will understand the fundamental principles of design of water retaining structures (tanks) and will be able to:

- ⦿ Explain the basic requirements of design of water retaining structures
- ⦿ Describe different types of water retaining structures (tanks) in different situations
- ⦿ Explain the approximate approaches of simple designs of water tanks
- ⦿ Design a circular tank of given capacity placed directly on the ground
- ⦿ Design a rectangular tank of given capacity placed directly on the ground
- ⦿ Design a rectangular tank of given capacity placed underground
- ⦿ Design a circular overhead tank of given capacity and height
- ⦿ Explain the procedure of design of an Intze tank
- ⦿ Design an Intze tank of given capacity

11.1 INTRODUCTION

11.1.1 General

Water is essential for human life and is required for daily consumption. It is required to be stored for consumption in towns, cities, villages, residential colonies, hotels, industries, stations, military settlements, etc. for different purposes. Various types of water tanks are used to meet daily consumption requirements. These tanks used for various purposes can be classified according to their shape and location as follows:

    (a) Circular— placed on the ground or elevated on the staging
    (b) Rectangular— placed on the ground or underground
    (c) Intze tank— elevated (overhead)

    The circular tanks may have flat slab bottom or conical bottom. Rectangular tanks have flat bottom slab both in underground and on the ground.

The dimensions of the tank (height and diameter or length and breadth) are decided on the basis of its volume. Actual height (H) is kept about 200 mm (free board FB) more than the full supply level (FSL) height (h) as calculated from the volume capacity. Structural design is based on the maximum level (actual height H) to safeguard against the worst case of loading for failure.

11.1.2 Method of Analysis and Design

Actual behaviour of the tank components is based on three dimensional interaction of various forces and is quite complex. For accurate and economical design, advanced procedures may be adopted from *bending theory of cylinders* or *plate theory* with different edge conditions. We shall adopt simple approximate methods of design as stated in IS:3370 revised in 1999, wherein various coefficients for bending and shear are specified. IS:3370 (Part IV) provides tables for selecting shear and moment coefficients both for rectangular and cylindrical tank walls.

In approximate method of analysis in cylindrical tanks, it is assumed that the bottom 1 m or 1/3rd the height (whichever is greater) is predominantly under cantilever action. Similarly, in the case of rectangular tank wall panels bottom 1/4th height or 1 m, whichever is greater, is mainly under cantilever action. Rest of the wall shall be resisting water pressures by developing resistance in horizontal directions.

The approximate methods are simple, conservative on safer side and provide guidance of critical structural behaviour. These methods result in slightly uneconomical but safer designs. All engineers and engineering students must learn these approximate approaches for quick understanding of structural behaviour for design of tanks and must possess IS:3370 latest editions for their guidance and reference for design of tanks.

In case of tanks, apart from adequate strength requirement, tank walls must avoid wider cracks to achieve imperviousness. This may be achieved by:

(i) Using richer concrete (viz. M30, M35, etc.)
(ii) Provide a minimum clear cover of 25 mm
(iii) Provide smaller diameter bars at closer interval
(iv) Keep the tensile stresses in concrete low
(v) Adopt good construction practices, e.g. proper mixing, proper compaction and curing, etc.

To avoid leakage problems, limit state method of design should not be used in water tanks. IS:456-2000 is silent about permissible stresses in direct tension in concrete in water tanks.

11.1.3 Design Requirements of Tanks

Follow various parts of Indian standard code IS:3370-1967 revised in 1999. The code is available in four parts.

Part I :   General requirements
Part II :   Reinforced concrete structures
Part III : Prestressed concrete structures
Part IV : Design tables

It may be noted that working stress method is used in design of tanks and the design tables provided in Part IV, IS:3370-1967 are based on working stress method according to earlier version of IS:456 guidelines. Tables 11.1 and 11.2 specify permissible stresses in concrete and steel for use in the design of water tanks.

In design of water tanks to avoid wide cracks, certain minimum steel reinforcement is provided as stated below:

- For thickness up to 100 mm, minimum reinforcement should be 0.3%
- For thickness of 450 mm and above, minimum reinforcement should be 0.2%
- For thicknesses of 100–450 mm, it is reduced linearly from 0.3–0.2%.

Minimum reinforcement shall be ensured in both directions and provided on both faces if the thickness is more than 225 mm. If the total steel provided for bending, shear or tension is more than the minimum in each direction, no additional reinforcement is necessary for distribution of temperature.

For example, minimum steel required for a thickness of 300 mm will be

$$p_{min}(\text{for 300 mm}) = 0.3 - \frac{(300 - 100)}{(450 - 100)} \times 0.10 = (0.3 - 0.057143) = 0.2429\%$$

Table 11.1: Permissible stresses in concrete

Grade of concrete	Permissible stress in tension (N/mm²)		Permissible stress in shear (N/mm²)
	Direct tension	Bending tension	
M20	1.2	1.7	1.7
M25	1.3	1.8	1.9
M30	1.5	2.0	2.2
M35	1.6	2.2	2.5
M40	1.7	2.4	2.7

Table 11.2: Permissible stresses in steel reinforcements

Type of stress	Permissible stress in (N/mm²)	
	Mild steel	HYSD bars
1. Direct tensile stress	115	150
2. Tensile stress in bending		
(i) On liquid retaining face	115	150
(ii) On face away from liquid, if $t < 225$ mm	115	150
(iii) On face away from liquid, if $t \geq 225$ mm	125	190
3. Tensile stress in shear reinforcement		
(i) For members < 225 mm thick	115	150
(ii) For members ≥ 225 mm thick	125	175
4. Compressive stress in columns subjected to direct load	125	175

11.2 DESIGN OF CIRCULAR TANKS RESTING ON GROUND

Circular tanks resting on ground can have *flexible base joint* (Fig. 11.1(a)) or *rigid base joint* (Fig. 11.1(b))

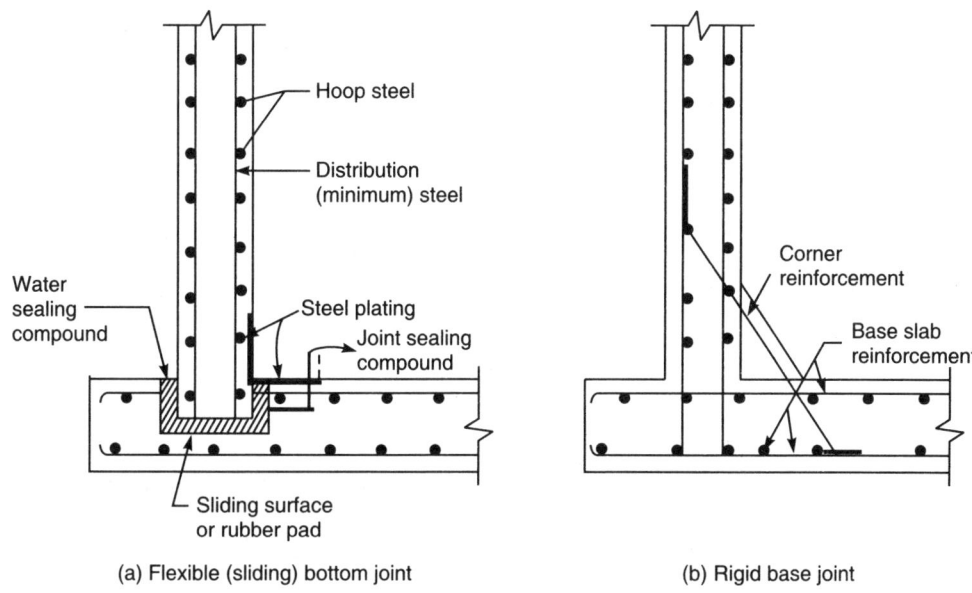

(a) Flexible (sliding) bottom joint   (b) Rigid base joint

Fig. 11.1: Typical circular tank joints

In case of flexible joint, the wall is free to move outward when water pressure acts from inside and the wall is subjected to hoop forces T.

$$\text{Hoop tension } T = \frac{pD}{2} = \frac{wHD}{2} \qquad \qquad \text{... Eq. (11.1)}$$

where, $p = wH = $ Water pressure
                                $w = $ Unit weight of water
                                $H = $ Height of tank
                                $D = $ Diameter of circular tank

The reinforcement for hoop tension is to be given in horizontal direction as ring bars. In vertical direction only minimum reinforcement is to be provided.

$$\text{Hoop tension steel area } A_{st} = \frac{T}{\text{Permissible stress}} = \frac{wHD}{2\sigma_{st}}$$

when concrete tension is neglected   ... Eq. (11.2)

where, $\sigma_{st} = $ Permissible tensile stress in steel (*see* Table 11.2).

In case of rigid joint, bottom 1 m or 1/3rd height acts predominantly as cantilever while the upper portion is subjected mainly to hoop tension. Figure 11.2 shows the approximate loading diagram for cantilever and hoop actions. Let BD = h, then cantilever moment at the base B will be

$$M = \frac{wHh}{2} \times \frac{h}{3} \text{ kN-m/m} \qquad \qquad \text{... Eq. (11.3)}$$

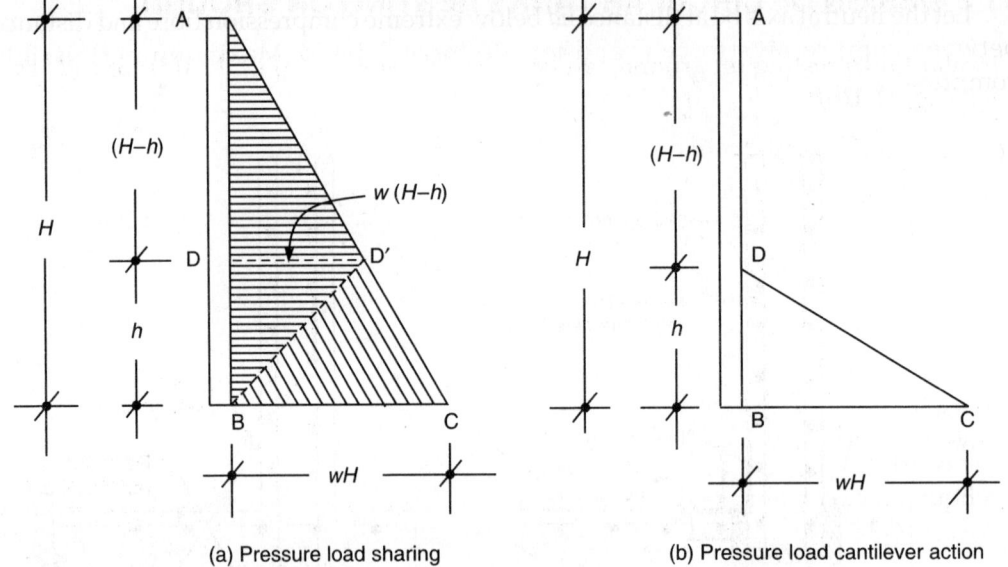

(a) Pressure load sharing   (b) Pressure load cantilever action

Fig. 11.2: Water pressure loading and its sharing

D is the critical point located at $1/3\,H$ or 1 m, whichever is greater.

At D, water pressure = $w\,(H-h)$

At B, water pressure = wH

∴ Max. hoop tension at D is

$$T = w\,(H-h)\,\frac{D}{2} \qquad\qquad \dots \text{Eq. (11.4)}$$

These aspects of design shall be explained in Examples 11.1 and 11.2.

Design Constants

Consider a rectangular section of width b, overall depth D, and effective depth d and steel reinforcement A_{st} as shown in Fig. 11.3.

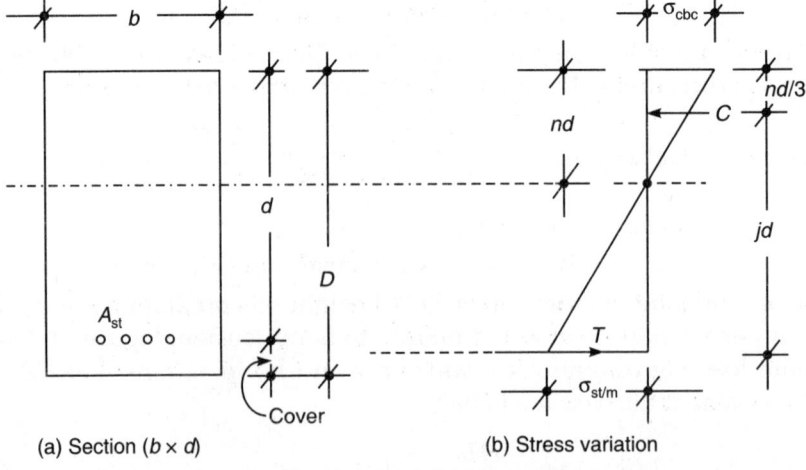

(a) Section ($b \times d$)   (b) Stress variation

Fig. 11.3: Rectangular section ($b \times d$) and stress variation

Let the neutral axis be at distance nd below extreme compression fibre and distance between compressive force C and tensile force T be jd. Maximum permissible compressive stress in concrete $= \sigma_{cbc}$ and maximum permissible stress in steel $= \sigma_{st}$.

$$\text{The critical NA factor } n_c = \frac{m\sigma_{cbc}}{m\sigma_{cbc} + \sigma_{st}} \qquad \qquad \text{... Eq. (11.5)}$$

$$\text{Lever arm} = jd, \text{ where } j = \left(1 - \frac{n_c}{3}\right) \qquad \qquad \text{... Eq. (11.6)}$$

Bending moment = moment of resistance (M_r) of the section, i.e .

$$M_r = kbd^2, \text{ where } k = \frac{\sigma_{cbc}}{2} \times j \times n \qquad \qquad \text{... Eq. (11.7)}$$

$$\text{or} \qquad M_r = \frac{\sigma_{cbc}}{2} b \cdot nd \cdot jd = \frac{\sigma_{cbc}}{2} njbd^2 \qquad \qquad \text{... Eq. (11.7a)}$$

Equivalent concrete section A_e will be

$$A_e = A_c + m\, A_{st} = A_g + (m-1)\, A_{st} \qquad \qquad \text{...Eq. (11.8)}$$

$$\text{where,} \qquad m = \left(\frac{E_s}{E_c}\right) = \frac{280}{3\sigma_{cbc}}, \text{ as per IS:456-2000} \qquad \qquad \text{... Eq. (11.9)}$$

In all designs for water tanks, we shall use above working stress data and equations.

Free Board

In all water tanks, the height of tank is always taken 200 mm more than the height of water in tank. This extra height of 200 mm above full supply level (FSL) is called *free board* (FB). Total height $(h + FB)$ is considered for the design since the water level may rise beyond FSL in free board (FB) depth also. This approach will now be explained through examples (Fig. 11.4).

Example 11.1

Design a circular tank with flexible base and resting on the ground to store 60,000 liters of water. The total depth of the tank may be kept 4.50 m including free board. Use M25 CC and Fe 415 steel.

Solution: $\sigma_{cbc} = 8.5\ \text{N/mm}^2$, $\sigma_{st} = 150\ \text{N/mm}^2$, capacity $V = 60,000\ \text{liters} = 60\ \text{m}^3$

Permissible tensile stress of CC $= 1.30\ \text{N/mm}^2$

$$V = \frac{\pi}{4} D^2 h = 60\ \text{m}^3, h = (4.50 - \text{FB}) = 4.30\ \text{m}$$

$$\frac{\pi}{4}(D^2)\, 4.30 = 60,\ D^2 = \frac{60 \times 4}{\pi \times 4.3} = 17.766,\ D = 4.215\ \text{m} \cong 4.22\ \text{m}$$

Unit weight of water $w = 9.8\ \text{kN/m}^3$

$$\text{Max. hoop tension } T = \frac{wH}{2} D = \frac{9.8 \times 4.5 \times 4.22}{2} = 93.051\ \text{kN/m height}$$

$$A_{st}\ \text{(reqd.)} = \frac{93.051 \times 1000}{150} = 621\ \text{mm}^2\ (12\phi\ @\ 180\ \text{c/c},\ A_{st} = 628\ \text{mm}^2/\text{m})$$

Increase the spacing of rings to $12\phi\ @\ 300\ \text{c/c}$ above 2 m height from bottom of tank.

Design of wall thickness:

Permissible tensile stress = $1.30 \, \text{N/mm}^2$,

$$m = \frac{280}{3 \times 8.5} = 11$$

If t is the thickness of wall, equivalent area of concrete per metre height is

$$A_c = 1000t + (m-1) A_{st} = 1000t + (10) \, 628$$

$$\frac{T}{A_c} \le 1.3 \quad \text{or} \quad \frac{93.051 \times 1000}{(1000t + 6280)} \le 1.30$$

or

$$(1000t + 6280) \ge \frac{93.051}{1.30} \times 1000, \text{ or } t \ge \frac{(71577.7 - 6280)}{1000} = 65.3 \text{ mm}$$

Provide minimum thickness of 100 mm.

Vertical steel:

$$\text{Minimum steel} = \frac{0.30}{100} \times 100 \times 1000 = 300 \text{ mm}^2 \, (8\phi \text{ bars @ 150 c/c}, A_{st} = 333 \text{ mm}^2/\text{m})$$

Design of base slab:

Since the load gets directly transferred to lean concrete base on the ground, a minimum thickness of 150 mm may be provided. Also, a minimum reinforcement of

$$\left\{ 0.30 - \frac{0.10\,(150 - 100)}{(450 - 100)} \right\} = 0.286\% \text{ in each direction may be provided,}$$

i.e. $A_{st} = \dfrac{0.286}{100} \times 150 \times 1000 = 429 \text{ mm}^2/\text{m}$ (half reinforcement near each face, i.e.

8ϕ 230 c/c near both faces (top and bottom) in each direction.

Details of reinforcement:

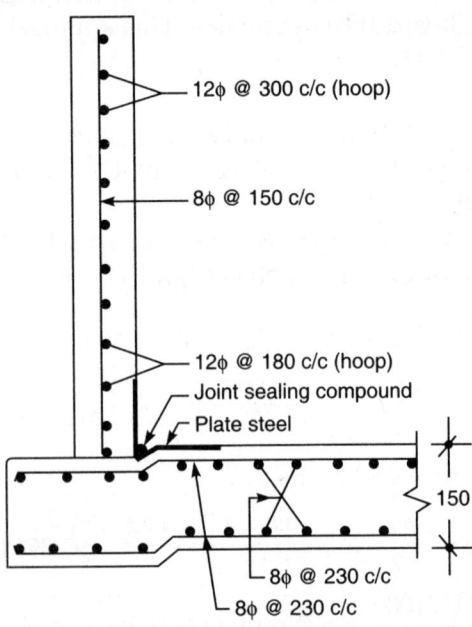

Fig. 11.4: Details of reinforcement (bottom flexible joint circular tank)

Example 11.2

Design the water tank for Example 11.1, with bottom joint rigidly connected to the base slab. All other data remains the same. Use approximate design method.

Solution: Diameter $D = 4.22$ m, $H = 4.5$ m, H' (max. water height) $= 4.30$ m, $\sigma_{cbc} = 8.5$ N/mm^2

$$\sigma_{st} = 150 \text{ N/mm}^2, \, m = \left(\frac{E_s}{E_c}\right) = 11$$

Design constants: $n = \dfrac{m\sigma_{cbc}}{m\sigma_{cbc} + \sigma_{st}} = \dfrac{11 \times 8.5}{11 \times 8.5 + 150} = 0.384, \, j = 1 - \dfrac{n}{3} = 0.872,$

$$k = \frac{1}{2} \times 8.5 \times 0.384 \times 0.872 = 1.423$$

Design for cantilever action in vertical wall:

$$h = \frac{4.5}{3} \quad \text{or} \quad 1 \text{ m whichever is greater} = 1.50 \text{ m}$$

$$\text{Cantilever moment} = \left(wH \cdot \frac{h^2}{6}\right) = \frac{9.8}{6} \times 4.5 \times 1.5 \times 1.5 = 16.54 \text{ kN-m/m}$$

$$d \text{ (balanced section)} = \sqrt{\frac{16.54 \times 10^6}{1.423 \times 1000}} = 108 \text{ mm}$$

To keep the section under reinforced, $D = 170$ mm, $d = 139$ mm $\left(170 - 25 - \dfrac{12}{2}\right)$

$$A_{st} = \frac{M}{\sigma_{st} jd} = \frac{16.54 \times 10^6}{150 \times 0.872 \times 139} = 910 \text{ mm}^2 \ (12\phi \ @ \ 110 \text{ c/c}, A_{st} = 1027 \text{ mm}^2/\text{m})$$

Near inner face keeping clear cover of 30 mm at least.

Provide vertical bars 12ϕ @ 110 c/c near inner face at a clear cover of 30 mm. Curtail these vertical bars beyond 1.50 m height from the base, i.e. beyond 1.50 m height provide 12ϕ 220 c/c (in top 3.0 m height).

Design of wall section for hoop tension:

Max. hoop tension T at a height of 1.50 m above base is

$$T = w \, (H - h) \cdot \frac{D}{2} = 9.8 \, (4.5 - 1.5) \frac{4.22}{2} = 62.03 \text{ kN}$$

$$A_{sto} = \frac{62.03 \times 1000}{150} = 414 \text{ mm}^2 \ (10\phi \text{ bars @ 180 c/c}, A_{sto} \text{ provided} = 436 \text{ mm}^2/\text{m})$$

Check for tensile stress in concrete:

$$\sigma_{ct} = \frac{62030}{A_e} = \frac{62030}{1000 \times 170 + (11-1)436}$$

$$= 0.36 \text{ N/mm}^2 < 1.3 \text{ N/mm}^2, \text{ permitted for M25 concrete.}$$

Thus, 10ϕ @ 150 mm c/c may be maintained up to bottom 1.50 m height and it may gradually be increased to 10ϕ @ 300 c/c in upper 3.0 m height.

Distribution steel in vertical wall in vertical direction:

$$\text{Minimum \% steel} = 0.3 - 0.10\,\frac{(170-100)}{(450-100)} = 0.30 - 0.02 = 0.28\%$$

$A_{st} = \dfrac{0.28}{100} \times 1000 \times 170 = 476 \text{ mm}^2$ (vertical steel provide for cantilever 12ϕ @

220 c/c = 514 mm^2 > minimum, hence OK).

Base slab:

Provide nominal thickness of 150 mm and nominal reinforcement 8ϕ 200 c/c in both directions. Also provide haunches of 150 mm × 150 mm along the junction of base and vertical wall to provide extra rigidity (Fig. 11.5). Minimum steel = 429 mm^2 (8ϕ 220 c/c near each face in both directions, $A_{st} = 2 \times 227$ mm^2).

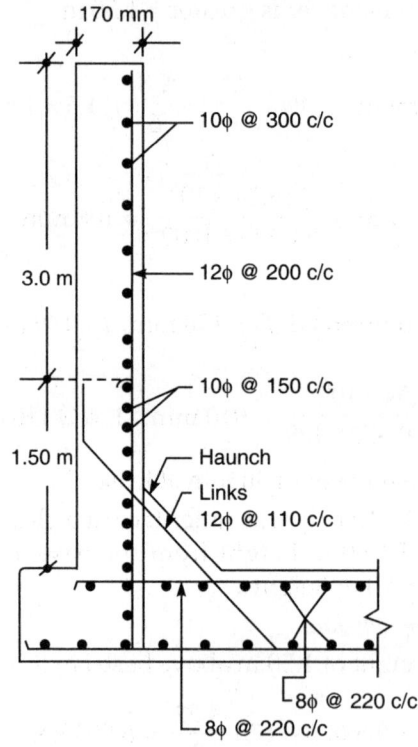

Fig. 11.5: Details of reinforcement of rigidly jointed circular tank

11.3 RECTANGULAR TANKS RESTING ON GROUND

Consider rectangular tanks of size $L \times B \times H$, where

L = Length of tank

B = Breadth of tank

H = Total height of tank including FB

For design of these tanks by approximate methods, the tanks are divided into two categories based on the ratio of L to B as follows:

(i) Design of Tanks with $L/B < 2$

The design of these tanks is similar to that of circular tanks. In this case lower portion of height, 1/4th the total height or 1 m, whichever is greater, is considered predominantly under cantilever action and upper portion to have resistance by horizontal action. The load taken by the two actions is shown in Fig. 11.6(a), where the point D is at a height $\dfrac{H}{4}$ or 1 m, whichever is more. Hence the maximum cantilever moment shall be

$$M = \frac{wHh}{2} \cdot \frac{h}{3} = \frac{wHh^2}{6} \quad \left(\text{where } h = \frac{H}{4} \text{ or 1 m, whichever is more}\right) \qquad \text{... Eq. (11.10)}$$

Horizontal action is caused at h metre above base, where water pressure loading is $P_h = w\,(H-h)$, as shown in Fig. 11.6.

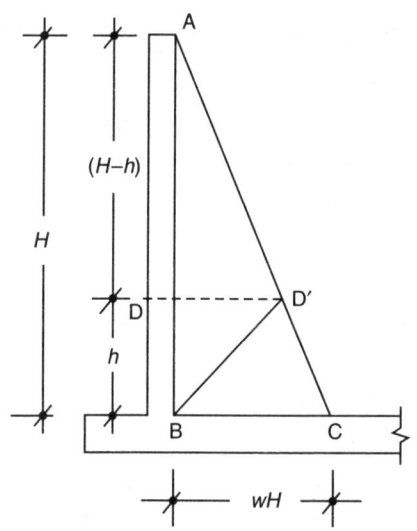

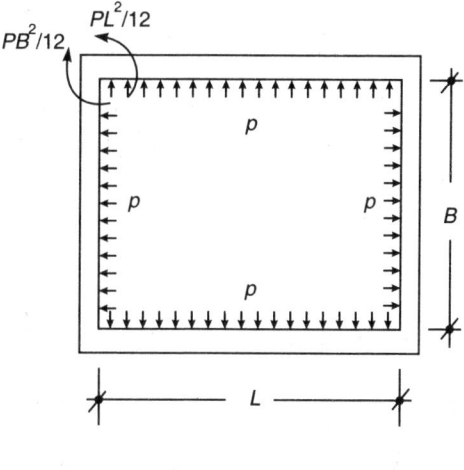

(a) Water pressure loading at critical section   (b) Pressure as lateral loading on tank walls (plan view)

Fig. 11.6: Rectangular tank under cantilever and frame action

The loading due to pressure is resisted by frame action. The frame being symmetrical, the analysis gives the fixed end moments as:

$$M = P_h \frac{B^2}{12} \quad \text{and} \quad P_h \frac{L^2}{12} \qquad \text{... Eq. (11.11)}$$

These moments are balanced by moment distribution.

Since long walls are supported on short walls and short walls are supported on long walls, these walls develop horizontal tensions as:

$$T_L = w\,(H-h)\frac{B}{2} \qquad\qquad \text{... Eq. (11.12)}$$

$$T_B = w\,(H-h)\frac{L}{2} \qquad\qquad \text{... Eq. (11.12a)}$$

The effect of horizontal tensile force is to reduce the net moment in walls such that net moment

$$M = (M_1 - T \cdot x) \qquad\qquad \text{... Eq. (11.13)}$$

where, x is the distance of tensile reinforcement from the centre of the wall.

The bending moment reduces towards top above the critical point D, and the reinforcement may be reduced up to the minimum reinforcement level. It may be ensured that minimum reinforcement is provided at all the sections.

(ii) Design of Tanks with $L/B \geq 2$

In case, where L is more than or equal to twice the width B, the long walls behave mainly like cantilevers of height H. Thickness of walls may be decided on the basis of cantilever moments in long walls. The horizontal steel required in the long wall is to resist direct tension $T_L = w\,(H-h)\dfrac{B}{2}$. Generally, the requirement of tension steel is satisfied with the minimum temperature steel of 0.30%.

Lower portion of short wall of height h is resisting the load by cantilever action and top $(H-h)$ height resist the load by horizontal frame action. Cantilever moment in short wall shall be

$$M = w\,(H-h)\frac{h^2}{6}$$

Due to horizontal frame action, bending moment may be taken equal to

$$M = \pm\,\frac{w(H-h)B^2}{16}\text{, both at ends and middle of the strip span B.}$$

At ends, tension due to moment is on inner face and at centre it is on the outer face. Though long walls are predominantly resisting the load by cantilever action, end 1 m may be considered as supported by short walls, $T_B = w(H-h) \times 1$.

This will reduce the design moment by $T_B.x$, where x is the distance of reinforcement from the centre of the section. The reinforcement is calculated for bending and direct tension separately and total steel is provided in the horizontal direction. The process of design shall be explained by solved examples.

Example 11.3

Design a rectangular tank of size 6 m × 4 m × 3 m depth and resting on firm ground. Use M25 concrete and HYSD steel bars.

Solution: Size 6 m × 4 m × 3 m, CC grade M25, $\sigma_{cbc} = 8.5$ N/mm^2, $m = 11$, $\sigma_{st} = 150$ N/mm^2

Design constants: $n = 0.384$, $j = 0.872$, $k = 1.423$

$\dfrac{L}{B} = \dfrac{6}{4} = 1.5 < 2$, hence both short and long walls resist the load by cantilever action up to $\dfrac{H}{4}$ or 1 m height from the bottom and resist by horizontal action the load in the top $(H - h) = 3 - 1 = 2$ m height. In these cases, moment due to horizontal action is considerable and it governs the design of thickness of walls. Therefore, the horizontal action is considered first.

Horizontal frame action:

The critical section is at a height of 1 m $\left(\text{greater of } \dfrac{H}{4} \text{ and } 1 \text{ m}\right)$.

$P_h = w \, (H - h) = 9.8 \, (3 - 1) = 19.6 \text{ kN/m}^2$

Fixed end moments: $\dfrac{P_h L^2}{12} = \dfrac{19.6 \times 6^2}{12} = 58.8$ kN-m in long wall end, and

$\dfrac{P_h B^2}{12} = \dfrac{19.6 \times 4^2}{12} = 26.13$ kN-m in short wall end.

Since thickness of short and long wall will be same, the moment distribution coefficients shall be:

	Stiffness	Σk	*Distribution factors*
Short wall	$\dfrac{4EI}{L} = \dfrac{4EI}{4} = EI$	$1.667 \, EI$	0.60
Long wall	$\dfrac{4EI}{6} = EI \, (0.667)$		0.40

Moment distribution:

Short wall	Long wall
0.60	0.40
– 26.133	58.80
– 19.60	– 13.07
– 45.73	+ 45.73

Corner moment = 45.73 kN-m

Effective depth $d = \sqrt{\dfrac{45.73 \times 10^6}{1.423 \times 1000}} = 179.3 \text{ mm} \left(D = 180 + 30 + \dfrac{10}{2} = 215 \text{ mm}\right)$

Adopt overall thickness 230 mm (d = 195 mm with effective cover of 35 mm)

Direct pull: $T_L = P_h \dfrac{B}{2} = 19.6 \times \dfrac{4}{2} = 39.2$ kN, and

$T_B = P_h \dfrac{L}{2} = 19.6 \times \dfrac{6}{2} = 58.8$ kN

Distance of tensile steel from the centre $x = \dfrac{230}{2} - 35 = 80$ mm (0.080 m)

Long wall:

Design moment (horizontal) $= 45.73 - 39.2 \left(\dfrac{80}{1000}\right) = (45.73 - 3.14) = 42.60$ kN-m

At corner horizontal reinforcement required is

$$A_{st1} = \frac{42.60 \times 10^6}{\sigma_{st} jd} = \frac{42.60 \times 10^6}{150 \times 0.872 \times 195} = 1670 \text{ mm}^2/\text{m}$$

$$A_{st2} \text{ (direct tension)} = \frac{39200}{150} = 262 \text{ mm}^2/\text{m}$$

Total $A_{st} = 1670 + 262 = 1932 \text{ mm}^2$

(20ϕ 150 c/c, $A_{st} = 2093 \text{ mm}^2/\text{m}$ on inner water face)

Middle span of long wall moment $= \left(\dfrac{p_h L^2}{8} - 45.73 \right)$

$$= \left(\frac{19.6 \times 6^2}{8} - 45.73 \right) = 42.47 \text{ kN-m}$$

Design moment (mid span) $= (42.47 - 39.2 \times 0.080) = 39.33 \text{ kN-m}$

$$A_{st1} = \frac{39.33 \times 10^6}{150 \times 0.872 \times 195} = 1542 \text{ mm}^2$$

$$A_{st2} \text{ (direct tension)} = \frac{39200}{150} = 261.33 \text{ mm}^2$$

Total steel $A_{st} = 1542 + 261.33 = 1803.3 = 1804 \text{ mm}^2/\text{m}$

(20ϕ 150 c/c, $A_{st} = 2093 \text{ mm}^2$ on outer face of wall)

Short wall:

End junction

$M = 45.73 - T_B x = (45.73 - 58.8 \times 0.08) = 41.03 \text{ kN-m/m}$

$$A_{st1} = \frac{41.03 \times 10^6}{150 \times 0.872 \times 195} = 1609 \text{ mm}^2$$

$$A_{st2} \text{ (direct tension)} = \frac{58.8 \times 1000}{150} = 392 \text{ mm}^2$$

Total $A_{st} = 1609 + 392 = 2001 \text{ mm}^2$ (20ϕ @ 150 c/c, $A_{st} = 2093 \text{ mm}^2/\text{m}$)

BM at centre of wall $= \dfrac{9.80(3-1) \times 4^2}{8} - 45.73 = 39.2 - 45.73 = -6.53 \text{ kN-m}$

$$A_{st} = \frac{6.53 \times 10^6}{150 \times 0.872 \times 195} = 256 \text{ mm}^2$$

This is small and the minimum steel takes care of this. Bend alternate bar provided at end at a distance of $\dfrac{B}{4} = 1$ m from the end. Hence at the centre, reinforcement consists of 20ϕ @ 300 c/c on both faces.

Reinforcement in vertical direction:

$$\text{Minimum steel } \% = 0.30 - \frac{0.1(230 - 100)}{350} = 0.263\%$$

A_{st} (minimum) $= 0.263 \times 2300 = 605$ mm²

(10ϕ 250 c/c each face, $A_{st} = 2 \times 314 = 628$ mm²)

$$\text{Cantilever moment} = wH \frac{h^2}{6} = \frac{9.8 \times 3 \times 1^2}{6} = 4.9 \text{ kN-m}$$

$$A_{st} = \frac{4.9 \times 10^6}{150 \times 0.872 \times 195} = 192 \text{ mm}^2 \left(< \text{minimum} \frac{605}{2} \text{ mm}^2 \right)$$

Provide 10ϕ 250 c/c on each face vertical bars (both outer and water faces)

Total A_{st} provided $= 2 \times 314 = 628$ mm²

Base slab:

Reinforcement in joint should be provided as shown below in Fig. 11.7:

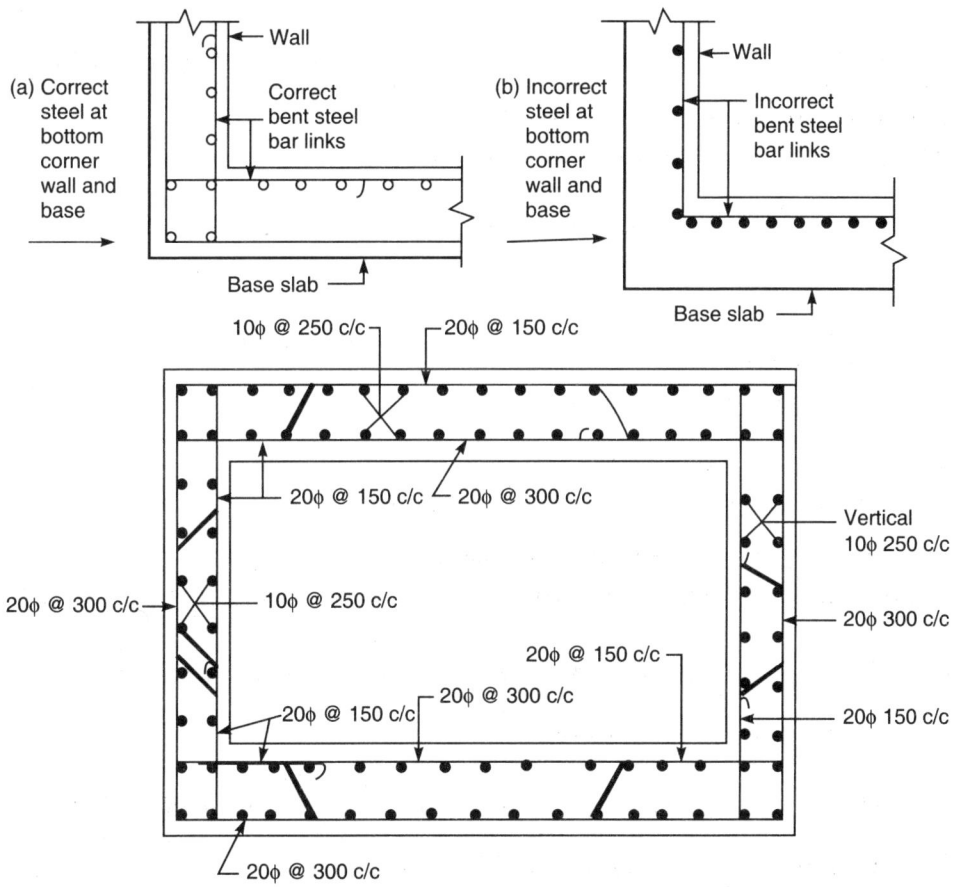

(a) Correct steel at bottom corner wall and base

(b) Incorrect steel at bottom corner wall and base

(c) Plan showing details of reinforcement (junction of walls)

Fig. 11.7: Details of reinforcement of rectangular tank

Provide nominal base slab of thickness 150 mm with 8 mm bars at 230 mm c/c (A_{st} = 435 mm²). For 150 mm thickness, the minimum steel percentage

$$= \left[0.3 - 0.10\frac{(150-100)}{350}\right] = 0.286\%, \text{ i.e. } A_{st} = 0.286 \times 1500 = 430 \text{ mm}^2 \text{ provide on each}$$

face 8ϕ 230 c/c in both direction (A_{st} = 2 × 217.4 = 435 mm²)

Example 11.4

Design an open rectangular tank of size 3 m × 8 m × 4.50 m deep and resting on firm ground. Use M30 grade concrete and Fe 415 HYSD steel.

Solution: Size B = 3 m, L = 8 m, H = 4.50 m, σ_{cbc} = 10 N/mm², σ_{st} = 150 N/mm²

$$\therefore \quad m = \frac{280}{3\sigma_{cbc}} = \frac{28}{3} = 9.33 \text{ N} \cong 9$$

Coefficients:
$$n = \frac{m\sigma_{cbc}}{(m\sigma_{cbc} + \sigma_{st})} = \frac{90}{(90+150)} = 0.375$$

$$j = \left(1-\frac{n}{3}\right) = 0.875, k = \frac{1}{2}\sigma_{cbc} \times n \times j = 1.64, \frac{L}{B} = \frac{8}{3} = 2.67 > 2, \text{ hence the long wall}$$

predominantly behaves as cantilever of 4.50 m height.

Design of long wall: Cantilever moment = M

$$M = \frac{wH^3}{6} = \frac{9.8(4.5)^3}{6} = 148.84 \text{ kN-m/m}$$

$$d = \sqrt{\frac{M}{kb}} = \sqrt{\frac{148.84 \times 10^6}{1.64 \times 1000}} = 301.25 \text{ mm, assume effective cover} = 35 \text{ mm}$$

D = 301.25 + 35 ≅ 345 mm, d = 310 mm

$$A_{st} = \frac{M}{\sigma_{st} \times jd} = \frac{148.84 \times 10^6}{150 \times 0.875 \times 310} = 3658 \text{ mm}^2$$

(22ϕ 100 c/c in vertical direction on inner face)

A_{st} provided = 3800 mm²/m, curtail half bars at a height of h m below top level,

hence $\dfrac{A_{st1}}{A_{st0}} = \left(\dfrac{h}{H}\right)^3 = \dfrac{1}{2}$, $h^3 = \dfrac{1}{2}$ (4.5)³ or h = (40.563)$^{1/3}$ or h = 3.44 m from top, i.e.

(4.50 – 3.44) m = 1.06 m above bottom.

Actual curtailment of alternate bars = 1.06 m + $\dfrac{12 \times 22}{1000}$ = 1.324 m = 1.35 m (say),

i.e. curtail alternate bars at height of 1.35 m above base.

Provide on outer face vertical bars 8ϕ @ 120 c/c (A_{st} = 417 mm²/m) as minimum temperature steel.

Horizontal bending in long walls:

$$\left(As \frac{L}{B} > 2, \text{ bending in vertical direction is main}\right)$$

Direct horizontal tensile force transferred from short wall at the critical point,

$$T_L = w(H - h) \frac{B}{2}, \text{ where } h = \frac{H}{3} \text{ or 1 m, whichever is more, } h = 1.5 \text{ m}$$

$$\therefore T_L = \frac{9.8 \times (4.5 - 1.5) \times 3}{2} = 44.1 \text{ kN}, A_{st} = \frac{44.1 \times 1000}{150} = 294 \text{ mm}^2/\text{m} < \text{minimum.}$$

$$\text{Minimum temperature steel} = 0.30 - \frac{0.1(245)}{350} = 0.23\% \text{ MS and}$$

$$0.23 \times 0.8 \text{ HYSD} = 0.184\%,$$

i.e. on two faces (HYSD) $A_{st} = \dfrac{0.184}{100} \times 345 \times 1000 = 635 \text{ mm}^2$ (each face 318 mm^2/m)

Provide minimum 8ϕ @ 120 c/c horizontally on each face (inner and outer) and 8ϕ @ 120 c/c vertically on outer face only ($A_{st} = 417 \text{ mm}^2/\text{m}$) as inner face has main vertical reinforcement.

Design of short wall:

Reinforcement in vertical direction (cantilever up to critical point 1.50 m)

$$M = \frac{wHh^2}{6} = \frac{9.8 \times 4.5 \times (1.5)^2}{6} = 16.54 \text{ kN-m}$$

$$A_{st} = \frac{16.54 \times 10^6}{150 \times 0.875 \times 310} = 406 \text{ mm}^2 \le 417 \text{ mm}^2/\text{m each face}$$

Provide vertically on each face 8ϕ 120 c/c ($A_{st} = 417 \text{ mm}^2/\text{m}$)

Horizontal reinforcement in short wall (B = 3 m):

P_h(1.5 m above base) $= w(H - h) = 9.8 (4.5 - 1.5) = 29.4 \text{ kN/m}^2$

$$\text{Bending moment at ends} = \frac{P_h \cdot B^2}{12} = \frac{29.4 \times 3^2}{12} = 22.05 \text{ kN-m/m at ends.}$$

Actual tension (direct) from long wall $T_B = w(H - h) \times 1 \text{ m} = 29.4 \text{ kN}$
(only 1 m length considered to transfer direct tension)

$$A_{st1} \text{ (direct)} = \frac{29.4 \times 1000}{150} = 196 \text{ mm}^2/\text{m}$$

$$A_{st2} \text{ (bending)} = \frac{22.05 \times 10^6}{150 \times 0.875 \times 310} = 542 \text{ mm}^2/\text{m}$$

Total A_{st} reqd. $= 542 + 196 = 738 \text{ mm}^2/\text{m}$ (12ϕ @ 150 c/c, $A_{st} = 753 \text{ mm}^2$)
Provide 12ϕ 150 c/c horizontally on inner face near ends.

Middle span:

$$\text{Moment at mid span} = w(H-h)\,\frac{B^2}{24} = 9.8\,(4.5-1.5)\times\frac{3^2}{24} = 11.03\ \text{kN-m/m}$$

$$A_{st} = \frac{11.03\times10^6}{0.875\times310\times150} = 271\ \text{mm}^2$$

Total $= 271 + 196 = 467\ \text{mm}^2/\text{m}$ (12ϕ 240 c/c, $A_{st} = 471\ \text{mm}^2/\text{m}$ on outer face) or (12ϕ 300 c/c by bending from inner face $+ 8\phi$ 300 c/c additional,

$$A_{st} = 376 + 167 = 543\ \text{mm}^2,\ \text{OK}).$$

Thus provide:
- Vertically 8ϕ 120 c/c near each face
- Horizontally 12ϕ 150 c/c near inner face near ends. Alternate bars bent towards outer face at 1/4th length from ends
- Horizontally 12ϕ 300 c/c bent plus 8ϕ 300 c/c on outer face in middle portion.

Base slab:

Since base slab is directly resting on ground, hence provide nominal thickness of 150 mm and minimum temperature steel in each direction.

$$\text{Minimum temp. steel \% } = 0.30 - \frac{0.10(150-100)}{350} = 0.286\%$$

$$A_{st} = \frac{0.286}{100}\times1000\times150 = 429\ \text{mm}^2/\text{m}\ (8\phi\ 230\ \text{c/c},\ A_{st} = 434\ \text{mm}^2/\text{m}\ \text{on both faces}$$

and in both directions as mesh).

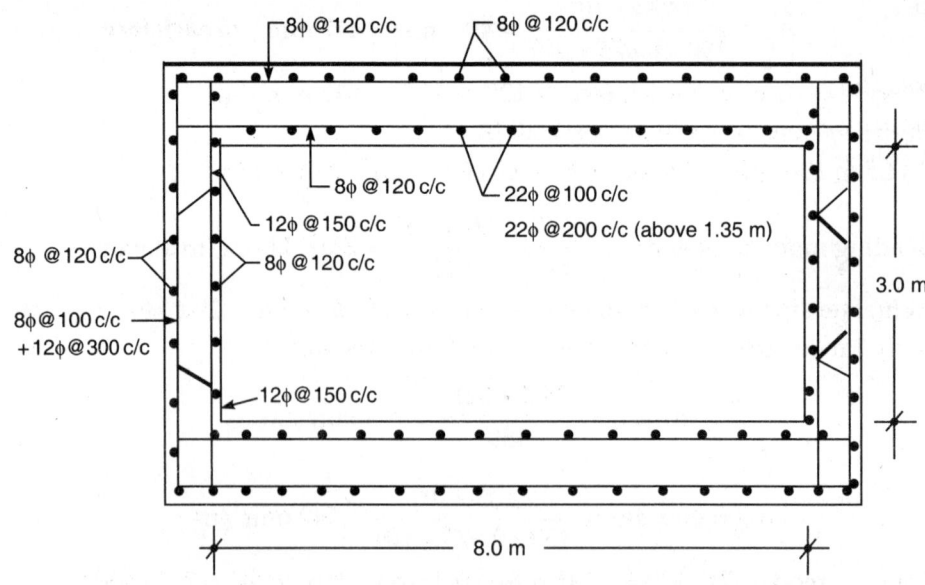

(a) Plan view of a rectangular tank with reinforcement details

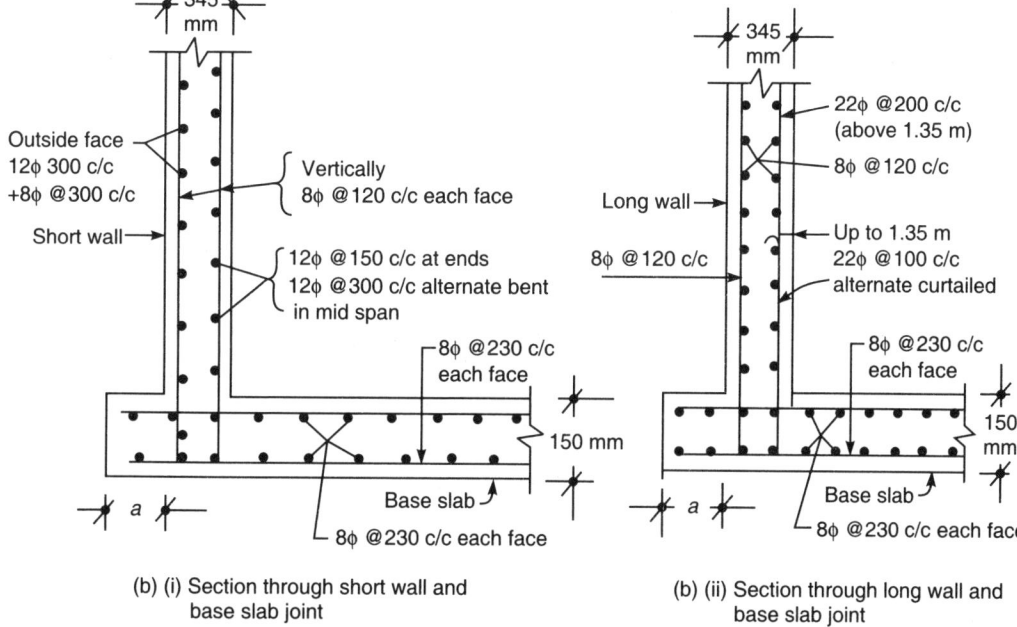

(b) (i) Section through short wall and
base slab joint

(b) (ii) Section through long wall and
base slab joint

Fig. 11.8: Details of reinforcement of a tank resting on ground

11.4 UNDERGROUND TANKS

Underground tanks are used to store water for water supply system of towns, cities, human settlements, or industries, etc. Circular tanks are economical for large capacities, but for lesser capacities, rectangular tanks are better. The design of underground tanks is based on the following two cases of critical loading:

Case I: Tank full and no earthfill on outside.

This is same as a tanks placed on the ground which is already dealt.

Case II: Tank empty and active earth pressure acting from outside.

The active earth pressure on outside may be caused by dry earth or saturated earth or wet cohesionless soil (Fig. 11.9).

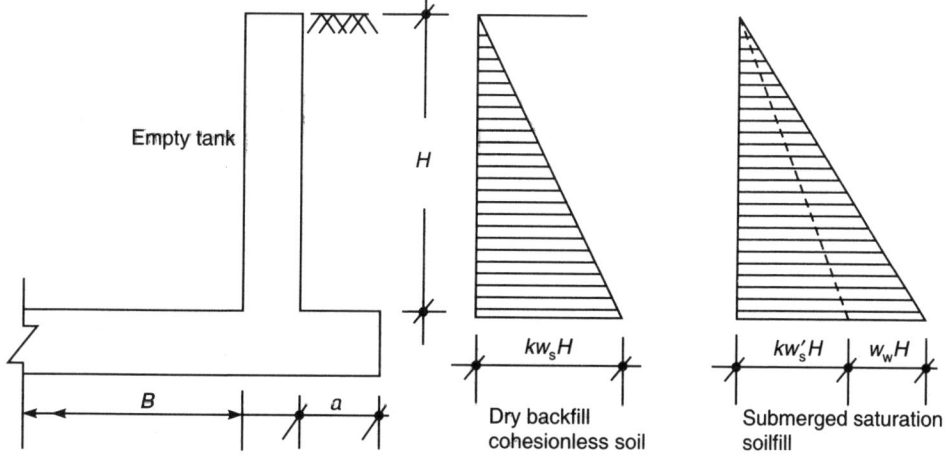

Fig. 11.9: Outside pressure on the empty tank

The maximum active earth pressure due to dry backfill earth:

$$p = k w_s H \qquad \text{... Eq. (11.14)}$$

where, $\quad k = \text{Rankine's coefficient of earth pressure} = \left(\dfrac{1-\sin\phi}{1+\sin\phi}\right)$,

w_s = Unit weight of soil

H = Total height

ϕ = Angle of repose of soil

The maximum active earth pressure due to saturated or wet earthfill will be the sum of pressures due to saturated soil and water pressure from outside.

$$p = k w'_s H + w_w H \qquad \text{... Eq. (11.15)}$$

where, $\quad w'_s$ = unit weight of saturated soil

$w_w = 9.8 \text{ kN/m}^3$ is the unit weight of water.

The designed reinforcement shall be efficiently adjusted for the two types of loading cases. The bottom base slab also needs to be designed and checked for uplift force. To avoid lifting of the tank under uplift force, the bottom slab shall be projected on outside along the periphery so as to increase the weight of empty tank as shown in Figs 11.8 and 11.9. This projection a increases the weight of earth on the tank to resist upward water pressure on the bottom slab.

Underground tanks also require roof slab to cover the tank for keeping water clean. The design of roof slab is similar to the design of roof slabs in building under gravity forces (dead and live).

The design of rectangular tanks is by considering appropriate forces exerted by earth pressure and horizontal frame action. The circular tanks develop hoop forces in horizontal planes. The procedure of design is explained by an illustrative example that follows.

Example 11.5

Design an underground tank of size 3 m × 8 m × 4.50 m for the following data: Earth-submerged sandy soil with unit weight $w_s = 16 \text{ kN/m}^3$, $\phi = 30°$ water table can rise up to GL. Use M30 grade CC for the tank and M20 CC for the roof slab and use steel Fe 415 HYSD grade. Unit weight of water = 9.8 kN/m³, live load on the roof slab = 2.0 kN/m², $H = 4.50$ m.

Solution: *Design of roof slab (M20, Fe 415) with ultimate method:*

Let roof slab be simply supported on the tank walls. The roof slab is 8 m × 3 m in size. As $\dfrac{8}{3} = 2.67 > 2$, hence the roof slab will behave as one way slab.

Span = 3.0 m, $d = \dfrac{3000}{25} = 120$ mm, $D = 120 + 25 + \dfrac{10}{2} = 150$ mm

Permissible stresses: $f_{ck} = 20 \text{ N/mm}^2$ (7 N/mm²), $f_y = 415 \text{ N/mm}^2$ ($\sigma_{st} = 230 \text{ N/mm}^2$), self-wt. = 0.150 × 25 = 3.75 kN/m², finishing plaster = 0.50 kN/mm²,

$w_l = 2.0 \text{ kN/m}^2$, total $w = 3.75 + 0.5 + 2 = 6.25 \text{ kN/m}^2$, w_u (ultimate) = 1.5 × 6.25 =

9.375 kN/m², $M_u = \dfrac{9.375 \times 3^2}{8} = 10.55$ kN-m/m

$M_{u \, lim}$ (limit state design) $= 0.138 \times 20 \times bd^2 = 2.76 \times 1000 \times (120)^2 = 39.45 \times 10^6$ N-mm

$M_{u \, lim} = 39.45 > 10.55$ kN-m, hence the section shall be under reinforced.

$$A_{st} = \frac{0.5 f_{ck} bd}{f_y} \left[1 - \sqrt{1 - \frac{4.6 \times M_u}{f_{ck} \cdot b(d)^2}} \right] = 255 \text{ mm}^2/\text{m}$$

(10ϕ 300 c/c, $A_{st} = 262$ mm^2/m) main in short direction.

Min. temp. steel $= \dfrac{0.12}{100} \times 1000 \times 150 = 180$ mm^2/m (8ϕ 270 c/c, $A_{st} = 185$ mm^2/m)

Design of walls (working stress method):
M30 CC, $\sigma_{cbc} = 10$ N/mm^2, $\sigma_{st} = 150$ N/mm^2
$m = 9$, $n = 0.375$, $j = 0.875$, $k = 1.64$.

Design of long walls: $c_a = \left(\dfrac{1 - \sin 30°}{1 + \sin 30°} \right) = \dfrac{1}{3}$.

With tank empty, $\quad p = c_a w_s H + w_w H = \dfrac{1}{3} \, (16 - 9.8) \, 4.5 + 9.8 \times 4.5, \, w_s' = (w_s - w_w)$

$$= 6.2 \times 1.5 + 9.8 \times 4.5 = 53.4 \text{ kN/m}^3$$

$$M \text{ (cantilever)} = \frac{1}{2} \times 53.4 \times 4.5 \times \frac{4.5}{3} = 180.225 \text{ kN-m}$$

$$d = \sqrt{\frac{180.225 \times 10^6}{1.64 \times 1000}} = 331.5 \text{ mm} = 335 \text{ mm}, D = 335 + 25 + \frac{20}{2} = 370 \text{ mm}$$

$$A_{st} = \frac{180.225 \times 10^6}{150 \times 0.875 \times 335} = 4099 \text{ mm}^2/\text{m} \ (22\phi @ 90 \text{ c/c}, A_{st} = 4222 \text{ mm}^2/\text{m})$$

Curtailing alternate bars at a depth x below top and hence

$$\frac{x^3}{H^3} = \frac{1}{2}, x = (0.5)^{1/3} H, x = 3.57 \text{ m, from bottom} = 4.5 - 3.57 = 0.93$$

Point of theoretical curtailment $= 0.93 + \dfrac{12 \times 22}{1000} = 1.20$ m

Actual curtailment may be done at 1.50 m from the bottom
(i.e. above 1.5 m: 22ϕ @ 180 c/c)
Minimum temperature steel for 370 mm thickness will be

$$= \left[0.30 - \frac{0.10(0.370 - 0.10)}{0.350} \right] 0.80 = 0.223 \times 0.8 = 0.178\% \text{ HYSD bars}$$

Minimum temperature steel $A_{st} = \dfrac{0.223}{100} \times 0.80 \times 370000 = 660$ mm^2 (each face

330 mm^2).

Provide horizontally 8ϕ 150 c/c (A_{st} = 333 mm²) on each face in bottom 2.00 m height and gradually increase the spacing to 8ϕ 250 c/c in top 2.50 m height.

Design when the tank is full and outside earth is not existing:

$P = w_w\, H = 9.8 \times 4.5 = 44.1$ kN/m² (with no outside earth)

Vertical cantilever moment $M = \dfrac{44.1 \times 4.5^2}{6} = 148.8$ kN-m/m in long walls.

Vertical steel $A_{st} = \dfrac{148.8 \times 10^6}{150 \times 0.875 \times 335} = 3395$ mm² (working stress design)

Provide on inner face 22ϕ 110 c/c vertical bars (A_{st} = 3455 mm²/m) for bottom 1.5 m and curtail alternate bars above 1.50 m.

Check for direct tension in long wall:

$T_L = w_w\,(H-h) = \dfrac{B}{2} = \dfrac{9.8 \times 3}{2}\ (4.5 - 1.5) = 44.1$ kN (since $h = $ or 1 m = 1.5 m)

A_{st} reqd. $= \dfrac{44.1 \times 1000}{150} = 294$ mm²/m ht. (<min. temp. steel 660 mm²/m), OK.

Thus, horizontal bars of 8ϕ 150 c/c on each face will be quite adequate and A_{st} = 667 mm²/m).

In upper portion, 8ϕ 250 c/c (A_{st} = 200 × 2 = 400 mm²/m > 294 mm²/m is OK).

Also tapper the wall from 370 mm thickness at 1.50 m height from the bottom to 200 mm thickness at top.

Design of short wall:

Lower $\dfrac{H}{4}$ or 1 m shall be designed for cantilever action whichever is greater, i.e.

$\dfrac{4.5}{4} = 1.125$ m bottom height.

Case I: *When tank is empty and earth is saturated*

p (as in long wall) $= 53.4$ kN/m²,

Cantilever moment $M = \dfrac{53.4 \times 1.125^2}{6} = 11.264$ kN-m/m

Vertical steel (earth face) $= \dfrac{11.264 \times 10^6}{150 \times 0.875 \times 335} = 256$ mm²/m

(<minimum temp. steel 8ϕ @ 150 c/c each face, A_{st} = 2 × 333 mm²/m on 2 faces).

Thus, provide 8ϕ 150 c/c each face (earth and water faces) in vertical direction.

Max. direct compression due to 1 m wide end strip of long wall (at average thrust from outside)

$$P = 53.4 \times 1 \times (4.5 - 1.125) = 180.225 \text{ kN}$$

Stress in concrete $= \dfrac{180.225 \times 1000}{1000 \times 335} = 0.538$ N/mm² (< 10 N/mm²), OK.

Case II: *When the tank is full and no earthfill on outside*:
$$p_h = 9.8 \times 4.5 = 44.1 \text{ kN/m}^2$$

Cantilever moment $M = 44.1 \times \dfrac{1.125^2}{6} \times 1 = 9.30 \text{ kN-m/m}$

$$A_{st} = \dfrac{9.30 \times 10^6}{150 \times 0.875 \times 335} = 242 \text{ mm}^2$$

(Minimum temp. 8ϕ @ 150 c/c, $A_{st} = 333 \times 2$ on 2 faces)

Provided 8ϕ 150 c/c is OK.

Design of top portion $(4.5 - 1.125) = 3.375$ m, $p_h = 9.8 \times 3.375 = 33.1 \text{ kN/m}^2$

Moment in horizontal strip $= \dfrac{33.1 \times 3^2}{12} = 24.8 \text{ kN-m/m}$

$A_{st} = \dfrac{24.8 \times 10^6}{150 \times 0.875 \times 335} = 564 \text{ mm}^2$ (10ϕ 130 c/c, $A_{st} = 604 \text{ mm}^2 > 564 \text{ mm}^2/\text{m}$), OK.

Thus, provide horizontal bars 10ϕ @ 130 on inside face (water face) at ends and increase spacing gradually to 10ϕ 230 c/c at top.

At mid span, the BM is half of 24.8 kN-m, hence provide half steel 10ϕ 260 c/c outer face in mid span point by bending alternate bars on outer face in the mid span points from ends.

Bottom slab:

Assume bottom slab thickness = 0.30 m (300 mm).

H at bottom of slab = 4.5 + 0.3 = 4.8 m

Upward pressure of soil when saturated = $9.8 \times 4.8 = 47.04 \text{ kN/m}^2$

Base slab shall be projected outside the wall so that uplift pressure can be compensated by weight of the earth on the projection (Fig. 11.10).

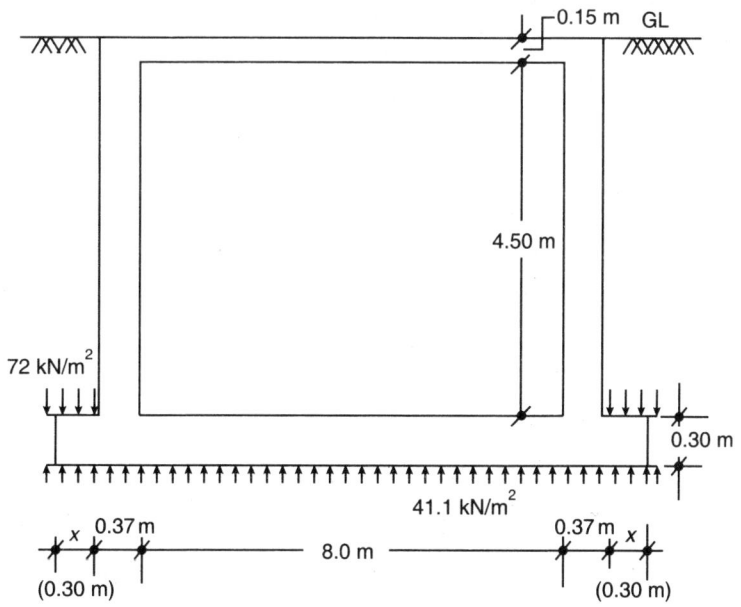

Fig. 11.10: Section of tank with projections (*x*)

Let the base slab be extended by x beyond the outer face of walls in each side. Downward loads and upward lifting water force and to avoid uplift of tank, total weight should be equal or more than the uplift force caused by saturation of soil (Fig. 11.11).

Total weights:

   (i) Weight of top slab $= 0.15 \times 1 \times 25 \, (8 + 2 \times 0.37) \, (3 + 2 \times 0.37) \cong 123$ kN

  (ii) Weight of long walls $= 0.37 \times 25 \times 2 \, (8.74) \, (4.5) = 728$ kN

 (iii) Weight of short walls $= 0.37 \times 25 \times 2 \, (3.0) \, (4.5) = 250$ kN

 (iv) Weight of bottom slab (assume $t = 0.3$ m) $= (8.74 + 2x) \, (3.74 + 2x) \, (0.30 \times 25)$
$$= 245.0 + 187.2x + 30x^2$$

  (v) Weight of earth on the projection $= [(8.74 + 2x) \, (3.74 + 2x) - 8.74 \times 3.74] \, 4.5 \times 16$
$$= 1797.12x + 288x^2$$

$$\text{Total weight} = 1346.0 + 1984.32x + 318x^2$$

$$\text{Uplift force on the bottom} = 9.8 \, (4.5 + 0.30) \, [(8.74 + 2x) \, (3.74 + 2x)]$$
$$= 1537.6 + 1174.12x + 188.2x^2$$

Therefore, $1346 + 1984.32x + 318x^2 \geq 1537.6 + 1174.12x + 188.2x^2$

or $129.8x^2 + 810.2x - 191.6 \geq 0$ or $x^2 + 6.242x - 1.480 = 0$

$$x = \frac{-6.242}{2} \pm \frac{\sqrt{38.96 + 5.92}}{2} = 0.23 \text{ m, say } 0.30 \text{ m}$$

Provide projection of 0.30 m on all the sides, thus base slab will be 9.34 m × 4.34 m.

Net upward pressure $=$ total weight/total area of base

$$= \frac{1346 + 595.3 + 28.6}{(9.34 \times 4.34)} = 48.6 \text{ kN/m}^2$$

Weight of base slab (0.30 m thick assumed) $= 0.3 \times 25 = 7.5$ kN/m

Net upward pressure (empty tank) $= 48.6 - 7.5 = 41.1$ kN/m^2

Weight of earth on projection $= 16 \times 4.5 = 72.0$ kN/m^2

Total downward pressure $= 72 + 7.5 = 79.5$ kN/m^2

Pressure due to submergence on walls $= \dfrac{wh^2}{2} = \dfrac{9.8 \times 4.5^2}{2} = 99.23$ kN/m acts at

$(1.5 + 0.3) = 1.80$ m above base

Cantilever moment at the outer face of wall

$$M = \frac{41.1 \times 0.3^2}{2} + 99.23 \times 1.8 - \frac{79.5 \times 0.3^2}{2}$$

or $M = 176.88$ kN-m/m

Moment at the centre of bottom slab:

Weight of long wall/m wide strip of base slab + half the weight of top slab

$$= 0.37 \times 4.5 \times 1 \times 25 + \frac{1}{2} \, (3 + 0.74) \, 0.15 \times 25$$

$$= 41.63 + 7.01 = 48.64 \text{ kN at centre of long wall.}$$

The moment at the centre of the bottom slab resting on the ground shall be maximum when the tank is full because the weight of water shall be transferred directly on the ground and the net uplift shall not change. The lateral pressure (p) shall cause moment at the face of wall about the bottom of base slab.

$$p = \frac{wH^2}{2} = \frac{9.8 \times 4.5^2}{2} = 99.23 \text{ kN/m width, acting at} \left(\frac{4.5}{3} + 0.3\right) = 1.8 \text{ m}$$

Moment at the centre of base slab (fully saturated and tank empty) M_1:

$$= 99.23 \times 1.8 + \frac{41.1}{2} \times (1.5 + 0.37 + 0.3)^2 - 72 \times 0.3 \left(1.5 + 0.37 + \frac{0.30}{2}\right)$$

$$- 48.64 \times \left(1.5 + \frac{0.30}{2}\right)$$

$$= 178.61 + 96.77 - 43.63 - 81.96 = 149.8 \text{ kN-m/m}$$

M_2 (moment at the centre when tank is full and no earth on outside

$$= \frac{41.1}{2} \times (1.5 + 0.37 + 0.3)^2 - 48.64 \left(1.5 + \frac{0.37}{2}\right) - 99.23 \, (1.8) = 96.77 - 81.96 - 178.61$$

$$= -163.8 \text{ kN-m/m}$$

Greater of 176.88 kN-m, 149.8 kN-m, and $-$ 163.8 kN-m will be 176.88 kN-m (M30 grade $k = 1.64$)

$$d = \sqrt{\frac{176.88 \times 10^6}{1000 \times 1.64}} = 328.41 \text{ mm, thus provide } d = 335 \text{ mm}, D = 370 \text{ mm}$$

$$A_{st} \text{ (cantilever)} = \frac{176.88 \times 10^6}{150 \times 0.875 \times 335} = 4023 \text{ mm}^2/\text{m}$$

(Provide 22ϕ 90 c/c vertically, A_{st} = 4222 mm^2) near bottom.

$$A_{st} \text{ (middle)} = \frac{163.8 \times 10^6}{150 \times 0.875 \times 335} = 3725 \text{ mm}^2/\text{m}$$

(Provide 22ϕ 90 c/c, throughout on water face, A_{st} = 4222 mm^2/m)

$$A_{st} \text{ (middle) bottom face} = \frac{149.8 \times 10^6}{150 \times 0.875 \times 335} = 3407 \text{ mm}^2/\text{m}$$

(Provide 22ϕ 90 c/c in bottom face, A_{st} = 3489 mm^2/m) throughout.

Distribution steel $p_t = 0.3 - \dfrac{(370 - 100)}{(450 - 100)} \times 0.1 = 0.223\%$ MS bars or

(0.223 $\times$ 0.8 = 0.1784% HYSD or 10ϕ 230 c/c)

$$A_{st} = \frac{0.1784}{100} \times 370 \times 1000 = 660 \text{ mm}^2 \text{ (on each face } \frac{660}{2} = 330 \text{ mm}^2 \text{ (HYSD)}$$

(12ϕ 300 c/c HYSD bars, A_{st} = 377 mm^2/m or 10ϕ 230 c/c, A_{st} = 341 mm^2/m)

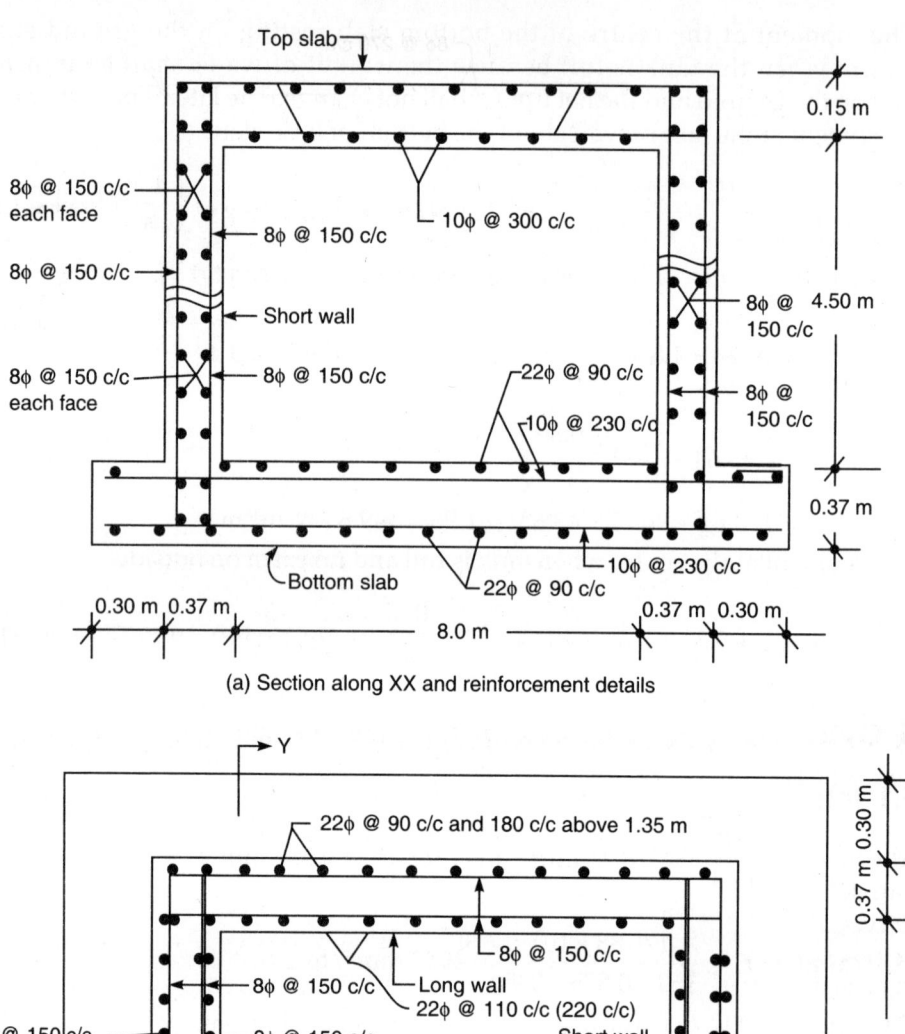

Top slab

0.15 m

8φ @ 150 c/c
each face

10φ @ 300 c/c

8φ @ 150 c/c

8φ @ 150 c/c

Short wall

8φ @ 4.50 m
150 c/c

8φ @ 150 c/c
each face

8φ @ 150 c/c

22φ @ 90 c/c

10φ @ 230 c/c

8φ @
150 c/c

0.37 m

Bottom slab

10φ @ 230 c/c

22φ @ 90 c/c

0.30 m 0.37 m

0.37 m 0.30 m

8.0 m

(a) Section along XX and reinforcement details

Y

22φ @ 90 c/c and 180 c/c above 1.35 m

0.37 m 0.30 m

8φ @ 150 c/c

8φ @ 150 c/c

Long wall

22φ @ 110 c/c (220 c/c)

Short wall

8φ @ 150 c/c

8φ @ 150 c/c

3.0 m

X

X

8φ @ 150 c/c

22φ @ 110 c/c (220 c/c)

0.30 m 0.37 m

22φ @ 90 c/c and 180 c/c above 1.35 m height

Y

0.30 m 0.37 m

0.37 m 0.30 m

8.0 m

(b) Sectional plan and reinforcement details

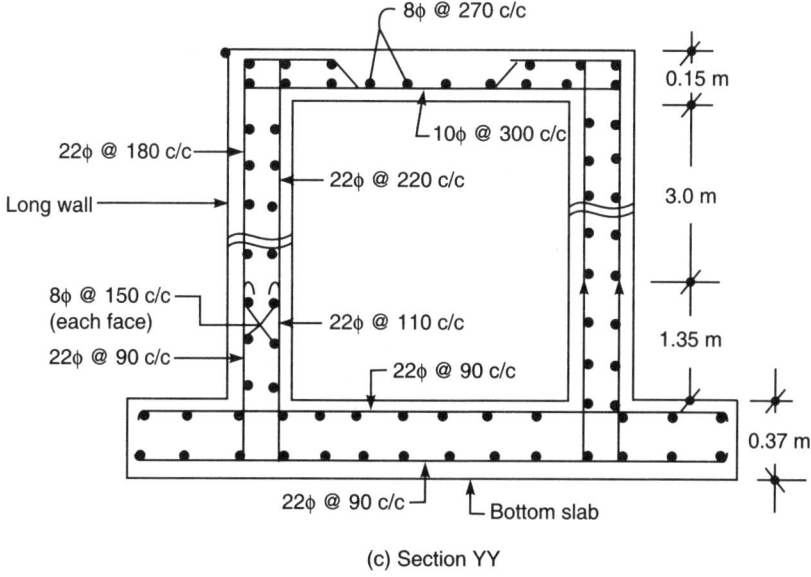

(c) Section YY

Fig. 11.11 (a) – (c): Details of reinforcement in underground tank

11.5 OVERHEAD SERVICE RESERVOIRS (OHSR)

11.5.1 Introduction

Overhead tanks are used to store water for water supply in cities, towns, villages, human settlements, military establishments, industries, etc. Overhead tanks are of three types: *rectangular*, *circular*, and *Intze*. Rectangular tanks are used for smaller capacities up to 50,000 litres. Circular tanks are used for moderate capacities (75,000 to 7,50,000 litres). The diameter of circular tanks generally varies from 5 m to 15 m and height varies from 3 m to 4.50 m. Intze tanks are most economical for capacities more than 7,50,000 litres.

All overhead tanks are covered with top slab and these tanks are supported on staging comprising beams and columns or cylindrical shaft of suitable thickness and diameter offering adequate safety against failure. Top cover slab is generally made domical for light weight and economy.

The joint of top cover slab and tank wall shall be generally considered as hinged for analysis. Tank walls are monolithic with the bottom slab, and hence the tank walls are considered fixed at the base and hinged at the top. Sometimes the bottom slab is also made domical instead of flat.

The exact analysis of the tank elements is quite complex and hence generally approximate approaches are adopted. These approximate methods are more conservative and provide adequate safety margins. Exact designs based on finite element analysis provide more economical sections. IS:3370 (Part IV) reaffirmed in 1999 provides table of coefficients for shear and bending moments at various critical points. Sample Tables 10.1 and 10.2 of coefficients are reproduced for rectangular tank walls. Base slabs of tanks are most heavily loaded when the tanks are full.

Tank elements are designed by *working stress method* to avoid leakage problem. Approximate method shall be explained through solved examples. Rectangular tanks are designed in a similar manner as those placed on the ground with tank full and treating the edges as fixed. Bottom slab carries full load of water when tank is full in addition to the edge moments from wall bottom.

Stresses in working stress method may be taken as σ_{st} = 230 N/mm² for steel Fe 415 and σ_{st} = 190 N/mm² for mild steel. Live load on tanks may be taken as 2 kN/m². Shear and moment coefficients given in IS:3370 (Part IV) may be used for the design of tanks.

11.5.2 Design of Circular Overhead Tanks

Circular tanks are used for a maximum capacity up to 7,50,000 litres and usually covered with domical slab. Researchers found that the optimum rise shall be $\dfrac{1}{7}$ th of the diameter as shown in Fig. 11.12.

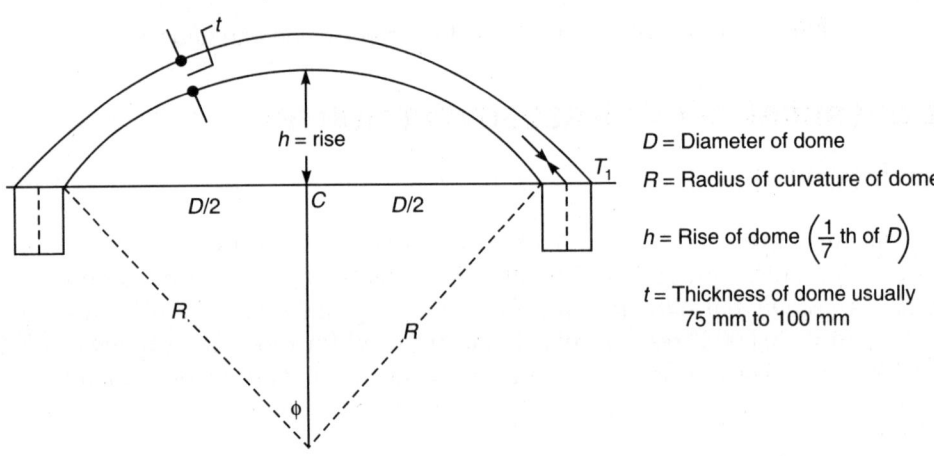

D = Diameter of dome

R = Radius of curvature of dome

h = Rise of dome $\left(\dfrac{1}{7}\text{ th of } D\right)$

t = Thickness of dome usually 75 mm to 100 mm

Fig. 11.12: Dome (cover slab)

Loads on the dome shall be taken as self-weight plus live load of about 1.5 kN/m². Finishing load may be added further to get the total load w on the dome per unit area. Considering total u.d.l. = w kN/m²

According to membrane theory of shells,

meridional thrust: $T_1 = \dfrac{wR}{(1 + \cos\phi)}$ per unit length ... Eq. (11.16)

Circumferential force: $T_2 = wR \left[\cos\phi - \dfrac{1}{1 + \cos\phi}\right]$ per unit length ... Eq. (11.17)

These forces are maximum when $\phi = \theta$, i.e. at junction with top ring beam. The reinforcement is provided for meridional thrust along meridional direction and for circumferential force along the circumference of the dome.

The dome rests on top ring beam and the meridional thrust T_1 causes hoop tension (T_p) in top ring beam.

$$T_p \text{ (Hoop tension in ring beam)} = T_1 \cos\theta \; \frac{D}{2} = \frac{wRD\cos\theta}{2(1+\cos\theta)} \qquad \text{... Eq. (11.18)}$$

Direct tensile stress in concrete should not exceed the values specified in IS code (*see* Tables 11.1 and 11.2). The size of the ring beam shall be found on the basis of these permitted tensile stresses. The moments are found by using various coefficients as specified in Table 11.3.

Table 11.3: Coefficients for bending moment, torsional moment and location of point of maximum torsion in ring beams

No. of column supports (n)	$\phi = \dfrac{360}{n}$	K	K′	K″	α for maximum torsion
4	90°	0.137	0.070	0.021	19.25
5	72°	0.108	0.054	0.150	15.25
6	60°	0.089	0.045	0.009	12.75
8	45°	0.066	0.030	0.005	9.33
10	36°	0.054	0.023	0.003	7.50
12	30°	0.045	0.017	0.002	6.25

Cylindrical tank wall shall be designed for cantilever action in lower h height $\left(\text{greater of } \dfrac{H}{3} \text{ or } 1\,\text{m}\right)$ and for hoop tension in upper portion of tank wall when the tank is full. Because of continuity of wall with the bottom slab, the lower edge cannot be treated as fixed. Along with the slab, the cylindrical wall also rotates and this results in decrease of cantilever moment and increase in the required depth of tank for hoop tension. The exact analysis of slab and tank wall involves cylindrical shell analysis of wall and plate analysis of slab to find the rotations and ensure same values to get continuity. Finite element analysis of tank can also be applied for accurate analysis and design. In approximate analysis lower h height $\left(\text{greater of } \dfrac{H}{3} \text{ or } 1\,\text{m}\right)$ shall be designed for cantilever moment and upper portion for hoop tension. Consider the entire height H for hoop tension. Maximum tension in wall $= \dfrac{wHD}{2}$, M (cantilever)

$$= \frac{wHh^2}{6}$$

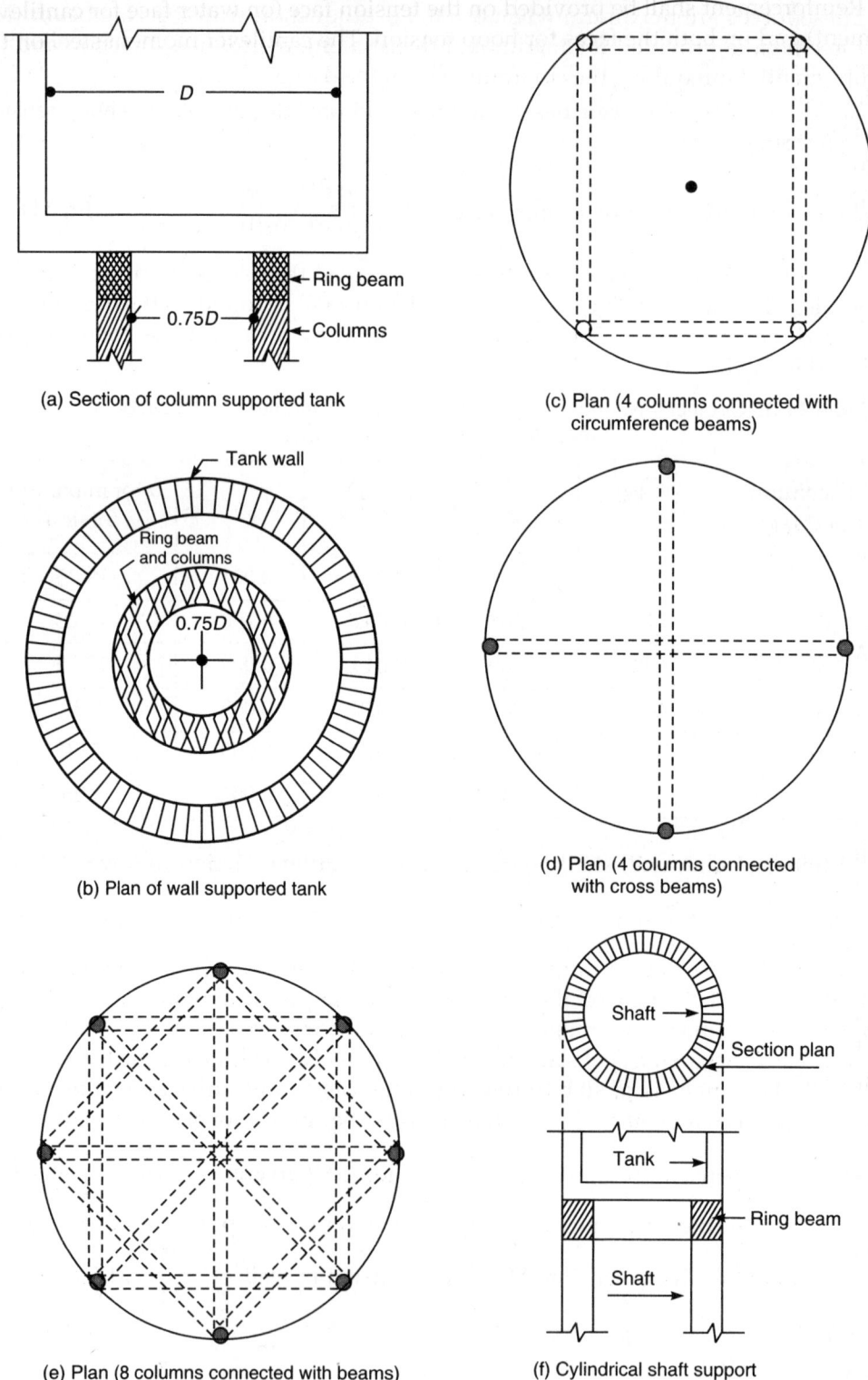

(a) Section of column supported tank

(b) Plan of wall supported tank

(c) Plan (4 columns connected with circumference beams)

(d) Plan (4 columns connected with cross beams)

(e) Plan (8 columns connected with beams)

(f) Cylindrical shaft support

Fig. 11.13 (a) – (f): Different types of supports for circular OHSR

Reinforcement shall be provided on the tension face (on water face for cantilever moment) and on both the faces for hoop tension. The cantilever moment steel on the water face shall be looped in such a way that adequate anchorage length is ensured both in tank wall and bottom slab on water face.

The design of base slab depends on how it is supported, i.e. supported on number of columns and beams joining the columns. Figure 11.13 shows different types of supports for circular OHSR.

Reinforcements shall be not less than the minimum temperature steel (0.12% of cross-section).

(i) Tank supported on wall

Small capacity tanks, required at lesser heights, are supported on circular walls along the periphery of tank. The bottom slabs of these tanks are designed as circular plate subjected to water pressure and self-weight. The end condition may be assumed as simply supported on supporting wall. From circular plate theory, radial and circumferential moments are given by the expressions:

$$M_r = \frac{q}{16} (3 + \mu) (a^2 - r^2) \cong \frac{3q}{16} (a^2 - r^2), \text{neglecting } \mu \qquad \text{... Eq. (11.19)}$$

$$M_\theta = q \left[\frac{a^2}{16}(3 + \mu) - \frac{r^2}{16}(1 + 3\mu) \right] \cong \frac{q}{16} (3a^2 - r^2), \text{neglecting } \mu \qquad \text{... Eq. (11.20)}$$

where, q = Load per unit area = $(w_w H + \text{self-weight})$

a = Radius of bottom slab

μ = Poisson's ratio of concrete slab

r = Radial distance where values are computed

Radial and circumferential reinforcements shall be designed for respective moments. The additional steel is provided at top and bottom of tank wall.

(ii) Tank supported on ring beam

In case of larger tanks, it is economical to support circular tank on base slab cast with a ring beam of diameter equal to 0.75 of the tank diameter as shown in Fig. 11.13a. The ring beam shall be supported on a number of columns spaced at regular intervals. The angle ϕ subtended by the arc between any two consecutive columns at the centre of the ring beam shall be $\phi = \dfrac{360}{n}$, where n is the number of columns supporting the ring beam.

Let w the load per unit run of beam, the shear force at the support will be

$$F_v = \frac{wR}{2} \phi \qquad \text{... Eq. (11.21)}$$

$$\text{Let support moment} = KwR^2\phi \qquad \text{... Eq. (11.22a)}$$

$$\text{Mid span moment} - K'wR^2\phi \qquad \text{... Eq. (11.22b)}$$

and   maximum torsional moment = $K''wR^2\phi$ \qquad ... Eq. (11.22c)

α = the angle at which maximum twisting moment occurs.

The values of K, K', K'', ϕ and α are already given in Table 11.3.

Bottom slab may be analysed by using plate theory. Figure 11.14 shows various loads on the slab. These loads comprise of total weight of dome, top ring beam and wall transferred at the edge of the base slab and uniformly distributed load of $w_w H$ plus self-weight of the slab. The base slab rests on ring beam of radius b. The total load on ring beam from slab may be computed from the total load on the base slab.

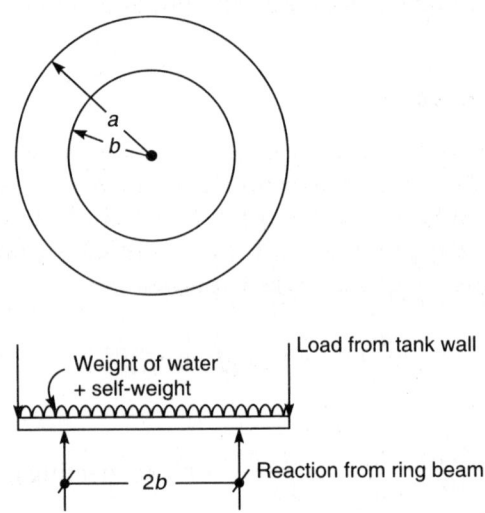

Fig. 11.14: Circular tank with flat bottom and ring beam

Let the total weight on the slab be W. Let the slab be simply supported at its outer edge along the periphery of wall and supports the weight of water ($w_w H$ plus self-weight).

Radius of wall support is b.

Tank wall load gets transferred along the periphery radius a.

According to plate theory the moments are given as follows:

Case I: Circular slab simply supported along outer edge and subjected to load ($w_w H$ plus self-weight of slab)

$$M_r = \frac{q}{16}(3 + \mu)(a^2 - r^2) \cong \frac{q}{16}(a^2 - r^2) \qquad \text{... Eq. (11.23a)}$$

$$M_\theta = \frac{qa^2}{16}(3 + \mu) - \frac{qr^2}{16}(1 + 3\mu) \cong \frac{q}{16}(3a^2 - r^2) \qquad \text{... Eq. (11.23b)}$$

Value of μ is small and hence in approximate analysis it is neglected.

Case II:

$r < b$:

$$M_r = M_\theta = \frac{W}{8\pi}\left[2\log\frac{a}{b} + 1 - \left(\frac{b^2}{a^2}\right)\right] \qquad \text{... Eq. (11.24)}$$

$r > b$:
$$M_r = \frac{W}{8\pi}\left[2\log\frac{a}{r} - \left(\frac{b^2}{a^2}\right) + \left(\frac{b^2}{r^2}\right)\right] \qquad \dots \text{Eq. (11.25)}$$

and
$$M_\theta = \frac{W}{8\pi}\left[2\log\frac{a}{r} - \left(\frac{b^2}{a^2}\right) + 2 - \left(\frac{b^2}{r^2}\right)\right] \qquad \dots \text{Eq. (11.26)}$$

when $r = a, \ M_r = 0$.

(iii) Tank bottom slab supported on four beams (Fig. 11.13c)

Slab between the beams has size $L = 2a \sin 45° = \sqrt{2}\,a$

Square slab of size $\sqrt{2}\,a \times \sqrt{2}\,a = 2a^2$ and it is designed as two way slab.

Beams are adequately reinforced for negative moment near column support.

If W is the total weight of water and self-weight of slab, each beam carries a load of $\frac{W}{4}$. The load is triangular in shape with maximum ordinate at mid span of beam.

Hence, maximum bending moment in the beam is $\frac{W}{4} \times \frac{L}{6} = \frac{WL}{24}$. The beam is designed as T-beam.

The tank wall shall be provided additional reinforcement so as to act as a beam to support own weight. Provide additional horizontal hoop reinforcement at top and bottom of the tank walls.

(iv) Slab supported by two crossed beams (Fig. 11.13d)

In this case, the slab is designed as two way reinforced for a span of 0.45 times the diameter. The beam is designed to carry half the total load. The span of the beam is equal to the diameter of the slab and it acts as a T-beam.

The tank wall acts as a curvedbeam and hence needs additional steel at top and bottom of the tank walls.

(v) Slab supported on a number of beams (Fig. 11.13e)

When the slab is of large diameter (12 m to 15 m), the slab must be supported from several beams joining several columns. The arrangement of beams and columns is shown is Fig. 11.13e. Each panel of slab between the beams is designed as continuous slab. The beams are also designed as continuous beams subjected to triangular loading. The tank wall is provided with additional steel to act as a beam.

Example 11.6

Design a circular OHSR of 9.60 m diameter and total height of 4.2 m. The tank is to be supported on a ring beam of 7.2 m diameter. The ring beam is to be supported by six columns equally spaced along the periphery of beam. Use M30 CC and Fe 415 steel. Design the following components of water tank:

   (i) Top cover dome   (ii) Top ring beam

 (iii) Cylindrical tank wall   (iv) Bottom slab of tank

  (v) Bottom ring beam.

Solution: $D = 9.6$ m, Radius $a = 4.8$ m, $H = 4.20$ m, diameter of bottom ring $= 7.20$ m, radius $b = 3.60$ m, M30 CC, $\sigma_{cbc} = 10$ N/mm², Fe 415 steel, $\sigma_{st} = 190$ N/mm²

(i) *Design of top dome* (refer to Fig. 11.12):

$$D = 9.60 \text{ m}, a = 4.8 \text{ m, rise } h = \frac{D}{7} = \frac{9.6}{7} = 1.37 \text{ m, say } 1.40 \text{ m}$$

$$R = \text{radius of dome, } (2R - h) h = \frac{D^2}{2^2} = 4.80^2$$

$$(2R - 1.4)\, 1.4 = 4.80^2, \text{ or } (2.80\, R - 1.96) = 23.04, \text{ or } R = \frac{(23.04 + 1.96)}{2.80} = 8.93 \text{ m}$$

$$\text{Semi central angle } \theta = \cos^{-1}\left(\frac{8.93 - 1.4}{8.93}\right) = \cos^{-1}(0.8432) = 32.52°$$

Assume thickness of dome = 75 mm

Self wt. of dome = $0.075 \times 25 = 1.875$ kN/m²

Live load $\qquad\qquad = 1.50$ kN/m²

Finishing $\qquad\qquad = 0.50$ kN/m²

$$\overline{w = 3.875 \text{ kN/m}^2 \ (3.90 \text{ kN/m}^2)}$$

$$\text{Max. meridional thrust } T_1 = \frac{wR}{(1 + \cos\theta)} = \frac{3.9 \times 8.93}{(1 + \cos 32.52°)} = \frac{34.83}{1.8432} = 18.90 \text{ kN/m}$$

$$\text{Max. circumferential force } T_2 = wR\left[\cos\theta - \frac{1}{1 + \cos\theta}\right]$$

$$= 3.9 \times 8.93\left[0.8432 - \frac{1}{1.8432}\right] = 10.47 \text{ kN/m}$$

$$\text{Max. stress} = \frac{18.90 \times 1000}{1000 \times 75} = 0.252 \text{ N/mm}^2$$

(< max. permissible compressive for M30 = 10 N/mm²), OK.

$$\text{Nominal reinforcement} = \frac{0.12}{100} \times 1000 \times 75 = 90 \text{ mm}^2/\text{m}$$

(8ϕ 300 c/c, $A_{st} = 167$ mm²/m in both circumferential and meridional directions).

(ii) *Design of top ring beam*

$$\text{Hoop tension} = (T_1 \cos\theta)\,\frac{D}{2} = 18.90 \times 0.8432 \times \frac{9.6}{2} = 76.5 \text{ kN/m}$$

$$A_{st} = \frac{76.5 \times 1000}{190} = 403 \text{ mm}^2 \text{(provide 6 bars of 10}\phi, A_{st} = 471 \text{ mm}^2)$$

Considering $\sigma_{st} = 190 \text{ N/mm}^2$

$$m \text{ (modular ratio)} = \frac{280}{3 \times 10} = 9.3 \cong 9$$

Area of concrete required $(A_c + 9 \times 471) = \dfrac{76.5 \times 1000}{1.5} = 51000$

(Permissible direct tension for M30 concrete $= 1.5 \text{ N/mm}^2$)

$A_c = 51000 - 4239 = 46761 \text{ mm}^2$ (250×300 mm) with 6 bars 10ϕ main reinforcement and 6 mm stirrups @ 225 mm c/c.

(iii) *Design of tank wall*

Depth of water $= 4.20$ m, diameter of tank $= 9.60$ m

Max. hoop tension in the tank wall $= wH\dfrac{D}{2} = (9.8 \times 4.2) \times \dfrac{9.6}{2} = 197.6 \text{ kN/m}$

$A_{st} = \dfrac{197.6 \times 1000}{190} = 1040 \text{ mm}^2/\text{m}$ ($520 \text{ mm}^2/\text{m}$ on each face)

12ϕ @ 200 c/c on each face ($A_{st} = 2 \times 565 = 1130 \text{ mm}^2/\text{m}$), OK.

This is provided near the bottom and then the spacing may be gradually increased towards the top to 12ϕ @ 300 c/c on each face.

Let the thickness of wall $= t$, then

$$\frac{197600}{1000t + (9-1) \times 1130} \leq 1.50, \qquad t \geq 122.7 \text{ mm (provide } t = 150 \text{ mm say)}$$

Cantilever moment:

$$h = \frac{H}{3} \text{ or } 1 \text{ m} = \frac{4.2}{3} = 1.40 \text{ m}$$

$$M \text{ (cantilever moment)} = \frac{wHh^2}{6} = \frac{9.8 \times 4.2 \times 1.4^2}{6} = 13.45 \text{ kN-m/m}$$

For M30 CC and Fe 415 steel,

$\sigma_{cbc} = 10 \text{ N/mm}^2$, $m = 9$, $\sigma_{st} = 190 \text{ N/mm}^2$, $n = 0.321$, $j = 0.893$, $K = 1.433$

$d = (150 - 35) = 115 \text{ mm}$, $A_{st} = \dfrac{13.45 \times 10^6}{190 \times 0.893 \times 115} = 689 \text{ mm}^2$

(12ϕ @ 150 c/c, $A_{st} = 753 \text{ mm}^2$) on inner face, ($12\phi$ 300 c/c on inner face beyond 1.50 m height)

$A_{st} \text{ (min.)} \cong \dfrac{0.30 \times 0.8}{100} \times 150 \times 1000 = 360 \text{ mm}^2$ (180 mm^2 each face)

(8ϕ 270 c/c on outer face vertically)

Inner face: Vertical bars 12ϕ @ 150 c/c near bottom 1.50 m and 12ϕ 300 c/c beyond 1.50 m.

Outer face: Vertical bars 8ϕ @ 270 c/c throughout.

Hoop steel on each face: 12@ 200 c/c on each face near bottom and 12ϕ 300 c/c in upper portion beyond 1.50 m above base (Fig. 11.15).

(iv) *Design of base slab*

$$\text{Total load from cover dome} = T_1 \sin\theta \ 2\pi \times \frac{D}{2}$$

$$= 18.90 \sin 32.52° \times \pi \times 9.6 = 306.5 \text{ kN}$$

$$\text{Weight of ring beam} = 0.25 \times 0.30 \times \pi (9.6) \times 25 = 56.55 \text{ kN}$$

$$\text{Weight of wall} = 0.15 \times (4.2 - 0.30) \ 2\pi \ (4.8 + 0.15) \times 25 = 454.90 \text{ kN}$$

$$\text{Total weight of empty tank} = 306.5 + 56.55 + 454.9 = 817.91 = 818 \text{ kN}$$

$$\text{Weight of water} = wH\frac{\pi}{4}D^2 = 9.8 \times 4.2 \times \frac{\pi}{4}(9.6)^2 = 2980 \text{ kN}$$

Edge of slab:

$$\text{Self-wt. of slab: Assuming slab thickness } t = \frac{D}{35} = \frac{9600}{35} = 275 \text{ mm, say 300 mm}$$

Self-wt. of slab: $0.3 \times 1 \times 1 \times 25 = 7.5 \text{ kN/m}^2$

Total slab diameter $= 9.6 + 0.15 + 0.15 = 9.90 \text{ m}$

$$\text{Total self-wt. of slab} = 0.3 \times \frac{\pi}{4}(9.9)^2 \times 25 = 578 \text{ kN}$$

$$\text{Finishing load} = 0.6 \times \frac{\pi}{4}(9.6)^2 = 44.0 \text{ kN}$$

Total downward load $= 818 + 2980 + 578 + 44.0 = 4420 \text{ kN}$

Total reaction of ring beam $= 4420 \text{ kN}$

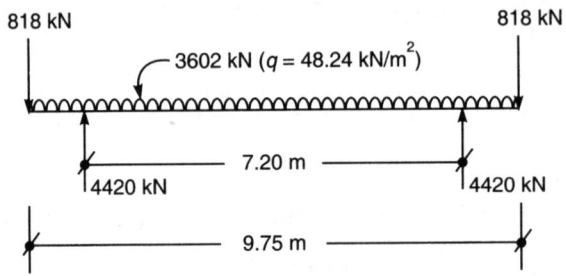

Fig. 11.15: Total load on the base slab and ring beam

The base slab is freely supported by walls and subjected to following loads:

- u.d.l. downward $q = 48.24$, $\left\{ 3602 \times \dfrac{4}{\pi(9.75)^2} \right\}$

- upward ring load $= 4420 \text{ kN}$ (total)

For u.d.l.: $\qquad M_r = \dfrac{3q}{16}(a^2 - r^2)$ and $M_\theta = \dfrac{q}{16}(3a^2 - r^2)$

$$a = \frac{9.75}{2} = 4.875 \text{ m}, r = 0, 1.80 \text{ m}, 3.60 \text{ m}, 4.875 \text{ m}$$

Moment at critical points $\left\{ \dfrac{q}{16} = \dfrac{48.24}{16} = 3.015 \right\}$, $a^2 = 4.875^2 = 23.77$, due to u.d.l. q.

r (in m)	0.0	1.80	3.60	4.875
M_r (in kN-m) $\dfrac{3q}{16}(a^2 - r^2)$	215	186.0	98.0	0.0
M_θ (in kN-m) $\dfrac{q}{16}(3a^2 - r^2)$	215	205.0	176.0	143.0

Due to upward ring load of $W = 4420$ kN, $\{a = 4.875$ m, $b = 3.60$ m, $\dfrac{W}{8\pi} = 175.87$ kN$\}$

For $r \leq 3.60$ m   *For $r > 3.60$ m*

$$M_r = \frac{W}{8\pi}\left\{ 2\log_e \frac{a}{b} + 1 - \frac{b^2}{a^2} \right\} \qquad M_r = \frac{W}{8\pi}\left[2\log_e \frac{a}{r} - \left(\frac{b^2}{a^2}\right) + \left(\frac{b^2}{r^2}\right) \right]$$

0.0	1.8	3.60	4.875
−107.0	−107.0	−107.0	0.0

$$M_\theta = \frac{W}{8\pi}\left[2\log_e \frac{a}{b} + 1 - \left(\frac{b^2}{a^2}\right) \right] \qquad M_\theta = \frac{W}{8\pi}\left[2\log_e \frac{a}{r} - \left(\frac{b^2}{a^2}\right) + 2 - \left(\frac{b^2}{r^2}\right) \right]$$

−107.0	−107.0	−107.0	−160.0

Net moments in the base slab due to downward u.d.l. and upward reaction W:

r (in m)	0.0	1.80	3.60	4.875	Max. value
M_r (in kN-m)	+108.0	+79.0	−9.0	0.0	+ 108.0 at the centre
M_θ (in kN-m)	+108.0	+98.0	+69.0	−17.0	+ 108.0 at the centre

Design moments: $M_r = M_\theta = +108$ kN-m at the centre

$M_r = -9$ kN-m, $M_\theta = -17$ kN-m at 3.60 m and 4.875 m

$$d = \sqrt{\frac{108 \times 10^6}{1.64 \times 1000}} = 257 \text{ mm, provide } d = 265 \text{ mm, } D = 300 \text{ mm}$$

$$A_{st} = \frac{108 \times 10^6}{150 \times 0.875 \times 265} = 3105 \text{ mm}^2/\text{m } (25\phi \text{ 150 c/c,}$$

$A_{st} = 3272$ mm^2/m), considering $\sigma_{st} = 150$ N/mm^2

Thus, provide 25ϕ @ 150 radially plus 25ϕ 150 c/c circumferentially in bottom face in middle portion.

Alternatively, provide a mesh of 25ϕ @ 150 c/c in mutually perpendicular directions in bottom face in middle portion.

$$A_{st} \text{ (– ve) at edges} = \frac{17 \times 10^6}{150 \times 0.875 \times 265} = 489 \text{ mm}^2/\text{m} \text{ (12}\phi \text{ 230 c/c on water face at}$$

corners both radially and circumferentially in overhanging portion 3.60 m to 4.875 m radii, $A_{st} = 491 \text{ mm}^2/\text{m}$.

(v) *Design of bottom ring beam*

Radius $R = 3.6$ m, total load on the ring beam $W = 4420$ kN

$$\text{Load per metre run} = \frac{4420}{2\pi R} = \frac{4420}{2 \times \pi \times (3.6)} = 195.4 \text{ kN/m}$$

Since the ring beam is subjected to torsion, a wider beam shall be preferred (say width = 350 mm).

Consider depth of beam approximately equal to $\frac{1}{15}$ th of diameter,

$$D = \frac{9750}{15} = 650 \text{ mm}, d = 600 \text{ mm, (say)}$$

Self-wt. of beam = $0.35 \times 0.65 \times 1 \times 25 = 5.6875 = 5.7$ kN/m + 0.90 kN/m finishing.

Total load on the beam = $195.4 + 6.6 = 202$ kN/m

Let there be 6 columns supporting the ring beam.

$$\phi = \frac{360°}{6} = 60° \left(\frac{\pi}{3} \text{rad} \right)$$

$$\text{Max. shear at support} = \frac{wR \cdot \phi}{2} = \frac{202 \times 3.6}{2} \times \frac{\pi}{3} = 381 \text{ kN}$$

Refer to Table 11.3 for K, K', K'' and α.

$$\text{Support moment} = Kw \, R^2\phi = 0.089 \times 202 \times 3.6^2 \times \frac{\pi}{3} = 244 \text{ kN-m}$$

$$\text{Mid span moment} = K'w \, R^2\phi = 0.045 \times 202 \times 3.6^2 \times \frac{\pi}{3} = 123.37 \text{ kN-m}$$

$$\text{Max. torsional moment} = K''w \, R^2 = 0.009 \times 202 \times 3.6^2 \times \frac{\pi}{3} = 24.7 \text{ kNm}$$

Max. torsion occurs at α (= 12.75°) joining the columns (consecutive columns). Since this is not in contact with water, the limit state design may be adopted.

Use SP – 16 for design. Let $\frac{d'}{d} = 0.10, d' \cong 0.10 \times 600 = 60$ mm (cover)

$$\frac{M_u}{bd^2} = \frac{1.5(244 \times 10^6)}{350 \times 600^2} = 2.905, \text{ [from Table 4 of SP – 16 (M30, Fe 415), page 50].}$$

$p_t = 0.922\%$, p_c = nominal two anchor bars as the beam with singly reinforced is sufficient.

$$A_{st} = \frac{0.922}{100} \times 350 \times 600 = 1936 \text{ mm}^2 \text{ (provide } 22\phi \text{ 6 bars, } A_{st} = 2280 \text{ mm}^2\text{)}$$

Provide 6 bars 22 mm as tensile steel and 2 bars 22 mm as compression steel.

A_{st} provided at top near supports = 2280 mm^2

A_{sc} provided at bottom near support = 628 mm^2 (nominal anchor bars)

A_f mid span: Provide in bottom 4 bars of 22ϕ and top 2 bars of 22ϕ continuous throughout.

Check for torsion at $\alpha = 12.75° \left(0.2225 \text{ radians}, \frac{12.75 \times \pi}{180} \right)$

Distance from the support = $0.2225 \times R = 0.2225 \times 3.6 = 0.800$ m

Torsional moment $T = 24.7$ kN-m

$T_u = 1.5 \times 24.7 = 37.5$ kN-m

$$\text{Bending moment} = \left(244 - 202 \times \frac{0.80^2}{2} \right) = 179.40 \text{ kN-m}$$

$M_u = 1.5 \times 179.4 = 269.10$ kN-m

$$M_e = 269.1 + \frac{37.05}{1.70} \left(1 + \frac{650}{350} \right) = 269.1 + 62.3 = 331.4 \text{ kN-m}$$

At support $M_u = 1.5 \times 244 = 366$ kN-m (> 331.4 kN-m). Hence, continue support reinforcement to this point and 12ϕ beyond this point.

Shear force $V = 381$ kN, $V_u = 1.5 (381) = 572$ kN

$$\tau_v = \frac{572000}{350 \times 600} = 2.724 \text{ N/mm}^2 \text{ (< 3.5 N/mm}^2, \tau_{c\,max}\text{) hence size is OK.}$$

$$> \tau_c \text{ design of } 0.63 \text{ N/mm}^2$$

Provide shear reinforcement (12ϕ 2 legged vertical stirrups)

$$S_v \text{ (spacing)} = \frac{0.87 \times 415 \times 226 \times 600}{(V_u - \tau_c.bd)} = \frac{48958380}{(572000 - 0.63 \times 350 \times 600)} = \frac{48958380}{439700}$$

$$= 111.35 \text{ mm c/c (provide } 12\phi \text{ 2 legged @ 110 c/c)}$$

Shear force reduces towards mid span and hence increase the spacing of 12ϕ 2 legged stirrups from 110ϕ c/c to 200 c/c at the mid span point.

Since the beam is very deep, provide two bars 16ϕ in the middle of depth (one bar on each face).

The details of reinforcement are given in Fig. 11.16.

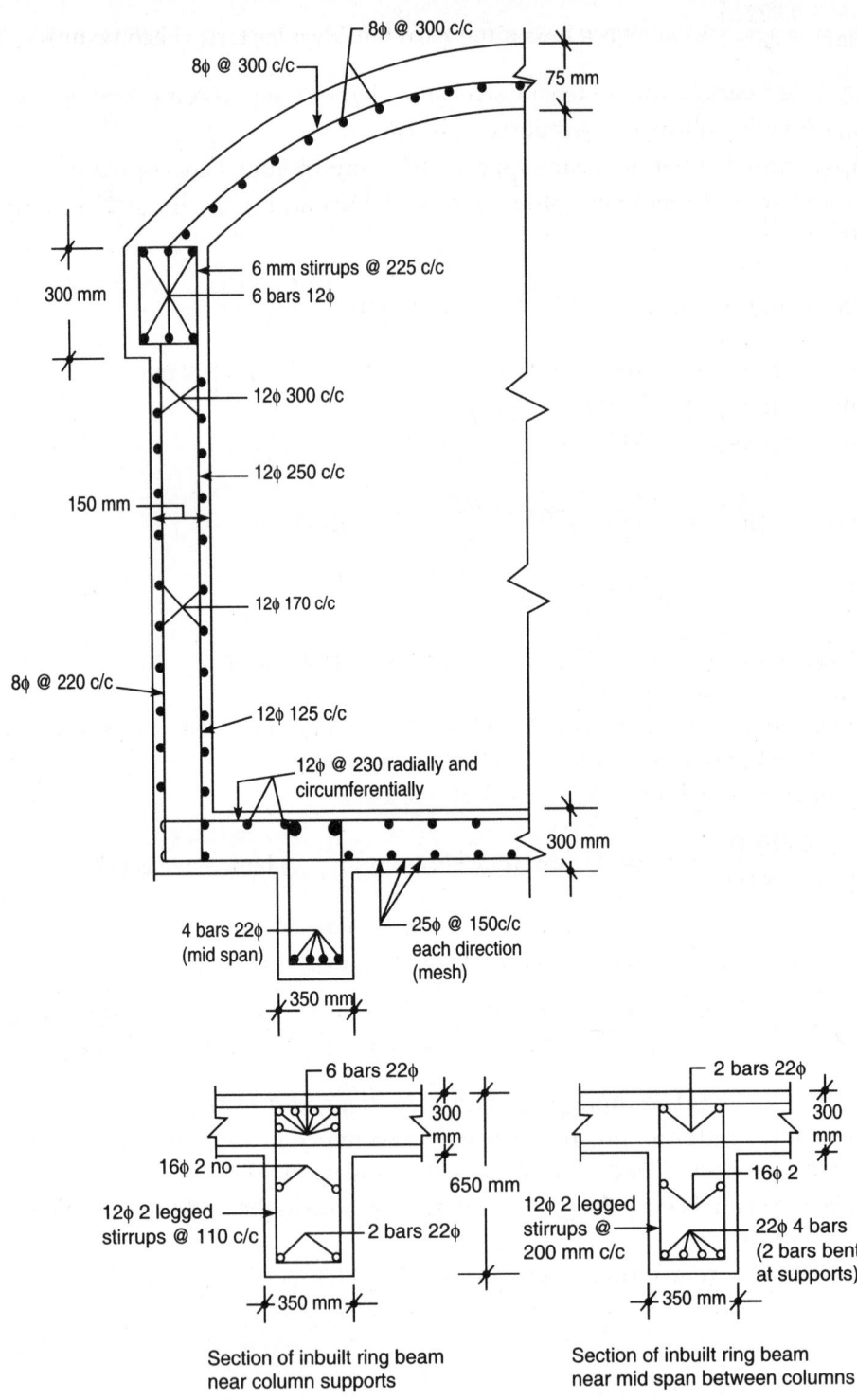

Fig. 11.16: Details of reinforcement of OHSR tank

11.5.3 Design of Intze Overhead Tanks

For larger capacity of OHSR tanks, flat base slab circular tanks become uneconomical due to thick flat base slab. To reduce the thickness of bottom base slab, it is changed to thin spherical dome (shell). Most economical proportions of Intze tank components are suggested by various researchers in relation to the diameter D of the tank. Various components of Intze tank (Fig. 11.17) are:

  (i) Top spherical dome
 (ii) Top circular ring beam (beam A–A)
(iii) Main cylindrical tank wall (A–B)
 (iv) Bottom circular ring beam (beam B–B) (with maintenance plateform)
  (v) Conical shell wall (B–C)
 (vi) Bottom spherical dome
(vii) Bottom circular girder beam (C–C)
(viii) Supporting stage of columns (or cylindrical shaft)

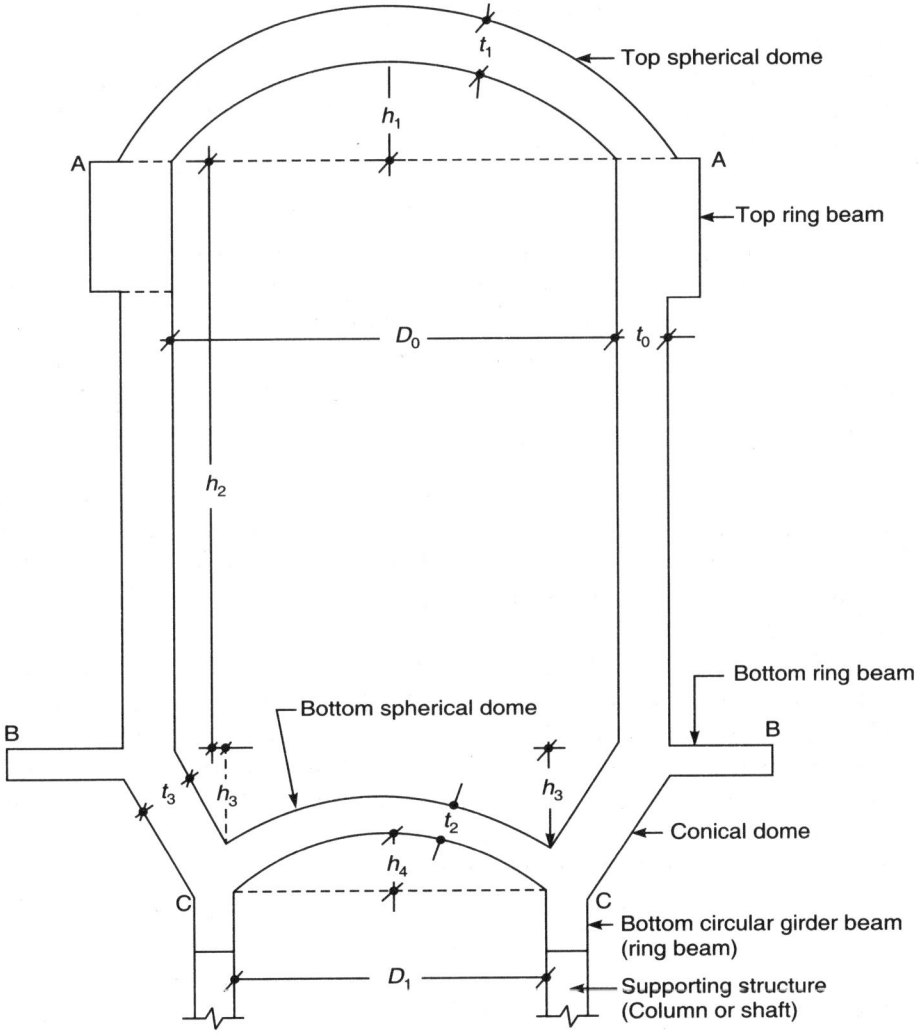

Fig. 11.17: Intze tank

Economical Proportions of Intze Tank Elements

Considering cost of materials and operational cost, economical proportions of various elements shall be:

- Rise of top spherical dome $h_1 = \dfrac{1}{7} D_0$

- Height of cylindrical tank, $h_2 = 0.4 \, D_0$
- Height of conical dome (angle with vertical $\theta = 45°$), $h_3 = 0.2 D_0$

- Rise of bottom spherical dome $h_4 = \dfrac{1}{7} D_0$

- Diameter of bottom circular girder beam $D_1 = 0.6 D_0$

Design Principles

Membrane theory being simplest, shall be adopted for design of elements. The joints between various members are rigid and will cause error in membrane analysis but this error will be limited in the vicinity of joints only. The effect of rigidity shall be accounted by extending all reinforcements of the member into the next member element through the joint. This will take care of rigidity and continuity of the joint between the connecting elements.

(i) *Design of top spherical dome*: The rise h_1 shall be taken as $\dfrac{1}{7} D_0$ ($D_0 =$ diameter of the tank). Generally, the thickness of top dome shall be taken as 75 mm. In case of severe extreme environment, this thickness may be taken as 100 mm. The design is based on meridional thrust and circumferential force.

(ii) *Design of top ring beam (A–A)*: Ring beam is subjected to a force from meridional thrust T_1 as ($T_1 \cos \theta$). This force will result in hoop tension in the ring beam as calculated tension (T_{hp}) = ($T_1 \cos \theta$) $\dfrac{D_0}{2}$, as done in previous case.

(iii) *Cylindrical tank wall*: Cylindrical tank wall shall be designed for hoop tension and shall be provided with minimum of 0.30% reinforcement in vertical direction for some cantilever effect. Hoop tension shall be resisted by reinforcement alone and the section shall be designed for maximum permissible tensile stress in concrete.

(iv) *Bottom ring beam (B – B)*: Total weight of top dome, top ring beam A–A and the cylindrical tank wall shall be acting on the bottom ring beam B–B (Fig. 11.15). This load V_1 per metre run shall try to push out beam B–B by rotating it about bottom girder ring beam C–C. The beam B–B shall develop a horizontal resistance H_1 to balance this load. The resultant force R_1 of V_1 and H_1 at the joint of beam B–B, tank wall and conical shell shall be in equilibrium. If the conical wall is inclined at an angle θ with the vertical, then we have:

$$R_1 \sin \theta = H_1$$
$$R_1 \sin \theta = V_1$$

and $\qquad\qquad \tan \theta = \dfrac{H_1}{V_1}$ or $H_1 = V_1 \tan \theta$ $\qquad\qquad$ ... Eq. (11.27)

Tension in the ring beam B–B due to vertical load from the wall shall be equal to

$$T_1 = H_1 \cdot \frac{D}{2} = V_1 \tan \theta \cdot \frac{D}{2} \qquad \text{... Eq. (11.28)}$$

Apart from this, the ring beam shall also be subjected to hoop tension due to horizontal water pressure wh_2 (total hoop tension in beam B–B should be) is given by

$$T_B = (V_1 \tan \theta + wh_2) \frac{D}{2} \qquad \text{... Eq. (11.29)}$$

The reinforcement in the ring beam B – B shall be designed to resist total tension (T_B) while the section shall be designed to limit the tensile stress in concrete within permissible value of the concrete (Fig. 11.19).

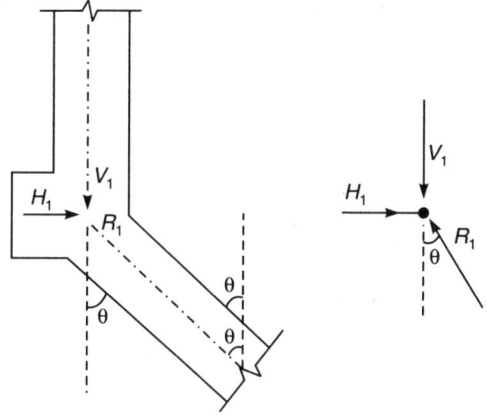

Fig. 11.18: Joint at bottom ring beam

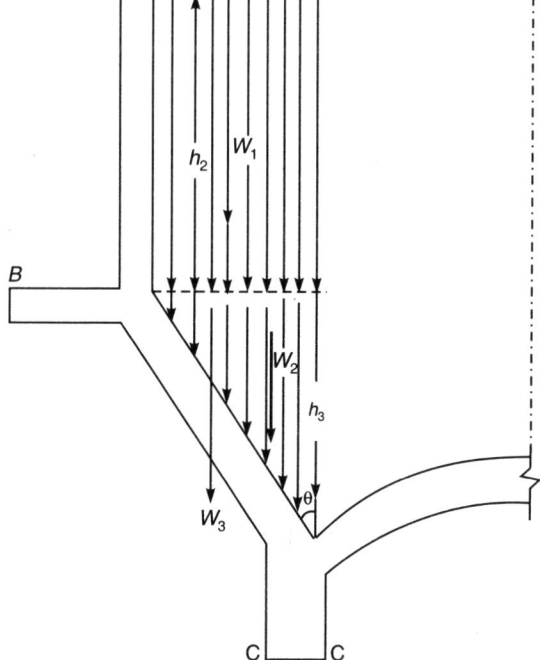

Fig. 11.19: Loads on conical dome

(v) *Conical dome:*

Conical dome is subjected to meridional thrust as well as hoop tension. Figs 11.19 and 11.20.

(a) Meridional thrust:

$$V_1 = \text{Total load from tank wall/m run}$$

$$\text{Total load } W = V_1 \times \pi D_0$$

$$W_w = \text{Wt. of water on conical shell}$$

$$W_w = W_1 + W_2$$

$$W_w = w \frac{\pi}{4} (D_0^2 - D_1^2) h_2 + w (D_0^2 + D_1^2 + D_0 \cdot D_1) h_3 - w \frac{\pi}{4} D^2 \cdot h_3$$

Self-wt. of conical shell of thickness 't'

$$= \pi \left(\frac{D_0 + D_1}{2} \right) l w_c, \text{ where } w_c = 25 \text{ kN/m}^3$$

$$\text{Total vertical load} = V_1 \pi D_0 + W_w + W_s$$

Total vertical load per unit length on ring beam C–C

$$V_2 = \left(\frac{\pi D_0 V_1 + W_w + W_s}{\pi D_1} \right)$$

Meridional thrust per unit length in the conical shell wall at bottom near

$$\text{C–C} = V_2 \cos \theta = \left(\frac{\pi D_0 V_1 + W_w + W_s}{\pi D_1} \right) \cos \theta \qquad \dots \text{ Eq. (11.30)}$$

where θ is the angle of conical wall with vertical.

(b) Hoop tension: Figure 11.20 shows conical shell subjected to water pressure p at a depth h below top, and self-weight q per unit area. If D' is diameter at this level, then hoop tension $T = (p \cos \theta + q \tan \theta) \dfrac{D'}{2}$

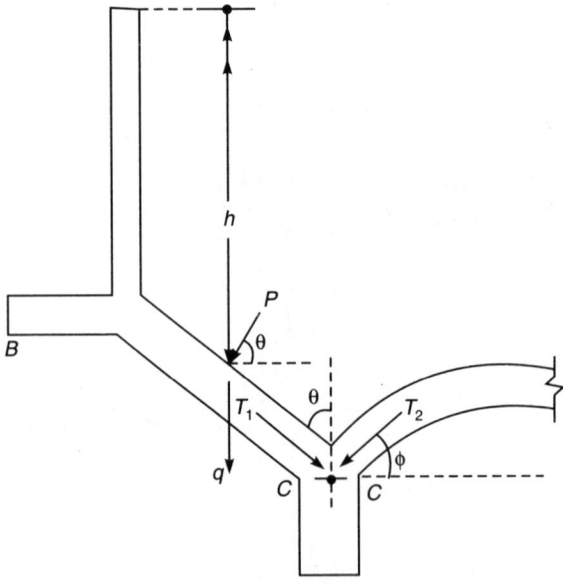

Fig. 11.20: Forces on conical and bottom spherical shells

Hoop tension may be calculated at top, middle and bottom by taking appropriate values of D', p, and q of conical shell and steel reinforcement is provided to take full hoop tension. Thickness of conical dome is to be checked for permissible stress in concrete for direct tension by considering equivalent area of concrete.

(vi) *Bottom spherical dome*: Bottom dome is designed for meridional thrust (T_2) and circumferential forces caused due to weight of water plus self-weight (Figs 11.19 and 11.20).

(vii) *Bottom ring girder C–C*: Figure 11.19 shows the forces acting on bottom ring girder beam C–C are thrust T_1 from the conical shell and thrust T_2 from the spherical bottom dome.

Net horizontal thrust on the ring girder $p_h = (T_1 \sin \theta - T_2 \cos \phi)$

Hence, hoop compression

$$T_p = p_h \cdot \frac{D_1}{2} = (T_1 \sin \theta - T_2 \cos \phi) \frac{D_1}{2} \qquad \text{... Eq. (11.31)}$$

The vertical pressure on the girder $P_v = (T_1 \cos \theta + T_2 \sin \phi)$ ... Eq. (11.32)

per unit length

The girder is to be designed for these forces (P_v and T_p) from the adjoining conical and spherical shells. These forces are based on membrane analysis and checked for continuity effect at rigid joints by extending reinforcement of one element to the other element making the joint rigid.

Effect of Continuity (rigidity) of Joints

Membrane analysis holds good only if the elements meeting at the joints are free to deflect and rotate. In actual practice, all these joints are rigid and result in moments and forces near the joints causing edge disturbances in members. Reinforcement of one element should be extended to the other elements connected through the joint to provide rigidity and continuity at the joint. However, the errors introduced due to rigidity are small and generally on the safer side. Finite element analysis may be carried out for economical and accurate design. We shall discuss and demonstrate only membrane analysis approach by solved example.

Example 11.7

Design an Intze type OHSR water tank with a capacity of 8,00,000 litres supported on 8 columns symmetrically placed. Use M25 CC and Fe 415 steel.

Solution: Capacity (V) = 800 m³, f_{ck} = 25 N/mm², σ_{cbc} = 8.5 N/mm², σ_{st} = 150 N/mm²

$$m = \frac{280}{3\sigma_{cbc}} = 11, n = 0.384, j = 0.872, K = 1.423 \text{ N-mm}$$

Let the diameter of cylindrical tank = D

Diameter of bottom ring girder = $0.6D$

Height of conical shell, $h_3 = 0.2D$, rise of bottom dome = $\dfrac{D}{7}$

Free board in top spherical dome portion.

V = Volume of (cylindrical tank + conical shell – bottom spherical dome)

$$V = \frac{\pi}{4} D^2. h_2 + \frac{\pi}{12} (D^2 + D^2_1 + D_1.D)h_3 - \frac{\pi}{3} h^2_4 (3R_2 - h_4)$$

Assume $D_1 = 0.6D$, $h_2 = 0.4D$, $h_3 = 0.2D$, $h_4 = \dfrac{D}{7}$

$$\left(2R_2 - \frac{D}{7}\right)\frac{D}{7} = \frac{(0.6D)^2}{2^2} = \frac{0.36D^2}{4}, R_2 = 0.386D$$

$$V = \frac{\pi}{4} D^2 \times 0.4D + \frac{\pi}{12} \times 0.2D (D^2 + 0.36D^2 + 0.6D^2) - \frac{\pi}{3} (0.2D)^2\left\{3 \times 0.386D - \frac{D}{7}\right\}$$

$$800 \text{ m}^3 = 0.314D^3 + \frac{3.14D^3}{60} (1.96) - \frac{3.14 \times 0.04}{3} (1.158 - 0.143) D^3 = 0.3745D^3$$

$$D = (2136.3)^{1/3} = 12.88 \text{ m}$$

Assume $D = 13.0$ m, say $D_1 = 0.6 \times 13 = 7.8$ m $\cong 8$ m, $h_2 = 0.4 \times 13 = 5.2$ m,

$$h_3 = 0.2 \times 13 = 2.60 \text{ m}$$

$$h_4 = \frac{D}{7} = \frac{13}{7} = 1.85 \text{ m},$$

$$(2R_2 - 1.85)1.85 = 4 \times 4, R_2 = 5.25 \text{ m}$$

Actual volume:

$$V = \frac{\pi}{4} (13)^2. (5.2) + \frac{\pi}{12} 2.6 (13^2 + 8^2 + 13 \times 8) - \frac{\pi}{3} (1.85)^2 (3 \times 5.25 - 1.85)$$

$V = \{690.2 + 229.4 - 49.8\} = 869.80$ m^3 (870 m^3) (little more than 800 m^3, OK)

Design of top cover dome:

$$D_0 = 13 \text{ m, rise } h_1 = \frac{D_0}{7} = \frac{13}{7} = 1.85, \quad \therefore R_1 = \frac{1}{2}\left(\frac{6.5^2}{1.85} + 1.85\right) = 12.34 \text{ m}$$

Semicentral angle $\phi_1 = \cos^{-1}\left(\dfrac{12.34 - 1.85}{12.34}\right) = 31.7°$

Let the thickness of top cover dome = 75 mm, (self-wt. = $0.75 \times 25 = 1.875$ kN/m^2)
Live load = 1.50 kN/m^2 (say), weight of finishing (say) = 0.625 kN/m^2

Total load $w = 4.0$ kN/m^2

Meridional thrust:

$$T_1 = wR_1 \frac{1}{(1 + \cos \phi)} = \frac{4 \times 12.34}{(1 + \cos 31.7)} = \frac{4 \times 12.34}{1.8508} = 26.67 \text{ kN}$$

Meridional stress (in effective section) = $\dfrac{26.67 \times 1000}{1000 \times 50} = 0.53$ N/mm^2

($< 6 \, \text{N/mm}^2$ direct permissible compressive stress)

Circumferential force:

$$T_2 = \left(\cos\phi - \frac{1}{1+\cos\phi} \right) = 4 \times 12.34 \left(0.8508 - \frac{1}{1.8508} \right) = 15.33 \, \text{kN}$$

Circumferential stress in effective section (MS) = $\dfrac{15.33 \times 1000}{1000 \times 50} = 0.3065 \, \text{N/mm}^2$

($< 6 \, \text{N/mm}^2$ direct compressive stress)

Nominal temperature steel = $\dfrac{0.3}{100} \times 75000 = 225 \, \text{mm}^2$ (MS), 180 mm^2 (HYSD bars)

(8ϕ @ 200 c/c MS radially and circumferentially), alternatively 8ϕ HYSD @ 270 c/c, A_{st} provided HYSD 185 mm^2.

A_{st} (MS) provided = 250 mm^2 (OK).

Design of top ring beam (A–A):

Hoop tension in the ring beam = $T_1 \cos\phi_1 \cdot \dfrac{D}{2} = 26.67 \times 0.8508 \times \dfrac{13}{2} = 147.5 \, \text{kN}$

$$A_{st} \, (\text{reqd.}) = \frac{147.5 \times 1000}{150} = 983 \, \text{mm}^2$$

(Provide 6 bars 16ϕ, 3 in top and 3 in bottom), A_{st} (provided) = 1206 mm^2

Direct tensile stress in concrete (allowed) = 1.30 N/mm^2

Area of concrete required (gross) = $\dfrac{147500}{1.3} = 113462 \, \text{mm}^2$

$$A_g = 113462 \, (350 \, \text{mm} \times 350 \, \text{mm})$$

Nominal stirrups $A_{st} = \dfrac{0.240}{100} \times 350 \times 350 = 294 \, \text{mm}^2$ (HYSD steel)

(8ϕ 2 legged 300 c/c, A_{st} = 333 mm^2/m)

Design of cylindrical tank wall (H = 5.20 m, D_0 = 13 m):

Hoop tension at the base of tank = $\{9.8 \times 5.2\} \times \dfrac{13}{2} = 331.24 \, \text{kN/m}$ height

A_{st} required = $\dfrac{331.24 \times 1000}{150} = 2209 \, \text{mm}^2$ (using 16ϕ bars, spacing = 90.99 mm c/c,

providing 16ϕ rings @ 180 c/c on each face) for resisting hoop tension.

Hoop steel (ring bars) spacing may be reduced gradually to 16ϕ rings @ 300 c/c near top (3 m) on each face (A_{st} reqd. at 3 m below top = 1274 mm^2/m, 16ϕ @ 300 c/c gives A_{st} = 1340 mm^2).

A_{st} (provided) = 2233 mm^2/m height on both faces to 2 × 670 mm^2/m (top 3.00 m height)

Since hoop tension at 3.00 m depth = 9.80 × 3.0 × 6.5 = 191.0 kN/m

$$A_{st} = 1237 \text{ mm}^2/\text{m (total)}$$

Area of concrete A_c by restricting direct tensile stress to 1.3 N/mm^2, thus

$$\frac{331240}{A_c + (m-1)2233} \leq 1.30 \text{ or } \{A_c + 10 \times 2233\} \geq \frac{331240}{1.30} \geq 254800,$$

$$A_c = 232470 \text{ mm}^2, t \geq \frac{232470}{1000} = 232.47 \text{ mm}$$

Thus, provide uniform thickness of 250 mm.

Distribution (temperature steel = 0.3%)

$$p = \left\{ 0.3 - 0.1 \frac{(250 - 100)}{(450 - 100)} \right\} = 0.257\% \text{ (MS), HYSD} = 0.2056\%,$$

$A_{st} = 643 \text{ mm}^2/\text{m}, (514 \text{ mm}^2 \text{ HYSD/m}), (\text{each face } 257 \text{ mm}^2/\text{m})$

8ϕ bars, $S_v = \dfrac{50000}{322} = 155 \text{ mm c/c (MS) or } 8\phi \text{ 190 c/c HYSD on each face. (Provide }$ 8ϕ vertical bars at 190 c/c on each face, A_{st} (HYSD) = 526 mm²/m width).

Extend vertical bars 8ϕ 190 mm c/c in outer face to conical shell outer face to take care of continuity moments through the lower ring beam B–B.

Design of bottom ring beam (B–B):

Vertical loads on bottom beam (B–B) are:

(i) Load from top dome = $T_1 \sin\phi = 26.67 \sin 31.7° = 14.02 \text{ kN/m}$

(ii) Wt. of top ring beam = $0.35 \times 0.35 \times 1 \times 25 = 3.0625 \text{ kN/m}$

(iii) Wt. of tank wall $= 0.25 \times 1 \times 5.2 \times 25 = 32.5 \text{ kN/m}$

(iv) Self-wt. of bottom ring beam (B–B) width kept large for maintenance (width = 1200 mm) and depth say 500 mm, wt. = $1.20 \times 0.50 \times 1 \times 25 = 15 \text{ kN/m}$

Total load $V_1 = 14.02 + 3.06 + 32.5 + 15 = 64.55 \cong 65 \text{ kN/m}$

Slope ϕ_2 of conical dome with vertical $= \tan^{-1}\left(\dfrac{D - D_1}{2h_3}\right) = \tan^{-1}\left(\dfrac{13 - 8}{2 \times 2.6}\right) = 43.88°$

Hoop tension due to vertical load $V_1 = V_1 \tan\phi \cdot \dfrac{D}{2} = 65 (\tan 43.88) \dfrac{13}{2} = 406.3 \text{ kN}$

Hoop tension due to water pressure (top of beam B–B) = $9.8 \times 5.2 \times \dfrac{13}{2} = 331.24 \text{ kN}$

Total hoop tension in the beam B–B = (406.3 + 331.24) kN = 737.54 kN

A_{st} reqd. $= \dfrac{737.54 \times 1000}{150} = 4917 \text{ mm}^2$ (provide 16 ring bars of 20 mm, 8 in top face and 8 in bottom face, A_{st} = 5024 mm in 1200 mm width).

Area of concrete reqd. $A_g + (m-1) 5024 = \dfrac{737.540 \times 1000}{1.30}$

or $A_g = (567339 - 50240) = 517099$ mm^2 (1200 mm × 500 mm), $d = 460$ mm.
Provide nominal 12ϕ 2 legged stirrups @ 100 c/c ($A_{st} = 1130$ mm^2/m width, $p = 0.25\%$).
τ_c permissible stress M25 = 0.23 N/mm^2
Cantilever moment and shear force due to projecting plate form (u.d.l. $w = 15$ kN/m)
Projection = $1.20 - 0.25 = 0.95$ m, inspection team 5 kN/m at end.

$$V = 15 \times 0.95 + 5.0 = 19.25 \text{ kN/m}, \; M_{max} = \frac{14.25 \times 0.95}{2} + 5 \times 0.95 = 11.52 \text{ kN/m},$$

$$A_{st} = \frac{11.52 \times 10^6}{0.872(450) \times 190} = 155 \text{ mm}^2 \text{ (beam outside water contact)}$$

Thus, 12ϕ 2 legged radial stirrups @ 100 c/c shall be quite adequate
($A_{st} = 1130$ mm^2/m) in top face.

$$\text{Shear stress} = \frac{19250}{1000 \times 460} = 0.042 \text{ N/mm}^2 \text{ (very small), OK.}$$

Design of conical shell:

$$h_3 = 2.60 \text{ m}, D_0 = 13 \text{ m}, D_1 = 8 \text{ m}, h_4 = 1.85 \text{ m}, \phi = \tan^{-1}\left(\frac{2.5}{2.6}\right) = 43.88°$$

(i) Total load from the cylindrical wall = $V_1 \pi D = 65 \times \pi \times 6.50 \times 2 = 2655$ kN
(ii) Weight of water = $w \times$ volume = $9.8 \times 870 = 8526$ kN

Slant height of conical shell = $\sqrt{(2.60^2 + 2.5^2)} = 3.607$ m

Mid height diameter = $\dfrac{13 + 8}{2} = 10.5$ m,

Assuming thickness 350 mm
(iii) Self-wt. conical dome = $(\pi \times 10.5) \, 3.607 \times 0.35 \times 25 = 1041$ kN
Total wt. = $2655 + 8526 + 1041 = 12222$ kN

$$\therefore \qquad\qquad V_2 = \frac{12222}{(\pi 8)} = 486.3 \text{ kN/m}$$

Meridional thrust = $V_2 \cos \phi = 486.3 \cos 43.88° = 375.2$ kN/m

$$\text{Meridional stress} = \frac{375.21 \times 1000}{1000 \times 350} = 1.0715 \text{ N/mm}^2 < 6 \text{ N/mm}^2 \text{ permissible limit.}$$

Hoop tension at top of conical shell:
$q = 0.35 \times 1 \times 1 \times 25 = 8.75$ kN/m^2, p(water pressure) $= 9.8 \times 5.2 = 50.96 = 51$ kN/m^2
$\phi = 43.90°$

Hoop tension near top:

$$T_p = (p \cos \phi + q \tan \phi) \frac{D}{2}$$

$$= (51 \cos 43.90° + 8.75 \tan 43.90°) \frac{13}{2} = (36.75 + 8.42) \frac{13}{2} = 294.0 \text{ kN}$$

$$A_{st} = \frac{294 \times 1000}{150} = 1960 \text{ mm}^2 \ (A_{st} \text{ on each face} = 980 \text{ mm}^2/\text{m})$$

$(12\phi @ 110 \text{ mm c/c}, A_{st} = 1027 \text{ mm}^2/\text{m})$,
section to limit hoop tensile stress to 1.3 N/mm^2

$$A_g + (11-1) 2 \times 1027 = \frac{(293.634 \times 1000)}{(1.3)}$$

$$A_g = (225872 - 20540) = 205332 \text{ mm}^2/\text{m}$$

$$\text{Thickness} = \frac{205332}{1000} = 205.332 \text{ mm (provide 300 mm thickness instead of 350 mm}$$

assumed).

$$\theta \cong 43.9°$$

Hoop tension at lower edge:

$$p = 9.8 (5.2 + 2.6) = 76.44$$

$$T_{bott} = (76.44 \cos 43.9 + 8.75 \tan 43.9) \frac{8}{2}$$

$$= (55.097 + 8.414) 4 = 254.00 \text{ kN} < 294.00 \text{ kN (at top end), OK.}$$

Therefore, provide $12\phi @ 110 \text{ c/c}$ on each face throughout.

$$\text{Minimum temperature steel (MS)} = \left\{0.3 - \frac{0.1(300-100)}{(350)}\right\} 300 \times \frac{1000}{100}$$

$$= 0.243 \times 3000 = 729 \text{ mm}^2/\text{m (MS bars)}$$

$$(584 \text{ mm}^2/\text{m} - \text{HYSD bars})$$

Provide on each face $8\phi 160 \text{ c/c}$, $A_{st} = 313 \text{ mm}^2/\text{m}$ HYSD bars
or $\qquad\qquad 8\phi 130 \text{ c/c}$, $A_{st} = 385 \text{ mm}^2/\text{m}$ MS bars.

Design of bottom spherical dome:

Diameter $D_1 = 8.0 \text{ m}$, h_4 (rise) = 1.85 m, $R_2 = \frac{1}{2}\left(\frac{4 \times 4}{1.85} + 1.85\right) = 5.25 \text{ m}$

$$\theta = \cos^{-1}\left(\frac{R_2 - h_4}{R_2}\right) = \cos^{-1}\left(\frac{3.40}{5.25}\right) = 49.64°, \text{ let thickness of dome} = 200 \text{ mm}$$

Self-wt. of dome = $2\pi R_2 h_4 t \times 25 = 2\pi (5.25 \times 1.85 \times 0.2 \times 25) = 305 \text{ kN}$

Wt. of water = $\left[\frac{\pi}{4} 8^2 (5.2 + 2.6) - \frac{\pi}{3} 1.85^2 (3 \times 5.25 - 1.85)\right] 9.8 = 3354.00 \text{ kN}$

Total load = $(3354.00 + 305) = 3659 \text{ kN}$

$$\text{Load/unit area } (w) = \frac{3659}{\dfrac{\pi}{4}(8)^2} = 72.8 \text{ kN/m}^2, \cos 49.64° = 0.6476$$

$$\text{Meridional thrust} = wR_2 = \left(\frac{1}{1+\cos\theta}\right) = \frac{72.8 \times 5.25}{(1+\cos 49.64)} = 232.0 \text{ kN}$$

$$\text{Meridional stress} = \frac{232.0 \times 1000}{200 \times 1000} = 1.16 \text{ N/mm}^2 < 6 \text{ N/mm}^2, \text{ OK.}$$

$$\text{Circumferential force} = wR_2\left[1 - \frac{1}{(1+\cos\theta)}\right]$$

$$= 72.8 \times 5.25\left[1 - \frac{1}{(1.6476)}\right] = 150.23 \text{ kN}$$

$$\text{Hoop stress} = \frac{150.23 \times 1000}{200 \times 1000}$$

$$= 0.751 \text{ N/mm}^2 \text{ (safe)} (<6 \text{ N/mm}^2 \text{–direct comp. stress})$$

$$\text{Min. reinforcement (MS) } p = \left\{0.3 - 0.1\frac{(200-100)}{350}\right\} = 0.2714\%$$

$$A_{st} = \frac{0.2714}{100} \times 200 \times 1000 = 543 \text{ mm}^2 \text{ (272} \times 2) \text{ (MS) or (435 HYSD steel)}$$

Provide MS bars 8 mm ϕ @ 180 c/c on each face ($A_{st} = 2 \times 278 = 556$ mm^2/m) both as circumferential and meridional reinforcement or HYSD bars 8ϕ @ 220 c/c each face ($A_{st} = 2 \times 227$ mm^2/m).

Design of ring girder beam C–C:

Bottom ring beam C–C is subjected to inward thrust from conical wall and outward thrust from the spherical dome.

$T_1 = 254$ kN, vertical load $= T_1\cos\theta = 254 \cos 43.9° = 183.08$ kN/m ($\downarrow$)

Dome $T_2 = 232$ kN, vertical load $= 232.0 \sin 49.6° = 177$ kN/m ($\downarrow$)

Vertical downward load $w' = 183.08 + 177 = 360.08$ kN/m ($\downarrow$)

Assuming section of ring beam (C–C) as 600 mm $\times$ 1000 mm

Self-wt. $= 0.6 \times 1 \times 25 = 15.0$ kN/m

Let the circular girder (ring beam) be supported on 8 columns,

$$\phi = \frac{360}{8} = 45° = \left(\frac{\pi}{4}\right) \text{radians}$$

Total downward vertical load $w' = 360.08 + 15 = 375.1$ kN/m $\cong 375.1$ kN/m

$$\text{Max. SF near the support} = \frac{1}{2} \times 375.1 \times \frac{\pi}{4} \times 8 = 1179 \text{ kN}$$

From Table 11.3, moment coefficients, $K = 0.066$, $K' = 0.030$, $K'' = 0.005$ (torsion)

$$\text{Support moment} = -0.066 \times 375.1 \times \frac{\pi}{4} \times \left(\frac{8}{2}\right)^2 = -311 \text{ kN-m}$$

$$\text{Mid span moment} = +0.030 \times 375.1 \times \frac{\pi}{4} \times \left(\frac{8}{2}\right)^2 = +142 \text{ kN-m}$$

$$\text{Max. torsional moment} = 0.005 \times 375.1 \times \frac{\pi}{4} \times \left(\frac{8}{2}\right)^2 = 24 \text{ kN-m}$$

Max. torsional moment shall be at $\alpha = 9.33°$, with the line joining the centre of circle and the column.

Max. design moment = 311 kN-m/m. Since this girder is not in contact with water, the limit state design can be adopted. Using above dimension of beam C–C (600 × 1000), steel reinforcement for moment can be determined.

Depth required for shear force $V_u = 1.5 \times 1197 = 1796$ kN

$$\tau_{uv} = \frac{1796 \times 1000}{600 \times 1000} = 2.993 \text{ N/mm}^2 < 3.1 \text{ N/mm}^2 \text{ (design shear reinforcement)}.$$

Reinforcement for moment (ring beam):

$$M_u = 1.5 \times 311 = 467 \text{ kN-m}, \, d = 1000 - 50 = 950 \text{ mm},$$

M_u (mid span) $= 1.5 \times 142 = 213$ kN-m

$M_{u \lim}$ (capacity) $= 0.138 \times 600 \times 950^2 \times 25 = 1868.18 \times 10^6$ N-mm

$> 474 \times 10^6$ N-mm and hence singly reinforced section.

$$A_{st} = \frac{0.5 f_{ck} b \cdot d}{f_y}\left(1 - \sqrt{1 - \frac{4.6 \times M_u}{f_{ck} \, bd^2}}\right)$$

or

$$A_{st} = \frac{0.5 \times 25 \times 600 \times 950}{415}\left(1 - \sqrt{1 - \frac{4.6 \times 467 \times 10^6}{25 \times 600 \times 950^2}}\right) = 1423 \text{ mm}^2$$

(20ϕ 5 bars, $A_{st} = 1570$ mm^2) in top face near column supports.

$$p_t = \frac{1570 \times 100}{600 \times 950} = 0.275, \, \tau_c = 0.373 \text{ N/mm}^2$$

$$\text{Minimum } A_{st} = \frac{0.2}{100} \times 600 \times 1000 = 1200 \text{ mm}^2 < 1423 \text{ mm}^2$$

Provide 12ϕ 4 legged vertical stirrups @ spacing S_v,

$$S_v = \frac{0.87 \times 415 \times 452 \times 950}{(1179000 - 0.275 \times 600 \times 950)} = 152.0 \text{ mm c/c (140 mm c/c)}$$

Spacing of stirrups be increased in middle of span to 250 c/c.

Torsion moment is maximum at $\alpha = 9.33°$, $l = R_2 \alpha = \frac{\pi}{180}(9.33) \times 4.0 = 0.65$ m

BM reduces at the point of max. torsion (T_u):

$$M = \left(311 - 381 \times \frac{0.65^2}{2}\right) = (311 - 80.5) = 230.5 \text{ kN-m/m}$$

$$M_u = 1.5 \times 230.5 = 346 \text{ kN-m/m}, \ T_u = 1.5 \times 24.0 = 36.0 \text{ kN-m}$$

$$M_e \text{ (equivalent } M) = 346.0 + \frac{36.0}{1.7} \times \left(1 + \frac{1000}{600}\right) = 346 + 56.50$$

$$= 403 \text{ kN-m}, (A_{st} \text{ reqd.} = 1270 \text{ mm}^2)$$

Hence, continue 5 bars of 20ϕ beyond this point. A_{st} provided will be 1570 mm². Shear force also gets reduced towards middle span and hence reduce the shear reinforcement in the middle half span, i.e. increase spacing of 4 legged 12ϕ vertical stirrups from 140 mm c/c to 250 mm c/c for the middle half span (Figs 11.21 and 11.22).

Since beam C–C is very deep, provide side face reinforcement (0.1%).

$$\text{Side face reinforcement} = \frac{0.10}{100} \times 600 \times 1000 = 600 \text{ mm}^2$$

Provide 3 × 2 of 12ϕ bars 3 on each face, $A_{st} = 678$ mm²

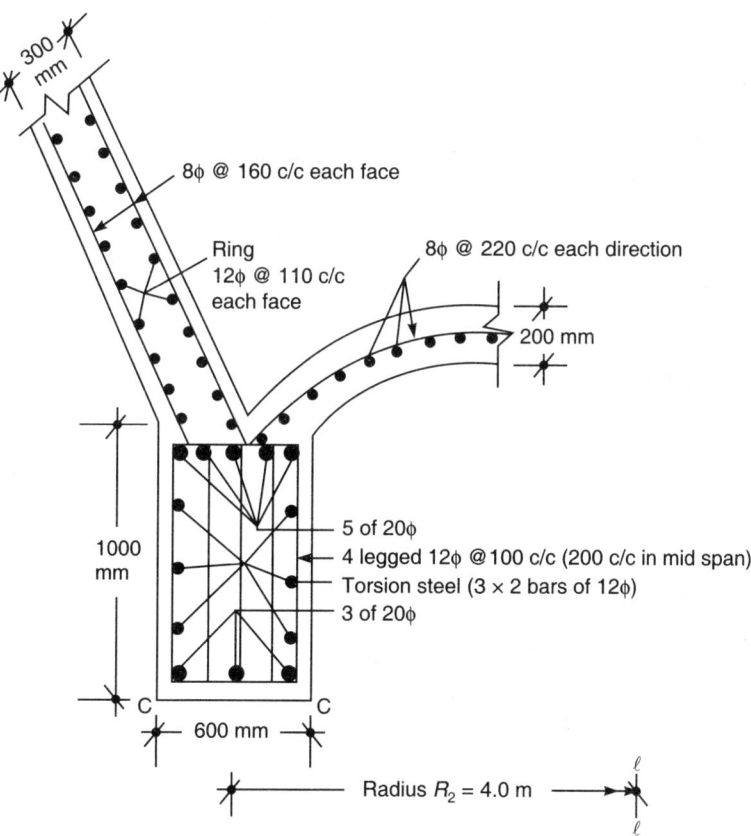

Fig. 11.21: Details of reinforcement of ring beam girder

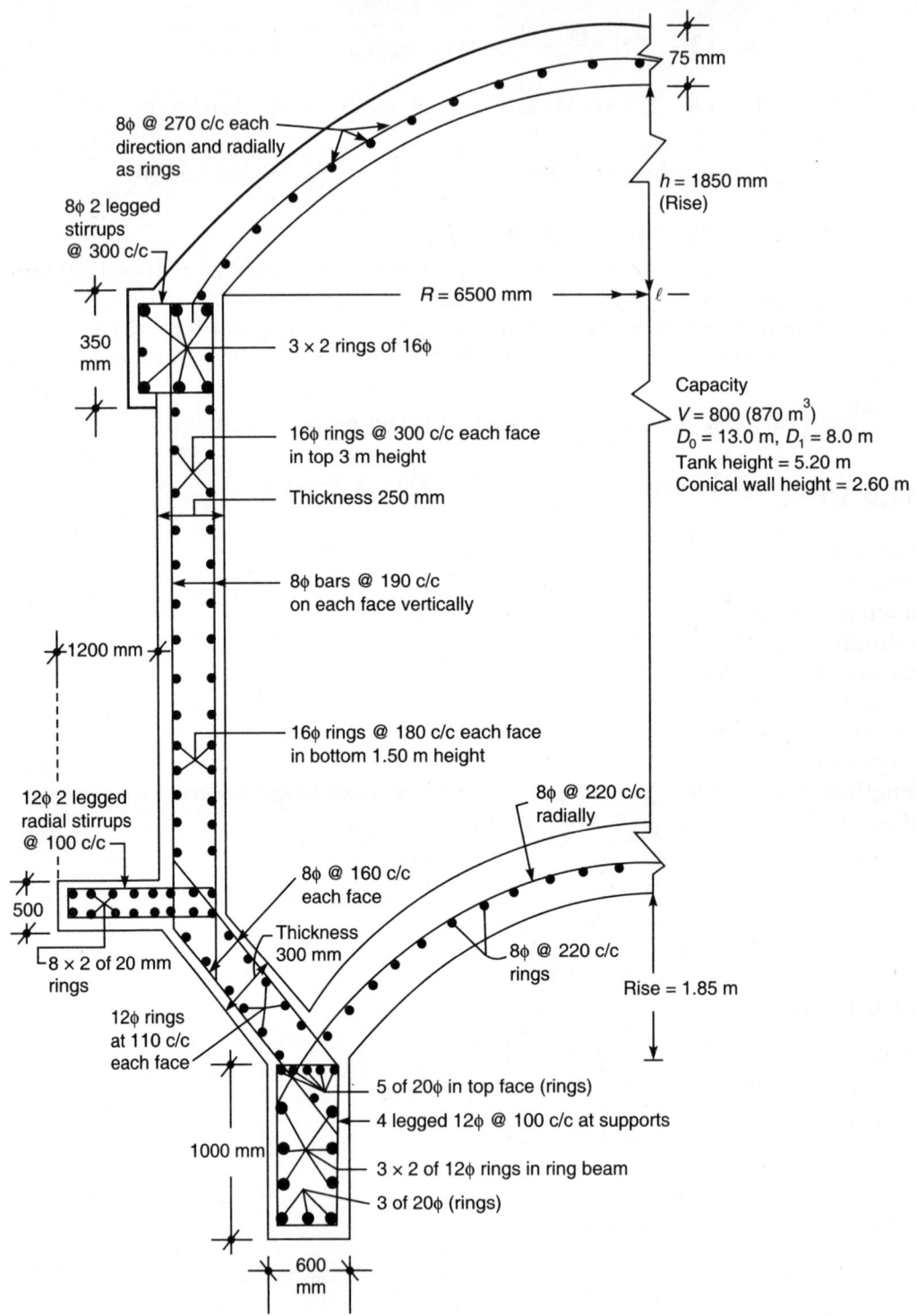

8ϕ @ 270 c/c each direction and radially as rings

8ϕ 2 legged stirrups @ 300 c/c

350 mm

3 × 2 rings of 16ϕ

16ϕ rings @ 300 c/c each face in top 3 m height

Thickness 250 mm

8ϕ bars @ 190 c/c on each face vertically

1200 mm

16ϕ rings @ 180 c/c each face in bottom 1.50 m height

12ϕ 2 legged radial stirrups @ 100 c/c

500

8 × 2 of 20 mm rings

8ϕ @ 160 c/c each face

Thickness 300 mm

12ϕ rings at 110 c/c each face

1000 mm

5 of 20ϕ in top face (rings)

4 legged 12ϕ @ 100 c/c at supports

3 × 2 of 12ϕ rings in ring beam

3 of 20ϕ (rings)

600 mm

75 mm

h = 1850 mm (Rise)

R = 6500 mm

ℓ

Capacity
$V = 800 (870 \text{ m}^3)$
$D_0 = 13.0$ m, $D_1 = 8.0$ m
Tank height = 5.20 m
Conical wall height = 2.60 m

8ϕ @ 220 c/c radially

8ϕ @ 220 c/c rings

Rise = 1.85 m

Fig. 11.22: Details of reinforcement of a Intze tank

Bottom face reinforcement ($M = 1.5 \times 142 = 215$ kN-m):

$$A_{st2} = \frac{0.50 \times 25 \times 600 \times 950}{415}\left[1 - \sqrt{1 - \frac{4.6 \times 215 \times 10^6}{25 \times 600(950)^2}}\right] = 642 \text{ mm}^2$$

(3 bars 20 mm), $A_{st} = 942$ mm^2, OK.

Net horizontal force = $T_c \sin\phi - T_2\cos\theta$, where $\phi = 43.9°$, $\theta = 49.64°$

$$= 254 \sin 43.9° - 232 \cos 49.64° = 176.06 - 150.24$$

$$= 25.82 \text{ kN/m}$$

Hoop compression $= 25.82 \times \dfrac{8}{2} = 103.3$ kN,

direct hoop comp. stress $= \dfrac{103.3 \times 1000}{600 \times 1000} = 0.172$ N/mm$^2 < 6$ N/mm^2, OK.

11.6 DESIGN OF TOWER/ SHAFT FOR OHSR

For water supply schemes, water tanks are placed at elevated places by constructing towers or staging at an elevation of 10 to 20 m depending on the local requirements of town water supply schemes. The bottom ring beam girder of tank rests on a number of columns placed at equal intervals. More number of columns are necessary to support tanks of large base diameter. Number of columns are decided in such a way that the curved distance between columns is generally not more than 4 to 5 m. Columns of tower are braced appropriately, at suitable intervals so as to make columns of economical size. Columns should be braced at 3 to 4 m height to limit the effective length and slenderness ratio to achieve sufficient economy and load carrying capacity (Fig. 11.23).

The supporting structures are generally of two types:

(i) Group of columns properly braced at appropriate intervals

(ii) Hollow cylindrical shaft and staircase on inner face

11.6.1 Analysis of Braced Column Tower

Total vertical load of tank shall be shared equally by columns. Tank and tower shall also be subjected to wind pressure and earthquake forces and safety must be checked against these forces also. For wind forces, certain assumptions are made for analysis by approximate method of frames:

- Point of contraflexure ($M = 0$) occurs at mid points of columns and braces.
- The vertical forces in the columns are proportional to the distance of the column from the axis of bending (with the foundation slab rigid).

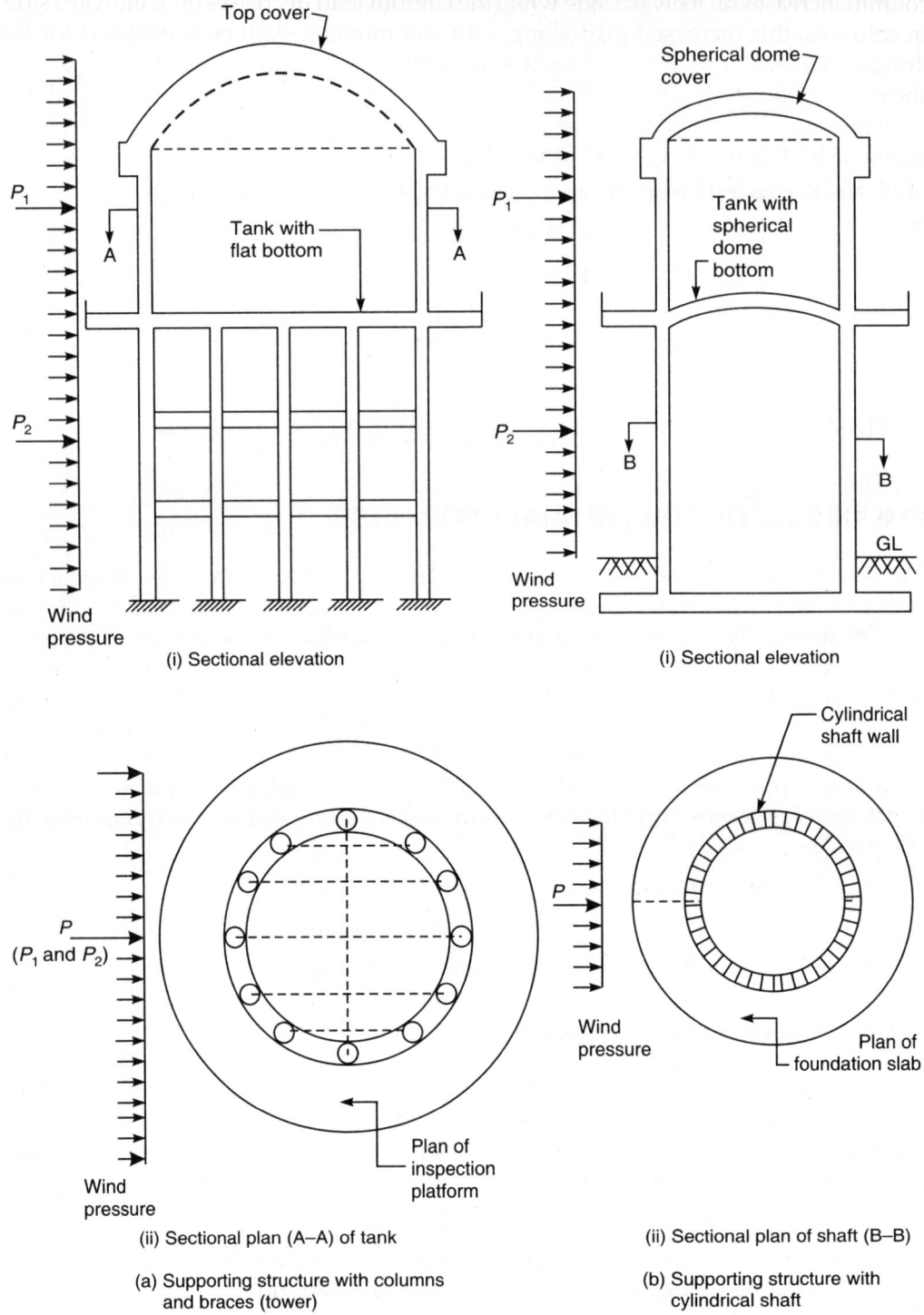

(i) Sectional elevation

(i) Sectional elevation

(ii) Sectional plan (A–A) of tank

(ii) Sectional plan of shaft (B–B)

(a) Supporting structure with columns and braces (tower)

(b) Supporting structure with cylindrical shaft

Fig. 11.23: Supporting structure with columns and shaft

Further, it may be noted that wind may blow in any direction, hence the analysis shall be carried out by considering wind in most critical direction. Axial loads on column increases on leeward side while the column load decreases on windward side. In columns, this increased load along with any moment shall be considered for the design. Moments also develop in columns and braces. The braces are also subjected to shear force (Fig. 11.24).

The moments at mid height of columns (points O_1, O_2, O_3) can be determined after finding wind loads on various elements (Fig. 11.24). Let the moment at level 1 be equal to M_1 and let the SF in extreme column located at a distance r from the axis of bending be V. The resisting moment of ith column at a distance of a from the axis of bending $= V_i \cdot a$,

but
$$V_i = \frac{a \cdot V}{r}, M = V_1 \cdot a = \frac{a \cdot V}{r} \cdot a = \frac{Va^2}{r}$$

Resisting moment of all the columns = external moment

Hence
$$M_1 = V\Sigma \frac{a^2}{r} = \frac{V}{r}\,\Sigma a^2 \qquad \qquad \dots \text{Eq. (11.33)}$$

Thus, maximum compression developed in column due to wind can be found at level 1, 2, or 3.

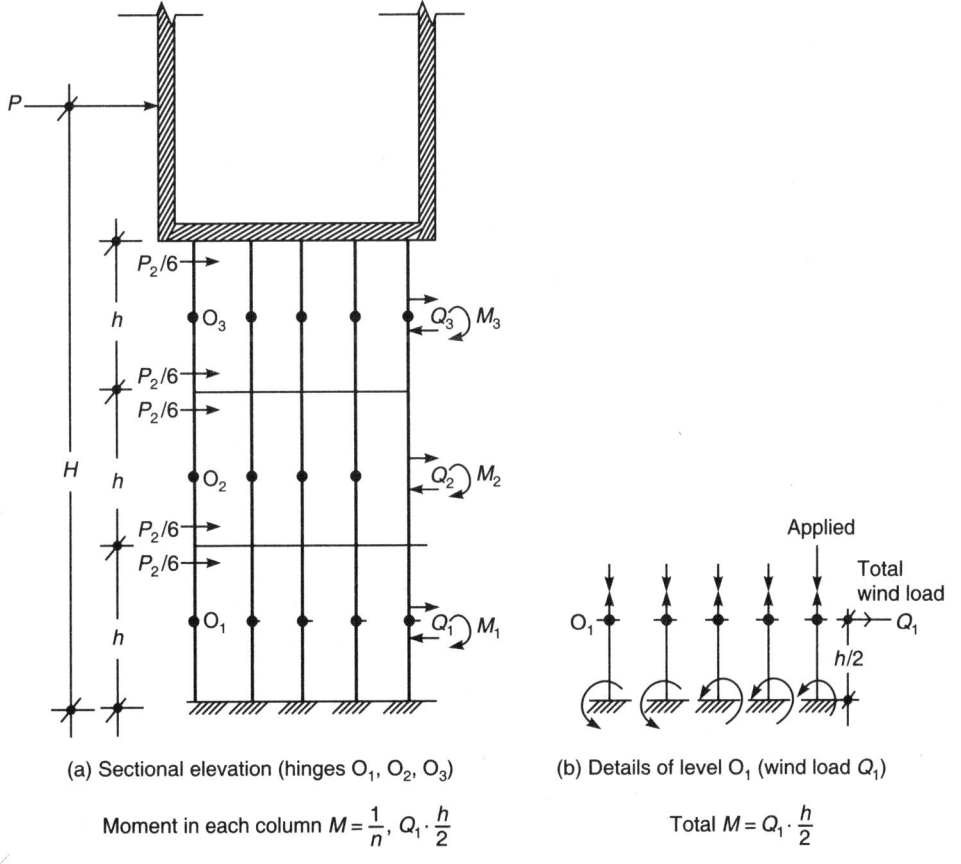

(a) Sectional elevation (hinges O_1, O_2, O_3)

Moment in each column $M = \dfrac{1}{n}, Q_1 \cdot \dfrac{h}{2}$

(b) Details of level O_1 (wind load Q_1)

Total $M = Q_1 \cdot \dfrac{h}{2}$

Fig. 11.24: Details of forces on columns due to wind

Hinges are formed at point of contra flexure at level O_1, O_2, O_3 (middle of column heights).

$$\text{Moment in columns} = Q_1 \frac{h}{2}$$

$$\therefore \text{Moment in each column} = \frac{1}{n} Q_1 \frac{h}{2} \qquad \qquad \text{... Eq. (11.34)}$$

11.6.2 Design Force in Braces

Consider joint A as shown in Fig. 11.25a. Joint moment due to wind forces Q_1 and Q_2 in level 1 and level 2 at joint A (braces and column) is given as

$$M'_1 + M'_2 = \left(Q_1 \frac{h}{2}\right)\frac{1}{n} + \left(Q_2 \frac{h}{2}\right)\frac{1}{n}$$

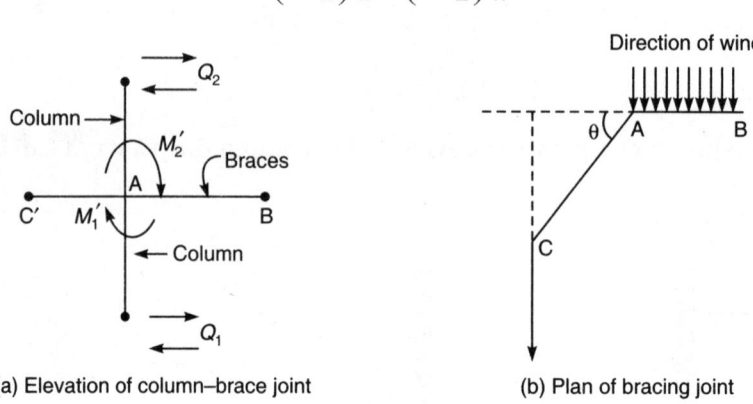

(a) Elevation of column–brace joint   (b) Plan of bracing joint

11.25: Moment in braces

Braces meeting at the joint are required to develop more or equal to above moment for equilibrium condition,

i.e. $$\qquad\qquad M_A = (Q_1 + Q_2)\frac{h}{2}\frac{1}{n} \qquad\qquad \text{... Eq. (11.35)}$$

To be on safer side $M'_2 = M'_1$, and moment resisted by the braces may be taken as

$$2M'_1 = 2\frac{Q_1 h}{2n} = \frac{Q_1 h}{n} \qquad\qquad \text{... Eq. (11.36)}$$

When wind is blowing normal to one of the braces as shown in Fig. 11.25b, the brace AB will not resist any moment while brace AC will be subjected to maximum moment. If θ is the angle of deviation of the two braces, then

$$M_{AC} \cos \theta = 2M' \text{ or } M_{AC} = 2M' \sec \theta \qquad\qquad \text{... Eq. (11.37)}$$

All braces must be designed for above moments where as the point of contra flexure is assumed to be at the midpoint of the brace. Hence, shear in the brace is given by

$$S \times \frac{l}{2} = M_{AC} \text{ or } S = \frac{2M_{AC}}{l} \qquad\qquad \text{... Eq. (11.38)}$$

where, l is length of the brace. Brace shall be designed for moment M_{AC} and SF S.

The process of design of braces and columns shall be explained by the following example.

Example 11.8

Design a tower of height 12 m to support the Intze tank of Example 11.7. Assume wind pressure of 1.5 kN/m². Use M25 grade CC and Fe 415 grade steel. Provide 8 columns in the supporting structure. Let the bracing be provided at an interval of height 4 m c/c (Fig. 11.26).

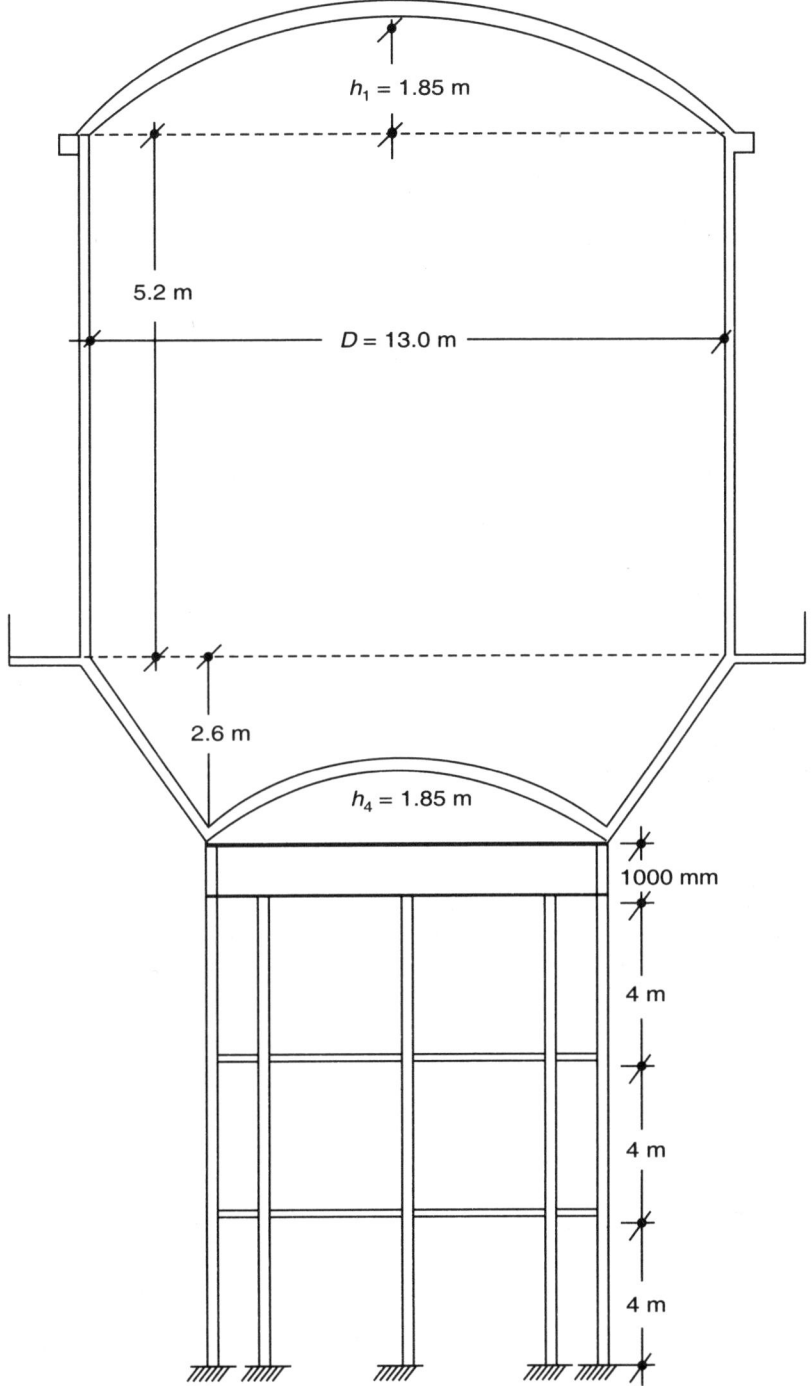

$h_1 = 1.85$ m

5.2 m

$D = 13.0$ m

2.6 m

$h_4 = 1.85$ m

1000 mm

4 m

4 m

4 m

Fig. 11.26: Intze tank with supporting structure (columns and braces)

Solution: *Vertical load on each column:*

Load from tank = 360.1 kN/m × length of ring beam between two columns

$$= 360.1 \times \left(4 \times \frac{\pi}{4}\right) = 1131.3 \text{ kN}$$

Self-weight of column (assuming 600 mm diameter and 12 m height)

$$= \frac{\pi}{4} (0.6)^2 \times 12 \times 25 = 85 \text{ kN each}$$

Wt. of braces (assuming size of 300 mm × 500 mm) and length

$$\text{length} = \frac{2\pi R}{8} = \frac{\pi \times 4}{4} = 3.14 \text{ m (c/c column)}$$

Wt. of 3 braces between 2 columns = 0.3 × 0.5 × 3.14 × 3 × 25 = 35.3 kN

Total vertical load on each column = 85 + 35.3 + 1131.3 = 1251.6 ≅ 1252 kN

Wind loads:

Intensity of wind = 1.5 kN/m²

Shape factor for dome, cylindrical wall, conical dome,bottom ring beam, etc. = 0.70

Wind load on dome, cylindrical wall and conical dome,

$$P_u = 1.5 \times 0.7 \left[\frac{2}{3} \times 13.08 \times 1.85 + 13.5 \times 5.2 + \left(\frac{13.7 + 8 + 2 \times 0.35}{2}\right) \times 3.607\right]$$

$$= 1.05[16.13 + 70.2 + 40.32] = 133.07 \cong 133.0 \text{ kN}$$

This may be assumed to act at about 0.52D from the top of bottom ring, i.e. 0.52 × 13 = 6.76 ≅ 6.80 m from the top of bottom ring.

Hence, total height from the bottom bay of columns, i.e. 6.8 +1.0 + (12 – 2) = 17.8 m above the point of contraflexure = 17.8 m above point O_1.

Wind load on bottom girder = 1.5 × 0.7 × 1.0 × (8.0 + 2 × 0.6) = 9.66 kN acting at (12 – 2 + 0.5) = 10.5 m from O_1.

Wind load on column = 0.7 × 1.5 × 0.6 × 12 × 8 = 60.48 kN

Wind load on bracing = 1.5 × 0.3 × 0.5 × 8 × 2.0 = 3.60 kN

Wind force (column and bracing) = 60.48 + 3.60 = 64.08 kN

(This load acts at middle height from bottom = $\frac{12}{2}$ = 6 m or 4 m above point O_1.

Wind load moment about O_1:

M = 133 × 17.8 + 9.66 × (12 + 0.5 – 2) + 64.08 × 4 = (2367.4 + 101.43 + 256.32) m

$$M = 2725.15 \text{ kN} = \frac{V}{\gamma}\Sigma a^2 = \frac{V}{(8 \times 0.5)}\left[2 \times 4^2 + 4\left(\frac{4}{\sqrt{2}}\right)^2 + 2 \times 0\right] = \frac{V}{4}[32 + 32] = 16V$$

$$V = \frac{2725.15}{16} = 170.32 \text{ kN, max. load on column} = 1252 + 170.32 = 1422.32 \text{ kN}$$

Q_1 (total wind load) = $133 + 9.66 + 64.08 = 206.74$ kN

Total moment at base of the column = $206.74 \times 2 = 413.48$ kN-m

Moment at base of each column = $\dfrac{413.48}{8} = 51.7$ kN-m

Forces in braces:

Moment at joint from lower column = $\dfrac{Q_1}{n} \dfrac{h}{2} = \dfrac{206.74}{8} \times \dfrac{4}{2} = 51.7$ kN-m

Moment from upper column = $\dfrac{413.48}{8} = 51.7$ kN-m each.

(Considering $M'_2 = M'_1 = 413.48$ kN-m)

Total joint moment from column = $(51.7 + 51.7) = 103.4$ kN-m

Braces will be subjected to this moment = 103.4 kN-m

If the wind pressure is normal to the brace at the joint, the joint moment in the brace

$$\dfrac{103.4}{\cos \theta} = \dfrac{103.4}{\cos 45°} = 103.4\sqrt{2} = 146.23 \text{ kN-m}$$

SF in the brace ($l = 3.14$ m) = $\dfrac{146.23}{(\text{length of brace})} = \dfrac{146.23}{3.14} = 46.57$ kN

Design of column: (limit state design $f_{ck} = 25$ N/mm^2, $f_y = 415$ N/mm^2)

Column shall be designed for an axial load of 1422 kN and moment of 51.7 kN-m {$1252 + 170.32$ (wind effect) $= 1422.32 \cong 1422$ kN}

e (eccentricity of load) = $\dfrac{51.7 \times 1000}{1422.32} = 36.35$ mm > 28 mm (e_{min})

$$\left\{ e_{min} = \dfrac{L_e}{500} + \dfrac{600}{30} = \left(\dfrac{4000}{500} + \dfrac{600}{30} \right) = 28 \text{ mm} \right\}$$

The column shall be designed for critical axial load of 1422 kN and a moment of 51.7 kN-m.

$P_u = 1.5(1422) = 2133$ kN each column, $M_u = 1.5 \times 51.7 = 77.55$ kN-m each column (78 kN-m, say)

Using 1% steel (0.8 to 6%), $A_{sc} = \dfrac{A_c}{100}$,

2133000 N $= 0.4f_{ck} \cdot A_c + 0.67f_y \left(\dfrac{A_c}{100} \right) = 0.4 \times 25\,(0.99A_g) + 0.67 \times 415 \left(\dfrac{0.99A_g}{100} \right)$

or $2133000 = 12.6527\, A_g$, or $A_g = \dfrac{2133000}{12.6527} = 168581$ mm^2

$\dfrac{\pi}{4}(D)^2 = 168581$, $\therefore D = 463.3$, provide $D = 500$ mm, clear cover = 40 mm, core = 420 mm

$$A_{sc} = \frac{A_c}{100} = \frac{0.99}{100} \frac{\pi}{4} \times (500)^2 = 1944 \text{ mm}^2 \text{ (provide } 20\phi \text{ 8 bars, } A_{sc} = 2512 \text{ mm}^2) \text{ or}$$

$$\text{check stress due to BM} = \frac{77550000}{\dfrac{\pi}{32}(500)^3} = \pm 6.32 \text{ N/mm}^2$$

$$A_e = (m-1) A_{sc} + A_c = (11-1) 2512 + 196350$$
$$A_e = 25120 + 196350 = 221470 \text{ mm}^2$$

$$\text{Direct stress in concrete} = \frac{2133000}{221470} = 9.63 \text{ N/mm}^2,$$

$$p_{max} = 9.63 \pm 6.32 = +15.95 < \frac{25}{1.5} \text{ N/mm}^2 = 16.67, \text{ OK or}$$

$$+ 3.31 \text{ N/mm}^2, \text{ OK.}$$

Design of bracing (limit state method):

$$M = 146.23 \text{ kN-m}, M_u = 1.50 \times 146.23 = 220 \text{ kN-m } (220 \times 10^6 \text{ N-mm})$$
$$\text{SF } V = 46.57 \text{ kN}, V_u = 1.5 \times 46.57 = 70.0 \text{ kN } (70000 \text{ N})$$

$M_{u \text{ lim}}$(for bracing 0.30 m × 0.50 m) = $0.1380 \times 25 \times 300 \times 450^2$ ($R_u \cdot f_{ck} \cdot bd^2$)

$$= 209.6 \times 10^6 \text{ N-mm} < M_u \text{ of } 220 \times 10^6 \text{ N-mm, increase size,}$$

provide bracing 300 mm × 600 mm size

$$(M_{u \text{ lim}} = 0.138 \times 25 \times 300 \times 550^2 = 313 \times 10^6 \text{ N-mm}), \text{ OK.}$$

$$A_{st} \text{ reqd.} = \frac{0.50 \times 25 \times 300 \times 550}{415} \left(1 - \sqrt{1 - \frac{4.6 \times 220 \times 10^6}{25 \times 300 \times 550^2}}\right) = 1271 \text{ mm}^2$$

Provide 5 bars 20 mm (A_{st} = 1570 mm²), continue 3 bars 20ϕ in top face and in bottom face throughout. At junction provide 2 bars 20ϕ diameter extra in top face and 2 bars 20ϕ extra in bottom face, i.e. total 5 bars 20ϕ both in bottom and top each, at the column–bracing junction.

$$\% \text{ steel } p_t = \frac{1570 \times 100}{300 \times 550} = 0.95\%, \qquad \tau_c \text{ (design shear)} = 0.63 \text{ N/mm}^2$$

$$\text{Minimum temperature steel} = \frac{0.12}{100} \times 300 \times 600 = 216 \text{ mm}^2 < 1570 \text{ mm}^2, \text{ OK.}$$

(12ϕ 2 bars = 226 mm²)

Since, the depth of 600 mm is more than 450 mm, provide in middle 0.10% steel

$$= \frac{0.10}{100} \times 300 \times 600 = 180 \text{ mm}^2 \text{ (12}\phi \text{ 2 bars} = 226 \text{ mm}^2)$$

Check for shear:

$$\text{Shear stress } \tau_v = \frac{70000}{300 \times 550} = 0.424 \text{ N/mm}^2 \text{ (<0.63 N/mm}^2, \text{ i.e. } \tau_c \text{ design shear}$$

stress).

Provide nominal vertical stirrups 8ϕ 2 legged @ 300 c/c (A_{st} = 333 mm^2/m) throughout.

Alternative system of supporting in the form of cylindrical shaft shall be discussed.

Design of staging shaft:

For supporting OHSR, cylindrical shaft (hollow type) is provided as modern design instead of column and bracing type of staging. The shaft construction is easier and economical specially for heavy capacity tanks and high elevations. Simple design of shaft is based on provision of minimum shell thickness and minimum vertical and horizontal steel reinforcements and limiting compressive and tensile stresses.

Compressive or tensile stresses are developed due to direct vertical loads and bending moments due to wind loads or earthquake loads. The wind pressures and earthquake acceleration coefficients are selected from the respective Indian standards relevant to the geographical location.

(i) (a) Minimum thickness of shell (t_1) = 125 mm

(b) Thickness based on permissible stresses (working stress design) and on design stresses (limit state design)t_2.

Total load on the shaft base (W_f) = W_o (empty tank) + wt. of water (W_w)

$$+ W_{sh} \text{ (wt. of shaft)} = W_o + W_w + W_{sh}$$

Moment due to wind pressure at the shaft base,

$M_1 \text{ (wind)} = P_1 \times h_1 + P_2 \times h_2$

where, P_1 = Wind pressure on the tank

h_1 = CG of tank from the shaft base

P_2 = Wind pressure on the shaft

h_2 = CG of shaft wind pressure from the shaft base.

$M_2 \text{ (EQ)} = \alpha_h (W_o + W_w) \cdot h_3 + \alpha_h (W_{sh}) h_4$

where, α_h = Earthquake coefficient for the place and h_3 and h_4 are distances of CG of loads from the shaft base.

(ii) (a) Minimum vertical steel = 0.24% in case of HYSD and 0.30% in case of MS

(b) Minimum horizontal steel = 0.20% in case of HYSD and 0.25% in case of MS

(iii) Equivalent cross-section of shaft

(a) t_e (equivalent thickness) = $t_o + \dfrac{0.24}{100} t_o \times (m-1) = \left\{ 1 + \dfrac{0.24}{100}(m-1) \right\} t_o$

where, t_o = Assumed thickness (minimum 125 mm)

m = Modular ratio

$\left(m = \dfrac{280}{3\sigma_{cbc}} , \sigma_{cbc} = \text{Permissible stress in concrete} \right)$

For example, m = 13 for M20 grade of concrete

and t_e(for minimum thickness and M20 CC)

$$= \left(125 + \dfrac{12 \times 0.24 \times 125}{100} \right) = 128.6 \text{ mm}$$

(b) A_e (equivalent area) $= \pi D_o \times t_e$

where, D_o = Centreline diameter of shaft and t_e = equivalent thickness.

(iv) Equivalent section modulus $Z_e = \pi R_o^3 t_e / R'$

where, R_o = Centreline radius of the shaft

t_e = Equivalent thickness and

R' = Outer radius

(v) Check of stresses w.r.t. equivalent cross-section.

$$\text{Max. stress (wind and tank full)} = \left(\frac{W_F}{A_e} \pm \frac{M_1}{Z_e} \right), \qquad \dots \text{Eq. (11.39)}$$

(not more than permissible stress)

where, W'_F = Total load (tank full)

M_1 = Total wind pressure moment (tank full)

A_e and Z_e are equivalent areas and section modulus respectively.

$$\text{Max. stress (EQ and tank full)} = \left(\frac{W'_F}{A_e} \pm \frac{M_2}{Z_e} \right) \qquad \dots \text{Eq. (11.40)}$$

(not more than permissible stress)

where, W'_F = Total load (tank full with earthquake effect),

M_2 = Total earthquake moment due to horizontal acceleration

A_e and Z_e are equivalent cross-sectional area and section modulus respectively.

Similarly stresses are also checked with tank empty.

(vi) Stresses are further checked by interaction formula in all the cases of wind pressure and earthquake as follows:

$$\frac{\text{Direct stress}}{\text{Direct permissible stress}} + \frac{\text{Bending stress due to wind or EQ}}{\text{Permissible bending stress}} \not> 1.33 \quad \dots \text{(11.41)}$$

(vii) A check for buckling stress can also be done by calculating reduction factor as:

$$C_r = \left[1 + \frac{f_{ck}}{0.2 E_c} \times \frac{R}{t} \right]^{-1}, E_c = 5000 \sqrt{f_{ck}} \text{ N/mm}^2$$

For example, for M20 CC and $R = 2300$ mm, $t = 125$ mm, $E_c = 0.22361 \times 10^5$ N/mm²

$$\left[C_r = 1 + \frac{20}{(0.2 \times 22361)} \times \frac{2300}{125} \right]^{-1} = 0.924$$

Permissible stress under vertical load $= 5.0 \times 0.924 = 4.62$ N/mm²

[M20 CC, f_c (direct) = 5 N/mm²]

Check direct stress under full tank case $= \dfrac{W_F}{A_e} \le 4.62$ N/mm² (for M20 CC)

(viii) *Maximum shear stress*: Total wind pressure or horizontal EQ force divided by effective area A_e.

Max. shear stress $\le$ (1.33 $\times$ permissible shear stress for the given grade of concrete).

Consider the following example for understanding the design of staging shaft.

Design of column–bracing, staging and foundation

Example 11.9

Design the shaft staging for the tank, following the data from Example 11.8:

Bearing capacity (SBC) = 120 kN/m^2

Ultimate (DL + LL) combination = 1.50 (1152) $\times$ 8 = 15024 kN (all the columns)

Ultimate (DL + LL + WL) combination = 1.2 (1252 + 170) $\times$ 8 = 13651 kN (all the columns)

Solution: Thus, (DL + LL) combination is critical = 15024 kN

Self-weight of columns = 1502 kN. Thus, total load at the base of columns = 16526.

Self-weight of foundation $\dfrac{10}{100}$ (16526) = 1653 kN.

Total load on foundation soil = 16526 + 1653 = 18179 kN

Let there be circular slab base for columns and projection on outside.

Diameter of foundation on slab = D

$$\frac{\pi}{4}(D)^2 = \frac{18179}{1.5 \times \text{SBC}} = \frac{18179}{1.5 \times 120} = 101 \text{ m}^2$$

$$D = \sqrt{\frac{101 \times 4}{\pi}} = 11.34 \text{ m, provide foundation slab diameter} = 12 \text{ m}$$

Projection beyond column – strap beam = $\dfrac{(12-8)}{2} - 0.25 = 1.75$ m

Upward soil pressure = $\dfrac{18179 \times 4}{\pi(12)^2}$ = 161 kN/m^2

Assuming base slab 500 mm, w = 0.50 $\times$ 25 = 12.5 kN/m^2 $\times$ 1.5 = 18.75 kN/m^2

Thus, net ultimate upward loading on the base slab = 161 – 18.75 = 142.25 kN/m^2

Cantilever projection 1.75 m, upward loading w_u = 142.25 kN/m^2

For approximate design:

SF at the face of strap beam V_u = 142.25 $\times$ 1.75 = 249 kN/m

(projection M = 218 kN-m/m).

BM at the face of strap beam $M_u = \dfrac{142.25 \times 1.75^2}{2}$ = 218 kN-m/m.

Assuming % steel as 0.25%, τ_c= 0.36 N/mm^2

Depth d (based on BM) = $\sqrt{\dfrac{218 \times 10^6}{0.138 \times 25 \times 1000}}$ = 251.4 mm,

provide D = 500 mm, d = 450 mm (say)

SF at d from face (at 0.45 from the face), $V'_u = 142.25\ (1.75 - 0.45) = 185.0\ \text{kN/m}$

Depth d (based on SF) $= \dfrac{185000}{1000(0.36)} = 514\ \text{mm}$ (provide $d = 550\ \text{mm}$, $D = 600\ \text{mm}$)

A_{st} (bottom face) $= \dfrac{0.5 \times 25 \times 1000 \times 550}{415} \left[1 - \sqrt{1 - \dfrac{4.6 \times 218 \times 10^6}{25 \times 1000\ (550)^2}}\right] = 1138\ \text{mm}^2/\text{m}$

(16ϕ 170 c/c, $A_{st} = 1182\ \text{mm}^2/\text{m}$)

Min. temp. steel $= \dfrac{0.12}{100} \times 1000 \times 600 = 720\ \text{mm}^2$ (each face 360 mm^2), $\%\ p_t = 0.215\%$

(12ϕ 300 c/c, A_{st} each face $= 377\ \text{mm}^2/\text{m}$, OK).

provide in bottom face 16ϕ @ 170 c/c radially. Also provide in bottom ring bars 12ϕ 300 c/c and radial bars 12ϕ 300 c/c in top and 12ϕ 300 c/c ring bars in top face.

Weight of empty tank (W_o):

Top dome, $W_1 = \left(4\pi\, r^2\, \dfrac{2\theta}{360}\right) t \times 25$

$\quad = \dfrac{4\pi \times 2 \times 31.7}{360} \times (12.38)^2 \times 0.08 \times 25 \qquad\qquad = 678\ \text{kN}$

Tank wall, $W_2\ (\pi\, Dh.t \times 25) = \pi\ (13.25)\ 5.2 \times 0.25 \times 25 \qquad = 1353\ \text{kN}$

Top beam A, W_3 Add $l = (\pi \times 13.6 \times 0.1 \times 0.35 \times 25) = 13.78 \qquad = 37\ \text{kN}$

Middle beam, (0.90 m $\times$ 0.50 m), $W_4 = \pi\ (13.5 + 0.95)0.5 \times 0.95 \times 25 \qquad = 539.08$

$\qquad\qquad\qquad\qquad\qquad\qquad\qquad\qquad\qquad\qquad\qquad\qquad\qquad = 539\ \text{kN}$

Concial wall, ($L = 3.61$ m), $W_5 = \pi(10.5 + 0.35)0.35 \times 3.61 \times 25 = 1077 \quad = 1077\ \text{kN}$

Lower dome, $W_6 = 4\pi\ (5.25)^2 \left(\dfrac{2 \times 49.64}{360}\right) \times 0.2 \times 25 = 478 \qquad = 478\ \text{kN}$

Lower ring beam, (0.6 m $\times$ 1.0 m), $W_7 = \pi(8.6)1.0 \times 0.60 \times 25 = 405 \qquad = 405\ \text{kN}$

Weight of empty tank (W_o) $\qquad\qquad\qquad\qquad\qquad\qquad\qquad\qquad\qquad = 4567\ \text{kN}$

Weight of tank full (W_{full}) $= (4567\ \text{kN} + 8700\ \text{KN}) = 13267\ \text{kN}$

Approximate average CG above bottom of ring beam $= 5.0$ m (say)

(Exact CG can be found by $\tilde{y} = \dfrac{\Sigma M}{\Sigma W}$

Let the shaft be 0.20 m thick and c/c diameter 8.0 m, $D_o = 8.2$ m, $D_i = 7.8$ m

Weight of staging shaft (0.20 m) $= \pi \cdot D \cdot L \cdot t \times 25 = \pi\ 8 \times 12 \times 0.2 \times 25 = 1508\ \text{kN}$

Weight of foundation 10% of $(13267 + 1508) = 1478\ \text{kN}$

Weight of full tank and staging $= 14775\ \text{kN}$

Weight of full tank, staging and foundation $= 16253\ \text{kN}$

Weight of empty tank, staging and foundation $= 7553\ \text{kN}$

Direct stress (shaft bottom), tank full $= \dfrac{14775 \times 10^3}{\pi(8)(0.20)10^6} = 2.94\ \text{N/mm}^2$ ($<6\ \text{N/mm}^2$,

OK).

$$\text{Area of foundation required} = \frac{16253 \times 1.5}{120 \times 1.5} = 135.44 \text{ m}^2$$

$$\frac{\pi}{4} D^2 = 135.44, D = \sqrt{\frac{135.44 \times 4}{\pi}} = \sqrt{172.75} = 13.13 \text{ m, provide } D = 13.4 \text{ m}$$

$$\text{Thus, upward pressure of soil} = \frac{162525}{\left(\frac{\pi}{4} \times 13.4^2\right)} = 115.25 \text{ kN/m}^2 < (120 \times 1.5), \text{ OK.}$$

Check stress in shaft:

Ultimate direct stress in shaft base = 2.94 N/mm² × 1.5

$$= 4.41 \text{ N/mm}^2 \text{ (ultimate stress)}$$

$$\text{Ultimate bending stress} = \pm \frac{3274.27 \times 10^6 \times 1.5}{Z_s} = \frac{3274.27 \times 10^6 \times 1.5}{9814.03 \times 10^6}$$

$$= \pm 0.334 \text{ N/mm}^2 \times 1.5 = 0.501 \text{ N/mm}^2 \text{ (ultimate stress)}$$

Ultimate net stress = 1.5 (2.94 ± 0.334) = (3.274)1.5 N/mm² and min.

$$= +(2.606) \times 1.5 \text{ N/mm}^2, \text{ i.e. } 4.911 \text{ and } 3.91 \text{ N/mm}^2 < f_{ck}, \text{ OK.}$$

Wind load on tank elements:

(wind pressure = 1.50 kN/m², shape factor = 0.70, approximate CG of wind pressure 5.0 m above) bottom of lower ring beam of the tank.

Tank elements:

$$= 1.5 \times 0.7 \left[\frac{2}{3} \times 13.16 \times (1.85 + 0.08) + 13.50 \times 5.2 + \left(\frac{13+8}{2} + 0.70 \right) 2.6 + 8.6 \times 1 \right]$$

P_1 = 148.87 kN at 5.0 m above bottom of bottom ring beam CC.

Staging (circular shaft thickness 0.20 m, height = 12 m above GL)

P_2 = 1.5 × 0.7 × (8 + 0.2) 12 = 103.32 kN at 6 m above GL

Let foundation be 2 m deep (below GL).

Moment of wind forces about bottom of shaft,

M_1 = 148.87 × (12 + 5.00) + 103.32 × 6 = 2530.79 + 619.92 = 3150.71 kN-m

Moment of wind forces about base of foundation (2 m below GL)

M_2 = 148.87 (5.0 + 12.0 + 2) + 103.32(6 + 2) = 2828.53 + 826.56 = 3655.1 kN-m

$$Z_s \text{ of shaft base} = \frac{\frac{\pi}{64}(8.2^4 - 7.8^4)1000^4}{\frac{8.2}{2} \times 1000} = \frac{\pi \times 2 \times 10^9}{64 \times 8.2} \times (8.2^4 - 7.8^4) = 9814.03 \times$$

10^6 mm^3

$$(Z_s \text{ also} \cong \frac{\pi R^3 \cdot t}{R_o} = 9808 \times 10^6 \text{ mm}^2)$$

Foundation slab is 13.4 m in diameter and *e* (say) 0.60 m thick.

$$A_f = \frac{\pi}{4} (13.4)^2 \times 10^6 = 141.03 \times 10^6 \text{ mm}^2 \text{ (area of foundation} = A_f)$$

Z_f (foundation slab) = $\dfrac{\pi}{32}(d^3)$ = $\dfrac{\pi}{32}(13.4)^3 \times 10^9 = 236.22 \times 10^9\,\text{mm}^3$

Stresses in foundation:

(a) Tank full: $\dfrac{16253 \times 1000}{141.03 \times 10^6} \pm \dfrac{3655.10 \times 10^6}{236.22 \times 10^9} = +0.115245 \pm 0.0155$

$= +0.1307\,\text{N/mm}^2 < 6.0\,\text{N/mm}^2$

(b) Tank empty: $\dfrac{(4567 + 1508 + 1478)10^3}{141.03 \times 10^6} \pm \dfrac{3655.1 \times 10^6}{236.22 \times 10^6} = 0.0536 \pm 0.0155$

$= +0.0691\,\text{and} + 0.0381\,\text{kN/m}^2, \text{OK.}$

Similarly under earthquake ($\alpha = 0.075\,\text{ms}^{-2}$)

P_{H_1} = Tank force (earthquake) = $0.075 \times 13267 = 995$ kN at $(5 + 12 + 2)$ 19 m above

P_{H_2} = Shaft force (earthquake) = $0.075 \times 1508 = 113.1$ kN at $(6 + 2)$ 8 m above foundation base.

Also vertical force (earthquake) = $\dfrac{1}{2} \times 995 = 498$ kN and 56.0 kN (total = 554 kN)

Moment = $995 \times 19 + 113.1 \times 8 = 18905 + 905 = 19810$ kN-m

Foundation stress (max) = $\pm \dfrac{19810 \times 10^6}{236.22 \times 10^9} + \dfrac{(554 + 16253)10^3}{141.03 \times 10^6} = \pm 0.084 + 0.1192$

$= +0.2032\,\text{N/mm}^2, +0.0352\,\text{N/mm}^2$

Check of stresses in foundation:

Shaft $A_s = 5.027 \times 10^6$, $Z_s = 9814 \times 10^6\,\text{mm}^3$

Foundation $A_{F_2} = 141.03 \times 10^6$, $Z_f = 236.22 \times 10^9\,\text{mm}^3$

Case I: Only gravity forces

(a) Tank empty weight W_o (total) = 4567 +1508 +1478 = 7553 kN, M (total) = 0

(b) Tank full weight W_1 (total) = 16253 kN, $M_1 = 0$

Case II: Wind force ($p = 1.5\,\text{kN/m}^2$)

(a) Tank empty weight, W_o (total) = 7553 kN (foundation), 6075 kN (shaft)

Moment (wind) shaft, $M_1 = 3150.71$ kN-m

Moment (wind) foundation, $M_2 = 3655.1$ kN-m

(b) Tank full weight, W_1 (total) = 16253 kN

Moment (wind) shaft, $M_1 = 3150.71$ kN-m

Moment (wind) foundation, $M_2 = 3655.1$ kN-m

Case III: Earthquake force ($\alpha = 0.075\,\text{m/sec}^2$), $M_1 = 6075 \times 0.075 \times 17 = 7746$ kN-m

Check stresses in shaft bottom:

Case I: (a) Stress = $\dfrac{(4567 + 1508)10^3}{5.027 \times 10^6} \pm 0 = 1.21\,\text{N/mm}^2, \text{OK.}$

(b) Stress = $\dfrac{16253 \times 10^3}{5.027 \times 10^6} = +3.233\,\text{N/mm}^2, \text{OK} < 6.0\,\text{N/mm}^2$

Case II: (a) Stress $= \pm \dfrac{6075 \times 10^3}{5.027 \times 10^6} \pm \dfrac{3150.71 \times 10^6}{9814 \times 10^6} = 1.21 \pm 0.321 = 1.531 \, \text{N/mm}^2$

$$(< 6 \times 1.33 = 8)$$

(b) Stress $= \dfrac{16253 \times 10^3}{5.027 \times 10^6} \pm \dfrac{3150.71 \times 10^6}{9814 \times 10^6} = 3.233 \pm 0.321 = +3.554 \, \text{N/mm}^2$

$$(< 6 \times 1.33)$$

Case III: (a) Stress $= \dfrac{6075 \times 10^3}{5.027 \times 10^3} \pm \left\{ \dfrac{4567 \times 0.075 \times 17 + 1508 \times 0.075 \times 6}{9814 \times 10^6} \right\} 10^3$

$$= 1.21 \pm 0.000662 = 1.211 \, \text{N/mm}^2$$

(b) Stress $= \dfrac{16253 \times 10^3}{5.027 \times 10^3} \pm \left(\dfrac{(13267 \times 17 + 1508 \times 6)0.075 \times 10^3}{9814 \times 10^6} \right)$

$$= 3.233 \pm 0.001793 = +3.235 \, \text{N/mm}^2$$

Stress check in foundation slab:

Foundation slab diameter $D = 13.4$ m, $r = 6.7$ m, $A_f = 141.03 \times 10^6 \, \text{mm}^2$, $Z_f = 236.22 \times 10^9 \, \text{mm}^3$

Case I: Gravity loads only:

(a) Tank empty: $\dfrac{(4567 + 1508 + 1478)10^3}{141.03 \times 10^6} = + \dfrac{7553 \times 10^3}{141.03 \times 10^6} = 0.054 \, \text{N/mm}^2$

(b) Tank full: $\dfrac{16253 \times 10^3}{141.03 \times 10^6} = 0.1152 \, \text{N/mm}^2$

Case II: Wind pressure

(a) $0.054 \pm \dfrac{3655.1 \times 10^6}{236.22 \times 10^9} = 0.054 \pm 0.0155 = 0.0695, 0.0385 \, \text{N/mm}^2$

(b) $0.1152 \pm \dfrac{3655.1 \times 10^6}{236.22 \times 10^9} = 0.1152 \pm 0.0155 = 0.1307, 0.10 \, \text{N/mm}^2$

Case III: Earthquake

(a) $0.054 \pm \dfrac{19810 \times 10^6}{236.22 \times 10^9} = 0.054 \pm 0.084 + \dfrac{554 \times 1000}{141.03 \times 10^6} = 0.142 \, \text{N/mm}^2$

(b) $0.1152 \pm 0.0039 \pm 0.084 = 0.203 \, \text{N/mm}^2 < 1.33 \times 6$, OK.

It may be noted that the tank components are designed only as a sample and in real design the sections are so designed that the safety margin reduces, just equal to permitted value so as to make the section economical. For making the section economical, it may require number of cycles and computer design shall be very useful.

Example 11.10

Design a cylindrical shaft staging for a 50,000 litres capacity OHSR with staging height of 20 m above GL and water depth 3.50 m with 0.30 m free board. Shaft centre line diameter is 3.40 m (3.55 m outer and 3.25 m internal). Thickness of tank walls is 125 mm. Depth of foundation 1.52 m below GL. Permissible stress increased by 33 $\frac{1}{3}$ % in case of wind and earthquake forces. Permissible SBC = 70 kN/m² (vertical loads) while 105 kN/m² for earth quake effects and 99 kN/m² for wind effects. Seismic coefficients at the place are α_h = 0.075 ms⁻², α_v = 0.0375 ms⁻².

Assume tank wall thickness = 125 mm, top slab 100 mm thick, diameter of tank = 4.40 m, weight of empty tank (W_1) = 303.00 kN, weight of tank water (W_2) = 503 kN, average CG above bottom of base = 2 m, shaft wall thickness = 150 mm, inner radius = 1.625 m, outer radius = 1.775 m, centre line radius = 1.70 m, height of water in tank = 3.50 m. Design the shaft using M20 CC and Fe 415 HYSD bars by working stress method.

Solution: f_{ck} = 20 N/mm², f_y = 415 N/mm², σ_{cbc}= 7 N/mm², σ_c = 5 N/mm², σ_{st} = 230 N/mm², SBC = 70 kN/m², (with wind effect = 99 kN/m², with EQ effect = 105 kN/m²), $W_w \cong$ 5,00,000 N = 503 kN (given), W_o (empty) = 303 kN, CL rad. shaft = 1.70 m,

height of tank CG from foundation slab top = 20 + 0.30 − $\dfrac{3.80 + 0.2}{2}$ + (1.52 − 0.30)

= 19.52 m, (assuming thickness of foundation slab = 300 mm)

Height of shaft = (20 + 1.12 + 0.3 − 4) = 17.42 m, t_{shaft} = 0.15 m

Weight of shaft = $\pi D t$ × 17.42 × 25 $\cong$ 694 kN, CG 8.71 m above base

Weight of tank (full) = 303 + 503 = 806 kN, CG (average) = 17.42 + 2.1 = 19.52 m

Weight of tank (empty) = 303 kN, CG (average) = 19.52 m from base

Wind force (tank) = 0.7 × 4.52 × 4.1 × 1.5 = 19.56 kN, CG = 19.52 m from base.

(Diameter = 4.52 m, height = 4.1 m, coeff. = 0.7, wind pressure = 1.5 kN/m²)

Wind force shaft = 0.7 (17.42 − 1.12) × (2 × 1.84) 1.50 = 63 kN, CG = 9.27 m (above GL)

Moment wind force (tank) = 19.56 × 19.52 = 382 kN-m

Moment wind force (shaft) = 63 × 9.27 = 584 kN-m

Total moment (wind) about base = 382 + 584 = 966 kN-m (966 × 10⁶ N-mm)

Similarly, earthquake forces (IS:1893)

Given α_H = 0.075, α_v = 0.0375, weight of tank full = 806 kN

Horizontal EQ force (tank full) = 806 × 0.075 = 64.450 kN, CG (av.) = 19.52 m

Vertical force (EQ) full = 806 × 0.0375 = 30.22 kN

Total max. (EQ) wt. (tank) = 806 ± 30.23 = 836.23 kN, CG 0.0 (min. 775.78 kN)

Horizontal EQ force (shaft) = 694 × 0.075 = 52.05 kN, CG = 9.27 m

Max. (EQ) wt. (shaft) = 694 × 0.0375 = 26.03 kN, CG 0.0

Total max. vertical load (shaft base)= 806 + 30.23 + 694 + 26.03 = 1556.263 kN

Moment (EQ) about base (M_2) = 60.45 × 19.52 + 52.05 × 9.27 = 1180 + 482.5 = 1662.5 kN-m

Thickness of shaft t (given) = 150 mm, A_v radius = 1700 mm, outer R = 1850 mm

Equivalent thickness t_e for 0.24% vertical steel $= 150 + \dfrac{0.24}{100}(13-1)150 = 154.32$ mm

Equivalent cross–section area $A_e = \dfrac{154.32}{1000} \times \pi\,(2 \times 1.70) = 1.648$ m^2

or $(1.650 \times 10^6\,\text{mm}^2)$

Equivalent section modulus $Z_e = \dfrac{\pi \cdot R^3 \cdot t_e}{R_0} = \dfrac{\pi \cdot (1700)^3 \times 154.32}{1850} = 1287.5 \times 10^6\,\text{mm}^3$

Check stresses in staging shaft:

(i) Direct stress:

Total weight (full) $= 806 + 694 = 1500$ kN.

Total weight (empty) $= 303 + 694 = 997$ kN

$$\text{Direct stress in base of the shaft (full)} = \frac{1500 \times 10^3}{1.65 \times 10^6} = 0.910\,\text{N/mm}^2$$

$$\text{(very small} < 5.0\,\text{N/mm}^2\,\text{permissible)}$$

$$\text{Direct stress in base of the shaft (full and EQ)} = \frac{1556.26 \times 10^3}{1.65 \times 10^6}$$

$$= 0.94\,\text{N/mm}^2\,\text{(very small), OK.}$$

$$\text{Direct stress in base of shaft (empty)} = \frac{(303 + 694) \times 10^3}{1.65 \times 10^6}$$

$$= 0.604\,\text{N/mm}^2\,\text{(very small), OK.}$$

(ii) Wind action:

$$\text{Max. stress (empty tank)} = 0.604 \pm \frac{966 \times 10^6}{Z_e} = 0.604 \pm \frac{966 \times 10^6}{1287.5 \times 10^6}$$

$$= 0.604 \pm 0.7503 = -0.146\,\text{N/mm}^2, + 1.354\,\text{N/mm}^2$$
$$(< 5 \times 1.33\,\text{N/mm}^2),\,\text{OK.}$$
$$(-0.146\,\text{also small less than permitted), OK.}$$

$$\text{Max. stress (tank full)} = \frac{1500 \times 10^3}{1.65 \times 10^6} \pm \frac{966 \times 10^6}{Z_e} = (0.910 \pm 0.75)$$

$$= + 1.660\,(< 1.33 \times 5\,\text{N/mm}^2),\,\text{and}$$
$$+ 0.160\,\text{N/mm}^2,\,\text{(small), OK.}$$

(iii) Earthquake:

$$\text{Max. stress (tank full)} = + 0.94 \pm \frac{1662.5 \times 10^6}{1287.5 \times 10^6} = + 0.94 \pm 1.2913$$

$$= + 2.2313\,\text{N/mm}^2,\,\text{and}$$

minimum stress (tank full) $= 0.94 - 1.2913 = -0.3513\,\text{N/mm}^2$

Thus, earthquake condition is most critical and gives rise to maximum compressive stress of 2.231 N/mm² $(< 5 \times \frac{4}{3}, \text{OK.})$

Maximum tensile stress of 0.3513 N/mm² $\left(< \frac{4}{3} \times 1.7 \text{ or } \frac{1}{3} \times 2.2313, \text{OK.} \right)$

Interaction check

Moment (EQ): $\dfrac{0.94}{(1.33 \times 5)} + \dfrac{1.2913}{(7 \times 1.33)} = 0.1414 + 0.1387 = 0.2801 < 1.0, \text{OK.}$

Permissible tensile stress $= \dfrac{1}{3}$ (1.2913) or 1.33 (1.7) N/mm², whichever is less,

i.e. 0.4304 or 2.261, i.e. 0.430 N/mm² > 0.3513 N/mm², actual, OK.
Check for buckling stress (M20 grade CC)

Modulus of elasticity for M20 concrete $= 5000 \sqrt{f_{ck}} = 5000 \sqrt{20} = 2.24 \times 10^4$ N/mm²

Reduction factor $\left(1 + \dfrac{f_{ck}}{0.2E} \cdot \dfrac{R}{t}\right)^{-1} = \left(1 + \dfrac{20 \times 1700}{0.2 \times 2.24 \times 10^4 \times 150}\right)^{-1} = (1.0506)^{-1} = 0.9518$

Thus max. permissible comp. stress = 0.9518 × 5 = 4.76 N/mm²
Actual maximum comp. stress = 2.2313 N/mm² (< 4.76 N/mm², OK.)

Check for Shear Stress

Max. shear force = max. horizontal force (wind) = (19.56 + 63 = 82.56 kN) or
Max. shear force = max. horizontal (EQ) = (60.45 + 52.05 = 112.5 kN)

Max. shear stress (EQ effect) $= \dfrac{112.5 \times 10^3 \times 4}{\pi D \cdot t \times 3}$

$= 0.0942 \text{ N/mm}^2 < \dfrac{4}{3} \times 0.22 \text{ N/mm}^2, \text{(OK)}$

Provide thickening at the base, i.e. from 150 mm to 300 mm, i.e. outer diameter of the shaft from 3.53 m to 3.68 m in bottom 0.60 m height within GL.

Thus, shaft is 3400 mm centerline diameter and has 150 mm thickness throughout except bottom 0.60 m height where thickness is splayed to 300 mm from 150 mm at 0.60m above base (i.e. external diameter splayed from 3550 mm to 3850 mm) (Fig. 11.27).

Vertical steel $= \dfrac{0.24}{100} \times 150 \times 1000 = 360$ mm² (180 mm² each face 8ϕ @ 250 c/c vertical bars each face), $(A_{st} = 200 \text{ mm}^2/\text{m each face})$

Horizontal steel rings $= \dfrac{0.20}{100} \times 150 \times 1000 = 300$ mm² (150 mm² each face 8ϕ 300 c/c rings on each face)

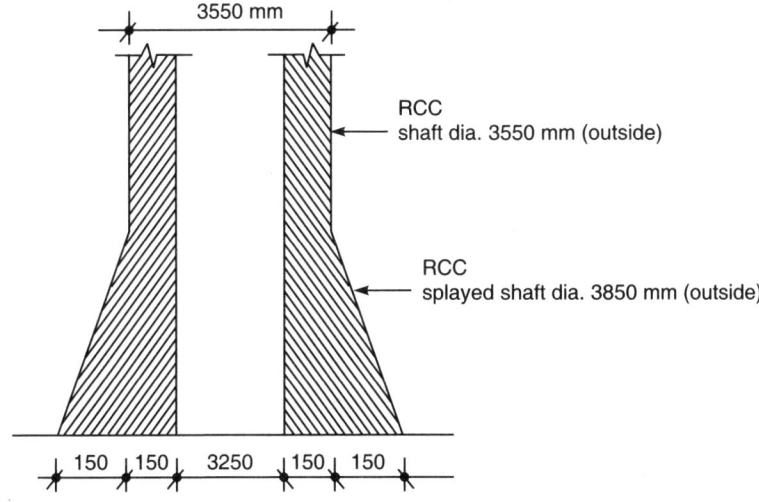

3550 mm

RCC
shaft dia. 3550 mm (outside)

RCC
splayed shaft dia. 3850 mm (outside)

150   150   3250   150   150

Fig. 11.27: Splayed shaft bottom (3550 mm to 3850 mm)

Uniform thickness 150 mm, diameter (centerline) = 1690 mm

Bottom 0.60 m height from the base will be thickened (splayed) on outside by 150 mm, i.e. outside diameter is 3400 + 150 + 300 = 3850 mm and inner diameter = 3250 mm (also provide extra vertical dowel bars 8ϕ 250 c/c and extra rings 8ϕ @ 300 c/c in 600 mm height).

Example 11.11

Design a circular slab foundation for a tank and supporting shaft of Example 11.10 on the soil having SBC of 70 kN/m² (normal), seismic (SBC) = 105 kN/m², and wind conditions of 87.5 kN/m². In bottom 0.60 m bottom height of shaft is splayed from 150 mm to 300 m thickness at the junction with raft slab. At the junction with the raft slab the diameter of the shaft shell vary from 3250 mm (inner) to 3850 mm (outer). Speciy gravity of soil = 16 kN/m³. M20 CC and Fe 415 steel (σ_{st} = 230 N/mm²) shall be used.

Solution: Superimposed loads on the foundation slab (W_e) = 303 + 694 = 997 kN

Superimposed loads on the foundation slab (tank full = W_F) = 503 + 997 = 1500 kN

0.1152 ± 0.0039 ± 0.084 = 0.203 N/mm² < 1.33 × 6, OK

- Wind pressure (horizontal load) moment = 966 kN-m about shaft base
- Earthquake load (horizontal load) moment = 1662.5 kN-m about shaft base

    Assume foundation raft slab thickness = 0.40 m for moment calculation.

    Assuming base slab diameter = 7.0 m (say)

- Area = $\dfrac{\pi}{4}$ (7)² = 38.485 m²

- Section modulus $Z_e = \dfrac{\pi}{32}$ (7)³ = 33.674 m³

Soil pressure (direct) = $\dfrac{\left(1500 + \dfrac{1500}{10}\right)}{38.485}$ = 42.87 kN/m²

Additional weight of concrete raft replacing soil = $38.485 \times (25 - 16) = 346.0$ kN

- Total actual vertical load (tank full and EQ condition)
$$= 1500 + (30.22\ EQ + 26.03) + 346.0 = 1903.00\ kN$$
- Total load (empty tank) = $(303 + 694) + 346.0 = 1343.0$ kN
- Total moment (tank full and EQ forces) about foundation bottom
$$= 1662.5 + 0.40\ (60.45 + 52.05) = 1662.50 + 45.5 = 1708.0\ kN\text{-}m$$
- Total moment (tank empty and wind) about foundation bottom
$$= 966 + (19.5 + 63)\ 0.40 = 966 + 33.0 = 999.00\ kN\text{-}m$$

- Max. soil pressure (tank full and EQ) $= \dfrac{1903}{38.485} \pm \dfrac{1708}{33.674} = 49.45 \pm 50.71$

$$= +\ 100.16\ kN/m^2\ \text{and} - 1.26\ kN/mm^2 < 105\ kN/m^2,\ (OK)$$
and $<$ wt. of surcharge soil for 1.12 m depth, OK.
Max. soil pressure (tank empty and wind)

$$= \dfrac{(997 + 346.0)}{38.485} \pm \dfrac{999.00}{33.674} \pm = 34.90 \pm 29.70 = +\ 64.60\ kN/m^2 < 87.5\ kN/m^2,$$

$$+\ 5.20\ kN/m^2, OK.$$

Thus, 7 m diameter is OK.

Safety against overturning

(i) Tank full (earthquake) condition, gravity forces (earthquake weight acts upward or downward) $= 1500 \pm \dfrac{60.45}{2} \pm \dfrac{52.05}{2} + 346.4 = 1902.65$ kN, 1790.0 kN

Min. restoring moment about foundation edge $= 1790.0 \times \dfrac{\text{Diameter}}{2}$

or min. restoring moment = $1790.0 \times 3.5 = 6265$ kN-m (radius of base = 3.5 m).
Overturning moment due to EQ = $1662.5 + 0.40\ (60.45 + 52.05) = 1708$ kN-m (about bottom of base slab)

(a) FOS on overturning (EQ conditions)

$$= \dfrac{\text{Restoring moment due to weight}}{\text{overturning moment about bottom edge}} = \dfrac{6265}{1708} = 3.67\ (\text{more than 1.5, OK.})$$

(b) Tank empty and wind pressure
Restoring moment = $(997 + 346.4)\ 3.5 = 4702$ kN-m
Max. overturning moment (wind) = $966 + 0.40\ (19.56 + 63) = 966 + 33$
= 999 kN-m
FOS in overturning (tank empty and wind blowing)

$$FOS = \dfrac{4702}{999} = 4.70\ (> 1.5),\ OK.$$

Foundation slab (assumed diameter = 7.0 m) and radius = 3.50 m outerline of splayed cylindrical shaft diameter = 3850 mm, centreline of splayed shaft bottom diameter – 3550 mm, radius – 1775 mm = 1.78 m

Net vertical load soil reaction = 1500 kN (upward)

Net pressure intensity on raft slab $p = \dfrac{1500}{\pi(3.5)^2} = 39 \text{ kN/m}^2$ (direct loading)

Design for circular raft slab (diameter assumed 7.0 m)

Case I : Direct loads

$p = 39 \text{ kN/m}^2$, $W = 1500 \text{ kN}$, $\dfrac{W}{8\pi} = 59.7 \text{ kN}$, $\dfrac{p}{16} = 2.44$

Radius of raft slab $a = 3.50$ m, $b = $ CL of support $= 1.78$ m,

$a^2 = 12.25 \text{ m}^2$, $3a^2 = 36.75 \text{ m}^2$, $b^2 = 3.15 \text{ m}^2$, $\left(\dfrac{a}{b}\right) = 1.972$, $\left(\dfrac{b^2}{a^2}\right) = 0.2572$

$2 \log_e \left(\dfrac{a}{b}\right) = 2 \times 0.67905 = 1.3581$

These values shall be used in various equations for calculations of moments at various critical points (centre, near support of shaft, edges, etc.)

Moments in circular raft slab using plate theory:

For r < b

$$M_r = \dfrac{W}{8\pi}\left[2\log_e\left(\dfrac{a}{b}\right)+1-\left(\dfrac{b^2}{a^2}\right)\right]-\dfrac{p}{16}(3a^2-3r^2) = 59.7 \ 1.3581 + 1 - 0.2572$$

$$- 2.44 \ (36.75 - 3r^2)$$

or $M_r = 125.43 - 89.67 + 7.32r^2 = (35.76 + 7.32r^2)$

$$M_\theta = \dfrac{W}{8\pi}\left[2\log_e\left(\dfrac{a}{b}\right)+1-\left(\dfrac{b^2}{a^2}\right)\right]-\dfrac{p}{16}(3a^2-r^2) = 59.7 \ (2.101) - 2.44 \ (36.75 - r^2)$$

or $M_\theta = 125.43 - 89.67 + 2.44r^2 = (35.75 + 2.44r^2)$

For r > b

$$M_r = \dfrac{W}{8\pi}\left[2\log_e\dfrac{a}{r}-\left(\dfrac{b^2}{a^2}\right)+\left(\dfrac{b^2}{r^2}\right)\right]-\dfrac{p}{16}(3a^2-3r^2)$$

or $M_r = 59.7\left[2\log_e\dfrac{3.5}{r}-(0.2572)+\left(\dfrac{3.151}{r^2}\right)\right]-2.44\ (36.75 - 3r^2)$

$$= 119.4 \log_e \dfrac{3.5}{r} - 15.355 + \dfrac{188.115}{r^2} - 89.67 + 7.32r^2$$

$$= 119.4 \log_e \dfrac{3.5}{r} + \dfrac{188.115}{r^2} + 7.32r^2 - 105.025$$

$$M_\theta = \frac{W}{8\pi}\left[2\log_e\frac{a}{r}+2-\left(\frac{b^2}{a^2}\right)-\left(\frac{b^2}{r^2}\right)\right]-\frac{p}{16}\,(3a^2-r^2)$$

$$= 59.7\left[2\log_e\frac{3.5}{r}+2-(0.2572)-\left(\frac{3.151}{r^2}\right)\right]-2.44\,(36.75-r^2)$$

$$= 119.4\log_e\frac{3.5}{r}+104.05-\frac{188.115}{r^2}-89.67+2.44r^2$$

$$= 119.4\log_e\frac{3.5}{r}-\frac{188.115}{r^2}+14.38+2.44r^2$$

BM at various radii

r (m)	0.0	0.80	1.63	1.78	1.93	2.50	3.50
M_r (kN-m)	+ 35.76	+ 40.45	+ 55.2	–	+ 50.24	+ 10.78	0
M_θ (kN-m)	+ 35.76	+ 37.32	+ 42.243	–	+ 44.043	+ 39.71	+ 28.91

Case II: *Tank full and earthquake moments*
$$a = 3.50 \text{ m}, b = 1.78 \text{ (CL)},$$

$$p'\text{ (variable intensity moment–EQ)} = \frac{1708}{Z_e} = \frac{1708}{33.674} = \pm\,50.71 \text{ kN/m}^2, a^2 = 12.25$$

$$q = \frac{1708}{\pi(1.78)^2} = 171.6 \text{ kN/m (CL of shaft support)}, \beta = \frac{h}{a} = \frac{1.78}{3.5} = 0.509$$

$$\beta^2 = 0.259, \beta^4 = 0.0669, \beta^4+2\beta^2-3 = -2.4151, \beta^4+2\beta^2 = 0.5849, k = \frac{r}{3.5}$$

For r < b

$$M_r = \frac{5}{48}\,p'\,a^2\,(k-k^3)+\frac{qa}{8}\,k\,(\beta^4+2\beta^2-3)$$

$$= \frac{5}{48}\times 50.71\times 12.25\,(k-k^3)-2.4151\times 171.6\times\frac{3.5}{8}\,k$$

$$= -116.6k-64.71\,k^3$$

$$M_\theta = \frac{p'a^2}{48}\left(\frac{5}{3}k-k^3\right)+\frac{qa}{8}\cdot\frac{k}{3}\,(\beta^4+2\beta^2-3)$$

$$= \frac{50.71\times 12.25}{48}\left(\frac{5}{3}k-k^3\right)+\frac{171.6\times 3.5k}{24}\times(-2.4151)$$

$$= -12.942\,k^3-38.87\,k$$

For r > b

$$M_r = \frac{5}{48} p' a^2 (k - k^3) + \frac{qa}{8} \cdot \frac{1}{k^3} \{k^4 (\beta^4 + 2\beta^2) - 2\beta^2 k^2 - \beta^4\}$$

$$= \frac{5}{48} \times 50.71 \times 12.25 (k - k^3) + \frac{171.6 \times 3.5}{8} \cdot \frac{1}{k^3} k^4 (0.5849) - 0.518k^2 - 0.0669$$

$$= 64.71 (k - k^3) + 43.911k - \frac{38.9}{k} - \frac{5.023}{k^3}$$

$$= 108.62k - 64.71k^3 - \frac{38.9}{k} - \frac{5.023}{k^3}$$

$$M_\theta = \frac{p'a^2}{48} \left(\frac{5}{3}k - k^3 \right) + \frac{qa}{8k^3} \left\{ \frac{k^4}{3} (\beta^4 + 2\beta^2) - 2\beta^2 k^2 - \beta^4 \right\}$$

$$= \frac{+ 50.71 \times 12.25}{48} \left(\frac{5}{3}k - k^3 \right) + \frac{171.6 \times 3.5}{8k^3} \left\{ \frac{k^4}{3} (\beta^4 + 2\beta^2) - 2\beta^2 k^2 - \beta^4 \right\}$$

$$= - 12.942k^3 + 36.21k - \frac{38.9}{k} + \frac{5.023}{k^3}$$

Thus, moments at various points at various radii are given as under:

r (m)	0.0	0.80	1.63	1.78	1.93	2.50	3.50	Remarks
M_r (kN-m)	0	∓27.50	∓60.9	–	∓49.9	∓14.23	0.0	Max. $M_r = 60.9$
M_θ (kN-m)	0	∓9.06	∓19.42	–	∓22.8	∓19.53	∓10.61	Max. $M_\theta = 22.8$
$k = \dfrac{r}{3.5}$	0	0.229	0.466	–	0.5514	0.7143	1.0	–
k^3	0	0.01201	0.1011	–	0.1677	0.3644	1.0	–
Net design moments (M_r) kN-m	+35.76	+67.95	+116.1 −5.7	–	+100.14 −0.34	+25.01 −3.45	0.0	+116.1
Net design moments (M_θ) kN-m	+35.76	+46.38	+61.66 +22.82	–	+66.84 +21.24	+59.24 +20.18	+39.52 18.30	+66.84

Thus, maximum design moments are $M_r = 116.1$ kN-m, $M_\theta = 66.84$ kN-m, permissible stresses shall be $33\frac{1}{3}\%$ more when earthquake forces are included.

For M20 CC and steel Fe 415

$$\sigma_{cbc} = 7 \, \text{N/mm}^2, \sigma_{st} = 190 \, \text{N/mm}^2 \text{ (say), coefficients (m)} = 13$$

$$\sigma_c = 5 \, \text{N/mm}^2, \sigma_{st} = 190 \, \text{N/mm}^2$$

$$\text{Critical NA coefficient } (N_c) = \frac{13 \times 7}{13 \times 7 + 190} = 0.324$$

$$j \text{ (lever)} = 0.892$$

$$k = 1.0115 \text{ (moment)}$$

$$\text{Effective } d = \sqrt{\frac{M}{bk \times 1.33}}$$

$$d = \sqrt{\frac{116.1 \times 10^6}{1000 \times 1.0115 \times 1.33}} = 294 \text{ mm } (D = 375 \text{ mm}, d = 300 \text{ mm}), \text{ cover 75 mm for raft.}$$

$$A_{str} = \frac{116.1 \times 10^6}{1900 \times 0.892 \times 300} = 230 \text{ mm}^2/\text{m}$$

(provide 8ϕ radial bars @ 200 c/c on each face, A_{st} = 250 mm²/m each face)

Also provide circumferential bars as rings 8ϕ @ 200 on each face since reversal of stresses can occur due to reversal of earthquake or wind forces. In the central portion of cylindrical shaft, provide a mesh of 8ϕ @ 200 c/c in each direction on each face.

$$\text{Minimum temp. steel} = 0.24\% = \frac{0.24}{100} \times 1000 \times 375 = 900 \text{ mm}^2 \text{ (450 mm}^2 \text{ each face)}$$

in each direction. (0.24% steel similar to cylindrical shell). Since minimum temp. Steel is more than required for max. M_r, and hence provide this min. steel (12ϕ 250 c/c).

Minimum radial bars in projected portion of slab 12ϕ @ 250 c/c each face (A_{st} = 452 mm²/m). Also provide rings 12ϕ @ 250 c/c on each face (top and bottom).

In the central portion provide a mesh of 12ϕ @ 250 c/c on each face.

SUMMARY

Water is required for human consumption, and hence it should be stored for regular supply. Water is stored in tanks — underground, on ground and overhead at elevated positions for various purposes. These tanks may be rectangular, circular or Intze in case of elevated tanks. Dimensions (L, B, H, etc.) are computed from the volume capacity, and other design parameters (thickness, reinforcement, etc.) are computed on the basis material characteristics (grade of concrete and steel) and environmental factors (earthquake, wind pressures, safe bearing capacities of soils,etc.).

Generally water retaining structures are designed by working stress method. In this book, only approximate method of analysis (membrane analysis) shall be used while exact method of analysis (considering continuity analysis) may be referred to in advanced books. For design guidance, IS:3370 shall be referred. To achieve imperviousness and longer life (durability), the following guidelines shall be adopted:

(i)   Use richer mix of CC (M30, M40, etc.)

(ii) Provide a minimum clear cover of 25 mm

(iii) Provide smaller diameter bars at closer intervals

(iv) Keep tensile stresses low in concrete

(v) Adopt good concreting practices (mixing, compacting, curing, etc.)

(vi) Use working stress method

(vii) Minimum temperature steel shall be

0.30% steel up to 100 mm thickness and

0.20% steel for 450 mm thickness and above

(viii) For thickness above 225 mm, provide minimum steel in both directions and on both the faces

(ix) Stresses in concrete and steel shall be restricted to permissible limits.

Design data and equations:

- Circular elements of diameter D subjected to water pressure p at a water depth of H shall be developing hoop tension $T = p \cdot \dfrac{D}{2} = wH\dfrac{D}{2}$

- Hoop steel required shall be for the total tension.

$$A_{st} = \frac{\text{Hoop tension }(T)}{\text{Permissible tensile stress}} = \frac{wHD}{2\sigma_{st}}$$

- In case of rigid joints of tank wall and base slab,

cantilever moment M in wall $= wH \cdot \dfrac{h}{2} \cdot \dfrac{h}{3} = \dfrac{wHh^2}{6}$

where, $\qquad\qquad\qquad H = $ Depth of water and

$\qquad\qquad\qquad h = \dfrac{H}{3}$ or 1 m whichever is greater for circular and

for rectangular $\qquad h = \dfrac{H}{4}$ or 1 m, whichever is greater

- Max. hoop tension h above rigid base

$$T = \frac{D}{2}\, w\,(H - h)$$

- Circular tanks on the ground has flexible or rigid joints with the base slab. Rigid joints of tank wall and base slab are generally thickened at bottom.

- Reinforcement in rigid joints of tank wall with base slab are provided with appropriate links (corner) so that the concrete in cover is not thrown out as shown in Fig. 11.28.

- In rectangular tank, joints between long and short wall are also provided with special horizontal links along the four corner joints as shown in Fig. 11.28.

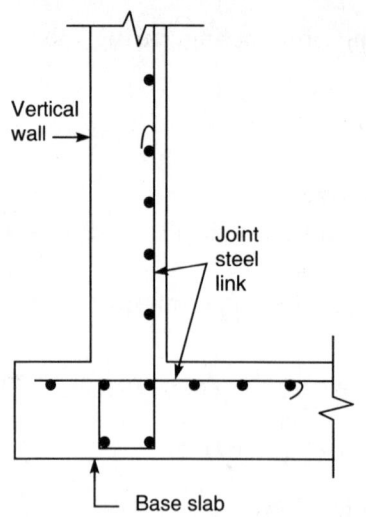
(a) Vertical wall and base slab junction steel link

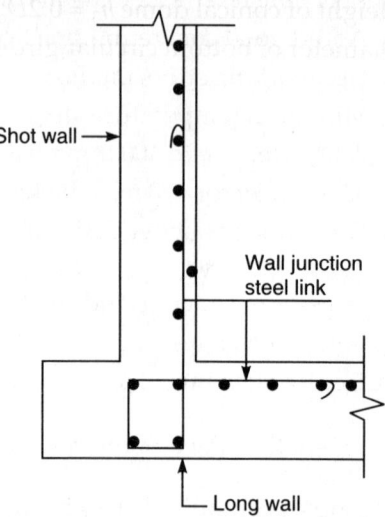
(b) Short and long wall's junction steel link

Fig. 11.28: Rigid joints of wall with wall or with base slab

- Max. liquid (water) pressure on walls (above rigid joint)

$$p_h = w\,(H - h)$$

- Moment due to horizontal pressure, h m above base

$$M = p_h \cdot \frac{B^2}{12} \text{ and } p_h \cdot \frac{L^2}{12} \quad \text{or} \quad \left\{ w(H-h) \cdot \frac{B^2}{12} \right\} \quad \text{and} \quad \left\{ w(H-h) \cdot \frac{L^2}{12} \right\}$$

- Tension in walls

$$T_B = p_h \cdot \frac{L}{2} \text{ or } T_B = w(H-h)\frac{L}{2}, \quad T_L = p_h \cdot \frac{B}{2} = w(H-h) \cdot \frac{B}{2}$$

- Net moment due to horizontal tension:

$$M = (M_1 - T \cdot x)$$

where x is the distance from the centre of the wall.

- Cantilever moment in short walls

$$M = w(H-h)\frac{h^2}{6}$$

Overhead tanks are generally circular or Intze type. OHSR generally comprises spherical dome (shell), top ring beam, cylindrical tank wall, middle ring beam, conical shell, bottom ring beam, supporting stage, foundation, etc. Complete design consists of 3 parts: tank elements, staging, and foundation. Simple design fundamentals are dealt here for understanding basic requirements in tank design.

Economical proportions of Intze tank elements shall be assumed as:

- Rise of spherical dome $h_1 = \dfrac{D_o}{7} = h_4$ (bottom dome)

- Height of cylindrical tank $h_2 = 0.4 D_o$

- Height of conical dome $h_3 = 0.2D_o$
- Diameter of bottom circular girder $= 0.60D_o$

Top spherical cover dome:

Meridional thrust $T_1 = \dfrac{wR}{(1 + \cos\phi)}$ per unit length

Circumferential force $T_2 = wR\left[\cos\phi - \dfrac{1}{(1 + \cos\phi)}\right]$

Appropriate radial steel bars and circumferential ring bars are provided to resist meridional thrust and circumferential force alternatively mesh of steel bars may be provided.

Meridional thrust T_1 causes hoop tension in top ring beam.

Hoop tension in ring beam $T_p = T_1 \cos\theta \cdot \dfrac{D}{2} = \dfrac{wRD \cos\theta}{2(1 + \cos\theta)}$

where R is the radius of spherical shell.

Tensile stresses in ring beam of top cover shell should be restricted to permissible limits as given in Tables 11.1 and 11.2.

Cylindrical shell (tank wall) shall be designed on the basis of hoop tension in upper portion and cantilever moment in the bottom $\dfrac{H}{3}$ portion when the tank is full (depth of water H). Tank is supported on staging comprising of n columns and bracings or cylindrical shaft. Flat bottom slabs of circular tanks are designed by using plate theory with edges assumed as simply supported.

Radial moment (neglecting μ) $M_r = \dfrac{q}{16}(3 + \mu)(a^2 - r^2) \cong \dfrac{3q}{16}(a^2 - r^2)$

Circumferential moment (neglecting μ) $M_\theta = \dfrac{qa^2}{16}(3 + \mu) - \dfrac{qr^2}{16}(1 + 3\mu)$

$$\cong \dfrac{q}{16}(3a^2 - r^2)$$

where,

q = Load per unit area ($q = wH$ + self-wt.)

a = Radius of the circular slab

μ = Poisson's ratio of concrete (neglected being small in approximate method)

r = Radial distance of point from the centre where moments are found.

The forces and moments in a ring beam of radius R and supported on the column–bracing staging at column supports and mid span points between the columns can be found by using various coefficients (K, K', K'', etc.) as given in the Table 11.3.

Shear force $F_v = \dfrac{wR}{2}\phi$, where $= \phi = \dfrac{360}{n}$

Support moment = $KwR^2w\,\phi$, w is the load on the ring beam

Mid span moment = $K'wR^2\,\phi$

Torsional moment = $K''wR^2\,\phi$

Columns and bracings are designed in usual manner for maximum load and moment carried under worst combination of loading.

Design of shaft staging: Modern design of supporting stage of OHSR comprises circular shaft (cylindrical shell). Design of shaft is based on certain minimum dimensions and worst combinations of loading when the tank is empty or full under wind or earthquake forces. The most critical situation arises when the tank is in full load condition and combines with earthquake or when the tank is in empty condition and subjected to severe wind forces.

Minimum thickness of the cylindrical shaft shell shall be 125 mm.

Minimum steel in shaft:

Vertically 0.24% HYSD (0.3% for MS) and *horizontally* 0.20% HYSD (0.25% for MS).

Max. moment due to wind at the bottom of shaft shall be

$$M_1 \text{ (wind)} = P_1h_1 + P_2h_2$$

where, $P_1 = 0.70\,D_0h \times$ (wind pressure), D_0 = Outer diameter of tank

             h_1 = Height of CG of wind force on the tank from shaft base

             $P_2 = 0.70\,D_s \cdot h_s \times$ (wind pressure), D_s = Outer diameter of shaft

             h_2 = Distance of CG shaft wind force from the shaft base

M_2 (earthquake) $= \alpha_H W_1 h_1' + \alpha_H W_2 h_2')$

where, W_1 = Weight of tank full

             h_1' = Height of tank's CG from shaft bottom

             W_2 = Weight of shaft, α_H = Coeff. of EQ acceleration (refer to IS code)

             h_2' = CG of shaft from shaft base

Equivalent thickness of shaft (t_e) when provided with minimum steel (vertical) = 0.24% steel.

$$t_e = t + \frac{0.24}{100}(m-1)t$$

A_e (effective area) $= \pi D \cdot t_e$

$$Z_e\text{(section modulus)} = \frac{\pi R^3 \cdot t_e}{R_o}, \text{ where } R \text{ is centre line radius of shaft and } R_o \text{ is}$$

outside radius of shaft.

$$\text{Max. stress} = \left[\frac{(W_1 + W_2)(1 \pm \alpha_v)}{A_e} \pm \frac{M_2}{Z_e} \right]$$

where, α_v = Vertical EQ coefficient

             A_e = Effective shaft area

             Z_e = Equivalent section modulus of shaft

$$\text{Max. stress (wind)} = \left[\frac{(W_1' + W_2)}{A_e} \pm \frac{M_1}{Z_e} \right]$$

where, $\quad W_1' =$ Weight of empty tank

$W_1 =$ Weight of tank full

$W_2 =$ Weight of shaft

Maximum stress must be less than 1.33 permissible in case of wind and earthquake.

Design of circular raft slab: The diameter of foundation slab shall be assumed more than that required for limiting the maximum bearing pressure on soil less than permissible SBC under the effect of wind or EQ moments. Safety against overturning and direct maximum soil bearing pressure shall be checked for the assumed diameter and revised for optimum safe economical values. For check against overturning safety factor (at least 1.5), restraining moment due to weight of tank, shaft and additional weight of foundation shall be calculated and compared.

$$\text{Restraining moment (tank empty and tank full)} = \text{Total weight} \times \text{assumed } \frac{D}{2}$$

$$\text{FOS against overturning} = \frac{\text{Restraining moment}}{\text{Overturning moment(due to wind or EQ)}} > 1.5.$$

Wind pressure moment $(M_1) = [0.7D_o \times h_{t1} \times h_1 + 0.7 D_s \times h_{t2} \times h_2]$ about foundation base

EQ moment $(M_2) = [W_f.\alpha_H h'_1 + W_s\, \alpha_H h'_2]$ about foundation base

where,

$D_o =$ External diameter of tank,

$h_{t1} =$ External height. of tank

$h_1 =$ CG of tank wind force from the foundation base

$h_2 =$ CG of shaft wind force from the foundation base

h'_1 and h'_2 are respective distances of EQ forces on tank and shaft

α_H, α_V are the EQ acceleration coefficients at the place (refer IS code).

Check for maximum soil pressure:

$$\text{Max. soil pressure} = \frac{\text{Total vertical load}}{A(\text{base slab area})} \pm \frac{M_1(\text{wind})}{Z \text{ of base slab}}$$

$$\text{or} \quad \frac{\text{Effective verticalf load(EQ)}}{A(\text{base slab area})} \pm \frac{M_2(\text{EQ})}{Z \text{ of base slab}}$$

Maximum soil pressure on foundation base should not be more than SBC.

After all these foundation safety checks, the slab is designed for M_r and computed by plate theory with usual equations. Thickness is calculated for the maximum BM (M_r or M_θ). Radial reinforcement for M_r and circumferential steel for are calculated with reference to adopted value of thickness. Outer portion of slab reinforcement is provided as radial and circumferential while in the central portion reinforcement is provided as mesh.

PRACTICE QUESTIONS

(I) Objective Questions

Q. 11.1 Select the correct response given after each statement to complete it correctly and fill in the response sheet provided.

(i) What is the type of lower portion of Intze tank?
   (a) rectangular shape wall and circular flat base
   (b) circular shape wall and flat slab base
   (c) conical shell wall and spherical shell base
   (d) cylindrical shell wall and conical shell base

(ii) What value of free board is assumed in design of tanks?
   (a) 100 mm   (b) 200 mm
   (c) 300 mm   (d) 500 mm

(iii) What minimum clear concrete cover is provided to achieve good impermeability and durability in design of tanks?
   (a) 10 mm   (b) 15 mm
   (c) 20 mm   (d) 25 mm

(iv) What is the minimum %age of (MS) temperature steel provided in 275 mm thick water retaining structural element?
   (a) 0.12%   (b) 0.20%
   (c) 0.25%   (d) 0.30%

(v) What is the maximum hoop tension (T) in case of circular tank of 4 m diameter having flexible joint and containing liquid to a depth of 3 m (sp. gr. = 10 kN/m³)?
   (a) 120 kN   (b) 60 kN
   (c) 40 kN   (d) 20 kN

(vi) What is maximum hoop tension (T) in case of a circular tank of 4 m diameter containing liquid to a depth of 3 m above base if the joint of the wall and base slab is rigid (sp. gr. = 10 kN/m³)?
   (a) 40 kN   (b) 60 kN
   (c) 80 kN   (d) 180 kN

(vii) What is the maximum cantilever moment in the wall of a circular tank having wall–base joint rigid and tank has 4 m diameter? The tank is filled with liquid having sp. gr. 10 kN/m³ to a maximum height of 3 m above the base.
   (a) 45 kN-m/m   (b) 30 kN-m/m
   (c) 13.33 kN-m/m   (d) 5 kN-m/m

(viii) What is the maximum horizontal tension T_B developed in short wall of breadth $B = 3$ m when the length $L = 5$ m of the rectangular tank with rigid base joint and water is filled in the tank up to 4 m ?
   (a) 45 kN/m   (b) 60 kN/m
   (c) 75 kN/m   (d) 100 kN/m

(ix) What will be the maximum horizontal end support moments in a junction of long and short walls of a rectangular tank of size 10 m × 6 m filled with water (sp. gr. 10 kN/m³) before balancing the joint moments if the tank is filled with water to 4 m height and the tank is placed on ground?

(a) 90, 250 kN-m/m (b) 135, 375 kN-m/m

(c) 120, 333.3 kN-m/m (d) 180, 500 kN-m/m

(x) What will be the maximum horizontal end moment at a junction of long and short walls of a rectangular tank of size 15 m × 9 m placed on ground after balancing the joint if both short and long walls are of 200 mm thickness when end moments are 180 kN-m and 500 kN-m before balancing on either side of junction ?

(a) 300, 300 kN-m (b) 320, 320 kN-m

(c) 340, 340 kN-m (d) 380, 380 kN-m

(xi) A circular OHSR tank has spherical shell cover of radius 5 m supporting a total u.d.l. of 3 kN/m² including its own weight. The spherical dome is monolithically connected to a circular ring beam of 6 m diameter. The spherical dome subtends semi angle of 45° at its centre. Find the meridional thrust T_1 on the edge ring beam.

(a) 1.818 kN/m (b) 8.787 kN/m

(c) 19.392 kN/m (d) 51.195 kN/m

(xii) Horizontal edge beam of 6 m diameter is subjected to a radial meridional thrust of 20 kN/m at an inclination of 30° with the horizontal. Determine the hoop tension in the ring beam.

(a) 60 kN/m (b) 51.96 kN/m

(c) 30 kN/m (d) 20 kN/m

(xiii) Bottom circular slab of external diameter 4 m is supported along a circular shaft of diameter 3.0 m. If Poisson's ratio μ of concrete is neglected, determine the maximum approximate radial moment at the centre using simple plate theory when the slab is subjected to an uniformly distributed load of 10 kN/m².

(a) 7.5 kN-m/m (b) 22.5 kN-m/m

(c) 30.0 kN-m (d) 90.0 kN-m

(xiv) In Intze tank for economical proportions, what could be the height h of cylindrical tank portion considering the diameter of the cylindrical tank as D_o?

(a) $h = \dfrac{D_o}{7}$ (b) $h = 0.2\, D_o$

(c) $h = 0.40\, D_o$ (d) $h = 0.60\, D_o$

(xv) What type of forces are mainly considered for approximate design of conical shell of an Intze tank when the tank is full?

(a) Only hoop tension

(b) Only meridional thrust

(c) Meridional thrust and radial moment

(d) Meridional thrust and hoop tension

(xvi) What forces are mainly considered for the approximate design of bottom spherical dome?

(a) Only meridional thrust

(b) Only circumferential forces

(c) Both circumferential and meridional forces

(d) Both edge moments and radial moments

(xvii) Approximate method of designing Intze tank elements results in ....

(a) better economy

(b) better safety but lacks in optimum economy

(c) both better economy and safety

(d) lack of economy as well as proper safety

(xviii) Supporting structure for OHSR is provided by staging shaft which shall have minimum thickness t = ...

(a) 100 mm   (b) 125 mm

(c) 150 mm   (d) 200 mm

(xix) Foundation raft slab (circular) diameter D shall be assumed more than that required for direct load safe soil bearing pressure to safeguard against worst soil bearing pressures caused by ....

(a) maximum vertical loads when tank is full

(b) minimum vertical loads when tank is full

(c) maximum vertical loads along with moments due to EQ or wind

(d) sliding of structure along with foundation

(xx) Factor of safety against overturning is found as FOS = ...

(a) $\dfrac{\left(\text{Total minimum vertical loads}\times\text{Raft slab diameter}\right)}{2\left(\text{Maximum moments caused about bottom edge of raft slab due to EQ or wind}\right)}$

(b) $\dfrac{\text{Maximum moments caused about edge of raft slab due to EQ or wind}}{2\left(\text{Total minimum vertical loads}\times\text{Raft slab radius}\right)}$

(c) $\dfrac{\left(\text{Total maximum vertical loads}\times\text{Raft slab diameter}\right)}{\text{Total moments caused due to}\left(\text{EQ}+\text{wind}\right)\text{about the base bottom}}$

(d) $\dfrac{\text{Total minimum moments about the raft slab edge use to}\left(\text{EQ}+\text{wind}\right)}{\left(\text{Total vertical loads}\right)\times\left(\text{Raft slab diameter}\right)}$

Response sheet to Q. 11.1 (i to xx)

Question	(i)	(ii)	(iii)	(iv)	(v)	(vi)	(vii)	(viii)	(ix)	(x)
Response (a/b/c/d)										

Question	(xi)	(xii)	(xiii)	(xiv)	(xv)	(xvi)	(xvii)	(xviii)	(xix)	(xx)
Response (a/b/c/d)										

(II) Numerical Questions

Q. 11.2 Explain at least 5 design and construction guidelines for water tanks.

Q. 11.3 Compute hoop tension, thickness of wall and steel reinforcement in a circular tank with flexible wall–base joint and show arrangement of steel reinforcement. Diameter of the tank $D = 5$ m, height of water above base $= 3$ m. Use M20 CC and Fe 415 grade of steel.

[**Hint:** $T = 75$ kN, $A_{st} = 10\phi$ @ 190 mm c/c, $t = 125$ mm, stress $= 0.577$ N/mm$^2 < 1.2$]

Q. 11.4 Design a circular tank wall placed on the ground. The tank has diameter 6 m and total depth of water 4.5 m. Tank walls are fixed monolithically with the base slab of 200 mm thickness. Use M20 CC and Fe 415 HYSD bars.

[**Hint:** $T = 90$ kN, $M = 16.88$ kN-m/m, $A_{st} = 474$ mm^2/m (10ϕ @ 300 c/c on each of 2 faces), thickness min. $= 125$ mm, vertical steel inner face 12ϕ @ 100 c/c, $A_{st} = 1130$ mm^2, min. temp. steel $= 375$ mm^2 (8ϕ @ 250 c/c vertically on outer face only).]

Q. 11.5 Design a circular water tank placed on the ground for 60,000 litres capacity and total height of 4.5 m including FB with the fixed base joint. Use M20 CC and Fe 415 HYSD bars.

[**Hint:** $D = 4.22$ m, $h = 1.50$ m, $T = 63.3$ kN, $A_{st} =$ rings 8ϕ @ 250 c/c each face, min. thickness $= 125$ mm, M(vert.) $= 16.88$ kN-m, A_{st}(vert.) $= 1048$ mm^2 12ϕ

100 c/c inner face), min. temp. steel $= \dfrac{0.24}{100} \times 125 \times 1000 = 300$ mm^2 (outer

face vert. 8ϕ 300 c/c, $A_{st} = 167$ mm^2/m outer face).

Q. 11.6 Design a rectangular water tank of size 5 m $\times$ 4 m $\times$ 3 m (ht) resting on ground. Use M25 concrete and Fe 415 HYSD bars.

[**Hint:** $\sigma_{cbc} = 8.5$ N/mm^2, $m = 11$, $\sigma_{st} = 140$ N/mm^2, $n = 0.40$, $j = 0.867$, $k = 1.474$, $\dfrac{L}{B}$

$< 2 = 1.25$, cantilever moment action $h = 1$ m, horizontal action 2 m top,

$P_h = 10(2) = 20$ kN/m^2, fixed, M_1(ends) $= \dfrac{20 \times 5^2}{12} = 41.7$ kN-m/m, $\dfrac{20 \times 4^2}{12}$

$= 26.7$ kN-m/m. Balanced moments $41.7 - 15 \times 0.444 = 35.04$ and $26.7 + 15 \times$

$0.556 = 35.04$ kN-m d (eff. depth) $= \sqrt{\dfrac{35.04 \times 10^6}{1.474 \times 1000}} = 154$ mm, $D = 200$ mm,

$d = 165$ mm (say).

Long walls:

$$T_s = \frac{20 \times 4}{2} = 40 \text{ kN}, \ T_B = 20 \times \frac{5}{2} = 50 \text{ kN}$$

Design moment = $35.04 - 40 \times 0.065 = 32.44$ kN-m

$$A_{st} = \frac{32.44 \times 10^6}{140 \times 0.86 \times 165} = 1620 \text{ mm}^2, \text{ for direct tension} = \frac{50000}{140} = 357 \text{ mm}^2$$

Total = 1977 mm^2.

Ends inner face 20ϕ @ 150 c/c (2093 mm^2/m).

In mid span = $20 \times \frac{5^2}{8} - 35.04 = 27.46$ kN-m.

Design moment = $27.46 - 40 \times 0.065 = 24.86$ kN-m

$$A_{st1} = \frac{24.86 \times 10^6}{140 \times 0.867 \times 165} = 1241 \text{ mm}^2, \ A_{st2} = \text{(direct)} = \frac{40000}{140} = 286 \text{ mm}^2$$

Total $A_{st} = 1241 + 286 = 1527$ mm^2 (20ϕ @ 200 c/c on outer face),
provide 20 @ 150 c/c on outer and bend alternately at ends on inner face

Short wall:

Ends: $M = 35.04 - 50 \times 0.065 = 31.8$ kN-m, $A_{st1} = \dfrac{31.8 \times 10^6}{140 \times 0.867 \times 165} = 1588$ mm^2

$$A_{st2} = \frac{50000}{140} = 357 \text{ mm}^2, \text{ total } A_{st} = 1945 \text{ mm}^2 (20\phi \ 160 \text{ c/c inner corners})$$

Provide similar to long wall 20ϕ 150 c/c on inner face at corners.

Mid span: Horizontally outer face, 10ϕ @ 300 c/c ($A_{st} = 262$ mm^2) OK.

Mid span $M = \left\{ 20 \times \dfrac{4^2}{8} - 35.04 \right\} = 4.96$ kN-m (very small and can be taken

care of by min. temp. steel)

Cantilever moment = $10 \times 3 \times \dfrac{1}{6} = 5.0$ kN-m/m, $A_{st} = \dfrac{5 \times 10^6}{140 \times 0.867 \times 165}$
= 250 mm^2

Min. temp. steel = $0.24 - \left(0.08 \times \dfrac{100}{350} \right) = 0.217\% = \dfrac{0.217 \times 200 \times 1000}{100}$
= 434 mm^2

Thus, provide vertically 250 mm^2 on each face 10ϕ 300 c/c ($A_{st} = 262 \times 2$
= 524 mm^2)

Also horizontally on outer and inner face 10ϕ 300 c/c ($A_{st} = 262$ mm^2/m)

Bottom base slab is resting directly on the ground.

Provide nominal thickness of base slab = 150 mm and provide 8ϕ @ 230 c/c
on each face and in both directions.

Q. 11.7 Design an open rectangular tank of size 8 m × 3 m × 4.5 m total height and resting on rigid ground. Use M30 grade CC and Fe 415 HYSD steel bars (sp. gr. of water = 9.8).

[Hint: Design constants: σ_{st} = 150 N/mm², σ_{cbc} = 10 N/mm², m = 9, n = 0.375,

j = 0.875, k = 1.64 N-mm, $\dfrac{L}{B}$ > 2, long wall mainly as cantilever, bottom slab

= 150 mm, m (vert.) = 148.84 kN-m/m, D = 350 mm, d = 310 mm (say), A_{st} (vert. inner face) = 3660 mm²/m, 22ϕ @ 100 c/c inner face vertically (A_{st} = 3800 mm²/m), half curtailed at 1.40 m above base.

Min. steel = 0.185% (HYSD), A_{st} (each face) = 318 mm² (8ϕ @ 120 c/c horizontally each face).

Short wall M = 16.54 kN-m, A_{st} (reqd.) less than min. temp. Provide vertically each face 8ϕ @ 120 c/c. Horizontal bars 12ϕ @ 150 c/c on inner face near ends. Mid. span: Provide12ϕ 240 c/c outer face or 12ϕ 300 c/c alternate bent inside at ends and provide 8ϕ 300 c/c additional bars.]

Q. 11.8 Design an underground tank 8 m × 3 m × 4.5 m (depth) for following data:

Unit weight of submerged soil w = 16 kN/m³, ϕ = 30°, water table rises up to ground level.

Use main tank M30 grade CC and Fe 415 grade HYSD steel. Unit weight of water is 9.8 kN/m³.

Live load on the cover roof slab = 2.0 kN/m². Cover roof slab M20 CC, Fe 415 steel.

[Hint: Roof slab one way, D = 150 mm, d = 120 mm, A_{st} = (10ϕ 300 c/c) = 262 mm²/m. min. temp. 8ϕ 270 c/c, A_{st} = 185 mm². Tank empty M (cantilever) submerged = 181 kN-m/m, d (wall) = 335 mm, D = 370 mm, A_{st} (22ϕ @ 90 c/c) = 4222 mm²/m,

Half curtailed 1.50 m above base (i.e. 22ϕ @ 180 c/c).

Tank full and no earth on outside:

Long wall: p = 9.8 × 4.5 = 44.1 kN/m², cantilever moment M = $\dfrac{44.1 \times 4.5^2}{6}$ =

149 kN-m/m

Vert. steel A_{st} = $\dfrac{149 \times 10^6}{150 \times 0.875 \times 335}$ $\cong$ 3400 mm²/m (22ϕ @ 110 c/c vert. inner

face in bottom 1.50 m and then 22ϕ @ 220 c/c).

Walls may be tapered in bottom 1.50 m height from 370 mm to 200 mm thickness.

Direct tension long wall T_L = 44.1 kN (h = 1.5 m)

A_{st} (direct) = $\dfrac{44100}{150}$ = 294 mm²/m < min.

Provide min. 8ϕ 150 c/c (A_{st} = 333 mm²/m) on each face horizontally and 8ϕ 250 c/c on upper part.

Short wall: Lower $\dfrac{H}{4} = 1.125$ m as cantilever, earth saturated, tank empty $p = 53.4$ kN/m^2

Moment (cantilever) $= \dfrac{5.4 \times 1.125^2}{6} = 11.26$ kN-m/m, vert. steel (earth face)

$= 256$ mm^2/m

Provide 8ϕ 150 c/c on each face and 8ϕ 300 c/c on top portion.

$p_h = 44.1$ kN/m^2, $M = \dfrac{44.1 \times 1.125^2}{6} = 9.3$ kN-m/m

A_{st} (vert.) $= \dfrac{9.31 \times 10^6}{150 \times 0.875 \times 335} = 242$ mm^2/m (<min.), provide 8ϕ 150 c/c on both faces.

Horizontal moment $= \dfrac{40.05 \times 3^2}{12} = 30.04$ kN-m/m, $A_{st} = \dfrac{30.04 \times 10^6}{150 \times 0.875 \times 335}$

$= 684$ mm^2

8ϕ 140 c/c (2 × 357 mm^2/m > 684, OK), provide horizontal bars on each face 8ϕ 140 c/c and increase spacing gradually 8ϕ 280 c/c near top portion.

Bottom slab:

Nominal say 370 mm thickness, provide projection 0.30 m on outside so that soil bearing pressure is less than permissible SBC. Provide steel bars 22ϕ 90 mm c/c on water face and 20ϕ @ 90 c/c in bottom face. Distribution bars on each face 12ϕ 300 c/c.

Q. 11.9 Design a cylindrical tank with flat bottom slab and spherical dome cover. The tank is resting on a ring beam of centre line diameter 7.20 m. Height of water should be limited to 4 m. The ring beam is resting on a cylindrical shaft shell of 7.20 m centre line diameter. Use M30 CC and Fe 415 steel bars. The capacity of the tank is 290000 litres. Find the diameter and design the (i) spherical cover dome, (ii) top ring beam, (iii) tank wall, (iv) base slab and (v) bottom ring beam.

[**Hint:** $D = 9.60$ m, $H_{max} = 4 + 0.2 = 4.20$ m

Dome: Rise $\dfrac{D}{7}$, $R = 8.93$ m, ϕ (semi angle) $= 32.52°$, min. thickness $= 75$ mm.

T (meridional thrust) $= 18.90$ kN/m, $T_2 = 10.5$ kN/m, nominal reinforcement 0.12%, 8ϕ @ 300 c/c radial as well as circumferential. Central portion mesh 8ϕ 300 c/c.

Top ring beam: 250 mm × 300 mm, hoop tension $= 76.5$ kN/m, $A_{st} = 12\phi$ 6 ring bars, 6 mm 2 legged stirrups @ 225 mm c/c.

Tank wall: $H_{max} = 4.20$ m, $D = 9.60$ m, hoop tension $= 198$ kN

A_{st} rings 12ϕ 170 c/c on each face near bottom and gradually spacing increased in top.

Cantilever moment (bottom 1 m) = 13.5 kN-m/m, vertically 12ϕ 125 c/c inner face, outer 8ϕ 220 c/c.

Base slab: u.d.l. q = 48.24 kN/m^2 (3602 kN total, ring load tank wall centre line = 818 kN reactions ring beam 7.20 m = 4420 kN (upward).

$$a = \frac{9.75}{2}, M_r = \frac{3q}{16}(a^2 - r^2), M_\theta = \frac{q}{16}(3a^2 - r^2), M_r\,(\text{max}) = M_{\theta\,(\text{max})} = 215 \text{ kN-m}$$

(u.d.l.)

Net $M_{r\,\text{max}}$ = 108 kN-m = $M_{\theta\,\text{max}}$, D = 300 mm, d = 265 mm, A_{str} = 3105 mm^2/m, 25ϕ @ 150 c/c radial bars and circumferentially 25ϕ @ 150 c/c in bottom portion or bottom mesh 25ϕ @ 150 c/c in central portion. Negative on water face over ring beam 12ϕ 230 c/c both radially and circumferentially.

Bottom ring beam: $\dfrac{\text{load}}{m}$ = 196.0 kN/m, b = 350 mm, D = 650 mm,

SF = $wR\phi$ = 381 kN.

Support moment = $KwR^2\phi$ = 244 kN-m, mid span moment = $K'wR^2\phi$ =123.4 kN-m

Limit state design: $\dfrac{M_u}{bd^2}$ = 2.91, p_t = 0.922%, A_{st} = 1936 mm^2 (6 bars 22ϕ) and 2 anchor bars 22ϕ

Shear stress τ_v = 2.724 N/mm^2, provide shear reinforcement

S_v (spacing) 12ϕ 2 legged @ 110 mm c/c near supports to 12ϕ 2 legged @ 200 c/c near mid span points.

Q. 11.10 Design various dimensions of an Intze tank of 800 m^3 capacity.

[**Hint:** Optimum proportions: Diameter of cylindrical tank = D, rise of bottom dome

$= \dfrac{D}{7}$, height of conical shell = 0.2D, diameter of bottom ring girder = 0.60D,

FB in top dome, height of cylindrical tank = 0.40D

$$V = 800 = \frac{\pi D^2}{4} \times 0.4D + \frac{\pi}{12} \times 0.2D\,(D^2 + 0.36D^2 + 0.6D^2) - \frac{\pi}{3}(0.2D)^2$$

$$\left\{3 \times 0.386D - \frac{D}{7}\right\}$$

D = 12.88 m (say) D = 13 m, D $\cong$ 8 m, h_2 = 5.2 m, h_3 = 2.60 m, $h_4 = \dfrac{D}{7}$ = 1.85 m,

R_2 = 5.25 m, V (actual) = 870 m^3, OK.

Q. 11.11 Design top cover spherical dome and top ring beam of an Intze tank of 13 m diameter. Use M25 CC and Fe 415 HYSD.

[**Hint:** Rise = $\dfrac{D}{7} = \dfrac{13}{7}$ = 1.85 m, radius $R_1 = \dfrac{1}{2}\left(\dfrac{6.5^2}{1.85} + 1.85\right)$ = 12.34 m

Semicentral angle $\phi = \cos^{-1}\left(\dfrac{12.34 - 1.85}{12.34}\right) = 31.7°$, thickness $t = 75$ mm

Total load $w = 4.0$ kN/m^2. Meridional thrust $T_1 = \dfrac{wR_1}{(1+\cos\theta)} = 26.7$ kN

Circumferential force $T_2 = wR_1\left(\cos\theta - \dfrac{1}{1+\cos\theta}\right) = 15.33$ kN.

Min. temp. steel, MS = 225 mm^2/m or (HYSD = 180 mm^2/m)
Top ring beam (8ϕ @ 200 c/c MS) or (8ϕ 250 c/c HYSD bars)

Hoop tension (ring beam) $= T_1 \cdot \cos\theta \cdot \dfrac{D}{2} = 26.7 \cos 31.7°\left(\dfrac{13}{2}\right) = 148$ kN

A_{st} (reqd.) ($\sigma_{st} = 150$ N/mm^2) $= \dfrac{148000}{150} = 987$ mm^2

(provide 16ϕ 6 ring bars, 3 top + 3 bottom), $A_{st} = 1206$ mm^2

A gross of concrete ($\sigma_{cbt} = 1.3$ N/mm^2) $= \dfrac{148000}{1.30} = 113846$ mm^2, (350 mm × 350 mm)

Nominal stirrups $A_{st} = \dfrac{0.24}{100} \times 350 \times 350 = 294$ mm^2 (8ϕ 2 legged stirrups @ 300 c/c, $A_{st} = 333$ mm^2/m)

Q. 11.12 Design cylindrical tank wall of an Intze tank with diameter = 13 m and height of water = 0.4D = 5.20 m, Use M25 CC and Fe 415 HYSD.

[Hint: Hoop tension T (base of tank) $= (9.8 \times 5.2)\dfrac{13}{2} = 331$ kN/m

$t = 232.5$ mm ($t = 250$ mm), min. $A_{st} = 0.206\%$ HYSD
vertical bars 8ϕ @ 190 c/c ($A_{st} = 526$ mm^2/m)

A_{st} (reqd.) $= \dfrac{331 \times 1000}{150} = 2207$ mm^2/m

16ϕ rings 180 c/c each face in bottom 1.5 m
16ϕ rings 300 c/c each face in top.

Q. 11.13 Design conical shell wall and middle ring beam of an Intze tank of following data: Upper end of conical shell diameter = 13 m, lower end of conical shell diameter = 8.0 m, height of conical shell = 2.6 m. Lower end of conical shell connected to a spherical shell of 1.85 m rise and having radius of 5.25 m, total weight of tank wall, cover shell, top beam and self-weight of ring beam = 65 kN. Slope of conical shell wall = 43.9° with the vertical. Use M25 CC and Fe 415 HYSD steel.

Assume: $m = 11$, $\sigma_{ct} = 1.3$ N/mm^2, $\sigma_{st} = 150$ N/mm^2, depth of water at middle beam level = 5.20 m, cylindrical wall = 250 mm thick.

[**Hint:** (Fig. 11.29) Hoop tension due to vertical loads at the middle beam joint

$$= (W \tan 43.9°) \frac{D}{2} = 407 \text{ kN}$$

Hoop tension due to water pressure at middle beam level ($h = 5.2$ m)

$$= \frac{(9.8 \times 5.2)13}{2} = 331 \text{ kN}$$

Total hoop tension at middle beam level = $407 + 331 = 738$ kN

Hoop steel reqd. in the middle beam = $\frac{738000}{150} = 4920 \text{ mm}^2$

Provide 16 rings of 20 mm diameter, 8 in top face and 8 in bottom face.
A_{st} (provided) $= 16 \times 314 = 5024 \text{ mm}^2$

Area of concrete reqd. $= \frac{738000}{1.3} = 567690 \text{ mm}^2$,

thus, $A_g + (m - 1) A_{st} = 567690 \text{ mm}^2$, $A_g = 517450 \text{ mm}^2$ (provide $1200 \times 500 \text{ mm}^2$)
$d = 460$ mm
Projection beyond wall = $1200 - 250 = 950$ mm = 0.95 m.
The projection is used as inspection gallery with 15 kN/m^2 u.d.l. and 5 kN/m edge live load.
SF $V_a = 15 \times 0.95 + 5 = 1925$ kN/m

BM $M = 14.25 \times \frac{0.95}{2} + 5 \times 0.95 = 11.55$ kN-m/m

A_{st} (reqd.) as cantilever in top face $= \frac{11.55 \times 10^6}{150 \times 0.872 \times 460} = 192 \text{ mm}^2$

Thus, providing 12ϕ 2 legged radial bars in top face extending in cylindrical wall and conical wall at 150 mm c/c will be quite safe.

Top $A_{st} = \frac{113000}{150} = 753 \text{ mm}^2/\text{m}$ (> 192 mm^2/m, OK).

Design of conical wall (tank capacity actual = 8,70,000 litres):
Total vertical load from top side = $65\pi D = 65\pi (13) = 2655$ kN
Weight of water = $870 \times 9.8 = 8526$ kN

Weight of slant height of conical wall $= \pi \left(\frac{13 + 8}{2} \right) \times \left(\sqrt{2.6^2 + 2.5^2} \right) \times 0.35 \times 25$

$= 1041$ kN
($t = 0.35$ m, assumed)
Total wt. (bottom end) = $(2655 + 8526 + 1041) = 12222$ kN,

$V_2/\text{m} = \frac{12222}{\pi(8)} = 486.3$ kN/m

Meridional thrust $= V_2 \cos \theta = 486.3 \cos 43.9° = 375.2 \text{ kN/m}$

Meridional stress ($t = 350$ mm, assumed) $= \dfrac{375.2 \times 1000}{1000 \times 350}$

$$= 1.072 \text{ N/mm}^2 \text{(comp.)}$$
$$(< \text{permitted } 6 \text{ N/mm}^2)$$

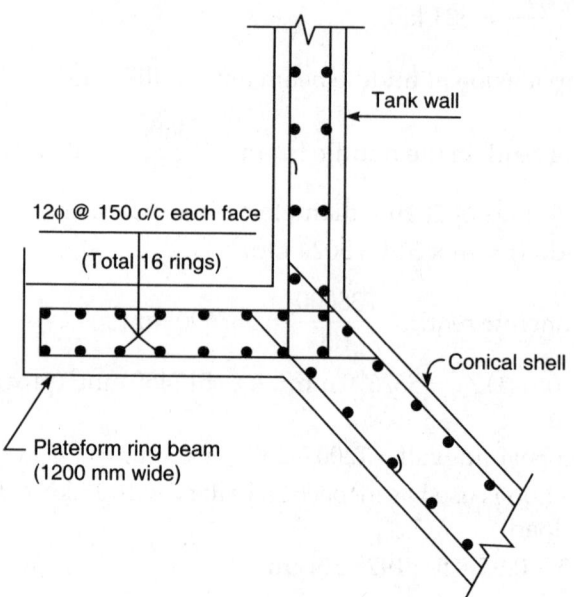

12φ @ 150 c/c each face

(Total 16 rings)

Tank wall

Conical shell

Plateform ring beam
(1200 mm wide)

Fig. 11.29: Plateform ring beam junction

Provide min. steel.

Weight q of conical wall $= 0.35 \times 1 \times 1 \times 25 = 8.75 \text{ kN/m}^2$

Top end water pressure $p = 9.8 \times 5.2 = 51 \text{ kN/m}^2$

$\theta = 43.9°$ with the vertical,

hoop tension near top end of conical wall $= (p \cos \theta + q \tan \theta)\dfrac{D}{2}$

$= 302.66 \text{ kN/m}$

A_{st} (hoop steel) $= \dfrac{302.66 \times 1000}{150} = 2018 \text{ mm}^2$ (each face $= \dfrac{2018}{2} = 1009 \text{ mm}^2\text{/m}$)

Provide 12φ @ 110 c/c rings on each face $A_{st} = 1027 \text{ mm}^2\text{/m}$ on each face

Section dimensions of conical wall:

Effective area of section $= A_g + (m - 1) A_{st} = (1000t + 10 \times 2 \times 1027)$

Permissible tensile stress in concrete $= 1.3 \text{ N/mm}^2$

Hence, $(1000 t + 20540) = \dfrac{302660}{1.3} = 232815$

$t = \dfrac{(232815 - 20540)}{1000} = 212.3 \text{ mm}$ (provide 300 mm thickness)

Hoop tension at lower edge of conical wall
$p = 9.8 (5.2 + 2.6) = 76.44$ kN/m^2.

Hoop tension (lower end) $T = (76.44 \cos 43.9° + 8.75 \tan 43.9°) \dfrac{D}{2}$

$$= (58.98 + 7.22) \frac{8}{2} = 265 \text{ kN/m} < 302.66 \text{ (top end)}$$

Hence design is OK.

Min. temp. steel for 300 mm thickness (HYSD steel) $= \left\{ 0.24 - \dfrac{0.08(200)}{350} \right\}$

$= 0.195\%$

$A_{st} = \dfrac{0.195}{1000} \times 300 \times 1000 = 585$ mm^2 (293 mm^2/m each face)

8ϕ 170 mm c/c, $A_{st} = 294$ mm^2/m.

Provide min. steel 8ϕ 170 c/c on each face along the slant length. Extend these bars in middle beam and vertical tank wall and on bottom side of bottom ring beam girder.

Q. 11.14 Design bottom ring girder of an Intze tank supported on 8 columns with following data:

Diameter ring girder = 8.0 m, rise of spherical dome = 1.85 m, radius = 5.25 m

$$\phi = \cos^{-1}\left(\frac{3.40}{R_2} \right) = \cos^{-1}\left(\frac{3.4}{5.25} \right) = 49.64°, \text{ thickness of dome } t_0 = 0.20 \text{ m},$$

Total load of water and dome = 3660 kN, $\dfrac{\text{load}}{\text{area}} = \dfrac{3660}{\frac{\pi}{4}(8^2)} = 72.8$ kN/m^2

[Hint: Meridional thrust $= \dfrac{w \cdot R}{(1 + \cos\theta)} = \dfrac{72.8 \times 5.25}{(1 + 0.6476)} = 232$ kN/m

Circumferential force = 150.23 kN/m

Thrust from conical shell = 265 kN/m, vert. load from cone = 191 kN/m $\downarrow$

Meridional thrust = 232 kN/m, vert. load = 232 sin 55.16° = 177

Total vertical load on the ring beam = 191 + 177 = 368 kN/m

Assume ring beam size = 600 mm × 1000 mm deep, self-wt. = 0.6 × 25 = 15 kN/m

Total downward load/m = 368 + 15 = 383 kN/m

The ring beam is supported on 8 columns and can be designed by limit state design. Moments and torsion can be computed by using various coefficients given in Table 11.3

Total downward load/m = 383 kN.

Max. SF near column support $= \dfrac{1}{2} \times 383 \times \dfrac{\pi}{4} \times 8 = 1203$ kN

Various coefficients from Table: $K = 0.066$, $K' = 0.030$, $K'' = 0.005$ (torsion), $\alpha = 9.33°$

Support moment $= -0.066 \times 383 \dfrac{\pi}{4} (4)^2 = -318$ kN-m

Mid span moment $= +0.030 \times 383 \dfrac{\pi}{4} (4)^2 = +144.4$ kN-m

Max. torsional moment $= 0.005 \times 383 (4)^2 \dfrac{\pi}{4} = 24.07$ kN-m

Thus, max. design moment $= 318$ kN-m.

$M_u = 1.5 \times 318 = 477$ kN-m

$V_u = 1.5 \times 1203 = 1805$ kN, $\tau_v = \dfrac{1805 \times 1000}{600 \times 1000} = 3.01 \le 3.10$ N/mm^2, (max. τ_c)

Just Safe.

Increase the depth to 1200 mm, $\tau_v = \dfrac{1805 \times 1000}{600 \times 1200} = 2.51$ N/mm$^2 < 3.10$, $\tau_{c\,max}$

Very Safe.

Design shear reinforcement (vertical stirrups).

$D = 1200$ mm, $d = 1160$ mm (cover 40 mm),

$M_{u\,lim}$ (600 × 1160 mm) $= 0.138 \times 600 \times 1160^2 \times 25 = 278539 \times 10^4$ N-mm
$= 2785.39 \times 10^6$ N-mm $> M_{u\,max}$ (494 kNm), OK, under reinforced.

$$A_{st} = \dfrac{0.5 f_{ck} b \cdot d}{f_y} \left[1 - \sqrt{1 - \dfrac{4.6 \times M_u}{f_{ck} \cdot b \cdot d^2}} \right]$$

$$= \dfrac{0.50 \times 25 \times 600 \times 1160}{415} \left[1 - \sqrt{1 - \dfrac{4.6 \times 494 \times 10^6}{25 \times 600 (1160)^2}} \right] = 1216 \text{ mm}^2$$

(12ϕ 5, $A_{st} = 1570$ mm^2 in top face near columns)

Min. temp steel $= 0.20\%$ above 450 mm thickness $= \dfrac{0.20}{100} \times 1200 \times 600 = 1440$

<1570, OK

In bottom face provide 20ϕ 3 ring bars through out and add 2 bars 20ϕ over column supports in top face.

Provide 12ϕ 4 legged vertical stirrups @ $S_v = \dfrac{0.87 \times 415 \times 452 \times 1160}{(1871000 - 0.275 \times 600 \times 110)}$

$S_v = 113$ mm c/c (provide near columns, 12ϕ 4 legged @ 110 c/c and spacing gradually increased towards mid span to 200 c/c)

Torsion moment max. at $\alpha = 9.33°$, i.e. $l = R_z\alpha = \dfrac{\pi(9.33)}{180} \times 4 = 0.65$ m

BM reduces at the point of max. torsion (T_u)

$$M = \left(329.3 - 381.4 \times \frac{0.65^2}{2}\right) = 248.73 \text{ kN-m}, \; M_u = 1.5 \times 248.73 = 373.11 \text{ kN-m}$$

$T_u = 1.5 \times 24.07 = 36.11$ kN-m/m

$$M_e = 373.1 + \frac{36.11}{1.7}\left(1 + \frac{1200}{600}\right) = 373.1 + 63.73 = 436.83 \text{ kN-m} < 494 \text{ kN-m,}$$

OK.

Thus continue 20ϕ 5 bars throughout in top face and provide 3 bars 20 as

rings in lower face throughout. Also provide at least $\dfrac{0.10}{100} \times 600 \times 1200$

= 720 mm². Provide additional 16ϕ 2 × 2 = 4 bars (2 on each face) as ring bars. Provide 4ϕ legged 12ϕ stirrups @ 110 c/c near columns and spacing gradually increased to 200 c/c near the mid span.

Q. 11.15 Design a cylindrical shaft (staging) for a circular tank of 13.4 m outside diameter and 25 m FSL above ground. Total height of the tank may be taken as 5 m. Foundation raft slab may be assumed as 0.50 m thick. Raft slab is 1.50 m deep below GL. Weight of empty tank = 4570 kN, weight of water = 8700 kN, CG of tank (empty as well as full) may be assumed as 2 m above tank bottom. Total height of shaft above foundation slab top = 21 m. Use M20 CC and Fe 415 HYSD. CL diameter of shaft = 8.0 m

[Hint: $f_{ck} = 20, \sigma_{cbc} = 7.0$ N/mm², $\sigma_c = 5.0$ N/mm², $f_y = 415$ N/mm², $\sigma_{st} = 190$ N/mm², $m = 13$,

wind pressure $p = 1.5$ kN/m², EQ coefficient $\alpha_H = \pm 0.075$, $\alpha_V = \pm 0.0375$, assume thickness = 150 mm.

Wt. of shaft = $\pi(8)\, 0.15 \times 21 \times 25 = 1980$ kN, CG above base slab top = $\dfrac{21}{2}$

= 10.5 m.

Total weight (empty tank) on the shaft base $W_o = 4570 + 1980 = 6550$ kN

Total weight (tank full) on the shaft base $W_F = 6550 + 8700 = 15250$ kN

Effective thickness (0.24% vertical steel) = $150 + \dfrac{0.24}{100}\,(13 - 1)\,150 = 154.32$ mm

Effective area of shaft $A_e = \pi D \times 0.15432 = \pi(8) \times 0.15432 = 3.88 \times 10^6$ mm²

Effecrtive section modules of shaft $Z_e = \dfrac{\pi R_0^3 \cdot t}{R_1} = \dfrac{\pi(4)^3 \times 0.1543}{4.075} = 7.613 \times$

10^9 mm³

Moments: Wind force (P_{w1}) tank = $0.70 \times 1.5 \times 13.4 \times 5.0 = 70.35$ kN

CG from shaft bottom = $2 + 21 = 23$ m.

Wind force (P_{w2}) shaft $= 0.70 (8.15 \times 20) \times 1.5 = 171.15$ kN

$$CG = \left(\frac{20}{2} + 1GL\right) = 11 \text{ m above base of shaft}$$

Max. moment M_1 (due to wind) $= 70.35 \times 23 + 171.1 \times 11 = 1618 + 1882.7 = 3501$ kN/m.

Horizontal force (tank) (EQ) $= (8700 + 4570) \times 0.075 = 13270 \times 0.075 = 995.25$ kN at 23 m above shaft base.

Horizontal force (shaft) EQ $= 1980 \times 0.075 = 149$ kN at CG $\frac{21}{2} = 10.5$ m

Moment (EQ) about shaft base $= 22891 + 149 \times 10.5 = 1564 + 22891$
or $M_2 = 24455$ kN-m

Vertical load during EQ $= 0.0375 \times (13270 + 1980) = \pm 572$ kN

Check of stresses at the base of shaft (working stress method):

Case I: Wind (tank empty) $= \dfrac{6550 \times 1000}{3.88 \times 10^6} + \dfrac{3501 \times 10^6}{7.613 \times 10^9}$

$$= 1.6881 \pm 0.460 = +2.148 + 1.2281 \text{ N/mm}^2$$

Wind (tank full) $= \dfrac{15250 \times 10^3}{3.88 \times 10^6} \pm \dfrac{3501}{7613} = 3.93 \pm 0.46$

$$= +4.93 \text{ and } 3.47 \text{ N/mm}^2$$

$$\left(< \frac{4}{3} \times 7 = 9.33 \text{ N/mm}^2, \text{OK}\right).$$

Case II: EQ (tank empty) $= \dfrac{(15250 \pm 572)10^3}{3.88 \times 10^6} \pm \dfrac{24530 \times 10^6}{7.613 \times 10^9} = 4.078 \pm 3.222$

$$= +7.30 \text{ N/mm}^2 \text{and} + 0.854$$

$$\left(< \frac{4}{3} \times 7 = 9.33 \text{ N/mm}^2, \text{OK}\right).$$

Design of reinforcement:

$$t = 150 \text{ mm}$$

Min. vertical steel: $A_{st} = \dfrac{0.24}{100} \times 150 \times 1000 = 360 \text{ mm}^2$

(180 mm^2, each face 8ϕ 250 mm c/c)

Min. horizontal steel (ring) $A_{st} = \dfrac{0.20}{100} \times 150 \times 1000 = 300 \text{ mm}^2$

(150 mm^2, each face 8ϕ 300 c/c rings)

Thus, provide 8ϕ rings @ 300 c/c on each face and 8ϕ vertical bars @ 250 c/c throughout the staging shaft. Stresses are checked and found safe with respect to effective thickness.

Check of stresses by interaction formula:

In case of EQ stresses (worst case): Direct stress $= \dfrac{15822 \times 10^3}{3.88 \times 10^6} = 4.078\ \text{N/mm}^2$

Bending stress $(Z_e = 7.613 \times 10^9) = \dfrac{24530 \times 10^6}{7.613 \times 10^9} = \pm 3.222\ \text{N/mm}^2$

By interaction $= \left(\dfrac{4.078}{5.00} + \dfrac{3.222}{7.00}\right)(0.8156 + 0.4603) = 1.276,\ \text{OK}$

$$\left(< \frac{4}{3} \times 1 = 1.333\right)$$

Reduction of direct stress for buckling effect:

$C_r = \left[1 + \dfrac{f_{ck}}{0.2E_c} \times \dfrac{R}{t}\right]^{-1},\ E_c = 5000\ \sqrt{f_{ck}} = 5000\ \sqrt{20} = 2.236 \times 10^4\,\text{N/mm}^2$

$C_r = \left[1 + \dfrac{20}{0.2 \times 2.2361 \times 10^4} \times \dfrac{4000}{150}\right]^{-1} = (1.12)^{-1} = 0.8935$

Thus, maximum direct stress permitted shall be $5 \times 0.8935 = 4.47\ \text{N/mm}^2$
Actual maximum direct stress when tank full and EQ $= 4.078 < 4.47$. Further bottom 1 m of shaft may be splayed on outside from 150 mm to 300 mm thickness, i.e. outside diameter of junction with base $= 8.45$ m, inside diameter $= 7.85$ m.

Q. 11.16 Design a circular raft slab foundation for the tank given in Q 11.15. SBC for foundation soil $= 120\ \text{kN/m}^2$, wt. of empty tank $W_o = 4570\ \text{kN/m}$, wt. of shaft $= 1980$ kN, wt. of water $= 8700$ kN, area of shaft (effective) $= 3.88 \times 10^6$ mm^2, diameter of shaft (CL) $= 8.15$ m ($8.45 - 7.85$ m upper thickness $t = 150$ mm($t_e = 154.32$ mm), wind force on tank $= 70.35$ kN, CG $= 23$ m above base slab top (23.50 m from bottom of slab), wind force on shaft $= 171.15$ kN, CG $= 11$ m above top of base slab (11.5 m above bottom), EQ force (horizontal) tank full $= 995.25$KN, at CG 23 m above base slab, (23.50 m above the bottom of foundation slab, horizontal force due to EQ on shaft $= 149$ kN, CG at 10.5 m above base slab top (11.0 m above bottom of base).

[Hint: *A* of base reqd.

$$= \frac{\text{Max. vertical loads}}{\text{SBC}} = \frac{\{(4570 + 8700 + 1980) \pm 572\}\,\text{m}^2}{120}$$

$$A = \left(\frac{15822}{120}\right) = 132\ \text{m}^2, D = 12.96\ \text{m. Bearing pressures increase due}$$

to bending moments caused by EQ or wind pressures.
Assume $D = 16$ m so that the maximum bearing pressure on the edge of base slab must remain within limits.

Area of base $A = \dfrac{\pi}{4}\ (16)^2 = 201.06\ \text{m}^2$, section modulus $Z = \dfrac{\pi(16)^3}{32} = 402.12\ \text{m}^3$

Moments about bottom of base M_1 (wind) $= 70.35 \times 23.5 + 171.15 \times 11.5$
$$= 3622.0 \text{ kN-m}$$
Moments about bottom of base M_2 (EQ) $= 995.25 \times 23.5 + 149 \times 11.0$
$$= 25028 \text{ kN-m}$$
Thus, EQ forces are more critical,
assume weight of base slab = 10% of maximum load.

Max. bearing pressure $= \dfrac{(15822 + 1582)}{201.06} \pm \dfrac{25028}{402.12} = 86.56 \pm 62.24 = +148.8,$

$+24.32 \left(< \dfrac{4}{3} \times 120 = 160 \text{ kN/m}^2, \text{OK} \right)$

Thus, 16 m diameter is quite safe.

Design of foundation raft slab ($D = 16$ m, $a = 8$ m, $b = 4$ m)

loading upward $w = 148.8 \text{ kN/m}^2(\text{max}), < \dfrac{4}{3} \times 120 = 160 \text{ kN/m}^2, \text{OK}.$

Check for safety against overturning:

Net minimum load (tank full) EQ $= (15250 + 1582 \text{ raft}) - 572 \text{ EQ} = 16260 \text{ kN}$

Minimum restoring moment about edge of raft slab

Moment restoring $= 16260 \times \dfrac{16}{2} = 130080 \text{ kN-m}$

Max. overturning moment about bottom edge (EQ) $M_2 = 25028 \text{ kN-m}$

$\text{FOS} = \dfrac{130080}{25028} = 5.20 \text{ (more than 1.50), OK.}$

Tank empty and wind pressure:

Wt. of empty tank, shaft and raft $= (6550 + 1582) = 8132 \text{ kN}$

Moment due to wind when empty about bottom edge $= 3622 \text{ kN-m}$

$\text{FOS against (OT) empty tank} = \dfrac{8132 \times \dfrac{16}{2}}{3622} = 18.00 \gg 1.50, \text{OK}$

Thus, the structure is safe against overturning in all critical situations.

Design of raft foundation slab:

Outer diameter of slab $D = 16$ m, $R = 8.0$ m, assumed thickness = 0.500 m

Junction of staging and raft slab: outer diameter = 8.45 m, radius, 4.225 m inner diameter = 7.85 m, radius = 3.925 m, CL of shaft = 8.15 m, radius = 4.075 m

Using plate theory, the design of circular raft slab can be done.

Data:

$a = 8.0$ m, $a^2 = 64.0$, $3a^2 = 192$, $2 \log_e \left(\dfrac{a}{b} \right) = 2 \,(0.6746) = 1.3492$

$b = 4.075$ m (centre line radius), $b^2 = 16.606$, $\dfrac{a}{b} = 1.9632$, $\left(\dfrac{b}{a} \right)^2 = (0.5094)^2 = 0.2595$

Total downward load = net upward soil reaction on the slab = 15822 kN

$W = 15822$, $\dfrac{W}{8\pi} = 629.54 \text{ kN units}$, $p = \dfrac{15822}{\pi(8)^2} = 78.70$, $\dfrac{p}{16} = 4.92$

Most critical situation shall be under earthquake and tank full condition. Analysis shall be done for (i) direct vertical forces for $W = 15822$ kN (max.) and (ii) Variable forces due to EQ bending moments M_2 (tank full) $= 25028$ kN-m

$$p' \text{ (variable)} = \frac{M_2}{Z} = \frac{25028}{402.12} = \pm 62.24 \text{ kN/m}^2$$

$$q \text{ (support reaction – shaft CL)} = \frac{25028}{\pi(4.075)^2} = 479.80 \text{ kN/m}$$

$$\beta = \frac{b}{a} = \frac{4.075}{8.0} = 0.509, \ \beta^2 = 0.259, \ \beta^4 = 0.0671, \ (\beta^4 + 2\beta^2 - 3) = -2.4149$$

$$\beta^4 + 2\beta^2 = 0.5851, \ p = \frac{r}{8.0}$$

Radial and circumferential moments due to direct loading:
For $r < b$:

$$M_r = \frac{W}{8\pi}\left[2\log_e\left(\frac{a}{b}\right) + 1 - \left(\frac{b}{a}\right)^2\right] - \frac{p}{16}\,(3a^2 - 3r^2) = (370.90 + 14.76r^2)$$

$$M_\theta = \frac{W}{8\pi}\left[2\log_e\left(\frac{a}{b}\right) + 1 - \left(\frac{b}{a}\right)^2\right] - \frac{p}{16}\,(3a^2 - r^2) = (370.90 + 4.92r^2)$$

For $r > b$:

$$M_r = \frac{W}{8\pi}\left[2\log_e\frac{a}{r} - \left(\frac{b}{a}\right)^2 + \left(\frac{b}{r}\right)^2\right] - \frac{p}{16}\,(3a^2 - 3r^2)$$

$$= [1259.08\log_e\frac{8}{r} + \frac{10454.14}{r^2} + 14.76r^2 - 1108]$$

$$M_\theta = \frac{W}{8\pi}\left[2\log_e\frac{a}{r} + 2 - \left(\frac{b}{a}\right)^2 - \left(\frac{b}{r}\right)^2\right]\frac{p}{16}\,(3a^2 - r^2)$$

$$= [1259.08\log_e\frac{8}{r} - \frac{10454.14}{r^2} + 4.92r^2 + 151.07]$$

Radial and circumferential moments due to variable loading:
For $r < b, \ k = \dfrac{r}{8.0}$

$$M'_r = \frac{8p'a^2}{48}\,(k - k^3) + \frac{qak}{8}\,(\beta^4 + 2\beta^2 - 3) = -743.74\,k - 414.93\,k^3$$

$$M'_\theta = \frac{p'a^2}{48}\left(\frac{5}{3}k - k^3\right) + \frac{qak}{24}\,(\beta^4 + 2\beta^2 - 3) = -247.90\,k - 82.99k^3$$

For $r > b$, ($p' = 62.24, \ q = 479.8, \ a = 8$ m), $k = \dfrac{r}{8.0}$

$$M'_r = \frac{5p'a^2}{48}\,(k - k^3) + \frac{qa}{8}\frac{1}{k^3}\,[k^4\,(\beta^4 + 2\beta^2) - 2\beta^2 k^2 - \beta^4]$$

$$= 695.66k - 414.93k^3 - \frac{248.54}{k} - \frac{32.2}{k^3}$$

$$M'_\theta = \frac{p'a^2}{48}\left(\frac{5}{3}k - k^3\right) + \frac{qa}{8k^3}\left[\frac{k^4}{3}(\beta^4 + 2\beta^2) - 2\beta^2 k^2 - \beta^4\right]$$

$$= 231.90k - 82.99k^3 - \frac{248.63}{k} + \frac{32.2}{k^3}$$

Moments at various points are computed and tabulated as given below:

Calculations of moments at various radii substituting the values

r (m)	0.0	1.0	2.0	3.90	4.075	4.25	6.00	7.00	8.00	
r^2	0.0	1.0	4.0	15.2	C_L	18.06	36.00	49.00	64.00	
Moments due to direct loads										Max.
M_r (kN-m)	\multicolumn{3}{c}{$M_r = 370.90 + 14.76\,r^2$}			\multicolumn{4}{c}{$M_r = 1259.08 \log_e \frac{8}{r} = \frac{10454.14}{r^2}$ $+ 14.76r^2 - 1108$}	
	370.9	385.66	430	595.4		534	76.0	-3.3	0	595.4
M_θ (kN-m)	\multicolumn{3}{c}{$M_\theta = 370.90 + 4.92\,r^2$}			\multicolumn{4}{c}{$M_\theta = 1259.08 \log_e \frac{8}{r}\,\frac{10454.14}{r^2}$ $+ 4.92r^2 - 151$}	
	370.9	375.8	390.6	445.7		458	400	347.0	302.6	458.0
Moments due to variable loading caused by EQ/ wind moments										
$k = \frac{r}{8.0}$	0.0	0.125	0.25	0.4875		0.5313	0.75	0.875	1.0	
k^3	0.0	0.001953	0.01563	0.11586		0.1499	0.4219	0.6699	1.0	
M'_r	\multicolumn{3}{c}{$M'_r = -743.74\,k - 414.93\,k^3$}			\multicolumn{4}{c}{$M'_r = 695.66k - 414.93k^3 - \frac{248.54}{k}\,\frac{32.2}{k^3}$}	
(kN-m)	0.0	∓93.78	∓192.40	∓410.65		∓375.3	∓61.02	∓1.38	0.0	410.65
M'_θ	\multicolumn{3}{c}{$M'_\theta = -247.90k - 82.99k^3$}			\multicolumn{4}{c}{$M'_\theta = 231.90k - 82.99k^3 - \frac{248.54}{k} + \frac{32.2}{k^3}$}	
(kN-m)	0.0	∓31.15	∓63.28	∓130.47		∓142.3	∓116.16	∓88.68	∓67.4	∓142.3
Net moments (kN-m) – Worst condition under EQ/ wind										
M_r (kN-m)	−370.9	479.44	622.4	1006.0		909.3	137.02	−4.68	0	1006.0
M_θ (kN-m)	370.9	406.95	453.88	576.17		600.3	516.16	435.68	370.04	600.3

$$= 231.90k - 82.99k^3 - \frac{248.63}{k} + \frac{32.2}{k^3}$$

Max. Moments $M_r = + 1006$ kN-m/m (inner edge at 3.90 m radius),
$M_\theta = +600.3$ kN-m/m (at 4.25 m radius on outer edge)
Using M20 CC, Fe 415 HYSD and working stress approach
$\sigma_{cbc} = 7$ N/mm^2, $\sigma_{st} = 190$ N/mm^2, $\sigma_c = 5$ N/mm^2, $m = 13$, constants $n = 0.324$,
$j = 0.892$, $k = 1.0115$ for EQ/ wind, $k' = 1.33k$

Effective d required $= \sqrt{\dfrac{M}{b \cdot k'}}$

$$d = \sqrt{\frac{1006 \times 10^6}{1000 \times 1.33 \times 1.0115}} = 865 \text{ mm}, D = 865 + 50 = 915 \text{ mm}, d = 865 \text{ mm}$$

$$A_{st} \text{ (radial)} = \frac{1006 \times 10^6}{1900 \times 0.892 \times 865} = 686 \text{ mm}^2/\text{m}$$

(Provide 12ϕ radial bars @ 150 c/c, $A_{st} = 753$ mm^2/m, OK.)

Minimum temp. steel $= \dfrac{0.2}{100} \times 1000 \times 915 = 1830$ mm^2/m (each face = 915 mm^2/m)

Thus provide radial bars 16ϕ 200 c/c ($A_{st} = 1005$ mm^2/m on each face)
In the central portion as mesh of 16ϕ 200 c/c in each direction and on each top face and bottom faces.
At free ends provide circular ring bars 16ϕ 200 c/c and also radial bars 16ϕ 200 c/c on each face (length from radius 8 m to 3.50 m and also on both the faces. Half of the mesh bars can be bent and extended into the shaft wall while half of the bars are extended by about 500 mm in the raft slab beyond shaft wall. Ring bars are also provided from 3.50 m radius to 8.0 m radius (Fig. 11.30).

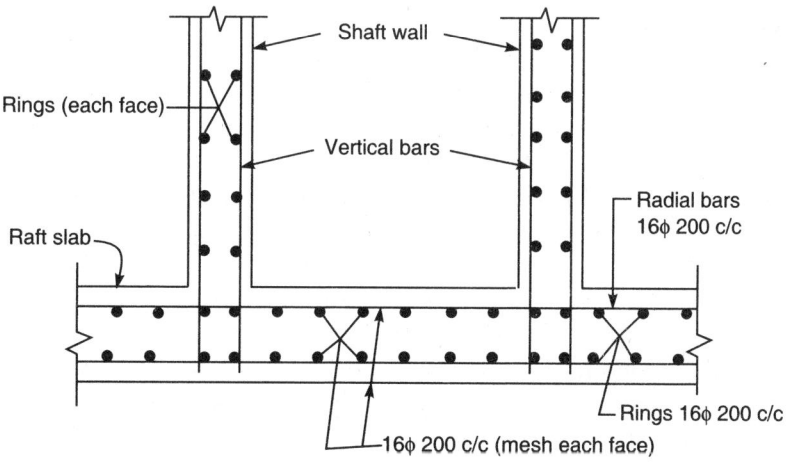

Fig. 11.30: Section of raft slab and shaft

Chapter 12

Introduction to Prestressed Structural Concrete

LEARNING OBJECTIVES

After the study of *introduction to prestressed structural concrete*, the learner with understand the basic concepts of prestressed concrete and shall be able to:

- ⊙ Explain basic concepts of prestressing cement concrete
- ⊙ Describe advantages of prestressed concrete elements
- ⊙ Explain the characteristics of materials of prestressed cement concrete
- ⊙ Explain different losses of prestress in prestressed cement concrete
- ⊙ Explain classification of prestressed cement concrete members
- ⊙ Explain common systems of prestressing
- ⊙ Calculate losses of prestress due to different causes in simple cases
- ⊙ Design simple section of prestressed concrete beams for a given loading and span details

12.1 INTRODUCTION

Cement concrete is weak in tension and it does not carry any tensile force in RCC members. In RCC members concrete resists compressive stress while tensile stresses are borne entirely by steel reinforcement.

The concrete surrounding reinforcement simply acts as bonding material but does not share any tension. Thus in flexural members, the concrete is effective only in compression zone, while it is totally ineffective in tension zone. Prestressed concrete was developed to utilize most of concrete by subjecting the concrete to compressive stresses at working loads by prestressing both steel and concrete prior to loading.

In prestressed concrete there is intentional introduction of internal stresses of magnitude and distribution so that the resulting stresses from the given external loadings are counteracted to a desired degree. The prestress is generally introduced by tensioning the steel reinforcement in RCC elements. Thus, prestressing is the intentional introduction of permanent compressive stress specially in tension zones for balancing the likely tensile stresses which may be caused by working loads for efficient use of the constituent materials of RCC. Prestressing makes possible to use high tensile steel and also using concrete efficiently ultimately making the members much lighter weight.

Earlier attempts of prestressing were made by prestressing tensile steel to working stress, i.e. in case of mild steel reinforcement, the strain

$$e = \frac{p\left(140 \text{ N} / \text{mm}^2\right)}{E\left(2.1 \times 10^5 \text{ N} / \text{mm}^2\right)} = 0.00067$$

While permanent negative strain induced due to shrinkage and creep in concrete is of the order of 0.0008. Thus the prestress induced in mild steel totally disappears in short period of time due to shrinkage and creep. Thus, earlier attempts of prestressing failed due to various causes of prestress losses. With development of new materials and new techniques, prestressing became a reality after First World War.

E Freyssinet (France) made a viable specific contribution to prestressed concrete in 1928 which is used even today with some modifications. High strength steel wires with ultimate tensile strength of 1750 N/mm^2 and yield strength of 1250 N/mm^2 are used in prestressing and stressed to about 1050 N/mm^2 with resulting strain $e = \dfrac{1050}{2.1 \times 10^5} = 0.005$. Allowing for total loss of about 0.0008 due to shrinkage and creep, the net strain remains = (0.005 – 0.0008) = 0.0042 and the final stress = 2.1 × 10^5 × 0.0042 = 882 N/mm^2 (about 84% of the original).

Thus, this system of prestressing using high tensile strength steel works.

In Germany, America, Russia, Belgium and many other countries, engineers and scientists developed these materials and techniques further and provided many systems of prestressing.

12.2   BASIC CONCEPTS

Prestressing is the process of prior introduction of stresses, controlled in magnitude and direction, in a structural element to counteract the undesirable stresses (viz. tension in concrete) which may be caused by working loads. Prestressing is primarily introduced by tensioning the steel reinforcement. The behaviour of prestressed concrete can be explained by 3 concepts, viz. (a) stress concept, (b) strength concept and (c) balanced load concept.

(a) Stress concept

Eugene Freyssinet visualized prestressed concrete as an elastic material due to introduction of precompression in concrete. To explain this concept consider a concrete beam of cross-section A and section modulus Z and reinforced with prestressed steel bar (called tendon) having tensile force T, located centrally (along the centroidal axis) or eccentrically with eccentricity e below the centroid in case of simply supported beam. When this prestressed simply supported beam is loaded with working loads developing a bending moment M, net compressivestress developed across the section at a distance y above the centroidal axis (maximum and minimum) shall be $\left(\dfrac{T}{A} \pm \dfrac{M}{I} \cdot y\right)$ (Fig. 12.1),

when prestress force T is along the centroidal axis.

or $$\frac{T}{A} \mp \frac{T\,e}{I} \cdot y \pm \frac{My}{I}$$

when prestress force T is eccentric and e is below the centroid.

The values of T and e shall be so arranged that the resulting stresses in concrete across the section $(+y \text{ to } -y)$ remain zero orcompressive. Because of concrete remaining under compressive stress, it does not crack anywhere and hence the prestressed concrete behaves as perfectly elastic under the given working loads.

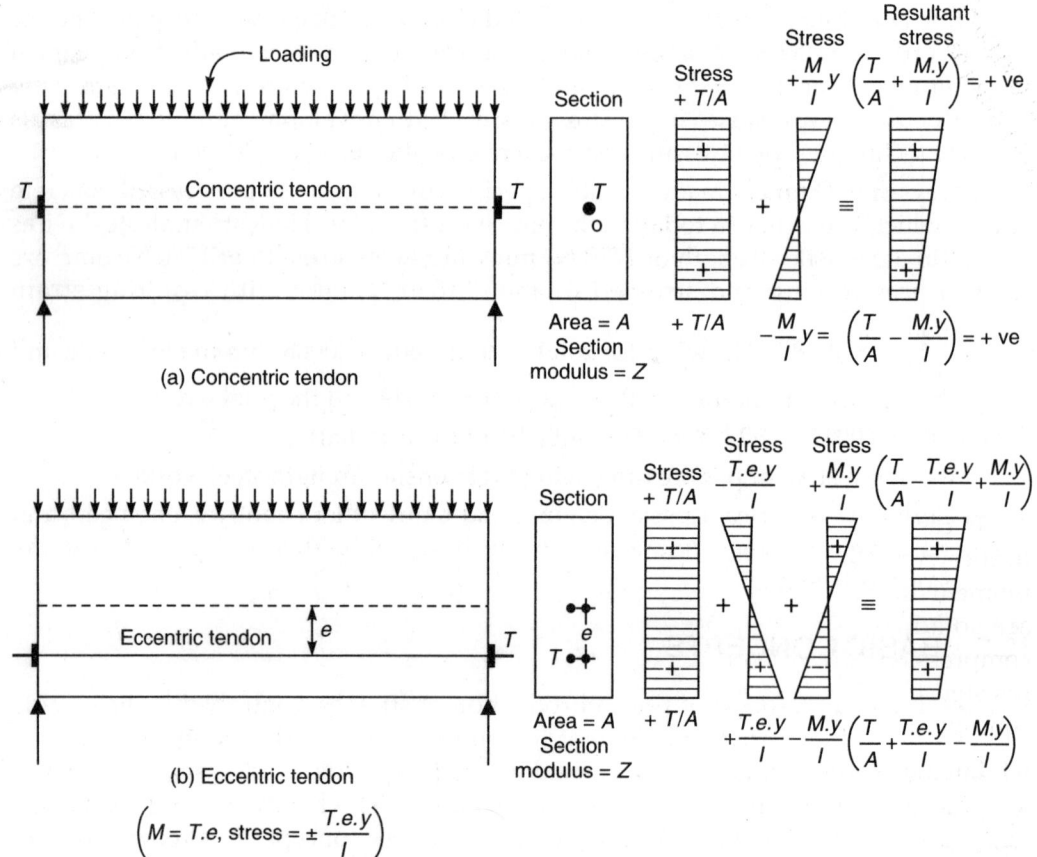

$$\left(M = T.e, \text{ stress} = \pm \frac{T.e.y}{I}\right)$$

Fig. 12.1: Prestressed concrete section developing compressive stress

(b) Strength concept

Prestressed concrete is a combination of high strength steel and high strength cement concrete. In RCC, steel takes tension in tension zone (T) while concrete takes compression (C) in compression zone forming a couple $(C \text{ or } T \times \text{distance between } C \text{ and } T)$ which resists external moment M (Fig. 12.2). The same concept can be applied to prestressed concrete section also.

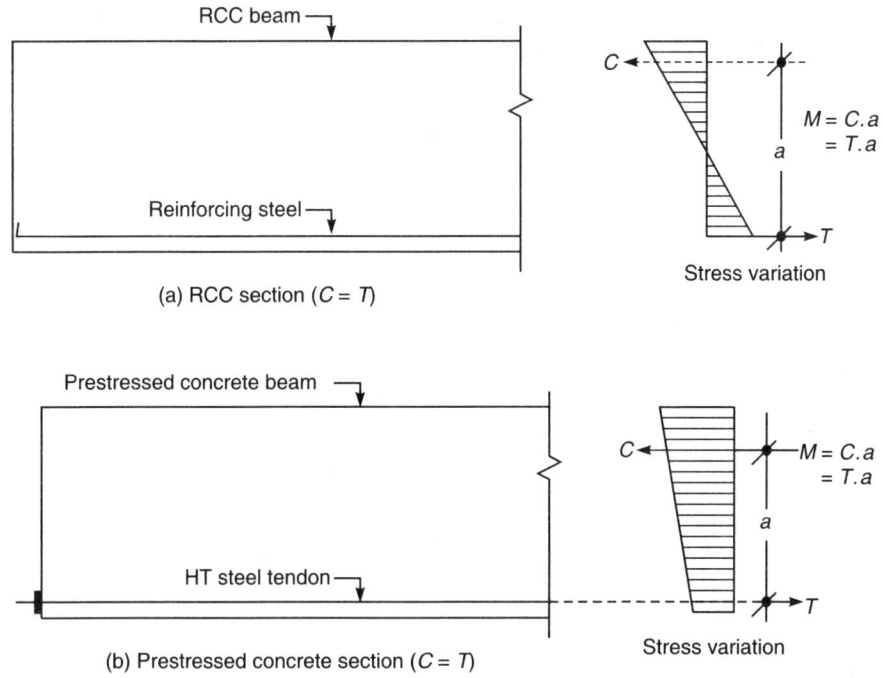

(a) RCC section ($C = T$)

(b) Prestressed concrete section ($C = T$)

Fig. 12.2: RCC and prestressed concrete beams

In case of prestressed concrete, the tendons are stressed to high tensile stress and induce compressive force C in the whole section while this force T also induces moment due to eccentricity of tendon force T. Eccentricity and magnitude of tension T are so adjusted that the whole concrete section remains in compression. Total compression of concrete section $= C$, and $C = T$ (tension in steel). External moment is resisted by the couple. The ultimate strength is computed by using the strength concept.

(c) Load balancing concept

According to Lin, the flexural members are transformed into members under direct stresses. The effect of prestressing is considered as balancing of gravity loads (external loads plus self-weight) with prestressing force so that the member under bending will not be subjected to flexural stresses under given transverse load conditions. This is possible by putting the tendons eccentrically. Flexural members such as beams, slabs etc. are normally subjected to variable bending moments along the span with maximum moment at the mid span and zero moment at the simple supports at ends. In such a case the eccentricity e must also vary accordingly and the tendons can be placed along the curved path. Concrete is considered as free body and tendons are represented by forces.

Consider the concrete beam as free body diagram and tendon as tensile force. Figure 12.3 shows the parabolic profile of tendon with central sag as h with respect to centre line of the section. Consider two points P and Q on the tendon at distances of x_1 and x_2 respectively from the centre line C. The tendon (cable) is pulled to a uniform tension of T throughout. If θ_1 and θ_2 are slopes of the cable with horizontal at P and Q respectively.

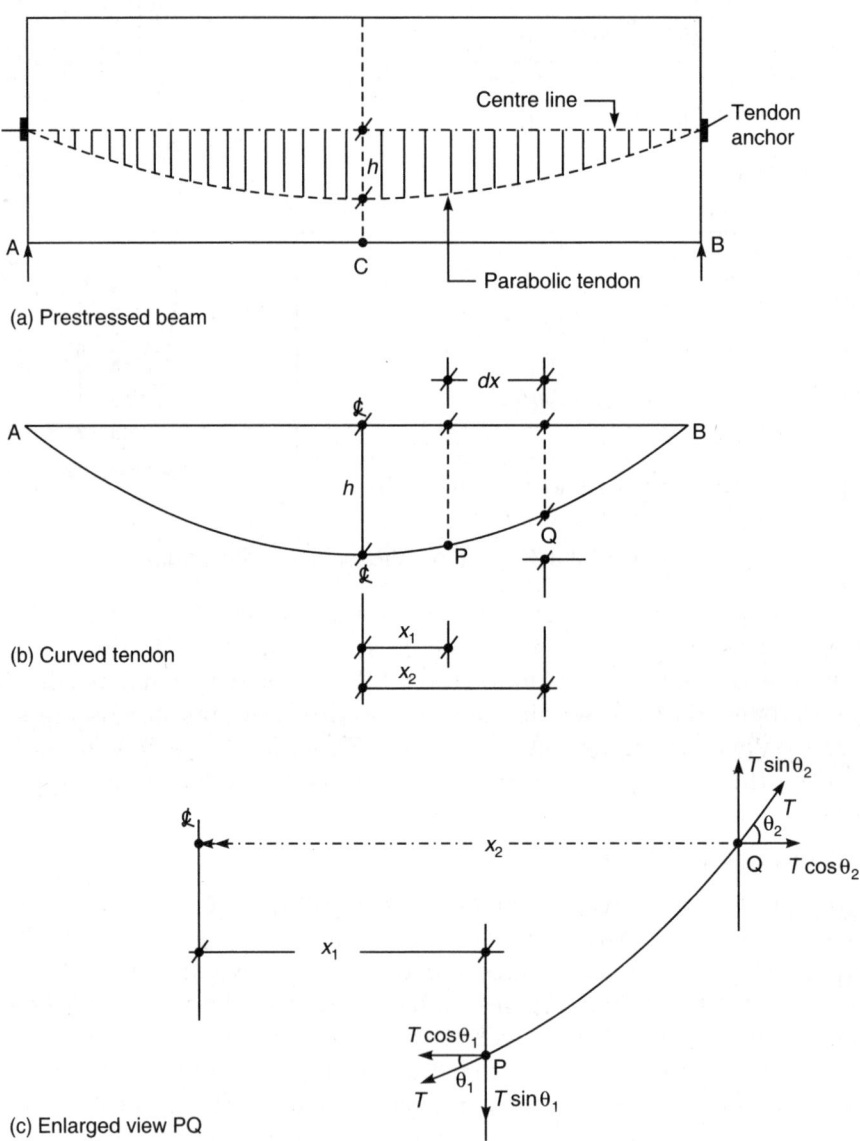

(a) Prestressed beam

(b) Curved tendon

(c) Enlarged view PQ

Fig. 12.3: Prestressed beam with parabolic tendon (cable)

we have upward force = $(T \sin \theta_2 - T \sin \theta_1)$, θ_1 and θ_2 are small since h is very small compared to span L.

Since θ_1 and θ_2 are small,

$$\tan \theta_1 = \sin \theta_1 \text{ and } \tan \theta_2 = \sin \theta_2$$

Hence upward force

$$= T(\tan\theta_2 - \tan\theta_1) \qquad\qquad \text{... Eq. (12.1)}$$

Equation of parabola

$$y = \frac{4h}{L^2}\left(x^2 - \frac{L^2}{4}\right) \qquad\qquad \text{... Eq. (12.2)}$$

or

$$\frac{dy}{dx} = \frac{4h}{L^2}(2x - 0) = \frac{8hx}{L^2},$$

or

$$\left(\frac{dy}{dx}\right)_1 = \frac{8hx_1}{L^2} = \tan\theta_1$$

or

$$\left(\frac{dy}{dx}\right)_2 = \frac{8hx_2}{L^2} = \tan\theta_2$$

Upward force on PQ $= T \cdot \dfrac{8h}{L^2}(x_2 - x_1)$

Upward force on PQ length $(x_2 - x_1)$ per unit length $= T\dfrac{8h}{L^2}$ $\qquad$ ... Eq. (12.3)

From Eq. (12.2) $\qquad \dfrac{d^2y}{dx^2} = \dfrac{8h}{L^2} \quad \therefore \quad p = T\dfrac{8h}{L^2} = T\dfrac{d^2y}{dx^2},$

or

$$p = T \cdot \frac{d^2y}{dx^2}, \qquad\qquad \text{... Eq. (12.4)}$$

where

$$\frac{8h}{L^2} = \frac{d^2y}{dx^2}$$

Thus, the parabolic cable supports a uniform distributed load p given by $p = T\dfrac{d^2y}{dx^2}$. If the external u.d.l. w is equal to p, transverse load on the beam is balanced by the axial force T which produces uniform stress in concrete, i.e. $(f) = \dfrac{T}{A}$. If the external load w is greater than p, the mid-span moment M for the remaining force $(w - p)$ can be computed and fibre stresses due to that moment would be $\dfrac{MY}{I}$. Therefore, resulting stress will be

$$\frac{T}{A} \pm \frac{MY}{I} \qquad\qquad \text{... Eq. (12.5)}$$

M = Moment due to unbalanced load of $(w - p)/m$.

The balancing force $V = 2T\sin\theta$, where V counter balances a point load W applied at mid-span (Fig. 12.4).

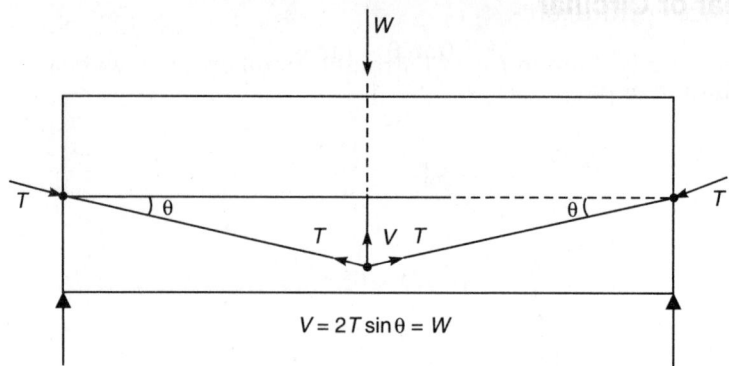

Fig. 12.4: Variable eccentricity to balance gravity loads

The uniform fibre stress in beam shall be given by

$$f = \frac{T \cos \theta}{A_c} = \frac{T}{A_c}, \text{(since } \theta \text{ is generally very small and } \cos \theta \cong 1)$$

or
$$f = \frac{T}{A_c}$$
... Eq. (12.6)

where, T is force in the cable and A_c = area at the central section

This method of *load balancing concept* is very simple and can be applied advantageously in case of indeterminate structures.

12.3 CLASSIFICATION AND TYPES OF PRESTRESSING

Classification of prestressed concrete depends on design, construction, type of structure and method of applying prestress as given below:

12.3.1 External or Internal

Prestress in concrete structure can be maintained by adjusting external reactions or by built in internal thrust with high tensile wires. External prestressing is achieved through reactions from external supports such as abutments (Fig. 12.5). External prestressing is practically not common as it requires very strong abutments and very elaborate arrangements of jacks. Figure 12.5 shows a simply supported beam externally prestressed with reactions from abutments using jacks.

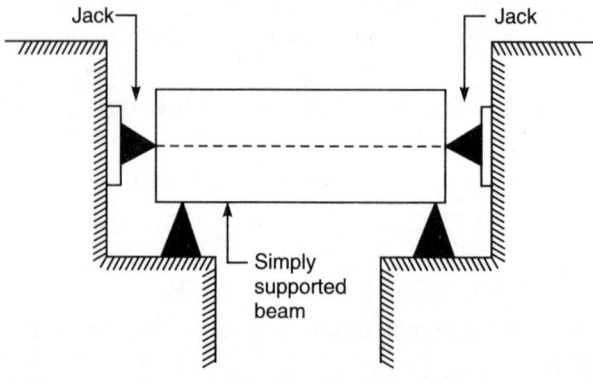

Fig. 12.5: External prestressing

12.3.2 Linear or Circular

Linear prestressing is done in case of straight members such as beams, slabs, piles, poles etc. In linear prestressing, the tendons may not be necessarily straight but can be along straight or curved path according to the type of external loading and design.

Circular prestressing is applicable to circular members such as pipes, bunkers, silos and tanks to counterbalance hoop tension.

12.4 PRETENSIONING AND POST-TENSIONING SYSTEMS

Internal prestressing has two approaches, viz. pretensioning and post tensioning depending upon to whether the steel bars (tendons) are stressed (or tensioned) before or after casting the concrete.

12.4.1 Pretensioning System (Fully Bonded Construction)

In this system of prestressing, the tendons are tensioned (stretched) before the concrete is placed. The tendons pass through the moulds and are temporarily anchored against some abutments. The tension is maintained when the concrete is placed (casted) in the mould. When the concrete is sufficiently hard, the ends of tendons are slowly released from anchorages, thereby transferring the prestress from steel tendons to concrete through bond. In factories, for mass production, the wires may be stretched the full length of casting yard (100 m to 150 m) between rigid anchor blocks. These wires pass through a whole series of moulds as shown in Fig. 12.6.

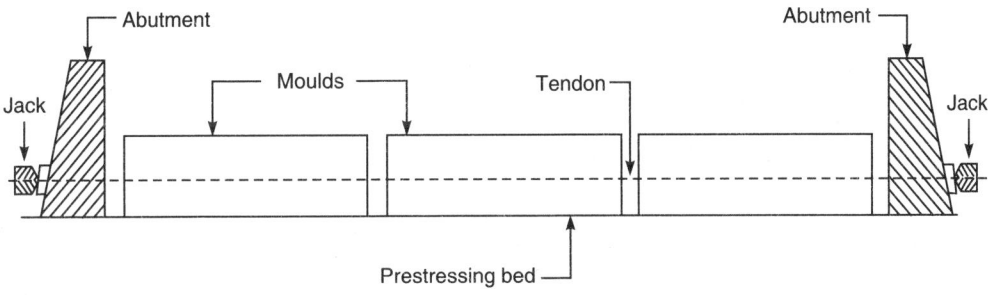

Fig. 12.6: Prestressing bed for pretensioning

Each mould may be mounted on rollers and spaced with small gaps at ends between adjacent moulds to allow the cutting of wires after hardening and curing of concrete to release individual members. After cutting the wires, the members get released from the end anchorages and thus transferring the prestress force from anchors to the concrete around these wires through bond. Since the prestress force is transferred through bond of steel wires to concrete, it is better to use smaller diameter wires. Generally, the diameter of these wires is 2 mm to 3 mm. Sometimes to improve the bond, these wires are twisted or notched. On cutting the wires, the bond length (transmission length) near ends loose some prestress and to minimize this loss, the diameter of wires at ends is increased initially for creating additional mechanical anchorage. The transmission length (or bond length) is generally considered as 85 times the diameter for 2 mm wires and about 120 times diameter for 5 mm wires.

12.4.2 Post-tensioning System (End Anchored Construction)

In post-tensioning, a duct with a given profile is formed in the member while casting concrete in the mould. When the concrete hardens and becomes adequately strong after curing, tendon consisting of either a bar, cable or strand is placed in the duct and then tensioned and anchored at ends of the concrete members. The prestress in the tendon is transferred to the member through appropriate anchorage wedges or anchor blocks provided at the end of the member as its integral part.

Anchoring in post-tensioning has to be done very carefully so that prestress is not lost by anchor slippage and hence anchoring has to be doubly secured. End of the wires are pulled up and anchored at member ends. There are many post tensioned systems developed under their patent names such as: Freyssinet, Magnel, Leonhardts, Lee McCall, Hobling, etc. After prestressing and anchoring, the ducts are filled with cement grout under pressure to doubly secure wires and also to protect the wires from weathering forces against corrosion. In case of unbonded cables (or wires) protection against corrosion is done by galvanizing or greasing.

In post-tensioned structure, complications arise due to friction of wires in the ducts at the anchorages and jacks during prestressing. In case of pretensioning, the major loss occurs due to elastic deformation of concrete when the prestress is first applied. Pretensioned structure also suffers severely from effects of shrinkage and creep. The total losses in pretensioned work may be about 25% while total losses in post tensioned work may be about 15%.

12.4.3 Full or Partial Prestressing

Concrete members are called fully prestressed, if there are no tensile stresses in the concrete member when these members are subjected to the working loads. If some tensile stresses are developed in the concrete under working loads, the members are said to be partially prestressed. Additional steel reinforcement may be provided in zones where tensile stresses are existing in partially prestressed members.

12.5 END ANCHORAGES AND PRESTRESSING SYSTEMS

Prestressing system essentially consists of method of stressing and method of anchoring to the concrete.

Pretensioning System and End Anchorages

The simplest prestressing system consists of two bulk heads anchored against the ends of casting yard. The tendons are pulled between these bulk heads. The bulk heads and the casting bed must be strong enough to resist the prestressing force and its eccentricity. Prestressing casting yard has advantage of intermediate bulk head making it possible to cast members of different size and also prestressing of bent cables.

Hoyer (Germany) is based on this method of mass production using construction yard with end anchoring blocks. Construction yard (100 m to 150 m long) is used for pulling the wires through series of moulds and stretched wires are anchored to end blocks.

After hardening and curing of concrete, wires are cut to transfer the prestressing force from end blocks to concrete through bond of concrete. Moulds are mounted on rollers to easily separate individual members.

The anchoring devices for holding pretensioning wires to the bulk heads are made on the wedge and friction principle. One common device consists of a split cone wedge, which is made from a tapered conical pin. The tapered conical pin is drilled axially and then cut in half longitudinally to form a pair of wedges (Fig. 12.7).

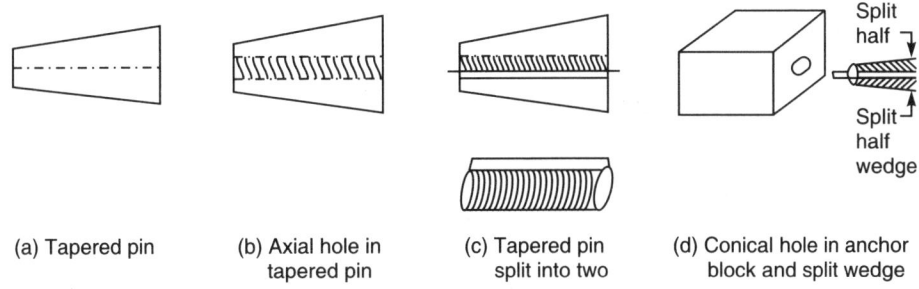

(a) Tapered pin   (b) Axial hole in   (c) Tapered pin   (d) Conical hole in anchor
                        tapered pin   split into two   block and split wedge

Fig. 12.7: Split wedge and anchor block for wire grip

The anchoring block has a conical hole, in which the split conical wedge holds the wire (or twisted wire strands). Anchorage consisting of clips can be gripped to the tendons under high pressure and the edges of the clips can then be welded together at several points. In such cases tendons of greater diameter can be used.

12.6 POST-TENSIONING SYSTEM AND ANCHORAGES

Post-tensioning systems are of four types : (i) Mechanical using jacks (ii) Electrical applying heat (iii) Chemical using expanding cement, (iv) Miscellaneous methods of using combination approach.

We shall discuss some of the important mechanical systems here.

Mechanical prestressing

This is the most common approach of prestressing by stretching wires using hydraulic jacks, screw jacks or levers. Hydraulic jacks of capacity ranging from 30 kN to 1000 kN are used depending on the requirement of design. Wires can be pulled one by one or in groups.

Magnel system use hydraulic jacks which pulls two wires at a time.

Freyssinet system uses double acting jacks which pulls 12 strands or 18 wires at a time. Pressure gauges are installed with jacks to indicate tensile force applied to the tendon or the pressure on the piston of jack.

When several wires are pulled, care should be taken to avoid excessive eccentricity of prestressing force during stretching.

Wires are anchored to the end of concrete member in their stretched condition by using wedging device. Following methods are commonly used to anchor stretched wires to the concrete:

- Wedge action producing a friction grip on the wires;
- Direct bearing from rivet or bolt heads formed at the end of wires, and
- Looping the wires around concrete.

Variations in jacking method, wire arrangement and anchoraging system give rise to different systems. Some of most common systems are explained briefly:

    (a) Freyssinet system

    (b) Magnel–Blaton system

    (c) Gifford–Udall system

    (d) Lee–McCall system

    (e) PSS Mono-wire prestressing system

(a) Freyssinet System

Freyssinet system is the first post tensioning system developed by a French engineer and used most widely in the world. It uses high tensile steel wires of 5 mm diameter in group of 8, 10, 12, or 18 and 7 mm wires in group of 12. The group of wires in each unit is called 'cable'. These cables are encased in flexible tube or sheething of 32 gauge metal sheet, with a helical spring inside to maintain proper spacing between the wires and forms a channel for cement grout. The wires are anchored by being held between two reinforced concrete cones male and female which fit one inside the other. The female cone is a conical steel wound lining heavily reinforced with high tensile steel spirals to resist bursting (Fig. 12.8).

The male cone is of mesh reinforced concrete fluted to space evenly the requisite number of wires. After pulling the wires, cement grout can be injected in the tube. In Freyssinet system all the wires are stretched simultaneously by double acting jack capable of pulling 18 wires at a time (Table 12.1). After completing the pulling of wires, a plug is pushed into the anchorage to secure wires. The pressure of the jack is released gradually. The anchoring unit is buried with the face of the concrete by injecting cement grout through central hole in the male cone.

Table 12.1: Group of wires used in Freyssinet system

Group of wires (cables)	Area (mm²)	Weight (N/m)	Metal hose	
			Outside dia. (mm)	Inside dia. (mm)
8 × 5 mm	157	12	28.6	25.4
10 × 5 mm	196	15	28.6	25.4
12 × 5 mm	235	18	31.8	28.6
18 × 5 mm	353	27	41.3	38.1

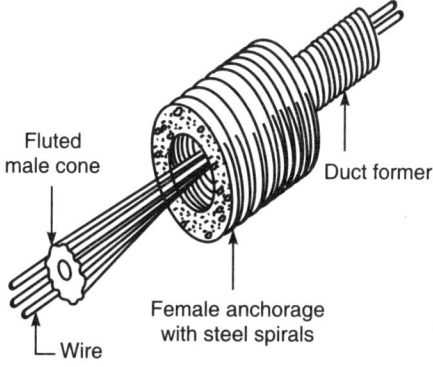

(a) Male–Female cones

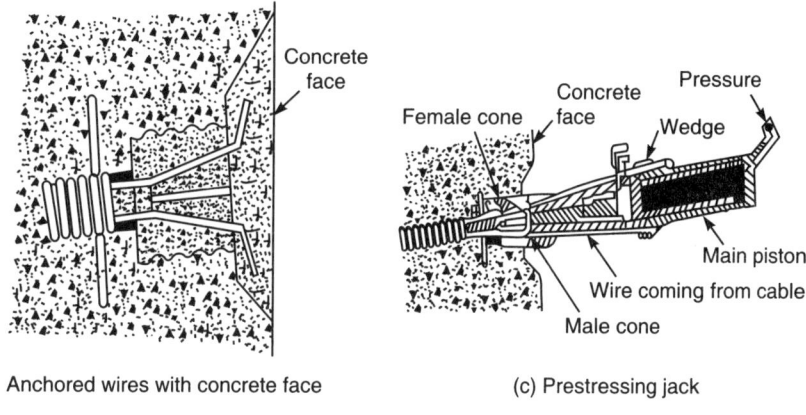

(b) Anchored wires with concrete face   (c) Prestressing jack

Fig. 12.8: Freyssinet system

(b) Magnel–Blaton system

Professor Magnel and contractor Blaton of Belgium developed this system and it consists of pulling two wires at a time (Fig. 12.9). In this a cable consists of even number up to a maximum of 64 wires of 5 mm or 7 mm diameter. The anchorage device consists of sandwich steel plates having grooves to hold wires and wedges. Each layer contains 4 wires (2 × 2) and the anchorage plates have 16 layers, thus making total 64 wires (4 × 16). The wires are maintained separately (with a spacing of 0.75 mm) throughout the length of the cable by using grills (spacers) at an interval of 1.50 m to 2.50 m along the length. The steel plate wedges are about 25.4 mm thick with two wedge shaped grooves in upper and lower faces. These sandwich plates are placed one above the other and held against the distribution plate by temporary bolting clamps during stressing. Regular ducts are cast along the length with rubber cores of about 30 mm initial diameter. These rubber ducts are pulled out after 6 to 8 hours of casting concrete to form ducts for introducing wires or cables at the time of pre-stressing (Table 12.2).

Hydraulic jacks are used to stress two wires at a time. The jack takes support against the end plates of suitable size. Two wedges are driven in each layer holding tensioned wires (2 × 2). After completion of each layer, the anchorage is covered with concrete by grouting.

Table 12.2: Details of Magnel wire system

No. of wires in cable	5 mm size wires				7 mm size wires			
	End plate		Duct hole		End plate		Duct hole	
	Length (mm)	Width (mm)	Width (mm)	Depth (mm)	Length (mm)	Width (mm)	Width (mm)	Depth (mm)
16	130	130	55	55	150	200	65	60
24	160	160	55	75	185	260	65	90
32	200	170	55	100	240	260	65	130
40	230	180	55	125	290	270	65	150
48	260	190	55	150	340	280	65	180
56	290	200	55	175	370	290	65	190
64	320	210	55	200	390	300	65	200

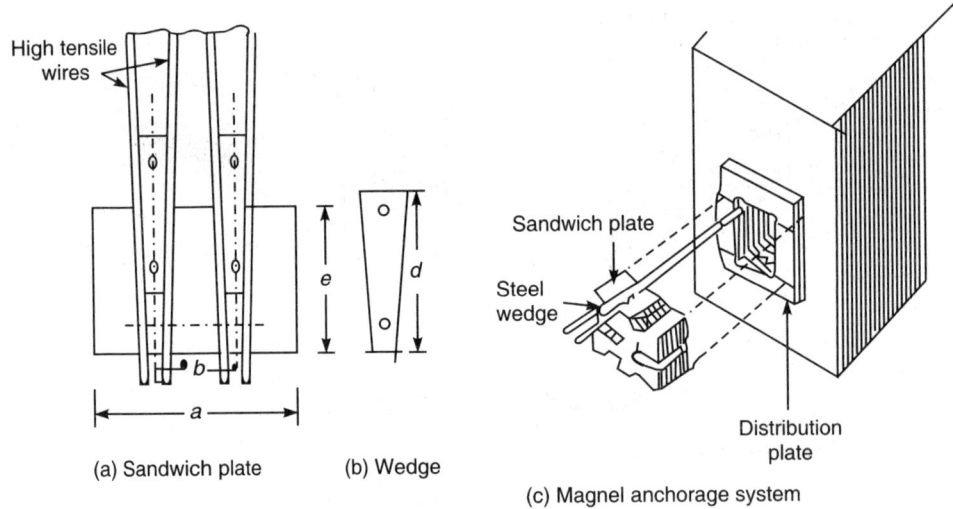

(a) Sandwich plate   (b) Wedge

(c) Magnel anchorage system

Fig. 12.9: Magnel–Blaton system

(c) Gifford–Udall system

Gifford–Udall system is one of the most widely used systems, using 7 mm wires. The cable comprises parallel wires which are separated by spacers placed at 1.0 m interval along the length. The diameter of spacers shall be:

39 mm for up to 8 wires, while 51 mm for up to 12 wires.

The system uses two types of anchorages, viz. plate and tube anchorage.

- **Plate anchorage:** The Gifford–Udall system essentially consists of a bearing plate, a thrust ring anchor grips and steel helix. The wires are stressed and anchored individually by small wedge type Udall grips seating against a bearing plate. The thrust ring is cast into the concrete. The bearing plate is located against the thrust ring. Theend of the duct is encircled by a helix. The anchor grips consist of an outer barrel and two semicircular wedges which fit in the tappered hole of the barrel. The wire is gripped between the serrated parallel faces of the wedges (Fig. 12.10).

- **Tube anchorage:** In this system the wedges fit directly into the tappered recesses in the bearing plate which is thick sufficiently. There is no need for separate anchor grips. The bearing plate includes a threaded hole for grouting purpose. The bearing plate seats against a tube unit incorporating the thrust ring and helix in a single element which is bolted to formwork and is cast into the concrete. The tube unit provides the necessary spreading of the wires in the cable so as to pass through the hole of the bearing plate. This type of anchorage is available for cables of 8 wires or 12 wires. The wires are pulled (stretched) one by one in this system (Fig. 12.11).

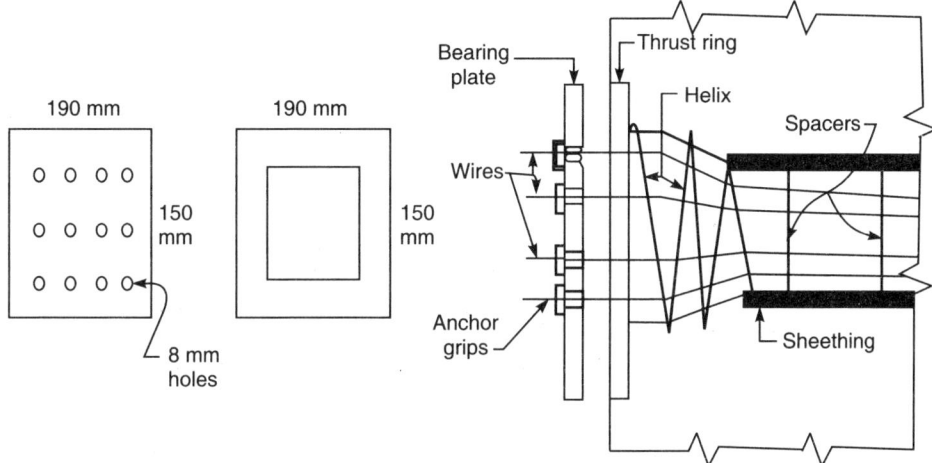

Fig. 12.10: Gifford–Udall system (plate anchorage)

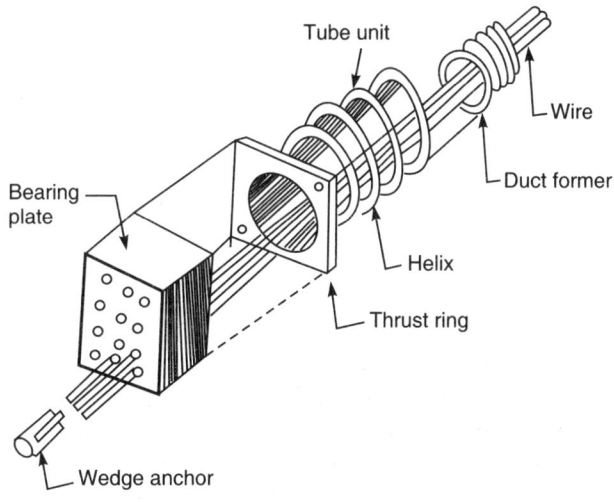

Fig. 12.11: Gifford–Udall system (tube anchorage)

(d) Lee–McCall system

In this system, prestressing tendons are used instead of prestressing wires. The tendons are made of special Macalloy steel containing silico-manganese and produced by hot rolled process. The diameter of these tendons vary from 12 mm to 28 mm and these rods are threaded at ends. Ducts are formed while casting the member with the help of rubber core. These tendons are introduced in the duct holes after completion of concrete curing and these tendons are stretched to the required tension. After stretching, these tendons are tightened with the help of nut and bolts provided at the ends. These tendons in the duct are either grouted for bond or kept unbonded (Fig. 12.12).

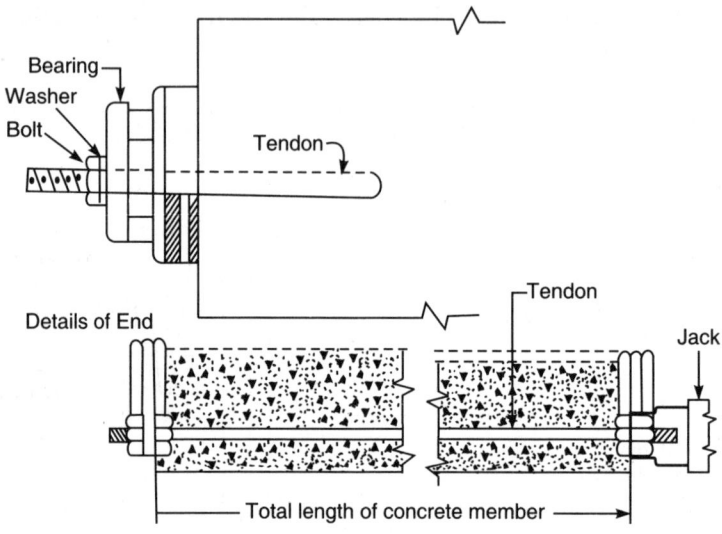

Fig. 12.12 : Lee–McCall system

(e) PSC mono-wire prestressing system

In this system wires are stretched individually, and consists of conical sleeves wedging in conical holes. The cable contains 1,2,4, 8 or 12 wires having size up to 7 mm. High strength plastic spacers are used to keep individual wires separate for stressing.

Electric prestressing

Electric prestressing is applicable in post-tensioning approach. In this method the bars are stretched by heating with high amperage current at low voltage. Here jacks are not used for stretching wires. The reinforcing bars are smooth and coated with thermoplastic material such as sulphur before embedding them in concrete. After the concrete hardening a low voltage high amperage current is passed through the bars to heat up to 170°C, when sulphur melts and allows steel bars to expand. At the end of member the protruding end of these bars are threaded which are tightened after expansion against heavy washers. On cooling, the sulphur solidifies and restores bond. This method is not practically accepted for high strength tensile wires.

12.7 LOSSES OF PRESTRESS

The prestressing force is applied to the member by stretching wire, but this does not remain constant. This prestress force gradually reduces due to various reasons. The amount of reduction in prestress force is known as loss of prestress which varies 15% to 20%. Various causes of prestress losses are:

(a) elastic deformation of concrete (on application of prestress)

(b) shrinkage of concrete

(c) creep of concrete under sustained loading

(d) relaxation of steel

(e) deformation and slip of anchorage, and

(f) friction between tendon and concrete

(a) Loss due to elastic deformation of concrete:

This loss occurs in pretensioned members when prestress force is transferred to the member. Due to prestress force, the concrete member undergoes elastic shortening along with shortening of steel giving corresponding reduction in prestress in steel.

Let P_i = Initial prestress force

A_e = Equivalent (transformed) area of concrete

A_{st} = Area of tensioned steel

$$\text{Shortening in unit length } e_s = e_c = \frac{\text{Force } P_i}{A_e \cdot E_c} = \frac{P_i}{(A_c + mA_{st}) E_c} = \frac{P_i}{(A_c \cdot E_c + E_s \cdot A_{st})}$$

$$\text{Loss of prestress in steel} = E_s \times e_s = \frac{E_s \cdot P_i}{A_c \times E_c + A_{st} \times E_s} = \left[\frac{P_i \times \dfrac{E_s}{E_c}}{A_c + \dfrac{E_s}{E_c} \times A_{st}} \right]$$

$$= \left[\frac{mP_i}{A_c + m \times A_{st}} \right] = \frac{mP_i}{A_e}$$

$$\frac{(\Delta f)\, \text{elasticity}}{(\text{Loss of stress})} = \frac{mP_i}{A_e} = \frac{m\left(f_{si} \times A_{st}\right)}{A_e} \qquad \qquad \text{... Eq. (12.7)}$$

$$P_i = f_{si} \times A_{st}$$

where, $$m = \frac{E_s}{E_c} = \frac{2.1 \times 10^5}{5000\sqrt{f_{ck}}} = \frac{42.0}{\sqrt{f_{ck}}},$$

f_{si} = Initial steel stress in N/mm²

f_{ck} is characteristic strength of concrete in N/mm², $E_s = 2.1 \times 10^5$ N/mm², $E_c = 500\sqrt{f_{ck}}$ N/mm²

There is no loss of prestress due to tensioning of wires/cables/tendons since the shortening of concrete is simultaneously adjusted during tensioning in case of post tensioned members.

(b) Loss due to shrinkage (Δf_{sh} of concrete):

Cement concrete shrinks on hardening and shrinkage depends on w/c ratio, size and grading of aggregate, amount of cement and atmospheric conditions, etc.

Shrinkage is time dependent and most of the shrinkage occurs during early days of the hardening of concrete. On release of prestress in pretensioned members, the concrete shrinks with time and hence shortening of wires also occur which results in loss of stress in steel wires. In post-tensioned member, most of the shrinkage takes place unhampered and hence loss due to shrinkage is relatively small. According to IS:1343-1980, the approximate shrinkage strain for design is assumed as follows:

Pretensioning members = 0.0003

$$\text{Post tensioning members} = \frac{0.0002}{\log_{10}(t+2)}$$

where, t = age of concrete in days at the time of transfer of prestress force.

In case of dry atmospheric conditions, shrinkage of the post-tensioned concrete may be increased by 50% but limited to the maximum value of 0.0003. Thus, if e_{sh} is the shrinkage strain, then the loss of prestress $(\Delta f)_{sh}$ due to shrinkage is given as:

$$\Delta f_{sh} = e_{sh} \cdot E_s \qquad \qquad \text{... Eq. (12.8)}$$

where, E_s = modulus of elasticity of high tension wires ($E_s = 2.1 \times 10^5\,N/mm^2$). The loss due to shrinkage may be assumed as 4% to 6% for pretensioned members and 3% to 4% for post tensioned members.

(c) Loss due to creep of concrete:

Under sustained loading, concrete undergoes permanent deformation (known as creep) which is also time-dependent. This occurs due to plastic flow of concrete under compressive force of prestress. The amount of creep strain also depends on the magnitude of stress and generally 2 to 3 times the elastic strain of concrete. The rate of creep is higher initially and reduces with time. The creep is higher with higher water/cement and also inversely proportional to the concrete strength. It depends on the strength of concrete at the time of loading.

The ratio of the final creep strain to the elastic strain in concrete is defined as creep coefficient (c_c), the relation of which is given as under:

$$e_c = (c_c - 1)\, e_{el}, \qquad \qquad \text{... Eq. (12.9)}$$

where,
e_c = strain due to creep

e_{el} = elastic strain of concrete

c_c = coefficient of creep

Values of creep coefficient are given in Table 12.3.

Table 12.3: Creep coefficient (c_c) for concrete

S. No	Condition of exposure of concrete	Final creep coefficient (c_c)
1	In water	$0.5\,k_c$ to k_c
2	In very humid atmosphere	$1.5\,k_c$ to $2.0\,k_c$
3	In average humidity	$2.0\,k_c$ to $3.0\,k_c$
4	In dry atmosphere (in dry internal rooms)	$2.5\,k_c$ to $4.0\,k_c$

Where k_c depends on the ratio of strength (f_{cr}) of concrete at transfer of prestress and cube crushing strength of concrete at 28 days as measured on 150 mm size (as given in Table 12.4).

Table 12.4: Values of k_c

S. No	(f_{cr}/f_{cu}) at the time of stressing	k_c
1	0.50	2.20
2	0.84	1.50
3	1.00	1.0
4	1.08	0.75
5	1.30	0.50

The loss of prestress due to creep of concrete under load should be determined for all the permanently applied loads including the prestress. The creep loss due to live loads, erection loads and other short duration minor loads must be ignored. Loss of prestress is obtained by multiplying modulus of elasticity (E_s) of prestress tendons/ wires and total comprehensive (ultimate) creep strain of concrete along the entire length (e_{cr}).

As per IS:1343-1980, the ultimate creep strain may be estimated from Table 12.5 (creep coefficients c_c), that is the ultimate creep strain e_{cr}, elastic strain e_{el} at the age of loading.

Table 12.5: Creep coefficients (c_c)

Age of loading	7 days	28 days	1 year and more
Creep coefficients (c_c)	2.2	1.6	1.0

About half the creep strain takes place in the first month after the loading and that about three quarters of the total creep takes place in the first six months after loading.

Total loss in prestress due to creep may range 5% to 10%.

(d) Loss due to relaxation of steel:

When the stress in steel is more than half the yield stress, the steel also undergoes creep. When the steel wires (or tendons) are kept under certain stress higher than half of its yield (proof) stress, some part of strain (or deformation) becomes permanent and this part of strain (deformation) is called relaxation of steel. This relaxation depends on the composition of steel. The creep (also called relaxation) is high under high stressing force. For low stressing force the creep is very small and can be neglected. Creep in high tension steel is time dependent and most of the creep in steel occurs at early stage. The magnitude of the loss of prestress due to relaxation in steel may be taken to vary from 2% to 8% of the average initial stress and depends on the order of the initial prestress. Overstressing to compensate the loss of prestress due to relaxation (creep) of steel may be permitted if the overstressing does not exceed 85% of the ultimate tensile strength (f_{su}) of the prestressing steel.

(e) Loss due to deformation and slip of anchorage:

This loss is applicable in post-tensioning system only. After stretching the wires, the jack is released to transfer prestress to concrete by bearing of the anchorage system. During the transfer, the anchorage system slips before wires or tendons are firmly gripped. The amount of slippage (δ) depends on the type of wedge and stress in the wires. The average value of slippage may be taken as 2 mm to 3 mm. For very heavy strands, the slippage may be 5 mm. Manufacturers of various systems specify this slippage (e.g. Freyssinet system for 7 mm cables the slippage is 9 mm).

Loss of slippage prestress $\Delta f \text{ (slip)} = E_s \dfrac{\delta}{L}$ ... Eq. (12.10)

where, L = Length of the cable/wires (mm)

E_s = Modulus of elasticity of steel (N/mm²)

δ = Slippage (mm)

The loss of prestress due to slippage in anchorage must be given proper consideration specially in short length members by additional elongation at the time of tensioning to compensate this loss.

(f) Loss due to friction:

During tensioning, tendons/wires slide relative to the surrounding duct or concrete surface friction is caused reducing the prestress force. As per IS:1343-1980, for straight or moderately curved cable profile, the prestressing force P_x at a distance x from the tensioning end may be given as:

$$P_x = P_0 e - (\mu\alpha + kx)$$ ... Eq. (12.11)

where, P_x = Prestressing force at a distance x from the tensioning end

P_0 = Prestressing force in prestressing steel at the tensioning end

For small values of $(\mu\alpha + kx)$, the above equation may be reduced to

$$P_x = P_0 (1 - \mu\alpha - kx)$$ ... Eq. (12.12)

where α = Cumulative angle in radian which the tangent makes to the cable profile between the two points

k = Coefficient of wave effect, varying from 15×10^{-4} to 50×10^{-4} per metre

μ = Coefficient of friction in curve (0.55 for steel moving on smooth curve, 0.30 for steel moving infixed duct and 0.25 for steel moving on lead).

The friction loss can be reduced by following methods:

- Passing the cables through metal tubes at bends
- Making the radius of curvature at bends as large as possible but not less than 800 times the diameter of cable/wire
- Supporting cables at closer intervals
- Avoid double curvature
- Using lubricants in unbonded concrete. For bonded concrete, water soluble oil should be used which should be flushed out after stressing
- Prestressing may be done from both ends
- Making the bends through as small an angle as possible
- Over tensioning to compensate friction.

Total amount of losses

The amount of total losses will be reduced from the initial prestress.

Apparent jacking stress: It is the force shown in the gauge of the jack.

Jacking stress: It is the actual pull in the wire/cable near the jack before the anchorage. This stress is always less than the apparent Jacking stress.

Initial prestress: The jacking stress minus the anchorage loss, i.e. stress at the anchorage after release of jack.

Effective or Design stress: It is stress equal to the initial prestress minus the losses.

The losses to be considered are: elastic shortening of concrete, creep and shrinkage of concrete and relaxation of steel. Friction losses away from the jack may also be considered. Losses may be expressed in terms of strains, stresses or percent of initial stress. Total losses are assumed as in Table 12.6:

Table 12.6: Total losses (% of initial)

S. No	Losses due to	%age losses	
	(cause of loss in prestress)	Pretension	Post-tension
1	Elastic shortening of concrete	3.0	1.0
2	Creep of concrete	6.0	5.0
3	Shrinkage of concrete	7.0	6.0
4	Creep in steel (relaxation)	2.0	3.0
	Total losses (all causes)	18.0	15.0

Total losses shown in Table 12.6 are assumed for overstressing (over tensioning) to compensate for various losses at the time of tensioning. Friction loss may be considered separately if not compensated during stressing. If the initial stress (P_0/A_c) is high (more than $7 \, N/mm^2$) the above losses shown in Table 12.6 should be increased to 25% for pretensioning and 20% for post tensioning. When the average prestress is low (less than $2 \, N/mm^2$), the losses should be decreased to 14% for pretensioned members and 12% for post tensioned members. Accurate calculations of losses and prestressing can be done after preliminary design of members. For preliminary design of members assumptions of losses may be made as shown in Table 12.6.

Example 12.1

A simply supported prestressed concrete beam of 5 m span and rectangular section of $300 \, mm \times 500 \, mm$ carries an u.d.l. of $24 \, kN/m$ (including self-wt.) over the entire span. If tensile force of $900 \, kN$ is applied at (i) centroid of the section (ii) 100 mm below centroid in longitudinal direction, find the stresses across the section at mid-span point and supports.

Solution: $A = 300 \times 500 = 15 \times 10^4 \, mm^2$ (Fig. 12.13)

$$I = \frac{300 \times (500)^3}{12} = 3125 \times 10^6 \, mm^4, \; Z = 12.5 \times 10^6 \, mm^3$$

$T = 900 \times 10^3 \, N$, (i) $e = O$, (ii) $e = 100 \, mm$ below CG.

$$M = \frac{24 \times 5 \times 5000}{8} = 75 \times 10^3 \, kN\text{-}mm = 75 \times 10^6 \, N\text{-}mm \; \text{(mid span)}$$

$M_0 = 0$ (end).

(i) *Centric load* $= 9 \times 10^5 \, N$, direct stress $\dfrac{9 \times 10^5}{15 \times 10^4} = +6 \, N/mm^2$ (throughout)

$$fb_1 = \pm \frac{M}{Z} = \pm \frac{75 \times 10^6}{12.5 \times 10^6} = \pm 6 \, N/mm^2, fb_0 = 0 \; \text{(ends)}$$

Net stress (mid-span) $= + 6 \pm 6 = + 12, 0, N/mm^2$ (top and bottom)

Net stress (ends) $= + 6 \pm 0 = + 6 \, N/mm^2$ (top and bottom)

(ii) *Eccentric prestress force*

$$T = 9 \times 10^5 \, N, \, e = 100 \text{ mm below}$$

Moment due to eccentric prestressing force:

$$M = T.e = 9 \times 10^5 \times 100 = 90 \times 10^6 \, \text{N-mm.}$$

Net stresses:

Mid-span: $\dfrac{9 \times 10^5}{15 \times 10^4} = +6 \, \text{N/mm}^2 \text{ (direct stress)}$

Eccentrict stress $= \mp \dfrac{9 \times 10^5 \times 100}{Z} = \mp \dfrac{90 \times 10^6}{12.5 \times 10^6} = \mp 7.2 \, \text{N/mm}^2$

Bending stress $= \mp \dfrac{75 \times 10^6}{12.5 \times 10^6} = \pm 6.0 \, \text{N/mm}^2$

Thus, net stress (mid-span) =

$$+ 6.0 \pm 6.0 \mp 7.2 = + 4.80 \, \text{N/mm}^2 \text{ (top fibres),}$$
$$+ 6.0 \mp 6.0 \pm 7.2 = + 7.20 \, \text{N/mm}^2 \text{ (bottom)}$$

Net stress end (support) =

$$+ 6.0 \pm 0 \mp 7.2 = - 1.2 \, \text{N/mm}^2 \text{ (top fibres)}$$
$$+ 6.0 + 7.2 + 0 = + 13.2 \, \text{N/mm}^2 \text{ (bottom)}$$

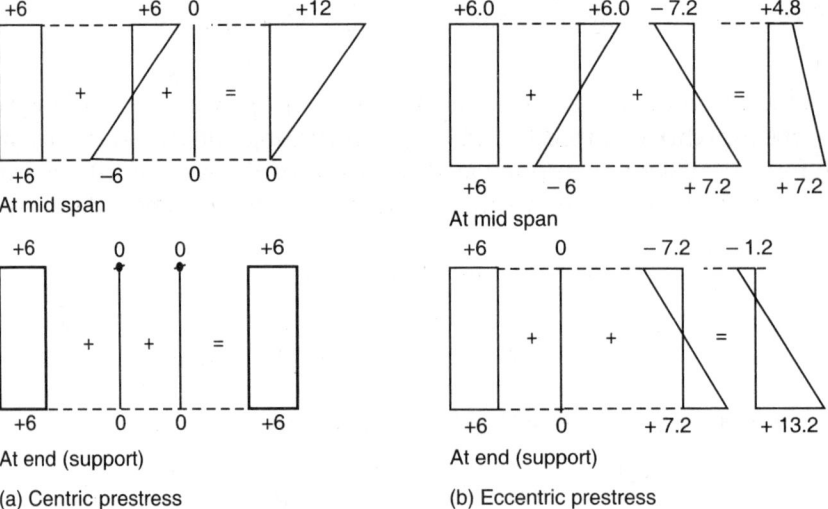

Fig. 12.13: Stress variation across the section

Example 12.2

A cantilever beam of rectangular section 300 mm × 500 mm and 3 m span carries an u.d.l. of 12 kN/m (including self-weight). If the section is prestressed with a force of 648 kN at 80 mm above the centroid, find the net maximum stress across the section by using the principle of internal force.

Solution: Prestress tension $T = 648$ kN, $Z = 12.5 \times 10^6 \, \text{mm}^3$ (Fig. 12.14)

Area $(A) = 150 \times 10^3 \, \text{mm}^2$, $I = \dfrac{300 \times (500)^3}{12} = 3125 \times 10^6 \, \text{mm}^4$

$M_{\text{max}} \text{ (BM)} = \dfrac{w.l^2}{2} = \dfrac{12 \times 3^2}{2} = 54 \, \text{kN-m} = 54 \times 10^6 \, \text{N-mm}$

Direct stress due to prestress force 648 kN

$f_0 = \dfrac{648000}{150 \times 10^3} = +4.32 \, \text{N/mm}^2 \text{ (uniform)}$

Moment due to eccentricity, $e = 80$ mm

$M' = T.e = 648 \times 80 = 51840 \, \text{kN-mm} = 51.84 \times 10^6 \, \text{N-mm}$.

Net stress is due to prestress force, bending moment due to external loading and moment caused by eccentricity.

Stress across the section at the point of maximum BM, i.e. fixed end.

Net stress (fixed end) :

$f_0 \pm fb_1 \mp fb_2 = +4.32 \pm \dfrac{54 \times 10^6}{Z} \mp \dfrac{M'}{Z}$

Top fibre $= +4.32 - \dfrac{54 \times 10^6}{12.5 \times 10^6} + \dfrac{51.84 \times 10^6}{12.5 \times 10^6} = +4.32 - 4.32 + 4.15 = +4.15 \, \text{N/mm}^2$

Bottom fibre $= +4.32 + 4.32 - 4.15 = +4.49 \, \text{N/mm}^2$

Net stress (free end):

Top fibre $= +4.32 \pm 0 \pm 4.15 = +8.47 \, \text{N/mm}^2$

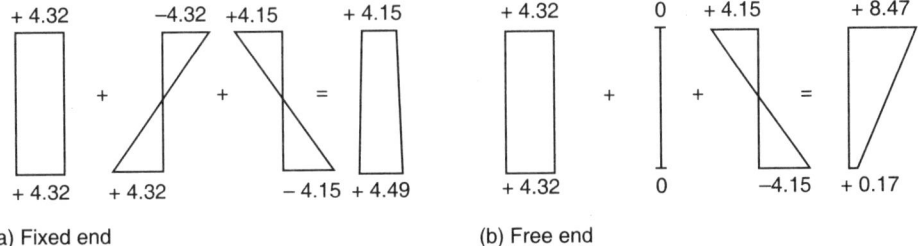

(a) Fixed end · (b) Free end

Fig. 12.14: Stress variation across the section

Bottom fibre $= +4.32 \pm 0 \mp 4.15 = +0.17 \, \text{N/mm}^2$

Thus, maximum stress $= +8.47 \, \text{N/mm}^2$ in top fibres at free end

At fixed end: stress varies $+4.15 \, \text{N/mm}^2$ in top fibre to $+4.49 \, \text{N/mm}^2$ in bottom fibre.

Example 12.3

A simply supported prestressed concrete beam of rectangular section 400 mm × 600 mm, is loaded with u.d.l. of 60 kN/m over the entire span of 6 m. If a prestress force of 1980 kN is applied through a parabolic tendon having 100 mm eccentricity at supports and 200 mm eccentricity at the mid-span point (i.e. sag of tendon = 100 mm).

Solution: Upward uniform load due to tension force of 1980 kN and sag of 100 mm

$$(p) = T\,\frac{8h}{l^2} = \frac{1980 \times 8 \times 0.1}{6 \times 6} = 44\ \text{kN/m}$$

Net downward load/m = $60 - 44 = 16$ kN/m

BM at mid span (net loading of 16 kN/m) = $\dfrac{w \cdot l^2}{8} = \dfrac{16 \times 6 \times 6}{8} = 72$ kN-m

Area of cross-section = $400 \times 600 = 24 \times 10^4\ \text{mm}^2$

$$Z = \frac{400 \times 600^2}{6} = 24 \times 10^6\ \text{mm}^3,\ I = \frac{400 \times 600^3}{12} = 72 \times 10^8\ \text{mm}^4$$

Bending stress due to unbalanced load = $\dfrac{M}{Z} = \dfrac{72 \times 10^6}{24 \times 10^6} = \pm 3.0\ \text{N/mm}^2$

At ends:

Bending stress = 0, direct prestress = $\dfrac{1980 \times 1000}{24 \times 10^4} = +8.25\ \text{N/mm}^2$

Bending due to prestress = $-\dfrac{(1980 \times 100)1000}{24 \times 10^6} = \mp 8.25\ \text{N/mm}^2$

Net stress: Top face = $+8.25 - 8.25 = 0$

Bottom face = $+8.25 + 8.25 = +16.5\ \text{N/mm}^2$ (Fig. 12.15)

Mid span:

Top face = $+8.25 - 8.25 + 3.0 = +3.0\ \text{N/mm}^2$

Bottom face = $+8.25 + 8.25 - 3.0 = +13.5\ \text{N/mm}^2$

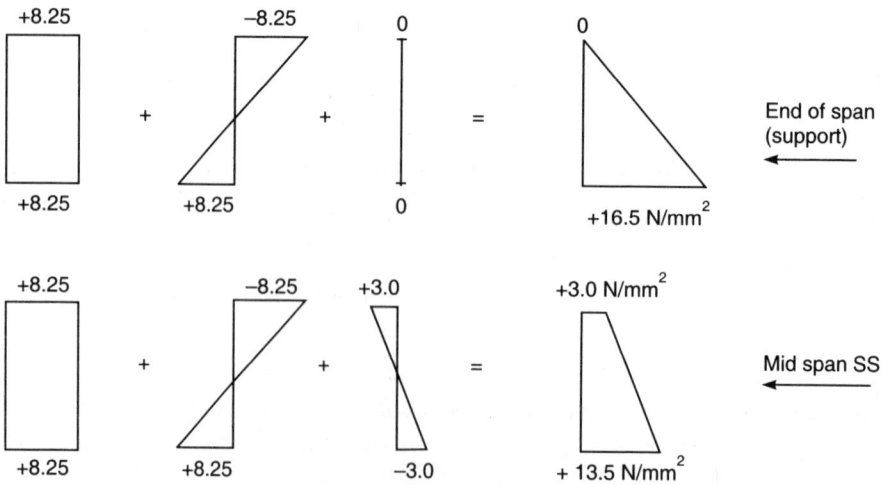

Fig. 12.15: Net stress distribution across the section

Example 12.4

A prestressed concrete beam of rectangular section 200 mm × 600 mm deep is prestressed by a parabolic cable located at an eccentricity of 100 mm at mid-span point and zero at the supports. If the beam is simply supported over 10 m span and carries uniformly distributed load of 4 kN/m in addition to self-load. Determine the effective tension required in the cable so that shear stress in the beam is zero. Weight of concrete is 25 kN/m^3.

Solution: Area of beam = $0.20 \times 0.60 = 0.12 \text{ m}^2$

Self-weight of beam = $0.20 \times 0.60 \times 1 \times 25 = 3.0 \text{ kN/m}$, live load = 4 kN/m

Total load on the beam = $3 + 4 = 7.0 \text{ kN/m}$

Let the tension force in the cable = T

In case of parabolic cable (upward loading), $w = \dfrac{T \cdot 8h}{l^2}$,

where h = total sag of cable = 0.10 m, l(span) = 10 m

For zero shear, upward loading should be equal to total downward loading, i.e

$$w = \frac{T \cdot 8h}{L^2} = 7.0, \ T = \frac{7 \times 10 \times 10}{8 \times 0.10} = 875 \text{ kN}$$

The stress across the section shall be uniform as downward loading is fully balanced. Uniform stress in concrete = $\dfrac{875 \times 1000}{200 \times 600} = 7.29 \text{ N/mm}^2$.

Example 12.5

A prestressed rectangular concrete beam 100 mm × 250 mm deep spanning 8 m is prestressed by a straight cable carrying an effective tension of 250 kN located at an eccentricity of 40 mm below the centroid. The beam supports a live load of 1.2 kN/m, determine the resultant stresses at the mid-span cross-section. Concrete weight is 25 kN/m^3. Also determine the prestressing force with an eccentricity of 40 mm so that the resultant stress in bottom fibre of the section at mid-span point is fully neutralized (zero) due to dead and live loads.

Solution: Dead load = $0.1 \times 0.25 \times 1 \times 25 = 0.625 \text{ kN/m}$, $L = 8$ m,

$$A = 100 \times 250 = 25000 \text{ mm}^2, Z = \frac{100 \times 250^2}{6} = 1041.67 \times 10^3 \text{ mm}^3,$$

DL of beam = 0.625 kN/m, LL = 1.2 kN/m (given),

Total weight = 1.825 kN/m, T = 250 kN at e = 0.04 m (40 mm)

Direct stress throughout = $\dfrac{T}{A} = \dfrac{250 \times 1000}{25000} = + 10 \text{ N/mm}^2$

Stress due to eccentricity = $\dfrac{250 \times 1000 \times 40}{1041.6 \times 10^3} = \mp 9.6 \text{ N/mm}^2$

SS BM = $\dfrac{1.825 \times 8^2}{8} = 14.6 \text{ kN-m} \ (14.6 \times 10^6 \text{ N-mm})$

Bending stress due to moment $= \dfrac{14.6 \times 10^6}{1041.67 \times 10^3} = \pm 14.01 \text{ N/mm}^2$

(i) final stress (mid-span): at top fibre $= +10 - 9.6 + 14.01 = +14.41 \text{ N/mm}^2$

at bottom fibre $= 10 + 9.6 - 14.01 = + 5.59 \text{ N/mm}^2$

(ii) final stress: at mid span $= 0$, in bottom fibre,

i.e. $\dfrac{T \times 1000}{25000} + \dfrac{T \times 40 \times 1000}{1041.67 \times 10^3} - 14.01 = 0,$

$T(0.04 + 0.0384) = 14.01, \quad T = 178.7 \text{ kN}$

Example 12.6

A prestressed beam with rectangular section 150 mm × 300 mm deep is used as simply supported beam over a span of 10 m. The cable is stressed to an effective tension of 500 kN and has zero eccentricity at the supports which increases to 50 mm at the mid span section. Concrete weighs 25 kN/m³, find the magnitude of a concentrated load W located at the mid span point so that (i) the load nullifies the effect of the prestressing force (neglecting the self-wt. of the beam). (ii) The pressure line passes through the upper kern of the section under the action of external load, self-weight and prestress.

Solution: Let inclination of cable $= \alpha$, area $= 150 \times 300 = 45000 \text{ mm}^2$ (Fig. 12.16)

$L = 10 \text{ m}, \sin \alpha = \tan \alpha = \dfrac{50}{5000} = \dfrac{1}{100}, \ Z = \dfrac{150 \times 300^2}{6} = 2250 \times 10^3 \text{ mm}^3,$

Self-wt. $= 0.15 \times 0.3 \times 25 = 1.125 \text{ kN}$

(i) Upward force component of cable tension $= 2T \sin \alpha = 2 \times 500 \times \dfrac{1}{100} = 10 \text{ kN}$

Hence $W = 10 \text{ kN}$ to balance tension of cable at mid span point (vertical component).

(ii) When the pressure line passes through the upper kern of the section under external load, self-weight and prestress. Thus, the stress is triangular and bottom stress is zero.

Direct stress $= \dfrac{T \cos \alpha}{A} = \dfrac{500 \times 10^3}{45000} \cos \alpha$ (since $\cos \alpha = 1.0$) $= 11.11,$

i.e. $f_0 = 11.11 \text{ N/mm}^2$

$\text{BM} = \dfrac{1.125 \times 10^2}{8} = 14.0625 \text{ kN-m}$

Upward component of prestress force $= 2T \sin \alpha = \dfrac{2 \times 500}{100}$

$= 10 \text{ kN} \left(\text{since, } \sin \alpha = \dfrac{1}{100} \right)$

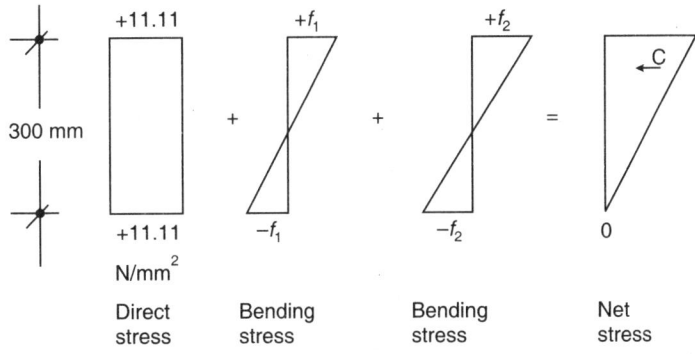

Fig. 12.16: Stress variation across the section (150 mm × 300 mm)

Net downward load at mid span = $(W - 10)$ kN

$$\text{Total BM at mid span} = \left\{ 14.0625 + \frac{(W-10)10}{4} \right\} = \{14.0625 + 2.5W - 25\}$$

$M = (2.5W - 10.9375)$ kN-m $= (2.5W - 10.9375) \cdot 10^6$ N-mm

$$\text{Bending stress} = \frac{M}{Z} = \pm \frac{(2.5W - 10.9375)10^6}{2250 \times 10^3} = \pm \frac{100}{225}(2.5W - 10.9375)$$

$$\text{Net stress in bottom fibre} = \left\{ 11.11 - \frac{100}{225}(2.5W - 10.9375) \right\} = 0$$

$$11.11 + \frac{1093.75}{225} - \frac{250W}{225} = 0, \quad W = \frac{225}{250}\{11.11 + 4.8611\} = 14.375 \text{ kN}$$

Thus, in this situation the point load required at mid span point will be 14.375 kN.

Example 12.7

A prestressed beam of rectangular section 250 mm × 600 mm deep supports an u.d.l. of 15 kN/m over an effective span of 12 m (simply supported). Calculate the minimum initial prestress force T and the corresponding eccentricity e for the mid-span section if the maximum tensile stress in concrete is limited to 1.4 N/mm² and the loss in prestress due to all factors at any time may be assumed as 20%. Weight of concrete = 25 kN/m³.

Solution: $A = 250 \times 600 = 150 \times 10^3 \text{ mm}^2$, $Z = \dfrac{250 \times 600^2}{6} = 15.0 \times 10^6 \text{ mm}^3$, $L = 12$ m,

$f_t = -1.4 \text{ N/mm}^2$, DL (self-wt.) $= 0.25 \times 0.60 \times 25 = 3.75 \text{ kN/m}$,

$$M_{dmax} \text{ due to self-weight} = \frac{3.75 \times 12^2}{8} = 67.5 \text{ kN-m}$$

$$\text{Max. bending stress due to self-wt.} f_d = \frac{M_d}{Z} = \frac{67.5 \times 10^6}{15 \times 10^6} = \pm 4.5 \text{ N/mm}^2$$

Let initial prestress = T_0 kN.

After transfer, any time the cable tension after loss = $0.80T_0$

live load moment $M_L = \dfrac{15 \times 12^2}{8} = 270$ kN-m (270×10^6 N-mm), eccentricity (m) = e

(i) At the time of transfer of tension, extreme stress will be

$$\dfrac{T_0 \times 10^3}{15 \times 10^4} \mp \dfrac{T_0 \times e \times 10^6}{15 \times 10^6} \mp \dfrac{67.5 \times 10^6}{15 \times 10^6} = + \dfrac{T_0}{150} \mp \dfrac{T_0 \times e}{15} \pm 4.5 = -1.4$$

or $\qquad\qquad (T_0 - 10\, T_0.e) = -5.9 \times 150 = -885$ ... Eq. (1)

(ii) After transfer at working condition (bottom reaches -1.4 N/mm^2)

$$\dfrac{0.80T_0 \times 10^3}{15 \times 10^4} + \dfrac{0.8T_0 \times e \times 10^6}{15 \times 10^6} \mp \dfrac{67.5 \times 10^6}{15 \times 10^6} \mp \dfrac{270 \times 10^6}{15 \times 10^6} = -1.4$$

or $\qquad\qquad (0.8T_0 + 8\,T_0.e) = (4.5 + 18 - 1.4)150 = 21.1 \times 150$

or $\qquad\qquad (T_0 + 10\, T_0.e) = 3956.25$ ... Eq. (2)

Adding Eq. (1) and Eq. (2), we get

$$T_0 = 3956.25 - 885 = 3071.25, \; T_0 = 1535.63 \text{ kN}$$

Substituting value of T_0 in Eq. (1), we get $1535.63 - 10\,(1535.63)\,e = -885$

or $\qquad\qquad \dfrac{(1535.63 + 885)}{10(1535.63)} = e$ or $e = 0.1576$ m (157.6 mm)

Example 12.8

A prestressed concrete I-beam supports live load of 4 kN/m over a simply supported span of 8 m. The beam has an overall depth of 400 mm, flanges of 200 mm × 60 mm and web of 80 mm × 280 mm. The beam is to be prestressed with an effective tension of 240 kN applied at suitable eccentricity so that the final stress at the bottom of beam at the mid span point is zero. Find (i) eccentricity needed (ii) what should be magnitude of prestressing force if the tendons are placed concentrically so that the final stress is zero at the bottom of the mid span section.

Solution: $A = 60 \times 200 \times 2 + 280 \times 80 = 46400$ mm^2 (Fig. 12.17)

$$I = \left[\dfrac{200 \times 60^3}{12} + 12000 \times 170^2 \right] \times 2 + \left[\dfrac{80 \times 280^3}{12} \right] = 847.15 \times 10^6 \text{ mm}^4$$

$Z = 4235.73 \times 10^3$ mm^3, $\; L = 8.0$ m, $\; T = 240$ kN

DL of beam = $(0.2 \times 0.06 \times 2 + 0.28 \times 0.08) \times 1 \times 25 = 1.16$ kN/m

Max. BM due to DL, $M_d = \dfrac{1.16 \times 8^2}{8} = +9.28$ kN-m (9.28×10^6 N-mm)

LL moment, $M_L = \dfrac{4 \times 8^2}{8} = +32.0$ kN-m (32×10^6 N-mm)

Case I: Tendon placed eccentricically (e)

Direct stress = $+ \dfrac{240 \times 10^3}{46400} = +5.17$ N/mm^2

Stress due to eccentricity $= + \dfrac{240 \times 10^3 \times e \times 10^3}{4235.73 \times 10^3} = + (56.66e) \text{ N/mm}^2$

Bending stress due to DL $= + \dfrac{M}{Z} = \dfrac{9.28 \times 10^6}{4235.73 \times 10^3} = \pm 2.191 \text{ N/mm}^2$

Bending stress due to LL $= \pm \dfrac{32 \times 10^6}{4235.73 \times 10^3} = \pm 7.555 \text{ N/mm}^2$

Stress in bottom face to be zero, we have

$+ 5.17 + 56.66e - 2.191 - 7.555 = 0, \ \text{or} \ e = \dfrac{(7.555 + 2.191 - 5.17)}{56.66} = 0.0808 \text{ m}$

$e = 80.8 \text{ mm}$

Case II: Centric prestress (T_0) kN

Bottom stress at mid span = 0,

i.e. $\dfrac{T_0 \times 10^3}{46400} - 2.191 - 7.555 = 0, \ T_0 = \dfrac{(2.191 + 7.555)46400}{10^3} = 452.2 \text{ kN}$

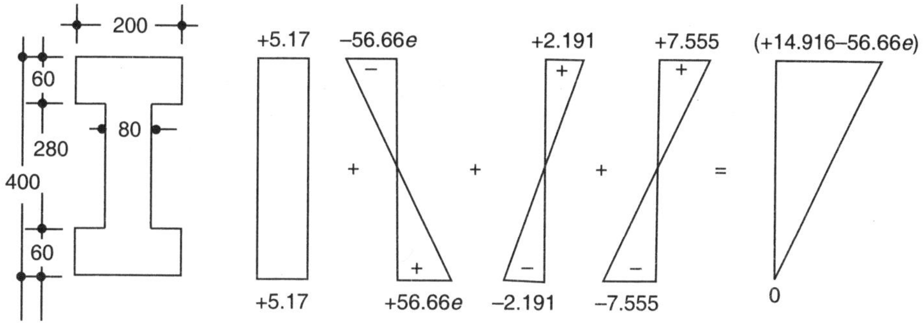

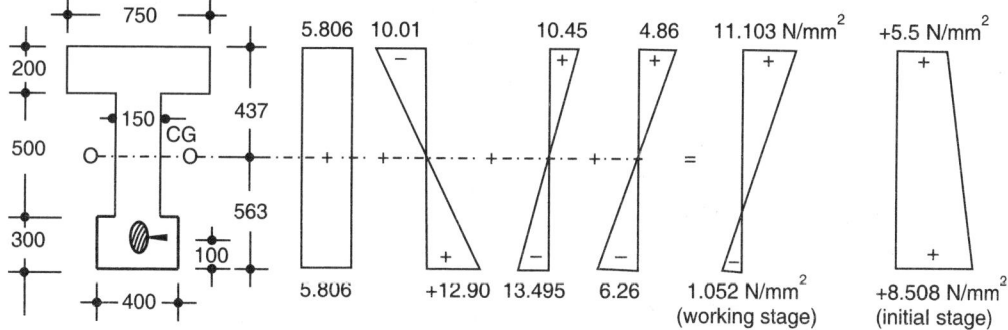

Fig. 12.17 : Stress variation across the section due to various loads

Example 12.9

A prestressed concrete I–beam has its upper flange 750 mm × 200 mm, lower flange 400 mm × 300 mm and connecting web 500 mm × 150 mm thick. The beam is supported over a span of 30 m and carries an u.d.l. of 4 kN/m excluding self-weight. It is prestressed with 120 wires of 5 mm diameter with their centroid 100 mm above the bottom edge and initially tensioned to 1000 N/mm². Assuming 15% of losses in prestress, determine the extreme fibre stresses at mid span at different stages (Fig. 12.18).

Fig 12.18: Stress variation at mid-span section (working stage)

Solution: $A = 750 \times 200 + 500 \times 150 + 400 \times 300 = 345 \times 10^3 \, \text{mm}^2,$

$y_t = 437.0 \, \text{mm}, \; y_b = 563.0 \, \text{mm}$

$$I = \frac{750 \times 200^3}{12} + 150000(437{-}100)^2 + \frac{150 \times 500^3}{12} + 75000\,(450-437)^2 + \frac{400 \times 300^3}{12}$$

$$+ \, 120000\,(563-150)^2 = 404.79 \times 10^8 \, \text{mm}^4$$

$$Z_t = \frac{404.79 \times 10^8}{437} = 92.63 \times 10^6 \, \text{mm}^3, \; Z_{\text{bottom}} = 71.90 \times 10^6 \, \text{mm}^3$$

$$e = 563 - 100 = 463 \, \text{mm}, \; A_{st} = 120 \; \pi \frac{(5)^2}{4} = 2356.2 \, \text{mm}^2, \; T_0 = 2356.2 \, \text{kN}$$

$$\text{Dead load } w_d = \frac{345 \times 10^3}{10^6} \times 1 \, \text{m} \times 25 \, \text{kN/m} = 8.625 \, \text{kN/m},$$

$$\text{BM } M_d = \frac{8.625 \times 30^2}{8} = 970.3125 \, \text{kN-m} \; (970.313 \times 10^6 \, \text{N-mm})$$

$$\text{Live load } w_1 = 4 \, \text{kN/m}, \; M_1 = \frac{4 \times 30^2}{8} = 450 \, \text{kN-m} \; (450 \times 10^6 \, \text{N-mm})$$

Case I: when prestress is initially transferred, $T_0 = 2356.2 \, \text{kN} \; (2356.2 \times 10^3 \, \text{N})$

$$\text{Direct stress } f_o = \frac{2356.2 \times 10^3}{345 \times 10^3} = +6.83 \, \text{N/mm}^2, \text{ eccentricity } e = 463 \, \text{mm}$$

$$\text{Stress due to eccentricity (top) } f_{bt} = - \frac{2356.2 \times 10^3 \times 463}{Z_t}$$

$$= - \frac{2356.2 \times 10^3 \times 463}{92.63 \times 10^6} = -11.78 \, \text{N/mm}^2$$

$$\text{Stress due to eccentricity bottom } f_{bb} = + \frac{2356.2 \times 10^3 \times 463}{Z_b}$$

$$= + \frac{2356.2 \times 10^3 \times 463}{71.90 \times 10^6} = +15.173 \, \text{N/mm}^2$$

$$\text{Stress in top due to DL} = f_{dt} = + \frac{970.3125 \times 10^6}{92.63 \times 10^6} = +10.475 \, \text{N/mm}^2$$

$$\text{Stress in bottom due to DL} = f_{db} = - \frac{970.3125 \times 10^6}{71.90 \times 10^6} = -13.495 \, \text{N/mm}^2$$

$$\text{Stress in top due to LL} = + \frac{450 \times 10^6}{92.63 \times 10^6} = +4.86 \, \text{N/mm}^2$$

Stress in bottom due to LL $= -\dfrac{450 \times 10^6}{71.90 \times 10^6} = -6.26 \, \text{N/mm}^2$

Case II: Final stress at the time of initial transfer

Top face: $+ 6.83 - 11.78 + 10.45 = + 5.5 \, \text{N/mm}^2$

Bottom face: $+ 6.83 + 15.173 - 13.495 = + 8.508 \, \text{N/mm}^2$

Final stress at the time of working loads transfer and after prestress losses

Top face: $(+ 0.85 \times 6.83 - 0.85 \times 11.78) + 10.45 + 4.86 = +11.103 \, \text{N/mm}^2$

Bottom face: $+0.85 \times 6.83 + 0.85 \times 15.173 - 13.495 - 6.26 = -1.052 \, \text{N/mm}^2$

Example 12.10

A straight pretensioned, concrete beam 15 m long and cross-section 400 mm × 400 mm, is concentrically prestressed with 900 mm² steel wires. The wires are anchored to the bulk heads with a stress of 1000 N/mm². If the modular ratio $m = 6$, determine the loss of prestress due to elastic shortening of concrete at the transfer of prestress.

Solution: $T_0 = 900 \times 1000 = 9 \times 10^5 \, \text{N}, \quad A = 400 \times 400 = 16 \times 10^4 \, \text{mm}^2$

$$\text{Loss of prestress} = \Delta f_{el} = m \, \frac{T_0}{A} = \frac{9 \times 10^5}{16 \times 10^4} \times 6 = 33.75 \, \text{N/mm}^2$$

$$\%\text{age loss} = \frac{33.75 \times 100}{1000} = 3.375\%$$

Example 12.11

A post-tensioned concrete member of 400 mm × 400 mm and 15 m long is prestressed with four tendons each of 225 mm² cross-section. The tendons are stressed to a stress of 1100 N/mm². Determine the loss of prestress in each tendon due to elastic shortening of concrete if the tendons are pulled one by one, find the total loss and % loss of prestress. If it is desired that a stress of 1100 N/mm² be maintained after stressing the last tendon, find the actual stress in each tendon to which these tendons are pulled. Take $m = 6$.

Solution: Since pretension is done one by one tendon, 1st tendon undergoes 3 times loss for each tendon force, 2nd tendon 2 times force, 3rd tendon one force of tendon at $m = 6$, Thus, total force causing elastic shortening $= 3 \times 225 \times 1100 = 742500 \, \text{N}$(1st tendon)

$$\text{Loss of prestress in 1st tendon} = \frac{m \times 742500}{A_c} = \frac{6 \times 742500}{400 \times 400} = 27.844 \, \text{N/mm}^2$$

$$\text{Loss of prestress in 2nd tendon} = \frac{6 \times (2 \times 225 \times 1100)}{400 \times 400} = 18.563 \, \text{N/mm}^2$$

$$\text{Loss of prestress in 3rd tendon} = \frac{6 \times 225 \times 1100}{400 \times 400} = 9.281 \, \text{N/mm}^2$$

Loss in 4th tendon $= 0$

$$\text{Average loss of prestress} = \frac{(27.844 + 18.563 + 9.281 + 0)}{4} = 13.92 \, \text{N/mm}^2$$

$$\text{Average \% loss} = \frac{13.92}{1100} \times 100 = 1.266\% = 1.27\%$$

Initial stress in 1st tendon = $1100 + 27.844 = 1127 \, \text{N}/\text{mm}^2$

Initial stress in 2nd tendon = $1100 + 18.563 = 1118.56 \, \text{N}/\text{mm}^2$

Initial stress in 3rd tendon = $1100 + 9.281 = 1109.281 \, \text{N}/\text{mm}^2$

Initial stress in 4th tendon = $1100 \, \text{N}/\text{mm}^2$

Example 12.12

Calculate the loss of prestress due to shrinkage of concrete in a post tensioned beam if the age of concrete at transfer is 15 days, $E_s = 2.1 \times 10^5 \, \text{N}/\text{mm}^2$.

Solution: Shrinkage strain = $\dfrac{2 \times 10^{-4}}{\log_{10}(T+2)} = \dfrac{2 \times 10^{-4}}{\log_{10}(17)} = \dfrac{2 \times 10^{-4}}{1.23045} = 1.625 \times 10^{-4}$

Loss of stress in steel = $E_s \times$ (shrinkage strain of concrete)

$$= 2.1 \times 10^5 \times 1.6254 \times 10^{-4} = 34.134 \, \text{N}/\text{mm}^2$$

∴   Loss of stress = $34.134 \, \text{N}/\text{mm}^2$

Example 12.13

A post-tensioned prestressed concrete beam spanning 10 m and having cross-section 200 mm × 300 mm deep is prestressed with a straight cable of 320 mm² cross-section located at a constant eccentricity of 50 mm. The initial stress in the cable is 1000 N/mm². Calculate the %age loss of stress in the cable using following data:

Relaxation of steel stress = 5% of initial

Shrinkage strain in concrete = 200×10^{-6} mm/mm

Ultimate creep strain in concrete = 20×10^{-6} mm/mm/ N/mm²

Slip at anchorage = 1 mm, $E_s = 2.1 \times 10^5 \, \text{N}/\text{mm}^2$, $E_c = 35 \times 10^3 \, \text{N}/\text{mm}^2$

Solution: $A_c = 200 \times 300 = 6 \times 10^4 \, \text{mm}^2$, $I = \dfrac{200 \times 300^3}{12} = 450 \times 10^6 \, \text{mm}^4$

$Z = \dfrac{200 \times 300^2}{6} = 3 \times 10^6 \, \text{mm}^3$, modular ratio $m = \dfrac{2.1 \times 10^5}{3.5 \times 10^4} = 6.0$

Initial prestress = $1000 \, \text{N}/\text{mm}^2$, force = $1000 \times 320 = 32 \times 10^4 \, \text{N}$

Initial stress in concrete at the level of tendon = $\dfrac{32 \times 10^4}{6 \times 10^4} = + \dfrac{16}{3} \, \text{N}/\text{mm}^2$

$= + 5.33 \, \text{N}/\text{mm}^2$

Due to eccentricity at the level of cable = $+ \dfrac{32 \times 10^4 \times 50 \times 50}{450 \times 10^6} = 1.78 \, \text{N}/\text{mm}^2$

Total stress = $5.33 + 1.78 = 7.11 \, \text{N}/\text{mm}^2$

Losses of prestress due to:

(a) Shortening of concrete = $m \,(5.33 + 1.78) = 6 \times 7.11 = 42.66 \, \text{N}/\text{mm}^2$

(b) Relaxation of steel = $\dfrac{5}{100}\,(1000) = 50 \, \text{N}/\text{mm}^2$

(c) Shrinkage of concrete $= 200 \times 10^{-6} \times E_s = 200 \times 2.1 \times 10^{-6+5} = 42 \text{ N/mm}^2$

(d) Creep in concrete $= (20.0 \times 10^{-6} \times 7) E_s = 14 \times 2.1 = 29.4 \text{ N/mm}^2$

(e) Friction $f_0 k \cdot x = 0.0015 \times$ length $\times$ cable stress

$$= 0.0015 \times 10 \times 1000 = 15 \text{ N/mm}^2$$

(f) slip $= \dfrac{\Delta \text{slip}}{l} \times E_s = \dfrac{0.001}{10} \times 2.1 \times 10^5 = 21 \text{ N/mm}^2$

Total loss of prestress $= 42.66 + 50 + 42 + 29.4 + 15 + 21 = 200.06 \text{ N/mm}^2$

$$\%\text{age loss} = \frac{200.06 \times 100}{1000} = 20\%$$

Example 12.14

A prestressed concrete pile with a square section 250 mm $\times$ 250 mm has 60 wires of 2 mm diameter each, uniformly distributed over the section. The wires are initially tensioned on the prestressing bed with a total force of 300 kN. Calculate the final stress in concrete after all the losses have occurred using the following data:

$E_s = 2.1 \times 10^5 \text{ N/mm}^2$, $E_c = 30 \times 10^3 \text{ N/mm}^2$, shortening due to creep $= 3 \times 10^{-5}$ per unit length per unit stress

Total shrinkage $= 200 \times 10^{-6}$ per unit length.

Relaxation of steel $= 5\%$ of initial stress.

Solution: Area of concrete $= 250 \times 250 = 62500 \text{ mm}^2$

Initial Load $= 300$ kN (T_0)

$$\text{Initial stress in wires} = \frac{300 \times 1000}{60 \times \dfrac{\pi}{4}(2.0)^2} = 1591.55 \text{ N/mm}^2$$

$$\text{Modular ratio } m = \frac{2.1 \times 10^5}{30 \times 10^3} = 7$$

Various losses are:

(i) Elastic shortening of concrete $= m \dfrac{T_0}{A} = \dfrac{300000 \times 7}{62500} = 33.6 \text{ N/mm}^2$

(ii) Creep strain $= (3 \times 10^{-5}) \times \dfrac{300000}{62500} = 1.44 \times 10^{-4}$

$\therefore$ Loss due to creep of concrete $= 1.44 \times 10^{-4} \times 2.1 \times 10^5 = 30.24 \text{ N/mm}^2$

(iii) Loss due to shrinkage $= (200 \times 10^{-6}) \times 2.1 \times 10^5 = 42.0 \text{ N/mm}^2$

(iv) Loss of stress due to steel relaxation $= 5\%$ initial $= \dfrac{5}{100} \times 1591.55$

$$= 79.58 \text{ N/mm}^2$$

Total loss $= 33.6 + 30.24 + 42.0 + 79.58 = 185.42 \text{ N/mm}^2$

Thus $\%\text{age loss} = \dfrac{185.42}{1591.55} \times 100 = 11.65\%$

$$\text{Net prestressing force} = \frac{(100-11.65)}{100} \times 300000 = 265050 \text{ N}$$

$$\therefore \text{ Stress in concrete} = \frac{265050}{62500} = 4.241 \text{ N/mm}^2$$

Example 12.15

A prestressed concrete is provided with a parabolic cable tensioned from both ends as shown in Fig. 12.19. If the stress in the cable at ends is 1050 N/mm², calculate the loss of prestress from the ends to the centre. Take $\mu = 0.35$ and $k_f = 0.0015$ per metre.

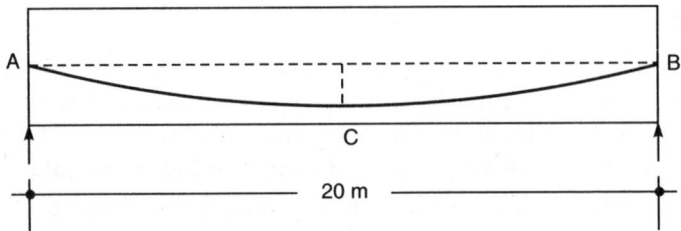

Fig 12.19 : Prestressed concrete beam–friction loss

Solution: Sag of parabola $(h) = 0.20$ m

$$\text{Equation of parabola profile } y = \frac{4h.x(L-x)}{L^2} = \frac{4 \times 0.2 x(20-x)}{20 \times 20}$$

$$y = 0.002\, x(20-x), \text{ where } x \text{ is in metre}$$

or

$$y = 0.002(20\, x - x^2)$$

$$\frac{dy}{dx} \text{ (slope)} = 0.002\,(20-2x)$$

Putting $x = 0$, slope at end A $= 0.002\,(20-0) = 0.04$ radian

Putting $x = \dfrac{L}{2} = 10$ m, slope at mid point C $= 0.002\,(20 - 2 \times 10) = 0$

Change of slope from A to C $= 0.04$ radian, $\therefore \alpha = 0.04$ radian,

Loss of stress A to C $= (\mu\alpha + k_f \cdot x) \times$ initial stress at A.

$$(\alpha = 0.04, x = 10 \text{ m}, \mu = 0.35, k_f = 0.0015)$$

or   loss of stress $= (0.35 \times 0.04 + 0.0015 \times 10) \times 1050 = 30.45 \text{ N/mm}^2$

Loss of prestress from end to mid span point $= 30.45 \text{ N/mm}^2$

Example 12.16

A simply supported post tensioned concrete beam of 15 m span has rectangular section 300 mm × 800 mm. The prestress at the ends is 1300 kN with eccentricity of zero at supports and an eccentricity of 250 mm at the mid span point. The cable profile is parabolic. Assume $k_f = 0.15$ per 100 m and $\mu = 0.35$, determine the loss %age due to friction at the centre of the beam.

Solution: $L = 15$ m, $h = 250$ mm, the equation of parabolic profile

$$y = \frac{4h}{l^2}x(l-x)$$

$$y = \frac{4h \cdot x(15-x)}{15^2}$$

Slope $\dfrac{dy}{dx} = \dfrac{4h(15-2x)}{15^2}$, slope at ends, $\alpha = \dfrac{4 \times 15h}{15^2} = \dfrac{4 \times 15 \times 0.25}{15 \times 15} = \dfrac{1}{15}$

Slope at centre $\alpha = 0$, $(x = 7.5$ m$)$

Angle of deviation $\alpha = \dfrac{1}{15}$ radian (since α is small, $\sin \alpha = \tan \alpha = \alpha$ radian)

k_f (wave friction) $= \dfrac{0.15}{100}$ per m, $\mu = 0.35$ (given), $\alpha = \dfrac{1}{15}$

$P_x = 1300 \left(1 - 0.35 \times \dfrac{1}{15} - \dfrac{15}{10^4} \times 7.5\right) = 1300\ (1 - 0.0233 - 0.01125) = 1255.04$ kN

$\therefore$ Loss of prestress $= 1300 - 1255 = 45$ kN

Percentage loss $= \dfrac{45}{1300} \times 1000 = 3.45$

Example 12.17

A prestressed concrete simply supported beam of uniform rectangular cross-section and 15 m span supports a total u.d.l. of 272 kN excluding self-weight. Determine the suitable dimensions of the beam and calculate the area of tendons and their position (eccentricity). The permissible stresses for concrete and the tendons are respectively 14 N/mm² and 1050 N/mm² respectively.

Solution: It may be noted that the prestressing force is so adjusted that the maximum stress with dead load and prestressing load (T_0) develops zero stress in top fibres and maximum compressive stress at bottom concrete fibres. Also after application of live load, the bottom fibre stress becomes zero and top fibre concrete stress becomes equal to maximum compressive stress of concrete. The design is approximately based on this principle.

$$\text{BM due to live load } (M_l) = \frac{272 \times 15}{8} \times 10^6 = \pm 510 \times 10^6 \text{ N-mm}$$

Max. compressive concrete stress in top fibre $= 14$ N/mm²

$$\therefore \qquad Z = \frac{M}{f} = \frac{510 \times 10^6}{14} = \frac{255 \times 10^6}{7} \text{ mm}^3$$

Depth required in simply supported beams generally assumed 1/20th to 1/25th of the span. Thus $d = \dfrac{15000}{20} = 750$ mm (say).

Thus, $\dfrac{bd^2}{6} = \dfrac{255 \times 10^6}{7}$ or $b\ \dfrac{6 \times 255 \times 10^6}{7 \times (750)^2} = 388.6$ mm ($b = 400$ mm, say)

The beam section of 400 mm × 750 mm may be adopted (A = 3 × 10⁵ mm²)

Self-wt. of beam = 0.4 × 0.75 × 25 = 7.5 kN/m,

total load = 7.5 × 15 = 112.5 kN on the entire span.

Assuming max. comp. stress = f_c(14 N/mm²) and zero at top we have total prestress

force $T = \dfrac{14 \times (400 \times 750)}{2} = 21 \times 10^5$ N (assuming full area under compression).

Compression (T) = Tension in tendons

$21 \times 10^5 = A_{st} \times 1050$, or A_{st} (prestress wires) = 2000 mm²

If wires are of 5 mm diameter, no of wires = $\dfrac{2000 \times 4}{\pi (5)^2} = 102$

BM due to self-wt. $(M_d) = \dfrac{112.5 \times 15}{8} = 210.94$ kN-m (210.94 × 10⁶) N-mm

BM due to LL $(M_l) = \pm 510 \times 10^6$ N-mm

BM due tension load eccentricity $= \mp (21 \times 10^5 e)$, (where e is in mm)

Before application of live load, we have zero stress in top face due to DL and T,

i.e. $\dfrac{T}{A} - \dfrac{T.e}{Z} + \dfrac{M_d}{Z} = 0$, or $\dfrac{T.e}{Z} = \dfrac{T}{A} + \dfrac{M_d}{Z}$ ... Eq. (1)

After live load application, we have zero stress in bottom fibres

i.e $\dfrac{T}{A} + \dfrac{T.e}{Z} - \dfrac{(M_d + M_l)}{Z} = 0$, or $\dfrac{T.e}{Z} = -\dfrac{T}{A} + \dfrac{(M_d + M_l)}{Z}$ ... Eq. (2)

Adding Eqs (1) and (2), we get $\dfrac{2T.e}{Z} = \dfrac{(2M_d + M_l)}{Z}$ or $e = \dfrac{(2M_d + M_l)}{2T}$ ... Eq. (3)

Substituting different values: $e = \dfrac{(2 \times 210.94 \times 10^6 + 510 \times 10^6)}{2 \times 21 \times 10^5} = 221.9$ mm

Thus, the beam section shall be 400 mm × 750 mm deep and tendons area (102 wires) shall be 2000 mm² and tendons of 5 mm located at 221.9 mm from centroid (below centroid).

Example 12.18

Design a prestressed concrete slab simply supported over a span of 15 m for following details:

Live loads (w) = 20 kN/m², safe stress in concrete = 14 N/mm²,

stress in steel wires = 900 N/mm²

Solution: Consider 1 m wide strip (Fig. 12.20).

$$\text{LL moment} = \dfrac{20 \times 15^2 \times 10^6}{8} = 562.5 \times 10^6 \text{ N-mm}$$

$$Z = \dfrac{562.5 \times 10^6}{14} = 40.18 \times 10^6 \text{ mm}^3$$

For rectangular section of slab

$$Z = \frac{1000 \times d^2}{6} = 40.18 \times 10^6$$

$$d^2 = 40.18 \times 6 \times 10^3$$

$d = \sqrt{241080} = 490.99$, (say) 500 mm, DL $= 1 \times 0.5 \times 1 \times 25 = 12.5$ kN/m

$$\text{BM } M_d = \frac{12.5 \times 15^2 \times 10^6}{8} = 351.5 \times 10^6 \text{ N-mm}$$

Area of slab/m width $= 1000 \times 500 = 5 \times 10^5$ mm^2

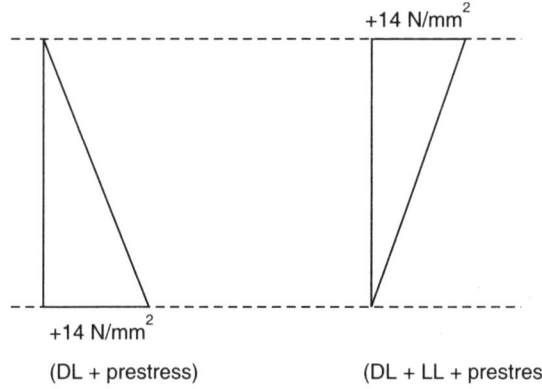

(DL + prestress)   (DL + LL + prestress)

Fig. 12.20

Prestressing force (T) average stress $\times$ area $= \dfrac{14}{2} \times 5 \times 10^5 = 35 \times 10^5$ N

Area of prestressing steel $A_{st}/\text{m} = \dfrac{35 \times 10^5}{900} = 3889$ mm^2

(200 wires of 5 mm) in 2 layers.
(i.e. 10 cables of 10 wires each @ 100 mm c$_c$ and in 2 layers)

$$K_t = K_b, \qquad \frac{d}{6} = \frac{500}{6} = 83.3 \text{ mm},$$

$$e = \frac{M_d}{T} + k_b = \frac{351.56 \times 10^6}{35 \times 10^5} = (100.45 \text{ mm} + 83.33 \text{ mm}) = 183.78 \text{ mm}$$

Example 12.19

Figure 12.21 shows the section of a prestressed concrete beam of span 9 m simply supported. The beam has to carry a superimposed u.d.l. of 20 kN/m. The prestressing force is transmitted by tendons in two cable ducts. Each tendon consists of 14 wires of 5 mm diameter subjected to an initial prestress of 1000 N/mm^2. Analyse the beam section at the mid span point, for the stresses induced before and after the application of superimposed load. Allow 20% loss of prestress.

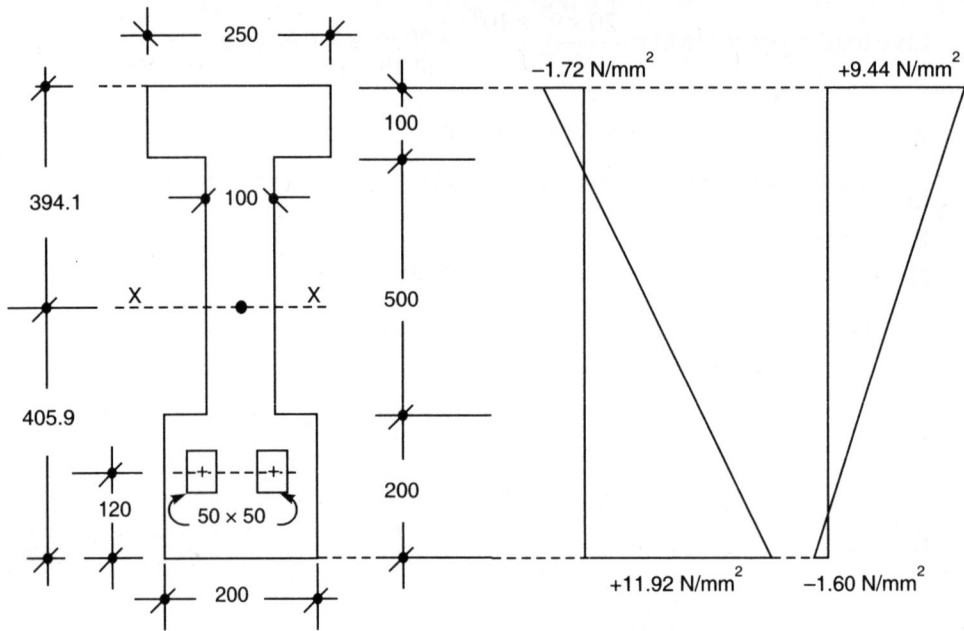

Fig. 12.21: PSC beam section and stress variation

Solution: Properties of the section

$A = 250 \times 100 + 500 \times 100 + 200 \times 200 - 50 \times 50 \times 2 = 11.0 \times 10^4 \, \text{mm}^2$

$y_t = 10^3 \left\{ \dfrac{25 \times 50 + 50 \times 350 + 40 \times 700 - 5 \times 680}{110 \times 10^3} \right\} = \dfrac{43350}{110} = 394.1 \, \text{mm}$

$y_b = (800 - 394.1) = 405.9 \, \text{mm}$

$I_{xx} = \dfrac{250 \times 100^3}{12} + 25000 + (344.1)^2 + \dfrac{100 \times 500^3}{12} + 50000 \, (44.1)^2 + \dfrac{200 \, (200)^3}{12} +$

$40000 \, (305.9)^2 - \left\{ \dfrac{100 \times 50^3}{12} + 5000 \, (285.9)^2 \right\}$

$\quad = 10^6 \, [20.833 + 1041.667 + 133.333 + 2960.12 + 97.2405 + 3 - 1.0417 + 408.694]$

$I_{xx} = 7586.46 \times 10^6 \, \text{mm}^4$

$Z_{top} = \dfrac{7586.46 \times 10^6}{394.1} = 19.25 \times 10^6 \, \text{mm}^3, \quad Z_{bott} = \dfrac{7586.46 \times 10^6}{405.9} = 18.69 \times 10^6 \, \text{mm}^3,$

$Z_{cable} = 26.5354 \times 10^6 \, \text{mm}^3$

Dead load of girder $= \dfrac{11 \times 10^4 \times 1 \times 25}{10^6} = 2.75 \, \text{kN/m}, \, LL = 20 \, \text{kN/m}$

Dead load moment $(M_d) = \dfrac{2.75 \times 9^2 \times 10^6}{8} = 27.844 \times 10^6 \, \text{N-mm}$

Live load moment $(M_l) = \dfrac{20 \times 9^2 \times 10^6}{8} = 202.5 \times 10^6$ N-mm

Area of steel wires $= 2 \times 14 \times \dfrac{\pi}{4} \, (5)^2 = 28 \times 19.635 = 549.8$ mm^2

Initial prestressing force $T = 549.8 \times 1000$ N $= 549.8$ kN

Eccentricity $e = 405.9 - 120 = 285.9$ mm

Direct pre-stress in concrete at the time of initial prestress force T_0

$f_0 = \dfrac{549.8 \times 1000}{110 \times 10^3} = +5.0$ N/mm^2 (area of concrete section $= 11$ mm)

f_{by} (top) (due to eccentricity) $= \dfrac{549.8 \times 1000 \times 285.9}{Z_t} = \mp 8.166$ N/mm^2

f_{bb} (bottom) (due to eccentricity) $= \dfrac{549.8 \times 1000 \times 285.9}{Z_b} = + 8.166$ N/mm^2

Bending stress (f_{bdt}) due to DL at top) $= \dfrac{27.844 \times 10^6}{19.25 \times 10^6} = + 1.45$ N/mm^2

Bending stress $(f_{bdb}) = -\dfrac{27.844 \times 10^6}{18.69 \times 10^6} = 1.49$ N/mm^2

Bending stress due to LL at top $(f_{blt}) = + \dfrac{202.5 \times 10^6}{19.25 \times 10^6} = + 10.52$ N/mm^2

Bending stress due to LL at bottom $(f_{blb}) = -\dfrac{202.5 \times 10^6}{18.69 \times 10^6} = 10.835$ N/mm^2

Case I: Stress before live load

In top face $= + 5.0 - 8.166 + 1.45 = - 1.716$ N/mm^2 $(- 1.72)$

In bottom face $= + 5.0 + 8.41 - 1.49 = + 11.92$ N/mm^2 $(+11.92)$

Case II: Stress after live load (when 20% losses occur in prestress)

In top face: $(+ 5 - 8.166)(1 - 0.2) + 1.45 + 10.52 = -2.533 + 1.45 + 10.52 = + 9.44$ N/mm^2

In bottom face: $(+ 5 + 8.41) \, 0.8 - 1.49 - 10.835 = - 1.60$ N/mm^2

SUMMARY

Cement concrete is strong in compression and resists compressive stresses quite effectively. Prestressed concrete develops mostly compressive stresses in total cross-sectional area. Compression is introduced in that portion of the concrete member where tension is likely to develop after application of working loads. Thus, in simply supported beam the lower portion is compressed by stretching high tension wires and creating opposite type of stresses. Stresses get neutralized after application of working loads. This makes the design much lighter and lighter loads on foundation. The steel wires are required to be of high yield strength. Concrete of high characteristic strengths are to be used in prestressed concrete elements. Shrinkage, creep and elastic strains are required to be compensated by additional tension in prestressing wires.

In a prestressed element stresses develop due to the following factors:

- direct stress and eccentricity bending due to prestressing force (T)
- bending stress due to dead loading (self-wt.) M_d and
- bending stress due to live loading (M_l)
- losses of prestressing force due to shrinkage, creep and elastic shortening, etc.

Final stress at any point across the section

$$\text{Top} f_t = \frac{T}{A_c} - \frac{(T.e)}{I_0} \cdot y_t + \frac{M_d}{I_0} \cdot y_t + \frac{M_l}{I_0} \cdot y_t$$

$$\text{Bottom} f_b = \frac{T}{A_c} + \frac{(T.e)}{I_0} \cdot y_b - \frac{M_d}{I_0} \cdot y_b - \frac{M_l}{I_0} \cdot y_b$$

where, T = prestress force

A_c = area of concrete section

e = eccentricity

y_t, y_b = distances of top and bottom fibres from the NA

I_0 = moment of inertia of concrete section about NA

Prestressing force also balances the external loading if the cable is provided in inclined or parabolic profile. If θ is inclination of cable with horizontal, we have vertical component $V = 2T \sin \theta$, V balances the vertical loading.

Horizontal component $H = T \cos \theta$, and direct stress f is given by

$$f_0 = \frac{T\cos \theta}{A_c} \simeq \frac{T}{A_c} \text{, when } \theta \text{ is small and } \cos \theta = 1$$

In case of parabolic profile with central sag h, we have vertical loading

$$p = \frac{T \cdot 8h}{l^2},$$

where, h = sag of cable, also $p = \dfrac{T \cdot d^2y}{dx^2}$,

l = span, T = tension in the cable

Net vertical loading = $(w - p)/m$ on the beam

Prestressing can be pretensioning of elements before casting while post-tensioning is done after hardening through cables in the duct provided while casting. Total losses due to shrinkage, creep and elastic shortening ranges from 20–25% in pretensioning while these losses range 12%–15% in case of post-tensioning.

The system of prestressing depends on the method of tensioning and anchoring the tensioned wires/cables/tendons. The common systems are: Freyssinet, Magnel–Blaton, Gifford–Udall, Lee McCall, etc.

Freyssinet, the most widely adopted system uses 5 mm wires in group of 8, 10, 12, or 18 and 7 mm wires in group of 12. Each group of HT wires are stretched simultaneously and anchored with the help of male and female cones at the ends.

Magnel–Blaton (Belgium) system comprises two wires pulling simultaneously up to 64 wires of 5 mm or 7 mm diameter. The anchorage device consists of sandwitch steel plates with grooves to hold wires and wedges with 2×2 wires in each layer. Sandwitch plates are placed one over the other and held against the distribution plate. After completing the prestressing process, the anchorage system with sandwitchand distribution plates are grouted with concrete at the end of element.

Gifford–Udall system uses 7 mm parallel wires separated with spacers and forming cable. This system uses either thrustring-bearing plate or tube type of anchoring system, 39 mm diameter spacers for cables of 8 wires while 51 mm diameter spacers for cables containing 12 wires.

Lee–McCall system uses tendons (12 to 28 mm diameter) instead of wires. These tendons are threaded at ends for anchoring with the help of nut-bolt system. The tendon nut bolt system is tightened after stretching and generally grouted with cement grout.

The prestressing force reduces on account of shortening of concrete element due to following reasons:

(i) loss due to elastic deformation of concrete on application of force

(ii) loss due to shrinkage of concrete after anchoring

(iii) loss due to creep of concrete because of sustained load of prestress

(iv) loss due to relaxation of steel (creep of steel)

(v) loss due to deformation and slip of anchorage

(vi) loss due to friction between tendon and concrete surface

Analysis:

(i) Loss due to elastic deformation $\Delta f_{el} = \dfrac{m\left(f_{si} \cdot A_{st}\right)}{A_e}$, $\quad m = \dfrac{E_s}{E_c}$

(ii) Loss due to shrinkage strains (0.0002–0.0003), $\Delta f_{sh} = e_{sh} \times E_s$

(iii) Loss due to creep strain in concrete (generally 5% to 10%)

$$e_c = (c_c - 1)\, e_{el}, \; c_c = 0.5\, k_c \text{ to } 4\, k_c, \; k_c \text{ depends on } f_{cr}/f_{cu} \text{ (2.2 to 0.50 for } \dfrac{f_{cr}}{f_{cu}} = 0.5$$

to 1.30)

(iv) Loss due to relaxation of steel (2%–8%) depending on initial prestressing. Overstressing to compensate relaxation is allowed.

(v) Loss due to deformation and slip of anchorage $\Delta f_{slip} = E_s \cdot \dfrac{\delta}{l}$

(vi) Loss due to friction: Prestress at X from end or $P_x = P_0(1 - \mu\alpha - kx)$

In designs of prestressed concrete members, these losses must be accounted for computing prestressing force.

PRACTICE QUESTIONS

(I) Objective Questions

Q. 12.1 Select the correct response given after each statement to complete it correctly and fill in the response sheet provided.

(i) A cantilever beam of span l and subjected to a u.d.l., the rectangular section shall be prestressed by placing the prestressing high tension steel wires/tendons in .... so that external moment is balanced to a great extent and no tension develop anywhere?

    (a) tension zone below NA   (b) tension zone above NA

    (c) compression zone below NA   (d) netural zone along NA

(ii) The advantage of prestressed concrete section shall be more ....

    (a) heavier   (b) costlier

    (c) efficient   (d) bulky

(iii) At what stage(s) the safety of prestressed beam section should be checked?

    (a) one stage only: at prestress application to concrete

    (b) two stages only: at prestress application and dead load application

    (c) three stages: at prestress application, live load application and after losses occur in prestress

    (d) four stages: at prestress application, dead load application, live load application and after anchoring tendons.

(iv) The Freyssinet system of post tensioning comprises specially................ ?

    (a) cable of 8 wires of 5 mm or 7 mm diameter

    (b) 5 mm wires in group of 20 and 7 mm in group of 12 wires

    (c) tendons of 64 wires of 5 mm or 7 mm diameter

    (d) 5 mm wires in group of 8, 10, 12, or 18 and 7 mm wires in group of 12

(v) Which system uses tendons of 12 mm to 28 mm diameter with ends threaded for anchoring with nut-bolt system?

    (a) Gifford–Udall system   (b) Magnel–Blaton system

    (c) Freyssinet system   (d) Lee McCall system

(vi) Elastic shortening of concrete may result in loss of prestress given as $\Delta f_{el} =$ ...., when shortening of concrete $= \delta$, modulus of elasticity of concrete $= E_c$, modulus of elasticity of steel $= E_s$, initial prestress $= T_i$,

t = age in days at prestressing, $\dfrac{E_s}{E_c} = m$ and $A_e = A_c + mA_{st}$

    (a) $\Delta f_{el} = \delta \times E_c$ (b) $\Delta f_{el} = m \cdot \dfrac{T_i}{A_e}$

    (c) $\Delta f_{el} = e_{sh} \cdot E_c$ (d) $\Delta f_{el} = \dfrac{0.0002}{\log_{10}(t+2)}$

(vii) Shrinkage of concrete after prestressing causes loss of pre stress given as Δf_{sh} =........................... with usual notations.

(a) $\Delta f_{sh} = P_0(1 - \mu\alpha - k_x)$ (b) $\Delta f_{sh} = \dfrac{E_s \delta}{E_c l}$

(c) $\Delta f_{sh} = e_{sh} \cdot E_s$ (d) $\Delta f_{sh} = (C_{sh} - 1) e_{sh} E_c$

(viii) For designing a SS rectangular concrete post tensioned (mid-span) beam section by approximate method, the stresses are assumed as.....................

    (a) zero in bottom fibres and maximum permissible stress in top fibres on application of initial prestress (T_0)

    (b) zero in top extreme fibres and maximum permissible in bottom fibres on application of initial prestress (T_0)

    (c) zero both in top and bottom fibres on application of initial prestress

    (d) maximum permissible compressive stress both in extreme top and bottom fibres on application of initial prestress (T_0)

(ix) For designing a SS prestressed rectangular concrete beam section at mid span by approximate method, the stresses are assumed as ....

    (a) zero both in top and bottom fibres after application of working loads and prestressing force

    (b) maximum permissible compressive stress both in top and bottom fibres after application of working load and prestressing force

    (c) maximum permissible compressive stress in top fibres and zero stress in bottom fibres after application of live loads and prestress force

    (d) zero in top and maximum permissible compressive stress in bottom fibres after application of final prestress force and working live loads.

(x) Total losses in prestressed members shall be......................................

    (a) more in post-tensioned members as compared to pretensioned in members of the same size

    (b) more in pretensioned members as compared to post-tensioned in members of the same size

    (c) equal in both pretensioned and post-tensioned members of the same size and placed in similar environment

    (d) none of the above statements are true.

Response sheet to Q. 12.1 (i to x)

Question	(i)	(ii)	(iii)	(iv)	(v)	(vi)	(vii)	(viii)	(ix)	(x)
Response (a/b/c/d)										

(II) Numerical Questions

Q. 12.2 A prestressed concrete beam has 400 mm × 600 mm section and 6 m simply supported span. The beam is subjected to anu. d.l. of 16 kN/m including self-weight of the beam. Prestressing tendons are located at lower one-third point of the section and provide an effective prestressing force of 1000 kN. Determine the extreme fibre stress in concrete at the mid span section.

[**Hint:** $A = 24 \times 10^4$ mm^2, $Z = 24 \times 10^6$ mm^3, max. BM $= 72 \times 10$ N-mm, $e = 100$ mm

Final stresses in top fibres $= +3.0$ N/mm^2, Bottom fibres $= + 5.34$ N/mm^2]

Q. 12.3 A concrete beam 400 mm × 600 mm deep is simply supported over 6 m span. The beam is prestressed with an inclined tendon having effective 1200 kN force. At the mid span point, tendon is located at 150 mm below the centroid and at ends it is located at the centroid of section. Determine the extreme stress at the mid span when the external load of 200 kN acts at the mid-span point.

[**Hint:** Slope tendon $= \dfrac{150}{3000} = \dfrac{1}{20} = \tan\theta$, upward component of force $= 2P\tan\theta$,

i.e. V, upward force $= 2 \times 1200 \times \dfrac{1}{20} = 120$ kN, net vertical load $= 200 - 120 =$

80 kN, self-weight $= 6$ kN/m, M_1 (BM) $= 80 \times \dfrac{6}{4} = 120 \times 10^6$ N-mm,

M_2 BM (self-weight) $= 6 \times \dfrac{6^2}{8} = 27 \times 10^6$ N-mm, $Z = 24 \times 10^6$ mm³,

Bending stress $= \pm \dfrac{(120 + 27)10^6}{24 \times 10^6} = \pm 6.125$ N/mm²

Direct stress $= + \dfrac{1200 \times 1000}{24 \times 10^4} = +5.0$ N/mm²

Resultant top $= 5 + 6.125 = 11.125$ N/mm²

Bottom $= 5 - 6.125 = -1.125$ N/mm²

Extreme Stress: $+11.125$ N/mm² (top) and -1.125 N/mm² (bottom)]

Q. 12.4 A prestressed SS concrete beam of 400 mm × 600 mm section is prestressed with a parabolic cable profile having 75 mm sag at the mid-span and passes through centroid at the end section. The cable is prestressed with an effective force of 1200 kN and the beam is subjected to an u.d.l. of 35 kN/m over the entire span of 6 m.

[**Hint:** Upward loading due to inclined parabolic cable $w_1 = \dfrac{8h \cdot T}{l^2} = 20$ kN/m

$h = 0.075$ m, $l = 6$ m, $T = 1200$ kN, net downward u.d.l. $= 35 - 20 = 15$ kN/m

Max. BM $= \dfrac{15 \times 6^2}{8} = 67.5$ kN-m (67.5×10^6) N-mm, $Z = 24 \times 10^6$ mm³

Extreme stress (mid span section): Top and bottom $= \dfrac{1200 \times 1000}{24 \times 10^4} \pm \dfrac{67.5 \times 10^6}{24 \times 10^6}$

$= 5 \pm 2.8125 = +7.8125, +2.1875$ N/mm²]

Q. 12.5 Find the dip(sag) h of a parabolic cable profile at the centre of a SS beam of 8 m span and carrying an u.d.l. of 40 kN/m including the self-weight of the beam, if the prestresssing force is 1600 kN and the resultant stress in extreme bottom fibre is zero. The size of the beam is 400 mm × 600 mm. Also find the maximum stress.

[Hint: For parabolic cable profile upward load $= \dfrac{8h \cdot P}{L^2} = 200h$, $P = 1600$ kN, $l = 8$ m,

$A = 24 \times 10t$ mm², $Z = 24 \times 10^6$ mm³, Net $w = (40 - 200h)$

$$\text{Bending stress} = \pm \frac{M}{Z} = \frac{(40 - 200h)8^2 \times 10^6}{8 \times 24 \times 106} = \pm \left(\frac{40}{3} - \frac{200h}{3} \right)$$

$$\text{Resultant stress} = \frac{1600 \times 1000}{24 \times 10^4} - \left(\frac{40}{3} - \frac{200h}{3} \right) = 0, h = 0.1 \text{ m}$$

$f_{max} = 13.33$ N/mm²]

Q. 12.6 Find the cable dip h for the beam so that the applied loads $W = 100$ kN are fully balanced by prestress vertical component. The cable tension is 1000 kN. The load, W act at $\dfrac{1}{3}$rd span points. Span of SS beam = 9 m. If the section of the beam is 400 mm × 750 mm, also find the maximum stress.

[Hint: $L = 9$ m, $\tan \theta = \dfrac{h}{3} = \sin \theta$, $T = 10^6$ N, $W = 10^5$ N

Upward tension component $= T \sin \theta = W$ (being balanced),

i.e. $10^6 \times \dfrac{h}{3} = 10^5$, $h = 0.30$ m, $M = 0$ (as load is balanced)

$$\text{Stress} = \frac{T}{A} \pm \frac{M}{Z} = \frac{10^6}{30 \times 10^4} = \frac{10}{3} \; 3.33 \text{ N/mm}^2$$

Q. 12.7 A prestressed concrete rectangular beam 300 mm × 600 mm is prestressed with a force of 1600 kN applied at 180 mm above the bottom. The prestress looses 200 kN force due to various reasons. The span of the simply supported beam is 12.5 m and carries two equal live loads of 45 kN each at distance of 4.5 m from the support. Find the extreme fibre stress at mid-span under:
  (i) initial prestress and no live loads, and
  (ii) final condition after application of two live loads.

[Hint: $A = 18 \times 10^4$ mm², $Z = 18 \times 10^6$, $T = 1600$ kN, $T_1 = 1400$ kN,

$$e = \frac{600}{2} - 180 = 120 \text{ mm}, M_d = (0.3 \times 0.6) \frac{25}{8} \times 12.5^2 = 88 \times 10^6 \text{ N-mm},$$

$M_l = 45 \times 4.5 \times 10^6 = 202.5 \times 10^5$ N-mm

$$\textbf{\textit{Case I:}} \text{ Stress} = \frac{1600 \times 10^3}{18 \times 10^4} \mp \frac{1600 \times 120 \times 10^3}{18 \times 10^6} \pm \frac{88 \times 10^6}{18 \times 10^6} = 8.89 \mp 10.67 \pm 4.89$$

$f_{top} = +3.11$ N/mm² , $f_{bottom} = +14.67$ N/mm²

Case II: Stress $= +7.78 + 9.34 \mp 4.89 \pm 11.25$

$f_{top} = + 14.58$ N/mm², $f_{bottom} = + 0.98$ N/mm²]

Q. 12.8 A pretensioned beam of rectangular section 80 mm × 120 mm is to be designed to support concentrated live load of 4 kN each at one third span points over an effective SS span of 3 metres. The permissible stress in concrete are limited to zero and 1.40 N/mm² working tension at transfer and working loads respectively. If 3 mm wires are initially stressed to 1400 N/mm², find the number of wires required and eccentricity of the prestressing force assuming 20% loss in prestress. Concrete weight is 25 kN/m³.

[Hint: $A = 9600$ mm², $Z = 192 \times 10^3$ mm³, M_d (u.d.l.) $= \dfrac{0.24 \times 3^2}{8} = 0.27 \times 10^6$ N-mm,

$M_l = 4 \times 10^6$ N-mm and T_0 is in N, (e) in mm.

$$\text{Initial stress} = \frac{T_0 \cdot e}{192 \times 10^3} \pm \frac{0.27 \times 10^6}{192 \times 10^3} = 0, \text{ at top}$$

or $\qquad 20T_0 - T_0 \cdot e = -270000$ ... Eq. (1)

After working loads $T' = 0.8 \, T_0$

Thus, at bottom edge stress $= +\dfrac{0.80T_0}{9600} + \dfrac{0.8T_0 e}{192 \times 10^3} - \dfrac{0.27 \times 10^6}{192 \times 10^3} - \dfrac{4 \times 10^6}{192 \times 10^3}$

$$= -1.40$$

or $\qquad 16T_0 + 0.8T_0 \cdot e = +270000 + 4000000 - 268800$

or $\qquad 20T_0 + T_0 \cdot e = 5001500$ ... Eq. (2)

Adding (1) and (2) $40T_0 = 4731500$, $T_o = 118.29$ kN, $e = 22.28$ mm

Area of each wire $= \dfrac{\pi}{4}(3)^2 = 7.069$ mm², $A_{st \, (reqd.)} = \dfrac{118290}{1400} = 84.5$ mm²,

$n = 12$ wires.

Q. 12.9 A prestressed concrete T-beam is to be designed to support a superimposed load of 4.4 kN/m over the entire SS span of 5.0 m. The T-beam has a flange of 400 mm × 40 mm thick and the rib of 200 mm × 100 mm thick. The stress in the concrete must not exceed 15 N/mm² in compression and zero in tension at any stage. Check for adequacy of the section provided and determine the minimum prestressing force necessary and the corresponding eccentricity.

Assume 20% losses of pre stress after transfer.

$n = 12$ wires.

[Hint: $A = (0.4 \times 0.04 + 0.2 \times 0.1) = 0.036$ m² $(36 \times 10^3$ mm²$)$, $w_d = 0.9$ kN/m,

$L = 5.0$ m, $w_l = 4.4$ kN/m, $M_d = 2.8125 \times 10^6$ N-mm, $M_l = 13.75 \times 10^6$ N-mm

$y_t = 86.67$ mm, $y_b = 153.33$ mm, $I_{xx} = 1967.8 \times 10^5$ mm⁴,

$Z_t = 22.7 \times 10^5$ mm³, $Z_b = 12.834 \times 10^5$ mm³.

Let T_0 be initial prestressforce (N) and let eccentricity $= e$(mm) below NA

$$\text{Initial stress at transfer} = \frac{T_0}{A} \mp \frac{T_0 \cdot e}{Z} \pm \frac{Md}{Z}$$

$$\text{Stress in top } f_t = \frac{T_0}{36000} - \frac{T_0 e}{22.7 \times 10^5} + \frac{2.8125 \times 10^6}{22.7 \times 10^5} = 0$$

$T_0 - 0.01586 \, T_0 \cdot e + 44604 = 0$ ... Eq. (1)

Stress in bottom $f_b = \dfrac{T_0}{36000} + \dfrac{T_0 \cdot e}{12.834 \times 10^5} - \dfrac{2.8125 \times 10^6}{12.834 \times 10^5} = +15.0$

$T_0 + 0.02805\ T_0 \cdot e - 618892.0 = 0$ ... Eq. (2)

Solving Eqs (1) and (2), we get $T_0 = 195046$ N (195.05 kN), $e = 77.50$ mm

Check for stress after application of working loads and 20% reduction in prestress force. These stresses are within the permissible limits.

Q. 12.10 A prestressed concrete beam is provided with a parabolic tendon as shown in Fig. 12.22. The tendon is tensioned from both ends. If the stress in tendon at the end is 1200 N/mm², compute the loss of prestress from the ends to the centre. $\mu = 0.30$, and $k_f = 0.0015$ per metre.

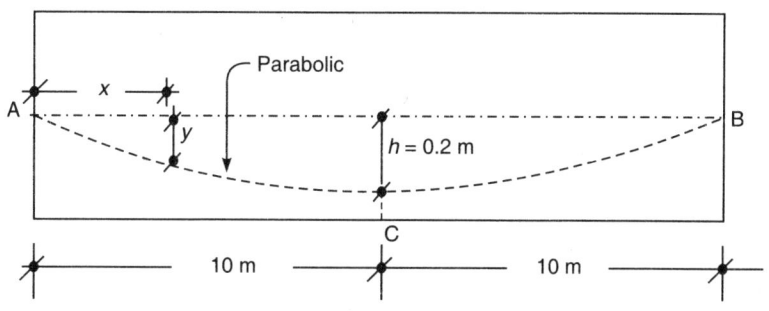

Fig. 12.22

[Hint: $y = \dfrac{4h \cdot x(l-x)}{l^2} = \dfrac{0.8x(20-x)}{20 \times 20}$

$\dfrac{d_y}{d_x}$ (Slope), (when $x = 0$) = 0.04 radians

Angle of deviation A to C, $\alpha = 0.04$ radians

Loss of stress from A to C = $(\mu\alpha + k_f \cdot x)T_i$,

i.e. loss of stress = $(0.30 \times 0.04 + 0.0015 \times 10)1200 = 32.4$ N/mm²

Q. 12.11 Calculate the loss of prestress due to shrinkage of concrete in a post tensioned beam if the wires are tensioned 21 days after the casting. $E_s = 2.1 \times 10^5$ N/mm².

[Hint: Shrinkage strain = $\dfrac{2 \times 10^{-4}}{\log_{10}(T+2)} = \dfrac{2 \times 10^{-4}}{\log_{10} 23} = \dfrac{2 \times 10^{-4}}{1.36173} = 1.47 \times 10^{-4}$

Loss of stress = $1.47 \times 10^{-4} \times 2.1 \times 10^5 = 30.87$ N/mm²]

Q. 12.12 A post-tensioned prestressed concrete beam spanning 12 m and having cross-section 300 mm × 450 mm deep is prestressed with a straight cable having an area of 500 mm² located at a constant eccentricity of 75 mm. The initial stress in the cable is 1000 N/mm². Calculate the percentage loss of stress in the cable with following data:
- relaxation of steel stress = 5% of initial
- shrinkage strain of concrete – 200 × 10⁻⁶ mm/mm
- ultimate creep strain of concrete = 2 × 10⁻⁵ mm/mm per N/mm²
- friction coefficient for wave effect = 0.0015/metre

- slip at anchorage = 1 mm
- modulus of elasticity of steel = $2.1 \times 10^{-5}\,\text{N/mm}^2$
- modulus of elasticity of concrete = $35 \times 10^3\,\text{N/mm}^2$

[**Hint:** $A = 1.35 \times 10^5\,\text{mm}^2$, $I = 22.78 \times 10^8\,\text{mm}^4$, $Z = 10.125 \times 10^6\,\text{mm}^3$

Initial stress in concrete $(f_c) = \dfrac{5 \times 10^5}{1.35 \times 10^5} + \dfrac{5 \times 10^5 \times e^2}{I} = 4.94\,\text{N/mm}^2$

Losses due to various causes:

(i) Elastic shortening = mf = $29.63\,\text{N/mm}^2$

(ii) Steel relaxation = $50\,\text{N/mm}^2$

(iii) Shrinkage in concrete = $200 \times 10^{-6} \times 2.1 \times 10^5 = 42\,\text{N/mm}^2$

(iv) Creep in concrete = creep strain $\times E_s = 20.75\,\text{N/mm}^2$

(v) Friction of cable = $f_0 \times 0.0015 \times 12 = 18\,\text{N/mm}^2$

(vi) Slip of anchorage = $\dfrac{1 \times 2.1 \times 10^5}{12000} = 17.5\,\text{N/mm}^2$, total loss = $\dfrac{177.88 \times 100}{1000}$

= 17.8%]

Q. 12.13 Design a prestressed concrete beam for the following requirements: Span = 15 m, simply supported, superimposed load 30 kN/m(u.d.l.), (f_{cu}) concrete cube strength 35 N/mm² (M35), safe stress in concrete $f_r = 0.5 f_{cu}$ at 28 days, safe stress in concrete due to final prestress (transfer) = $0.4f_{cu}$, total loss of prestress = 20%, allowable tensile stress in concrete = 1.0 N/mm², ultimate stress in steel = 1500 N/mm², safe stress in steel = 60% of ultimate stress.

[**Hint:** Permissible stress (concrete) = +17.5 N/mm², $f_c = 14.0$ N/mm², $f_t = 1.0$ N/mm² Permissible tensile stress (steel) = 900 N/mm²,

Max BM (LL)$M_l = \dfrac{30 \times 15^2}{8} = 843.75\,\text{kN-m}$

Assume self-wt. = 20% of LL, $M_d = 168.75 \times 10^6$ N-mm, $M_l = 843.7 \times 10^6$ N-mm.

Total BM = $(843.75 + 168.75)10^6 = 1012.5 \times 10^6$ N-mm

Assume overall depth = 1200 mm.

Prestressing force $T_0 = \dfrac{M_t}{0.65h} = \dfrac{1012.5 \times 10^6}{0.65 \times 1200} = 1298077\,\text{N}\ (1298.07)\,\text{kN}$

Area of concrete required = $\dfrac{1298077}{0.5 f_c} = \dfrac{1298077}{0.5 \times 14} = 185440\,\text{mm}^2$

Assume I–section: top and bottom flange 500 mm × 120 mm thick and connecting web = 960 mm × 120 mm thick.

Area of assumed section

$A = (500 \times 120)2 + (960 \times 120) = 235200\,\text{mm}^2 > 185440\,\text{mm}^2$, OK.

$I = \dfrac{1}{12}(500 \times 1200^3 - 380 \times 960^3) = 10^6(72000 - 28016)$

$I = 10^6 (43984)$ mm^4

Section modulus $Z = 73.306 \times 10^6$ mm^3

Actual self-wt. $= 0.2352 \times 1 \times 25 = 5.88$ kN/m

DL $M_d = \dfrac{5.88}{8} \times 15^2 = 165.375$ kN-m

$(165.375 \times 10^6$ N-mm$) < 168.75 \times 10^6$ N-mm

Area of steel required $= \dfrac{T_0}{900} = \dfrac{1298077}{900} = 1442.3$ mm^2

Number of wires $= \dfrac{1442.3}{28.30} = 51$ wires. Using 6 mm wires (28.3 mm^2),

Provide 6 cables each containing 10 wires of 6 mm diameter (60) wires
Provide 3 cables in flange and 3 cables in web as shown.

Final prestressing force $= 60 \times 28.3 \times 900 = 1528200$ N (1528.2 kN).

Therefore, initial prestress required (20% loss) $= \dfrac{1528.2}{(1-0.2)} = 1910.25$ kN

CG of cables from the CG of bottom row $= \dfrac{1}{6}(1 \times 120 + 240 \times 1 + 360 \times 1) =$

$\dfrac{720}{6} = 120$ mm, i.e. from bottom $60 + 120 = 180$ mm.

Eccentricity of prestress force $= \dfrac{1200}{2} - 180 = 420$ mm (e)

Check of stress in concrete

Case I: Initial stage

Stress at top: $\dfrac{1910250}{235200} - \dfrac{1910250 \times 420}{73.306 \times 10^6} + \dfrac{165.3 \times 10^6}{73.30 \times 10^6} = 8.122 - 10.945 + 2.257$

$f_t = -0.568$ N/mm^2 $(< -1.0$ N/mm^2), OK

Stress at bottom $f_b = \dfrac{1910250}{235200} + \dfrac{1910250 \times 420}{73.306 \times 10^6} - \dfrac{165.3 \times 10^6}{73.30 \times 10^6} = 8.122 + 10.945$

$- 2.257 = +16.712$ N/mm^2 $(< +17.5$ N/mm^2), OK.

Case II: Final stage stress

Stress top (20% loss) $= 0.8(8.122) - 0.8(10.945) + 2.255 + \dfrac{843.75 \times 10^6}{73.30 \times 10^6}$

$= 6.50 - 8.76 + 2.255 + 11.51 = +11.505$ N/mm^2 $(<17.5$ N/mm^2, OK.)

Stress in bottom $= 6.50 + 8.76 - 2.255 - 11.51 = +1.50$ N/mm^2, OK.

Beam section is safe at initial transfer and final stage after working stress stage (Fig. 12.23).

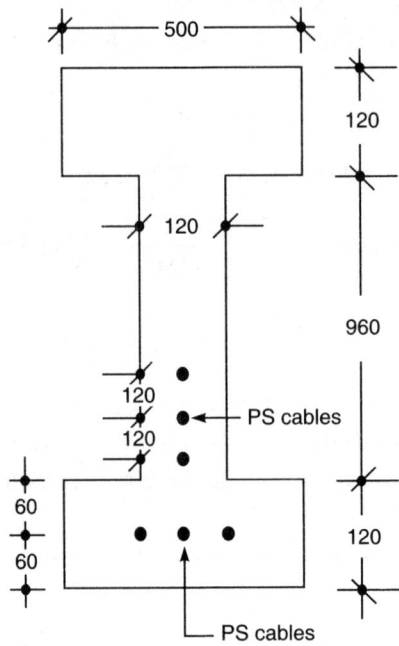

Fig. 12.23

Chapter 13

Design of RCC Staircases

LEARNING OBJECTIVES

After the study of *design of RCC staircases*, the learner will understand the design of different types of staircases and shall be able to:

- ⊙ Explain different types of RCC staircases and specific design considerations
- ⊙ State specific loading pattern on different type of RCC staircases
- ⊙ Determine the maximum bending moment and shear forces in different type of staircases
- ⊙ Design different type of RCC staircases.

13.1 INTRODUCTION

All the buildings are provided with staircases for access to different floors. A flight is the length of staircase between the two consecutive landings. The staircase comprises flight of steps generally with one or more intermediate landings provided between the floor levels. The structural components of a flight of steps consists of:

(a) *Tread* which formsthe horizontal portion of the step. The tread is usually 250 mm to 300 mm depending upon the type of building.

(b) *Riser* is the vertical distance between the adjacent treads (or vertical projection or portion of the step). Rise generally varies in the range of 150 mm to 200 mm depending upon the type of building.

(c) The *width* of stairs vary in the range of 1000 mm to 2000 mm with a minimum value of 850 mm. Generally public buildings shall be provided with larger widths of 2 m for free flow of public.

(d) *Going* is the horizontal plan projection of an inclined flight of steps between the first and the last riser between the consecutive landings of a flight.

Staircases are classified into following types:

- Dog-legged
- Open well staircase
- Tread–riser without any waist slab

- Isolated cantilever comprising of only the horizontal tread slab projecting from the wall or inclined beam serving as a fixed end with open risers.
- Double cantilever precast tread slab projecting on either side of a central inclined beam.
- Geometrical and spiral, etc.

13.2 DESIGN OF STAIRCASE

The design of staircase involves two aspects:
- (i) Proportioning of its different components (geometrical design).
- (ii) Design of section and reinforcement detailing to satisfy both the serviceability and strength requirements (structural design).

Geometrical proportioning of staircase depends on the available room space and common proportioning practices and empirical rules.

The width of staircase depends on the type of building such as residential, public building, offices, commercial, etc.

- Residential building

  Generally width is kept 1 m to 1.20 m (minimum 0.85 m)
- Public building, offices, etc.

  Generally width is kept 1.20 m to 2.0 m

Each flight has 3 to 12 steps in general. The tread of steps is usually kept 200 mm to 300 mm and rise 150 mm to 200 mm. Following guidelines may be followed to decide tread and rise:

- (rise × tread) = 40,000 mm^2 to 42,000 mm^2
- (tread + 2 rise) = 600 mm

The types and width of staircase are decided on the available room space for staircase, type of building and other service considerations. We shall discuss the structural design for different type of staircases.

13.3 LOADS ON STAIRCASE

The various types of loads acting on the staircase are grouped under dead and live loads.

- Dead loads including self-weight of the stair slab (waist slab), tread and risers and weight of finishes, hand rail etc.
- Live loads are to be considered as specified in IS:875-1987 (Part II).

For residential buildings, a uniformly distributed live load of 2 to 3 kN/m^2 depending on the users, and for public buildings, a uniformly distributed live load of 5 kN/m^2 is specified in the IS code. The loads are placed in such a critical position so that the moments and shear forces are most critical. In case of open well staircase, where the flights are at right angles, the load in each direction for common landing area is considered half of the actual load (IS:456-2000, Clause 33.2).

13.4 EFFECTIVE SPAN OF STAIRS

The effective length of the staircase flight depends on its support conditions. When the stair flight is supported at ends by landing beams, the effective span is the projected horizontal distance between the centre lines of landing beams. The effective span of stairs without stringer beam shall be taken as the following horizontal distances as per IS:456-2000, Clause 33.1:

(a) Where supported at top and bottom risers by beams spanning parallel with the risers, the distance centre to centre of beams;

(b) Where spanning on the edge of a landing slab, which spans parallel with the risers (Fig. 13.1), a distance equal to the going of the stairs plus at each end either half the width of the landing or one metre, whichever is smaller, and

(c) Where the landing slab spans in the same direction as the stairs, they shall be considered as acting together to form a single slab and the span determined as the distance centre to centre of the supporting beams or walls, the going being measured horizontally.

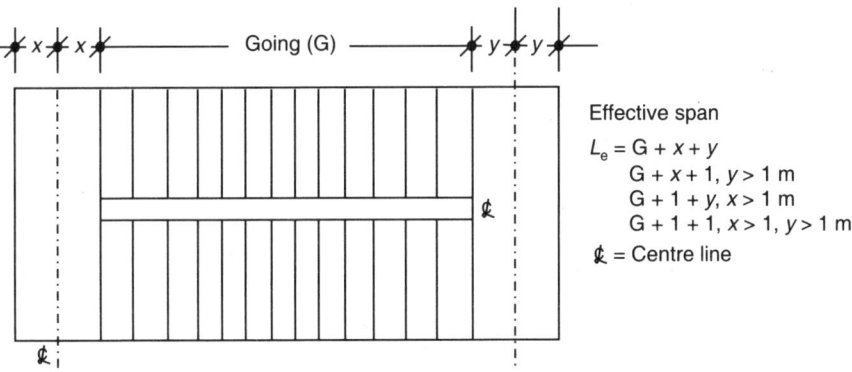

Fig. 13.1: Dog-legged staircase

Example 13.1

Design of dog-legged staircase with the steps on waist slab for floor to floor height of 3.30 m and width of flight equal to 1.20 m and carrying a superimposed load of 3 kN/m². Flight has the following support conditions:

(i) In the direction of flight

(ii) In the transverse direction

Use M20 grade CC and steel grade Fe 415, weight of finishing 0.50 kN/m and weight of concrete 25 kN/m³.

Solution: Assume tread = 250 mm, rise = $\dfrac{(600-250)}{2}$ = 175 mm,

$$\text{No. of rises} = \frac{3300}{175} = 19 \text{ steps, also rise} = \frac{40000 \text{ to } 42000}{250} = 160 \text{ to } 168 \text{ mm}$$

$$\text{No. of steps (rise)} = \frac{3300}{160} \text{ to } \frac{3300}{168} = 20.6 \text{ to } 19.6$$

Thus, provide 20 (steps); actual rise adopted $= \dfrac{3300}{20} = 165$ mm, tread $= 250$ mm,

$$\text{Steps (tread)} = \frac{20}{2} - 1 = 9 \text{ in each flight.}$$

In dog-legged type of staircase: each flight going $= 9 \times 250 = 2250$ mm (Figs 13.2 and 13.3).

Width of stair hall $= 2 \times 1200 + 100$, if landing is supported in the transverse direction $= 2500$ mm

- Effective span $L_e = 2250.0 + \dfrac{1200}{2} + \dfrac{1200}{2} = 3450$ mm, if landing is supported in the transverse direction of flight
- $(L_e) = 4650$ mm $= 4.650$ m (SS) if landing is supported in the direction of flight.
- Consider more critical case, $L_e = 4.65$ m.

$$\text{Thickness of waist slab} = \frac{L_e}{20} = \frac{4650}{20} = 235 \text{ mm}$$

$D = 235$ mm $(d = 210$ mm$)$

Loads (consider 1.0 m width)

$$\text{Weight of step (triangular)} = \frac{0.250 \times 0.165 \times 1 \times 25}{2 \times 0.250} = 2.0625 \text{ kN/m}$$

Weight of waist slab on 1 m slope $(w_s) = 0.235 \times 1 \times 25 = 5.875$ kN/m.

$$\text{Weight of slab on horizontal span } w = w_s \frac{\sqrt{R^2 + T^2}}{T} = \frac{5.875}{0.25} (0.2995) = 7.04 \text{ kN/m}$$

Finishing $= 0.5$ kN/m

Total DL $= 2.0625 + 7.04 + 0.5 = 9.602$ kN/m, say 9.6 kN/m

LL $w_1 = 3$ kN/m$^2 = 3$ kN/m (for 1 m width)

Total service load $w = 9.6 + 3 = 12.6$ kN/m

Factored load $w_u = 1.5 \times 12.6 = 18.9$ N $\simeq 19$ kN/m

$$\text{BM} = \frac{w_u}{8} \times L^2 = \frac{19 \times 4.65^2}{8} = 51.35 \text{ kN-m } (51.35 \times 10^6 \text{ N-mm})$$

$$\text{Depth required for BM } (d) = \sqrt{\frac{M_u}{0.138 f_{ck} \cdot b}}$$

$$= \sqrt{\frac{51.35 \times 10^6}{0.138 \times 20 \times 1000}} = 136.4 \text{ mm } (< \text{ actual 210 mm), OK}$$

Main reinforcement $(b = 1000$ mm, $d = 210$ mm, $f_{ck} = 20$ N/mm^2, $f_y = 415$ N/mm^2).

$$A_{st} = \frac{0.5 \times f_{ck} \cdot b \cdot d}{f_y} \left[1 - \sqrt{1 - \frac{4.6 M_u}{f_{ck} \cdot b \left(d^2\right)}} \right] = 731 \text{ mm}^2/\text{m}$$

or 12ϕ 150 c/c, (i.e. in 1.2 m of 8 bars.)

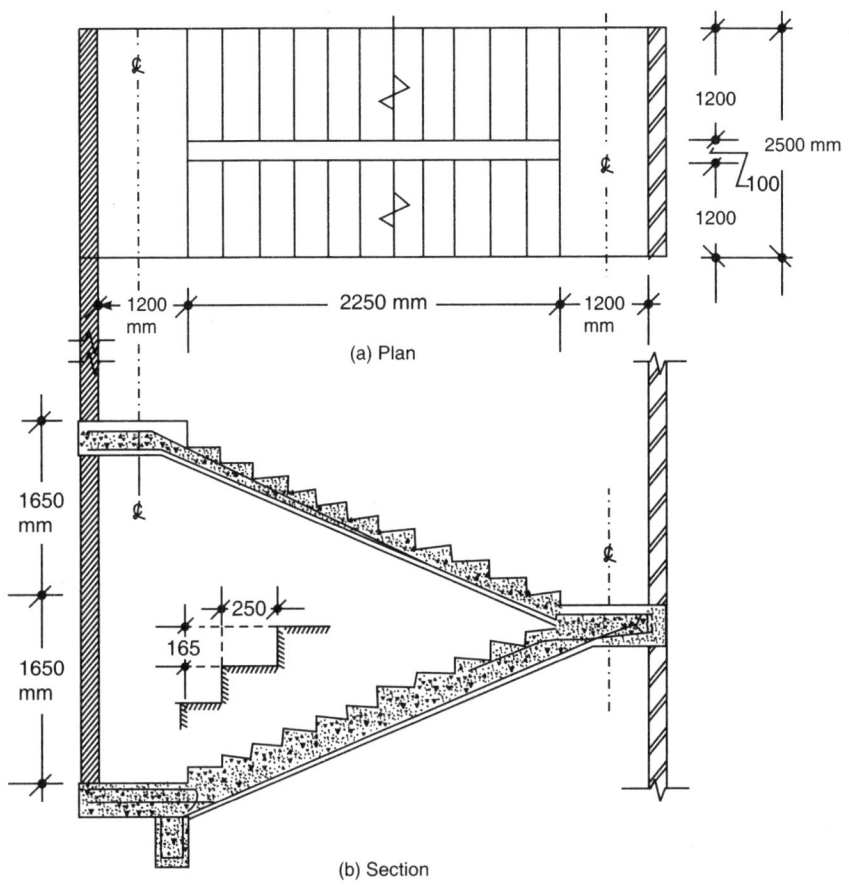

Fig. 13.2: Dog-legged staircase ($R = 165$ mm, $T = 250$ mm, $N = 10 \times 2 = 20$ rises)

Distribution Reinforcement

$$A_{st} = \frac{0.12}{100} \times 1000 \times 235 = 282 \text{ mm}^2/\text{m, provide 8 mm } \phi \text{ 170 c/c } (A_{st} = 294 \text{ mm}^2/\text{m})$$

Design as per SP-16 design chart

$$\frac{M_u}{bd^2} = \frac{51.35 \times 10^6}{1000 \times 210 \times 210} = 1.164, f_{ck} = 20, f_y = 415 \text{ N/mm}^2$$

$$p_t \text{ (main steel)} = 0.348\%, d = 210 \text{ mm}$$

$$\text{(Chart 2), } A_{st} = \frac{0.348}{100} \times 1000 \times 210 = 731 \text{ mm}^2/\text{m}$$

$(12\phi \text{ 150 c/c, } A_{st} = 753 \text{ mm}^2/\text{m})$. In 1.20 m, number of bars = 8
Same as obtained by analytical method.

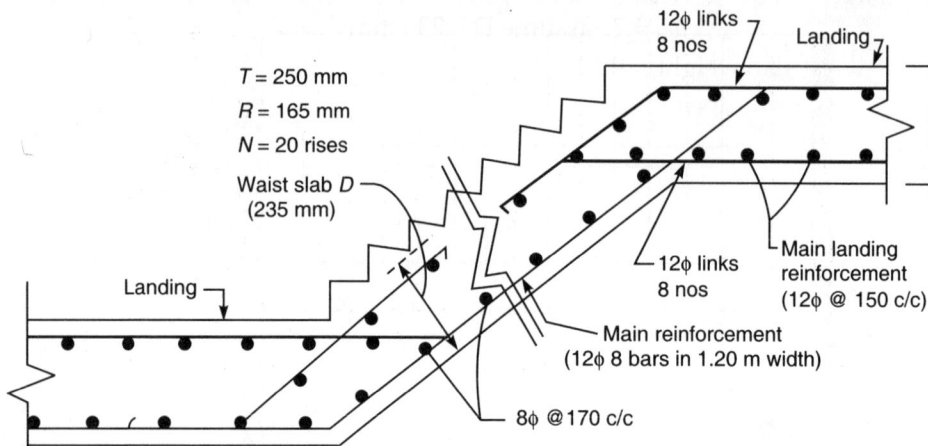

Fig. 13.3: Details of reinforcement of one flight of dog-legged staircase

Example 13.2

A staircase hall 4.5 m 4.5 m has a staircase as shown in Fig.13.4(a). The stair flight 1-1 is supported on the RCC beams and walls. Each flight has 6 steps having tread of 250 mm and rise of 165 mm. Floor to floor height is 3465 mm. The front side is level and the staircase floor is subjected to superimposed load of 5 kN/m² excluding self-weight. Take M20 CC and Fe 415 grade of steel reinforcement. Unit weight of concrete = 25 kN/m³.

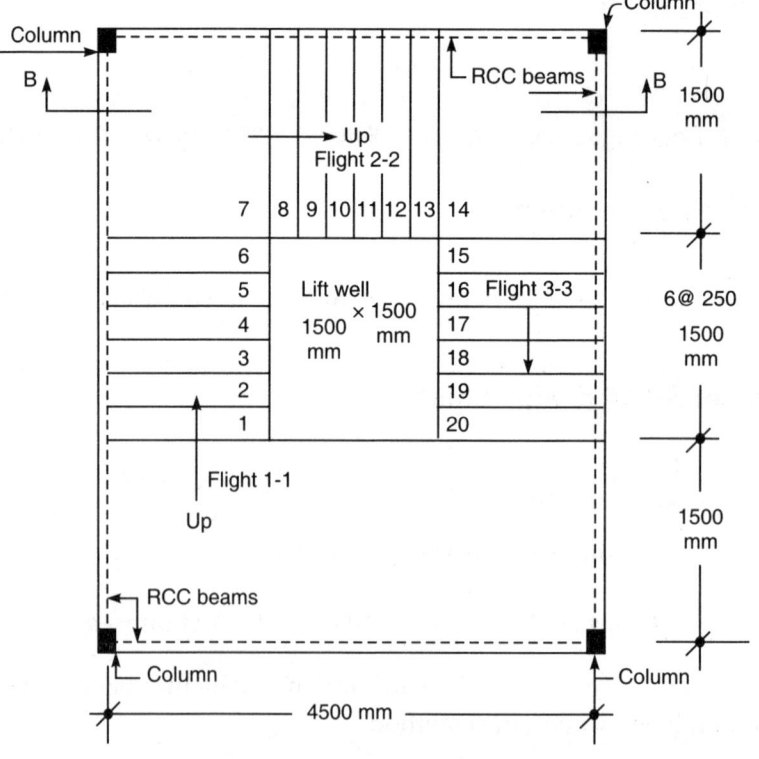

Fig. 13.4 (a): Stair hall (plan)

Solution: $f_{ck} = 20 \text{ N/mm}^2$, $f_y = 415 \text{ N/mm}^2$, $w_1 = 5 \text{ kN/m}^2$, assume $b = 1.0 \text{ m}$,

$$w_d = ?, \text{ assume } D = 230 \text{ mm},$$

waist slab weight $= 0.23 \times 25 \times 1$

on slope $w_s = 5.75 \text{ kN/m}$

Weight of finishing $= 1.25 \text{ kN/m}$ (assumed)

Weight of step $= 0.5 \times 0.25 \times 0.165 \times 25 = 0.516 \text{ kN}$

Weight of step/m $= \dfrac{0.516}{0.25} = 2.063 \text{ kN/m}$

Weight of slab/m horizontal span $w = w_s \left(\dfrac{\sqrt{R^2 + T^2}}{T} \right)$

or $\qquad\qquad w = \dfrac{5.75}{0.25} \times 0.29954 = 6.89 \text{ kN/m}$ (horizontal)

Total DL along the sloping portion

$$w_d = 6.89 + 2.063 + 1.25 = 10.2 \text{ kN/m}$$

w_{d_2} (landing) $= 5.75 + 1.25 + 0 = 7.0 \text{ kN/m}$ (no steps on landing)

Half the load on landing in each perpendicular direction
(i.e. 3.5 kN/m in each direction).

Effective span L_e, if simply supported at ends $= (4.50 + d) = 4.5 + 0.20 = 4.70 \text{ m}$
(beam to beam).

If fixed with RCC beams at lower end and upper end with RCC beams or walls,
span L (fixed) $= 4.50 \text{ m}$

Case I: Simply supported span 4.70 m Fig. 13.4 (b)

w_d (inclined portion) $= 10.2 \text{ kN/m}$, w_{d_2}(landing) $= 3.5 \text{ kN/m}$, $w_1 = 5.0 \text{ kN/m}$,
ultimate loading $w_{ud_1} = 15.3 \text{ kN/m}$, $w_{ud_2} = 5.25 \text{ kN/m}$, $w_{ul} = 7.5 \text{ kN/m}$

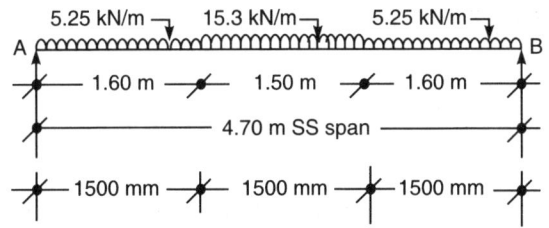

Fig. 13.4 (b): Loading on one flight

Analysis: $R_A = 19.875 \text{ kN}$, $F_{max} = 19.875 \text{ kN}$

$$M_{max}(\text{centre}) = 19.875 \times 2.35 - 5.25 \times 1.6 \,(1.55) - 15.3 \times 0.75 \times \dfrac{0.75}{2}$$

$$= (29.383 \text{ kN-m})$$

$$M_u = +29.383 \times 10^6 \text{ N-mm}, \ V_u = 19875 \text{ N}, \ b = 1000 \text{ mm}$$

$$d_{reqd.} = \sqrt{\dfrac{M_u}{0.138 f_{ck} \cdot b}} = \sqrt{\dfrac{29.383 \times 10^6}{0.138 \times 1000 \times 20}} = 103.2 \text{ mm} < 200 \text{ mm (assumed)}$$

Main reinforcement

$$A_{st} = \frac{0.5 f_{ck}.b.d}{f_y}\left[1 - \sqrt{1 - \frac{4.6 M_u}{f_{ck}.b.d^2}}\right] = 426 \text{ mm}^2/\text{m}$$

(12ϕ 250 c/c longitudinal main reinforcement).
No. of bars in 1.50 m = 6 bars [(Fig. 13.4(c)]

$$p_t = \frac{452 \times 100}{1000 \times 200} = 0.23\%$$

Shear check

$$\tau_v = \frac{19875 \text{ N}}{1000 \times 200} = 0.0994 \text{ N}/\text{mm}^2$$

$(< \tau_c = 0.345 \text{ N}/\text{mm}^2 \text{ for } p_t \text{ of } 0.23\%)$

Deflection

$$\frac{L_e}{d} = \frac{4700}{200} = 23.5 \text{ (for simply supported } \frac{l}{d} = 20 \times 1.85 \times 1 \times 1 \text{ for slabs)} = 37.0, \text{OK.}$$

Temp steel $= \dfrac{0.12}{100} \times 230 \times 1000 = 276 \text{ mm}^2/\text{m}$ (8ϕ 180 c/c, $A_{st} = 278 \text{ mm}^2/\text{m}$)

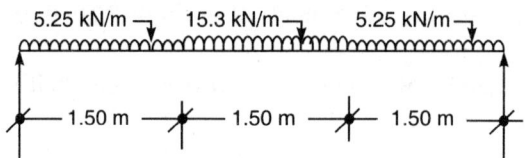

Fig. 13.4(c): Loading

Case II: Fixed slabs, $L_e = 4.50$ m.
Total u.d.l. = 5.25 × 1.5 + 15.3 × 1.5 + 5.25 × 1.5
$W_1 = 38.7$ kN

$$\text{BM}_{\text{max.}} \text{ (ends)} = -\frac{WL}{12} = \frac{38.7 \times 4.5}{12} = -14.51 \text{ kN-m } (14.51 \times 10^6 \text{ N-mm})$$

$$\text{BM}_{\text{max.}} \text{ (mid span)} = +\frac{WL}{16} = +10.9 \text{ kN-m } (10.9 \times 10^6 \text{ N-mm})$$

Depth is quite safe. Find A_{st} corresponding to both BMS.

$$A_{st} \text{ (ends)} = \frac{10 \times 1000 \times 200}{415}\left[1 - \sqrt{1 - \frac{4.6 \times 14.5 \times 10^6}{20 \times 1000\,(200)^2}}\right] = 198 \text{ mm}^2$$

$$\text{Minimum temp steel} = \frac{0.12 \times 1000 \times 230}{100} = 276 \text{ mm}^2/\text{m}$$

(8ϕ @ 150 c/c, $A_{st} = 333 \text{ mm}^2/\text{m}$), OK.

Thus, provide 8ϕ @ 150 c/c at ends in top face and 8ϕ 150 c/c in bottom at mid span section. Also provide 8ϕ @ 150 c/c distribution bars laterally (Fig. 13.5).

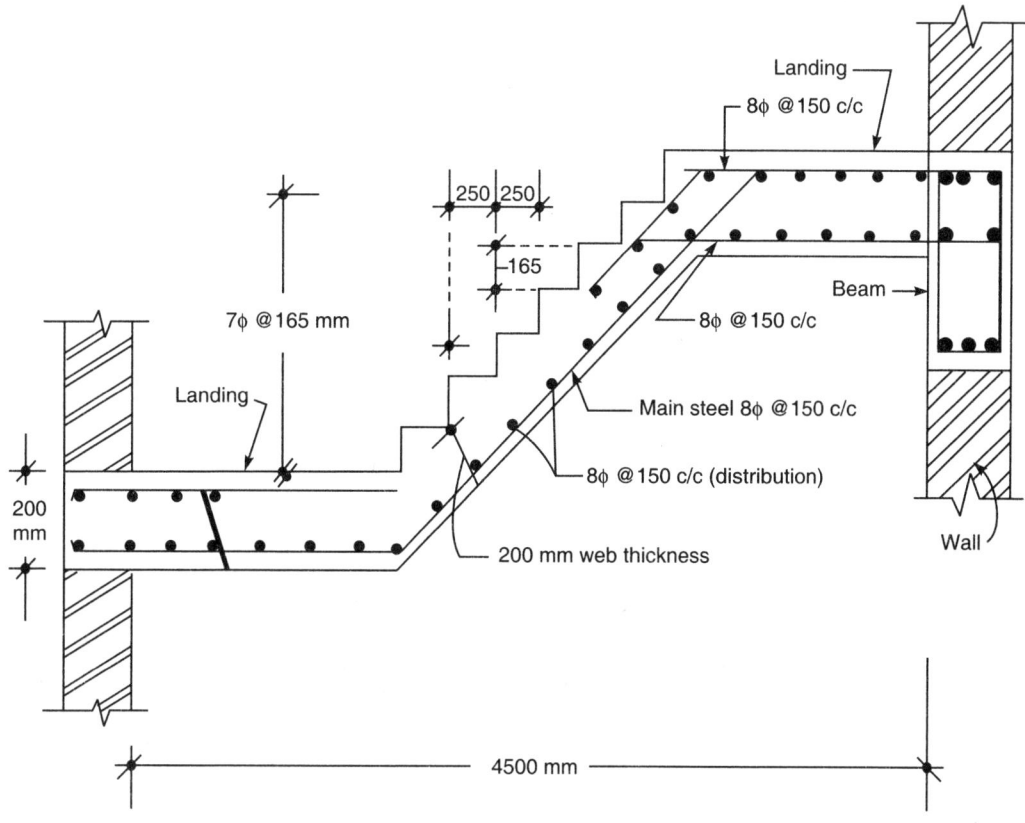

Fig. 13.5: Details of reinforcement (section BB)

Example 13.3

Design the waist slab B′ C′ and stringer beam B′ C′ for a staircase shown in Fig. 13.6. The stringer beam is supported at ends as shown on columns of 300 mm × 300 mm at ends. The staircase is 1.20 m wide and steps have rise 160 mm and tread 250 mm. Used M20 grade of CC and Fe 415 grade of steel reinforcement. The width of beam is 300 mm. The slab is supported on the central stringer beam. The staircase slab may be required to support a live load of 5 kN/m².

Solution: $R = 160$ mm, $T = 250$ mm, $B = 1.20$ m (width)

$f_{ck} = 20$ N/mm², $f_y = 415$ N/mm², assume $D_s = 150$ mm and $d = 120$ mm.

$$\text{weight of step/ m horizontal length} = 1.2 \times 25 \times 0.15 \times \sqrt{\frac{0.25^2 + 0.16^2}{0.25^2}}$$

$$= 4.5 \times 1.1873 = 5.343 \text{ kN/m}$$

$$\text{weight of steps/m} = \frac{1}{2} \times 25 \times 1.2 \times \frac{0.25 \times 0.16}{0.25} = 2.40 \text{ kN/m}$$

Weight of finishing $= 0.05 \times 1.2 \times 1 \times 25 = 1.5$ kN/m

Beam below slab along the slope/m $(300 \times 200) = 0.3 \times 0.2 \times 1 \times 25 = 1.5$ kN/m

Beam along the slope/m horizontal projection $= 1.5 \times 1.1873 = 1.781$ kN/m

Total DL along the horizontal and along the slope

W_{d1} (landing) = 4.5 + 1.5 + 1.5 = 7.5 kN/m

W_{d2} (sloping) = 5.343 + 2.4 + 1.5 + 1.78 = 11.023 kN/m

Live load (w_l) = 5 kN/m^2 × 1.20 = 6.0 kN/m

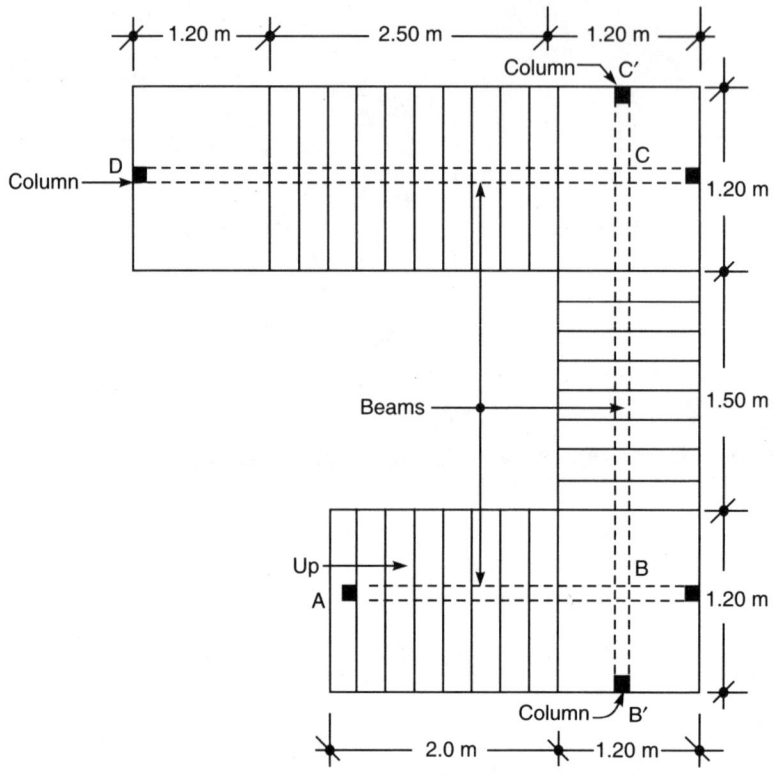

Fig. 13.6: Staircase plan

Ultimate Loading

$$w_{u1} = 1.50\ (7.5 + 6.0) = 20.25\ \text{kN/m (landing)}$$
$$w_{u2} = 1.50\ (11.023 + 6.0) = 25.54\ \text{kN/m (sloping)}$$

Since the landing portion is supported on beams in two directions, half load may be considered in each direction. Beam AB rests at B and beam CD rests at C, these reactions will be as point load (Fig. 13.7).

Consider beam CD, span = $0.15 + 1.2 + 2.5 + \dfrac{1.2}{2} = 4.45$ m

Total load = 1.35 × 20.25 + 2.5 × 25.54 + 0.6 × 20.25 = 103.34 kN

Reaction approx. = $\dfrac{1}{2}$ total load = $\dfrac{103.34}{2}$ = 51.67 kN

Slab is also supported on stringer beam as cantilever on either side.

Projection = $\dfrac{1}{2}$ (1.20 – 0.30)= 0.45 m.

Reaction at B = $\dfrac{1}{2}$ (2 × 25.54 + 0.6 × 20.25) = 31.62 kN as SS beam AB on the beam BC at B.

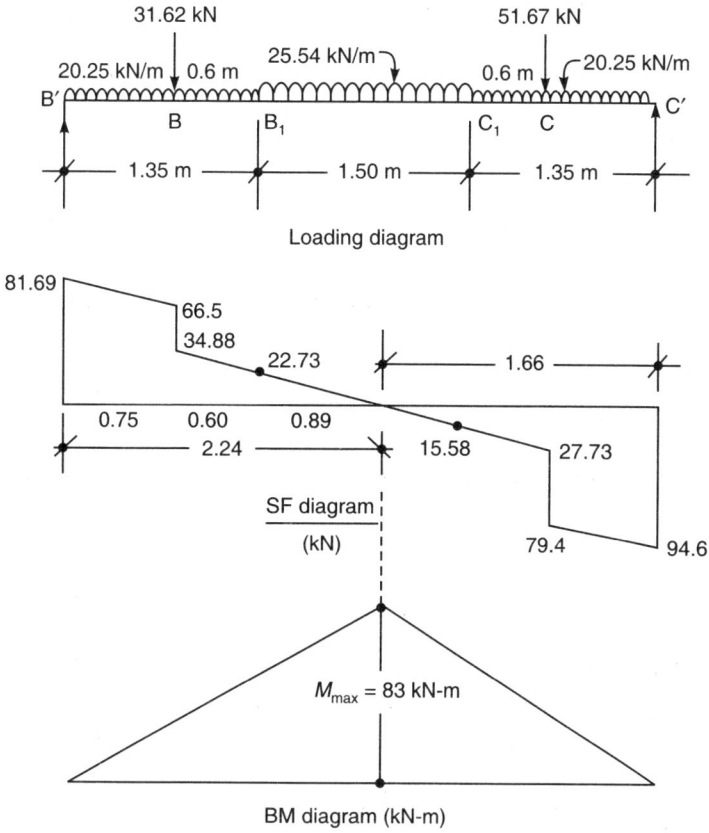

Fig. 13.7: Loading, SFD and BMD for beam B'BCC'

Reactions of beam B'C' (L_e = 4.20 m c/c of columns) (consider Fig. 13.7 for loading)

$$R_B' = \frac{1}{4.2}\left[1.35 \times 20.25 \times \frac{1.35}{2} + 51.67 \times 0.75 + 1.5 \times 25.54 \times (1.35 + 0.75) + 31.62 \times \right.$$

$$(0.6 + 2.85) + 20.25 \times 1.35\left(2.85 + \frac{1.35}{2}\right)$$

$$= \frac{1}{4.2}[18.45 + 38.75 + 80.45 + 109.1 + 96.365] = 81.70 \text{ kN}$$

$$R_C' = \frac{1}{4.2}[31.62 \times 0.75 + 20.25 \times 1.35 \times \frac{1.35}{2} + 1.5 \times 25.54 \times 2.1$$

$$+ 20.25 \times 1.35 \times 3.525 + 51.67 \times 3.45]$$

$$= \frac{1}{4.2}[23.72 + 18.45 + 80.45 + 96.37 + 178.26] = \frac{397.25}{4.2} = 94.60 \text{ kN}$$

Max. SF at C'(V_C') = 94.60 kN

SF at C = 94.60 – 20.25 × 0.75 = 79.4 kN(V_{uc})

SF zero at x = 94.60 – 20.25 × 1.35 – 51.67 – 25.54 × x = 0,

x = 0.61 from C_1 or x = 0.61+1.35 = 1.96 m from C_1

BM will be maximum at 1.96 m from C'

$$M_{max} = 94.60 \times 1.96 - 51.67 \times 1.21 - 20.25 \times 1.35 \times \left(0.61 + \frac{1.35}{2}\right) \times \frac{0.61^2}{2}$$

$$= 185.42 - 62.52 - 35.13 - 4.75 = 83.00 \text{ kN-m/m}$$

Design of beam B'BCC' (as T-beam) (Fig. 13.8)
$M_{max} = 83.00$ kN-m, $V_u = 94.6$ kN, $b = 300$ mm.

$$d \text{ (rectangular)} = \sqrt{\frac{83 \times 10^6}{300 \times 2.766}} = 317 \text{ mm (provide } D = 360 \text{ mm, } d = 320 \text{ mm)}$$

Design of flange slab

Projection = 0.45 m, load $w_u = 25.54$ kN/m,

$$V_u = 25.54 \times 0.45 \text{ kN} = 11.5 \text{ kN/m}, \quad M_u = 25.54 \times \frac{0.452}{2} = 2.59 \text{ kN-m}$$

$$\text{Eff. depth } (d_{reqd}) = \sqrt{\frac{2.59 \times 10^6}{1000 \times 2.766}} = 31 \text{ mm, provide } D = 100 \text{ mm, } d = 70 \text{ mm}$$

Minimum temp. steel of 8ϕ mm @ 300 c/c near top face with step reinforcement.
$$d(\text{web}) = 360 - 40 = 320 \text{ mm.}$$
Thus, depth of web below stair slab will be $360 - 100 = 260$ mm

$$A_{st} \text{ (main in beam)} = \frac{0.5 \times 20 \times 300 \times 320}{415} \left[1 - \sqrt{1 - \frac{4.6 \times 83 \times 10^6}{20 \times 300 (320)^2}}\right] = 890 \text{ mm}^2$$

Provide 3 bars of 20 mm diameter in beam near bottom face
($A_{st} = 942$ mm²), $p_t = 0.29\%$ steel, $\tau_c = 0.38$ N/mm²

$$\text{Nominal shear stress} = \frac{94600}{300 \times 320} = 0.99 \text{ N/mm}^2 \, (> 0.38 \text{ N/mm}^2 \text{ but } < 2.8 \text{ N/mm}^2)$$

$$\text{Provide 8}\phi \text{ 2 legged vertical stirrups @ } S_v = \frac{0.87 \times 415 \times (2 \times 50) \times 320}{94600 - 0.38 \times 300 \times 320} = 198 \text{ mm c/c}$$

Provide 8ϕ 2 legged vertical stirrups @ 150 c/c for shear resistance of stair beam.

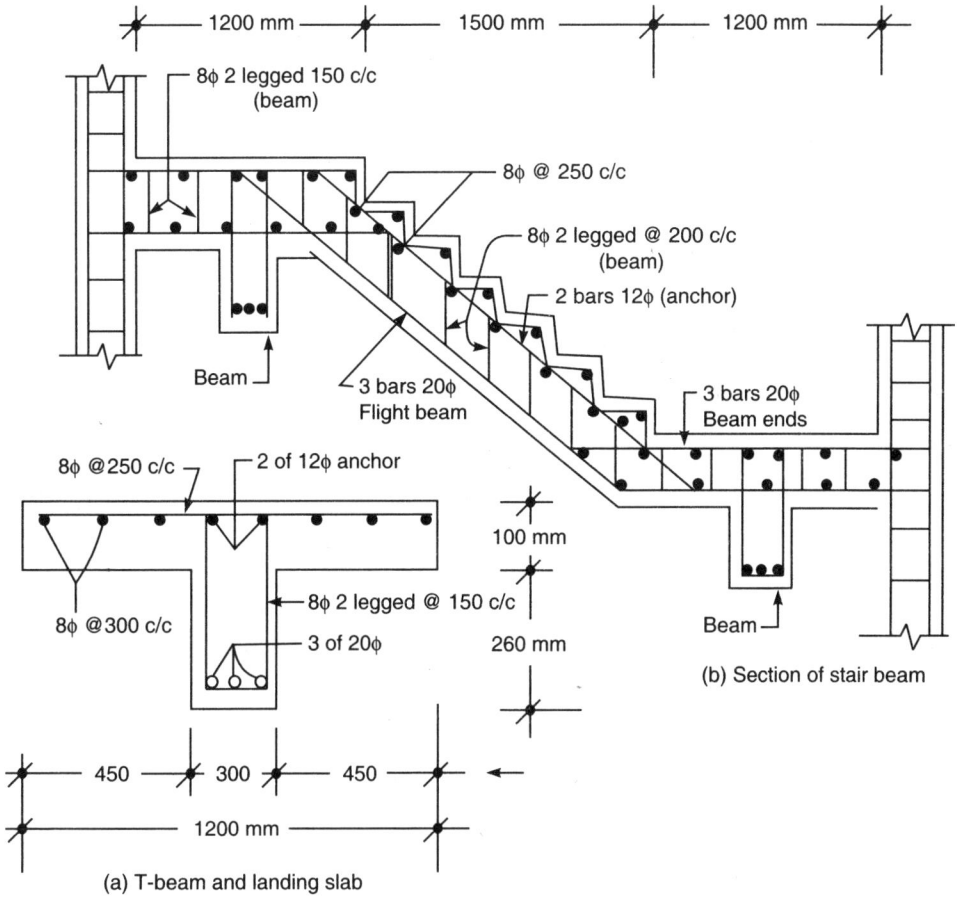

Fig 13.8: Details of reinforcement in beam and slab

SUMMARY

Staircases are required to provide access to different floors in a building. The staircase may be provided with simply supported or cantilever slab or beam. Staircase may be dog-legged, open well, isolated cantilever, double cantilever with central beam and geometrical (spiral type) etc.

Stair width ranges from 1 m to 1.20 m for residential and 1.2 m to 2.0 m for public buildings. Generally, steps have rise 150 mm to 200 mm and tread 200 mm to 300 mm.

A stair flight is subjected to different type of distributed and concentrated loads. Staircase slab or beam basically acts as a flexural member and it shall be designed for maximum BM and SF under worst placement of loads. Worst type of loads shall be adopted from IS:875 latest to compute maximum design moments and shear forces. A live load of 3 kN/m² for residential and 5 kN/m² for public buildings shall be considered.

Effective span of flight may be considered according to support conditions. If the flight slab is connected monolithically to perpendicular landing slabs, the effective span shall be from centre to centre of supporting landings. Details of effective spans shall be considered in accordance with IS:456-2000. Effective span, loads and placement of loads, etc. shall be considered in accordance to support conditions which give rise to most critical bending moments and shear forces.

PRACTICE QUESTIONS

(I) Objective Questions

Q. 13.1 Select the correct response given after each statement to complete it correctly and fill in the response sheet provided.

   (i) Flight in a staircase is defined as the distance between ....
     (a) two consecutive floors of a building
     (b) two consecutive steps in a staircase
     (c) two consecutive treads in a staircase
     (d) two consecutive landing in a staircase.

  (ii) Tread in a staircase forms the ....
     (a) vertical distance between the adjacent treads of step
     (b) horizontal distance between the adjacent risers of step
     (c) vertical distance of 150 mm to 200 mm
     (d) longitudinal distance of flight

 (iii) What is the range of number of steps generally in each flight?
     (a) 0 to 3   (b) 3 to 8
     (c) 3 to 12   (d) 8 to 20

  (iv) What is the general live load on the stair slab for public buildings?
     (a) $10 \, kN/m^2$ (b) $8 \, kN/m^2$
     (c) $5 \, kN/m^2$ (d) $1.25 \, kN/m^2$

   (v) What will be the effective span of each flight of 2.50 m horizontal length in case of dog-legged staircase if connected to transverse landings of 1.40 m width on each end?
     (a) 3.90 m   (b) 4.50 m
     (c) 4.90 m   (d) 5.30 m

  (vi) What will be dead weight of concrete waist slab of 1.20 m wide staircase per metre horizontal length along the inclined flight with thickness D and steps of tread T and rise R if the density of concrete $= 25 \, kN/m^3$?

     (a) $\left\{ 30D\sqrt{R^2 + T^2} + 15\dfrac{T}{R} \right\}$ (b) $\left\{ \dfrac{30D}{R}\sqrt{R^2 + T^2} + 15T \right\}$

     (c) $\left\{ \dfrac{D\sqrt{R^2 + T^2}}{T \times 25} + 15TR \right\}$ (d) $\left\{ \dfrac{30D\sqrt{R^2 + T^2}}{T} + 15R \right\}$

  (vii) What is the most accurate way of placing tensile reinforcement at the junction of landing and flight slab shown in Fig. a/b/c/d?

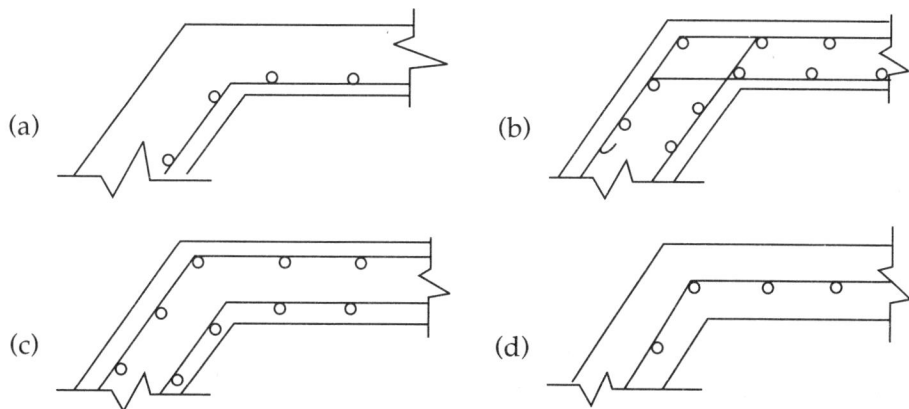

(a) (b) (c) (d)

(viii) In a staircase with cantilever steps, the worst case of loading for design shall
be when ....
   (a) point load lies at the middle of width of step
   (b) point load lies at the free end of the step
   (c) point load lies at the fixed end of the step
   (d) u.d.l. spreads over the entire span
(ix) Dead loading on the stair slab for the flight portion shall be ....
that on the landing portion.
   (a) equal to   (b) less than
   (c) more than   (d) sometimes equal or less than
(x) Staircase slab is basically flexural member and designed for ....
   (a) maximum bending moment only
   (b) maximum bending moment and maximum shear force
   (c) minimum shear force and maximum shear force
   (d) minimum deflection and minimum bending moment.

Response sheet to Q. 13.1: (i to x)

Question	(i)	(ii)	(iii)	(iv)	(v)	(vi)	(vii)	(viii)	(ix)	(x)
Response (a/b/c/d)										

(II) Numerical Questions

Q. 13.2 Figure 13.9 shows the geometrical arrangement of staircase for a public
building. The tread is 300 mm and rise is 150 mm. The stair is built in the side
wall along the flights. (a) Design the staircase and flight BC for a live load of
3 kN/m² taking the span in the direction of flight. Use M20 CC and Fe 415 steel,
width of staircase may be taken as 1.40 m (b) design flight AB.

[**Hint:** Limit state design constant: $\dfrac{x_u}{d}$ max $= 0.48$, $R_u - 2.761$, $f_{ck} = 20 \text{ N/mm}^2$

Assume thickness $D = 200$ mm, span AB $= 3 + 1.4 + 0.1 = 4.50$ m

(a) Effective span BC (L_e) = 0.10 + 1.40 + 1.50 + 1.4 + 0.1 = 4.50 m

Loading on BC (b = 1.40 m)

Steps: $\dfrac{1}{2} \times \dfrac{0.30 \times 0.15 \times 1.4 \times 25}{0.30}$ = 2.625 kN/m, Finishing = 0.75 × 1.4

= 1.05 kN/m

Waist slab = $\dfrac{0.20 \times \sqrt{0.30^2 + 0.15^2}}{0.30}$ × 1.4 × 25 = 7.830 kN/m

Loading in flight = 7.83 + 2.62 + 1.05 = 11.50 kN/m

Floor finish landing = 0.75 × 1.4 × 1.4 = 1.47 kN (total) in both directions

Landing slab = 0.20 × 1.4 × 1.4 × 25 = 9.8 kN (total) in both directions

Load on landing in each direction = $\dfrac{1}{2}$ (9.8 + 1.47) × $\dfrac{1}{1.4}$ = 4.00 kN/m

Total u.d.l. in flight = (11.50 + 3 × 1.4) = 15.7 kN/m

Total u.d.l. in landing = (4.00 + 3 × 1.4) = 8.2 kN/m

Ultimate loads w_{u_1} (flight slab) = 1.5 × 15.7 = 23.55 kN/m

Ultimate loads w_{u_2} (landing) = 1.5 × 8.2 = 12.30 kN/m

In case of upper landing since the landing is supported in one direction, the load will be in one direction only, i.e. u.d.l. w_{u3} = 2 × 12.3 = 24.6 kN/m for a width of 1.40 m. Considering the upper flight BC, approximate SF

R_B = 38.300 kN,

R_C = 52.300 kN, zero SF at 2.13 m from C and maximum

BM (M_u) = 55.550 kN-m, b = 1400 mm

$$d = \sqrt{\dfrac{55550000}{1400 \times 2.761}} = 120 \text{ mm (provide } D = 180 \text{ mm}, d = 150, \text{ cover 30 mm)}$$

Shear stress = $\dfrac{52300}{1400 \times 150}$ = 0.25 N/mm² (< 0.28 N/mm²), OK

$$A_{st} = \dfrac{0.5 f_{ck} b.d}{f_y}\left[1 - \sqrt{1 - \dfrac{4.6 \times 55.55 \times 10^6}{f_{ck}.b(150)^2}}\right] = 1160 \text{ mm}^2$$

(provide 12φ 11 bars in 1400 mm) 11 × 113 = 1243 mm² > 1160 mm², OK.

(b) Since flight AB has effective span of 4.50 m, the same design as that of flight BC may be adopted. There may be slight change in loading pattern for which SF and BM may be recalculated and if the difference is large, then redesign the section. Calculate the distribution steel, and show the details of reinforcement also.

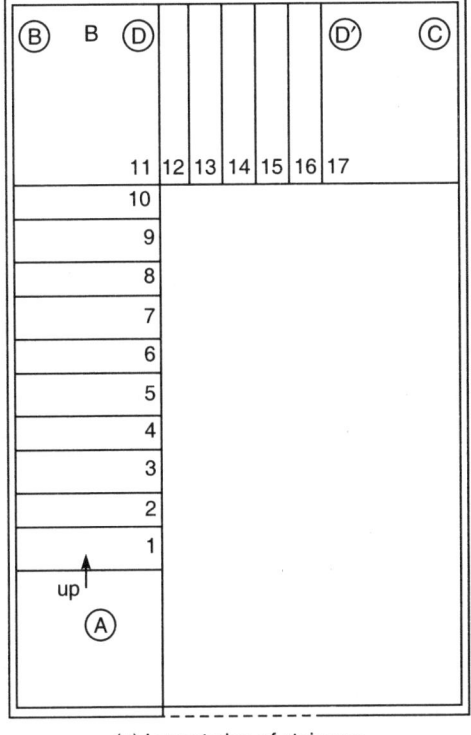

(a) Layout plan of staircase

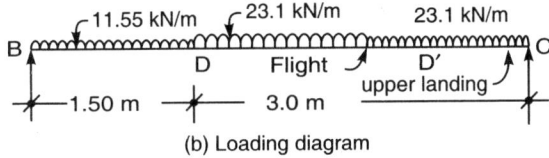

(b) Loading diagram

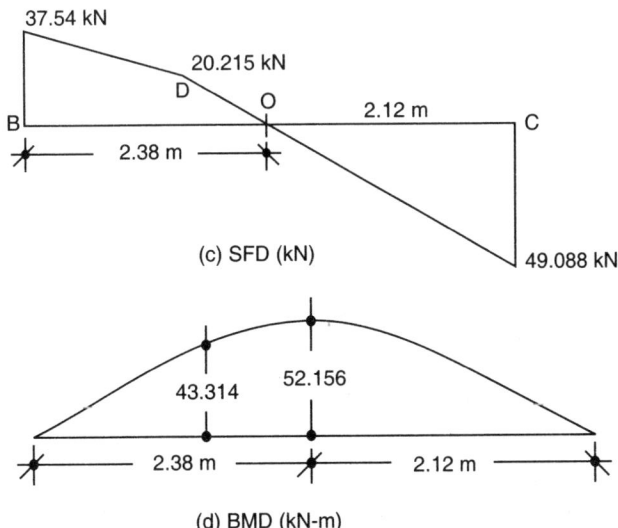

(c) SFD (kN)

(d) BMD (kN-m)

Fig. 13.9: Staircase layout and loading pattern

Q. 13.3 Design a dog-legged staircase for a building in which the vertical distance between the floors is 3.60 m. The stair hall measures 2.5 m × 5 m having 1.20 m width of stair. Assume rise of 150 mm and tread of 250 mm. Use M20 CC and Fe 415 steel. Take live load of 2.5 kN/m².

[Hint: No. of rises in each flight = 12, tread = 250 mm,

w_u (flight load) = 15.460 kN/m, landing portion w_{u1} = 13.208 kN/m

Assume width 1 m, also consider u.d.l. w_u =15.46 kN/m throughout

M_u = 32.485 × 10⁶ N-mm, d = 108.5 mm (adopt d = 125 mm, D = 150 mm)

A_{st} (main) = 837 mm² (10 mm @ 13 nos, distribution 8 mm bars @ 250 (c/c)

Check for shear stress. Show reinforcement placement and layout.]

Q. 13.4 Design a stair slab supported on the side wall on one side and stringer beam on the other side. Horizontal width of the staircase may be taken as1.2 m. Steps have risers of 150 mm and treads of 250 mm. Design the steps waist slab using M20 CC and Fe 415 steel.

[Hint: Design constants: $\dfrac{X_u}{d}$ max = 0.48, R_u = 2.761,

Steps: R = 150 mm, T = 250 mm, $l = \sqrt{R^2 + T^2}$ = 292 mm, waist slab = 80 mm

$$D_{total} = \left(80 + \frac{R.T}{l}\right) = \left(80 + \frac{150 \times 250}{292}\right) = 208 \text{ mm, equivalent } d = \frac{208}{2} = 104 \text{ mm}$$

Width of step along the flight = 292 mm and depth = 104 mm (say)

Span between the stringer beam and wall = $1.20 + \dfrac{0.01}{2}$ = 1.25 m

DL of each step = $25 \times \dfrac{0.15 \times 0.25}{2} \times 1$ = 0.47 kN/m width.

DL of waist slab = 25 × 0.080 × 0.292 kN = 0.58 kN/m width

DL of finishing = 0.070 kN/m   Total W_d= 1.12 kN/m

Live load w_1= 3 × 0.25 ×1 = 0.75 kN/m

Total load W =1.12 + 0.75 = 1.87 kN/m

W_u = 1.5 × 1.87 = 2.81 kN/m,

$$M_u = \frac{2.81 \times 1.25^2}{8} = 0.5488 \text{ kN-m } (54.88 \times 10^4 \text{ N-mm}) \text{ (assuming SS)}$$

$$d = \sqrt{\frac{54.88 \times 10^4}{2.761 \times 292}} = 26.1 \text{mm, provided } d = 104 \text{ mm, OK}$$

$$A_{st} = \frac{10 \times 292 \times 104}{415}\left[1 - \sqrt{1 - \frac{4.6 \times 0.5488 \times 10^6}{20 \times 292 (104)^2}}\right] = 14.8 \text{ mm}^2/\text{m}$$

Minimum = $\dfrac{0.12 \times 292 \times 208}{100}$ = 72.9 mm²

Provide 8 mm ϕ @ 300 c/c, A_{st} = 76.4 mm²/ in 292 mm.]

Q. 13.5 Design a two flight slab less tread-riser staircase for a floor height of 3.25 m and superimposed (*ll*) load of 3 kN/m². The staircase is supported at the end of landing slabs spanning in the direction of flight. Use concrete of M20 grade and steel of Fe 415 grade. Assume surface finish of 0.50 kN/m².

[**Hint:** No of rises = 10, rise = 162.5 mm (Fig. 13.10),

No. of treads in each flight = 10 – 1= 9, going = 9 × 250 = 2250 mm,

tread = 250 mm,

landing 1125 mm

$$L_e = 2 \times 1125 + 2250 + \frac{0.20}{2} \times 2 = 4700 \text{ mm, constant } m_{ft} = 1.30 \text{ (say)}$$

$$d = \frac{L_e}{20 \times m_{ft}} = \frac{4700}{20 \times 1.3} = 180.8 = 180 \text{ mm}, D = 180 + 15 + \frac{10}{2} = 200 \text{ mm}$$

Calculate load for each step, and landing at an interval of tread.

$P = 3.4425$ kN on landing

$P_q =$ Riser load at the end of landing = 4.077 kN above 1st riser,

$Q = 4.40625$ kN at the end of tread and above riser

Ultimate maximum moment $M_u = 30.89$ kN-m/m

$d = 180$ mm, $A_{st} = 517$ mm²/m,

Provide 10ϕ @ 150 mm c/c ($A_{st} = 524$ mm²/m > 517 mm²/m) OK

$$\text{Check for } d = \frac{L_e}{20 \times m_{ft}} = \frac{4700}{20 \times 1.52} = 155 \text{ mm } (< 180 \text{ mm}), \text{ OK}$$

$$\text{Minimum temp. steel} = \frac{0.12 \times b.D}{100} = \frac{0.12}{100} \times 1000 \times 200 = 240 \text{ mm}^2/\text{m}$$

∵ < 524 mm²/m, provided.

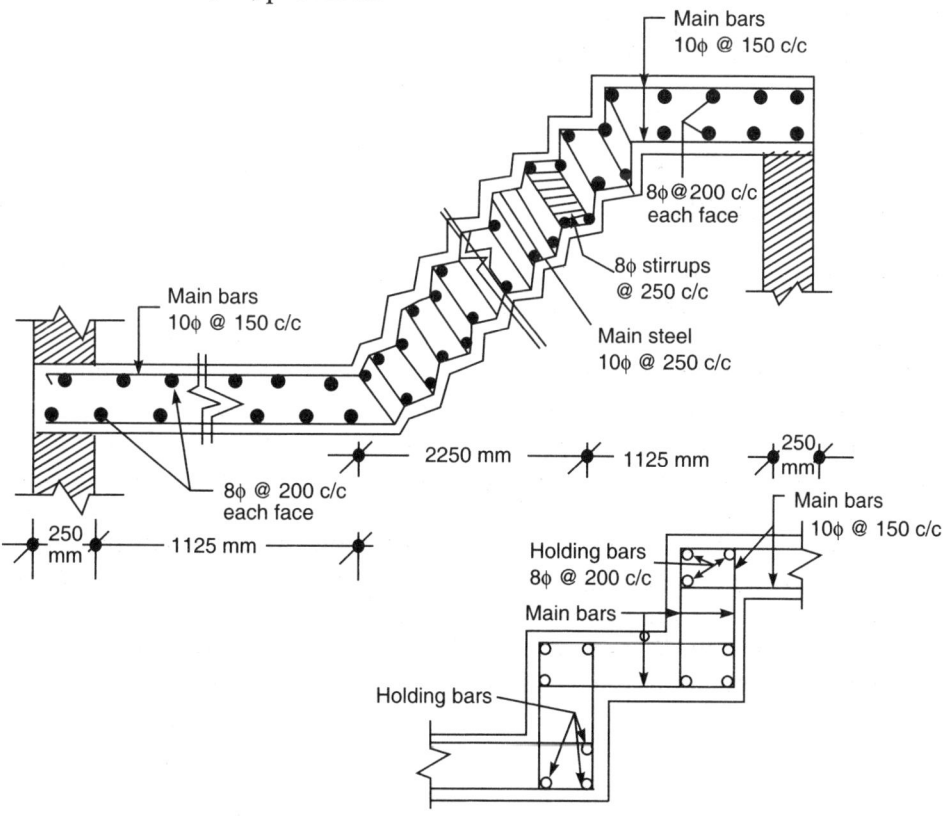

Fig. 13.10: Details of reinforcement cross-sections

Chapter 14

Design of RCC Culverts

LEARNING OBJECTIVES

After the study of *design of RCC culverts*, the learner will understand the design of RCC culverts and shall be able to:

- Differentiate between slab and box type of culverts
- Explain different types of IRC class of loading
- Compute different types of design loads on the culvert slab
- Explain the effect of impact load on the design of culvert slab
- Explain the effect of centrifugal force caused by moving vehicles
- Design the RCC slab of a culvert under given loads
- Design the RCC box culvert under given conditions and loads

14.1 INTRODUCTION

Culverts and bridges are needed to allow cross flow of drains, rivers, or traffic at different levels (Fig. 14.1). When the spans are small (up to 8.0 m), it is known as culvert and for large spans more than 8 m, it is known as bridge. Solid slabs are provided for culverts (small span up to 8.0 m) while beam-slabs are used for deck-slab in case of larger spans for bridges.

Culverts provided for road traffic only shall be of width generally not less than 4 m for single lane and increased by a minimum of 3 m for every additional lane. For railways, width of 4 m for single track and 7.6 m for double track trainway shall be provided. For any footway bridge width of not less than 1.50 m shall be provided.

Culvert slabs are basically flexural members and shall be designed in the same manner as oneway supported slabs outlined earlier except that the loading pattern is different and based on the loads of moving traffic. IRC loading of different classes shall be considered for worst or most critical placement along the slab span.

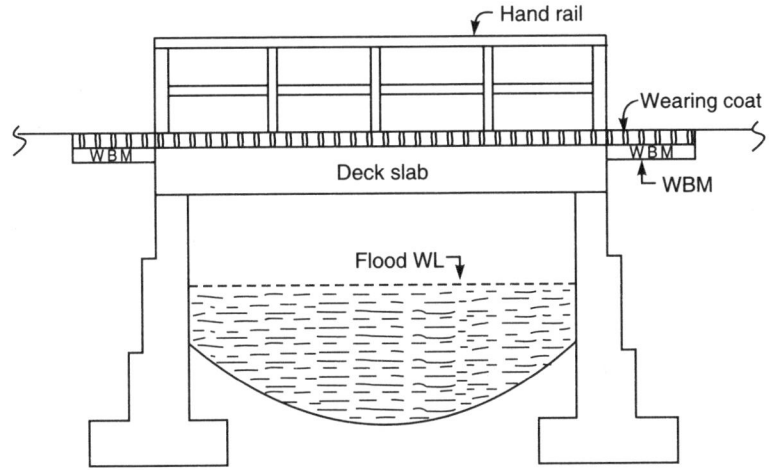

(a) Longitudinal section of a culvert

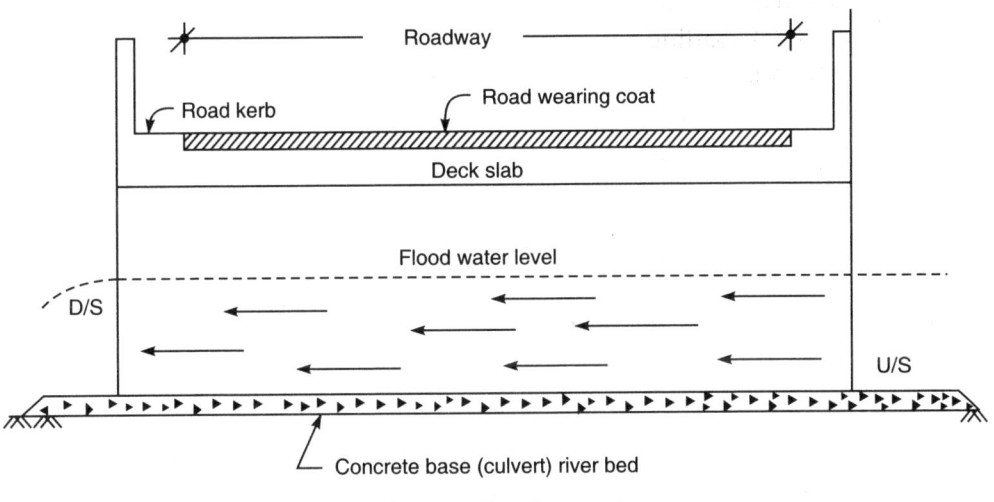

(b) Cross-section of a culvert

Fig. 14.1: A typical culvert

14.2 LOADS

The culverts and bridges are subjected to following type of loads:
1. Dead loads (own load)
2. Live loads (moving traffic of different types)
3. Impact loads (due to speed of vehicles)
4. Tractive forces (due to speed and braking effect)
5. Centrifugal forces (due to speed along curved path)
6. Wind forces (due to blowing of wind velocity)
7. Temperature stresses (due to variation in temperature)
8. Seismic forces (due to earthquake acceleration)

9. Erection forces (due to machinery weight and vibration) these loads shall be considered as specified in IRC (Indian road congress) Clauses.

1. Dead Loads

Dead loads are those loads which act permanently on the structure and these may include – self weight, weight of finishing, weight of wearing coat, kerb, handrails, etc. which act permanently on the slab. Unit weight of wearing coat may be taken as 22 kN/m^3 and that of RCC may be taken as 25 kN/m^3. Depth of culvert slab may be assumed as 80 mm for every one metre span.

2. Live Loads

Live loads on footway and road kerbs are caused by pedestrians and animals only as vehicular traffic is not allowed on footways. Live loads of 4 kN/m^2 may be taken under normal conditions while a live load of 5 kN/m^2 may be taken of heavily crowded areas near towns.

For design of deck slab, various vehicular live loads shall be considered as train of concentrated loads as specified in IRC code. IRC code loads are specified in 3 categories to avoid any confusion on loading variation of variety of vehicles. These are:

(a) IRC class AA loading

(b) IRC class A loading

(c) IRC class B loading

IRC class AA loading shall be adopted within certain municipal limits in industrial and other specified areas along highways. Bridges designed for IRC class 2A loading should also be checked for IRC class Aloading, as under certain conditions, class A loading may develop more serious stresses.

IRC class A loading shall be adopted for all roads on which permanent bridges and culverts are provided.

IRC class B loading shall normally be adopted for temporary structures and bridges constructed in certain specified (military and industrial) areas for specified projects.

Figures 14.2 and 14.3 show wheel loads, position and clear distances for IRC class AA and IRC class A and B loading details. Following notes are specified in IRC loadings.

(1) The nose to tail spacing between two successive vehicles shall not be less than 90 m.

(2) For multilane bridges and culverts one train of class AA tracked or wheeled vehicles whichever creates severe conditions should be considered for every two traffic lane width. When this type of vehicles are crossing the bridge or culvert, no other live load should be considered on any part of these two lanes wide carriageway.

(3) The maximum loads for the wheeled vehicles should be 20 tonnes (about 200 kN) for single axle or 40 tonnes (about 400 kN) for a bogie of two axles spaced not more than 1.20 m centres.

(4) Theminimum clearance between the road face of the kerb and the outer edge of the wheel or track C, shall be as given below in Table 14.1.

Table 14.1: Minimum clearance 'C' from the kerb face

Carriageway width	Minimum value of C
Single lane bridge – 3.8 m and above	0.30 m
Multilane bridge – less than 5.5 m	0.60 m
5.5 m and above	1.20 m

The details of wheel loads, their positions and clear distances to be considered for class A and class B are shown in Fig. 14.3.

Following points should be noted for these loadings:

(i) The nose to tail distance between successive trains shall not be less than 18.4 m.

(ii) No other live load shall cover any part of the carriageway where a train of vehicles (or train of vehicles in a multiple lane bridge) is crossing the bridge or culvert

(iii) The ground contact area of the wheels shall be as under (Table 14.2).

Table 14.2: Ground contact area

Axle load in tonnes	Ground contact area in mm	
	B	w
11.4	250	500 mm
6.8	200	380 mm
4.1	150	300 mm
2.7	150	200 mm
1.6	125	175 mm

The minimum clearance f, between outer edge of the wheel and the road way face of the kerb and the minimum clearance g, between the outer edge of the passing or crossing vehicle on multilane bridge shall be given in the Table 14.3.

Table 14.3: Values of minimum clearance, f, and g, outer edges of vehicles

Clear carriageway width	g	f
5.5 m to 7.5 m	Uniformly increasing from 0.4 m to 1.20 m	150 mm
Above 7.5 m	1.20 m	150 mm

Following points should be considered in IRC loading while designing the bridge:

(i) The trailer attached to the driving unit are not to be considered as detachable (i.e. considered as nondetachable).

(ii) Within the kerb to kerb width, the roadway, the standard train or vehicle shall be assumed to travel along the length of the bridge, and to occupy any position which will produce maximum stress provided that the minimum clearances between a vehicle and the roadway face of the kerb and between two crossing vehicles are not enchroached upon (i.e. minimum clearances are maintained).

(iii) All the axles of a train or standard vehicle shall be considered to act simultaneously in a position causing maximum stresses.

(iv) Vehicles in adjacent lanes shall be taken as headed in the direction producing maximum stresses.

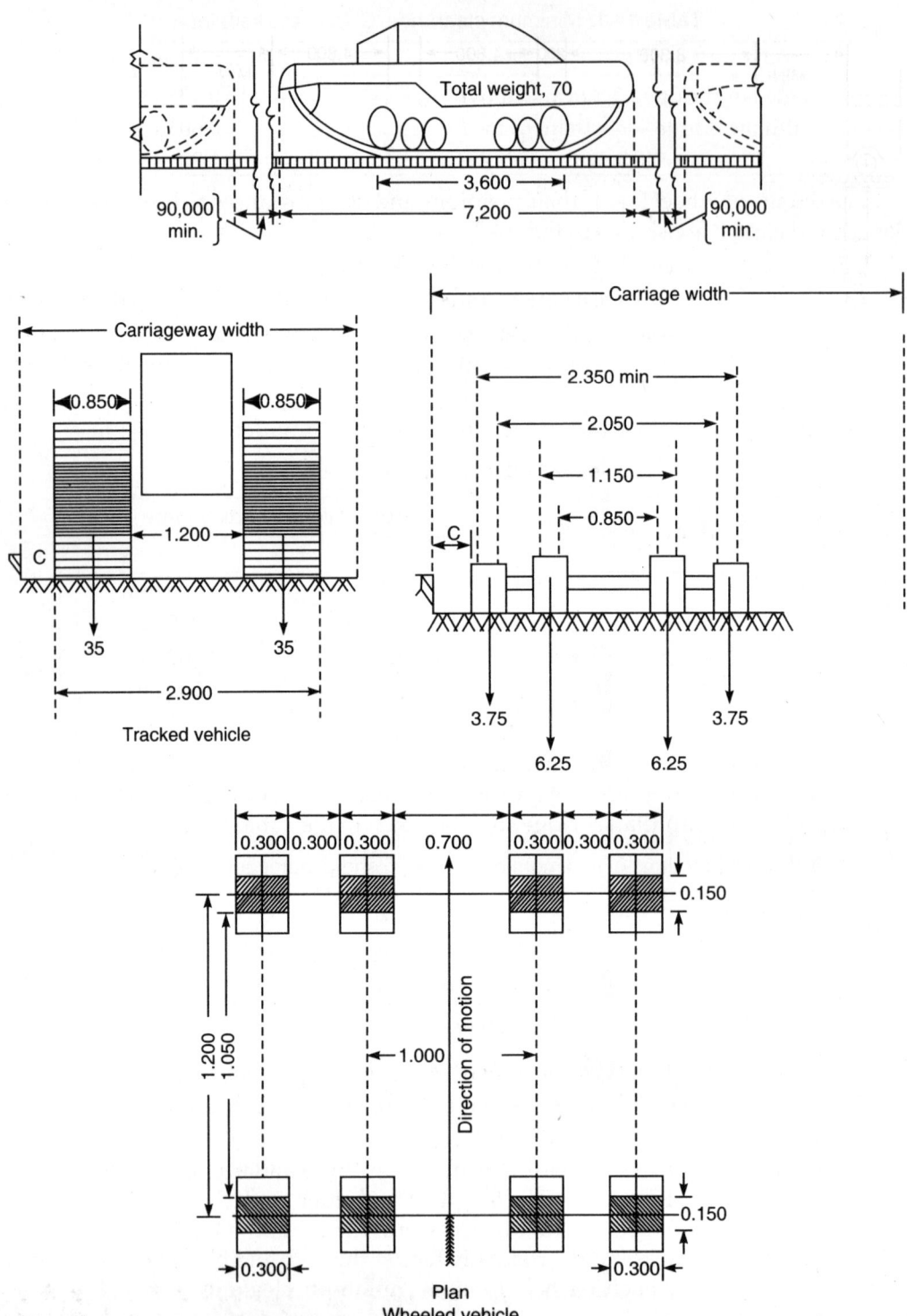

Fig. 14.2: IRC loading class AA tracked and wheeled vehicles (Clause 207.1)

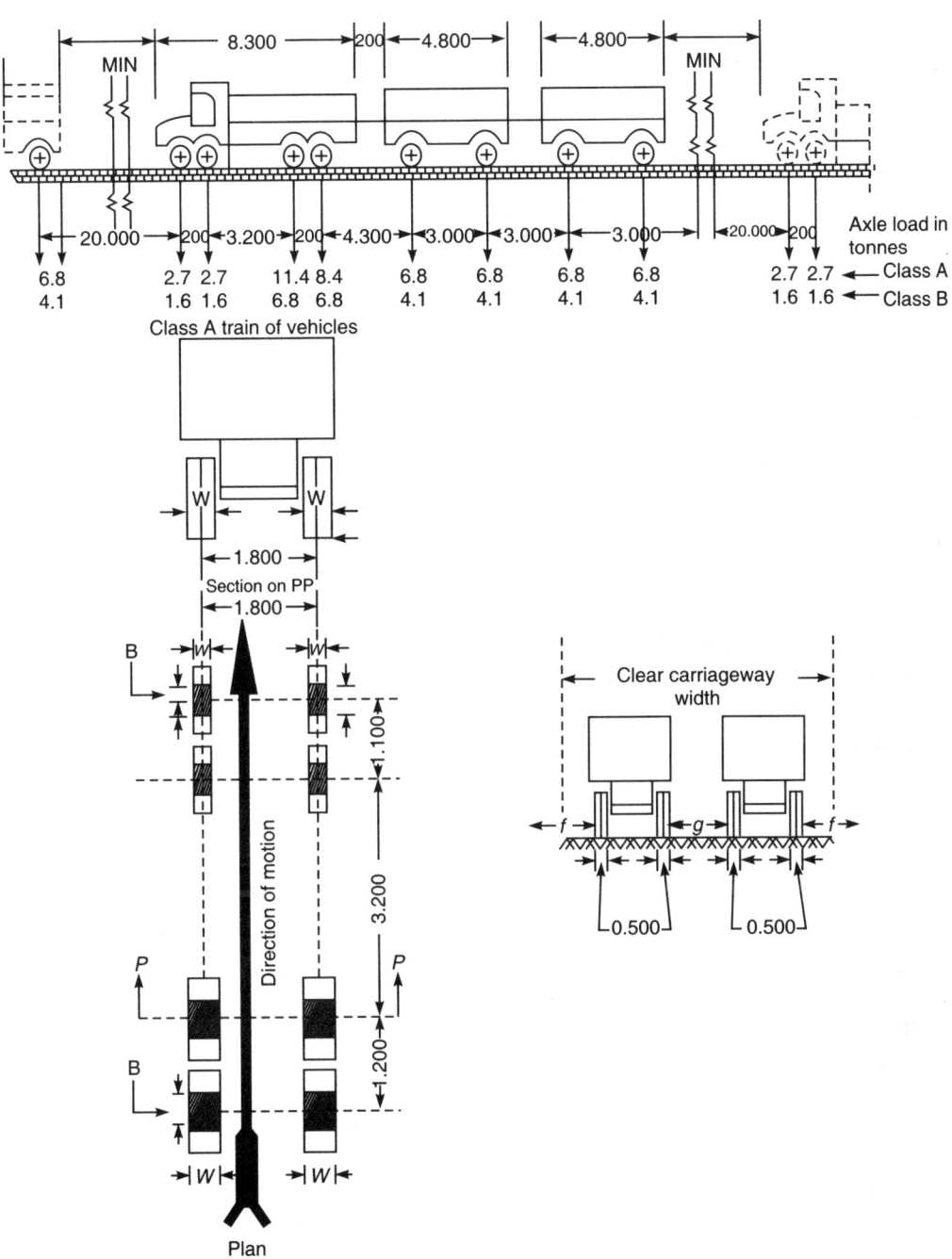

Fig. 14.3: IRC class A and class B loadings

(v) The space on the carriageway left uncovered by the standard train of vehicles shall not be assumed as subject to any additional live load.

3. Impact Load

For considering the effect of impact, the live loads or moving loads are increased by a certain percentage (known as impact allowance).

- Class A or class B loading

Impact factor fraction for reinforced concrete bridge $I_f = \dfrac{4.5}{(6+L)}$, (for spans 3 m to 45 m).

i.e. live loads are increased in the ratio of $(1 + I_f) : 1$

for example, if span $L = 10$ m (say), and live load is 5 kN/m,

$$\text{design load} = 5\left(1 + \frac{4.5}{6+10}\right) = 5\,(1.28125) = 6.406 \text{ kN/m}$$

- For class AA loading RCC bridges

  (a) *For spans less than 9 m*

  (i) For tracked vehicle: 25% for spans up to 5 m linearly reducing to 10% for spans of 9 m.

  (ii) For wheeled vehicles : 25%

  (b) *For spans 9 m or more*

  Reinforced concrete bridges

  (i) Tracked vehicles: 10% up to a span of 40 m and as per curves (for more than 40 m span)

  (ii) Wheeled vehicles: 25% for spans up to 12 m and as per curves (for spans more than 12 m).

Note:

  (i) No impact factor is allowed for foot bridge.

  (ii) Impact factor is reduced to one-half, if the slab carries filling of sand 0.60 m or more causing dampening of impact.

4. Tractive Forces

Bridge slabs or beams are subjected to longitudinal forces due to any one or more of the following causes:

  (i) Through application of brakes on wheel.

  (ii) Through acceleration of the driving wheels.

  (iii) Frictional resistance offered to the movement.

5. Centrifugal Forces

Centrifugal Forces are caused by moving vehicles along the curved road. Bridge structures must carry (or resist) the additional force/stress due to centrifugal force created by moving vehicles.

Centrifugal force $C = \dfrac{W \cdot V^2}{1.27R}$

where, C = centrifugal force acting normally to the traffic at the wheel contact point or u.d.l. over every metre length.

  W = live load of wheel loads (N) on the ground over contact length.

V = the design speed of the vehicles moving on the bridge km/hour

R = the radius of curvature in metre

Generally the centrifugal force C is considered to act at 1.20 m above the ground level of carriageway.

6. Wind Forces

Generally the wind forces are not critical for the design of culvert slab or beams. Wind forces may become critical for the design of the pier, abutments, etc.

7. Temperature Stresses, Seismic Forces and Erection Forces

Since the span of the slab culverts shall be small, the effect of these forces shall be small and may not influence the design of the slab culvert. However, in design of long bridges, these effects are accounted for planning the design.

Dead loads, live loads and impact loads directly influence the design of slab culvert and shall be considered for main design of the slab culvert and box culvert.

14.3 DESIGN OF SLAB CULVERT

Slab of culvert are supported on two opposite edges on walls or abutments and these slabs shall be spanning in one direction only (Fig. 14.4). The moment caused by a wheel load may be assumed to act on effective width of slab, given by b_{eff} :

$$b_{eff} = K_s\left(1-\frac{x}{1}\right)+b_w$$

where b_{eff} = effective width of slab

l = effective span

x = the distance of centre of gravity of load from the nearer support

b_w = breadth of concentrated load surface area

K = a constant depending upon the ratio $\dfrac{b}{l}$ as given in Table 14.4

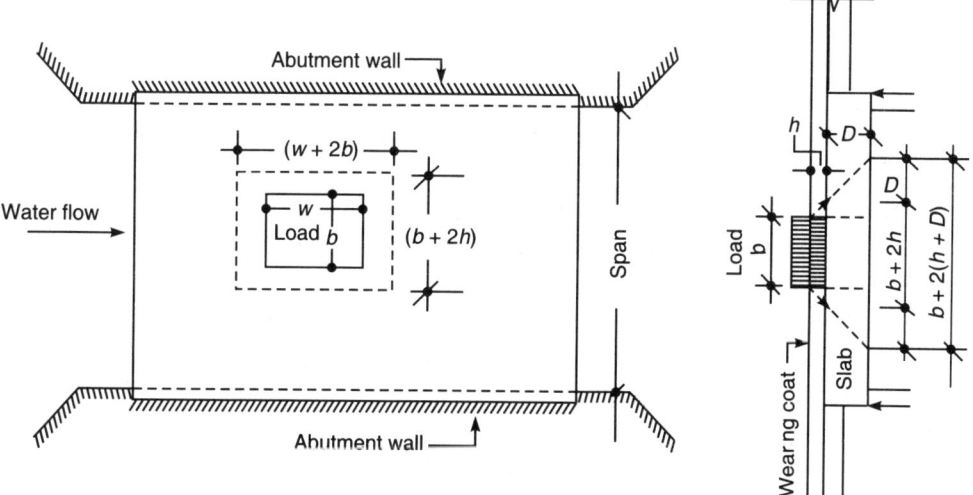

Fig. 14.4: Slab culvert span and load area ($b \times w$)

Table 14.4: Load dispersion width constant wrt $\left(\dfrac{b}{l_0}\right)$

$\dfrac{b}{l_0}$	K for simply supported	K for continuous slab	$\dfrac{b}{l_0}$	K for simply supported	K for continuous slab
0.10	0.40	0.40	1.10	2.60	2.28
0.20	0.80	0.80	1.20	2.64	2.36
0.30	1.16	1.16	1.30	2.72	2.40
0.40	1.48	1.44	1.40	2.80	2.48
0.50	1.72	1.68	1.50	2.84	2.48
0.60	1.96	1.84	1.60	2.88	2.52
0.70	2.12	1.96	1.70	2.92	2.56
0.80	2.24	2.08	1.80	2.96	2.60
0.90	2.36	2.16	1.90	3.00	2.60
1.00	2.48	2.24	2.00	3.00	2.60

Dispersion of Loads Along the Span

The effect of contact of track or wheel load in the direction of span length shall be taken as equal to the dimension of the tyre contact area b over the wearing surface of the slab in the direction of the span plus twice the overall depth of the slab inclusive of the thickness of wearing surface (IRC clause), illustrated in Fig. 14.4.

Dispersion of load through fills and wearing coat: The dispersion of load through fills and wearing coat shall be assumed at 45° in both, along and perpendicular to the span (as per IRC code). This is illustrated in Fig. 14.4. Illustrative examples given below will explain the various clauses.

Example 14.1

Design a reinforced concrete slab culvert for the following data:

Clear span = 5 m, width of support = 400 mm, clear width of roadways = 6.8 m, width of kerbs = 600 mm, thickness of wearing coat = 80 mm, loading Class A, grade of CC = M20, grade of steel = Fe 415 (Fig. 14.5).

Solution: Overall width of culvert = 6.80 + 2 (0.60) = 8.0 m

Assume thickness of slab = 80 mm/m × (5 m) = 400 mm (say)

Effective cover = 50 mm (say, d = 400 – 50 = 350 mm

Or clear span + support thickness = 5.0 + 0.400 = 5.40 m

L_e = lesser of 5.35 and 5.40 m, l_e = 5.35 m

$$F_{ck} = 20 \text{ N/mm}^2, f_y = 415 \text{ N/mm}^2, m = 13, \frac{x_u}{d} = 0.42, r_u = 2.71$$

Dead load

$$= 0.4 \times 25 + 0.08 \times 20 = 11.6 \text{ kN/m}^2$$

$$M_{ul} \text{ (DL)} = \frac{11.6 \times 1.5 \times 5.35^2}{8} = 62.254 \text{ kN-m/m}$$

Live load

(a) Since the width of carriage way is more than 6.80 m, the culvert will be two lane. Minimum clear distance between two adjacent vehicles varies from 0.4 m to 1.2 m linearly for carriage width of 5.5 m to 7.5 m. Hence for 6.80 m carriage width, minimum clearance between two vehicles shall be

$$g = 0.40 + \frac{(1.2 - 0.4)}{(7.5 - 5.5)} (6.8 - 5.5) = 0.92 \text{ m}$$

Width of contract area of wheels $w = 500$ mm $= 0.50$ m

Distance between centres of two vehicles $= 0.92 + 0.50 = 1.42$ m

Minimum distance of wheel from the kerb $= 0.15$ m

Minimum distance of centre of wheel from the kerb $= 0.15 + \dfrac{0.50}{2} = 0.40$ m

Hence, position of wheel loads in lateral direction for max. BM condition is shown in Fig. 14.6.

(b) Position of wheel loads in the longitudinal direction

For the maximum BM condition, heavy load of 11.4 tonnes (114 kN) are to be placed near mid span point. For such situation, other wheels are off the culvert.

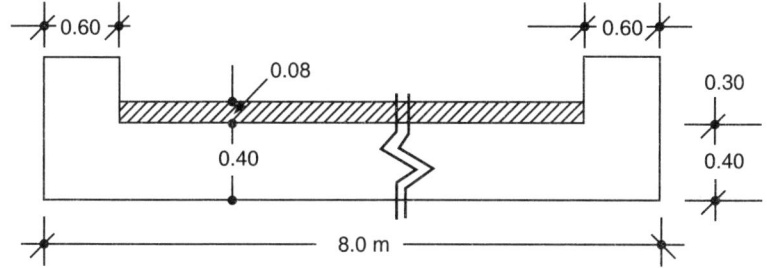

Fig. 14.5: Cross-section

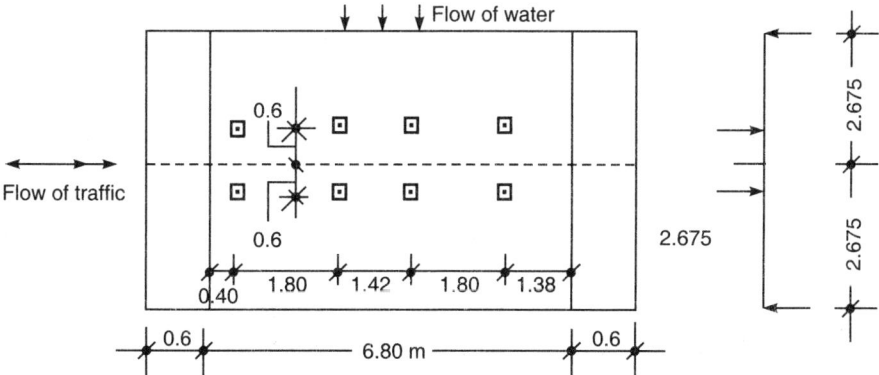

Fig. 14.6: Wheel position for max. BM

Since the moving load can be treated as wheel of length smaller than the span, the maximum bending moment will occur at mid span point, it will be placed symmetrically about centre of span. The position for maximum BM, in the longitudinal direction is shown in Fig. 14.6.

Effective width

For single concentrated load, the effective width is given by

$$B_{eff} = K \cdot x \left(1 - \frac{x}{l}\right) + b_w, \, l = 5.35,$$

x (distance from nearer support) = 1.775 m

$$K = \text{constant depending upon the ratio } \frac{b}{l}, \qquad \therefore \frac{b}{l} = \frac{8}{5.35} = 1.5 \text{ m}$$

b_w = (breadth of concentrated load area) = width of tyre + 2 (0.08)

$$= 0.5 + 0.16 = 0.66 \text{ m}$$

$$b_{eff} = 2.84 \times 1.775 \left(1 - \frac{1.775}{5.35}\right) + 0.66 = 4.03 \text{ m}$$

Hence, the effective width for heavy load vehicles overlap.

Therefore, all the four heavy loads will be considered as acting of effective width.

Total clear distance = 0.6 + 0.4 + 1.80 + 1.42 + 1.80 + 1.38 + 0.6 = 8.0 m

$$\text{Load/m width of slab} = \frac{2 \times 11.4}{8} = 2.85 \text{ tonnes (28.5 kN)}$$

Description of load along span: Width of wheel load

$$= 250 + 2000 (0.08 + 0.40) = 1210 \text{ mm (1.21 m)}$$

Load per metre width of slab shall be taken as shown in Fig. 14.7

$$\text{Reaction } R_A = R_B = \frac{1}{2} \text{ (total load)} = 28.5 \text{ kN}$$

Equivalent intensity of u.d.l.

$$w = \frac{28.5 + 28.5}{2.40} = 23.75 \text{ kN/m}$$

$$\text{Maximum BM} = 28.5 \times \left(\frac{5.35}{2} - \frac{1.2}{2}\right) = 59.14$$

$$\text{Impact factor} = \frac{4.5}{(6 + 5.35)} = 0.4$$

Additional BM due to impact 0.4 × 59.14 = 23.45 kN-m/m

Total BM (LL) = 59.14 + 23.45 = 82.59 kN-m/m

Total ultimate load BM (M_u)=1.5 (82.59) + 62.15 = 186.04 kN-m/m.

Design of section

$$M_u = 2.761 \, bd^2, d = \sqrt{1 - \frac{186.04 \times 10^6}{2.761 \times 1000}} = 259.6 \text{ mm}$$

Assume $d = 300$ mm, $D = 340$ mm (say)

Area of steel

$$A_{st} = \frac{0.5 f_{ck} \times 1000 \times 300}{415} \left[1 - \sqrt{1 - \frac{4.6 \times 186.04 \times 10^6}{20 \times 1000 (300)^2}} \right] = 1993.24 \text{ mm}^2/\text{m}$$

Provide 20 mm ϕ @ $\dfrac{314000}{1993.24} = 157.5$ mm c/c,

Provide 20ϕ @ 150 c/c ($A_{st} = 2093$ mm², i.e. 0.70%)

As per IRC distribution steel is to be provided to resist 0.3 times LL BM including impact BM and 0.20 times the DL BM.

$M_{u_2} = 0.3(82.59 \times 1.5) + 0.2(62.15) = 0.3 \times 123.9 + 0.2(62.15) = 49.6$ kN-m/m.

d(for distribution steel) $= 300 - \dfrac{20}{2} - \dfrac{12}{2} = 284$ mm, assuming distribution bars 12 mm ϕ

$$A_{st_2} = \frac{10 \times 1000 \times 284}{415} \left[1 - \sqrt{1 - \frac{4.6 \times 49.6 \times 10^6}{20 \times 1000 (284)^2}} \right] = 503 \text{ mm}^2/\text{m}$$

Provide 12ϕ @ 200 c/c, $A_{st} = 565$ mm²/m.

Check for shear

τ_c (design shear) = 0.46 N/mm² for 0.7% steel (Fig. 14.8)

$$\text{SF at support} = \frac{1}{2}(28.5 + 28.5) = 28.5 \text{ kN}$$

$$V_{u_1} = 1.5 \times 28.5 = 42.75 \text{ kN, SF (impact)} = 0.4 \times 42.75 = 17.1 \text{ kN}$$

$$\text{SF due to LL for worst location} = 28.5 + \frac{28.5 \times (5.35 - 1.2)}{5.35} = 50.6 \text{ kN (assuming}$$

one wheel on support)

$$\text{SF due to DL} = \frac{11.6 \times 5.35}{2} = 31.03 \text{ kN}$$

Total SF (ultimate) = 1.5(50.6 + 31.03) = 122.45 kN

$$\tau_c = \frac{122.45 \times 1000}{1000 \times 300} = 0.41 \text{ N/mm}^2 \ (< \tau_c \text{ of } 0.46 \text{ N/mm}^2, \text{ design shear), OK}$$

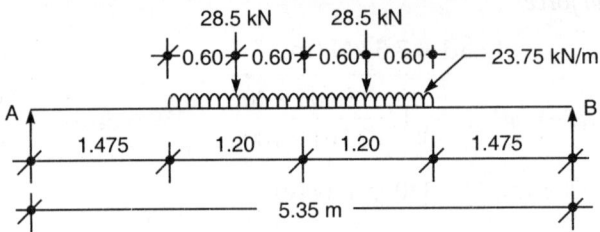

Fig. 14.7: Slab with loading

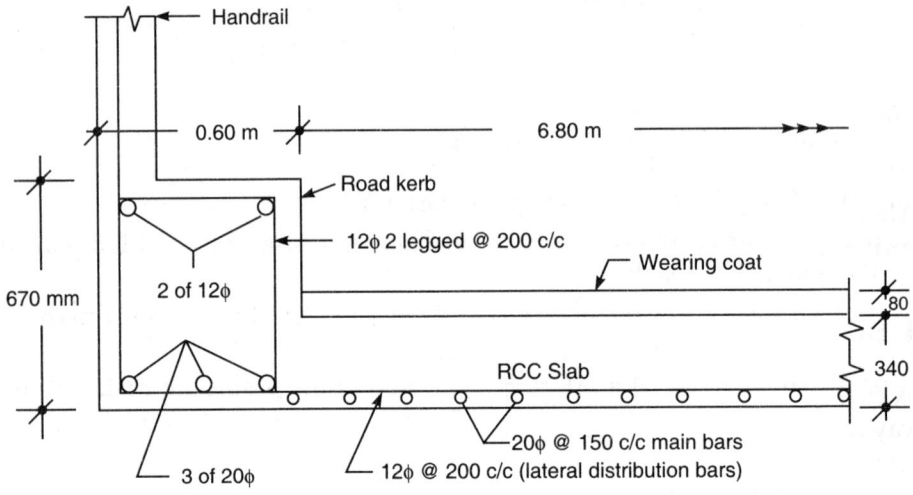

Fig. 14.8: Details of reinforcement in slab of culvert

Design of kerbs

$B = 0.60$ m, LL $= 4$ kN/m², horizontal load of 7.5 kN/m applied at the top of kerb.
Let the height of kerb above road $= 250$ mm

$$\text{Total height of kerb} = 340 + 80 + 250 = 670 \text{ mm}$$
$$\text{Effective depth } d = 670 - 50 = 620 \text{ mm}$$

Loads: DL $= 0.60 \times 0.67 \times 25 = 10.05$ kN/m,

LL $= 0.60 \times 1 \times 4 = 2.40$ kN/m

Weight of railing $= 0.6$ kN/m (assumed), total weight $= 13.05$ kN/m
Ultimate load $= 1.5 \times 13.05 = 19.58$ kN/m.

$$M_u = \frac{19.58 \times 5.35^2}{8} = 70.04 \text{ kN-m } (70.04 \times 10^6 \text{ N-mm})$$

$$V_u = \frac{19.58 \times 5.35}{2} = 52.40 \text{ kN}$$

$$A_{st} \ (d = 620 \text{ mm}) = \frac{10 \times 600 \times 620}{415}\left[1 - \sqrt{1 - \frac{4.6 \times 70.04 \times 10^6}{20 \times 600(620)^2}}\right] = 319 \text{ mm}^2$$

(3 of 20ϕ, $A_{st} = 942$ mm²)

Design for horizontal force

$$\text{Load} = 7.5 \text{ kN/m,}$$

$$\text{Cantilever span} = 670 - 340 = 330 \text{ mm (above road slab surface)}$$

$$b = 1000 \text{ mm (longitudinally)}$$

$$M_u = (1.5 \times 7.5) \times 330 \times 1000 \text{ N-mm} = 371.25 \times 10^4 \text{ N-mm}$$

$$d = 600 - 50 = 550 \text{ mm}$$

$$A_{st} = \frac{10 \times 1000 \times 550}{415}\left[1 - \sqrt{1 - \frac{4.6 \times 371.25 \times 10^4}{20 \times 1000(550)^2}}\right] = 19 \text{ mm}^2$$

$$\text{Minimum temp. steel} = \frac{0.12}{100} \times 1000 \times 600 = 720 \text{ mm}^2/\text{m} \ (12\phi \ 200 \text{ c/c 2 legged}$$

bent from distribution bars of main slab).

Alternatively provide 12 vertical bars on road face of kerb at 200 c/c by properly anchoring these bars in lateral direction in bottom of kerb slab.

14.4 BOX CULVERT

A box culvert is a square pipe of RCC is provided for crossing a drain across a road or railway with high embankment. A box culvert is a continuous rigid frame consisting of rectangular section in which the two side abutments, the top and the bottom slabs are cast monolithically to provide rigid joints. A box culvert is suitable in case of poor bearing capacity of soil of embankment. In case of high embankment of soil for road, an ordinary culvert shall be quite costly due to heavy sections of abutments, base slab and top slab. Heavy sections also transmit heavy loads to the foundations. Hence, in such high embankment, box culvert shall be economical and most suited.

A box culvert shall be subjected to soil pressures from outside and water pressures from inside. The bottom slab shall be subjected to upward soil pressure from outside and downward water pressurefrom inside. The top slab shall be subjected to downward weight of soil in embankment above the top slab plus traffic load from top. The vertical side walls (abutments) shall be subjected to variable soil pressure from outside and variable water pressure from inside.

The box culvert shall therefore be designed for the following conditions:

(a) Box culvert subjected to soil pressures from outside and there is no water flow inside the box

(b) Box culvert subjected to soil pressures from outside and there is water flow (water pressure) from inside the box.

The analysis of moments shall be done by considering the box culvert as a continuous rigid frame using moment distribution method. The method is illustrated with the following example.

Example 14.2

Design a box culvert having inside dimensions 3.65 m × 3.65 m clear. Thickness of box shell may be taken as 350 mm uniformly prior to design. The box culvert is subjected to a superimposed dead load of 12.00 kN/m² and live load of 45 kN/m² from the top. Assume unit weight of soil as18 kN/m³ and angle of repose of 30°. Use M20 CC and Fe 415 steel.

Solution: Consider 1 m length of the box.

Design analysis shall be done for (a) live load, dead load and soil pressure with no water in the drain, (b) live load, dead load and soil pressure from outside and water pressure from inside the drain, with no live load on the sides, (c) dead load and soil pressure from outside, and water pressure from inside with no live loads.

Considering the thickness as 0.35 m, the centre lines $= 3.65 + \dfrac{0.35}{2} \times 2 = 4.0$ m. For analysis, consider uniform thickness of 350 m throughout.

Case I: Live load, dead load and soil pressure from outside and no water pressure from inside.

$$\text{Self-weight of top slab} = 0.35 \times 1 \times 1 \times 25 = 8.75 \text{ kN/m}^2$$

$$\text{External LL and DL} = 45 + 12 = 57 \text{ kN/m}^2$$

$$\text{Total load on top slab} = 8.75 + 57 = 65.75 \text{ kN/m}^2$$

$$\text{Weight of each side wall} = 4.0 \times 0.35 \times 1 \times 25 = 35 \text{ kN/m}$$

$$\text{Net upward soil pressure reaction on base slab} = \frac{\left(65.75 \times 4 + 35 \times 2\right)}{4 \times 1} = 83.25 \text{ kN/m}^2$$

$$\text{Constant active soil pressure } K_a = \frac{1 - \sin 30^\circ}{1 + \sin 30^\circ} = \frac{1}{3}$$

$$\text{Lateral pressure due to external (DL and LL)} = K_a(57.0) = \frac{1}{3}(57) = 19 \text{ kN/m}^2$$

$$\text{Lateral pressure due to soil} = K_a wh = \frac{18}{3} h = 6h \text{ kN/m}^2$$

Hence total lateral pressure at depth $h = (19 + 6h)$, triangular

Lateral pressure at top $(h = 0)$, $p_1 = 19$ kN/m²,

and at bottom $p_2 = (19 + 6 \times 4) = 43$ kN/m².

Figure 14.9 shows the loading diagram on the box culvert

Due to symmetry, half of the frame (AEFD) of box culvert is considered for moment distribution (Fig. 14.10). All members have uniform section (thickness), the relative stiffness 'K' for AD shall be equal to 1 (say) while the relative stiffness for AE and DF shall be $\dfrac{1}{2}$.

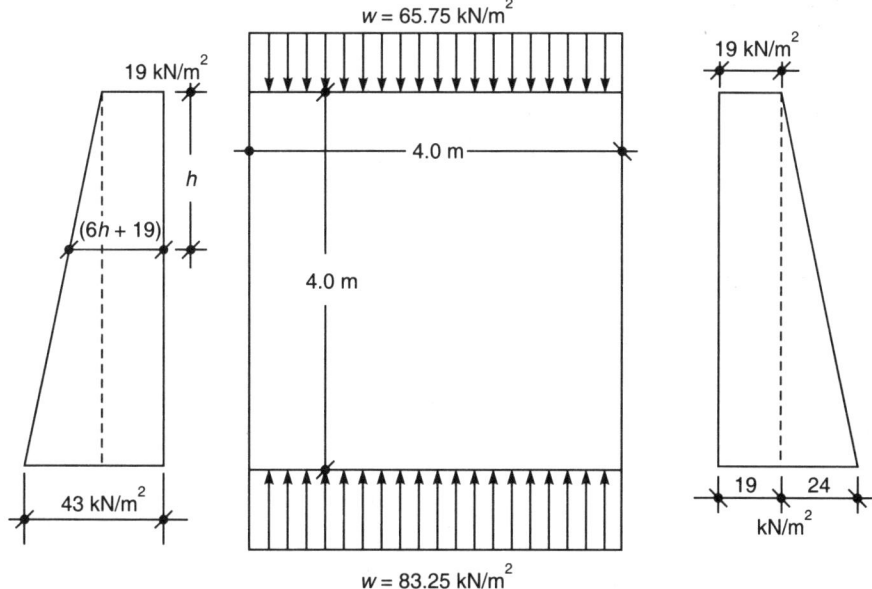

Fig. 14.9: A box culvert with various forces from outside

Distribution factors and AD and DA shall be $= \dfrac{1}{1+\dfrac{1}{2}} = \dfrac{2}{3}$ at A and D

Distribution factors for AB and DC at A and D shall be $= \dfrac{1/2}{1+\dfrac{1}{2}} = \dfrac{1}{3}$

The fixed end moments will be as under:

$$M_{FAB} = -\frac{w_1 l^2}{12} = -\frac{65.75 \times 4^2}{12} = -87.67 \text{ kN-m}$$

$$M_{FDC} = +\frac{w_2 l^2}{12} = +\frac{83.25 \times 4^2}{12} = +111.0 \text{ kN-m}$$

$$M_{FAD} = \frac{pl^2}{12} + \frac{Wl}{15}$$

where,

$$W = \text{total triangular load} = \frac{24 \times 4}{2} = 48 \text{ kN}$$

$$p = \text{uniform pressure intensity} = 19 \text{ kN/m}^2$$

$$M_{FAD} = \left(\frac{19 \times 4^2}{12} + \frac{48 \times 4}{15}\right) = +(25.33 + 12.8) = +38.13 \text{ kN-m}$$

$$M_{FDA} = -\frac{pl^2}{12} - \frac{Wl}{10} = -\left(\frac{19 \times 4^2}{12} + \frac{48 \times 4}{10}\right) = -44.53 \text{ kN-m}$$

The moment distribution is done as shown in Table 14.5

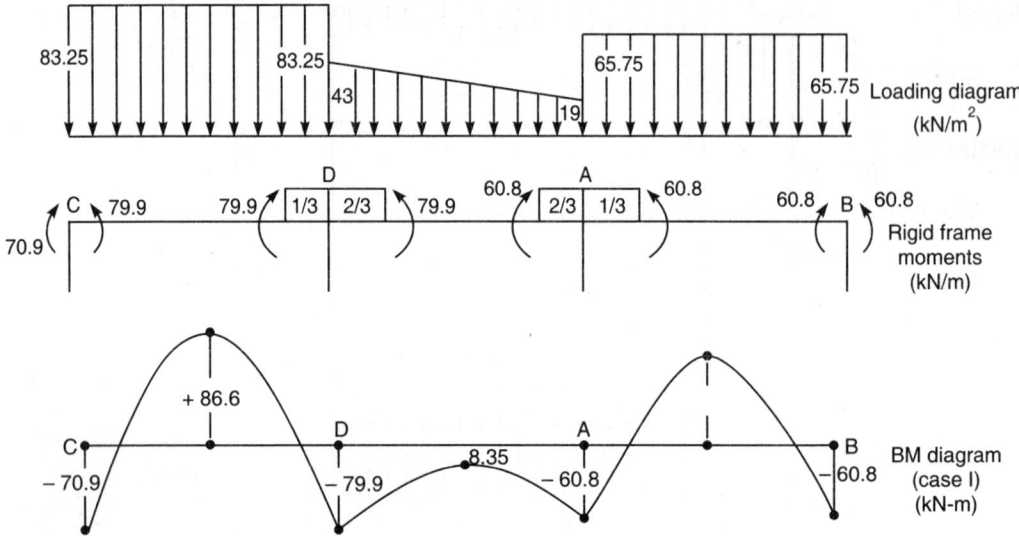

Fig. 14.10: Rigid frame, loading and BM diagram

Table 14.5: Moment distribution — *case I* (no water in drain)

Joint	D		A	
Member	**DC**	**DA**	**AD**	**AB**
Distribution factors	1/3	2/3	2/3	1/3
FEM	+111.00	−44.53	+38.13	−87.67
Balance	−22.16	−44.31	+33.03	+16.51
Carry over	–	+16.51	−22.16	–
Balance	−5.50	−11.01	+14.77	+7.39
Carry over	–	+7.39	−5.50	–
Balance	−2.46	−4.93	+3.67	+1.83
Carry over	–	+1.83	−2.46	–
Balance	−0.61	−1.22	+1.64	+0.82
Carry over	–	+0.82	−0.61	–
Balance	−0.27	−0.55	+0.41	+0.20
Carry over	–	+0.21	−0.27	–
Balance	−0.07	−0.14	+0.18	+0.09
Carry over	–	+0.09	−0.07	–
Balance	−0.03	−0.06	+0.05	+0.02
Carry over	–	+0.03	−0.03	–
Balance	−0.01	−0.02	+0.02	+0.01
Final moment (kN-m)	+79.90	−79.90	+60.80	−60.80

For the horizontal top slab carrying u.d.l. of 65.75 kN/m², the vertical reactions at

A and B shall be VA = VB = $\dfrac{65.75 \times 4}{2}$ = 131.50 kN upward.

Similarly, for the bottom slab CD carrying upward u.d.l. of 83.25 kN/m², the

vertical reactions at C and D shall be $V_C = V_D = \dfrac{83.25 \times 4}{2}$ = 166.50 kN downward.

The horizontal reaction at A shall be found by taking moments at D,

thus $(-h_a.4) + 60.80 - 79.90 + 19 \times 4 \times \dfrac{4}{2} + \dfrac{24 \times 4}{2} \times \dfrac{4}{3} = 0,$

$$h_a = \dfrac{(60.8 - 79.9)}{4} + 38 + 16 = 49.23 \text{ kN},$$

$$h_d = \dfrac{(19 + 43)}{2} \times 4 - 49.23 = 74.77 \text{ kN}$$

Free moment at midpoint (top slab) = $\dfrac{65.75 \times 4^2}{8}$ = 131.5 kN

Net moment at midpoint 'E' top slab = 131.5 − 60.8 = +70.7 kN-m

Similarly net moment at midpoint 'F' of bottom slab = $\dfrac{83.25 \times 4^2}{8} - 79.90$

= 86.6 kN-m

For vertical member AD,

SS BM at the mid span = $\dfrac{19 \times 4^2}{8} \times \dfrac{24 \times 4^2}{16}$ = 38 + 24 = 62 kN-m.

Net BM = $\dfrac{(79.9 + 68.8)}{2}$ 62 = (70.35 − 62) = +8.35 kN-m (mid height point)

Case II: In this case inner water pressure intensity at the top end A = 0 and at bottom end = 9.800 × 4 = 39.2 kN/m² in addition to forces considered in *case I*. The vertical walls shall thus be subjected to a net lateral pressure of intensity 19KN/m² at the top end A and (43 − 39.2) = 3.80 kN/m² at the lower end D. Net upward soil pressure will remain w = 83.25 kN/m² as additional water weight inside will be balanced by additional soil pressure (Table 14.6).

$$M_{FAB} = -\dfrac{wl^2}{12} = -\dfrac{65.75 \times 4^2}{12} = -87.67 \text{ kN-m},$$

Triangular loading will be 0 at bottom and 15.2 kN/m² at top.

$$M_{FDC} = \dfrac{w'l^2}{12} = \dfrac{83.25 \times 4^2}{12} = +111.0 \text{ kN-m}$$

Vertical member AD: Net pressure at top = 19.0 kN/m²,

at bottom (43 − 39.2) = 3.8 kN/m²

u.d.l. = 3.8 kN/m², triangular at bottom to (19 − 3.8) = 15.2 kN/m²

$$M_{FAD} = \frac{3.8 \times 4.0^2}{12} + \left(\frac{15.2 \times 4.0}{2}\right) \times \frac{4}{10} = 5.07 + 12.16 = + 17.23 \text{ kN-m}$$

$$M_{FDA} = \frac{3.8 \times 4.0^2}{12} - \left(\frac{15.2}{2}\right) \times \frac{4^2}{15} = -(5.07 + 8.11) = -13.18 \text{ kN-m}$$

Table 14.6: Moment distribution — *case II* (DL + LL from outside and water pressure inside)

Joint	D		A	
Member	DC	DA	AD	AB
Distribution factors	1/3	2/3	2/3	1/3
FEM	+111.00	−13.18	+17.20	−87.67
Balancing	−32.61	−65.21	+46.98	+23.49
Carry over	–	+23.49	−32.61	–
Balancing	−7.83	−15.66	+21.74	+10.87
Carry over	–	+10.87	−7.83	–
Balancing	−3.62	−7.25	+5.22	+2.61
Carry over	–	+2.61	−3.62	–
Balancing	−0.87	−1.74	+2.41	+1.21
Carry over	–	+1.21	−0.87	–
Balancing	−0.40	−0.81	+0.58	+0.29
Carry over	–	+0.29	−0.40	–
Balance	−0.10	−0.19	+0.27	+0.13
Carry over	–	+0.13	−0.10	–
Balancing	−0.04	−0.09	+0.07	+0.03
Carry over	–	+0.03	−0.04	–
Balancing	−0.01	−0.02	+0.03	+0.01
Final moment (kN-m)	+65.52	−65.52	+49.03	−49.03

For horizontal top slab AB, vertical reactions $V_A = V_B = \dfrac{65.75 \times 4.0}{2} = 131.5 \text{ kN} \uparrow$

(upward)

Similarly bottom slab CD, vertical reactions $V_D = V_C = \dfrac{83.25 \times 4.0}{2} = 166.5 \text{ kN} \downarrow$

(downward)

The horizontal reaction at A, by taking moments at D:

$$(h'a \times 4) + 49.03 - 65.52 + \frac{3.8 \times 4.0^2}{2} + \left(\frac{15.2 \times 4.0}{2}\right)\frac{2 \times 4}{3} = 0, h'a = 23.74 \text{ kN}$$

$$h'd = \frac{19 + 3.8}{2} \times 4 - 23.74 = 21.86 \text{ kN}$$

Net free moment mid span point E (top slab) = $\dfrac{65.75 \times 4.0^2}{8} - 49.03 = 82.47$ kN-m

Net free moment mid span point F (bottom slab) $= \dfrac{83.25 \times 4.0^2}{8} - 65.52$

$= 100.98$ kN-m

Net moment mid span point $= \dfrac{65.52 + 49.03}{2} - 22.8 = 34.50$ kN-m

Table 14.7: Moment distribution — *case III*

Joint	D		A	
Member	DC	DA	AD	AB
Distribution factors	1/3	2/3	2/3	1/3
FEM (kN-m)	+111.00	+7.47	−3.20	−87.67
Balancing	−39.49	−78.98	+60.58	+30.29
Carry over	–	+30.29	−39.49	–
Balancing	−10.10	−20.19	+26.33	+13.16
Carry over	–	+13.16	10.10	–
Balancing	−4.39	−8.77	+6.73	+3.37
Carry over	–	+3.37	−4.39	–
Balancing	−1.12	−2.25	+2.93	+1.46
Carry over	–	+1.46	−1.12	–
Balancing	−0.49	−0.97	+0.75	+0.37
Carry over	–	+0.37	−0.49	–
Balancing	−0.12	−0.25	+0.33	+0.16
Carry over	–	+0.16	−0.12	–
Balancing	−0.05	−0.11	+0.08	+0.04
Carry over	–	+0.04	−0.05	–
Balancing	−0.01	−0.03	+0.03	+0.02
Final moment (kN-m)	+55.23	−55.23	+38.80	−38.80

Case III: LL and DL on the top slab, water inside and no live load on sides

In this case there is no lateral pressure on vertical walls due to live loads and rest of forces are similar as earlier (Table 14.7).

Lateral pressure due to DL $= \dfrac{1}{3} \times 12 = 4$ kN/m²,

Lateral pressure due to soil $= \dfrac{18}{3} h = 6h$ kN/m².

Earth pressure at top of box $p_1 = 4$ kN/m²,

p_2(at bottom) $= 4 + 6 \times 4 = 28$ kN/m²

Water pressure from inside at top $(p'_1 = 0)$,

p'_2 (at bottom) $= 10 \times 4 = 40$ kN/m².

Net lateral pressure: Uniform 4 kN/m² from outside and 0 to (40 − 24) from inside.

$M_{FAB} = -87.67$ kN-m, $M_{FDC} = +111.00$ kN-m, as before,

$$M_{FAD} = \frac{4 \times 4.0^2}{12} - \left(\frac{1}{2} \times 16 \times 4\right)\frac{4}{15} = -3.2 \text{ kN-m},$$

$$M_{FDA} = -\frac{4 \times 4.0^2}{12} + \left(\frac{16}{2} \times 4\right)\frac{4}{10} = +7.47 \text{ kN-m}$$

Vertical reactions $V_A = V_B = \dfrac{65.75 \times 4.0}{2} = 131.5 \text{ kN} \uparrow$

$$V_C = V_{D \text{ (bott)}} = \frac{83.25 \times 4.0}{2} = 166.5 \text{ kN} \downarrow$$

Horizontal reaction $h_a = \dfrac{1}{4}\left\{(38.5 - 55.23) + \dfrac{4}{2} \times 4^2 - \dfrac{16 \times 4}{2} \times \dfrac{4}{3}\right\}$

$$= -6.77 \text{ kN} \rightarrow \text{(lateral)}$$

$$h_d = -9.23 \text{ kN} \rightarrow \text{(lateral)}$$

Midpoint E (top slab) and $M_E = \dfrac{65.75 \times 4.0^2}{8} - 38.8 = 92.70 \text{ kN-m}$

$$M_F = \frac{83.25 \times 4.0^2}{8} - 55.23 = 111.27 \text{ kN-m}$$

Mid point vertical (AD) $M_G = -\dfrac{4 \times 4.0^2}{8} + \dfrac{16 \times 4.0^2}{16} + \dfrac{(55.23 + 38.8)}{2} = 55.02 \text{ kN-m}$

Design of components

- *Top slab* (Table 14.8):

Table 14.8: Forces and design BM (details of Magnel wire system)

Forces	At centre		At ends		Direct force		Shear force	
		Mu_1	BM	Mu_2	P	P_u	V	V_u
Loading Conditions	BM	kN-m	kN-m	kN-m	kN	kN	kN	kN
Case I	70.70	106.05	60.80	91.20	49.23	73.85	131.5	197.25
Case II	82.47	123.71	49.03	73.55	23.74	35.61	131.5	197.25
Case III	92.70	139.10	38.80	58.20	-6.77	-10.16	131.5	197.25
Worst design value		139.10		91.20		73.85		197.25

$$d = \sqrt{\frac{M_u \times 10^6}{2.761b}} = 225 \text{ mm}, \quad \text{Adopt } d = 300 \text{ mm}, \quad D = 350 \text{ mm} \quad \text{(edges thickened)}$$

$$A_{st} = \frac{10000 \times 300}{415}\left[1 - \sqrt{1 - \frac{4.6M_u \times 10^6}{f_{ck}b(300)^2}}\right] = 1426 \text{ mm}^2 \text{ (mid span)},$$

At ends, $A_{st} = 899 \text{ mm}^2$, provide 20$\phi$ @ 200 c/c, alternate bent at ends plus 10ϕ @ 400 c/c

- *Bottom slab* (Table 14.9):

Table 14.9: Forces and design

Forces	At centre		At ends		Direct force		Shear force	
		M_{u_1}	BM	M_{u_2}	P	P_u	V	V_u
Loading conditions	BM	kN-m	kN-m	kN-m	kN	kN	kN	kN
Case I	86.6	130.0	79.9	120.0	74.77	112.20	166.5	250
Case II	100.98	151.50	65.52	98.3	21.86	32.80	166.5	250
Case III	111.27	167.0	55.93	82.9	9.23	13.9	166.5	250
Worst design value		167.0		120		112.20		250

$$d = \sqrt{\frac{167 \times 10^6}{2.761 \times 1000}} = 246 \text{ mm, Adopt } d = 300 \text{ mm, } D = 350 \text{ mm (edges thickened)}$$

$$A_{st} = \frac{3 \times 10^6}{415}\left[1 - \sqrt{1 - \frac{4.6 M_u \times 10^6}{20 \times 1000(300)^2}}\right] = 1756 \text{ mm}^2 \text{ (mid span), and } 1210 \text{ mm}^2,$$

at ends (1210 – 897 = 313 mm²) provide 20φ @ 175 c/c, alternate bent at ends plus 12φ @ 350 c/c additional.

- *Vertical walls* (Table 14.10):

Table 14.10: Forces and design

Forces	At centre		At ends		Direct force		Shear force	
		M_{u_1}	BM	M_{u_2}	P	P_u	V	V_u
Loading conditions	BM	kN-m	kN-m	kN-m	kN	kN	kN	kN
Case I	8.35	12.5	60.80	120	131.5	250	49.23	112.2
			79.90		166.5		74.77	
Case II	34.5	51.8	49.03	98.3	131.5	250	23.74	35.61
			65.52		166.5		21.86	
Case III	55.02	82.5	38.80	82.85	131.5	250	6.77	13.90
			55.23		166.5		9.23	
Worst design value		82.5		120		250		112.2

$$d = \sqrt{\frac{M_u \times 10^6}{2.76 \times 1000}} = 229 \text{ mm,} \quad \text{Adopt } d = 300 \text{ mm,} \quad D = 350 \text{ mm} \quad \text{(edges thickened)}$$

$$A_{st} = \frac{3 \times 10^6}{415}\left[1 - \sqrt{1 - \frac{4.6 \times 120 \times 10^6}{20 \times 1000(300)^2}}\right] - 1210 \text{ mm}^2 \text{ (at ends), and } 807 \text{ mm}^2 \text{ (mid span)}$$

Provide 12φ @140 c/c, alternate bent at ends. At ends 12φ @ 280 c/c plus 20φ @ 280 c/c between 12φ bent up bars.

Temp. steel $= \dfrac{0.12}{100} \times 1000 \times 350 = 420$ mm^2

(210 mm^2 each face—8ϕ @ 200 c/c, i.e. 250 × 2 = 500 mm^2/m)

Check the design for direct force on the vertical wall subject to eccentricity caused by bending moments. $M_{max} = 120$ kN-m, maximum load = 250 kN/m length.

Vertical walls as columns is 1000 mm × 350 mm, carrying maximum ultimate compressive load $P_u = 250$ kN and $M_{u\,max} = 120$ kN-m

Eccentricity $e = \dfrac{120000}{250000} = 0.48$ m (480 mm), more than the thickness.

Let the thickness of wall = 350 mm, $\dfrac{e}{D} = \dfrac{480}{350} = 1.37 < 1.50$ (less than 1.5)

$\dfrac{l}{b}$ for wall $= \dfrac{0.7 \times 4.0}{0.35} = 8.0$, wall may be considered short column.

(ends restrained $l_e = 0.7 \times 4 = 2.80$ m)

Use SP-16, interaction curves for axial force and uniaxial bending.

$f_{ck} = 20$ N/mm^2, $f_y = 415$ N/mm^2, $P_u = 250$ kN, $M_u = 120$ kN-m, steel on two sides

$D = 350$ mm, cover $d' = 50$ mm, $\dfrac{d'}{D} = \dfrac{50}{350} = 0.14$, $\left(\text{Use chart of } \dfrac{d'}{D} = 0.15\right)$

$\dfrac{P_u}{f_{ck}.b.D} = \dfrac{250000}{20 \times 1000 \times 350} = 0.036$, $\dfrac{M_u}{f_{ck}.b.D^2} = \dfrac{120 \times 10^6}{20 \times 1000 \times (350)^2} = 0.049$

Chart 33, P-118 of SP-16 design aids: $\dfrac{P}{f_{ck}} = 0.025$, $p = 0.025 \times 20 = 0.5\%$

$A_{sc} = \dfrac{0.5 \times 1000 \times 350}{100} = 1750$ mm^2 (875 mm^2 each face)

(20ϕ 300 c/c and $A_{st} = 1047 \times 2$ mm^2)

Already provided steel on each face is 20ϕ @ 280 c/c (>20ϕ @ 300 c/c, OK)

A_{st} available = 1121 mm^2 ($p = 0.32\%$ on each face)

Check for shear in wall: Maximum $V_u = 112.2$ kN (112200 N)

$\tau_v = \dfrac{112200}{1000 \times 300} = 0.374$ N/mm^2 (τ = for 0.32% steel, $\tau_{uc} = 0.394$ N/mm^2)

< 0.394 N/mm^2 design shear, OK (Refer to Table 19, P 73-IS: 456)

Refer to Fig.14.11 for reinforcement details.

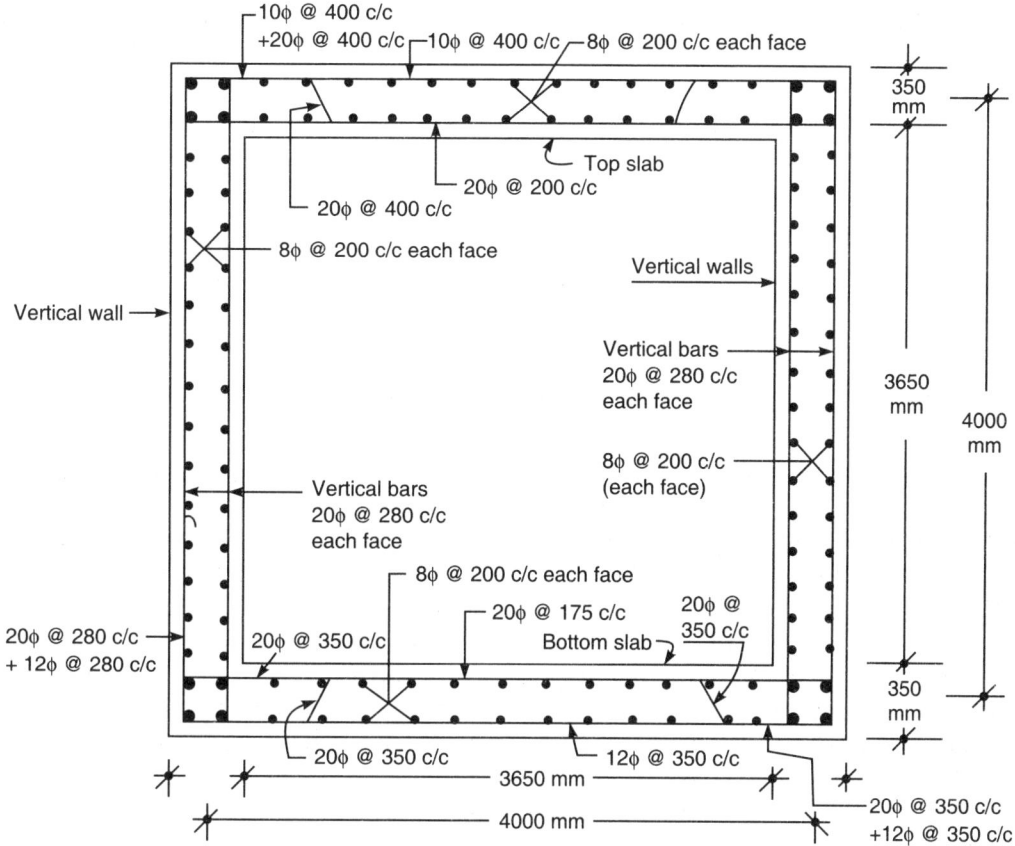

Fig. 14.11: Details of reinforcements of box culvert

SUMMARY

Culverts and bridges are needed to allow cross flow of drains, rivers or traffic at different levels. Culverts are provided for spans up to 8 m and bridges for spans more than 8 m. Generally Solid Slabs are provided for culverts or sometimes monolithically constructed RCC rigid box frame may be provided for the culvert. Culverts for road traffic are generally not less than 4 m span for single lane and 7.6 m span for double lane.

Culvert slabs are basically flexural members with design based on IRC (Indian road congress) loading of class AA, class A and class B. The culvert and bridge structures are subjected to dead, live, impact, tractive, centrifugal, wind, seismic, temperature and erection loads as specified in IRC clauses.

Dead loads comprises self-weight, finishing, wearing coat, kerbs, hand rail, etc. which act permanently. Live loads are caused by pedestrians, animals and vehicular traffic, etc. Normally live load of 4 kN/m² u.d.l. and 5 kN/m² u.d.l. for heavily crowded area may be assumed for design. For vehicular traffic IRC class AA, IRC class A and IRC class B may be considered as per given area and conditions. IRC loads may be placed in such a position which develops maximum BM, maximum shear or maximum deflection for the design purpose, i.e. worst condition of loading shall be used for the design. The structures designed for IRC class AA may also be checked for IRC class Aloading. Various IRC class AA, class A, or class B may be referred from IRC code.

No other live load shall be considered when IRC AA or IRC class A loads are acting on culvert or bridge structures. Details of IRC loadings may be taken from the IRC code. Impact factor I for spans 3 m to 45 m may be taken as $I = \left(\dfrac{4.5}{6 + L}\right)$ and impact load shall be considered in addition to live load (i.e. total load = $W(1 + I)$. Centrifugal force along curves may be taken as $C = \dfrac{W.V^2}{1.27R}$

where, W = wheel load (N), V = design speed (km/h)

R = radius of curvature in (m).

A box culvert is used to cross a drain across road or railway embankment. A box culvert is a continuous rigid frame consisting of two side abutments, top slab, and bottom slab cast monolithically to provide rigid joints. For high embankments with soil of low bearing capacity, the box culverts are economical compared to normal slab culverts.

Generally, the box culverts are subjected to soil pressure from outside and water pressure from inside the box. Bending moments and forces in various components (viz. top slab,bottom slab and two side walls may be computed for design. These components are basically flexural members. Vertical walls also act as a compression member subjected to axial compression along with uniaxial bending moment. When the eccentricity is more than the minimum, the design is done with the help of interaction curves provided in SP-16 design aids.

Moment analysis is done by the method of moment distribution considering half of the symmetrical frame. Relative stiffness of members may be considered at a joint and distribution factors may be computed for moment distribution of unbalanced moment at the joint.

PRACTICE QUESTIONS

(I) Objective Questions

Q. 14.1 Select the correct response given after each statement to complete it correctly and fill in the response sheet provided.

   (i) Maximum span of a culvert will be ....

     (a) 100 m   (b) 20 m

     (c) 8 m   (d) 5 m

  (ii) The slab of a slab culvert may be designed as a ....

     (a) flexural member for a general live load of 20 kN/m² u.d.l.

     (b) flexural member for a general live load of 4 kN/m²

     (c) compression member for an axial live load of 5000 kN/m²

     (d) tension member for an axial live load of 5 kN/m²

(iii) All permanent culverts or bridges on all type of roads must be normally designed for a loading of ....
   (a) IRC class AA and checked for IRC class B loading
   (b) IRC class B loading and checked for IRC class A loading
   (c) IRC class AA loading only
   (d) IRC class A loading

(iv) The maximum load of one train of the wheeled vehicles, IRC class AA loading shall be .... for single and for double axles respectively for bogie of two axles spaced not more than 1.20 m centres.
   (a) 200 kN and 400 kN
   (b) 200 kN/m² and 400 kN/m² u.d.l.
   (c) 4 kN/m² and 5 kN/m² u.d.l.
   (d) 54 kN and 114 kN

(v) The minimum clearance C between the road kerb face and multilane bridge of less than 5.5 m carriage way width shall be ....
   (a) 0.30 m   (b) 0.60 m
   (c) 1.20 m   (d) 2.10 m

(vi) The effect of impact of 10 kN/m moving u.d.l. in case of IRC class A loading for a concrete bridge of 9 m span (range 3 m to 45 m) will be I = ...

   (a) $\left(\dfrac{4.5}{9+10} \right)$ (b) $\left(\dfrac{4.5}{6+45} \right)$

   (c) $\left(\dfrac{4.5 \times 10}{9+6} \right)$ (d) $\left(\dfrac{4.5 \times 10}{3+6} \right)$

(vii) The value of impact coefficient for RCC bridge of 9 m span for IRC class AA loading for wheeled vehicles will be I = ...
   (a) 10%   (b) 15%
   (a) 20%   (d) 25%

(viii) Slab of a slab culvert rests on two abutments of 400 mm thickness and 6 m clear distance between the two abutments. The slab shall be about 300 mm thick excluding a cover of 40 mm. Determine the effective span of the slab
   (a) 6.0 m   (b) 6.30 m
   (c) 6.34 m   (d) 6.40 m

(ix) A moveable train of u.d.l. AB of 2 m length and 10 kN/m intensity of causing maximum BM in a slab CD of 6m simply supported span, the load AB shall be so placed that the ....
   (a) CG of load AB coincides with the mid span point
   (b) CG of load AB coincides with the 1/3rd span point
   (c) CG of load AB coincides with the quarter span point
   (d) tail end A of the load AB coincides with the simple support C

(x) A box culvert is a monolithically cast continuous rigid frame and analysed by moment distribution by considering ....

(a) worst condition of temperatur

(b) best condition of water flow

(c) worst condition of loading

(d) worst condition of weather

Response sheet to Q. 14.1 (i to x)

Question	(i)	(ii)	(iii)	(iv)	(v)	(vi)	(vii)	(viii)	(ix)	(x)
Response (a/b/c/d)										

(II) Numerical Questions

Q. 14.2 Design a reinforced slab culvert for the following requirement: clear span = 6 m, width of support = 400 mm, width of carriage way = 7.5 m, width of kerb = 600 mm, type of loading: IRC class AA or A whichever gives worst effect.

Grade of concrete M25, grade of steel = Fe 415

Sketch details of reinforcement.

[**Hint:** D slab = 510 mm, d = 460 mm (Fig. 14.12)

Main steel = 20 @ 125 c/c, alternate bars bent up at ends (A_{st} = 2512 mm²/m) (approximate design)

Road kerb 600 mm × 800 mm (say), bottom = 5 bars 20 and top 2 bars 20ϕ

Stirrups 8ϕ 2 legged @ 150 c/c

Lateral bars in slab (distribution) – 80ϕ @ 150 c/c on both faces.

(A_{st} = 2 × 333 = 666 mm², OK)

Moments are computed for worst placement of loading for IRC class AA or class A whichever gives maximum moments and shears.

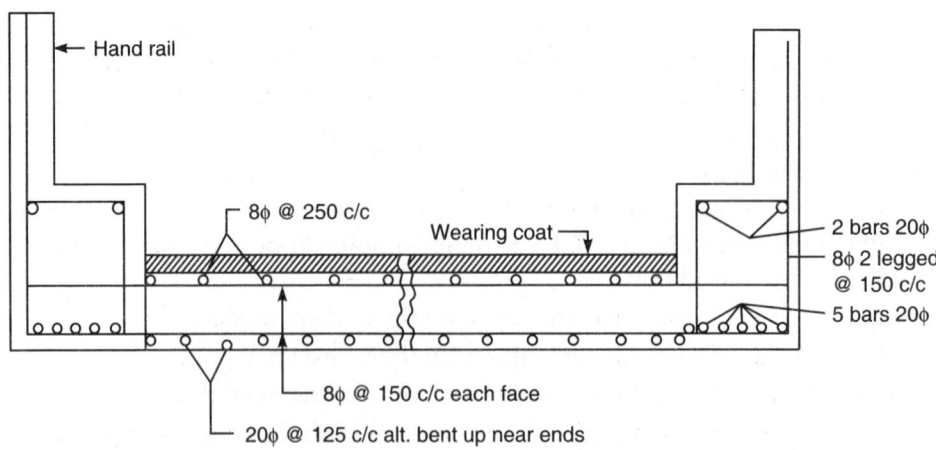

Fig. 14.12: Details of reinforcement in slab culvert

Q. 14.3 Design a box culvert having inside dimensions 3.5 m × 3.5 m. The box culvert is subjected to a superimposed dead load of 12 kN/m² and live load of 45 kN/m² from the road structure and traffic. Assume unit weight of soil as 18 kN/m³ and angle of repose 30°, use M20 cc and Fe 415 steel.

[**Hint:** C/C dimension of box = 3.80 m × 3.80 m, assuming D = 300 mm,

Total load on top slab AB (W) = 7.5 + 45.0 + 12 = 64.5 kN/m² $\downarrow$

Upward reaction (bottom slab CD) = $\dfrac{(64.5 \times 3.8 + 2 \times 28.5)}{3.8}$ = 79.5 kN/m $\uparrow$

$K_a = \dfrac{1}{3}$, lateral pressure on side walls = $\dfrac{1}{3}$ (45 + 12) = 19 kN/m²

Lateral pressure at 'h' m below (due to soil) = $\dfrac{1}{3}$ $w.h$ = $\dfrac{18h}{3}$ = $6h$ kN/m²

Total lateral pressure on walls = (19 + h) kN/m²

P_1 (at box top) = 19 kN/m²,

P_2 (at box bottom) = 19 + 6 × 3.8 = 41.8 kN/m².

Distribution factors: AB (top slab) = $\dfrac{1}{3}$ DC (bottom slab)

AD (vertical wall) = $\dfrac{2}{3}$ DA

Case I:

$$M_{FAB} = \frac{w.l^2}{12} = -\frac{64.5(3.8)^2}{12} = -77.61 \text{ kN-m},$$

$$M_{FAD} = \frac{19 \times (3.8)^2}{12} + \frac{1}{2} + (22.8 \times 3.8) \frac{3.8}{15} = 33.83 \text{ kN-m}$$

$$M_{FDC} = \frac{w.l^2}{12} = \frac{79.5(3.8)^2}{12} = +95.67 \text{ kN-m},$$

$$M_{FDA} = \frac{p.l^2}{12} - \frac{wl}{10} = -39.32 \text{ kN-m}$$

Final moments after moment distribution at the joint A and D, we have

M_{AD} = + 54.2 kN-m, M_{AB} = –54.2 kN-m

M_{DA} = – 69.01 kN-m, M_{DC} = + 69.01 kN-m

M_{UAD} = 1.5 × 54.2 = + 81.32 kN-m, M_{UAB} = –1.5 × 54.2 = – 81.32 kN-m

M_{UDA} = –1.5 × 69.07 = –103.52 kN-m, M_{UDC} = +1.5 × 69.07 = +103.52 kN-m

Vertical reactions at A and B = 122.55 kN (V_{u1} = 183.83 kN)

Vertical reactions at D and C = 151.05 kN (V_{u2} = 226.58 kN)

For vertical member AD, horizontal reactions $h_a = 46.62$kN, $h_d = 68.90$ kN.

Net ultimate moment at midpoint E top slab = 1.5 (116.42 – 54.2) = 93.33 kN-m

Net ultimate moment at midpoint F (bottom slab) = 1.5 (143.50 – 69.07)

$$= 111.65 \text{ kN-m}$$

Net ultimate moment at mid-point span of vertical wall

$$= 1.5 \left[\frac{(69.07 + 54.2)}{2} - 54.87 \right] = 10.12 \text{ kN-m}$$

Similarly for cases 2 and 3 loading the moment can be found and the design of components based on worst loading.

$$d = \sqrt{\frac{111.65 \times 10^6}{2.761 \times 1000}} = 201.1 \text{ mm, provide } D = 350 \text{ mm, } d = 300 \text{ mm}$$

$$A_{st} = \frac{1000 \times 300}{415} \left[1 - \sqrt{1 - \frac{4.6 \times 111.65 \times 10^6}{20000 (300)^2}} \right] = 1118 \text{ mm}^2$$

(20ϕ @ 225 c/c alternate bent down at ends).

At ends, provide lower face 12ϕ @ 450 c/c throughout, in addition to bent down bars of 20ϕ @ 450 c/c.

Chapter 15

Yield Line Analysis of Slabs

LEARNING OBJECTIVES

After the study of *yield line analysis of slabs*, the learner will understand the analysis of slabs by yield line theory and shall be able to:

- ⊙ Explain yield line analysis of slabs
- ⊙ State assumptions in yield line analysis
- ⊙ State specific characteristics of yield lines
- ⊙ Derive moment capacity across the yield line
- ⊙ Derive moment capacity across the yield line by virtual work method
- ⊙ Derive moment capacity of different shape of slabs with different supports
- ⊙ Design slabs using yield line theory

15.1 INTRODUCTION

Yield line theory is based on ultimate load for RCC slabs. A yield line is defined as a line in the plane of the slab along which all reinforcing steel bars have yielded. With the yielding of bars, the concrete cracks along the yield line as shown in Fig. 15.1

With the increase in loading, tensile stress in the steel bars also increases. Stress on reaching yield stress of steel, the bars start yielding and concrete starts cracking. The concrete reaches its maximum yield moment and stops offering resistance and hence excessive deflection occurs. Hinge is formed along the yield line and rotation of slab takes place about the yield line. At the time of failure a mechanism will be formed. Slabs with different type of supports may form different yield lines. Sometimes the formation of one yield line may not result in collapse mechanism. Redistribution of moments may occur and some more yield lines develop before formation of collapse mechanism. Figure 15.2 shows a typical yield line collapse mechanism for a two-way slab.

The yield line theory provides method of computing collapse load at the time of failure of RCC slab. It may be noted that the RCC slab must fail in tension and hence RCC slab has to be under reinforced. Modern codes recommend the design of under reinforced sections for adequate warning before collapse.

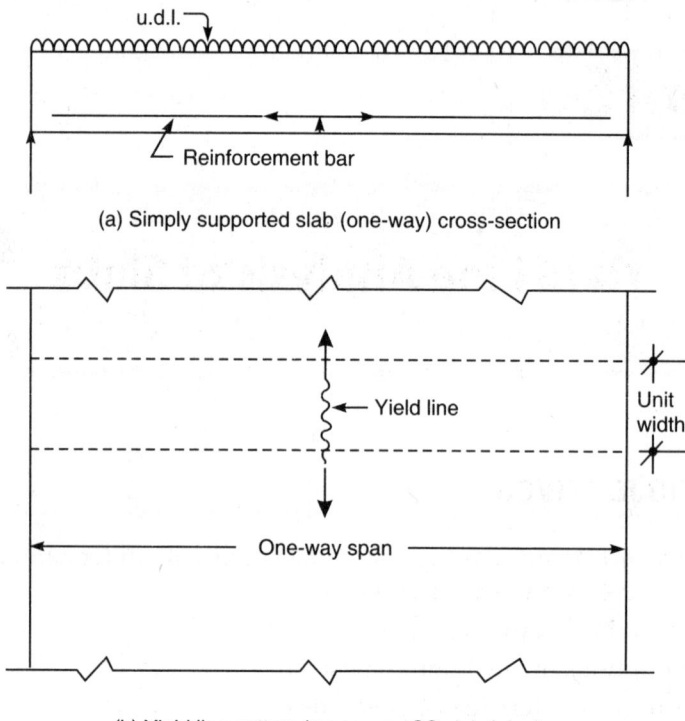

(a) Simply supported slab (one-way) cross-section

(b) Yield line pattern in one-way SS slab (plan)

Fig. 15.1: Yield line

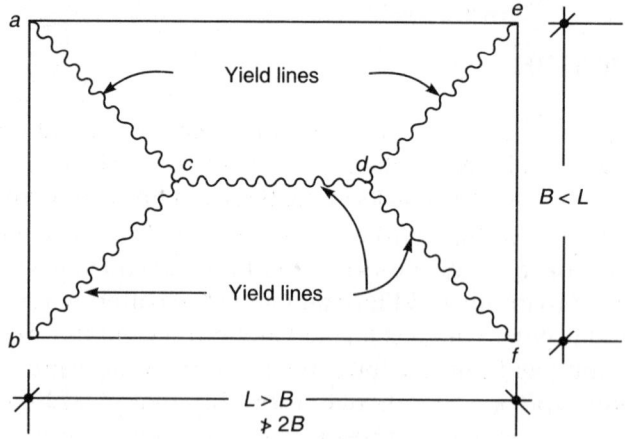

Fig. 15.2: Yield line in two-way SS slab

Different pattern of yield lines develop depending on support conditions and these are shown in Fig. 15.3. The calculation of ultimate collapse load is done by using virtual work method and equilibrium method. For yield line analysis of slabs, certain assumptions are made.

15.2 ASSUMPTIONS

Yield line analysis of slabs is based on the following assumptions:

(i) RCC slabs are under reinforced so as to develop tension failure.

(ii) The yield lines are straight lines.

(iii) Elastic deformation is considered negligible compared to plastic deformations.

(iv) After formation of collapse mechanism, each segment of the slab may be considered as rigid body. The entire rotation is assumed to occur along the yield line.

15.3 CHARACTERISTICS

In the yield line analysis of slab, it is necessary to predict the yield line pattern at the time of collapse. Yield lines will have the following characteristics (Fig. 15.3).

(i) Yield lines are straight lines and act as hinges of collapse mechanism.

(ii) Yield lines either terminate at the boundary of slab or at the intersection of other yield lines.

(iii) Yield lines act as axes of rotation of the segments of slab.

(iv) The axes of rotation lie along the lines of supports. Each segment of the slab will tend to rotate as a rigid body motion.

(v) A yield line may be formed along the support, if the edge is fixed or continuous.

(vi) The mechanism formed by the yield lines result in compatibility of deformations and the yield lines produced, pass through the intersection of the axes of rotation of adjacent slab segments.

15.4 SIGN CONVENTIONS AND YIELD LINE PATTERNS

Different yield lines pattern of collapse mechanisms are assumed as shown in Fig. 15.3(a). Different sign conventions are shown in Fig. 15.3(b).

15.5 MOMENT CAPACITY ACROSS A YIELD LINE

Yield line may occur at right angles to the direction of main reinforcement (x-direction) or the yield line may make an angle α with the direction of normal to main reinforcement.

According to IS:456-2000 code, we have ultimate moment capacity per unit width of slab, given by

(i) $M_u = 0.87 f_y \cdot A_{st} \cdot d \left[1 - \dfrac{A_{st}}{bd} \cdot \dfrac{f_y}{f_{ck}} \right]$

(ii) When the yield line makes an angle α with the normal to the main reinforcement direction (x-direction), i.e. the yield line makes an angle with the y-direction (normal to x-direction).

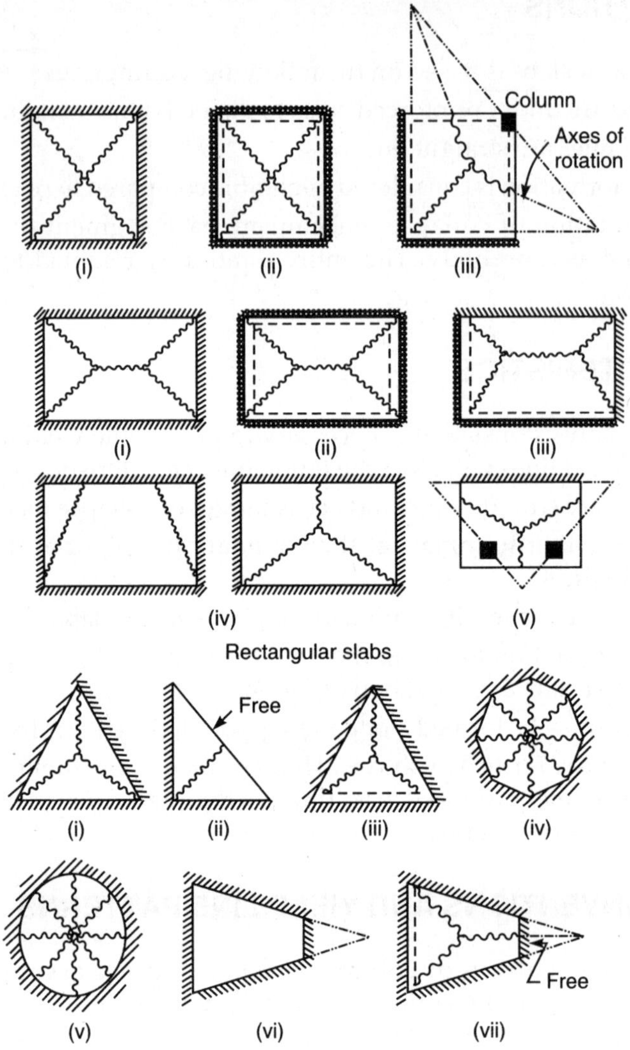
Column
Axes of rotation
(i)
(ii)
(iii)
(i)
(ii)
(iii)
(iv)
(v)
Rectangular slabs
Free
(i)
(ii)
(iii)
(iv)
Free
(v)
(vi)
(vii)
Slabs of other shapes

(a) Yield line patterns

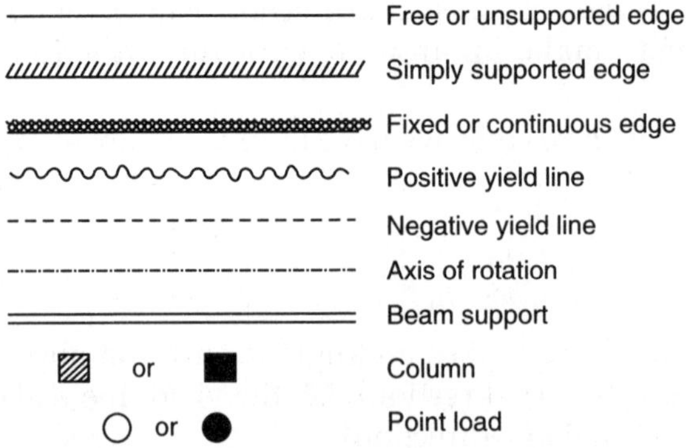
Free or unsupported edge
Simply supported edge
Fixed or continuous edge
Positive yield line
Negative yield line
Axis of rotation
Beam support
or
Column
or
Point load

(b) Symbols in yield line

Fig. 15.3: Different patterns in yield lines and symbols

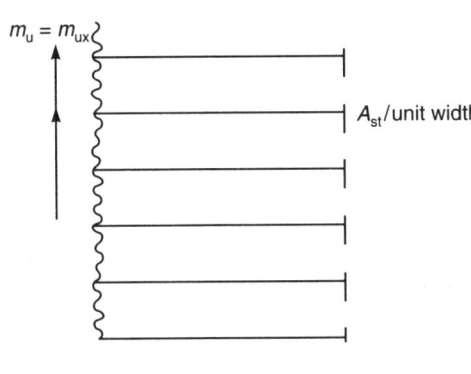

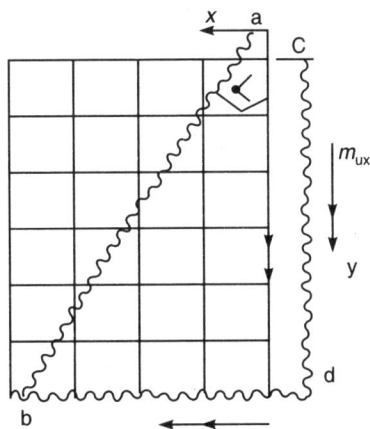

(a) Yield line normal to reinforcement   **(b) Yield line inclined to normal to the reinforcement**

Fig. 15.4: Yield line normal and inclined to normal

Let m_{ux} and m_{uy} be moment capacities per unit width in normal to x- and y-directions (Fig. 15.4), we have

$$M_{ux} = 0.87f_y.A_{stx}.d\left[1 - \frac{A_{stx}}{bd}\cdot\frac{f_y}{f_{ck}}\right]$$

$$M_{uy} = 0.87f_y.A_{sty}.d\left[1 - \frac{A_{sty}}{bd}\cdot\frac{f_y}{f_{ck}}\right]$$

Consider length ab of yield line. The projections of length ab along normal to x- and y-axes are cd and bd respectively. The component of m_{ux} to yield moment $m_{\alpha 1}$ is given as

$$m\alpha_1\,(ab) = m_{ux}.cd\,(\cos\alpha) = m_{ux}.(ab\cos\alpha).\cos\alpha = m_{ux}.ab\cos^2\alpha$$

or $$m\alpha_1 = m_{ux}.\cos^2\alpha$$

Similarly, the component of m_{uy} to yield moment $m\alpha_2$ is given as

$$m\alpha_2(ab) = m_{uy}.(bd)\,\alpha\cos(90°-\alpha) = m_{uy}\,(ab\sin\alpha).\sin\alpha = m_{uy}.ab\sin^2\alpha$$

or $$m\alpha_2 = m_{uy}.\sin^2\alpha$$

Total yield moment $m\alpha = m\alpha_1 + m\alpha_2 = m_{ux}.\cos^2\alpha + m_{uy}.\sin^2\alpha,$

i.e. $m_\alpha = m_{ux}.\cos^2\alpha + m_{uy}.\sin^2\alpha$ ... Eq. (15.1)

When the slab is reinforced isotropically (equal in x- and y-directions)

(i.e. $m_{ux} = m_{uy} = m\alpha$), then we have

$$m\alpha = m_u.\cos^2\alpha + m_u.\sin^2\alpha = m_u\,(\cos^2\alpha + \sin^2\alpha) = m_u,$$

i.e. incase of isotropically reinforced slab sections, yield moment is same in all directions.

If $m_{uy} = \mu m_{ux}$, i.e. the slab is orthotropically reinforced, the yield moment at an angle α with normal to x, is given as m_α and

$$m_{u\alpha} = m_u\,(\cos^2\alpha + \mu\sin^2\alpha) = m_u\,(\cos^2\alpha + \mu\sin^2\alpha)\quad\text{... Eq. (15.2)}$$

where, $m_{u\alpha}$ = ultimate moment of resistance per unit length of yield line, which makes an angle α with the normal of x-direction (i.e. makes an angle α with y-direction)

m_{ux} = ultimate moment of resistance per unit length in x-direction.

m_{uy} = ultimate moment of resistance per unit length in y-direction.

α = inclination of yield line with y-direction or normal of x-direction.

μ = ratio of ultimate unit moment of resistance about y-axis to ultimate unit moment of resistance about x-axis

$$\left(\text{i.e.} \, \mu = \frac{m_{uy}}{m_{ux}}\right)$$

$\therefore$ $\mu = 1$, in case of isotropically reinforced slabs.

The ultimate unit moment of resistance along a yield line at $\alpha = 45°$ shall be given as

(a) $m_{u\alpha} = m_{ux}$ in case of isotropically reinforced slab   ... Eq. (15.3a)

(b) $m_{u\alpha} = \dfrac{m_{ux}\left(1+\mu\right)}{2}$ in case of orthotropically reinforced slab   ... Eq. (15.3b)

15.6 ANALYSIS OF ULTIMATE LOAD CAPACITY OF SLABS

There are two methods for determining the ultimate load carrying capacity of slabs. These are:

(a) Virtual work method (also known as mechanism method): In virtual work method a collapse yield line pattern is assumed after the assumption of this collapse mechanism, the rotations of yield lines are quite large compared to elastic deformations. The elastic deformations are therefore neglected and virtual work done by moments to rotate the slab segments about the yield lines shall be computed and equated to the virtual work done by the loads on the slab (i.e. internal work done by the moments = external work done by the loads).

(b) Equilibrium method (also known as statical method): In this method, the segment of a slab, after the formation of the collapse mechanism, shall be considered in equilibrium and the equations of equilibrium shall be applied to determine the load.

It may be noted that both the methods give upper bound solutions, i.e. the maximum possible load carrying capacity. The predicted load carrying capacity is either equal or more than the actual load carrying capacity found by actual testing. The mechanism which gives the least value will be predicting the correct load carrying capacity. For simple cases it may be possible to easily choose the correct collapse mechanism. For complex cases several mechanisms may be tried to predict the collapse load by selecting the least of these values. It is found by tests that the actual collapse load is more than the predicted load carrying capacity with exact collapse mechanism. Hence the upper bound solutions obtained from the yield line theory can be used with reasonable degree of safety.

15.7 ISOTROPICALLY REINFORCED SS SQUARE SLAB WITH U.D.L

15.7.1 Virtual Work Method

The principle of virtual work states that:

"If a deformable structure in equilibrium under the action of a system of external forces is subjected to a virtual deformation compatible with its condition of support, the work done by these forces on the displacements associated with the virtual deformation is equal to the work done by the internal stresses on the strains associated with this deformation".

Hence, the work done during small motion of the collapse mechanism is equal to the work absorbed by the plastic hinges formed along the yield lines. The segments of the slab within the yield lines undergo through rigid body displacements with the collapse load on the structure.

Work done by external forces = work done by internal forces.

Work done by external loads on the slab = $W_{ext} = \iint W_u \cdot \delta xy \cdot dxdy = W_u \cdot \Delta \delta xy$

$$\Delta \delta xy = \text{virtual displacement}$$

where, W_u = resultant load on each segment

Δ = corresponding displacement at the centroid of load in each segment

m_u = ultimate moment across the yield line

l_0 = length of yield line

θ_n = relative rotation of two adjacent segments, perpendicular to the yield line

$W_{int} = \Sigma m_u \cdot \theta_n \cdot l_0$, hence we have

$W_{ext} = W_{int}$

or $\Sigma W_u \cdot \Delta = \Sigma m_u \cdot \theta_n \cdot l_0$. (work done by twisting moment is not considered)

This principle is illustrated with following examples:

i. *Isotropically reinforced SS square slab carrying u.d.l. on entire surface*

Typical yield lines in simply supported square slab is shown in Fig. 15.5.

Let m_u be the ultimate moment capacity of slab. The slab is reinforced isotropically and moment capacity will be the same in all directions.

Moment capacity across yield lines will also be m_u

Length of yield line 'ac' along the diagonal = $\sqrt{2}\, L$

Yield moment across 'ac' will be = $m_u \sqrt{2}\, L = \sqrt{2}\, m_u \cdot L$

Let the central point 'o' be given by virtual displacement δ

Resulting rotation $\theta = \dfrac{\delta}{\dfrac{1}{2}\left(\sqrt{2}L\right)}$

Total rotation of yield line 'ac' = $2\theta = 2\sqrt{2}\,\dfrac{\delta}{L}$

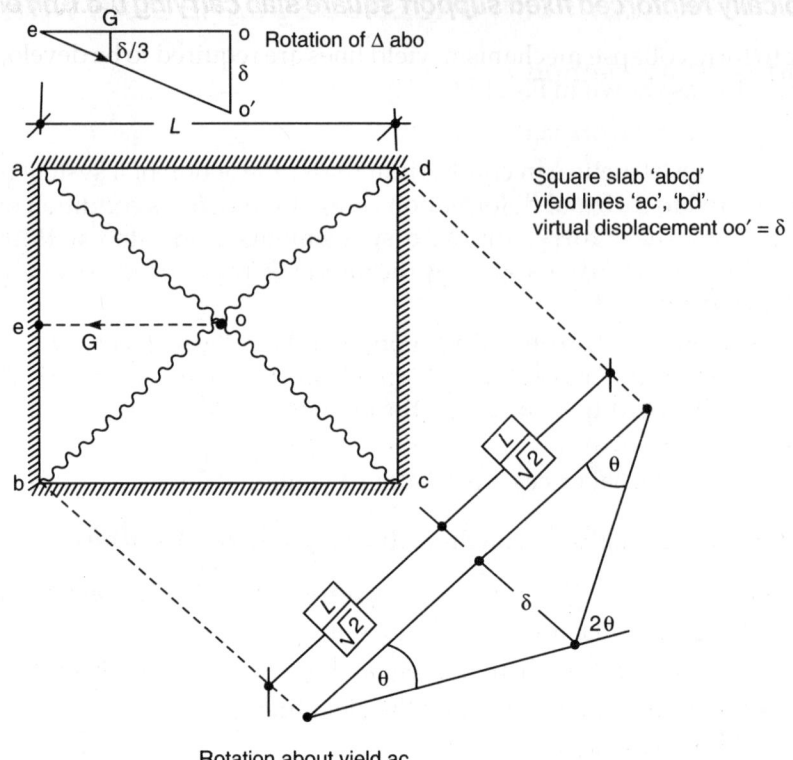

Rotation about yield ac

Fig. 15.5: SS square slab — yield lines aoc, bod, oo' = δ

Virtual work done by moment about yield line 'ac' = $\left(\sqrt{2}\, m_u . L \right) 2 \sqrt{2}\, \dfrac{\delta}{L}$

Virtual work done about yield line 'ac' = $4 m_u \cdot \delta$

Similarly virtual work done about yield line 'bd' = $4 m_u \cdot \delta$

Total internal virtual work done = $8 m_u \cdot \delta$

External work done by u.d.l. = w_u / m^2

There are 4 triangular segments, load on each segment = $\dfrac{w_u \cdot L^2}{4}$

Each triangular segment CG at $\dfrac{1}{3} \cdot \dfrac{L}{2}$ distance from base. Point o displaces by 'δ'

and CG displaces by $\dfrac{\delta}{3}$. Hence work done by external load = $\dfrac{w_u . L^2}{4} \cdot \dfrac{\delta}{3}$,

i.e. total external virtual work = $4 \times \dfrac{w_u \cdot L^2}{4} \cdot \dfrac{\delta}{3} = w_u \cdot L^2 \cdot \dfrac{\delta}{3} = \dfrac{w_u \cdot L^2}{3} \cdot \delta$

Total internal virtual work = total external virtual work,

i.e. $\qquad\qquad 8\, m_u \delta = w_u \cdot L^2 \cdot \dfrac{\delta}{3}$

∴ $\qquad\qquad\qquad m_u = \dfrac{w_u \cdot L^2}{24}$ $\qquad\qquad\qquad$ ... Eq. (15.4)

ii. *Isotropically reinforced fixed support square slab carrying u.d.l. on entire slab*

In this case to form collapse mechanism, yield lines are required to be developed along the supports also, as shown in Fig. 15.6

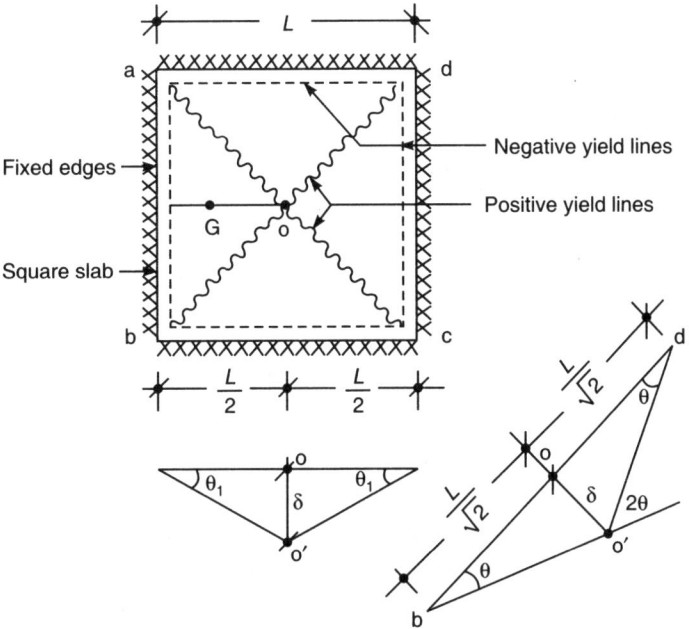

Fig. 15.6: Fixed support square slab with u.d.l. (w_u) on entire slab

If δ is the vertical displacement of centre point o, let the rotation about support yield lines be θ_1 (Fig. 15.6), we have $\theta_1 = \dfrac{2\delta}{L}$,

internal work done about edge yield lines $= 4 \times m_u.L.\theta_1 = 4\, m_u.L.\dfrac{2\delta}{L} = 8\, m_u.\delta$

Work done by +ve yield lines ac and bd $= (m_u.L).2.\dfrac{\delta\sqrt{2}}{L} \times 2 = 8\, m_u.\delta$

Total internal work done $= 8m_u.\delta + 8m_u.\delta = 16\, m_u.\delta$

External work done by load on 4 segments of slab shall be equal to

$$= 4\left[\frac{w_u\,L^2}{4} \times \frac{\delta}{3}\right]\frac{1}{3} = w_u.\delta.L^2$$

Equating total external virtual work to total internal virtual work,

i.e. $\qquad\qquad 16m_u.\delta = \dfrac{1}{3}w_u.\delta.L^2$

or $\qquad\qquad\qquad m_u = \dfrac{w_u \cdot L^2}{48}$ $\qquad\qquad\qquad$ ... Eq. (15.5)

iii. *Equilateral triangular isotropic slab (SS)*

Figure 15.7 shows an equilateral triangular isotropically reinforced slab with sides L and also the yield line pattern by lines OA, OB and OC. Since the slab is isotropic, moment about any yield line shall be m_u.

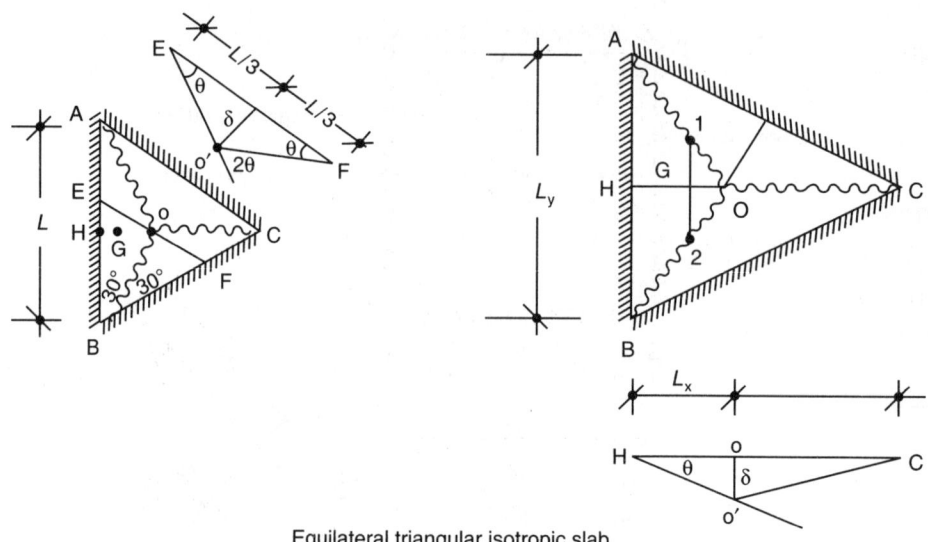

Equilateral triangular isotropic slab

Fig. 15.7: Equilateral SS triangular isotropic slab with u.d.l. on entire slab

Length of yield line OB = OA = OC = $\dfrac{L}{2} \cdot \dfrac{1}{\cos 30°} = \dfrac{L}{2} \times \dfrac{2}{\sqrt{3}} = \dfrac{L}{\sqrt{3}}$

Rotation about yield line may be computed by considering rotation of EOF,

$$OE = OF = \frac{1}{2} \times \frac{2L}{3} = \frac{L}{3}, \quad \left\{ \because EF = \frac{2L}{3}, OE = \frac{1}{2}, EF = \frac{L}{3} \right\}$$

Let δ be the displacement of the point o,

$\dfrac{L}{3} \cdot \theta = \delta,\, \theta = \dfrac{3\delta}{L}$, rotation about yield line = $2\theta = 2 \times \dfrac{3\delta}{L} = \dfrac{6\delta}{L}$

Internal work done about yield line OB,

$$m_u.l.2\theta = m_u. \frac{L}{\sqrt{3}}. \frac{6\delta}{L} = m_u. 2\sqrt{3}.\delta$$

Total internal work done = $3 \times m_u.2\sqrt{3}\,\delta = 6\sqrt{3}\,m_u.\delta$

External work done by loads:

consider segment AOB,

load on area AOB × virtual displacement at CG of segment

$$\text{External work done} = w_u \times \frac{L}{2} \times OH \times \frac{\delta}{3} = w_u . \frac{L}{2}\left(\frac{L}{2} \tan 30° \right) \frac{\delta}{3}.$$

$$= \frac{w_u.L^2.\delta}{12} \times \frac{1}{\sqrt{3}} = \frac{w_u.L^2.\delta}{12\sqrt{3}}$$

Total external work on 3 segments $= 3 \times \dfrac{w_u.L^2.\delta}{12\sqrt{3}} = \dfrac{w_u.L^2.\delta}{4\sqrt{3}}$

Equating external and internal work

$$6\sqrt{3}\, m_u.\delta = \frac{w_u.L^2.\delta}{4\sqrt{3}} \quad \text{or} \quad m_u = \frac{w_u.L^2}{72} \qquad \text{... Eq. (15.6)}$$

iv. Simply supported rectangular slab with orthogonal reinforcement and u.d.l

Figure 15.8 shows a simply supported slab on all the four edges with orthogonal reinforcement and carrying u.d.l. on the entire span. Let the length AB = CD = L and width AD = αL.

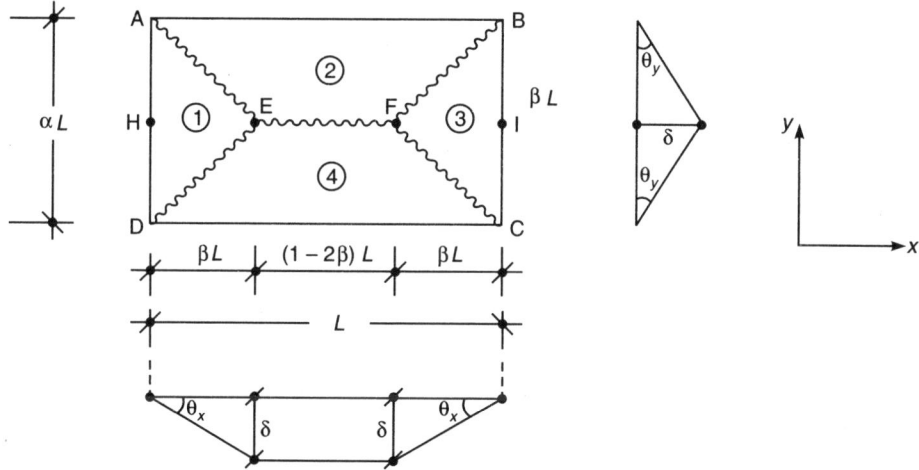

Fig. 15.8: SS Rectangular slab with orthogonal reinforcement

For part 1: $L_x = 2\beta L$ (for yield line AED) and $L_y = \alpha L$ (for yield line AED).
Let virtual displacement at E and F be = δ.
Let $\qquad\qquad\qquad m_{ux} = m_u,\ m_{uy} = \mu m_u$
Consider segment (1) and (3)
Internal work done $W_{int} = m_{ux}\,(2\,\beta L).\theta_x + m_{uy}.\alpha.L.\theta_y,\ \theta_{x1} = 0,\ \because$ level A, D and H remains the same.

$$\theta_y = \frac{\delta}{\beta L}, W_{int1} = m_u.2\,\beta L \times 0 + \mu\, m_u.\alpha L.\frac{\delta}{\beta L} = \frac{\mu\alpha\delta}{\beta}\, m_u$$

$$(W_{int1} + W_{int3}) = 2\,\frac{\mu\alpha\delta}{\beta}\, m_u$$

Consider segment (2) and (4), rigid body AEFB and DEFC:

L_x (projected yield line along x) = L, and $L_y = \dfrac{\alpha L}{2} \times 2$ (projection AE and BF)

$m_{ux} = m_u$ and $m_{uy} = \mu m_u$, θ_x (rotation of x-projection) $= \dfrac{\delta}{\dfrac{\alpha L}{2}} = \dfrac{2\delta}{\alpha L}$ $\theta_y = 0$

$\theta_y = 0$ ($\because$ any 2 points in x-direction are at the same level).

Work done internally in segment 2 = $W_{int2} = m_u.L.\dfrac{2\delta}{\alpha L} + \mu m_u \times \dfrac{\alpha L}{2} \times 0 = 2\,m_u \cdot \dfrac{\delta}{\alpha}$

Total work done in segment 2 and 4 = $2 \times 2\,m_u \dfrac{\delta}{\alpha} = 4\,m_u \cdot \dfrac{\delta}{\alpha}$

Total internal virtual work done = $\left(2\mu.m_u \cdot \delta \cdot \dfrac{\alpha}{\beta} + 4\,m_u \dfrac{\delta}{\alpha} \right)$

$$= 2\,m_u \cdot \delta \left(\dfrac{2}{\alpha} + \dfrac{\mu\alpha}{\beta} \right)$$

External work done on virtual displacement δ (total 1 + 2 + 3 + 4)

$$w_u \left(\dfrac{\beta L \cdot \alpha L}{2} \times \dfrac{\delta}{3} \right) \times 2 + \left(w_u \cdot \dfrac{1}{2}\alpha \cdot L \cdot \dfrac{\beta L}{2} \times \dfrac{\delta}{3} \times 2 \right) 2 + \left\{ w_u \cdot (L - 2\beta L)\dfrac{\alpha L}{2} \cdot \dfrac{\delta}{2} \right\} 2$$

$$\Sigma W_{ext} = w_u \dfrac{L^2 \cdot \alpha \cdot \delta}{6}(3 - 2\beta)$$

Equating $\quad \Sigma W_{int} = \Sigma W_{ext}$ or $2\,m_u.\delta \left(\dfrac{2}{\alpha} + \dfrac{\mu\alpha}{\beta} \right) = w_u \dfrac{L^2 \cdot \alpha \cdot \delta}{6}(3 - 2\beta)$

or $\quad m_u = w_u \dfrac{L^2.\alpha^2.\beta}{12} \cdot \dfrac{(3 - 2\beta)}{(2\beta + \mu\alpha^2)} = w_u \dfrac{L^2.\alpha^2}{12} \left(\dfrac{3\beta - 2\beta^2}{2\beta + \mu\alpha^2} \right)$ ... Eq. (15.7)

For m_u to be maximum (i.e. w_u to be minimum) with respect to variable β:

$$\dfrac{dm_u}{d\beta} = 0, \text{ or } \dfrac{(3 - 4\beta)(2\beta + \mu\alpha^2) - (3\beta - 2\beta^2) \times 2}{(2\beta + \mu\alpha^2)^2} = 0$$

or $\quad (3 - 4\beta)(2\beta + \mu\alpha^2) = (3\beta - 2\beta^2) \times 2$

or $\; 6\beta - 8\beta^2 + 3\mu\alpha^2 - 4\,\mu\alpha^2 \beta = 6\beta - 4\beta^2$, or $\; 4\beta^2 + 4\,\mu\alpha^2 \beta - 3\,\mu\alpha^2 = 0$

$$\beta = \dfrac{-4\,\mu\alpha^2 \pm \sqrt{16\mu^2\alpha^4 + 4 \times 4 \times 3\mu\alpha^2}}{8} = \dfrac{-\mu\alpha^2 \pm \sqrt{\mu^2\alpha^4 + 3\mu\alpha^2}}{2}$$

Considering +ve root value, we have

$$\beta = \frac{1}{2}\left(\sqrt{\mu^2\alpha^4 + 3\mu\alpha^2} - \mu\alpha^2\right)$$

Substituting the value of β in Eq. (15.7), we get

$$m_u = \frac{w_u \cdot \alpha^2 \cdot L^2}{24}\left(\sqrt{3 + \alpha^2 \cdot \mu} - \alpha\sqrt{\mu}\right)^2 \qquad \text{... Eq. (15.8a)}$$

Deriving specific case of SS square slab carrying u.d.l. over entire span, we have for square α = 1, for isotropically (μ = 1) reinforced slab

$$m_u = \frac{w_u \cdot L^2}{24}\left(\sqrt{4} - 1\right)^2 = \frac{w_u \cdot L^2}{24} \qquad \text{... same as Eq. (15.4)}$$

for SS rectangular slab with isotropically reinforced sections μ = 1

$$m_u = \frac{w_u \cdot \alpha^2 \cdot L^2}{24}\left[\sqrt{3 + \alpha^2} - \alpha\right]^2 \qquad \text{... Eq. (15.8b)}$$

v. *A polygonal slab isotropically reinforced and subjected to u.d.l. over the entire slab with edges fixed*

Let the length of side = L, and the perpendicular distance be r from the centre O. At collapse, the yield lines may develop as shown in Fig. 15.9. Let m_u be the yield line moment per unit length along the positive yield line AO, BO, ...etc. and m_{u1} along negative yield lines AB, BC, ...etc.

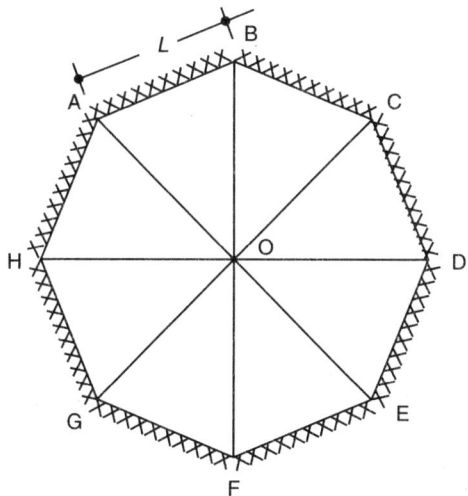

Fig. 15.9: Polygonal isotropically reinforced slab with and carrying u.d.l. on entire slab with edges fixed

Let there be virtual displacement δ at the centre point O. Consider rigid body rotation of ABO about O.

$$\theta = \frac{\delta}{r}, \text{moment of negative yield line AB} = m_{u1} \times L$$

Work done in rotating negative yield line $= m_{u_1} \times L \times \theta = m_{u_1} \times L \times \dfrac{\delta}{r}$

Work done along positive yield line $= m_u \cdot L \cdot \theta = m_u \cdot L \cdot \dfrac{\delta}{r}$

(projection of AO along AB $= L$ and rotation of AB $\theta = \dfrac{\delta}{r}$

Total internal virtual work of each segment $(W_{int}) = (m_{u_1} + m_u) L \cdot \dfrac{\delta}{r}$

Total internal work done $\Sigma W_{int} = n(m_{u_1} + m_u) L \dfrac{\delta}{r}$

External work done by each segment $= \left[w_u \times \dfrac{L \cdot r}{2} \right] \dfrac{\delta}{3} = \dfrac{w_u}{6} (L \cdot r \cdot \delta)$

Total external work done $\Sigma W_{ext} = \dfrac{n \cdot w_u}{6} (L \cdot r \cdot \delta)$

Equating total internal work and total external work, we have

$$(m_{u_1} + m_u) \dfrac{n \cdot L \cdot \delta}{r} = \dfrac{n \cdot w_u}{6} (L \cdot r \cdot \delta)$$

or $\qquad (m_{u_1} + m_u) = \dfrac{w_u}{6} \cdot r^2 \qquad\qquad$ ... Eq. (15.9)

(a) For regular hexagon: α at centre $= \dfrac{360°}{6} = 60°, \dfrac{\alpha}{2} = 30°, r = \dfrac{L}{2}$ cot $30° = \dfrac{L}{2}\sqrt{3}$

Thus $\qquad (m_{u_1} + m_u) = \dfrac{w_u}{6} \left(\dfrac{L\sqrt{3}}{2} \right)^2 = \dfrac{w_u}{6} L^2 \cdot \dfrac{3}{4} = \dfrac{w_u L^2}{8} \qquad$ ... Eq. (15.10)

(b) For regular octagon: $\alpha = \dfrac{360°}{8} = 45°, r = \dfrac{L}{2}$ cot $\dfrac{45°}{2} = 1.207L$

$\therefore \qquad (m_{u_1} + m_u) = \dfrac{w_u}{6} (1.207L)^2 = 0.24285 \, w_u \cdot L^2 \qquad$ ... Eq. (15.11)

In case of circular slab $r = R$ (radius of circular slab with edge fixed), the moment along the yield lines will be

$$(m_{u_1} + m_u) = \dfrac{w_u}{6} \cdot R^2 \qquad\qquad ... \text{Eq. (15.12)}$$

Here the circular slab is considered as polynomial and n is very large so that L tending to zero and r tending equal to radius R.

15.7.2 Yield Line Analysis by Equilibrium Method

In this method, moment equilibrium of segments is considered after collapse mechanism is formed and relationship between moment carrying capacity and load carrying capacity is determined. Different cases are illustrated below:

i. *Square slab isotropically reinforced*

(a) Simply supported

Figure 15.10(a) shows yield line pattern at the time of formation of collapse mechanism. Consider the element (segment) ABO with yield moment m_u per unit length of yield line. Since, slab is isotropically reinforced, m_u shall be same in any direction.

$$\Sigma \text{ moment about AB} = 0, \left(m_u \frac{L}{2}\right)2 = w_u \cdot \frac{1}{2}\left(L \times \frac{L}{2}\right) \cdot \frac{L}{6}, \ m_u = \frac{w_u \cdot L^2}{24},$$

same as ... Eq. (15.4)

(b) Fixed support [Fig. 15.10(b)]

$$m'_u L + \left(m_u \cdot \frac{L}{2}\right)2 = w_u \cdot \frac{1}{2}\left(L \times \frac{L}{2}\right) \times \frac{L}{6}$$

$$(m'_u + m_u) = \frac{w_u \cdot L^2}{24}, \qquad\qquad \textit{same as ...} \text{ Eq. (15.5)}$$

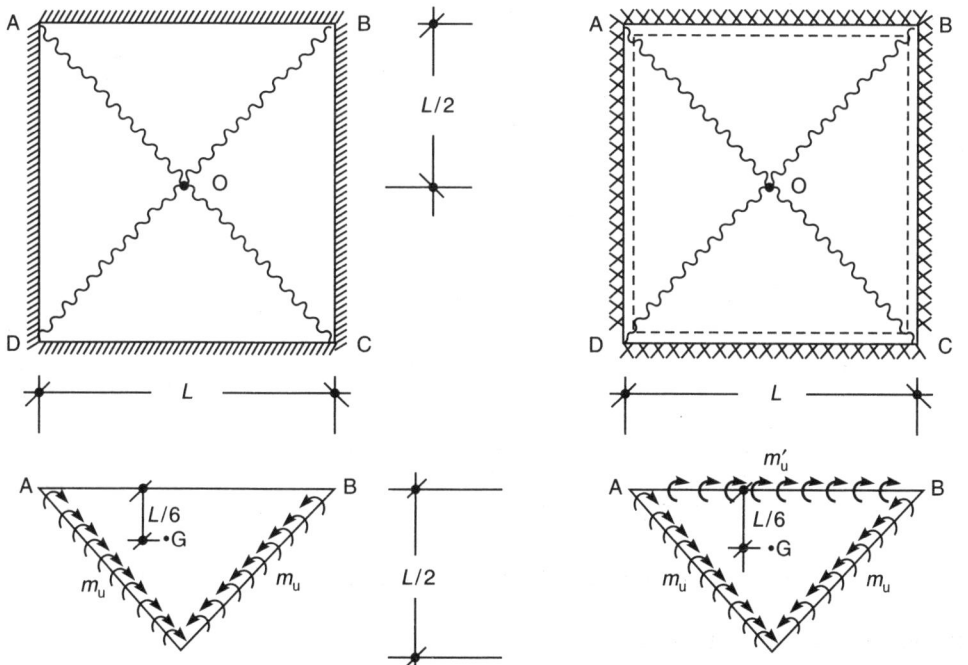

(a) Simply supported slab (isotropically reinforced)   (b) Fixed slab (isotropically reinforced)

Fig. 15.10: Square slab (SS and fixed)

For the same slab $m_u = m'_u$ at the time of mechanism formation for the same material.

Thus, $\qquad m'_u = m_u = \dfrac{1}{2}\dfrac{w_u \cdot L^2}{24} = \dfrac{w_u \cdot L^2}{48}$ $\qquad$ *same as Eq. ...(15.5)*

ii. *Orthotropically reinforced rectangular slab — simply supported*

Consider slab segments I (= III) and II (= IV) as shown in Fig. 15.11 carrying u.d.l. w_u/unit area (Fig. 15.11).

Consider equilibrium of segment I ($\triangle$ AEG, $\triangle$ BFH and $\square$ EFHG)

For equilibrium *I*: Σ moments about AB = 0, thus

$$m_u.(AG + GH + HB) = w_u \left[\frac{\alpha L \times \beta L}{2 \times 2} \times \frac{\alpha L}{6} \times 2 + (L - 2\beta L) \times \frac{\alpha L}{2} \times \frac{\alpha L}{4} \right]$$

or $\qquad m_u (L) = w_u \left\{ \frac{L^3}{6 \times 2} \cdot \alpha^2 \cdot \beta + \frac{L^3}{8}(\alpha^2 - 2\beta\alpha^2) \right\} = \frac{w_u \cdot L^3}{48} (4\alpha^2\beta + 6\alpha^2 - 12\beta\alpha^2)$

or $\qquad\qquad m_u = \dfrac{w_u \cdot L^2 \alpha^2}{48}(6 - 8\beta) = w_u \alpha^2 \cdot L^2 \left(\dfrac{1}{8} - \dfrac{\beta}{6} \right)$ $\qquad$ ... Eq. (15.13)

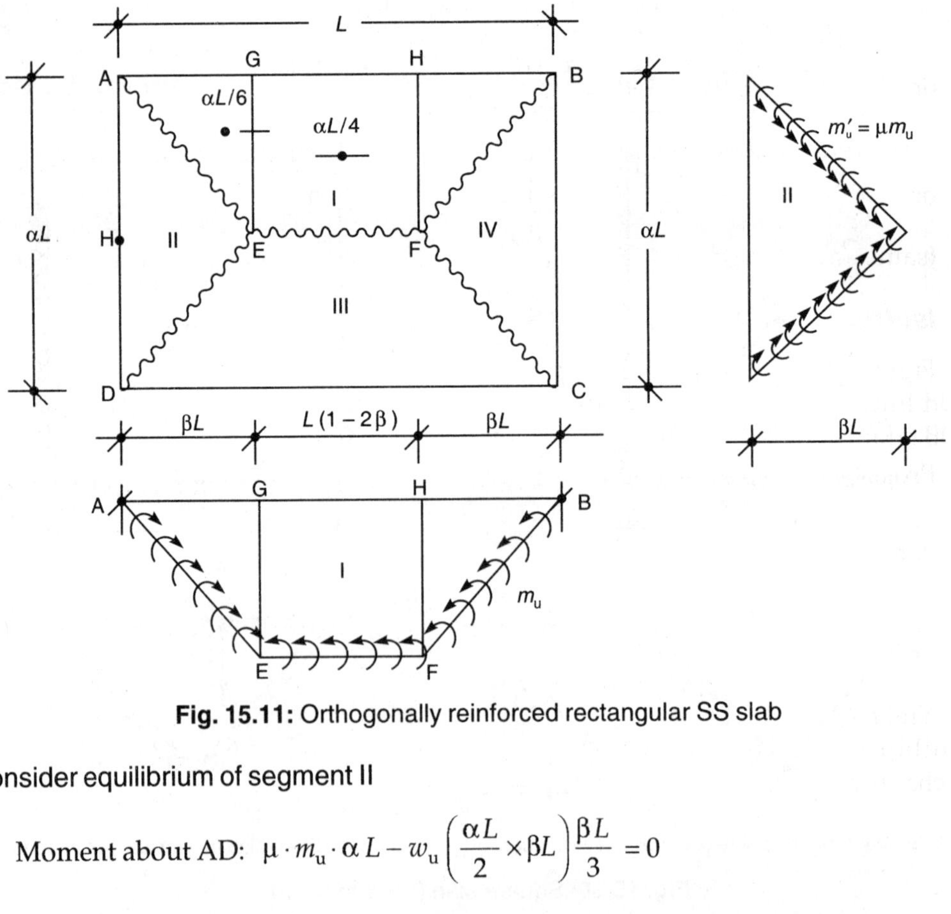

Fig. 15.11: Orthogonally reinforced rectangular SS slab

Consider equilibrium of segment II

Moment about AD: $\mu \cdot m_u \cdot \alpha L - w_u \left(\dfrac{\alpha L}{2} \times \beta L \right) \dfrac{\beta L}{3} = 0$

$$m_u = \frac{w_u L^2}{\mu}\left(\frac{\beta^2}{6}\right) = \frac{w_u \cdot \beta^2 \cdot L^2}{6\mu} \qquad \text{... Eq. (15.14)}$$

From Eqs (15.13) and (15.14), we get

$$m_u = \frac{w_u \cdot \beta^2 \cdot L^2}{6\mu} = w_u \, \alpha^2 \cdot L^2 \left(\frac{1}{8} - \frac{\beta}{6}\right) = w_u \, \alpha^2 \cdot L^2 \left(\frac{3 - 4\beta}{24}\right)$$

or $\qquad \dfrac{\beta^2}{6\mu} = \dfrac{\alpha^2}{24}(3 - 4\beta)$ or $4\beta^2 = \mu\alpha^2 \cdot 3 - 4\beta\,\mu\alpha^2$ $\qquad$ ... Eq. (15.15)

or $\qquad 4\beta^2 + 4\mu\alpha^2\beta - 3\mu\alpha^2 = 0, \quad \therefore \ \beta = \dfrac{-4\mu\alpha^2 \pm \sqrt{16\mu^2\alpha^4 + 48\mu\alpha^2}}{2 \times 4}$

or $\qquad \beta = \dfrac{1}{2}\left[-\mu\alpha^2 + \sqrt{\mu^2\alpha^4 + 3\,\mu\alpha^2}\right]$

Substituting the value of β in Eq. (15.14), we have

$$m_u = \frac{w_u \cdot L^2}{24\mu}\left(-\mu\alpha^2 + \sqrt{\mu^2\alpha^4 + 3\,\mu\alpha^2}\right)^2$$

or $\qquad m_u = \dfrac{w_u \cdot L^2}{24\mu} \times (\alpha^2 \cdot \mu)\left(-\alpha\sqrt{\mu} + \sqrt{\mu\alpha^2 + 3}\right)^2$

or $\qquad m_u = \dfrac{w_u \cdot \alpha^2 \cdot L^2}{24}\left(-\alpha\sqrt{\mu} + \sqrt{\mu\alpha^2 + 3}\right)^2 \qquad\qquad$ *same as* Eq. ... (15.8)

(same as derived by virtual work method)

iii. *Isotropically reinforced SS hexagonal slab carrying u.d.l. on entire slab*

Figure 15.12 shows a SS isotropically reinforced hexagonal slab with u.d.l. and yield lines at the time of collapse mechanism. Consider equilibrium of one segment AOB taking moment about AB.

Projection of yield line AO and BO on AB = L

$\therefore$ Moment about AB $= m_u \dfrac{L}{2} + m_u \dfrac{L}{2} - w_u \dfrac{L}{2} \times \dfrac{\sqrt{3}}{2} L \times \dfrac{1}{3} \times \dfrac{\sqrt{3}}{2} L = 0$ (by equilibrium)

or $\qquad m_u = \dfrac{w_u \cdot L^2}{8} \qquad$ *same as* (in virtual work method) Eq. ... (15.10)

Yield line method of design of slab shall be illustrated by solved examples. Equilibrium method appears to be simple but correct assumption of collapse mechanism of yield lines is most important for the correct solution.

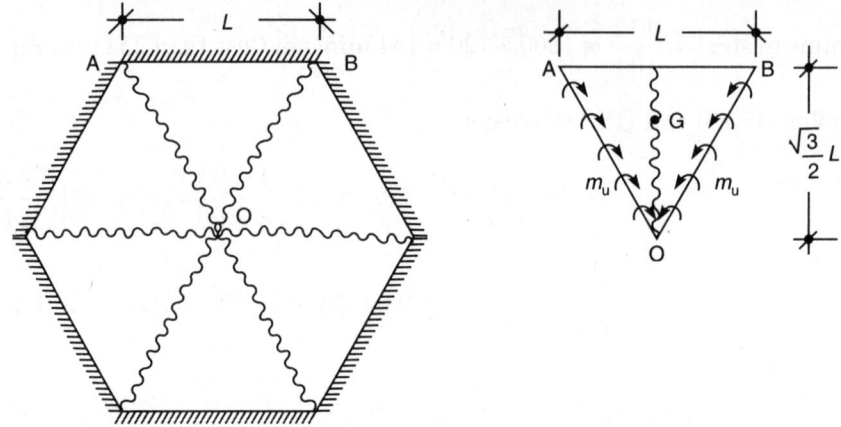

Fig. 15.12: SS isotropically reinforced hexagonal slab with u.d.l. on entire slab

Example 15.1

Design a simply supported square slab of 4 m side to carry a service u.d.l. of 5 kN/m². Use M25 grade concrete and Fe 415 grade steel.

Solution: $f_{ck} = 25\ \text{N/m}^2$, $f_y = 415\ \text{N/mm}^2$, two way slab span/depth = 35 × 0.8 = 28

$$\text{overall depth} = \frac{\text{span}}{28} = \frac{4000}{28} = 143\ \text{mm},$$

$$\text{Assume } D = 150\ \text{mm}, d = 150 - 25 - \frac{10}{2} = 120\ \text{mm}$$

$$\text{self-load} = 0.15 \times 25 = 3.75\ \text{kN/m}^2 + \text{finishing (say) } 1.25\ \text{kN/m}^2$$

$$\text{service load} = 5\ \text{kN/m}^2, \text{total u.d.l. } w = 3.75 + 1.25 + 5 = 10\ \text{kN/m}^2$$

$$w_u = 1.5 \times 10 = 15\ \text{kN/m}^2$$

Maximum moment from yield line theory for square slabs $m_u = \dfrac{w_u L^2}{24}$

$$\text{Effective span} = 4.0\ \text{m} + 0.12\ \text{m} = 4.12\ \text{m}$$

$$m_u = \frac{15(4.12)^2}{24} = 10.61\ \text{kN-m}$$

Limiting moment of the slab $(d = 120\ \text{mm}) = 0.138\,f_{ck}bd^2$

$$m_u = 0.138 \times 25 \times 1000 \times 120^2 = 49.680 \times 10^6\ \text{N-mm} \gg m_u$$

hence depth selected is OK.

Reinforcement

$$d = 120\ \text{mm}$$

$$A_{st} = \frac{0.5 f_{ck} \cdot bd}{f_y}\left[1 - \sqrt{1 - \frac{4.6 m_u}{f_{ck}\, b(d)^2}}\right]$$

$$= \frac{12.5 \times 120000}{415}\left(1 - \sqrt{1 - \frac{4.6 \times 10.61 \times 10^6}{25 \times 1000 \times 120^2}}\right) = 254\ \text{mm}^2/\text{m}$$

(10ϕ @ 300 c/c, $A_{st} = 262\ \text{mm}^2/\text{m}$, or use 8ϕ @ 180 c/c, $A_{st} = 278\ \text{mm}^2/\text{m}$)

Minimum steel $= \dfrac{0.12}{100} \times 1000 \times 120 = 144 \text{ mm}^2/\text{m}$ (less than 254 mm²/m)

Hence provide 8ϕ @ 180 c/c, ($A_{st} = 278 \text{ mm}^2/\text{m}$) in each direction to make the slab isotropically reinforced.

Example 15.2

A square slab of 6 m × 6 m is simply supported and reinforced with 10 bars of Fe 415 steel at a spacing of 180 mm c/c in both directions. The average depth of slab may be taken as 120 mm and overall depth equal to 150 mm. Determine the service load which can be permitted if the grade of concrete is M20.

Solution: $f_{ck} = 20 \text{ kN/mm}^2, f_y = 415 \text{ kN/mm}^2, A_{st} = \dfrac{\dfrac{\pi}{4}(10)^2 \times 1000}{180} = 436 \text{ mm}^2/\text{m}$

$$m_u = 0.87 f_y A_{st} d \left[1 - \dfrac{A_{st}}{bd} \times \dfrac{f_y}{f_{ck}} \right]$$

$$= 0.87 \times 415 \times 436 \times 120 \left[1 - \dfrac{436 \times 415}{120000 \times 20} \right] = 17.466 \times 10^6 \text{ N-mm}$$

$$m_{u \text{ lim}} = 0.138 f_{ck} \, bd^2 = 0.138 \times 20 \times 1000 \times (120)^2 = 39.744 \times 10^6 \text{ N-mm}$$

Section is under reinforced and hence $m_u = \dfrac{w_u \cdot L^2}{24}$, $L = 6$ m effective (say)

$$\therefore \quad \dfrac{w_u \times 6 \times 6 \times 10^6}{24} = 17.466 \times 10^6, \qquad \text{where, } w_u \text{ in kN/m}^2$$

$$w_u = \dfrac{17.466 \times 24}{36} = 11.64 \text{ kN/m}, \quad w \text{ (working load)}$$

$$w = \dfrac{11.64}{1.5} = 7.76 \text{ kN/m}$$

self-weight $= 0.15 \times 1 \times 25 = 3.75 \text{ kN/m}^2$, finishing (say) 1.25 kN/m²

Total dead load $= 3.75 + 1.25 = 5 \text{ kN/m}^2$

Service load $= 7.76 - 5.0 = 2.76 \text{ kN/m}^2$ (live load in addition to self-load)

Example 15.3

Design for BM: a rectangular slab of size 4 m × 6 m which is simply supported along the edges and has to carry a service load of 5 kN/m². Assume the coefficient of orthotropy $\mu = 0.80$. Use M20 CC and Fe 415 steel.

Solution: Assume $d = 120$ mm average (say), $f_{ck} = 20 \text{ N/mm}^2, f_y = 415 \text{ N/mm}^2$

$$L_y = 4.12 \text{ m}, \alpha_L = 4.12 \text{ m},$$

$$L_x = (6 + 0.12) = 6.12 \text{ m}, \ \alpha = 0.67, \mu = 0.80 \text{ (given)}$$

For two way slabs: $\dfrac{\text{span}}{\text{depth}} = 35 \times 0.8 = 28$

$$D = \dfrac{4120}{28} = 147, \text{ adopt } D = 150 \text{ mm}, d = 125 \text{ mm (say)}.$$

Self-load: $0.15 \times 1 \times 25 = 3.75 \text{ kN/m}^2$

Floor finish $= 1.25 \text{ kN/m}^2$, live load $= 5 \text{ kN/m}^2$

Total load $W = 3.75 + 1.25 + 5 = 10 \text{ kN/m}^2$, ultimate collapse $w_u = 1.5 \times 10 = 15 \text{ kN/m}^2$

m_u for orthotropically reinforced SS slab $= \dfrac{W_u \cdot \alpha^2 L^2}{24}\left[\sqrt{3 + \alpha^2 \mu} - \alpha\sqrt{\mu}\right]^2$

$$m_u = \frac{15.(0.67)^2(6.12)^2}{24}\left(\sqrt{3 + 0.67^2 \times 0.8} - 0.67\sqrt{0.80}\right)^2$$

$$= 10.5083 \, (1.8328 - 0.5993)^2$$

$$m_u = 15.99 \text{ kN-m}, \quad \mu m_u = 0.80 \times 15.99 = 12.8 \text{ kN-m}$$

$$m_{u\,lim} = 0.138 f_{ck}.b.(d)^2 = 0.138 \times 20 \times 1000 \times (125)^2 = 43.125 \times 10^6 \text{ N-mm}$$

$$= 43.125 \text{ kN-m}$$

$m_{u\,lim} > m_u$, the section is under reinforced.

Long span moment $= \mu m_u = 0.80 \times 15.99 = 12.8 \text{ kN-m}$

Main reinforcement in short span

$$A_{st} = \frac{10 \times 125000}{415}\left[1 - \sqrt{1 - \frac{4.6 \times 15.99 \times 10^6}{20 \times 1000 \times (125)^2}}\right]$$

$$= 379 \text{ mm}^2 \, (10\phi @ 200c/c, A_{st} = 393 \text{ mm}^2/\text{m})$$

Reinforcement in long direction $\mu m_u = 12.8 \text{ kN-m/m} \quad A_v.d = 125 \text{ mm}$

$$A_{st_2} = \frac{1250000}{415}\left[1 - \sqrt{1 - \frac{4.6 \times 12.8 \times 10^6}{20000 \times (125)^2}}\right]$$

$$= 299 \text{ mm}^2 \, (8\phi @ 160 \text{ c/c}, A_{st} = 313 \text{ mm}^2/\text{m})$$

Thus, in short direction provide main reinforcement $10\phi @ 200 \text{ c/c}$ and in long direction provide $8\phi @ 160 \text{ c/c}$.

Minimum temp steel $= \dfrac{0.12}{100} \times 1000 \times 150 = 180 \text{ mm}^2/\text{m}$ (< provided steel, OK)

Example 15.4

A rectangular slab of effective size 5 m × 6 m is simply supported at edges. The slab is reinforced with $10\phi @ 200 \text{ c/c}$ in short direction and $10\phi @ 225 \text{ c/c}$ in long direction. The slab has overall thickness $= 150 \text{ mm}$ and average effective depth may be taken as 120 mm. Use M20 CC and Fe 415 steel. Determine the superimposed load which the slab can carry safely if the finishing load for flooring is 1.25 kN/m^2.

Solution: $L = 6 \text{ m}, \alpha L = 5.0 \text{ m}, \alpha = \dfrac{5}{6} = 0.833$,

A_{stx} (reinforcement in x-direction, short span) $= \dfrac{78500}{200} = 393 \text{ mm}^2/\text{m}$

In short span

$$m_u = 0.87 f_y A_{st} d\left[1 - \frac{A_{st}}{bd} \times \frac{f_y}{f_{ck}}\right]$$

$$m_u = 0.87 \times 415 \times 393 \times 120 \left[1 - \frac{393}{120000} \times \frac{415}{20}\right]$$

$$= 17.027 \times 10^6 \ (0.932) = 15.87 \times 10^6 \ \text{N-mm}$$

$$= 15.87 \ \text{kN-m}$$

In long span

$$A_{st} = \frac{78500}{225} = 349 \ \text{mm}^2/\text{m}$$

$$\mu m_u = \text{BM} = 0.87 \times 415 \times 349 \times 120 \left[1 - \frac{349}{120 \times 1000} \times \frac{415}{20}\right]$$

$$= 15.12 \times 10^6 \ (0.939652) = 14.21 \times 10^6 \ \text{N-mm}$$

$$\mu = \frac{14.21}{15.87} = 0.8954$$

From yield line theory, we know that

$$m_u = \frac{w_u \cdot \alpha^2 L^2}{24} \left[\sqrt{3 + \alpha^2 \mu} - \alpha\sqrt{\mu}\right]^2$$

$$= \frac{w_u \cdot (0.833)^2 \times (5)^2}{24} \left[\sqrt{3 + 0.8954(0.833)^2} - 0.833\sqrt{0.8954}\right]^2$$

$$m_u = 0.7228 \ w_u \ [1.903 - 0.78823]^2 = 0.7228 \ w_u \times 1.1148 = 0.898 \ w_u$$

or $\qquad 15.87 = 0.898 \ w_u,$

$\therefore \qquad w_u = \dfrac{15.87}{0.898} = 17.67 \ \text{kN/m}^2, \ w = \dfrac{17.67}{1.5} = 11.78 \ \text{kN/m}^2$

Self-weight $= 0.15 \times 1 \times 25 = 3.75 \ \text{kN/m}^2, \ w_d = 3.75 + 1.25 = 5.0 \ \text{kN/m}^2$

Live load $= (11.78 + 5.0) = 6.78 \ \text{kN/m}^2.$

Example 15.5

A rectangular slab 3.50 m × 5.0 m in size (effective), simply supported at the edges. The slab is expected to carry a service live load of 3 kN/m² and a floor finish of 1.0 kN/m². Design the slab if (i) it is isotropically reinforced (ii) it is orthotropically reinforced with $\mu = 0.80$ for bending. Use M25 CC and Fe 415 steel.

Solution: $\qquad L = 5 \ \text{m}, \ \alpha L = 3.5 \ \text{m}, \ \alpha = \dfrac{3.5}{5.0} = 0.70$

For two way slabs:

$$\frac{\text{Span}}{\text{depth}} = 35 \ \text{(with MS steel), and take} \ \frac{\text{Span}}{\text{depth}} \ \text{(for HYSD)} = 0.8 \times 35 = 28$$

$$\text{Depth} = \frac{3500}{35 \times 0.8} = 125 \ \text{mm}, \ d = 125 - 15 - 10 = 100 \ \text{mm}$$

Dead load of slab $= 0.125 \times 1 \times 25 + \text{flooring} = 3.125 + 1 = 4.125 \ \text{kN/m}^2$

Live load $= 3.0 \ \text{kN/m}^2$, Total u.d.l. $w = 4.125 + 3.0 = 7.125 \ \text{kN/m}^2$

Ultimate load $w_u = 1.5 \times 7.125 = 10.69 \ \text{kN/m}^2$

(i) Isotropically reinforced slab ($\mu = 1$)

Ultimate moment by yield line theory, $m_u = \dfrac{w_u \cdot \alpha^2 L^2}{24} \left[\sqrt{3 + \alpha^2 \mu} - \alpha\sqrt{\mu} \right]^2$

$$m_u = \frac{10.69\,(0.7)^2\,(5)^2}{24} \left[\sqrt{3 + 0.7^2 \times 1} - 0.7\sqrt{1} \right]^2 = 7.44$$

$$A_{st} = \frac{12500\,(100)}{415} \left[1 - \sqrt{1 - \frac{4.6 \times 7.44 \times 10^6}{25 \times 1000 \times (75)^2}} \right] = 214 \text{ mm}^2/\text{m}$$

Use 8ϕ @ 225 c/c, $A_{st} = 222$ mm²/m, in both directions.

(ii) Orthotropically reinforced slab

$$\mu = 0.8,\ \alpha = 0.70,\ L = 5.0 \text{ m},\ w_u = 10.69 \text{ kN/m}^2$$

$$m_u \text{ (yield line theory)} = \frac{10.69 \cdot (0.70)^2\,(5.0)^2}{24} \left[\sqrt{3 + 0.7^2 \times 0.8} - 0.7\sqrt{0.80} \right]^2$$

or $\quad m_u = 5.4564 \left[\sqrt{3.392} - 0.6261 \right]^2 = 5.4564(1.478) = 8.06$ kN-m

$$A_{st_2} = \frac{0.5 f_{ck} bd}{f_y} \left[1 - \sqrt{1 - \frac{4.6 M_u}{f_{ck} b(d)^2}} \right] = \frac{12500\,(100)}{415} \left[1 - \sqrt{1 - \frac{4.6 \times 8.06 \times 10^6}{25 \times 1000(100)^2}} \right]$$

or $A_{st} = 233$ mm² (8ϕ 210 mm c/c, $A_{st} = 238$ mm²/m)

A_{st} in long direction 8$\phi = \dfrac{210}{0.80} = 260$ c/c, $A_{st} = 192$ mm²/m,

i.e. provide slab thickness of 125 mm with an effective cover of 21 mm for shorter (main) reinforcement of 8ϕ 210 c/c (A_{st}), and effective cover of $17 + 8 + 4 = 29$ mm (i.e. effective depth 96 mm) for secondary reinforcement (8ϕ 260 c/c)

Average effective depth $= \dfrac{96 + 104}{2} = 100$ mm (say)

Main reinforcement 8ϕ 210 c/c, distribution steel 8ϕ @ 260 c/c.

Example 15.6

A right angled triangular slab is simply supported along the two edges. The length of supported right angled triangular slab is simply supported along the two edges. The length of supported edges are 4.50 m × 6.0 m respectively. The slab is isotropically reinforced with 10ϕ bars spaced at 100 c/c both ways. The average effective depth of the slab may be taken as 110 mm and overall depth of 140 mm. Using M25 grade CC and Fe 415 grade steel, determine the safe permissible service live load on the slab. Assume the floor finish load of 1.5 kN/m² (Fig. 15.13).

Solution: $\quad L = 6$ m, $\alpha L = 4.5$ m, $\alpha = \dfrac{4.5}{6.0} = 0.75,\ d = 110$ mm $\ D = 140$ mm

$$A_{st} = \frac{78.5 \times 1000}{100} = 785 \text{ mm}^2\ /\text{m},\ \beta = 90°$$

The ultimate moment is given by $m_u = 0.87 f_y.A_{st}.d\left[1 - \dfrac{A_{st} f_y}{bd f_{ck}}\right]$

or $\qquad m_u = 0.87 \times 415 \times 785 \times 110 \left[1 - \dfrac{785 \times 415}{1000 \times 110 \times 25}\right] = 27.4834 \times 10^6$ N-mm

$\qquad m_u = 27.4834$ kN-m

According to yield line failure of triangular slabs

$$m_u = \frac{w_u \alpha L^2}{6} \sin^2 \frac{\beta}{2} \quad \text{or} \quad w_u = \frac{6 m_u}{\alpha L^2 \sin^2 \dfrac{90}{2}} = \frac{6 \times 27.4834}{0.75 \times 6^2 \left(\dfrac{1}{\sqrt{2}}\right)^2} = 12.215 \text{ kN/m}^2$$

$$\text{total service load } w = \frac{12.215}{1.5} = 8.14 \text{ kN/m}^2$$

$$\text{DL of slab} = 0.14 \times 25 + 1.5 = 5.0 \text{ kN/m}^2$$
$$\text{LL on slab} = \text{Total} - \text{DL} = 8.14 - 5.0 = 3.14 \text{ kN/m}^2$$

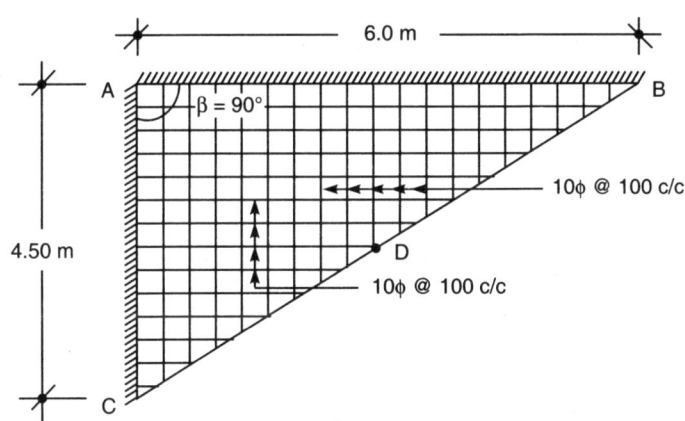

Fig. 15.13: SS triangular slab (isotropically reinforced)

Example 15.7

Design a reinforced circular slab of 5.0 m diameter and carrying a service live load of 5 kN/m² and floor finish of 1.5 kN/m². Use M20 grade CC and Fe 415 grade steel. The slab is simply supported along the edges (Fig. 15.14).

Solution: Diameter = 5.0 m, radius = 2.5 m, $f_{ck} = 20$ N/mm²

$$\text{Assume depth} = \frac{\text{span}}{35} \text{ for two-way slabs (with ms bars)},$$

$$\text{i.e. } D = \frac{5000}{0.8 \times 35} = 178.6 \text{ mm say (with HYSD)}$$

Assume $D = 180$ mm and average $d = 150$ mm (using HYSD steel Fe 415)

Self-load of slab (1 m²) = $0.18 \times 1 \times 25 = 4.5$ kN/m²

Floor finish (1m²) = 1.50 kN/m², service live load = 5 kN/m²

Total service load W = 4.5 + 1.5 + 5 = 11 kN/m²

Ultimate load w_u = 11 × 1.5 = 16.5 kN/m²

For circular slabs simply supported $= \dfrac{w_u \cdot r^2}{6} = \dfrac{16.5 \times 2.5^2}{6} = 17.19$ kN-m

$m_{u\,lim}$ = 0.138 $f_{ck}.b.d^2$ = 0.138 × 20 × 1000(150)² = 62.1 × 10⁶ N-mm (62.1 kN-m)

m_u = < $m_{u\,lim}$, i.e. slab is under reinforced.

m_u = 17.19 × 10⁶ N-mm

$$A_{st} = \frac{0.5 f_{ck}\, bd}{f_y}\left[1 - \sqrt{1 - \frac{4.6 m_u}{f_{ck}\, b(d)^2}}\right]$$

$$= \frac{10000(150)}{415}\left[1 - \sqrt{1 - \frac{4.6 \times 17.19 \times 10^6}{20 \times 1000 \times (150)^2}}\right] = 333 \text{ mm}^2$$

A_{st} = 335 mm² in face, 10ϕ @ 230 c/c, A_{st} = 341 mm²/m in both directions in bottom face.

At edges provide three 10ϕ ring bars along the circumference at edges @ 230 mm c/c near top face. Also provide radial bars of about 500 mm length, 10ϕ 230 c/c near top at ends.

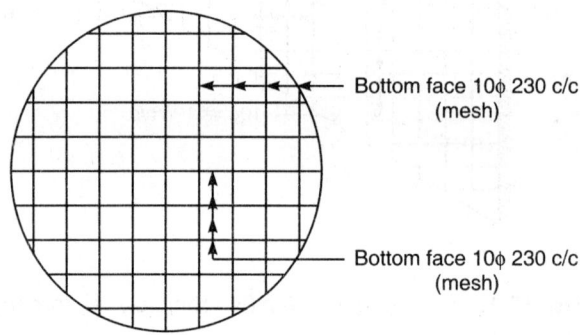

Bottom face 10ϕ 230 c/c (mesh)

Bottom face 10ϕ 230 c/c (mesh)

Fig. 15.14: Circular slab reinforcement

Example 15.8

A hexagonal slab is simply supported along the edges with a side length of 3.0 m each. It is isotropically reinforced with 10 mm diameter bars at 150 mm centre to centre both ways, at an average effective depth of 120 mm. The overall depth of the slab is 150 mm. Determine the ultimate load capacity of the slab and safe permissible service live load. Use M20 grade CC and Fe 415 HYSD bars.

Solution: f_{ck} = 20 N/mm², f_y = 415 N/mm², L = 3.0 m, (hexagonal), d = 120 mm

$$D = 150 \text{ mm}, A_{st} = 78.54 \times \frac{1000}{150} = 523.6 \text{ mm}^2/\text{m}$$

$$m_u = 0.87 f_y A_{st} d \left[1 - \frac{A_{st} f_y}{bd f_{ck}} \right]$$

$$= 0.87 \times 415 \times 523.6 \times 120 \left[1 - \frac{523.6 \times 415}{120000 \times 20} \right] = 20.632 \times 10^6 \text{ N-mm}$$

For hexagonal slab simply supported at edges, we have

$$m_u = \frac{w_u \cdot L^2}{8} \quad \text{or} \quad 20.632 = \frac{w_u \cdot (3)^2}{8}, \quad w_u = 18.34$$

∴ Total service load $w = \dfrac{18.34}{1.5} = 12.23 \text{ kN/m}$

Total DL of slab $= 0.15 \times 25 + 1.5$ (floor finish) $= 5.25 \text{ kN/m}^2$

Permissible live load $= 12.23 - 5.25 = 6.98 \text{ kN/m}^2$

SUMMARY

A yield line is defined as a line in the plane of the slab along which all reinforcing steel bars have yielded. After yielding of steel bars, the concrete also cracks along the yield line thus offering no further resistance to deformation or rotation of the slab segment. Initially with increase in loading, the tensile stress in steel bars also increase and reaches to the level of yield stress. The tensile steel bars undergo very large elongation at this yielding stage without any further increase of load. Thus, the slab segment rotates about the yield line(s) and a mechanism is formed.

The yield line analysis of slabs can be done by the following methods:

(i) Virtual work method (mechanism method)

In virtual work method, a collapse mechanism of yield lines is assumed. The elastic deformations are very small and hence yield line deformations (rotations) are considered in virtual work done by moments to rotate the slab segments about the yield lines and shall be equated to virtual work (external) done by loads.

(ii) Equilibrium method (statical method)

In this method, the slab segment after formation of collapse mechanism shall be considered in equilibrium under the yield line moments and other loads acting on the slab segment.

Both these methods predict upper bound values of load capacity. The real collapse load values by actual testing are always more than these predicted values of collapse loads. Hence, the values predicted by yield line theory are on a safer side with respect to the exact collapse capacity of slabs.

According to IS:456-2000 code, the ultimate load capacity per unit width of yield lines shall be computed from all reinforcement in that direction,

$$m_u = 0.87 f_y A_{st} d \left[1 - \frac{A_{st} f_y}{bd f_{ck}} \right] \quad \text{with usual notations}$$

When the slab is two-way reinforced then,

$$m_{ux} = 0.87 f_y A_{stx} d \left[1 - \frac{A_{stx} f_y}{bd f_{ck}} \right],$$

where, A_{stx} = reinforcement in x-direction

$$m_{uy} = 0.87 f_y A_{sty} d \left[1 - \frac{A_{sty} f_y}{bd f_{ck}} \right]$$

A_{sty} = reinforcement in y-direction

When the yield line is inclined at an angle α with the normal x-direction (i.e. α with y-direction), the resultant yield line moment $m_{u\alpha}$ shall be:

$$m_{u\alpha} = [m_{ux} \cos^2 \alpha + m_{uy} \sin^2 \alpha]$$

or

$$m_{u\alpha} = m_{ux} \left[\cos^2 \alpha + \frac{m_{uy}}{m_{ux}} \cdot \sin^2 \alpha \right]$$

or

$$m_{u\alpha} = m_{ux} [\cos^2 \alpha + \mu \sin^2 \alpha], \text{ where, } \mu = \frac{m_{uy}}{m_{ux}}$$

When the slab is reinforced isotropically $m_{ux} = m_{uy}$ and $\mu = 1$

We have

$$m_{u\alpha} = m_{ux} [\cos^2 \alpha + \sin^2 \alpha] = m_{ux} = m_u,$$

i.e. moment in all directions are the same.

If the yield line makes an angle of $\alpha = 45°$, the moment along any yield line at $45°$ for orthotropically reinforced slab shall be:

$$m_{u\alpha} = m_{ux} \left(\frac{1}{2} + \frac{\mu}{2} \right) = \frac{m_{ux}}{2} [1 + \mu]$$

For isotropically reinforced slabs $\mu = 1$, $\sin^2 45° = \cos^2 45° = \dfrac{1}{2}$

or

$$m_{u\alpha} = m_{ux} \left(\frac{1}{2} + \frac{1}{2} \right) = m_{ux}$$

(a) *Isotropically reinforced SS square slabs with u.d.l. on entire slab*

$$\boxed{m_u = \frac{w_u \cdot L^2}{24}}, \quad \text{where } L = \text{side of square}$$

(b) *Isotropically reinforced, fixed at edges square slab with u.d.l. on entire slab*

$$\boxed{m_u = \frac{w_u \cdot L^2}{48}}$$

(c) *Equilateral triangular, isotropically reinforced SS slab with u.d.l.*

$$\boxed{m_u = \frac{w_u \cdot L^2}{72}}, \quad \text{where } L = \text{side of triangle}$$

(d) *SS rectangular slab with orthogonal reinforcement with u.d.l.*

$$\beta = \alpha L, \, m_{uy} = \mu \, m_{ux}, \, m_{ux} = m_u,$$

$$m_u = w_u \frac{L^2 \cdot \alpha^2}{12} \left\{ \frac{(3\beta - 2\beta^2)}{(2\beta + \mu\alpha^2)} \right\}$$

For optimum value

$$\beta = \frac{1}{2} \left[-\mu\alpha^2 + \sqrt{\mu^2\alpha^4 + 3\mu\alpha^2} \right]$$

$$m_u = w_u \frac{\alpha^2 L^2}{24} \left[-\alpha\sqrt{\mu} + \sqrt{3 + \alpha^2\mu} \right]^2$$

(i) Isotropically reinforced polygonal slab with edges fixed and u.d.l.

$$(m_{u_1} + m_u) = \frac{w_u \cdot r^2}{6}, \quad \text{where } r = \text{distance from the centre}$$

(ii) Regular hexagon, $r = \dfrac{L}{2} \cot 30° = \dfrac{L\sqrt{3}}{2}$

$$(m_{u_1} + m_u) = \frac{w_u \cdot L^2}{8}$$

(iii) Regular octagon, $r = \dfrac{L}{2} \cot (22.5)° = 1.207L$

$$(m_{u_1} + m_u) = 0.24285 \, w_u L^2$$

(iv) Circular slab, $r = R$

$$(m_{u_1} + m_u) = \frac{w_u \cdot R^2}{6}$$

PRACTICE QUESTIONS

(I) Objective Questions

Q. 15.1 Select the correct response given after each statement to complete it correctly and fill in the response sheet provided.

(i) A yield line is defined as a line in the plane of the slab along which ....

    (a) all the reinforcing steel bars have resisted total moment

    (b) all the reinforcing steel bars have yielded

    (c) all the concrete shall have crushed by compression

    (d) all the concrete shall have sheared off

(ii) RCC slabs may fail in tension and yield line analysis requires the slabs to be only ....
(a) over reinforced for yield lines collapse mechanism
(b) balanced reinforced for accurate collapse mechanism
(c) under reinforced for formation of yield line collapse mechanism
(d) in straight line and supported at the same level

(iii) The slab analysis by yield line theory is not based on the assumption that ...
(a) RCC slabs are under reinforced
(b) yield lines are straight lines
(c) rotation of slab segments occur along the yield line
(d) elastic deformations are important compared to plastic one.

(iv) Which of the following does not represent the characteristic of a yield line ?
(a) mechanism formed by yield lines result in compatibility of deformations
(b) yield lines act as axes of rotation of the segment of slab
(c) yield lines are straight lines and form hinges of collapse mechanism
(d) yield lines are never formed along the fixed supports

(v) A yield line 'ab' is inclined at α with the y-axis along 'ac' normal to x-axis along 'ad'. If m_{ux} and m_{uy} are unit moment capacities of yield line along normal to x and normal to y, then component of m_{ux} and m_{uy} to total yield moment will be ....
(a) $m_{u\alpha} = m_{ux} \cdot \cos^2 \alpha + m_{uy} \sin^2 \alpha$
(b) $m_{u\alpha} = m_{ux} \cdot \sin^2 \alpha + m_{uy} \cos^2 \alpha$
(c) $m_{u\alpha} = m_{ux} \cdot \sin \alpha + m_{uy} \cos \alpha$
(d) $m_{u\alpha} = (m_{ux} + \mu m_{uy})$, where $\mu = \dfrac{\cos \alpha}{\sin \alpha}$

(vi) A two-way reinforced rectangular slab has yield lines at 45° with the shorter span (y-direction), the total ultimate moment capacity per unit length along the yield lines will be, $m_{u\alpha} = $ ...., if the slab is orthotropically reinforced.
(a) $m_{u\alpha} = m_{ux} \cdot \cos^2 \alpha$ (b) $m_{u\alpha} = \dfrac{m_{ux}}{2} (1 + \mu)$
(c) $m_{u\alpha} = m_{ux} + \mu m_{uy}$ (d) $m_{u\alpha} = m_{ux} + m_{uy}$

(vii) The unit moment capacities predicted by virtual work method or equilibrium method will be always .... the actual capacity obtained by real test on the slab.
(a) more than   (b) equal to
(c) less than   (d) 1.5 times

(viii) A simply supported isotropically reinforced square slab with side L carrying u.d.l. will have the ultimate moment capacity per unit length along the yield line will be ....
(a) $m_u = \dfrac{w_u \cdot L^2}{24}$ (b) $m_u = \dfrac{w_u \cdot L^2}{48}$
(c) $m_u = \dfrac{w_u \cdot L^2}{72}$ (d) $m_u = \dfrac{w_u \cdot L^2}{12}$

(ix) What will be ultimate moment per unit length of a yield line in case of a rectangular slab orthotropically reinforced ($m_{uy} = \mu m_{ux}$), if the yield lines are at 45° with Y axis and the slab is simply supported?

(a) $m_u = \dfrac{w_u . \alpha^2 L^2}{24}\left[-\alpha\sqrt{\mu} + \sqrt{\mu\alpha^2 + 3}\right]$

(b) $m_u = \dfrac{w_u . \alpha^2 . L^2}{24}\left[-\alpha\sqrt{\mu} + \sqrt{\mu\alpha^2 + 3}\right]^2$

(c) $m_u = \dfrac{w_u . \alpha^2 . L^2}{24}\left[\sqrt{3\mu\alpha^2}\right]$

(d) $m_u = \dfrac{w_u . \alpha^2 . L^2}{24}\sqrt{3 + \mu\alpha^2}$

(x) What will be the ultimate moment/unit length of a yield line in case of a circular slab of diameter D carrying u.d.l. on the entire slab and when edges are fixed ?

(a) $(m_{u_1} + m_u) = \dfrac{w_u D^2}{6}$ 　　　　　(b) $(m_{u_1} + m_u) = \dfrac{w_u D^2}{12}$

(c) $(m_{u_1} + m_u) = \dfrac{w_u D^2}{16}$ 　　　　　(d) $(m_{u_1} + m_u) = \dfrac{w_u D^2}{24}$

Response sheet to Q. 15.1 (i to x)

Question	(i)	(ii)	(iii)	(iv)	(v)	(vi)	(vii)	(viii)	(ix)	(x)
Response (a/b/c/d)										

(II) Numerical Questions

Q. 15.2 Design a simply supported square slab of 3.50 m to carry a service load of 4 kN/m² on the entire slab by using yield line theory. Use M20 CC and Fe 415 steel reinforcement.

[Hint: $D = \dfrac{3500}{(0.8 \times 35)} = 125$ mm, provide $D = 130$ mm, $d = 105$ mm.

$w = DL + LL = 8.25$ kN/m², $w_u = 12.375$ kN/m², $m_u = 6.32$ kN-m/m

$A_{st} = \dfrac{0.5 f_{ck} b.d}{2 f_y}\left[1 - \sqrt{1 - \dfrac{4.6 m_u}{f_{ck}.b(d)^2}}\right] = 173$ mm²/m (8ϕ @ 280 c/c, $A_{st} = 179$ mm²/m)

minimum temp. steel $= \dfrac{0.12 \times 1000 \times 130}{1000} = 156$ mm²/m (<173 mm²/m)]

Q. 15.3 Determine the permissible service load on a square slab of 6 m × 6 m size, if it is reinforced with 10ϕ @ 170 mm c/c in both directions and assume overall thickness of 150 mm. Average effective depth may be taken as 120 mm. Concrete grade of M25 and steel grade of Fe 415 HYSD bars are used in the slab.

[Hint: $f_{ck} = 25 \text{ N/mm}^2, f_y = 415 \text{ N/mm}^2, A_{st} = 461.8 \text{ mm}^2/\text{m}, M_{u \text{ lim}} = 0.138 f_{ck} bd^2$

$$\text{Actual } m_u = 0.87 \times f_y \times A_{st} \times d \left[1 - \frac{A_{sty} f_y}{bd f_{ck}} \right] = 18.73 \text{ kN-m} < M_{u \text{ lim}},$$

(under reinforcement)

$$18.73 \times 10^6 = \frac{w_u.(6)^2}{24} \times 10^6, w_u = 12.5 \text{ kN/m}^2, w = 8.33 \text{ kN/mm}^2$$

LL (superimposed) = 3.33 kN/m²]

Q. 15.4 A reinforced concrete slab 4 m × 6 m is reinforced with 10 mm bars at 150 mm c/c in short direction and 10 mm @ 200 mm c/c in longer direction. The slab is 120 mm thick with average effective depth of 90 mm. Find the ultimate unit moment (m_u), if the yield lines are inclined at 45° to either reinforcements. Use M20 CC and Fe 415 steel.

[Hint: $A_{sty} = 523 \text{ mm}^2, A_{stx} = 393 \text{ mm}^2, f_{ck} = 20 \text{ N/mm}^2, f_y = 415 \text{ N/mm}^2,$

$$m_{uy} = 0.87 f_y.A_{sty}.d \left[1 - \frac{A_{sty} f_y}{b.d.f_{ck}} \right] = 14.95 \times 10^6 \text{ N-mm}$$

$m_{ux} = 11.61 \times 10^6 \text{ N-mm.}$

$m_u = m_{ux} \cos^2 + m_{uy} \sin^2 \alpha, (\alpha = 45°) = 13.28 \times 10^6 \text{ N-mm}]$

Q. 15.5 A rectangular slab 4 m × 5 m is simply supported and reinforced isotropically with 10 mm bars at 200 c/c both-ways. The effective depth may be assumed as 90mm and overall depth of 120 mm. Determine the safe service load to be permitted on the slab if M25 CC and Fe415 steel are used. DL of finishing = 1 kN/m².

[Hint: $L = 5 \text{ m}, \alpha L = 4 \text{ m}, \alpha = 0.80, \mu = 1$ (isotropically reinforced),

$m_{ux} = m_{uy} = m_u, A_{stx} = A_{sty} = 393 \text{ mm}^2/\text{m}, d = 90 \text{ mm}$ (average)

$$\text{yield moment due to loading } m_u = \frac{w_u.\alpha^2.L^2}{24} \left[\sqrt{3 + \mu\alpha^2} - \alpha\sqrt{\mu} \right]^2$$

$$\frac{w_u.16}{24} = (1.1079)^2$$

$w_u = 14.48 \text{ kN/m}^2, w = 9.65 \text{ kN/m}^2$ (total)

LL $(w_L) = (9.65 - 4) = 5.65 \text{ kN/m}^2]$

Q. 15.6 A right angled triangular slab of 3 m × 4 m is simply supported along two edges and diagonal edge of 5 m is free. The slab is isotropically reinforced with 8 m diameter bars spaced at 100mm c/c in either direction. Determine the safe permissible live load on the slab, if M25 grade CC and Fe 415 grade steel are used. Overall depth of the slab is 120 mm and the average effective depth may be taken as 90 mm. DL of finishing may be taken as 1.15 kN/m².

[Hint: $L = 4 \text{ m}, \alpha L = 3 \text{ m}, \alpha = 0.75, \mu$ (isotropic) $= 1.0, A_{st} = 503 \text{ mm}^2/\text{m}, \beta = 90°$

$$m_u = 0.87 \times 415 \times 503 \times 90 \left[1 - \frac{503 \times 415}{90000 \times 25} \right] = 14.83 \times 10^6 \text{ N-mm}$$

$$m_u = \frac{w_u \cdot \alpha \cdot L^2}{6} \sin^2\left(\frac{90}{2}\right) = \frac{w_u}{6} \times 0.75 \times 4^2 \times \frac{1}{2}, w_u = \frac{14.83 \times 6 \times 2}{0.75 \times 4^2} = 14.83 \text{ kN/m}^2$$

$$LL = \left\{\frac{14.8}{1.5} - (3+1)\right\} = 5.89 \text{ kN/m}^2]$$

Q. 15.7 A hexagonal slab of 4 m long sides is simply supported along the edges. The slab is isotropically reinforced with 10ϕ bars at 100 mm c/c both-ways. Overall depth of 150 mm and an average effective depth of 120 mm may be adopted. Determine the ultimate load and permissible service live load on the slab if M20 CC and Fe 415 steel are used. Weight of finishing may be taken as 1.25 kN/m².

[**Hint:** $L = 4$ m, $A_{st} = \dfrac{78540}{100} = 785.4 \text{ mm}^2/\text{m}$,

$$m_u = 0.87 \times 415 \times 785.4 \times 120 \left[1 - \frac{785.4 \times 415}{120000 \times 20}\right] = 29.41 \times 10^6 \text{ N-mm}$$

$$m_u = \frac{w_u \cdot L^2}{8}, \quad w_u = \frac{8\, m_u}{L^2} = 14.7 \text{ kN/m}^2, w = 9.8 \text{ kN/m}^2, w \text{ (LL)} = 4.8 \text{ kN/m}^2]$$

Q. 15.8 Design a reinforced circular slab of 6.0 m diameter and fixed edges. The slab may carry a service load of 4 kN/m², and floor finish load of 1.25 kN/m². Use M20 CC and Fe 415 steel.

[**Hint:** Fixed slabs $D = \dfrac{\text{Span}}{(40 \times 0.8)} = \dfrac{6000}{32} = 187.5$ mm,

Say, $D = 190$ mm, effective depth (average) $= 160$ mm,
$w_u = 1.5 (0.19 \times 25 + 1.25 + 4.0) = 15 \text{ kN/m}^2$,

$$(m_{u_1} + m_u) = \frac{w_u \cdot r^2}{6} = \frac{15 \times 3^2}{6} = 22.5 \text{ kN-m, assume } m_{u_1} = m_u = 11.25 \text{ kN-m}$$

$d = 160$ mm, $m_{u\,\text{lim}} = 2.761 \times 1000\,(160)^2 = 70.68$ kN-m $> m_u$, under reinforced.
A_{st} (near bottom face of mesh)

$$= \frac{10 \times 1000 \times 160}{415}\left[1 - \sqrt{1 - \frac{4.6 \times 11.25 \times 10^6}{20 \times 1000\,(160)^2}}\right] = 200 \text{ mm}^2$$

$$\text{minimum steel} = \frac{0.12}{100} \times 1000 \times 190 = 228 \text{ mm}^2 > 200 \text{ mm}^2$$

Thus, provide in each direction (as mesh) 8ϕ 200 c/c ($A_{st} = 250 \text{ mm}^2/\text{m}$) at ends, and near top face provide 8ϕ @ 200 c/c rings, and 8ϕ @ 200 c/c radial bars for about 1.20 m radial distance from edges to resist fixity of edges.]

KEY TO
OBJECTIVE QUESTIONS
(Chapters 1 to 15)

Chapterwise Answer Key to Objective Questions

Chapter 1: Q. 1.1 (i to xv)

Question	i	ii	iii	iv	v	vi	vii	viii	ix	x	xi	xii	xiii	xiv	xv
Response	a	b	c	d	b	a	c	d	a	d	c	d	c	a	b

Chapter 2: Q. 2.1 (i to xxv)

Question	i	ii	iii	iv	v	vi	vii	viii	ix	x	xi	xii	xiii
Response	b	a	d	d	a	d	c	b	c	c	a	a	d

Question	xiv	xv	xvi	xvii	xviii	xix	xx	xxi	xxii	xxiii	xxiv	xxv
Response	b	c	c	d	a	a	b	b	a	a	b	b

Chapter 3: Q. 3.1 (i to xx)

Question	i	ii	iii	iv	v	vi	vii	viii	ix	x
Response	d	c	b	c	a	b	d	c	b	c

Question	xi	xii	xiii	xiv	xv	xvi	xvii	xviii	xix	xx
Response	d	c	a	c	d	a	c	b	c	a

Chapter 4: Q. 4.1 (i to xx)

Question	i	ii	iii	iv	v	vi	vii	viii	ix	x
Response	c	b	a	b	c	d	b	a	d	a

Question	xi	xii	xiii	xiv	xv	xvi	xvii	xviii	xix	xx
Response	c	c	b	a	b	b	d	a	d	c

Chapter 5: Q. 5.1 (i to xx)

Question	i	ii	iii	iv	v	vi	vii	viii	ix	x
Response	d	b	d	c	a	d	b	c	c	d

Question	xi	xii	xiii	xiv	xv	xvi	xvii	xviii	xix	xx
Response	b	d	a	d	c	b	a	c	c	c

Chapter 6: Q. 6.1 (i to x)

Question	i	ii	iii	iv	v	vi	vii	viii	ix	x
Response	a	b	d	c	b	c	c	c	b	a

Chapter 7: Q. 7.1 (i to xii)

Question	i	ii	iii	iv	v	vi	vii	viii	ix	x	xi	xii
Response	d	a	c	b	b	c	b	c	c	b	c	d

Chapter 8: Q. 8.1 (i to x)

Question	i	ii	iii	iv	v	vi	vii	viii	ix	x
Response	c	b	d	d	a	c	a	b	d	a

Chapter 9: Q. 9.1 (i to x)

Question	i	ii	iii	iv	v	vi	vii	viii	ix	x
Response	c	d	a	b	c	a	d	c	b	a

Chapter 10: Q. 10.1 (i to x)

Question	i	ii	iii	iv	v	vi	vii	viii	ix	x
Response	a	d	c	b	a	b	c	c	d	c

Chapter 11: Q. 11.1 (i to xx)

Question	i	ii	iii	iv	v	vi	vii	viii	ix	x
Response	c	b	d	c	b	a	d	c	a	d

Question	xi	xii	xiii	xiv	xv	xvi	xvii	xviii	xix	xx
Response	b	b	a	c	d	c	b	b	c	a

Chapter 12: Q. 12.1 (i to x)

Question	i	ii	iii	iv	v	vi	vii	viii	ix	x
Response	b	c	c	d	d	b	c	b	c	b

Chapter 13 : Q. 13.1 (i to x)

Question	i	ii	iii	iv	v	vi	vii	viii	ix	x
Response	d	b	c	c	a	d	b	b	c	b

Chapter 14: Q. 14.1 (i to x)

Question	i	ii	iii	iv	v	vi	vii	viii	ix	x
Response	c	b	d	a	b	c	d	b	a	c

Chapter 15: Q. 15.1 (i to x)

Question	i	ii	iii	iv	v	vi	vii	viii	ix	x
Response	b	c	d	d	a	b	c	a	b	d

GLOSSARY

Active earth pressure is the lateral pressure exerted by cohesionless soil particles on the retaining wall. The coefficient of active earth pressure $K_a = \dfrac{1-\sin\theta}{1+\sin\theta}$, where angle of repose of soil $= \theta$.

Actual neutral axis is the distance of the neutral axis from the extreme compressive fibre in the section where the moment of the compressive zone force is equal to the moment of the tensile zone force of the actual reinforcement.

Angle of repose (θ) of any cohesionless soil is the angle of the stable soil surface with the horizontal. It is also known as angle of friction.

Back fill is the material filled on the back side of the retaining wall.

Balanced section is the section where the distance of the critical neutral axis is equal to the distance of the actual neutral axis. The compressive stress in extreme concrete fibre and the tensile stress in the reinforcement develops simultaneously equal to the respective permissible limits.

Beams are structural elements having small widths and large depths and subjected to transverse forces causing bending in the element.

Bearing capacity is the safe load per unit area transferred to the soil below foundation.

Bond stress is the resistance caused by friction, mechanical adhesion of concrete per unit surface area of reinforcing bar along the interface of reinforcing bar with concrete.

Box culvert is a square pipe with four joints between horizontal and vertical sides constructed monolithically.

Centrifugal forces are caused by moving vehicles along the curved road which acts outward along the radius.

Characteristic strength (f_{ck}) is the mean strength of concrete in which not more than 5% samples fall below this strength. This strength of concrete (crushing) is measured on 150 mm cubes after 28 days of moist curing in N/mm^2.

Columns are axial structures mainly subjected to compressive forces (loads).

Compatibility represents equality of strains in concrete and steel reinforcement in RCC elements at the same point.

Concrete is the mixture of cement, aggregate (fine and coarse), water and appropriate admixture properly mixed, placed, compacted and cured.

Concrete cover is the distance of steel reinforcement from the outer surface of the concrete member.

Continuous slab is the slab supported on more than two supports and reinforcement is essential on both the faces according to variation of bending moment along the span.

Cracks develop in concrete in tension zone. The width of crack is kept within certain limits for serviceability and performance of the structural element.

Creep is the permanent strain in concrete due to permanent loading on the structure. Creep of concrete results in loss of prestress in prestressed concrete member.

Culvert is a small bridge structure to pass rain water or nallah across a road or railway.

Curing of concrete represents a condition of maintaining moisture and temperature of concrete during hardening process.

Curtailment is necessary for economy by curtailing steel bars wherever not required due to reduction in forces and moments. Bars are extended beyond the point of actual curtailment by certain distance for safety.

Cylindrical shaft is a thin circular supporting structure for overhead tanks. The minimum thickness of cylindrical supporting shaft is kept 125 mm.

Dead loads are permanent fixed loads of the structure.

Deflection of beams occur due to bending of original longitudinal axis under transverse loading. The vertical distance between the bent axis and original axis represents the deflection. The deflection in bending structure has to be kept within certain limits to maintain functionality and performance under service loads by suitably adjusting the ratio of (span to depth) of bending structures.

Design loads are sum of total working loads in working stress method and sum of total ultimate loads in limit state or ultimate load methods.

Development length of steel reinforcement is that which transfers the axial force in bar safely to the surrounding concrete by bond.

Doglegged staircase has two flights in opposite direction having a common horizontal landing.

Doubly reinforced section is reinforced in compression zone for enhancing the moment of resistance of the section in addition to reinforcing of tensile zone.

Ductile materials are those which undergo large deformation under the action of force prior to failure.

Durability of concrete represents its duration of satisfactory performance in the anticipated working environment and exposure conditions.

Earthquake loads are the result of horizontal and vertical components of accelerations caused by shaking of earth mass and depends on the earthquake zone of the country.

Economical depth of T–beam is that depth which gives minimum total cost of concrete and steel for the section. Economical depth depends on the ratio r of the cost of steel and concrete.

Effective length of the compressive member is the distance between the two end hinge supports or consecutive points of inflexion created by bending or buckling along the length.

Elastic deformation occurs when the structure is subjected to working service loads. When prestressed beam structure undergoes compressive deformation, loss of prestress occurs in steel wires.

Elastic design method (working stress method) of RCC members is based on linear properties of materials (steel and concrete) upto respective elastic limits. In this method, the materials (steel and concrete) are assumed to behave in linear elastic manner.

Equivalent concrete area of an axially reinforced compression member is that area of plain concrete column which can bear the same load as the reinforced column $[A_e = A + (m - 1) A_{sc}]$, where A = area of column, A_{sc} = area of compressive steel and m = modular ratio.

Erection forces are those which are caused due to weights and vibrations of machinery.

Flat slab is the slab directly supported on columns with or without capitals and cast monolithically with the support in both perpendicular directions. Flat slabs are designed by empirical or equivalent frame analysis.

Flexural tensile strength of concrete shall be equal to $0.70\sqrt{f_{ck}}$ N/mm^2.

Flexure means bending of structure under transverse loads.

Foundation is substructure below plinth to transfer the load of the superstructure to the ground soil.

Free board is the distance between the tank top and full supply level (FSL) of liquid (water).

Friction loss occurs in prestress due to friction between prestress wires and duct surface.

Grade of concrete is denoted by letter M followed by 28 days average compressive (crushing) strength specified in N/mm^2 as measured on 150 mm size cubes (M25, M40...).

Helical spiral reinforcement is the continuous circular lateral reinforcement encompassing main longitudinal bars in a circular section. Columns with helical spiral reinforcement gives higher strength as compared with tied columns of the same size.

High strength deformed steel bars have protruding ribs on the surface of the bar to develop superior bond with concrete (HYSD).

Hoop tension is the tensile force in a circular vessel caused by liquid pressure along the circumference of the vessel.

Impact load is load caused due to impact of the moving load. Impact load is more than the normal (average) load.

Imposed loads (live loads) are those loads which are applied externally on the structure and these loads are based on the type of occupancy and type of structure.

IRC loads are specified by Indian road congress for different type of highways according to the type of vehicles passing through that highway as class A, class B and class AA. According to Indian road congress specifications these loads are stated in terms of wheel loads and distances between various loads

Lever arm is the distance between the resultant compressive force in compression zone and the total tensile force in steel reinforcement in tension zone across the flexural member section.

Limiting concrete strain is the strain in the extreme compressive concrete fibre ($e_c = 0.0035$).

Limit state is a state of impending failure beyond which a structure ceases to perform the intended function.

Limit state design method is based on safety against various limiting conditions of strength, critical deformations and other limits.

Live loads (imposed loads) are those loads which are applied externally on the structure.

Long column (slender column) is that column which primarily fails by buckling (bending).

Minimum reinforcement in any RCC structure shall be provided to resist temperature strains (0.12% steel for HSD bars and 0.15% for MS plain bars).

Modular ratio of steel and concrete is ratio of modulii of elasticity of steel and concrete $\left(m = \dfrac{280}{3\sigma_{cbc}}, \text{with usual notations} \right)$.

Modulus of elasticity of concrete is normally related to its characteristic strength (f_{ck}) and is assumed as $E_c = 5000 \sqrt{f_{ck}}$ (N/mm^2).

Moment of resistance (M_r) of the structural element's section is the total moment of the couple formed by tensile force in steel reinforcement and the total resultant compressive force in compression zone of concrete across the section of flexural member.

Neutral axis in any flexural member is the line in the cross-section along which the bending stress will be zero (i.e. neither tensile nor compressive). The moment of the resultant compressive force and moment of the total tensile force will be equal about this neutral axis.

Nominal diameter of any deformed bar is the average diameter of an equivalent circle having the same area of cross-section.

One-way slab resists loading by bending mainly in one direction only and the main reinforcement is also provided in one direction. Distribution (temperature) steel shall be provided in the perpendicular direction.

Open well staircase has three flights with two quarter landings having open well in the middle. The quarter landings are at different levels and each flight moves around the open well.

Over reinforced section is the section reinforced with more than the required critical reinforcement for the balanced section. The compressive stress in the extreme concrete fibre will be equal to the maximum permissible compressive stress in concrete while the stress in the tensile steel reinforcement will be less than the maximum permissible tensile stress of the steel.

Partial safety factor is provided with respect to the material characteristics and with respect to the loading characteristics.

Partial safety factor for material is the ratio of characteristic strength to design strength of material.

Partial safety factor for loading is the ratio of design load to actual service load.

Passive earth pressure is the maximum lateral resistance offered by the soil mass against sliding of the retaining wall. Coefficient of passive earth pressure (k_p) = $\dfrac{1 + \sin \theta}{1 - \sin \theta}$, where θ = angle of repose.

Permissible stress is the maximum safe working stress permitted and obtained by dividing ultimate or proof stress or yield stress with the factor of safety of the material.

Post-tensioning is the system of stretching HYS steel wires after casting and hardening of concrete member through ducts provided.

Prestressed concrete is the special technique to utilize concrete with its maximum compressive stress by neutralizing tensile stresses developed by introducing internal compressive stress in concrete by stretching HYS steel wires in tensile zone of the flexural concrete member prior to loading of the member.

Prestress losses occur due to various causes such as elastic deformation, shrinkage and creep of concrete, relaxation of steel, slip of anchorage, and friction of tendon (wires) with concrete surface.

Pretensioning is the system of stretching HYS wires prior to casting of concrete member by suitable anchoring with bulk heads at ends.

Proof stress of steel is stress at 0.20% strain and is considered equivalent to yield stress.

Reinforced cement concrete (RCC) construction comprises cement concrete reinforced with steel bars to resist tensile forces caused by service loads in the structural member.

Retaining wall is a structure which retains safely earth or any other material having tendency to slide.

Riser is the vertical portion of each step of a staircase. Rise (height of riser) is generally 150 mm to 200 mm depending on the type of occupancy.

Seismic forces are caused due to earthquake acceleration of structural mass.

Serviceability of a structure is a condition which ensures deflection and cracking within certain limits under service loads so as to maintain its shape and size appropriately.

Shear force in a flexural structure represents algebraic sum of transverse forces on one side of the section under consideration.

Shear key is the protruding part of retaining structure below base slab of retaining wall to resist sliding of the wall against net active earth pressure.

Shear stress is transverse force (shear force) per unit area of section.

Short column is that compression member which primarily fails by crushing and has length/short dimension ratio quite small (less than 12).

Shrinkage is reduction in volume of concrete member during hardening process. Shrinkage results in loss of prestress in a prestressed member.

Simply supported structure is a structure which is supported on two simple supports at ends. Such a structure (beam or slab) bends in one direction and develops tension on one face only.

Singly reinforced section is one which is reinforced only on tensile zone/face.

Slenderness ratio is the ratio of effective length to the least lateral dimension of a compression member.

Slip loss occurs in a prestressed member due to slipping of anchors of stressed wires.

Span in any flexural element (beam/slab) will be the distance between two simple supports or two consecutive supports.

Stirrups are rings/ties provided to tie longitudinal bars to resist shear stresses along the section.

Strains are deformations per unit length when subjected to certain force. Strains occur both in steel and concrete in RCC section.

Stress block represents variation of stress in concrete/steel fibres along the depth of the section. IS:456-2000 specifies certain stress block for RCC members giving various parameters and constants.

Stress in any structure is caused due to internal resistance (force) per unit area of the structure.

Substructure is the foundation structure below plinth which interfaces with super structure. Sub structure transfers the loads of structure to sub base and to soil below foundation.

Super structure is the structure above plinth level.

Surcharge is the height of filled soil or any other material above the top of the retaining wall.

Target strength (f_t) of concrete mix is the design strength to achieve the characteristic strength (f_{ck}) for the desired concrete grade ($f_t = f_{ck} + ks$, where s = standard deviation and k = statistical constant).

T – beam section is a section with wide flange in the top and deep web and section is shaped as T.

Temperature stresses are caused due to variation in temperature in the structure. Expansion joints are necessary for more than 20 m – 30 m continuous length of concrete structure.

Tensile strength (flexural) of concrete (f_{cr}) is related to its characteristic strength (f_{ck}) as $f_{cr} = 0.70\sqrt{f_{ck}}$, (N/mm²).

Tensile stress in a structure is force per unit area tending to cause cracks in concrete. Tensile stress is resisted by steel reinforcement in RCC.

Ties are lateral rings encompassing main longitudinal steel bars in RCC columns.

Torsion is twisting moment caused by eccentric transverse loading. Torsional moment is resisted by providing longitudinal and lateral stirrup reinforcements.

Tractive force is the force caused by braking effect of moving wheel loads.

Tread is the horizontal portion of each step in a staircase. Tread is usually 250 mm to 300 mm.

Ultimate design method is the design approach based on ultimate strength of concrete and steel reinforcement along with respective safety factors.

Under reinforced section is the RCC section in which the actual tensile reinforcement provided is less than the required critical tensile reinforcement. In such a underreinforced section, the tensile stress in the reinforcement develops equal to the maximum permissible stress of steel while compressive stress in the extreme concrete fibre will remain less than the maximum permissible stress of concrete.

Width of stair is the breadth of the staircase and depends on the type of building. Width of stair case is usually 850 mm to 2000 mm.

Wind loads are loads on the surface of the structure caused by blowing of wind resulting in pressure on the surface.

Workability of fresh concrete is the property which represents the ease of compaction of concrete by available means.

Working stress design method is the design approach based on linear elastic behaviour of stress–strain of materials (concrete and steel).

Yield line is a line in the plane of the slab along which all reinforcing steel bars have yielded and concrete starts cracking after offering maximum moment of resistance.

Yield stress of steel is that stress beyond which the steel bar stretches at much higher rate without increase of any further force.

INDEX